Lecture Notes in Mechanical Engineering

Lecture Notes in Mechanical Engineering (LNME) publishes the latest developments in Mechanical Engineering—quickly, informally and with high quality. Original research reported in proceedings and post-proceedings represents the core of LNME. Volumes published in LNME embrace all aspects, subfields and new challenges of mechanical engineering.

To submit a proposal or request further information, please contact the Springer Editor of your location:

Europe, USA, Africa: Leontina Di Cecco at Leontina.dicecco@springer.com
China: Ella Zhang at ella.zhang@springer.com
India: Priya Vyas at priya.vyas@springer.com
Rest of Asia, Australia, New Zealand: Swati Meherishi at swati.meherishi@springer.com

Topics in the series include:

- Engineering Design
- Machinery and Machine Elements
- Mechanical Structures and Stress Analysis
- Automotive Engineering
- Engine Technology
- Aerospace Technology and Astronautics
- Nanotechnology and Microengineering
- Control, Robotics, Mechatronics
- MEMS
- Theoretical and Applied Mechanics
- Dynamical Systems, Control
- Fluid Mechanics
- Engineering Thermodynamics, Heat and Mass Transfer
- Manufacturing
- Precision Engineering, Instrumentation, Measurement
- Materials Engineering
- Tribology and Surface Technology

Indexed by SCOPUS, EI Compendex, and INSPEC

All books published in the series are evaluated by Web of Science for the Conference Proceedings Citation Index (CPCI)

To submit a proposal for a monograph, please check our Springer Tracts in Mechanical Engineering at https://link.springer.com/bookseries/11693

Katia Mocellin · Pierre-Olivier Bouchard ·
Régis Bigot · Tudor Balan
Editors

Proceedings of the 14th International Conference on the Technology of Plasticity - Current Trends in the Technology of Plasticity

ICTP 2023 - Volume 2

Springer

ICTP 2023
FRENCH RIVIERA

Editors
Katia Mocellin
CEMEF - MINES Paris
PSL Research University
Sophia Antipolis, France

Pierre-Olivier Bouchard
CEMEF - MINES Paris
PSL Research University
Sophia Antipolis, France

Régis Bigot
Arts et Metiers Institute of Technology
LCFC
Metz, France

Tudor Balan
Arts et Metiers Institute of Technology
LCFC
Metz, France

ISSN 2195-4356 ISSN 2195-4364 (electronic)
Lecture Notes in Mechanical Engineering
ISBN 978-3-031-40919-6 ISBN 978-3-031-40920-2 (eBook)
https://doi.org/10.1007/978-3-031-40920-2

This Springer imprint is published by the registered company Springer Nature Switzerland AG
The registered company address is: Gewerbestrasse 11, 6330 Cham, Switzerland

Contents

Sheet Forming

A New Type of CubeSat Structure Utilizing the Superplastic Forming Process

Yaqoob Alqassab[1](✉) and Firas Jarrar[2]

[1] National Space Science Agency, Hidd, Bahrain
yagoob.khalid@nssa.gov.bh

[2] Department of Mechanical Engineering, Al Hussein Technical University, Amman, Jordan
firas.jarrar@htu.edu.jo

Abstract. CubeSats are kind of small satellites that follow specific standards, and because of that, they are less costly to develop and launch compared to conventional satellites. The careful design of a CubeSat's structure allows for weight savings and increases the chances of a successful mission. In addition to being light in weight, the CubeSat structure must provide enough stiffness and strength to support the other subsystems. Superplastic forming (SPF) is a manufacturing process used to produce complex and lightweight components. It is known to significantly save manufacturing cost and time in certain niche applications. Up to our knowledge, this process has never been implemented in the CubeSat industry. This study considers the use of the superplastic forming process to produce a CubeSat enclosure that consists of six SPF panels. The Aluminum alloy AA5083 was used in the study, as it already has applications in the space industry. Several finite element simulations were carried out to ensure that the proposed structure passes the requirements of the CubeSat design specifications. At first, the SPF process was simulated to make sure that the panels' thickness profiles meet the industry requirements and that no excessive thinning is observed. The generated panels were integrated in a 1U CubeSat and then subjected to both quasi-static and modal analyses. The results show that the proposed structure has a mass that is 226g less than that of a certain commercially available off-the-shelf CubeSat structure. This is considered significant when compared with the typical mass of a complete 1U CubeSat, which is 2.0 kg. The results also demonstrate that the proposed structure satisfies all the CubeSat dimensional, stiffness, and strength requirements.

Keywords: Superplastic forming · CubeSat · Quasi-static analysis · Modal analysis · Finite element analysis

1 Introduction

CubeSats come in different sizes that are unit-based, for example, a 1U CubeSat, which consists of a one CubeSat standard unit, is a 10 cm cube with a mass of less than 2.0 kg [1]. In addition to being educational tools, whenever possible, CubeSats have been utilized to demonstrate new space technologies, whether that being for remote sensing,

K. Mocellin et al. (Eds.): ICTP 2023, LNME, pp. 3–11, 2024.
https://doi.org/10.1007/978-3-031-40920-2_1

communication, or other applications. The practice of developing large conventional satellites for the sake of testing new space technologies is now considered an inefficient approach since it is both expensive and it requires long development times. CubeSats on the other hand, due to their standardized design specifications and accordingly the availability of off-the-shelf components, offer a low cost and a short lead time approach for testing new technologies.

A CubeSat contains several subsystems such as the mechanical, onboard computer, electrical power, attitude determination and control, and communication systems. The mechanical subsystem is considered as one of the most critical subsystems in a CubeSat. It integrates and supports all the CubeSat's internal components. This is in addition to providing shielding against space radiation and atomic oxygen. As a mission requirement for any CubeSat, its structure needs to survive the harsh launch environment, which requires a high level of stiffness and strength. The structure shall follow certain specifications pertaining to its dimensions, mass, volume, and material properties. Moreover, different testing procedures are used to qualify a CubeSat for the space environment. This includes the vibration test that is used to simulate the environment in the launch vehicle. Thermal testing, in a thermal vacuum chamber, is also essential due to the significant thermal cycles the CubeSat will be subjected to while in orbit.

Superplastic forming (SPF) is an advanced industrial process used to produce thin and complex parts, especially in niche applications. A major advantage is that it can produce complex parts without the need for welding or joining operations. Thus, it is considered a near-net shape manufacturing process, which is highly desirable in industry [2]. Superplasticity becomes evident in certain fine grain size materials when such materials are deformed at low strain rates and at a temperature of half the absolute melting point of the material. Under these conditions, superplastic materials exhibit an elongation tens of times greater than that in traditional room temperature manufacturing processes. In general, the SPF process is based on applying a high-pressure inert gas, such as Argon, to slowly deform a sheet into the inner shape of a fixed die. The process can be completed within a single step with a significant reduction in cost and time when compared to other traditional processes.

On the other hand, one of the challenges in SPF is the non-uniformity of thickness of the produced component [3]. This is acceptable as long as the minimum thickness of the component is not very thin to the point it can jeopardize its structural integrity. In industrial practice, usually, if the minimum thickness is more than 50% of the original thickness of the sheet, the component is considered acceptable. However, from an application point of view, it is important to verify the strength of the component by subjecting it to actual testing and finite element simulations under the operational loading conditions.

This study is concerned with the implementation of the superplastic forming (SPF) manufacturing process in the CubeSat industry while at the same time considering the actual thickness profile of the produced panels. To the best knowledge of the authors, this has not been addressed in open literature. The purpose of this implementation is to reduce the mass and the number of parts while satisfying all the CubeSat and launch requirements. The procedure followed in this study takes into account the manufacturability and the design requirements.

2 Material Constitutive Model

The flow stress of materials undergoing the superplastic forming process is very sensitive to the strain rate, resulting in viscoplastic deformation. It is also considered an isothermal process as the thin sheet is introduced only after the die has reached the required forming temperature. Furthermore, the gas pressure is applied 10 min after placing the sheet [4, 5], giving enough time for thermal equilibrium. Equation (1) shows the constitutive relationship in the strain hardening form of the power law [6–9]:

$$\overline{\sigma} = k\overline{\varepsilon}^{n}\dot{\overline{\varepsilon}}^{m} \tag{1}$$

where $\overline{\sigma}$ is the effective flow stress, k is a material constant, $\overline{\varepsilon}$ is the effective strain, $\dot{\overline{\varepsilon}}$ is the effective strain rate, m is the strain rate sensitivity exponent and n is the strain hardening exponent. The material used for the sheet is Aluminum AA5083. This material has been considered in several SPF studies [6, 9]. Table 1. shows the experimentally calibrated values based on the extended power law model corresponding to a temperature of 500 °C [8].

Table 1. Aluminum AA5083 material values [8].

Parameter	Value
m	0.402
n	0.123
k	161.009

3 Methodology

In this study, a finite element analysis was done using Abaqus/Standard for simulating the SPF process. The SPF simulation was carried out to resemble the manufacturing of the original sheet into the die's desired shape (i.e., stiffened panel). The main objective of conducting the simulation is to ensure that the manufactured panel's thickness is within the acceptable range. As a general rule of thumb in industry, the minimum thickness in a manufactured part should be at least half the original thickness for it to be acceptable. After that, six identical Aluminum panels were assembled to form a 1U CubeSat that passes the CubeSat design requirements [1]. The formed CubeSat, which consists of the non-uniform thickness panels, was then imported to a new Abaqus/Standard model where quasi-static and modal analyses were carried out. Both simulations were conducted using 8-node, reduced integration, 3D stress elements.

4 Finite Element Analysis of the SPF Process

The model considered in the SPF simulation consists of two main parts, namely the original sheet and the die with protrusions. The original sheet has dimensions of 123.5 mm in length, 110 mm in width, and a thickness of 1.2 mm. The length and width of the die are 110 mm and 100 mm, respectively. According to the CubeSat Design Specification (CDS) document [1], the minimum width of the CubeSat rails should be greater than 8.5 mm. To meet this requirement, the depth of the die was selected to be 10 mm. The benefit of adding the protrusions on the die is to create stiffeners in the manufactured panels and thus increase their stiffness. When making the panel through traditional manufacturing processes, the stiffeners are usually joined to the panel by welding. The SPF process considered here will reduce the fabrication cost and time by manufacturing the panel with the stiffeners together in a single process. A no-lubrication, friction interaction between the sheet and the die geometries was considered in the SPF simulation with a coefficient of friction value of 0.8.

There were three boundary conditions used in the analysis, namely a fixed boundary condition for the reference point of the die, a fixed boundary condition at the stiffeners' reference point, and a pinned boundary condition at the sheet's edges, as shown in Fig. 1. The fixed boundary conditions ensure that the die and stiffeners will not move from the desired location. The pinned boundary condition is used to ensure that the sheet's edges remain in the exact location to resemble the blank holder's effect. The pressure loading consists of six steps applied in sequence, as shown in Fig. 2.

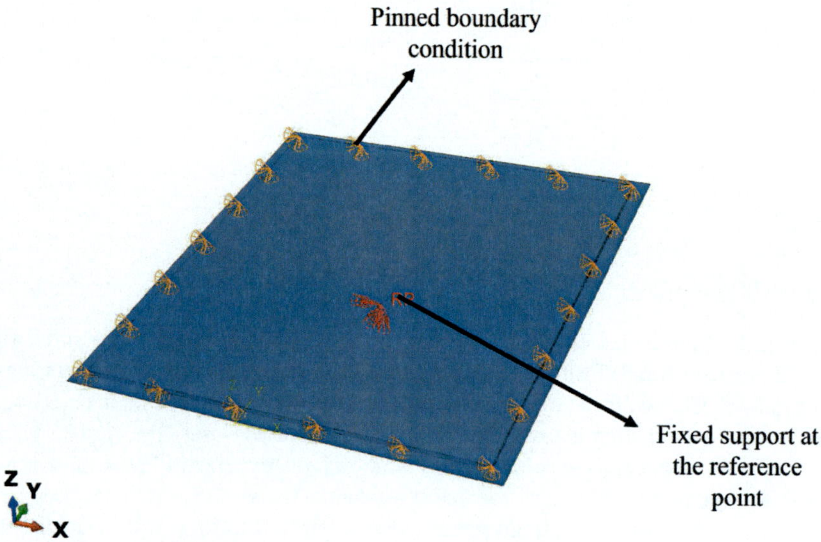

Fig. 1. The boundary conditions used in the SPF simulation

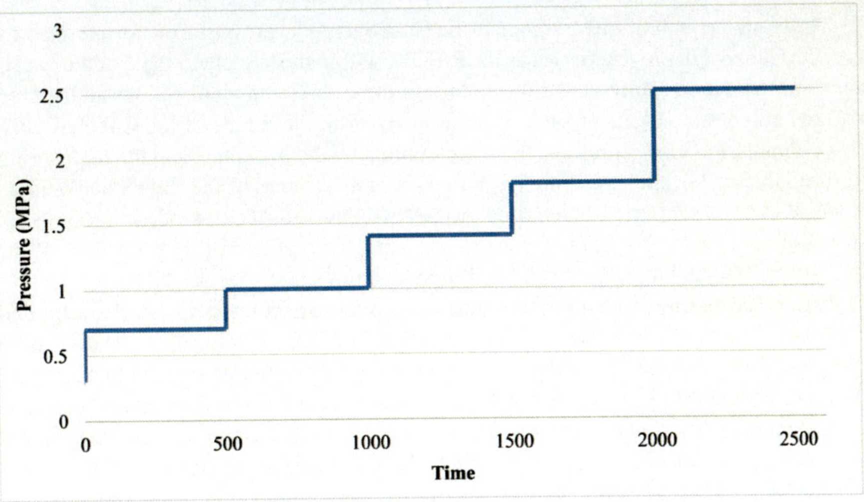

Fig. 2. Pressure load variation against time.

5 Finite Element Analysis of the 1U CubeSat

During launch, CubeSats are subjected to accelerations multiple times greater than the gravitational acceleration. This is also in addition to extreme dynamic loads, such as low and high-frequency random vibrations. CubeSats failures caused by vibrations during launch could result in damage to the deployer case and the CubeSats. Therefore, to prevent these potential failures, it is essential to estimate the response of the considered CubeSat to the loading environment of the launch vehicle. In simulations, this is usually done by quasi-static and modal analyses.

Using six of the panels produced from the SPF simulation, a 1U CubeSat was assembled, as shown in Fig. 3. The purpose of this second simulation is to test the assembled CubeSat performance under the conditions of the quasi-static and modal analyses.

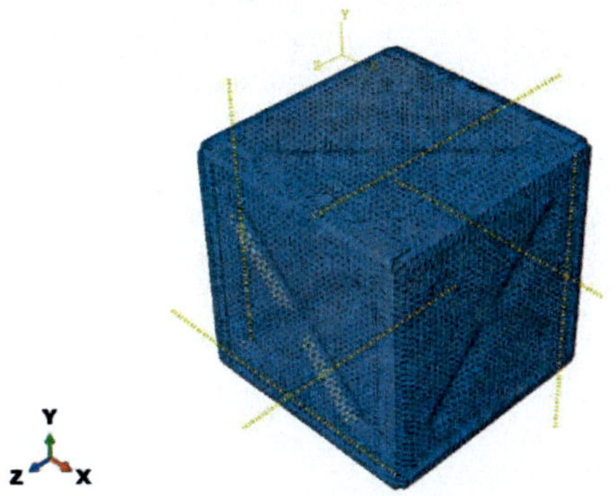

Fig. 3. 1U CubeSat SPF assembly.

The boundary conditions assigned to the CubeSat model are based on the interactions with the CubeSat deployer, known as the P-POD, within the launcher [10]. For the quasi-static analysis, it is required to fix the –z face of the CubeSat in all directions and adding a 46.6N compression load at each of the corners of the + z face of the CubeSat [10]. This compression load represents the force applied by the backplate and the mainspring on the CubeSat. For the modal analysis, the + z and −z faces of the CubeSat should be modeled as fixed supported as required by the JEM Payload Accommodation Handbook [10]. With no constraints at the CubeSat side rails, the surfaces of the CubeSat will be less constrained and thus the results will represent the worst-case scenario.

Liftoff can be simulated using the quasi-static analysis, where static loads are applied in a numerical analysis to simulate the spacecraft acceleration against gravity; these loads are typically applied in terms of g-force and differ depending on the launch vehicle used. Furthermore, since the CubeSat's actual orientation in the launch vehicle is unknown, three different loading scenarios, each in a certain direction, were considered. The loading was applied in each case along one of the axes (x, y, and z) directions.

The safety of the CubeSat was established by calculating the values of the yield and ultimate margins of safety and making sure that they are both greater than zero. The yield factor of safety was taken as 1.1 and the ultimate factor of safety was taken to be 1.4 [11].

The launch vehicle engine applies certain random vibrations on the CubeSat. These vibrations can cause significant damage to the CubeSat. Therefore, dynamic/modal analysis must be performed to demonstrate how the structure will respond when subjected to such dynamic loads. As stated in the CDS [1], the CubeSat's first mode must be higher than 100 Hz.

6 Results and Discussion

The quasi-static analysis was carried out on the proposed SPF 1U CubeSat enclosure to ensure that the yield margin of safety and the ultimate margin of safety are greater than zero. In addition, the modal analysis was carried out to ensure that the frequency in the first mode is greater than 100 Hz. Furthermore, the results were compared with an actual CubeSat structure, MYSAT-1. More details about the modeling and testing of MYSAT-1 can be found in the study by Alhammadi et al [12, 13]. The comparison made here was carried out without including the internal components.

Table 2. shows a comparison between the new SPF design and the MYSAT-1 structure in terms of mass, number of parts, and material. As shown in Table 2., the mass and the number of parts of the proposed structure are significantly lower, which makes it better for space applications.

Table 2. Comparison between SPF structure and MYSAT-1.

	SPF enclosure	MYSAT-1
Mass (g)	168	394
Material of the structure	Aluminum AA5083	Aluminum 6061
Number of parts	30	38

The quasi-static analysis was carried out for both structures, namely the proposed SPF structure and the MYSAT-1 structure. Table 3. shows the yield and ultimate margins calculation for the three considered scenarios and the results of the MYSAT-1 structure. It can be seen from Table 3. that for all cases, the yield and ultimate margins are greater than zero which satisfies the requirements. In addition, for the three loading cases, the stress in the SPF structure is less than that in the MYSAT-1 structure.

Table 3. The yield and ultimate margins of safety for the three analyses.

Analysis	Maximum stress results (MPa)	Yield margin of safety	Ultimate margin of safety	MYSAT-1 results (MPa)
Analysis A	32.74	4.41	5.00	45.08
Analysis B	38.67	3.58	4.08	39.62
Analysis C	30.54	4.80	5.43	43.46

The modal analysis was carried out to ensure that the first mode's frequency is significantly higher than 100 Hz, in order to satisfy the requirement in the CDS document [1]. Table 4. shows the results of the modal analysis conducted on the SPF structure compared with the MYSAT-1 results. It can be seen from Table 4. that the frequency for the SPF structure in the first mode is greater than 100 Hz which satisfies the requirement.

Table 4. The modal analysis results for MYSAT-1 and SPF CubeSat structures.

CubeSat	1st frequency mode	2nd frequency mode	3rd frequency mode
MYSAT-1	1098.9	1101.4	1556.9
SPF CubeSat	873.4	892.2	903.1

7 Conclusions

The main goal of this study is to show the benefits of applying the superplastic forming process in manufacturing CubeSats. To accomplish this, we had to address several challenges. The first has to do with the non-uniformity in thickness in the SPF produced parts. Therefore, a finite element simulation was carried out to simulate the SPF process, in order to ensure that the thickness of the formed part is at least more than 50% of the original sheet thickness. The second step was to investigate the effect of the produced panels with the non-uniform thickness on the CubeSat mechanical performance. Therefore, the exact (non-uniform) SPF panel, as produced from the manufacturing process, was transferred to Abaqus/Standard for the subsequent quasi-static and modal analyses.

A new 1U CubeSat design was proposed. It consists of 6 SPF sheets and 24 screws. The proposed SPF structure has less number of parts compared to the COTS MYSAT-1. Moreover, the SPF enclosure provided better results in the quasi-static analysis compared with MYSAT-1. The yield and ultimate margins of safety were all greater than zero. The first mode frequency is 873.39 Hz which is greater than 100 Hz and thus satisfies the CubeSat design requirements. The first three frequency modes were compared between the SPF CubeSat and MYSAT-1. The results show that the frequencies of MYSAT-1 in the three modes are slightly better than the SPF CubeSat.

In some cases, the launch cost could be in the same order, if not higher, than the components cost of the CubeSat. Therefore, it is essential to reduce the launch cost by reducing the weight of the CubeSat. The proposed SPF structure successfully achieves this objective as it has a mass of 168 g, which is lower by 226 g than the COTS satellite structure used in MYSAT-1.

References

1. The CubeSat Program (2022) CubeSat design specification rev. 14.1. California Polytechnic State University
2. Zaid, A., et al.: Forming of a super plastic sheet metal made of MgAZ31 alloy. In: IOP Conference Series: Materials Science and Engineering, vol. 146, p. 012012 (2016)
3. Jarrar, F., et al.: On the challenges and prospects of the superplastic forming process. Mater. Sci. Forum **941**, 2343–2348 (2018)
4. Sorgente, D., Palumbo, G., Scintilla, L.D., Tricarico, L.: Gas forming of an AZ31 magnesium alloy at elevated strain rates. Int. J. Adv. Manufact. Technol. **83**(5–8), 861–872 (2015). https://doi.org/10.1007/s00170-015-7614-0
5. Kaya, A., et al.: Evolution of microstructure and texture in AZ31 alloy subjected to gas forming. JOM **69**, 1041–1045 (2017). https://doi.org/10.1007/s11837-017-2350-6
6. Jarrar, F., et al.: Constitutive modeling for the simulation of the superplastic forming of AA5083. Mater. Sci. Forum **838**, 512–517 (2016)
7. Jarrar, F.S., Nazzal, M.A.: Inclination angle effect on the thickness distribution in a superplastic formed long rectangular pan. Mater. Sci. Forum **735**, 155–161 (2013)
8. Kappes, J., Liewald, M.: Evaluation of pneumatic bulge test experiments and corresponding numerical forming simulations. J. Mater. Sci. Eng. **B1**, 472–478 (2011)
9. Jarrar, F.: Designing gas pressure profiles for AA5083 superplastic forming. Procedia Eng. **81**, 1084–1089 (2014)
10. JEM Payload Accommodation Handbook (2013) Small satellite deployment interface control document rev 8. Japan Aerospace Exploration Agency
11. Raju, I., et al.: White paper on factors of safety. NASA/TM-2009–215732/REV1, NESC-PB-04–05 (2012)

12. Alhammadi, A., et al.: Quasi-static and dynamic response of a 1U nano-satellite during launching. In: The 8th European Conference for Aeronautics and Space Sciences (EUCASS) (2019)
13. Alhammadi, A.N., et al.: Effect of finite element model details in structural analysis of CubeSats. CEAS Space J. **13**(2), 231–246 (2020). https://doi.org/10.1007/s12567-020-00339-z

Flattening of Pyramidal Asperities Under Combined Normal Loading and In-Plane Biaxial Straining

Maximilian Zwicker, Úlfar Arinbjarnar, Maximilian Knoll, Niels Bay, and Chris V. Nielsen(✉)

Department of Civil and Mechanical Engineering, Technical University of Denmark, Kgs. Lyngby, Denmark
cvni@dtu.dk

Abstract. In metal forming processes, contact mechanics in the tool-workpiece interfaces determine the resulting friction and thus influence the quality and stability of the process. In models representing dry, thin film and boundary lubrication conditions in metal forming, friction may be assumed proportional to the ratio of the real contact area to the apparent contact area. It is therefore important to obtain a fundamental understanding of as many of factors influencing asperity flattening as possible. This paper presents asperity flattening of pyramidal model asperities loaded normally by a plane tool. The experiments are performed in a recently developed test setup, which allows for different in-plane strain ratios in combination with the normal load. The in-plane deformation is known, through the yield criterion, to influence the necessary yield pressure to flatten the asperities and hence the resulting real contact area. In-plane balanced biaxial stretching is shown to result in further asperity flattening than plane strain. When comparing to previous results, it is shown that in-plane strains reduce the effect of the asperity shape on the resulting real contact area ratio.

Keywords: Sheet metal forming · Friction · Asperity flattening · Pyramidal asperities

1 Introduction

Understanding of the influence of surface asperities on friction in sheet metal forming is important when attempting zero defect production. Friction emerges in contact points between asperity peaks on the sheet surface and the comparably smooth tool, referred to as the real contact area. The asperities flatten upon deformation until the area is large enough to carry the applied load. Friction between asperities and tools is largely determined by the real area of contact, and understanding its evolution throughout a metal forming process is, therefore, crucial to estimating friction. Asperity flattening as a determining mechanism for friction was first pointed out by Bowden and Tabor [1], Ernst and Merchant [2] and Holm [3]. They laid the foundation for the adhesion theory, which was the first quantitative theory of dry friction. Bowden and Tabor [1] pointed out

K. Mocellin et al. (Eds.): ICTP 2023, LNME, pp. 12–19, 2024.
https://doi.org/10.1007/978-3-031-40920-2_2

that friction is not a mere surface phenomenon but that bulk behavior of the subsurface material plays a role. The initial adhesion theory was only valid for low normal pressures and was later extended to high normal loads by Orowan [4] and Wanheim et al. [5, 6], among others as reviewed by Nielsen and Bay [7].

Since sheets are not only loaded normally but also experience sliding and the bulk material is strained, more research was conducted. Fogg [8] found that subsurface strains in the bulk when combined with normal load lead to further flattening, which is again explained by the reduction of yield pressure. Fogg's [8] research on subsurface deformation was continued by Sutcliffe [9], Wang et al. [10], Nielsen et al. [11], and Zwicker et al. [12] by means of model asperities in a fundamental manner. They confirmed the influence of the subsurface strains and varied parameters like normal load, asperity flank angle or strain hardening. Recently, Shisode et al. [13] conducted a study on asperity flattening on real surfaces, where asperities were exposed to normal loading and longitudinal elongation, and confirmed Fogg's [8] initial observations. Those studies have been conducted under uniaxial and plane strain conditions.

Zwicker et al. [14] recently presented a new test on sheet with model asperities, inspired by the Marciniak-Kuczynski test principle, which allows for different arbitrarily controlled in-plane strain ratios during asperity flattening due to normal loading. They found with concentric model asperities that balanced biaxial stretching led to more flattening than plane strain due to further reduction of the yield pressure by the presence of the in-plane stresses.

The work presented in this work expands the initial findings [14] to include results with pyramidal model asperities. Such model asperities were earlier studied by Childs [15] and Weidel et al. [16] without the effect of in-plane strains.

2 Experimental Set-Up

This work was carried out on the experimental set-up shown in Fig. 1, which was first presented by Zwicker et al. [14]. Figure 2 includes a detailed view of the deformation zone with the pyramidal asperities studied in the present work. The set-up shown in Fig. 1 combines the Marciniak-Kuczynski test principle with a center punch, which enables the simultaneous compression of the asperity test region and induction of in-plane strains. The compression of the specimen and the induced strains are independent of each other. The test employs a driving sheet and a test sheet as in the Marciniak-Kuczynski test.

The experimental set-up consisted of a forming unit and a compression unit. The forming unit was connected to a press, while the compression unit was controlled by an external hydraulic pump. The experimental procedure was the following:

1. The forming unit (1) was placed so that the test sheet and the driving sheet could be clamped by the clamping unit (2).
2. The forming unit was lowered to its zero point so that the sheet was aligned with the lower asperity flattening punch (3).
3. The hydraulic cylinder (4a) that was mounted onto the subpress (4b) was used to lower the upper asperity flattening punch (5) and clamp the test sheet against the lower asperity flattening punch (3).

4. The forming unit (1) was moved downwards until the desired in-plane strains were achieved.
5. The upper punch (5) was retracted and the formed specimen taken out of the set-up.

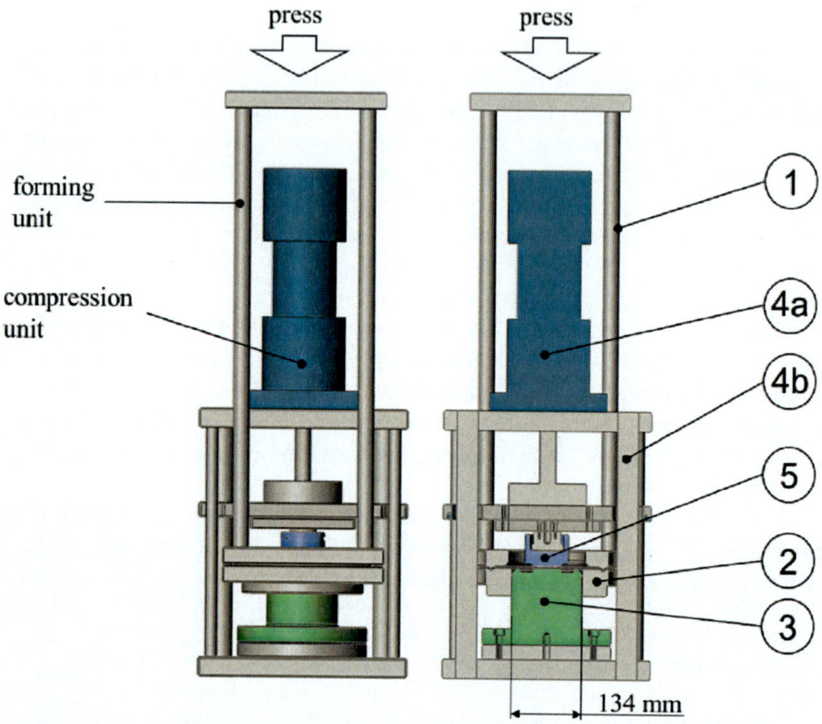

Fig. 1. Asperity flattening extended Marciniak-Kuczynski set-up (image adapted from [14]).

The test set-up was used inside a hydraulic *Mohr & Federhaff* press with a single-axis and up to 60 tons press capacity. The hydraulic cylinder (4a) inside the test set-up was from *Bosch* (*CDM1MF1/100/56/150A21/BB22CEUMWY*) and could, in combination with a hydraulic pump, provide up to 66 kN. During the experiments, the asperity region was placed under a normal load of 31.5 kN, resulting in a nominal pressure on the original test surface of 36.8 MPa. The level of normal pressure presented here was chosen such that the resulting real contact area ratio spans a large range within the possible range of subsurface strains.

The pyramidal asperities with a flank angle $\gamma = 10°$ and base 1.5 mm x 1.5 mm are presented in Fig. 2. They were machined into AA1050-H14 sheets with a thickness of 2.5 mm. Different specimen designs were used to achieve different in-plane strain ratios. Figure 3 presents the sheet geometries, which were designed to cover a wide range of strain ratios.

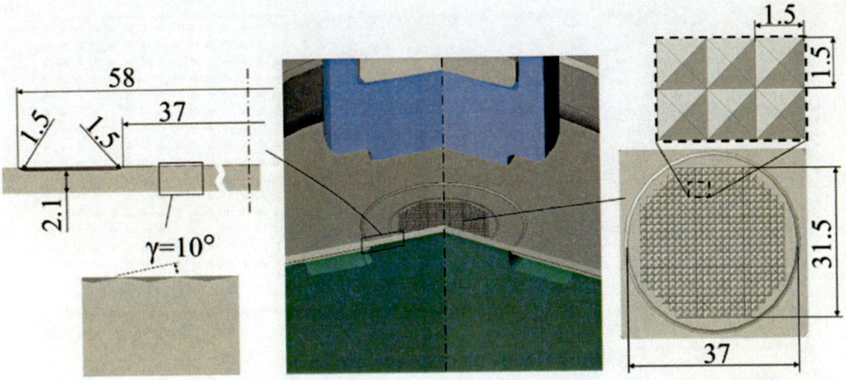

Fig. 2. Magnification of the test zone consisting of 381 pyramidal model asperities. Dimensions are in mm.

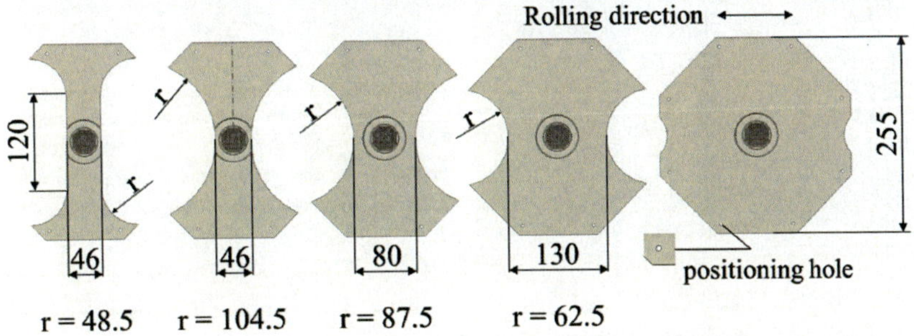

Fig. 3: Sheet geometries to cover a wide range of in-plane strain ratios. Dimensions are in mm.

After the experiments were performed, all asperity regions were scanned using an *Olympus LEXT OLS4100* confocal laser microscope. The whole asperity region was scanned and the 2D images were imported into *MATLAB 2020b*, where the region marked in Fig. 4a was cropped and rotated from the initial image and saved for further post-processing. Afterwards, the edges of the flattened asperities were automatically detected. Therefore, the existing image was first converted to grayscale (Fig. 4b), which was then smoothed using a Gaussian filter. After sharpening the contrast of the gray image, the image was transformed into a binary image (Fig. 4c), where the asperity plateaus were identified and colored in black and the valleys in white. A comparison of real contact surfaces determined by the presented procedure is shown in Fig. 4d-e by superposition of the edges to the grayscale images. With the final black and white image, it was possible to calculate the real contact area ratio α, as presented in Eq. (1). Under the assumption that a complete square of the 225 contact areas is detected incorrectly, a maximum error of 0.4% would occur.

$$\alpha = \frac{\text{real contact area}}{\text{nominal contact area}} = \frac{\text{number of black pixels}}{\text{total number of pixels}} \tag{1}$$

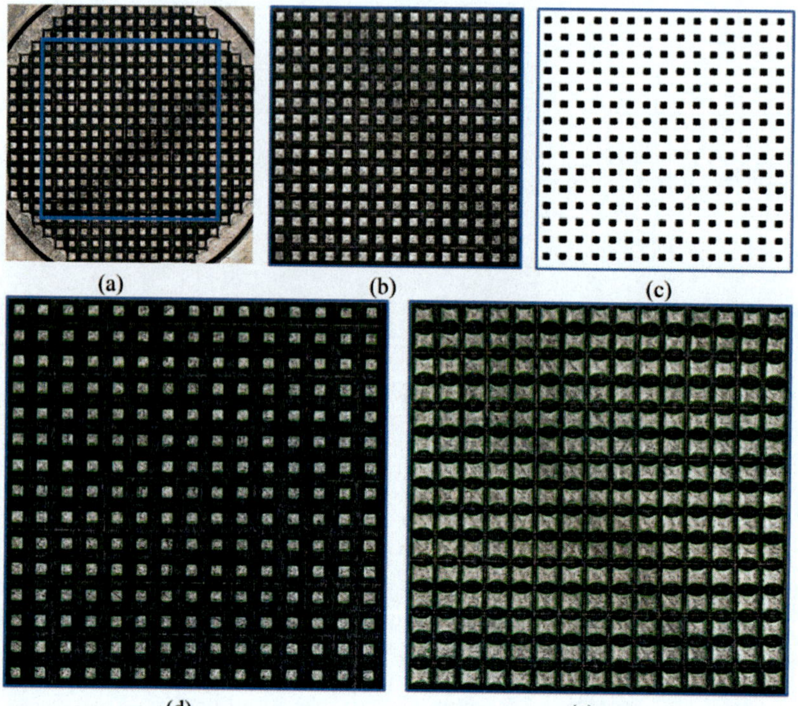

Fig. 4. Procedure to evaluate the real contact area ratio and the major and minor strains: (a) 2D image of real surface, (b) standardized part containing 225 pyramidal asperities cropped from each sample, (c) black and white image to calculate real contact area ratio, (d) identified edges of the flattened asperities superposed to the cropped, real surface image for the balanced biaxial specimen, and (e) similar for the middle specimen design of Fig. 3.

The in-plane strains were calculated from the same area as the real contact area ratio. Here, $\varepsilon_x = \ln(x_1/x_0)$ and $\varepsilon_y = \ln(y_1/y_0)$ where x_0 and y_0 correspond to 22.5 mm. The cropping of the images is subject to an error and thus influences the calculated strain. By repeated attempts, a maximum error of 10 pixels could be determined in the cropping process which corresponds to 0.064 mm. This is used for the calculation of the strain error. Following Gauss' law of error propagation,

$$\Delta f = \sqrt{\left(\frac{\partial f}{\partial x} \cdot \Delta x\right)^2 + \left(\frac{\partial f}{\partial y} \cdot \Delta y\right)^2 + \ldots} \tag{2}$$

and under the assumption that the initial length is not subject to any error, the strain errors in the two directions are

$$\Delta\varepsilon_x = \frac{\Delta x}{x_1},\ \Delta\varepsilon_y = \frac{\Delta y}{y_1} \tag{3}$$

This results in a maximum error of $\Delta\varepsilon = 0.0033$ for the strain estimation in both principal directions.

3 Results and Discussion

Figure 5 presents the subsurface major strain as a function of the subsurface minor strain for all experiments. It can be seen that the experimental results cluster into balanced biaxial stretching and plane strain. With reference to Fig. 3, the right most specimen results in balanced biaxial stretching, while all other specimen geometries approach plane strain. The difference across the remaining specimens is the amount of strain that can be achieved. Test specimens that would result in tensile in-plane strains in formability testing appear in plane strain due to a competition between the expected narrowing, widening due to the normal load and friction between the test workpiece and the tool surfaces compressing the asperity region.

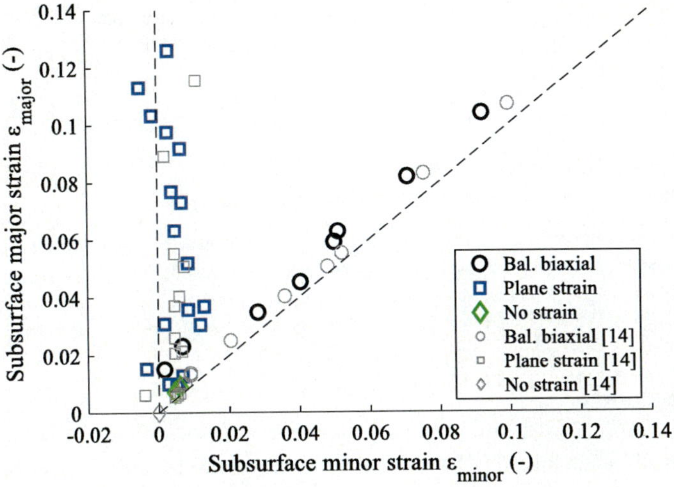

Fig. 5. Subsurface major strain ε_{major} as a function of the subsurface minor strain ε_{minor} similar to a forming limit diagram. Results are included for the pyramidal asperities and in grey for the concentric asperities presented in [14]. Data points labeled as no strain correspond to application of the normal pressure without further straining by the Marciniak-Kuczynski test principle.

Figure 6 presents the real contact area ratio α as a function of the subsurface major strain $\varepsilon_{\mathrm{major}}$ with labelling according to Fig. 5. The real contact area ratio grows in both cases with the subsurface strain. Furthermore, the real contact area grows at a higher rate $\alpha/\varepsilon_{\mathrm{major}}$ for balanced biaxial stretching than plane strain. This can be explained by the larger reduction of the yield pressure.

When comparing the resulting pyramidal asperity flattening to the flattening of concentric rings by Zwicker et al. [14] in Fig. 6, it is noticed that the behavior is very similar. This was not the case when only compressing ridge and pyramidal asperities without inducing in-plane strains [15]. The difference between the asperity flattening in the absence of in-plane strains might stem from different interactions between deformation fields from different asperity shapes, while the induced in-plane strains postpone interaction between the deformation fields as proven by Sutcliffe [9], and hence, the flattening will be governed by individual asperity flattening with less influence from the asperity shapes.

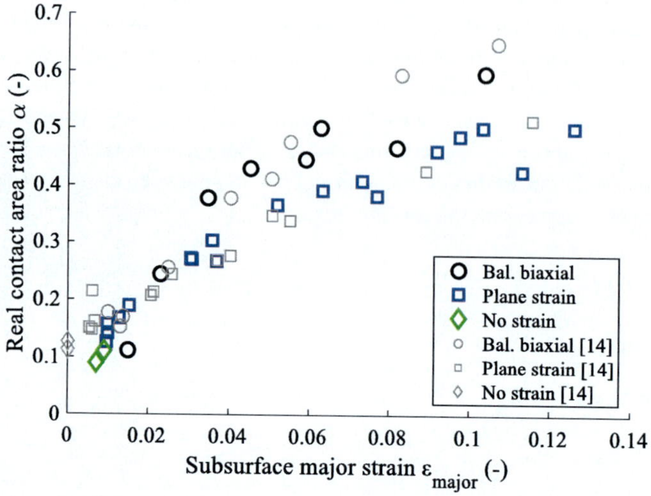

Fig. 6. The real contact area ratio α as a function of the subsurface major strain ε_{major} for the pyramidal asperities and in grey for the concentric asperities presented in [14].

4 Conclusions

The presented paper investigated the flattening of pyramidal model asperities due to combined normal loading and in-plane subsurface strains at strain ratios resembling plane strain and balanced biaxial stretching. The following can be concluded:

- As observed for concentric model asperities by Zwicker et al. [14], no strain ratios corresponding to tensile forming were achieved under the given normal load. This is attributed to the competition between narrowing of the width due to tension, widening due to the normal load and friction which acts as a damping of any of the effects.
- Balanced biaxial stretching leads to more asperity flattening than plane strain due to the further reduced yield pressure at the asperity tips.
- In-plane strains postpone interaction between deformation fields of individual asperities, and this seem to reduce the effect of asperity shape on the degree of flattening.

References

1. Bowden, F.P., Tabor, D.: Mechanism of metallic friction. Nature **50**, 197–199 (1942)
2. Ernst, H., Merchant, M.E.: Surface friction between metals – a basic factor in the metal cutting process. In: Special Summer Conference on Friction and Surface Finish, pp. 76–101 (1940)
3. Holm, R.: Über die auf die wirkliche Berührungsfläche bezogene Reibungs-kraft. Siemens-WerkeVeröff. Siemens-Werken **17**, 400–404 (1938). https://doi.org/10.1007/978-3-662-24676-4_4
4. Orowan, E.: The calculation of roll pressure in hot and cold flat rolling. Proc. Inst. Mech. Eng. **150**(1), 140–167 (1943)
5. Wanheim, T., Bay, N., Petersen, A.S.: A theoretical determined model for friction in metal working processes. Wear **28**, 251–258 (1974)
6. Wanheim, T., Bay, N.: A model for friction in metal forming processes. Ann. CIRP **27**(1), 189–194 (1978)
7. Nielsen, C.V., Bay, N.: Review of friction modeling in metal forming processes. J. Mater. Process. Technol. **255**, 234–241 (2018)
8. Fogg, B.: Preliminary study of the influence of stress and deformation in the substrate on junction growth and friction. Proc. Inst. Mech. Eng. Conf. Proc. **182**(11), 152–161 (1967)
9. Sutcliffe, M.P.F.: Surface asperity deformation in metal forming processes. Int. J. Mech. Sci. **30**(11), 847–868 (1988)
10. Wang, Z.G., Yoshikawa, Y., Suzuki, T., Osakada, K.: Determination of friction law in dry metal forming with DLC coated tool. Ann. CIRP **63**(1), 277–280 (2014)
11. Nielsen, C.V., Martins, P.A.F., Bay, N.: Modelling of real area of contact between tool and workpiece in metal forming processes including the influence of subsurface deformation. CIRP Ann. Manuf. Technol. **65**(1), 261–264 (2016)
12. Zwicker, M.F.R., Spangenberg, J., Bay, N., Martins, P.A.F., Nielsen, C.V.: The influence of strain hardening and surface flank angles on asperity flattening under subsurface deformation at low normal pressures. Tribol. Int. **167**, 107416 (2021)
13. Shisode, M., Hazrati, J., Mishra, T., de Rooij, M., van den Boogaard, T.: Evolution of real area of contact due to combined normal load and sub-surface straining in sheet metal. Friction **9**(4), 840–855 (2020). https://doi.org/10.1007/s40544-020-0444-6
14. Zwicker, M., Bay, N., Nielsen, C.V.: A new test for sheet metal asperity flattening under varying subsurface strain conditions. Tribol. Int. **108**, 108249 (2023)
15. Childs, T.H.C.: The persistence of asperities in indentation experiments. Wear **25**, 3–16 (1973)
16. Weidel, S., Engel, U., Merklein, M., Geiger, M.: Basic investigations on boundary lubrication in metal forming processes by in situ observation of the real contact area. Prod. Eng. Res. Devel. **4**(2), 107–114 (2010). https://doi.org/10.1007/s11740-009-0198-5

Concept for the Incorporation of Auxetics as Active Die Faces for Flexible Metal Forming Tools

Peter Frohn-Sörensen(✉), Jonas Reuter, and Bernd Engel

Chair of Forming Technology, Institute of Production Technologies, Breite Straße 11, 57072 Siegen, Germany
Peter.Frohn@Uni-Siegen.de

Abstract. Auxetic materials and cellular structures offer unique mechanical properties with respect to negative Poisson ratios, which essentially leads to elevated shear moduli and delivers superior mechanical resilience under specific mechanic and dynamic loading conditions. In the present study, we provide a literature review about auxetics, their specific properties and the resulting mechanical behavior. Systematically, we present concepts to apply these properties to the elementary functions of adjustable tools in groups of passive and active implementations of auxetic structures into forming tools. Conceptual application clusters are derived with respect to three scales of flexible forming tools: i) smallest adjustments of a surface layer to influence and control the material flow by pressure distribution, ii) intermediate adjustments to comply with manufacturing tolerances (e.g. elastic springback of the product) and iii) large tool surface adjustments by actuated substructures to produce variants. We present protype concepts of an elastically deformable cellular auxetic structure and a non-assembly mechanism with auxetic properties and quantify the benefits and limitations of both concepts. In future, we aim to apply these concepts to demonstrators used in sheet metal forming processes.

Keywords: Additive Manufacturing · Auxetics · Flexible Manufacturing · Lattice Structures

1 Introduction

Manufacturing flexibility has gained attention in the last decades to serve the needs for an individualized, decentralized and on-demand production [1]. Agility might either be achieved by increasing the level and speed of reconfigurability of a manufacturing system or by raising its flexibility level where the first defines how quickly a system might change and the latter shows the possibility of producing variants and largely changing the lot size of a production job [2, 3]. Forming technologies represent a major group of manufacturing systems according to technical standard DIN 8580 and traditionally serve these production margins where largest quantities are required with minimum material wastage [4]. This field of application however comes along with strongly shape-dependent tools with respect to the product geometry and often leads to complete inflexibility with respect to small changes in the product geometry and process fluctuations [5].

K. Mocellin et al. (Eds.): ICTP 2023, LNME, pp. 20–31, 2024.
https://doi.org/10.1007/978-3-031-40920-2_3

In order to overcome this limitation, various methodological approaches have been pursued. For instance, an indistinct tooling is utilized in kinematic forming processes, where product shape is defined by the tool kinematics rather than by tool geometry [6]. This strategy is often found in incremental forming processes [7, 8] and flexible bending methods [9–11]. For rotary draw bending, HEFTRICH et al. present a method to systematically remove those areas of the active tool faces which are not in significant contact with the workpiece during processing, hereby achieving new degrees of freedom (DoF) and, thus, increasing flexibility [12]. Segmentation is yet another approach to open up the geometric restrictions of a tool and, hence, increasing its universality. Amongst the processes in focus, drawing and stretch-drawing has underwent significant research to segmentation concepts with multipoint forming (MPF) [13]. Each of these segments "points" has its own actuator which is capable for a uniaxial adjustment of height and hereby providing a reconfigurable die [14]. Fundamentally, these discretized dies are investigated since the 20th century where the main goal is to achieve a pin design and density of closely packed pins to increase accuracy, smoothness and continuity of the obtained sheet metal product in stamping [15].

In general, transformable tool surfaces might either be achieved by an elastic deformation or the aforementioned discretization in to (partially) independent segments with respect to kinematics. Besides, transformation of tool surfaces might be categorized into active and passive [16] transformation where the passive case always requires a shape determined opposing tool as, for instance, in rubber pad forming [17, 18].

Considering active tool transformation, surfaces might be adjusted at different scales with different effects on the process. To this, the authors have introduced a concept of tool flexibility levels [19], see Fig. 1. In the first flexibility level, a surface adjustment of hundredths to tenths of millimeter enables a varying or a setting up of a defined contact pressure distribution. For instance, this adjustment might be necessary in blank holder contact areas for the purpose of controlling material flow [20, 21]. Surface adjustments of a few tenths of millimeter up to a few millimeters could be useful to compensate springback, either due to material changes or fluctuations, due to an initially compensated tool geometry or due to distinct surface adjustment and a second calibration step. In the third flexibility level, due to surface adjustment of several millimeters up to several centimeters, the adjustment of the whole tool geometry is achieved, which is crucial for the manufacturing of a variety of part geometries with a single tool. While for flexibility level 1 and 2 the adjustment by an actuator induced elastic deformation of a customized tool structure is conceivable, in flexibility level 3 a break-up of the conventionally closed tool surfaces and an independent adjustment of the tool parts is necessary, like it can be seen in MPF.

Nevertheless, the elastic and discretized approaches require any kind of actuation mechanisms below the active surface of the tool to achieve flexibility with respect to manufacturable variants. In the case of MPF, the actuation of pins is often achieved by hydraulic pistons which require an external pressure unit, a very extensive valve assembly and closed-loop control to assure position [22]. Since conventional mechanic actuation systems are often voluminous, rigid or might demand elaborate assemblies, therefore, novel actuation mechanisms and controlling systems are in demand to achieve actively transformable tools. Additive manufacturing (AM) opens up novel paths towards

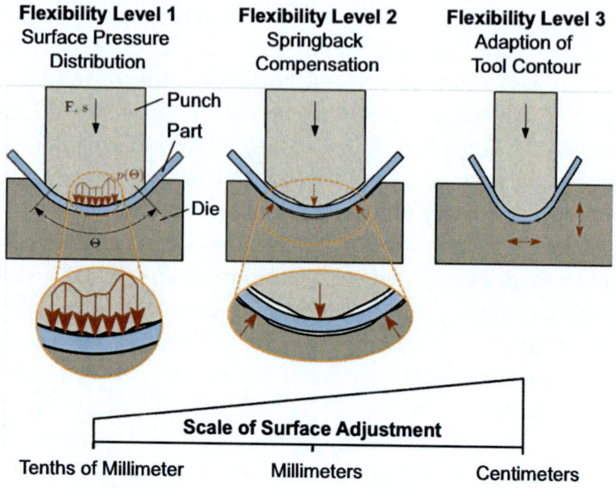

Fig. 1. Conceptual scheme for flexibility levels of adjustable tools for geometric compensations on successive logarithmic scale according to [19].

geometric complexity [23] and non-assembly mechanisms [24, 25] for the design of manufacturing tools. In particular, auxetic structures become manufacturable by AM methods which was possible only in very limited ranges by traditional manufacturing systems due to the intrinsic high geometric complexity of auxetics [26].

2 Functional Principles (and Literature Review) of Auxetics

In principle, auxetics are defined as such materials having a negative Poisson ratio ν [27]. Hence, under elastic deformation, transversal strains would possess the same algebraic sign as the applied principal strain. Practically, if a material is strained, it would expand laterally instead of contracting or, vice versa, would collapse/shrink transversely if compressed. Auxetic materials are usually applied where structures are demanded that provide a combination of high stiffness and strength such as in lightweight applications. From the general relation of elastic coefficients, Eq. (1), it is evident that negative Poisson ratios ν lead to highly increased shear moduli G if materials with the same elastic modulus E are considered:

$$G = \frac{\mathrm{E}}{2(1+\nu)} \tag{1}$$

According to ALDERSON and ALDERSON such a high shear modulus leads to specific advantages in technical application especially if approaching the infinite extreme of $\nu \to -1$: indentation resistance, thermal shock absorption, fracture toughness etc. [27]. Geometrically, auxetic materials offer a number of interesting deformation behaviors in addition to the uniaxial response discussed above, such as forming a dome shape (synclastic behavior) instead of a saddle shape (anticlastic behavior) when a plate is bent.

In nature, a number of materials are known to possess such an auxetic property, e.g. arsenic [28] and cadmium single crystals [29], α-cristobalite [30], iron pyrite [31] and some cubic metal elements [32]. Amongst biological materials, auxetics are found in certain forms of skins [33] and bones [34]. Artificial auxetic materials are being produced and investigated since the 1980s from the material classes of polymers, ceramics, metals and composites [27, 35].

Besides materials, a number of planar or spatial structures might possess auxetic properties. Cellular solids are amongst these and, as auxetic materials, based on the principle of elastic deformation. Specifically, lattice structures are open-celled structures of repeating cellular patterns which consist of struts connected in nodes according to MACONACHIE et al. [36] and might possess auxetic properties depending on the structure [37]. Because the main deformation of cellular solids leads to a proportional lateral deformation instead of anti-proportional, globally, a similar behavior like an auxetic material is achieved which has a negative Poisson ratio. Such cellular solids therefore might be considered as auxetic meta-materials. Similar to auxetic materials, the structures exhibit a very high shear stiffness [38] and therefore possess a number of advantages such as indentation resistance [39], fracture toughness [40], energy dissipation [41] and acoustic absorption [42].

In literature, a number of famous patterns are described to provide auxetic properties, with the so-called re-entrant honeycombs being the most frequently referenced patterns. YANG et al. provide a systematic parametric design with respect to the relevant geometric parameters for such a cell in spatial configuration [43]. Amongst the conclusions they state that the elastic deflection mainly bases on bending and shear loads. Moreover, they demonstrate, that the structure shows little sensitivity against the number of unit cells, if larger or equal to three. Additive manufacturing guidelines are provided for powder-based electron beam melting.

Next to the re-entrant honeycombs, other auxetic patterns are presented in literature. In the chiral honeycomb pattern, ligaments are curled around cylinders [44]. Under load, these ligaments unravel and cause a uniform planar expansion. The two-dimensional structure possesses an improved out of plane compressive strength due to the cylinders and a high out of plane shear strength given by the ligaments [45]. Star-shaped and double-arrow headed structures are both based on the principle of flexing or hinging ribs that open or close an elementary cell of the pattern [46, 47]. A detailed overview of all known two- and three-dimensional patterns investigated in research is reviewed by SAXENA, DAS & CALIUS [48] including further strut-based patterns, additional chiral types, rotating squares, triangles and parallelograms, foil and wire based examples as well as cut-out based auxetics.

Figure 2 provides a graphic summary of all reviewed auxetics and highlights groups with the potential to be manufactured by AM.

According to the review of JOSEPH, MAHESH and HARURSAMPATH [26], the application of additive manufacturing for auxetic structures is highly desirable since AM enables manufacturing of geometric complex parts. Since no known AM of an intrinsically auxetic material has been reported, the AM of auxetics currently involves only cellular solids and mechanisms. AM has been successfully applied in medical treatments by 3D printing tissues and organs [49]. Moreover, medical implants comprising

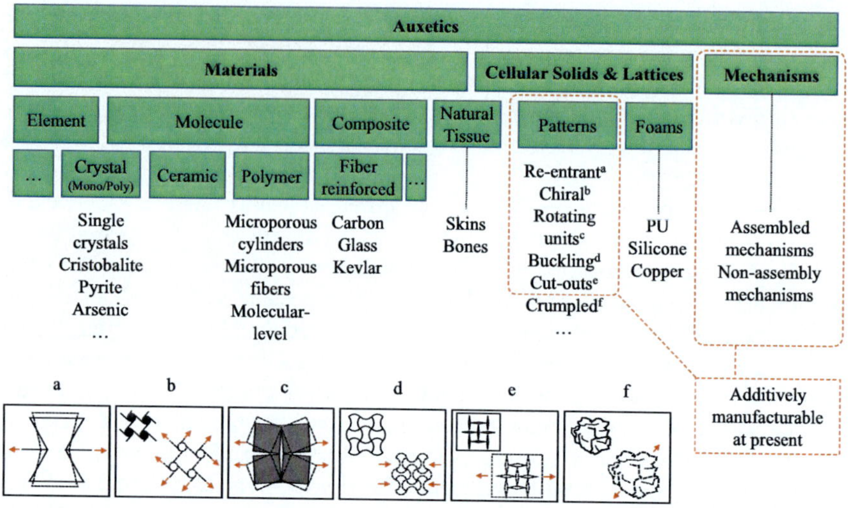

Fig. 2. Systematic overview of the reviewed auxetics, grouped into materials, cellular solids and mechanisms. Highlight of groups currently manufacturable by AM. Exemplary patterns of cellular solids are shown in the bottom for each type of cellular solids. Auxetic properties are obtained since transversal motion follows the main excitation (red arrows) of the patterns.

auxetic cellular structures from medical grade titanium were developed [50, 51] with the adjustability to the changing human body being among the unique advantages of using auxetics in implants. Yet, at present there is no AM method to directly manufacture natural tissue and bone with auxetic properties. In the following chapter, concepts are systematically derived to match the findings of additively manufacturable auxetics with the requirements and elementary functions of industrial manufacturing tools with particular regards towards sheet metal forming dies.

3 Application of Auxetics in Industrial Tooling

To regard the application of auxetics in an industrial tooling, the elementary functions of such a tool need to be defined. Afterwards, these functions are matched to the unique properties and potentials of auxetics of the preceding chapter. The functional analysis system technique, shortly referred to as FAST (-diagram), was developed in 1964 by Charles W. Bytheway as creativity technique [52]. In this methodology, a system is analyzed with respect to its fundamental functions. By asking "how" a specific function is achieved, the next subordinate level of functions is obtained (right direction) while the question "why" a specific function is necessary delivers the superordinate function level (left direction). Doing so, a structure of fundamental functions of a system is obtained with increasing degree of abstraction towards the analyzed system. In addition, permanent functions are listed above the interlinked functions.

In principle, the presented methodology is applicable for any manufacturing tool with changeable shape. With respect to forming, specific functions need to be regarded. Figure 3 shows the functional analysis for a forming tool with transformable surfaces.

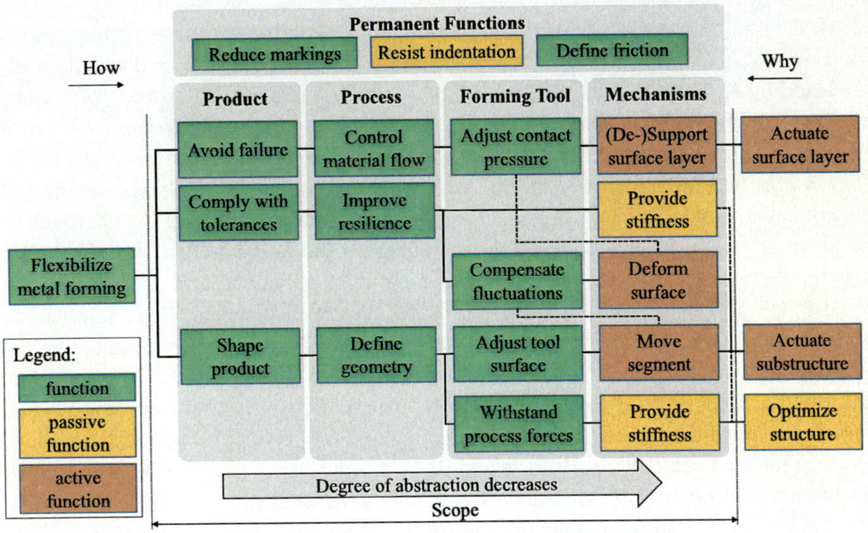

Fig. 3. Functional system analysis technique (FAST) of a forming tool with transformable surfaces. Passive application potentials for auxetics highlighted yellow and active implementation potentials highlighted red.

Within the scope of the system analysis in between both dashed lines, the functions are grouped in columns with decreasing degree of abstraction from left to right. Firstly, on the product level, three elementary functions are identified for flexible forming, i.e., shaping the product, to avoid failure (wrinkles, rupture, geometric collapse) during forming and to comply with tolerances. From these product-related functions, the principal functions of the system are derived on the level of the forming process and, subsequently, on tool level. In the fourth column, concrete application potentials for auxetics are identified on the level of mechanisms. These are grouped into passive and active potentials depending on if a specific mechanical behavior of the structure under load is intended or an active transformation.

Additive manufacturing has paved the way for new structural designs of mechanical components such as forming tools. According to the load collective, structural optimizations in the bulk of the tool become feasible by AM to save material, manufacturing time, energy and weight. By the method of successive volume reduction under stiffness maximization, the authors have demonstrated such an optimization for the additive tooling in a preceding study [18]. Using auxetic patterns might help to improve the stiffness obtained from a cellular substructure in a forming tool application due to the aforementioned beneficial mechanical properties of auxetics, in this context namely high shear stiffness [38], indentation resistance [39, 53], fracture toughness [40] or out of plane compressive strength [45], depending on the respective pattern. Hence, these particular properties might be used in the optimization process of forming tools. Doing so, stiffness optimized, lightweight tools with reduced costs in manufacture and material might be obtained [54] and made by metallic additive manufacture [55, 56].

Based on mechanisms, the application potentials of active auxetic structures derive from the FAST diagram in Fig. 3 according to the flexibility levels of adjustable tools predicted by KUHNHEN et al. [19] in Fig. 1. In detail, an active adjustment of surface pressure (tool flexibility level 1) might be achieved by actuating a (sub-) surface layer of an auxetic cellular structure. Based on a two-dimensional cellular structure, CHEN et al. demonstrate how auxetic metamaterials undergo different deformations when subjected to external stimuli [57] which might act as a principal starting point for the approach of an active auxetic surface layer in metal forming. WEI et al. utilize a chiral auxetic pattern to achieve small deformations by using shape-memory polymers [58].

Besides adjusting a certain surface pressure, larger displacements are required to serve the other two levels of geometrical adjustment. While the compensation of process fluctuations to comply within a given band of tolerance might be achieved by an elastic deformation of the forming tool surface, the geometric adjustment to shape a new product variant exceeds this margin. Therefore, a segmented tool concept is suggested such as multi point forming (MPF). Both levels of tool adjustment require the actuation of a sub-surface structure. If not driven by conventional actuators – such as individual hydraulic pistons in MPF – auxetic cellular structures might be applicable to achieve these adjustments. In contrast to the elastic deformation of such metamaterials, hinge-based mechanisms, inspired by such auxetic geometries, might be applied to achieve larger deflections of active tool surfaces. With respect to the identified groups in Fig. 2, these mechanisms might be assembled in conventional operations or use the non-assembly manufacturing principle known from additive manufacture [24]. Both concepts, the elastically deformable meta-material as well as the hinged mechanism are illustrated in Fig. 4 for the re-entrant sandglass-shaped honeycomb pattern.

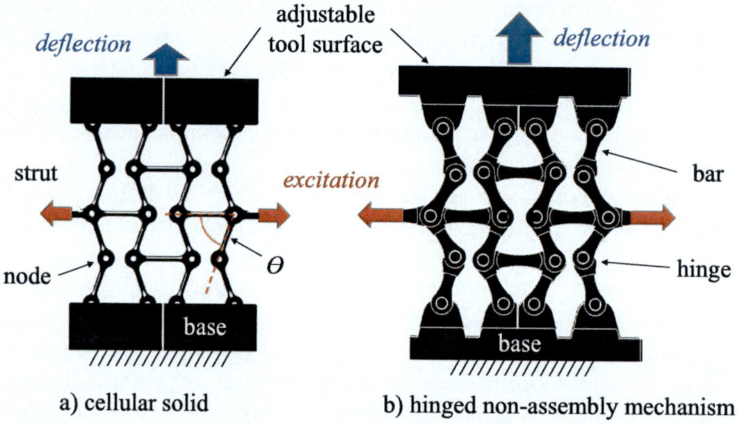

Fig. 4. Concept designs for actuated substructures of geometrical adjustable tools based on a) an elastically deformable cellular solid with mechanical properties of an auxetic metamaterial with the re-entrant angle Ø and b) a hinged non-assembly auxetic mechanism.

Both mechanisms were prototyped by direct polymer additive tooling (DPAT) according to the parameters used in a preceding study on DPAT [59]. In the fused filament fabrication (FFF) AM process, polylactide (PLA) polymer material was used to manufacture both. In qualitative comparison it is worth mentioning, that the concept of a cellular solid (a) offers a well-defined initial position and a zero-backlash deflection without any clearance. However, possible adjustments are always afflicted by the force necessary for elastic deformation of nodes and struts which leads to plastic deformation or failure of the structure at comparable few deflections of the adjustable surface. Hence the cellular solid might be favorable for the intermediate group of flexibility where smaller adjustments help to comply with tolerances. For the cellular solid, the ratio between lateral and vertical deflection is derived from the Poisson ratio of the auxetic structure which relates to the re-entrant angle Ø of the honeycomb according to the parametric study of YANG et al. [43].

In contrast, the concept of a hinged mechanism (b) allows for larger deflections until reaching singularities in the hinges. Ideally no deflection force is added by the mechanism. However, these mechanisms are more challenging to manufacture and possess clearances as given by the gaps in the hinges necessary to avoid fusion. The system is highly overdetermined by the large number of degrees of freedoms and must be guided carefully to restrict it to the desired direction of adjustment. Concludingly, the group of hinged auxetic mechanisms is suggested for the third group of adjustable tool segments. The relation between the lateral displacement and the resulting vertical deflection of the tool surface relates to the changing angle of the interconnected knee levers and thus follows a trigonometric function.

4 Conclusion and Outlook

Since AM enables the manufacture of geometrically complex shaped products [23], it is possible to use AM-manufactured forming tools with specific properties compared to conventional subtractively manufactured tools. These include defined compliance, mass reduction, integration of additional functions such as cooling channels in press hardening tools, creation of installation space for sensors and actuators and the integration of active as well as passive adjustment functions to increase flexibility. Due to their cellular structure, auxetic metamaterials are predestined for AM fabrication [26] and bring about unique mechanical behavior [53] which is reviewed in this paper in detail.

However, some challenges arise for the use of auxetics as load carrying structures in metal forming tools and should be addressed in further work:

- For the use as passive elements, suitable tooling areas must be identified which are expedient for the use of auxetics with regard to their unique elastic behavior.
- When using active elastic elements, the tool structure must be accurately designed and optimized to achieve the defined deformation behavior required for surface adjustments.
- When using active elements based on motion mechanisms, the challenging task of AM manufacturability of hinges must be considered [60].
- Generally, the structures need to be designed for strength and stiffness while being manufacturable by AM (e.g., by Selective Laser Melting). Therefore, manufacturing constraints must be considered in structural optimization.

In conclusion, auxetic metamaterials and auxetic motivated motion mechanisms offer potential applications in metal forming tools. Therefore, a literature review on auxetics, their specific properties and the resulting mechanical behavior is presented. Based on this, a generalized functional system analysis is utilized in the form of a FAST diagram showing conceptual application potentials of auxetics in forming tools. Finally, two design concepts for actuated substructures of geometrically adjustable forming tools based on cellular solids and based on a hinged motion mechanism are prototyped.

Acknowledgment. The authors acknowledge the financial support by the European Union and the Federal Ministry for Economic Affairs and Climate Action in the project SkaLaB (project number 13IK025B).

References

1. Yang, D.Y., et al.: Flexibility in metal forming. CIRP Ann. **67**, 743–765 (2018)
2. Andersen, A.-L., Brunoe, T.D., Nielsen, K.: Reconfigurable manufacturing on multiple levels: literature review and research directions. In: Umeda, S., Nakano, M., Mizuyama, H., Hibino, H., Kiritsis, D., von Cieminski, G. (eds.) APMS 2015. IAICT, vol. 459, pp. 266–273. Springer, Cham (2015). https://doi.org/10.1007/978-3-319-22756-6_33
3. ElMaraghy, H.A.: Flexible and reconfigurable manufacturing systems paradigms. Int. J. Flex. Manuf. Syst. **17**, 261–276 (2005). https://doi.org/10.1007/s10696-006-9028-7
4. DIN 8580:2022–12, Fertigungsverfahren_- Begriffe, Einteilung. https://doi.org/10.31030/3217872
5. Cao, J., et al.: Manufacturing of advanced smart tooling for metal forming. CIRP Ann. **68**, 605–628 (2019)
6. Hirt, G., Ames, J., Bambach, M., Kopp, R., Kopp, R.: Forming strategies and process modelling for CNC incremental sheet forming. CIRP Ann. **53**, 203–206 (2004)
7. Duflou, J.R., et al.: Single point incremental forming: state-of-the-art and prospects. Int. J. Mater. Form. **11**(6), 743–773 (2017). https://doi.org/10.1007/s12289-017-1387-y
8. Vorkov, V., Aerens, R., Vandepitte, D., Duflou, J.R.: Analytical prediction of large radius bending by circular approximation. J. Manuf. Sci. Eng. (2018). https://doi.org/10.1115/1.4041496
9. Chatti, S., Hermes, M., Tekkaya, A., Kleiner, M.: The new TSS bending process: 3D bending of profiles with arbitrary cross-sections. CIRP Ann. Manuf. Technol. **59**, 315–318 (2010)
10. Groth, S.: Methode zur Produktplanung beim Freiformbiegen. Shaker Verlag GmbH, Dissertation, Aachen (2020)
11. Groth, S., Frohn, P., Engel, B.: Product planning system for manufacture-oriented modeling of freeform bend tubes produced by three-roll-push-bending. Procedia Manuf. **34**, 10–18 (2019)
12. Heftrich, C., Steinheimer, R., Engel, B.: Rotary-draw-bending using tools with reduced geometries. Procedia Manuf. **15**, 804–811 (2018)
13. Zareh-Desari, B., Davoodi, B., Vedaei-Sabegh, A.: Investigation of deep drawing concept of multi-point forming process in terms of prevalent defects. Int. J. Mater. Form. **10**(2), 193–203 (2015). https://doi.org/10.1007/s12289-015-1268-1
14. Abosaf, M., Essa, K., Alghawail, A., Tolipov, A., Su, S., Pham, D.: Optimisation of multi-point forming process parameters. Int. J. Adv. Manuf. Technol. **92**(5–8), 1849–1859 (2017). https://doi.org/10.1007/s00170-017-0155-y
15. Walczyk, D.F., Hardt, D.E.: Design and analysis of reconfigurable discrete dies for sheet metal forming. J. Manuf. Syst. **17**, 436–454 (1998)

16. Elend, L-E.: Einsatz elastischer Niederhaltersysteme zur Erweiterung der Prozeßgrenzen beim Tiefziehen (2001). https://doi.org/10.15488/5808
17. Deladi, E.L.: Static friction in rubber-metal contacts with application to rubber pad forming processes (2006)
18. Frohn-Sörensen, P., Geueke, M., Belay, T.T., Kuhnhen, C., Manns, M., Engel, B.: 3D printed prototyping tools for flexible sheet metal drawing (2020). https://doi.org/10.31224/osf.io/w8rgb
19. Kuhnhen, C., Knoche, J., Reuter, J., Hassan Al-Maeeni, S.S., Engel, B.: Hybrid tool design for a bending machine. Procedia CIRP **99**, 370–374 (2021)
20. Doege, E., Elend, L.-E.: Design and application of pliable blank holder systems for the optimization of process conditions in sheet metal forming. J. Mater. Process. Technol. **111**, 182–187 (2001)
21. Yagami, T., Manabe, K., Yang, M., Koyama, H.: Intelligent sheet stamping process using segment blankholder modules. J. Mater. Process. Technol. **155–156**, 2099–2105 (2004)
22. Siegert, K., Hohnhaus, J., Wagner, S.: Combination of hydraulic multipoint cushion system and segment-elastic blankholders. SAE Trans. **107**, 35–43 (1998)
23. Gibson, I., Rosen, D., Stucker, B., Khorasani, M.: Design for Additive Manufacturing. In: Gibson, I., Rosen, D., Stucker, B., Khorasani, M., (eds).: Additive Manufacturing Technologies. Springer International Publishing, Cham, pp 555–607 (2021). https://doi.org/10.1007/978-1-4419-1120-9_11
24. Lussenburg, K., Sakes, A., Breedveld, P.: Design of non-assembly mechanisms: A state-of-the-art review. Addit. Manuf. **39**, 101846 (2021)
25. Reintjes, C., Reuter, J., Hartisch, M., Lorenz, U., Engel, B.: Towards CAD-Based mathematical optimization for additive manufacturing – Designing forming tools for tool-bound bending. In: Pelz, P.F., Groche, P. (eds.) ICUME 2021. LNME, pp. 12–22. Springer, Cham (2021). https://doi.org/10.1007/978-3-030-77256-7_2
26. Joseph, A., Mahesh, V., Harursampath, D.: On the application of additive manufacturing methods for auxetic structures: a review. Adv. Manuf. **9**(3), 342–368 (2021). https://doi.org/10.1007/s40436-021-00357-y
27. Alderson, A., Alderson, K.L.: Auxetic materials. Proc. Inst. Mech. Eng. Part G J. Aerosp. Eng. **221**, 565–575 (2007)
28. Gunton, D.J., Saunders, G.A.: The Young's modulus and Poisson's ratio of arsenic, antimony and bismuth. J. Mater. Sci. **7**, 1061–1068 (1972). https://doi.org/10.1007/BF00550070
29. Li, Y.: The anisotropic behavior of Poisson's ratio Young's modulus, and shear modulus in hexagonal materials. Phys. Status Solidi(a) **38**, 171–175 (1976)
30. Yeganeh-Haeri, A., Weidner, D.J., Parise, J.B.: Elasticity of α-cristobalite: A silicon dioxide with a negative poisson's ratio. Science **257**, 650–652 (1992)
31. Love, A.E.H.: A treatise on the mathematical theory of elasticity. Cambridge University Press (2013)
32. Baughman, R.H., Shacklette, J.M., Zakhidov, A.A., Stafström, S.: Negative Poisson's ratios as a common feature of cubic metals. Nature **392**, 362–365 (1998)
33. Veronda, D.R., Westmann, R.A.: Mechanical characterization of skin—Finite deformations. J. Biomech. **3**, 111–124 (1970)
34. Williams, J.L., Lewis, J.L.: Properties and an anisotropic model of cancellous bone from the proximal tibial epiphysis. J. Biomech. Eng. **104**, 50–56 (1982)
35. Lakes, R.: Foam structures with a negative Poisson's ratio. Science **235**, 1038–1040 (1987)
36. Maconachie, T., et al.: SLM lattice structures: Properties, performance, applications and challenges. Mater. Des. **183**, 108137 (2019)
37. Wang, S., Wang, J., Xu, Y., Zhang, W., Zhu, J.: Compressive behavior and energy absorption of polymeric lattice structures made by additive manufacturing. Front. Mech. Eng. **15**(2), 319–327 (2019). https://doi.org/10.1007/s11465-019-0549-7

38. Lakes, R.S., Elms, K.: Indentability of conventional and negative Poisson's ratio foams. J. Compos. Mater. **27**, 1193–1202 (1993)
39. Alderson, K.L., Simkins, V.R., Coenen, V.L., Davies, P.J., Alderson, A., Evans, K.E.: How to make auxetic fibre reinforced composites. Phys. Status Solidi(b) **242**, 509–518 (2005)
40. Bianchi, M., Scarpa, F.L., Smith, C.W.: Stiffness and energy dissipation in polyurethane auxetic foams. J. Mater. Sci. **43**(17), 5851–5860 (2008). https://doi.org/10.1007/s10853-008-2841-5
41. Bezazi, A., Scarpa, F.: Mechanical behaviour of conventional and negative Poisson's ratio thermoplastic polyurethane foams under compressive cyclic loading. Int. J. Fatigue **29**, 922–930 (2007)
42. Scarpa, F., Tomlin, P.J.: On the transverse shear modulus of negative Poisson's ratio honeycomb structures. Fatigue Fract. Eng. Mater. Struct. **23**, 717–720 (2000)
43. Yang, L., Harrysson, O., West, H., Cormier, D.: Mechanical properties of 3D re-entrant honeycomb auxetic structures realized via additive manufacturing. Int. J. Solids Struct. **69–70**, 475–490 (2015)
44. Prall, D., Lakes, R.S.: Properties of a chiral honeycomb with a poisson's ratio of — 1. Int. J. Mech. Sci. **39**, 305–314 (1997)
45. Spadoni, A., Ruzzene, M., Scarpa, F.: Global and local linear buckling behavior of a chiral cellular structure. Phys. Status Solidi (b) **242**, 695–709 (2005)
46. Theocaris, P.S., Stavroulakis, G.E., Panagiotopoulos, P.D.: Negative Poisson's ratios in composites with star-shaped inclusions: a numerical homogenization approach. Arch. Appl. Mech. **67**, 274–286 (1997). https://doi.org/10.1007/s004190050117
47. Larsen, U.D., Signund, O., Bouwsta, S.: Design and fabrication of compliant micromechanisms and structures with negative Poisson's ratio. J. Microelectromech. Syst. **6**, 99–106 (1997)
48. Saxena, K.K., Das, R., Calius, E.P.: Three decades of auxetics research – materials with negative Poisson's ratio: A review. Adv. Eng. Mater. **18**, 1847–1870 (2016)
49. Melchels, F.P.W., Domingos, M.A.N., Klein, T.J., Malda, J., Bartolo, P.J., Hutmacher, D.W.: Additive manufacturing of tissues and organs. Prog. Polym. Sci. **37**, 1079–1104 (2012)
50. Kolken, H.M.A., Janbaz, S., Leeflang, S.M.A., Lietaert, K., Weinans, H.H., Zadpoor, A.A.: Rationally designed meta-implants: a combination of auxetic and conventional meta-biomaterials. Mater. Horiz. **5**(1), 28–35 (2018)
51. Maciejewska, J., Jopek, H.: An analysis of the auxetic cranioplasty implant. Vibrations in Physical Systems (2020). https://doi.org/10.21008/j.0860-6897.2020.3.15
52. Bytheway, C.W.: FAST creativity and innovation: rapidly improving processes. Product Development and Solving Complex Problems J, Ross Publishing (2007)
53. Bronder, S., et al.: Hybrid auxetic structures: structural optimization and mechanical characterization. Adv. Eng. Mater. **23**, 2001393 (2021)
54. Burkart, M., Liewald, M., Wied, J., Todzy, T., Hartmann, M., Müller, M.: Optimization of a part holder design considering dynamic loads during return stroke of tool and ram. Procedia Manuf. **47**, 861–866 (2020)
55. Turk, C., Zunko, H., Aumayr, C., Leitner, H., Kapp, M.: Advances in maraging steels for additive manufacturing. Berg Huettenmaenn Monatsh **164**, 112–116 (2019)
56. Bajaj, P., Hariharan, A., Kini, A., Kürnsteiner, P., Raabe, D., Jägle, E.A.: Steels in additive manufacturing: A review of their microstructure and properties. Mater. Sci. Eng. A **772**, 138633 (2020)
57. Chen, Y., Ye, W., Xu, R., Sun, Y., Feng, J., Sareh, P.: A programmable auxetic metamaterial with tunable crystal symmetry. Int. J. Mech. Sci. **249**, 108249 (2023)
58. Wei, Y.-L., Yang, Q.-S., Tao, R.: SMP-based chiral auxetic mechanical metamaterial with tunable bandgap function. Int. J. Mech. Sci. **195**, 106267 (2021)

59. Frohn-Sörensen, P., et al.: Design for 3D printed tools: Mechanical material properties for direct polymer additive tooling. Polymers **14**, 1694 (2022)
60. Boschetto, A., Bottini, L.: Manufacturability of non-assembly joints fabricated in AlSi10Mg by selective laser melting. J. Manuf. Process. **37**, 425–437 (2019)

Investigation of Measures for Material Flow Control for a Backward Extrusion of Geared Components from Coil

Miriam Leicht(✉), Johannes Henneberg, and Marion Merklein

Institute of Manufacturing Technology (LFT), Friedrich-Alexander-Universität Erlangen-Nürnberg, Egerlandstraße 13, 91058 Erlangen, Germany
miriam.leicht@fau.de

Abstract. Increasing ecological and economic demands are setting new requirements for industrial manufacturing processes and components. The use of functional integration as a strategy of lightweight design offers the possibility to meet these requirements. However, a high functional integration implies an increasing component complexity, which challenges conventional forming processes to their limits regarding formability and efficiency. The novel process class of sheet-bulk metal forming (SBMF) offers the opportunity to combine the advantages of sheet metal and bulk forming by applying bulk forming operations to sheet metal semi-finished products. SBMF from coil offers further economic and ecological benefits through high output rates, short process times by eliminating costly handling tasks and a greater material efficiency. However, forming from coil also poses coil-specific challenges such as anisotropic material flow and reduced geometric part accuracy. Against this background, in this study, a backward extrusion process from coil for forming geared functional components is investigated on a high-speed press. The objective of this study is to analyze measures for a targeted material flow control in order to avoid unequal material flow and to improve part accuracy. Therefore, the mild deep drawing steel DC04 with an initial sheet thickness of 2.0 mm is analyzed in a combined numerical-experimental approach. Within these investigations, the influences of coil width, friction and feed width are analyzed and evaluated based on component and process parameters. Finally, measures for a purposive material flow are derived.

Keywords: Sheet-bulk metal forming · Extrusion · Cold forming

1 Introduction

In the course of advancing climate change and its effects on society and industry, the European Union has committed itself to achieving climate neutrality by 2050 as part of the Paris Climate Agreement [1]. This requires innovative solutions, particularly in the fields of industry and mobility, in order to achieve this global objective [1]. In the transport sector, the use of lightweight construction technologies, such as the application of functionally integrated components, is a particularly important measure. One key

K. Mocellin et al. (Eds.): ICTP 2023, LNME, pp. 32–42, 2024.
https://doi.org/10.1007/978-3-031-40920-2_4

approach in the industrial sector is the substitution of existing process technologies with more material- and energy-efficient processes [1]. Against this background, the new process class of sheet bulk metal forming offers the possibility of combining process-specific advantages of sheet metal forming and bulk forming by applying bulk forming operations to sheet metal, whereby complicated component geometries with locally adapted part properties can be produced [2]. In industrial application, SBMF is currently mainly used for pre-cut blanks, whose suitability for economical series production of high volumes is limited. This is caused by the part handling in multi-stage process chains, which is limited to an output quantities of 10 to 28 components per minute [3]. In the case of forming from coil in progressive die, output rates of more than 100 parts per minute are producible, which offers high economic potential. However, SBMF from coil is hardly established in industry due to low part accuracy. The occurrence of anisotropic material flow has been identified as a cause of unequal component forming [4]. For the extrusion of rotationally symmetrical functional elements, measures for a targeted material flow control have already been investigated in order to improve the part accuracy [5]. For a broad industrial application of SBMF from coil, the existing process knowledge must be transferred to more complicated geometries.

2 Objectives and Methodology

The central objective of this contribution is to obtain a profound understanding for the extrusion of geared components from coil. For this purpose, measures for a targeted material flow control are developed within the scope of this study. By influencing the material flow during forming, the anisotropic material flow, which is considered a coil-specific challenge, is to be reduced for improving the part accuracy. A combined experimental-numerical approach is used for the selection and investigation of measures. As a workpiece-side measure, the influence of the coil width is investigated. On the process side, the influence of the feed is evaluated. First, numerical investigations are carried out to evaluate the influence of the measures on the component geometry and the maximum process force using a validated simulation model. In the next step, recommendations for material flow control for SBMF from coil are derived. Finally, numerically identified effects are verified by experiments.

3 Process Layout

In this chapter, the process setup and kinematic as well as the resulting workpiece geometry of the backward extrusion process are described. Furthermore, the workpiece materials, the lubrication and the design of the virtual process model are explained.

3.1 Process Setup

The tool system consists of three active parts (Fig. 1a): A blankholder, an internally geared forming die and a counterholder. In this process a coil with a thickness of 2 mm and a width of 30 mm is used, which is formed by a high-speed press BSTA 510-125B2 of Bruderer AG with an output rate of 100 strokes per minute. The parts are formed in a single stroke with a punch penetration depth of 1 mm. The process kinematics are structured in four steps. In the first step, the sheet metal is clamped with a constant blankholder force of 130 kN. In the second step, the forming is carried out by the punch. After deforming the part, the blankholder is returned to its initial position and the workpiece is transported out of the forming zone with an automated feed of 20 mm. The resulting component geometry, which represents an abstracted functional element in the form of a pin with four external gearings, is illustrated in Fig. 1b.

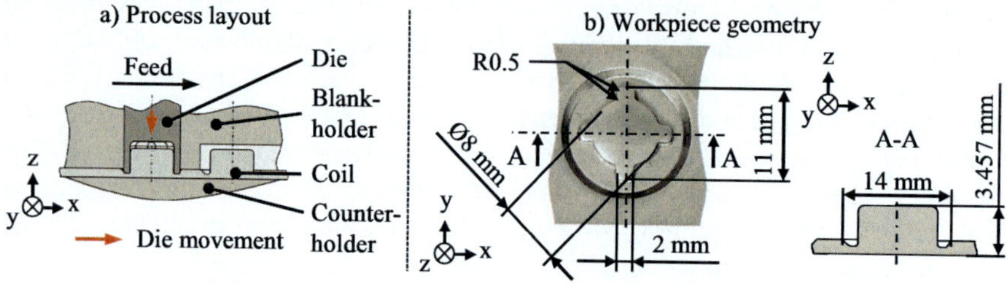

Fig. 1. Process layout (a) and workpiece geometry (b).

3.2 Materials and Lubrication

The workpiece material used is the mild deep-drawing steel DC04, which is frequently used in the automotive industry for structurally relevant components in car body constructions due to its advantageous forming properties [6]. DC04 shows a low anisotropic behavior [7], whereby a clear increase in yield stress and a decrease in vertical anisotropy can be seen at 45° to the rolling direction. In Table 1 the initial yield stress and the vertical anisotropy of DC04 with a sheet thickness of 2 mm is shown.

Table 1. Material characteristics of DC04 with a sheet thickness of 2 mm according to [8].

DC04, 2 mm	Angle	Initial yield stress k_{f0}	Vertical anisotropy r
	0°	169.09 ± 0.50 MPa	2.17 ± 0.02
	45°	181.23 ± 0.94 MPa	1.51 ± 0.02
	90°	175.75 ± 0.33 MPa	2.53 ± 0.15

The flow behavior of the used material in rolling direction is shown in Fig. 2a. The experimental values were determined in a layer compression test following DIN 50106 due to the similar load condition in the investigated process and due to the limited width of the coil.

To represent higher forming degrees, the extrapolation was carried out using the Hockett-Sherby approach. To reduce friction in the experiment, the lubricant Beruforge 150 DL from Carl Bechem GmbH, a water-based lubricant with a high solid wax content, was used.

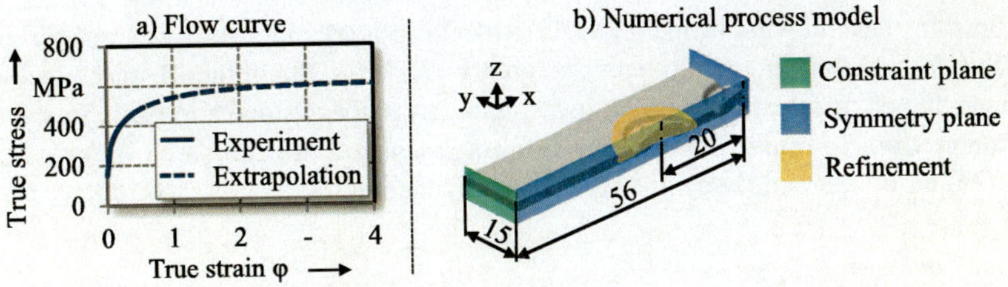

Fig. 2. Flow curve of DC04 (a) and setup of the numerical process model (b).

3.3 Numerical Process Model

The numerical process model was built using the simulation software Simufact.forming 14.01. The simulation model (Fig. 2b) was described in detail and validated by experimental results in [9] according to the recommendations of Tekkaya [10]. Hexahedral elements with an element size of 0.5 mm and a cylindrical refinement box in the area of the forming zone were used to mesh the semi-finished product. The tools are modeled rigid. The experimentally identified rotational symmetry of the components in single stroke [9] proves that no unequal component geometry for DC04 results due to possible anisotropic material properties. Therefore, an isotropic material model is used. In addition, the symmetry of the process along the X-axis is applied to reduce the computational time, resulting in the investigation of a 180°-segmented model. Furthermore, a global friction factor of 0.1 is used, which was determined in previous research work for the present tribologic system [11].

4 Numerical Analysis of Measures for Material Flow Control

Former process analysis identified an underfilling of the gearing elements and a low component accuracy as a reason for the coil-specific, asymmetrical material flow [9]. In this chapter, approaches to solve these challenges are investigated numerically to evaluate their potential as well as the effectiveness on the part geometry and the maximum force. For a better comparability of the forming of the cavity geometry as well as the gearing, relative measures of the radius r and the distance d, which relate the actual measurement to the target measure, are used for evaluation. The pin height is a parameter for the evaluation of the material flow into the bulk forming zone. The target parameters depicted in Fig. 3 are evaluated in + X, -X and Y directions in order to identify effects of the anisotropic material flow.

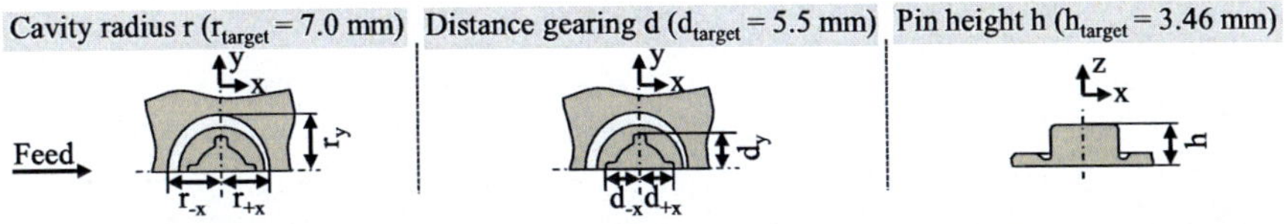

Fig. 3. Target parameters for the evaluation of the component geometry.

Forming several parts from coil leads to local pre-hardening from previous forming, which affects the accuracy of the part geometry [5]. Thus, the influence of the feed, the distance in between formed parts, is investigated as process-sided approach. It is varied in a range from 15 mm to 30 mm. The resulting measures are compared with results of the forming of one part (w/o). In Fig. 4, the influence of the feed is shown.

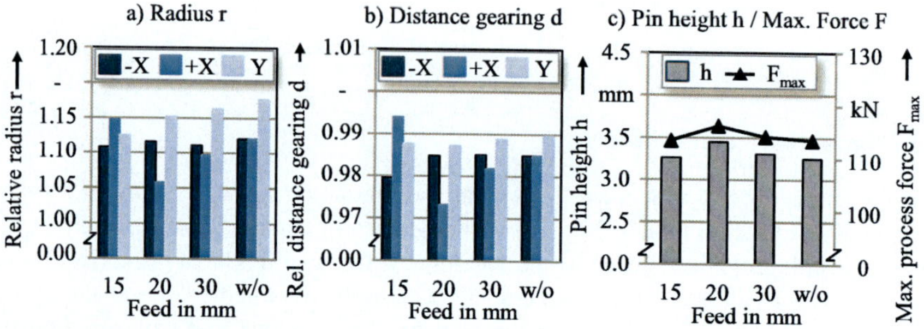

Fig. 4. Influence of the feed on part geometry – radius r (a), distance of the gearing d (b), and pin height h – and maximum force F_{max} (c).

The forming of the cavity geometry based on the measurements of r (Fig. 4a) shows no significant influence of the feed in −X direction due to relative differences of less than 1%. In +X direction, the radius is lowest for a feed of 20 mm and increases with an increasing feed by 4%. This behavior is caused by the higher distance between the components that decreases the influence of local strain hardening of the previous forming. This increases the outward material flow in +X direction. In Y direction, there is a continuous increase of the radius with the increase of the feed of 9%, whereby the greatest radius is achieved without feed with a value of 1.17. Furthermore, more isotropic component forming occurs, since the difference of radii in +X and −X directions decreases with increasing feed. At a feed of 30 mm, there is only a difference of 1.3% between + X and –X direction. A similar behavior can be observed for the forming of the gearing based on the measurement of d (Fig. 4b), whereby an underfilling of the gearing occurs regardless of the feed and direction. In addition, there is no clear influence of the feed on the pin height h (Fig. 4c). The highest pin with a value of 3.45 mm is achieved with a feed of 20 mm. The maximum process force F_{max} is analyzed as a process-side parameter (Fig. 4c). The maximum force is highest for a feed of 20 mm with 116 kN and decreases slightly to 114 kN with an increase and decrease of feed. The lower process force with a higher feed can be attributed to the decreasing influence of material pre-hardening of the coil, which is shown by the true strain distribution for different feeds in Fig. 5.

By increasing the feed, lower pre-hardening is introduced into the forming zone of the subsequent component. At a feed of 30 mm there is no pre-hardening, see Fig. 5. Lower pre-hardening of the coil in +X direction of the previous formed part causes more material to flow outward, resulting in a lower pin height. The lower material volume in the part center leads to lower true strain in the area of the cavity (Fig. 5) and thus to a decrease in local pre-hardening. For a more detailed analysis of the material flow, the

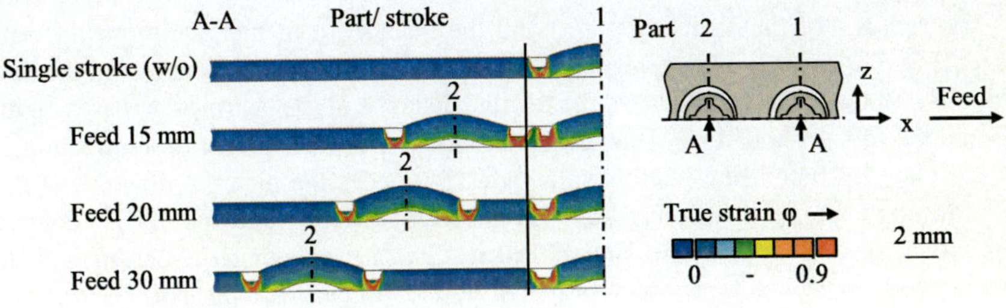

Fig. 5. Numerical prediction of the distribution of the true strain.

radial displacement for a feed of 20 mm and 30 mm are shown in Fig. 6. The material flow during coil forming has been described in [9].

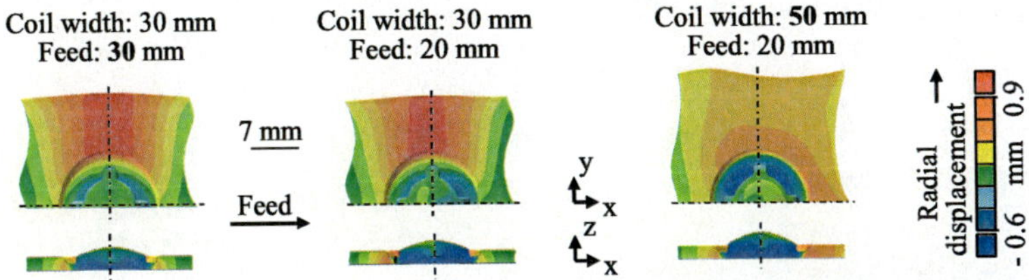

Fig. 6. Influence of the feed and the coil width on the radial displacement.

With a feed of 20 mm (Fig. 6), there is no symmetry of the material flow with respect to the Y- and Z-axis because of the pre-hardening of the coil in feed direction. By increasing the feed up to 30 mm, a symmetry of material flow with respect to the Y- and Z-axis can be identified, reducing the unequal forming in and against the feed direction. In [9], the sheet geometry was identified as a cause of the asymmetric material flow due to the smaller distance from the part center to the edges of the coil perpendicular than parallel to the feed direction. Thus, the coil width is varied from 20 mm to 50 mm as a workpiece-sided measure. Figure 7 shows the resulting part and process properties.

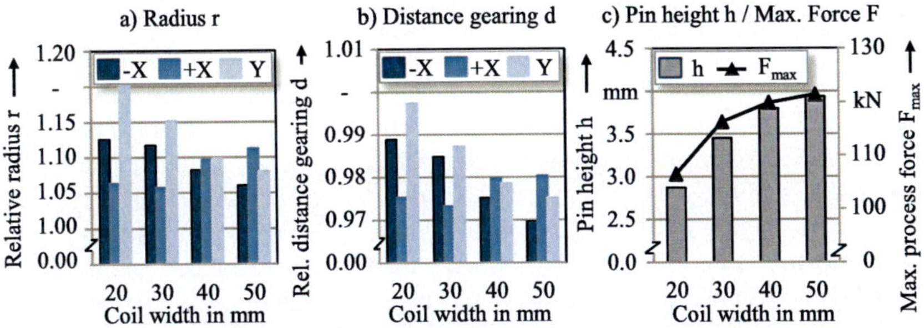

Fig. 7. Influence of the coil width on part geometry – radius r (a), distance of the gearing d (b), and pin height h – and maximum force F_{max} (c).

The radius r of the cavity (Fig. 7a) and the distance d of the gearing to the part center (Fig. 7b) show similar trends in +X, −X and Y directions depending on the coil width. By increasing the coil width, the relative distance of the gearing decreases about 3% and the radius about 14%. Due to the additional material volume in Y direction, a greater material supporting effect results, which impedes the material flow out of the bulk forming zone. Furthermore, there is an increase in friction due to the larger contact area between tool and workpiece. Both effects result in a greater material flow in the bulk forming zone in Y direction, which is shown by the larger blue color portion in Fig. 6 in the area of the forming zone and the yellow color portion in the sheet area for coil width of 50 mm compared to the coil width of 30 mm. In contrast, the higher material flow into the bulk forming zone leads to a lower die filling of the functional element in Y direction with a relative value of 0.968. This is a result of the higher material flow velocity into the die cavity, which is depicted in Fig. 8. The higher flow velocity causes a larger radius deflection of the material flow in the inlet area of the cavity, which reduces the dimensional accuracy of the gear element.

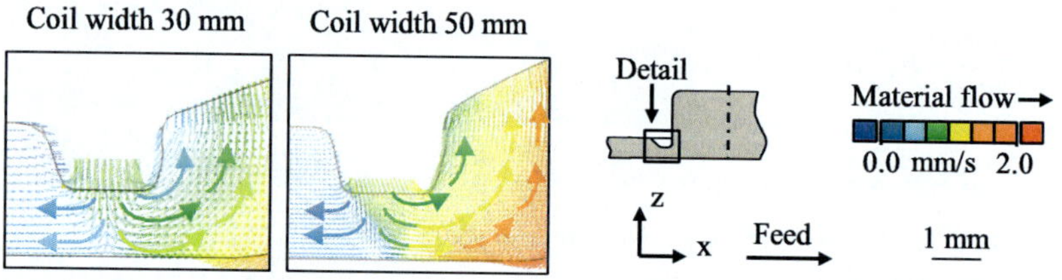

Fig. 8. Numerical prediction of material flow velocity for the coil width of 30 and 50 mm.

In addition, both geometrical parameters r (Fig. 7a) and d (Fig. 7b), also decrease in −X direction about 6% and 2% due to the increased material flow into the cavity. The reduced material flow out of the bulk forming zone with the higher coil width is confirmed by increase of the pin height (Fig. 7c) of 37%. Consequently, an adaptation of the coil width results in an improve of the accuracy of the cavity radii and a greater pin height, but in a lower accuracy of the gearing. By increasing the coil width, the maximum forming force (Fig. 7c) increases continuously from 106.5 kN to 121.4 kN about 14%. This is caused by the greater material flow into the forming zone, which leads to higher true strain in the cavity and pin area (Fig. 5).

Based on the numerical investigations, conclusions are drawn for the effectiveness of the measures investigated. By increasing the coil width, a decrease of the unequal material flow parallel and perpendicular to the feed direction occurs. It is a result of the greater supporting effect of the coil perpendicular to the feed direction. In addition to the higher supporting effect of the additional material, friction also increases due to the larger contact area between tool and workpiece, resulting in a higher volume of material flowing radially into the forming zone (Fig. 6), improving the cavity geometry and increasing the pin height as shown in Fig. 7. However, an underfilling of the gearing occurs simultaneously, because of the higher material flow velocity (Fig. 8). Consequently, there is a contrary functional relationship between the form filling of the gearing, of the pin as well as of the cavity. Finally, the challenge of unequal part forming, which is caused by the coil geometry [9], is counteracted. A greater coil width is recommended with regard to part forming, but it must be taken into account that the material efficiency of the process is negatively affected. By the adaption of the feed, no significant compensation of the asymmetric material flow parallel and perpendicular to the feed direction is possible. However, only a more equal forming in and against the feed direction is achieved, since the influence of local pre-hardening due to the previous forming is reduced. Thus, a higher feed should be selected to ensure that no pre-hardening is introduced into the forming area of the subsequent part, but material efficiency of the process is reduced.

5 Experimental Verification of Suitable Measures

The potential of the process- and workpiece-side measures for material flow control were numerically demonstrated. In the following, the numerical findings are verified by means of experiments. The part geometry was measured optically using the VHX-100K digital microscope from Keyence Deutschland GmbH with a 10x magnification. The influence of the feed as a process-side measure is evaluated by two process variants with a feed of 20 and 40 mm. In addition, a comparison is made with the forming without feed (w/o). Parameter for the evaluation of the part geometry and the maximum force are depicted in Fig. 9.

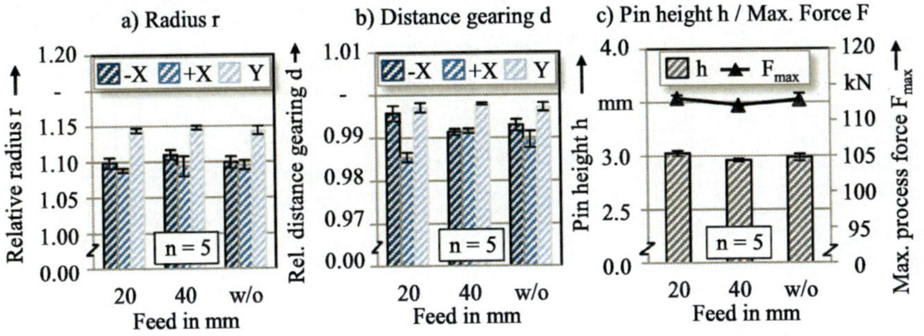

Fig. 9. Experimental influence of the feed on part geometry – radius r (a), distance of the gearing d (b), and pin height h – and maximum force F_{max} (c).

In the experiment, the adjustment of the feed has no significant influence on the forming of the radii r in all directions, as the differences are within the standard deviations. In contrast, by increasing the feed, the unequal form filling of the gearing in + X and –X direction is completely compensated. In +X and –X direction a value of 0.991 is achieved. For the feed of 20 mm, the difference of the values in +X and -X direction is a result of the local pre-hardening of the material in +X direction due to the previous forming, see Fig. 10.

As shown by the distribution of micro hardness in Fig. 10 comparing the forming of one and several components, the pre-hardening of the coil in feed direction ensures a higher resistance against material flow out of the bulk forming zone. This inhibits the flow of material radially outwards and increases the flow of material radially inwards.

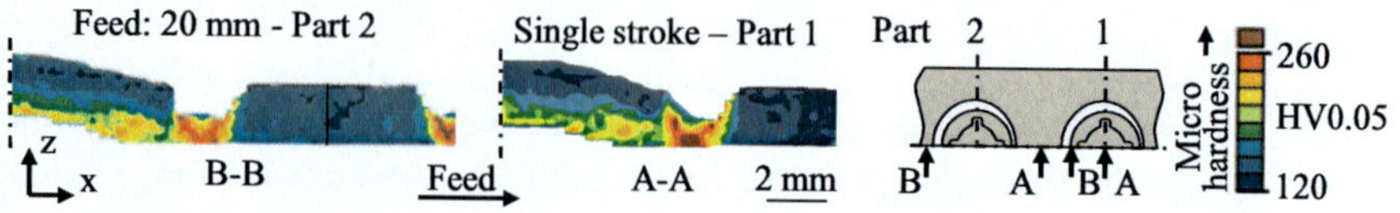

Fig. 10. Distribution of micro hardness – single stroke and multi stroke with a feed of 20 mm.

However, a deviation in X and Y direction still remains. In addition, the pin height as well as the maximum process force (Fig. 9) indicate no significant dependence on the feed with differences of less than 2%. By increasing the feed up to a value of 40 mm, an equal forming of the gearing in and against the feed direction is achieved, because there is no pre-hardening of the coil in feed direction due to the lager distance between the forming zone and the previous formed part (Fig. 5). Thus, similar material properties are present in the area of the forming zone in and opposite to the feed direction. Consequently, in the experiment mainly the forming of the gearing in and against the feed direction is influenced by the feed. To verify the influence of the coil width, an experimental study with coil widths of 20 and 30 mm is carried out (Fig. 11).

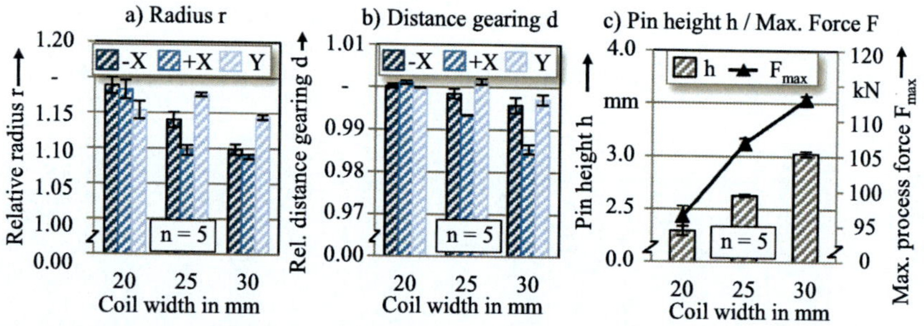

Fig. 11. Experimental influence of the coil width on part geometry – radius r (a), distance of the gearing d (b), and pin height h – and maximum force F_{max} (c).

The cavity radii r in + X and –X direction show a decrease by 8% as the coil width increases and approach the nominal value, thus improving dimensional accuracy. Hence, by adapting the coil width, the too large cavity radii are overall reduced. In contrast to the numerical results, no representative trend of the radii in Y direction can be observed. In contrast to the numerical results, a complete and uniform forming of the gearing is achieved experimentally with the coil width of 20 mm. This difference is caused by changed tribological conditions, which occur due to the experimentally greater expansion of the coil as a result of the low supporting effect of the narrow sheet metal strip. With the increase of the coil width, a higher and unequal underfilling occurs in the +X and –X direction with a difference of 1%. As in the numerical study, the pin height rises about 32% and the maximum force about 17%. Hence, as in numerical investigation, the increase of the coil width results in a greater material flow into the part center increasing the pin height and improving the dimensional accuracy of the cavity. In contrast to the feed, the coil width influences the forming of the component parallel and perpendicular to the feed direction. This measure has, however, a negative effect on the form filling of the functional elements, which requires a compromise between the geometrical accuracy of the cavity or pin height and the gearing. Based on the experimental results, it can be summarized that a high dimensional accuracy of the components, was achieved with a coil width of 20 mm or a feed of 40 mm.

6 Summary and Outlook

In this research work, measures for material flow control were investigated in order to meet the characteristic challenges of SMBF from coil including non-uniform and incomplete die filling due to unequal material flow. To address these challenges, the efficiency of material flow control measures coil width and feed were investigated numerically and verified experimentally. It was shown that the adaption of the coil width reduces the asymmetrical material flow in and perpendicular to the feed direction, resulting in uniform part forming, which proved that the coil geometry has a great influence on the dimensional accuracy. Simultaneously, a contrary interaction between the material flow into the bulk forming zone and the form filling of the gearing was determined. With a higher material flow into the bulk forming zone, the forming of the cavity and the pin is improved, but an underfilling of the gearing is caused by the higher material flow velocity.

Further research work is required to identify additional suitable measures for material flow control. In this context, local friction adjustment on the tool as well as adaptation of the geometry of the punch and the coil like sheet thickness and coil layout offer great potential. Furthermore, the transferability of these results to geared products of various sizes and to processes with other main material flow directions should be evaluated.

Acknowledgement. This work was funded by the German Research Foundation (DFG) within the scope of the project 'Fundamental investigations on the production of functional components from coil' (project number 471835637).

References

1. Bundesministerium für Umwelt, Naturschutz, Bau und Reaktorsicherheit (2016) Klimaschutzplan 2050, Klimaschutzpolitische Grundsätze und Ziele der Bundesregierung. Druck- und Verlagshaus Zarbock GmbH & Co. KG, Frankfurt
2. Merklein, M., et al.: Bulk forming of sheet metal. CIRP Ann. **61**, 725–745 (2012)
3. Altan, T.: Metal Forming Handbook. Springer, Berlin (1998)
4. Tajul, L., Maeno, T., Mori, K.-I.: Successive forging of long plate having inclined cross-section. Procedia Eng. **81**, 2361–2366 (2014)
5. Henneberg, J., Merklein, M.: Measures for controlling the material flow when extruding sheet-bulk metal forming parts from coil. Manuf. Rev. **7**, 36 (2020)
6. Birkert, A.R., Haage, S., Straub, M.: Umformtechnische Herstellung komplexer Karosserieteile: Auslegung von Ziehanlagen. Springer, Berlin/Heidelberg (2013)
7. Ghennai, W., Boussaid, O., Bendjama, H., Haddag, B., Nouari, M.: Experimental and numerical study of DC04 sheet metal behaviour—plastic anisotropy identification and application to deep drawing. Int. J. Adv. Manuf. Technol. **100**(1–4), 361–371 (2018). https://doi.org/10.1007/s00170-018-2700-8

8. Hildenbrand, P.: Entwicklung einer Methodik zur Herstellung von Tailored Blanks mit definierten Halbzeugeigenschaften durch einen Taumelprozess. PhD-Thesis. FAU (2019)
9. Leicht, M., Henneberg, J., Merklein, M.: Numerical and experimental investigation of a backward extrusion process for forming geared components from coil. In: Proceedings of the 26th International ESAFORM Conference on Material Forming (accepted, 2023)
10. Tekkaya, A.E.: A guide for validation of FE-simulations in bulk metal forming. Arab. J. Sci. Eng. **30**, 113–136 (2005)
11. Gröbel, D.: Herstellung von Nebenformelementen unterschiedlicher Geometrie an Blechen mittels Fließpressverfahren der Blechmassivumformung. PhD-Thesis. FAU (2018)

Research on the 5A06 Aluminum Alloy Thin-Walled Elbow in Cold Stamping Condition

Xiaomin Huang[1], Ben Guan[1,2,3](✉), and Yong Zang[1,2,3]

[1] School of Mechanical Engineering, University of Science and Technology Beijing, Beijing 100083, China
b20200250@xs.ustb.edu.cn, guanben@ust.edu.cn

[2] Beijing Key Laboratory of Lightweight Metal Forming, Beijing 100083, China

[3] Shunde Innovation School, University of Science and Technology Beijing, Beijing 528399, Guangdong, China

Abstract. 5A06 aluminum alloy exhibits low density, high strength, and excellent corrosion resistance, which is widely used in areas of the aerospace, aviation, missile, ship and other fields. However, the defects, such as springback and thinning will occur during the process of stamping. In this research, the formability of 5A06 aluminum alloy thin-walled elbow is studied by experiments and FE simulation. Thinning and springback of 5A06 aluminum alloy thin-walled elbow is analyzed when the BHF is 10 kN, 15 kN and 20 kN, and the friction coefficient is 0.15 and 0.3, respectively. The result shows that the maximum thinning rate increases with increasing of BHF and friction coefficient. Thinning occurs on the short edge and reaches the maximum value around the curved edge. When BHF and friction coefficient are 10 kN and 0.15, respectively, the maximum thinning rate is 5.22%. A large BHF contributes to control springback. When BHF is 20 kN, the maximum springback is 3.01 mm. The simulation results exhibit good consistency with the experiment results.

Keywords: Thin-walled elbow · Springback · Thinning

1 Introduction

5xxx series aluminum alloy exhibits low density and excellent corrosion resistance, which is widely used in the aerospace, aviation, missile, ship and other fields [1]. Hot stamping [2], cold stamping [3] and superplastic forming [4] are often used to produce aluminum alloy parts. Zhou et al. [5] analyzed the influence of processing parameters on springback during hot stamping. Guo et al. [6] investigated the springback of arched aluminum alloy gusset, and the result showed that the springback increased with decreasing of elastic modulus. Yakovtseva et al. [7] analyzed the effect of microstructure parameters and deformation regime on superplastic deformation of aluminum alloy. The above research showed that the formability of materials could be improved by hot stamping and superplastic forming, while, the above two forming methods would consume a large amount of time and energy. However, the formability of the alloy at room temperature is

K. Mocellin et al. (Eds.): ICTP 2023, LNME, pp. 43–50, 2024.
https://doi.org/10.1007/978-3-031-40920-2_5

relatively imperfect, and the defects of wrinkling [8], springback [9], thinning and even crack [10] were prone to occur during the stamping process. The current research on cold stamping mainly focused on steel and magnesium alloy [11, 12], and there was no systematic research on aluminum alloy.

In the study, the forming mechanism of defects such as springback and thinning is analyzed, and the influence of BHF and friction coefficient on defects is discussed.

2 Materials and Experiments

2.1 Material

In the test, the material used is a 5A06-O aluminum alloy with a thickness of 1.2 mm. The chemical composition of the alloy is shown in Table 1.

Table 1. Chemical composition of 5A06-O aluminum alloy (wt.%).

Mg	Si	Cu	Zn	Mn	Ti	Be	Fe	Al
6.2	≤ 0.4	≤ 0.1	≤ 0.2	0.6	0.06	0.002	0.3	Base

2.2 Uniaxial Tensile Test

Uniaxial tensile tests are conducted on a universal material testing machine to obtain the performance of 5A06-O aluminum. The specimen with a gauge distance of 32 mm is designed according to GB/T228.1-2010, which would be cut along the rolling direction at 0°, 45°, and 90°. The dimensions of the specimen, and force-distance curves are shown in Fig. 1. Each experiment is repeat three times to ensure accuracy of result.

Lankford ration of 5A06 aluminum can be calculated by the following formula:

$$r_\alpha = \frac{d\varepsilon_w^p}{d\varepsilon_T^p} = -\frac{d\varepsilon_w^p}{d\varepsilon_L^p + d\varepsilon_w^p}$$

where $d\varepsilon_w^p$, $d\varepsilon_T^p$ and $d\varepsilon_L^p$ represent the stress of material in the direction of width, thickness and length, respectively.

The mechanical properties of 5A06 aluminum obtained through experiments are shown in Table 2. The 5A06 used in this article does not exhibit significant anisotropy, the Mises yield function is used in this article.

2.3 Stamping Tests

The forming process of thin-walled elbow includes four main steps: blanking, stamping, cutting and welding, which are shown as Fig. 2.

The formability tests of thin-walled elbow are carried out on a 60 ton hydraulic press. The stamping die sets are shown in Fig. 3. The tool set consists of four parts: die,

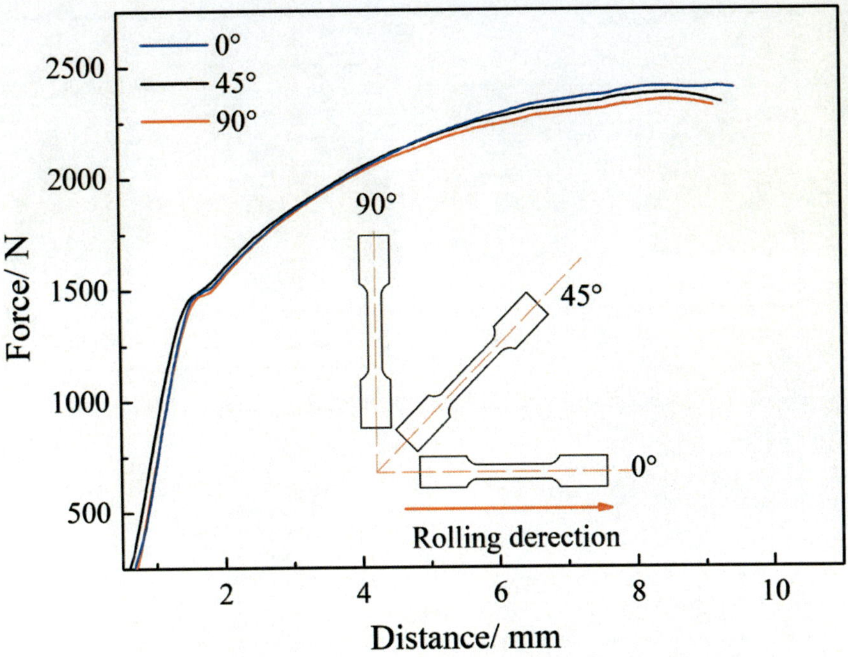

Fig. 1. The force-distance curves of 5A06.

Table 2. The mechanical properties of 5A06 aluminum

E/GPa	σ_s/MPa			σ_b/MPa			Lankford ration		
	σ_{s0}	σ_{s45}	σ_{s90}	σ_{b0}	σ_{b45}	σ_{b90}	r_0	r_{45}	r_{90}
68.9	198.97	197.65	189.37	324.17	321.54	319.61	1.03	0.99	1.11

blank holder, punch, and gas springs. Different blank holder force (BHF) are provided by controlling the number of gas springs used. First, the sheet is taken on the blank holder. Second, the die moves downward to make the sheet contact with the plane of die under the action of hydraulic press. Third, the die, sheet and the blank holder move down together, and the sheet deforms into the thin-walled elbow. Finally, the constraint of the deformed part is removed, the formed specimen goes on springback. The friction of the contact surface is controlled by whether to apply lubrication. The coefficient of friction is 0.15 and 0.3, the BHF is set as 10 kN, 15 kN and 20 kN, respectively.

2.4 Simulation Model

The FE model of thin-walled elbow was established by the finite element simulation software Abaqus, which is shown in Fig. 4. The material of tools was AISI-H13. The tools and the sheet are mashed by C3D10M element and S4R shell element, respectively,

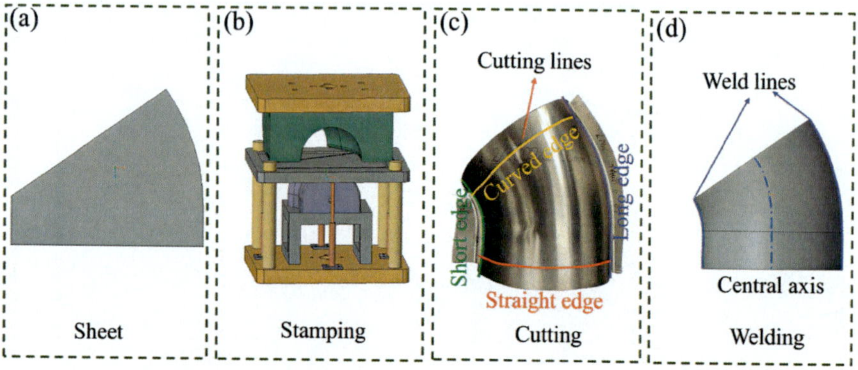

Fig. 2. The forming process: (a) blanking, (b) stamping, (c) cutting and (d) welding.

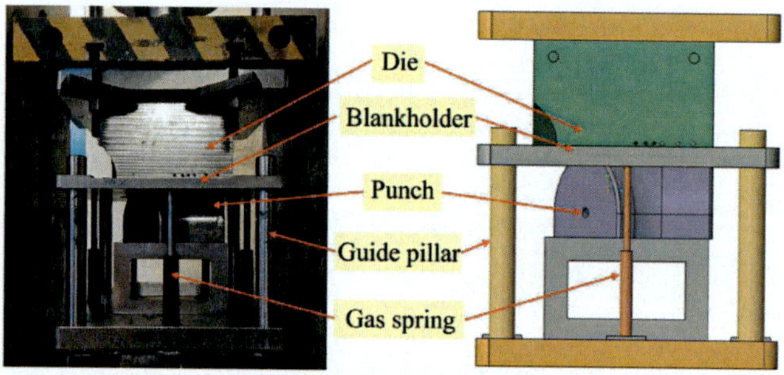

Fig. 3. The schematic diagram of the stamping die sets

and the element size of sheet is 0.5 mm, the total number shell element is 18096. Coulomb friction model is set in the contact between sheet and tools.

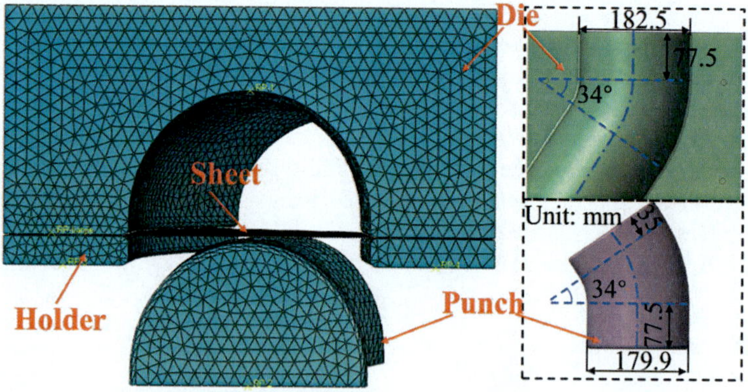

Fig. 4. Finite element model of stamping.

3 Results and Discussion

3.1 Thickness Variation

Excessive thinning affects the integral performance of the formed specimen. Therefore, it is significant to select the appropriate parameters of stamping according to the thickness distribution. The influence of stamping parameters on thinning is shown in Fig. 5, when the BHF and friction coefficient are 15 kN and 0.15, respectively. According to the results of experiment and simulation, thinning occurs on the short edge and reaches the maximum value around the curved edge, which is shown as Fig. 5(a), (b). The reason for this phenomenon is that during the stamping process, the sheet is mainly fed from the short edge with the action of BHF and punch. So the strain state of the material near the short edge is tension-tension, which leads to thinning. However, as for the material near the long edge, which is in a state of tension-compression, and the local thickness of this section is thickened and has a little change.

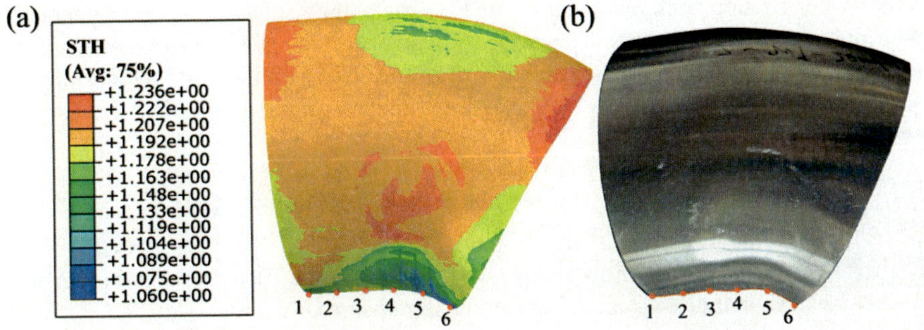

Fig. 5. The comparation of experiment and simulation: (a) simulation result, (b) experimental result.

In order to accurately measure the thickness along the short edge of the thin-walled elbow, six different points are selected along the short edge to quantify the difference of thinning, the position of points is shown in Fig. 6(a), (b). When friction coefficient is 0.15,the variation of thickness along the short edge of the thin-walled elbow under different BHF is shown as Fig. 6. Through the observation of Fig. 6(a), it is found that the thinning of specimen at different positions is different, and the maximum thinning occurs near the bending edge, and a small BHF is beneficial to control the maximum thinning rate of the specimen. When BHF is 10 kN, the maximum thinning rate of the specimen is minimum, which is 5.22%. When BHF is 20 kN, the minimum thickness of the specimen is 1.045 mm, that is, the maximum thinning rate is 12.91%. Under different BHF, the comparison of maximum thinning between experiment and simulation is shown in Fig. 6(b), it can be seen that the simulation results and the experiment have high consistence.

During the stamping, the lubrication state between the tool set and the sheet has a great influence on the thickness of the formed specimen. In this study, friction coefficient between the sheet and the tool set is controlled by applying butter (2#) on the surface

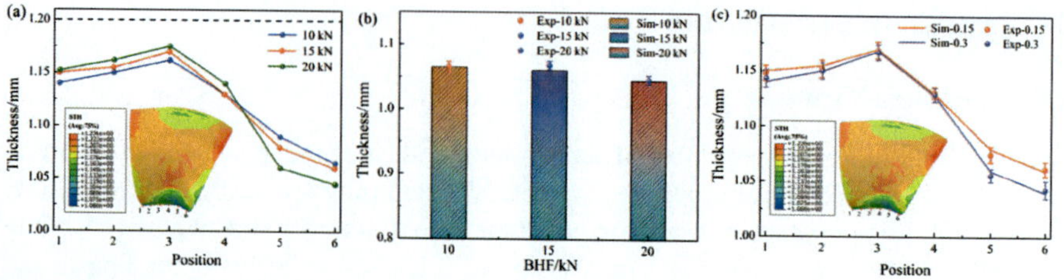

Fig. 6. The variation of thickness of short edge: (a) the variation of thickness under different BHF through simulation, (b) comparison of minimum thickness between experiment and simulation, (c) the influence of friction coefficient on thickness

of the tool set. When applying butter, friction coefficient is 0.15, otherwise it is 0.3 [3, 14]. When the BHF is 10 kN, the influence of the friction coefficient on the thickness of the formed specimen is shown in Fig. 6(c). A small friction coefficient is beneficial to control the thinning rate of the specimen, and the maximum thinning rate is up to is 13.3% when friction coefficient is 0.3. The FE simulation results are in good agreement with the experimental results. The reason for this phenomenon is that when friction coefficient is large, the movement resistance of the sheet is relatively high during the stamping process, which makes the metal sheet undergoes greater plastic deformation, resulting in more serious thinning.

3.2 Springback

Springback seriously affect the forming quality of specimen. The reason for springback is that the residual stress generating during the stamping releases with the removal of constraints, which leads to a dimensional deviation between the formed specimen and the target specimen. The thin-walled elbow studied in this paper has a complex shape, and the springback at different positions is different, which is different to measure springback adopting conventional method. Therefore, this paper selects a representative section-straight edge (cross section) to analyze the springback of thin-walled elbow, and the deviation of springback is applied to quantitatively describe the effect of BHF on springback, which are shown as Fig. 7. It can be seen from the Fig. 7 that there are differences in springback at different positions of the section. The maximum deviation of springback concentrates in the edge area of the straight edge, and a large BHF is beneficial to control springback. When the BHF is 10 kN, the maximum deviation of springback is 7.5 mm; when the BHF increases to 20 kN, it decreases to 3.01 mm, which indicates that the deviation of springback decreased by 59.9%. A larger BHF is beneficial to reduce the springback of the specimen. When the blankholder force is 15 kN, springback of simulation are in good agreement with that of the experiments.

The deviation of springback increases with increasing friction coefficient, however, when friction coefficient is large, the surface of the formed specimen has scratches, which seriously affects the quality of the formed specimen. In this section, the influence of friction coefficient on the springback is not studied in detail.

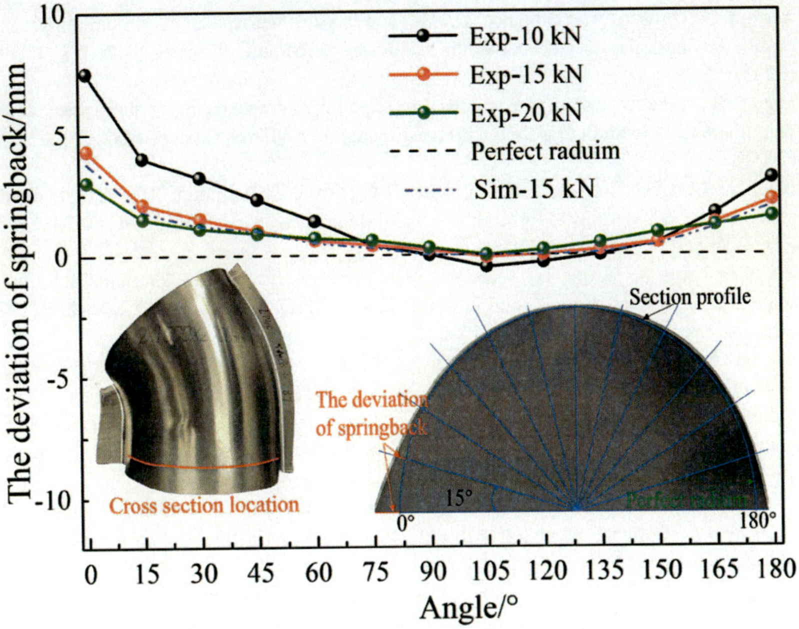

Fig. 7. The deviation of springback under different BHF

4 Conclusions

The formability of 5A06 aluminum alloy thin-walled elbow is studied by experiment and FE simulation at room temperature. The influence of BHF and coefficient of friction on thickness and springback is analyzed. The following conclusions are drawn.

(1) The FE model can well reflect the stamping process and defects of the thin-walled elbow.
(2) There is a significant thinning in the short edge of the thin-walled elbow, especially near the curved edge. The thickness change of the thin-walled elbow can be effectively controlled by reducing the BHF and the friction coefficient.
(3) The springback of different regions of the thin-walled elbow after stamping is quite different due to the asymmetry of specimen. A larger BHF and friction coefficient are beneficial to reduce the springback of the specimen.

Acknowledgements. This work is supported by the Innovation Talent Training Fund of Beijing University of Science and Technology.

References

1. Duan, T., et al.: Long-term localized corrosion behaviors of 5A06 aluminum alloys exposed in the natural deep-sea environment of South China Sea. J. Market. Res. **20**(2022), 4597–4607 (2022)

2. Atxaga, G., Arroyo, A., Canflanca, B.: Hot stamping of aerospace aluminium alloys: automotive technologies for the aeronautics industry. J. Manuf. Process, **81**(2022), 817–827 (2022)
3. Huang, X., Guan, B., Zang, Y., Wang, B.: Investigation of defect behavior during the stamping of a thin-walled semicircular shell with bending angle. J. Manuf. Process. **87**(2023), 231–244 (2023)
4. Brien, M.J.O., von Bremen, H.F., Furukawa, M., Horita, Z., Langdon, T.G.: A finite element analysis of the superplastic forming of an aluminum alloy processed by ECAP. Mater. Sci. Eng. A **456**(2007), 236–242 (2007)
5. Zhou, J., Yang, X., Wang, B., Xiao, W.: Springback prediction of 7075 aluminum alloy V-shaped parts in cold and hot stamping. Int. J. Adv. Manuf. Technol. **119**, 203–216 (2021). https://doi.org/10.1007/s00170-021-08204-x
6. Guo, X., Xu, H., Zeng, Q., Pet, T.: Springback characteristics of arched aluminum alloy gusset plate after stamping forming. Thin. Wall. Struct. **159**(2021), 107294 (2021)
7. Yakovtseva, O.A., Sitkina, M.N., Kotov, A.D., Rofman, O.V., Mikhaylovskaya, A.V.: Experimental study of the superplastic deformation mechanisms of high-strength aluminum-based alloy. Mater. Sci. Eng., A **788**(2020), 139639 (2020)
8. Yuan, S., Cui, X., Wang, X.: Investigation into wrinkling behavior of thin-walled 5A02 aluminum alloy tubes under internal and external pressure. Int J Mech Sci **92**, 245–258 (2015)
9. Liu, X., Cao, J., Chai, X., Liu, J., Zhao, R., Kong, N.: Investigation of forming parameters on springback for ultra high strength steel considering Young's modulus variation in cold roll forming. J. Manuf. Process. **29,** 289–297 (2017)
10. Wang, Y., et al.: Infuence of blank holder type on drawability of 5182-O aluminum sheet at room temperature. T Nonferr. Metal. Soc. **26**, 1251–1258 (2016)
11. Chang, Y., et al.: Prediction of bending springback of the medium-Mn steel considering elastic modulus attenuation. J. Manuf. Process. **67**, 345–355 (2021)
12. Wang, Z., Gu, R., Chen, S., Wang, W., Wei, X.: Effect of upper-die temperature on the formability of AZ31B magnesium alloy sheet in stamping. J. Mater. Process. Tech. **257**, 180–190 (2018)
13. Yuan, B.: Research on forming process of aluminum alloy skin for high speed EMU. Jilin University (2020)

Optimization of Stamping Process Parameters for Small Curved Beam Parts of Rail Train Based on GA-PSO-BP Algorithm

Hongchao Ji[1,2](✉), Mengmeng Li[1], Ran Yao[1], and Weichi Pei[1]

[1] College of Mechanical Engineering, North China University of Science and Technology, Tangshan 063210, China
jihongchao@ncst.edu.cn

[2] School of Mechanical Engineering, University of Science and Technology Beijing, Beijing 100083, China

Abstract. 6005A aluminum alloy is a medium-strength alloy widely used in the aerospace industry and throughout railway applications. The forming limit curve of 6005A aluminum alloy at 350–450 °C is obtained by forming limit test and the microstructure was analyzed. Particle swarm optimization (PSO)-BP neural network combined with genetic algorithm optimization method is used to obtain the best hot stamping process parameters for small curved beam parts of high-speed trains, and used to optimize the hot stamping forming parameters of the 6005A aluminum alloy track train small curved beam parts. The optimization results from PSO-BP neural network and genetic algorithm are verified by finite element simulation analysis and forming limit diagram. The results show that the deformation temperature has a great influence on the forming limit of aluminum alloy. The optimal process parameters for hot stamping of 6005A aluminum alloy small curved beam parts were determined: blank holder force is 4.051KN, stamping speed is 141 mm/s, friction coefficient is 0.05, and maximum thinning rate is 21.2%.

Keywords: 6005A aluminum alloy · BP neural network · Particle swarm optimization (PSO) · Genetic algorithm · Small curved beam · Forming limit

1 Introduction

6005A aluminum alloy is a medium strength alloy, widely used in panel components in the aerospace industry and throughout railway applications [1]. Small curved beam parts are important structural parts in the rail vehicle body, and are usually produced in batches by means of stamping. In order to ensure the forming quality of small curved beams, it is necessary to reasonably determine the forming process. Different stamping process parameters make the quality of aluminum alloy stamping distinct. Therefore, studying the stamping performance of aluminum alloys under different process parameters is of great significance to production [2].

K. Mocellin et al. (Eds.): ICTP 2023, LNME, pp. 51–58, 2024.
https://doi.org/10.1007/978-3-031-40920-2_6

The finite element method is used to simulate the forming process of stamping parts, which can predict the wrinkle, fracture and other trends of stamping parts as well as the shape of the parts after forming. It is helpful for designers to investigate the influence of different process parameters on forming quality, provides an important basis for correcting process parameters, which greatly reduces time and cost [3]. Therefore, this research mainly uses the finite element method to numerically simulate the hot stamping process of 6005A small curved beam parts, and combines optimization algorithms to obtain the best parameters.

2 Materials and Experiments

2.1 Materials

The experimental material is 6005A aluminum alloy, and its chemical composition is shown in Table 1 [4].

Table 1. Chemical composition of 6005A

Element	Si	Mg	Fe	Cu	Mn	Cr	Zn	Ti	Other
wt.%	0.0–0.9	0.4–0.6	≤0.35	≤0.1	≤0.1	≤0.1	≤0.1	≤0.1	0.15

2.2 High Temperature Tensile Test

The thermal simulation high temperature tensile test of 6005A aluminum alloy is carried out on the Gleebl-3800 thermal simulation testing machine. The tensile sample adopts a typical rectangular section, the sample size is shown in Fig. 1. According to the deformation temperature and strain rate range of 6005A aluminum alloy, the high temperature tensile test plan of aluminum alloy is formulated, under different conditions of deformation temperature of 350, 400 and 450 °C and strain rate of 0.01, 0.1 and 1 s^{-1}, the specimen is stretched and deformed until it breaks. After the deformation, the sample is quickly air-cooled to room temperature.

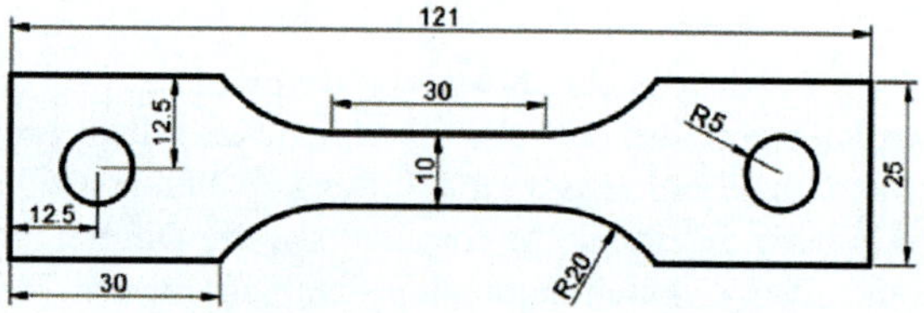

Fig. 1. The shape and size of the test piece

2.3 Forming Limit Diagram

The hot stamping experiment of 6005A aluminum alloy was carried out on the EC600H sheet forming test machine. And the heating process of the material was carried out on the HD-30KW high frequency electric induction heating machine. When obtaining the forming limit of 6005A aluminum alloy, five specimens with different widths were selected to simulate the forming limit under different strain paths. Taking into account the loss of temperature, the heating temperature is set 10 °C higher than the experimental temperature. The speed of the punch in this experiment is 100 mm/s, and the blank holder force is 20KN. The forming limit tests at 350, 400, and 450 °C forming temperatures were carried out.

2.4 Numerical Simulation of Small Curved Beam Forming Process

In order to reduce the assembly workload when establishing the finite element model, a gap of 1.1 times the thickness of the material is generally left between the molds. In the small bending beam simulation model, the mold gap is 2.2 mm, and the mold fillet is 5 mm. Figure 2 shows the finite element assembly model of the small bending beam forming simulation.

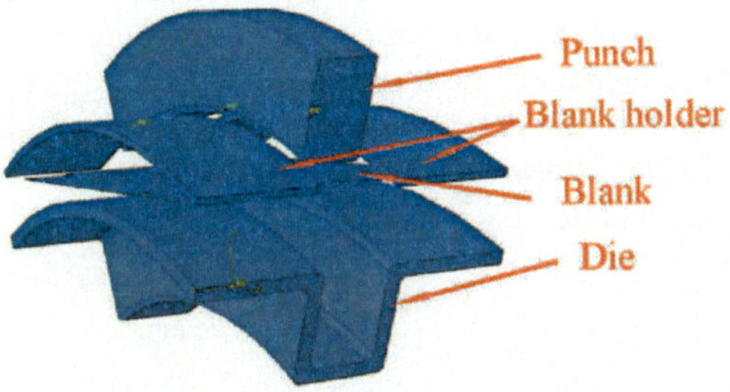

Fig. 2. Simulation assembly model of small bending beam forming

For the convenience of analysis, the optimized variable range is divided into five levels, as shown in Table 2.

Table 2. Orthogonal factors level table

Level	Blank holder force (KN)	Punching speed(mm/s)	Friction coefficient
1	4	80	0.05
2	5.5	100	0.075
3	7	120	0.1
4	8.5	140	0.125
5	10	160	0.15

2.5 Genetic Algorithm Parameter Optimization

Particle swarm optimization (PSO) [5] is used to optimize the BP neural network, thereby improving the stability and accuracy of the BP neural network [6]. Genetic Algorithm (GA) is an algorithm that can be globally optimized. It is widely used in various fields.

The process of parameter optimization of genetic algorithm based on Matlab software. The process parameters that need to be optimized are blank holder force, stamping speed and friction coefficient.

The single-point crossover method is used to reorganize the selected individuals. The crossover probability is set to 0.6, the mutation probability is set to 0.01, and the maximum number of iterations are set to 300 generations. After the maximum number of iterations is reached, the optimal solution obtained by the genetic algorithm is acquired.

3 Results and Discussion

3.1 The Stress-Strain Curves

The true stress-strain curves of the 6005A aluminum alloy obtained at the deformation temperatures of 350 °C, 400 °C and 450 °C, and the strain rates of 0.01 s^{-1}, 0.1 s^{-1} and 1 s^{-1} are shown in Fig. 3. It can be seen from the figure that the flow stress of the alloy is greatly affected by the deformation temperature and strain rate. At the same strain rate, as the temperature increases, the true stress decreases significantly. At the same deformation temperature, as the deformation rate increases, the stress level increases, which shows that the 6005A aluminum alloy is sensitive to the strain rate. It can also be seen that 6005A aluminum alloy has better plasticity at 400 °C.

3.2 Forming Limit Diagram

Figure 4 is the forming limit diagram of 6005A aluminum alloy sheet at 350–450 °C. It can be seen from the figure that the deformation temperature has a great influence on the forming limit of aluminum alloy; The temperature is between 350 and 400 °C, with the increase of deformation temperature, the forming limit rises significantly, but when the temperature is 450 °C, the plasticity decreases, and the forming limit decreases instead. This phenomenon is consistent with the results of high-temperature tensile experiments. Aluminum alloy has the lowest forming limit strain value at 450 °C, and its plasticity is poor; when the deformation temperature is 400 °C, the forming limit strain value is the highest, and its plasticity is also the best.

3.3 Fracture Morphology Analysis

The fracture morphology of the 6005A aluminum alloy sample with a width of 30mm at different temperatures was observed by scanning electron microscope, as shown in Fig. 5. The existence of small dimples at the fracture indicates that void nucleation has always been dominant in the ductile fracture mechanism of 6005A aluminum alloy.

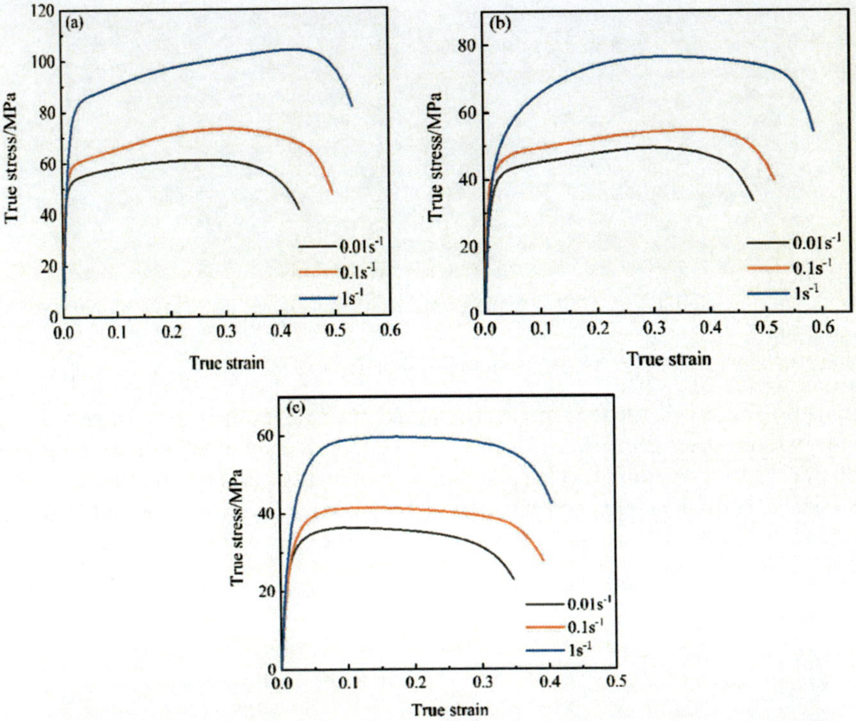

Fig. 3. The true stress-strain curves of 6005A aluminum alloy: (a) 350 °C; (b) 400 °C; (c) 450 °C

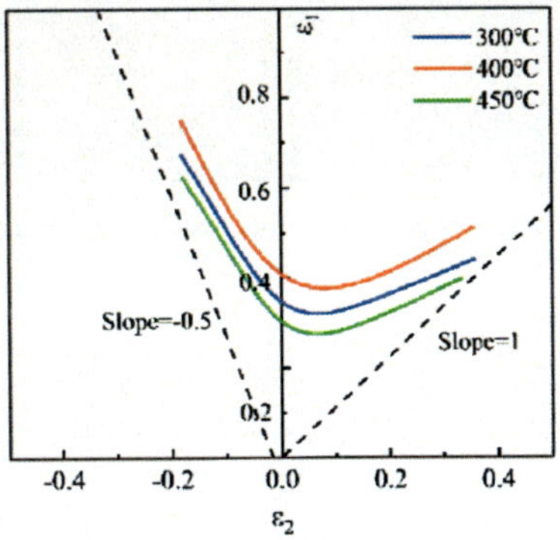

Fig. 4. The forming limit diagram of 6005A aluminum alloy

The fracture morphology of 6005A aluminum alloy samples of different widths at 400°C was observed by scanning electron microscope. The fracture morphologies under two different strain paths from single tensile stress state to equal double tensile state are shown, respectively. As shown in Fig. 6, Fig. 6(a) is the fracture morphology of the sample with w = 30 mm. It can be seen that the dimples are elongated in the transverse

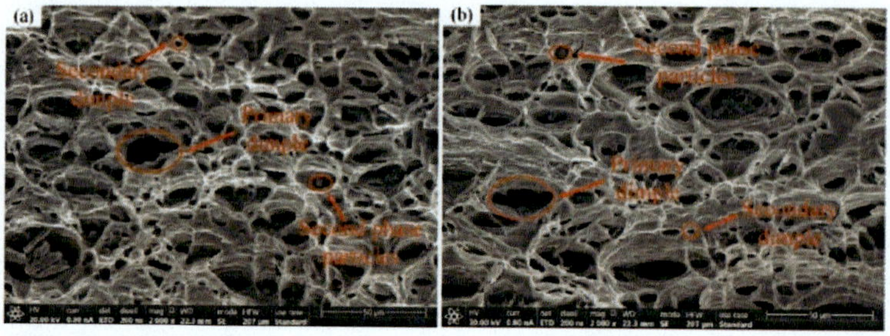

Fig. 5. Fracture morphology of the sample at different temperatures when w = 30 mm: (a) 350 °C; (b) 400 °C

direction. This is because the sample is in a state of single tensile stress. Large force in the lateral direction; Fig. 6(b) shows the fracture morphology of the sample with w = 200 mm. Most of the dimples are approximately circular. This is because the sample is in a state of equal double tension at this time. The force received in the vertical direction is equal. At the same time, the nucleation, growth and aggregation of the voids can be seen from both pictures.

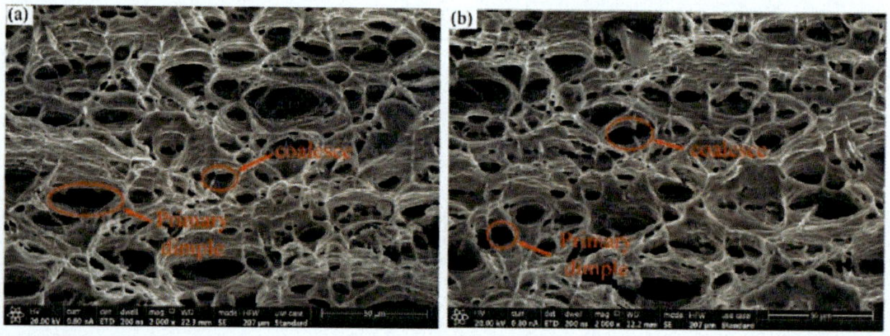

Fig. 6. The fracture morphology of the sample under different widths at 400 °C: (a) w = 30 mm; (b) w = 200 mm

3.4 Optimization Results

The final PSO-BP neural network prediction combined with genetic algorithm optimization to obtain the optimal process parameters of the hot stamping of 6005A aluminum alloy small curved beam parts are rounded to: blank holder force 4.051KN, stamping speed 141 mm/s, friction coefficient 0.05.

3.5 Validation of Optimization Results

The process parameters obtained by PSO-BP neural network and genetic algorithm optimization are substituted into the ABAQUS software to run, and the thickness cloud diagram of the small curved beam parts obtained is shown in Fig. 7. As can be seen from the figure, the minimum thickness of the small curved beam part is 1.588 mm. The maximum thinning rate of small curved beam parts is 20.6%, and the error of the maximum thinning rate of 21.2% obtained by combining with the PSO-BP neural network and the genetic algorithm is only 2.83%. This indicates the accuracy of PSO-BP neural network combined with genetic algorithm for optimization. Figure 8 shows the stress and strain distribution diagram of small bending beam parts. It can be seen from the figure that under the above process conditions, the stamping effect of the small curved beam is ideal, and most of the area has complete plastic deformation. The stress in the arc area at the bottom of the small curved beam is larger, and the strain on the left side of the part is larger.

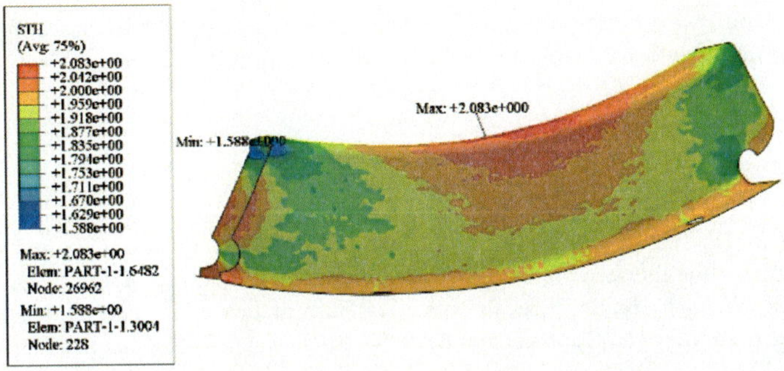

Fig. 7. Thickness cloud diagram of small curved beam parts

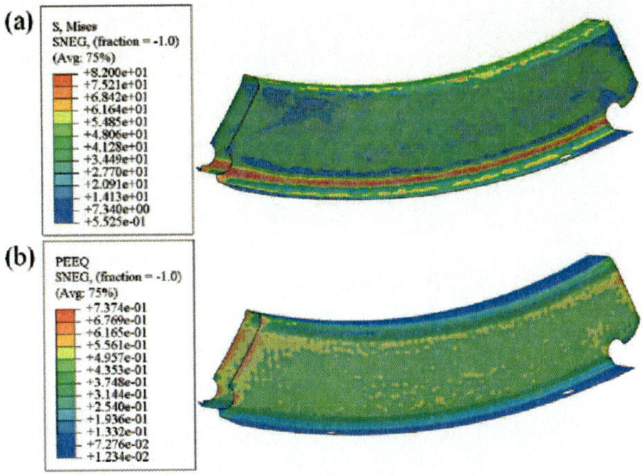

Fig. 8. Small bending beam parts: (a) stress distribution; (b) strain distribution

4 Conclusions

The specific conclusions are as follows:

(1) The forming limit diagram of 6005A aluminum alloy sheet at 350 ~ 450°C is obtained through forming limit experiments. The deformation temperature has a great influence on the forming limit of aluminum alloy. The temperature is between 350 and 400 °C, as the deformation temperature increases, the forming limit rises significantly.

(2) The PSO-BP neural network combined with the Genetic Algorithm is used to optimize the hot stamping forming parameters of the 6005A aluminum alloy track train small curved beam parts. Parameters were optimized through genetic algorithm, and the optimal process parameters for hot stamping of 6005A aluminum alloy small curved beam parts were determined: blank holder force is 4.051KN, stamping speed is 141mm/s, friction coefficient is 0.05, and maximum thinning rate is 21.2%.

References

1. Peng, D., Li, H., Sun, D., et al.: Effects of welding speed on the microstructure and hardness in friction stir welding joints of 6005A–T6 aluminum alloy. Mater. Des. **45**(3), 524–531 (2013)
2. Fakir, O.E., Wang, L., Balint, D., et al.: Numerical study of the solution heat treatment, forming, and in-die quenching (HFQ) process on AA5754. Int. J. Mach. Tools Manuf. **87**, 39–48 (2014)
3. Guo, X., Xu, H., Zeng, Q., et al.: Springback characteristics of arched aluminum alloy gusset plate after stamping forming. Thin-Walled Struct. **159**, 107294–107308 (2020)
4. Simar, A., Bréchet, Y., Meester, B.D., et al.: Microstructure, local and global mechanical properties of friction stir welds in aluminium alloy 6005A–T6. Mater. Sci. Eng. A **486**(1–2), 85–95 (2008)
5. Xia, X., Gui, L., He, G., et al.: An expanded particle swarm optimization based on multi-exemplar and forgetting ability. Inf. Sci. **508**(6184), 105–120 (2019)
6. Wan, P., Zou, H., Wang, K., et al.: Research on hot deformation behavior of Zr-4 alloy based on PSO-BP artificial neural network. J. Alloy. Compd. **826**, 154047–154056 (2020)

Novel Fabrication of Ultra-thin Copper/SS304L Composite Microchannels

Mengyuan Ren[1], Haibo Xie[1], Fei Lin[1], Fanghui Jia[1], Mingshuai Huo[1], Hui Wu[1], Shengnan Yuan[1], Ming Yang[2], Ken-ichi Manabe[3], and Zhengyi Jiang[1](✉)

[1] School of Mechanical, Materials, Mechatronic and Biomedical Engineering, University of Wollongong, Wollongong, NSW 2522, Australia
jiang@uow.edu.au

[2] Graduate School of System Design, Tokyo Metropolitan University, Hino-Shi, Tokyo 191-0065, Japan

[3] Department of Mechanical Engineering, Tokyo Metropolitan University, Hachioji, Tokyo 192-0397, Japan

Abstract. Microchannels on metallic substrates are essential parts of microreactors, micro heat exchangers, micro heat sinks, and fuel cell bipolar plates. Because of superior properties, such as good corrosion resistance, excellent thermal conductivity, high strength, excellent ductility and structural stability, composite metal foils are gaining popularity as substrates for microchannels. With the increasing demands of low cost, high efficiency and accuracy, novel micro forming technologies must be capable of creating intricate and precise micro features of microchannels at a low cost. An ultra-thin metal foil rolling for the fabrication of micro composite channels is proposed in this study, which can improve product function and reduce operational costs. Copper/ stainless steel 304L (SS304L) composite foils with a thickness of 0.41 mm after annealing treatment at 800 °C and holding for 0.5 h, 1.0 h and 2.0 h are used to fabricate microchannels under the rolling reduction of 60.9%. The hardness of composite foils has been characterised before and after different annealing processes, in which the hardness decreases due to the synergistic effects of decreased dislocation density, grain coarsening and phase transformation. Further, the formability of composite foils after annealing is discussed. In addition, the processing characteristics, material deformation behaviours and mechanisms during micro rolling are also investigated.

Keywords: Copper/SS304L composite foil · Roll forming · Microchannels

1 Introduction

Microchannels with widths of less than 1.0 mm have gained considerable interest in various applications such as microreactors, micro heat exchangers, micro heat sinks, and fuel cell bipolar plates in recent years [1, 2]. Generally, metallic substrates such as copper, stainless steel, titanium, and aluminium are widely used to fabricate microchannels for electronics and mechanical engineering-related applications. In order to achieve miniaturization, both the depth and thickness of microchannels should become smaller,

K. Mocellin et al. (Eds.): ICTP 2023, LNME, pp. 59–69, 2024.
https://doi.org/10.1007/978-3-031-40920-2_7

and the structural stability of the device must be maintained. Consequently, composite metal sheets have gained popularity owing to their ability to provide a combination of superior properties from each constituent material [3].

Traditional forming methods for the fabrication of metallic microchannels have been widely studied, including micro stamping, hydroforming and rubber pad forming [4, 5]. However, mould fabrication and maintenance require a high extra cost, and a large force is needed when forming a large area of microchannels, which limits the applicability of these forming methods. As a result, there is a growing demand for developing alternative technologies capable of creating intricate and precise features of microchannels with high efficiency and low cost.

In the field of micro forming, the micro rolling process has emerged as a promising method due to its high manufacturing efficiency, low forming force requirement, and excellent consistency. Buddhika et al. [6] proposed a novel multi-step micro rolling method with a feeding station and four forming stations and manufactured microchannels on the SS304L sheet for the bipolar plates. Huang et al. [7] fabricated microchannels on SS316L sheet by micro rolling process, with an aspect ratio up to 1.0 and thickness reductions lower than 18.7%. However, limited research focused on the microstructure and forming mechanisms of micro composite channels. Therefore, it is full of essentialities and significance to conduct further investigation on the material deformation behaviours and forming mechanisms in the fabrication of micro composite channels.

In this study, a novel micro rolling process for the fabrication of copper/SS304L composite microchannels has been proposed [8]. Annealing treatment at 800 °C has been conducted to improve the formability of copper/SS304L composite foil and further investigate the effect of holding times (0.5, 1.0 and 2.0 h) on the fabrication of microchannels during the micro rolling process. In addition, the microstructure and mechanical properties of composite foils after annealing treatments have been characterized, and the corresponding forming mechanism for the fabrication of composite microchannels has been discussed. This study provides a theoretical reference for the design of ultra-thin copper/SS304L composite microchannels and a novel method for manufacturing composite microchannels.

2 Experimental

The commercial copper/stainless steel 304L (SS304L) composite foils utilized in this study were prepared by explosive welding and subsequently machined to the uniform size of 10 mm × 10 mm × 0.41 mm, where the ratio of thickness for the copper and SS304L is ~ 3:1. The composition of SS304L layer is listed in Table 1, which was determined by optical emission spectroscopy. A heating furnace KTL1400 was utilized for the annealing treatment on the copper/SS304L composite foils at 800 °C for 0.5, 1.0 and 2.0 h, followed by the furnace cooling.

For the micro rolling process, the roll gap was adjusted by a screw knob, as shown in Fig. 1 [9], and a small amount of WD-40 lubricant was sprayed on the rolls before micro rolling. After the micro rolling process, the sample was cleaned for the following characterisation. Samples with different rolling reductions were prepared and nominated in Table 2.

Microhardness of copper/SS304L composite foils were measured under 100 gf on a Matsuzawa Via-F automatic Vickers Hardness Tester. The microstructure, element distribution and phase constituents were analysed by a JEOL JSM-7001F scanning electron microscopy (SEM) equipped with an energy dispersion spectrometer (EDS), and the electron backscatter diffraction (EBSD) was conducted with a step size of 4.0 μm, and 0.125 μm for characterising the microstructural evolution of copper and SS304L, respectively. 3D microstructure observations of microchannels were carried out on a VKX100K/X200K 3D laser scanning microscope.

Table 1. Chemical composition (wt. %) of SS304L.

Composition (wt. %)	Fe	Cr	Ni	Mn	Si	Mo	C
SS304L	71.62	17.87	8.28	1.14	0.50	0.06	0.026

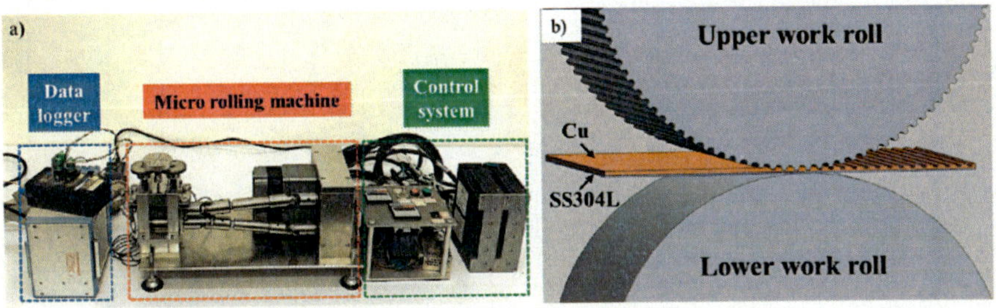

Fig. 1. a) Micro rolling experimental setup; b) Schematic diagram of the micro rolling process on copper/SS304L composite foils.

Table 2. Nomination of samples with different annealing process.

No	MA0		MA1	MA2	MA3
Annealing Temp (°C)	-		800	800	800
Holding hour (h)	-		0.5	1.0	2.0
Set reduction	-	60.9%		60.9%	60.9%
Rotation speed (rpm)	-	15.00		15.00	15.00
Linear velocity (mm/s)	-	21.21		21.21	21.21

3 Results and Discussion

3.1 Microstructure After Annealing Treatment

Figure 2 shows the XRD patterns for MA0, MA1, MA2 and MA3. Compared to MA0, the phase transformation from BCC to FCC occurs when the annealing temperature reaches 800 °C. However, there is no apparent transformation when the holding time increases from 0.5 to 2.0 h. After an annealing treatment of 800 °C for 2.0 h, the intimate bonding interface can still be observed in MA3, and the distribution of elements is shown in Fig. 3.

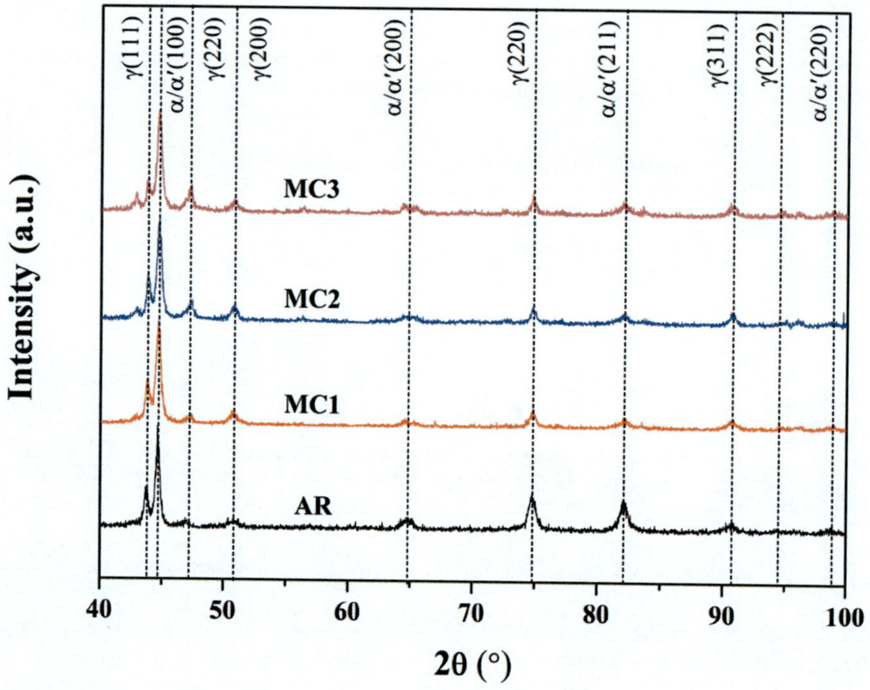

Fig. 2. X ray diffraction of copper/SS304L composite samples before and after different heat treatments.

From the previous study [10], the EBSD results of as-received sample have been reported. The average grain size of copper and SS304L is 1.67 μm and 1.41 μm, respectively. Figures 4 and 5 show the EBSD results of the copper layer for MA1 and MA3, in which the inverse pole figure (IPF) maps, grain size distribution histograms, Kernal average misorientation (KAM) maps and grain boundary maps are presented. Figures 4(e) and (f) show the grain size distributions of Cu in MA1 and MA3. It is evident that the average grain size of Cu significantly increases from 1.67 μm for MA0 to 25.98 μm and 33.43 μm for MA1 and MA3, respectively. The KAM values were calculated for the second nearest neighbour with a maximum misorientation of 5°. A higher KAM value indicates the formation of more geometrically necessary dislocations to accommodate the strain gradient between the two grains [11]. Compared to MA0, the KAM value of the copper layer decreases after annealing and keeps dropping when the holding time increases from 0.5 to 2.0 h. It indicates the decreased dislocation density with the increase in holding time.

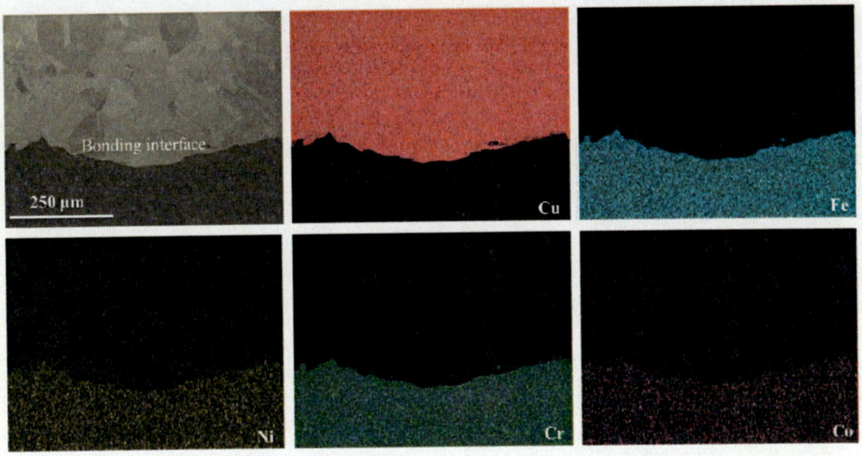

Fig. 3. SEM image and distributions of elements on MA3.

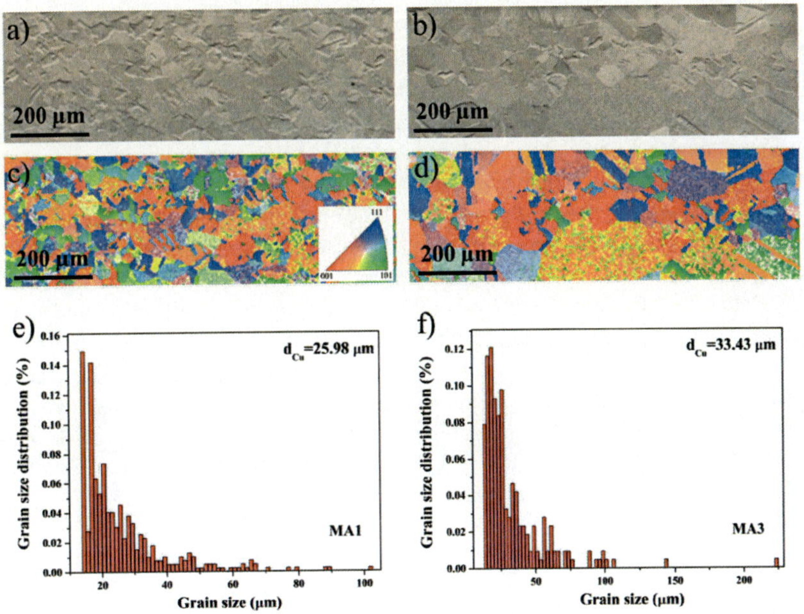

Fig. 4. SEM images for a) MA1, and b) MA3; IPFs for c) MA1, and d) MA3; Grain size distributions for e) MA1, and f) MA3.

Figures 6, 7 and 8 show the EBSD results of the SS304L layer for MA1 and MA3 (the tested areas are marked in red square), in which the IPF maps, grain size distribution histograms, phase distribution maps, KAM maps and grain boundary maps are presented. The grain size of MA3 is slightly smaller than that of MA1, and both are larger than MA0, as shown in Figs. 7(c) and (d). The phase maps, KAM maps, and

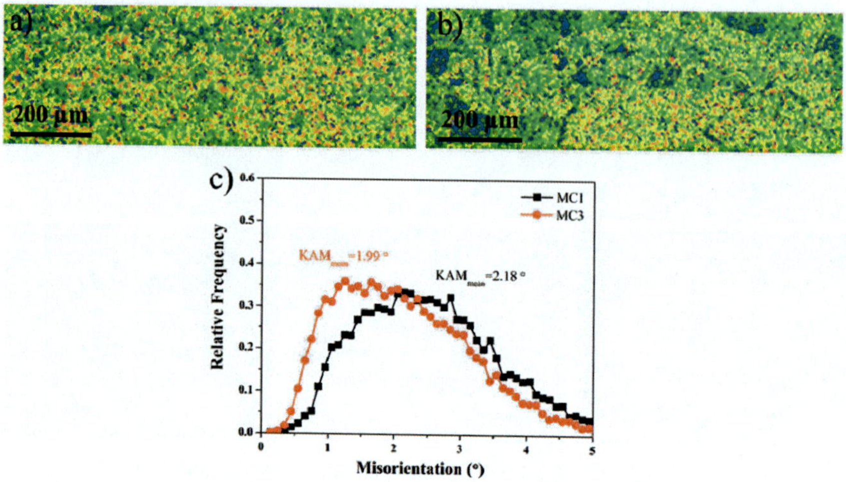

Fig. 5. KAM maps for a) MA1 and b) MA3; c) KAM values for MA1 and MA3.

KAM distributions of MA1 and MA3 have been presented in Fig. 7. From the previous study, the microstructure of the SS304L layer should be BCC-dominant. After annealing treatment, the transformation from BCC to FCC occurs, and the ratio of BCC become much lower than that of FCC in MA1 and MA3. However, the holding time increases from 1.0 to 2.0 h have almost no effect on the formation of FCC.

In addition, the KAM value decreases after the annealing treatment at 800 °C, but no obvious change in KAM value can be observed when the holding time increases from 0.5 to 2.0 h. The decreased KAM value represents the decreased dislocation density, which is attributed to the dislocation annihilation during recrystallisation under high-temperature annealing. During the micro rolling process, dislocation accumulation and grain rotation were accompanied by the formation of low angle grain boundaries (LAGs), and the orientation change of high grain boundaries (HAGs). Compared to MA1, the fraction of LAGs (shown as the green lines) in MA3 slightly decreases. The decreased fraction of LAGs on the SS304L layer can be attributed to the mechanism of dislocation annihilation after annealing treatment.

After annealing, the decreased hardness on the copper layer, bonding interface and SS304L layer can be observed. The hardness values of copper, bonding interface and SS304L in the as-received and annealed samples are shown in Fig. 9, which are 476.8 HV and 99.3 HV for the copper and SS304L layer in MA0, and 222.1 HV and 53.1 HV for copper and SS304L layer in MA3, respectively. In addition, the hardness value of bonding interface decreases from 120.5 HV for MA0 to 64.4 HV for MA3. The decreased hardness for the copper layer is mainly attributed to the dislocation annihilation and the grain coarsening after annealing treatment. Additionally, the formation of austenite, dislocation annihilation and recrystallisation have a synergistic effect on the decreased hardness of the SS304L layer.

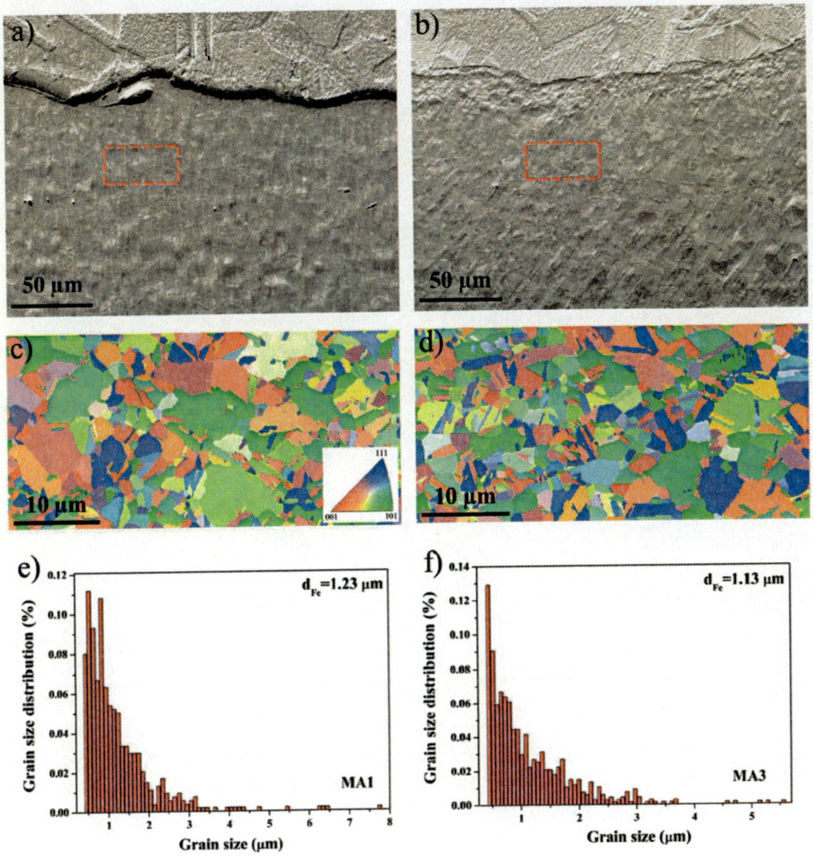

Fig. 6. SEM images for a) MA1, and b) MA3; IPFs for c) MA1, and d) MA3; Grain size distributions for e) MA1, and f) MA3.

3.2 Fabrication of Microchannels

After improving the formability of copper/SS304L composite foil by the annealing heat treatment, three annealed samples with different heat treatments were employed to form microchannels with a rolling reduction of 60.9% and a velocity of 21.21 mm/s). The cross-sectional and geometrical morphologies of formed composite channels are shown in Figs. 10 and 11, which indicate that stable and continuous microchannels were formed on all the samples. In terms of the forming shape of samples, the ridges and channels become more asymmetrical when the holding time increases from 0.5 to 2.0 h due to the decreased hardness and more significant material flow behaviour [12].

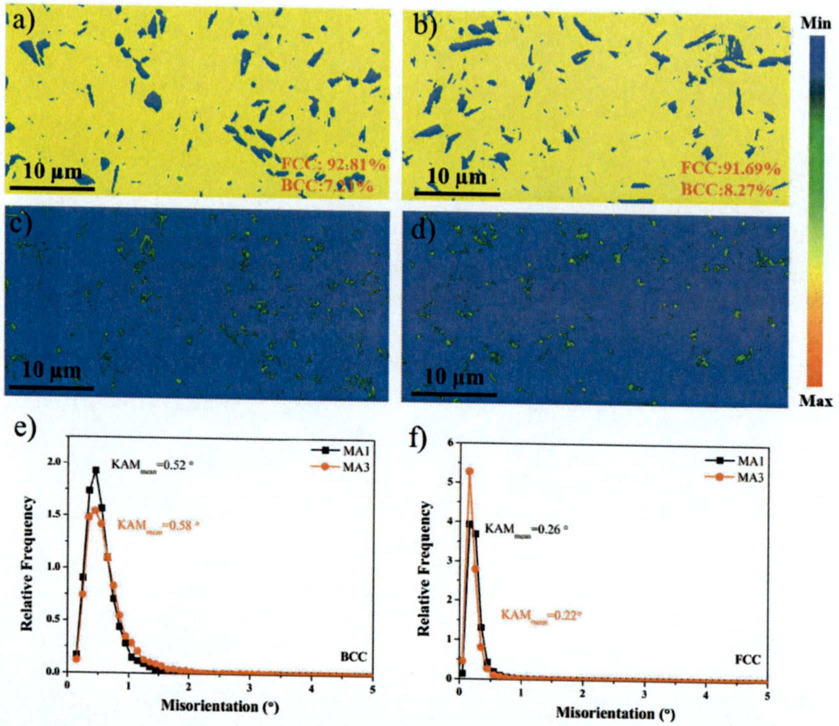

Fig. 7. Phase distribution maps for a) MA1 and b) MA3 (the yellow colour displayed in map indicates FCC phase, whereas the blue colour represents BCC phase); KAM maps for c) MA1 and d) MA3; KAM values for e) MA1 and f) MA3.

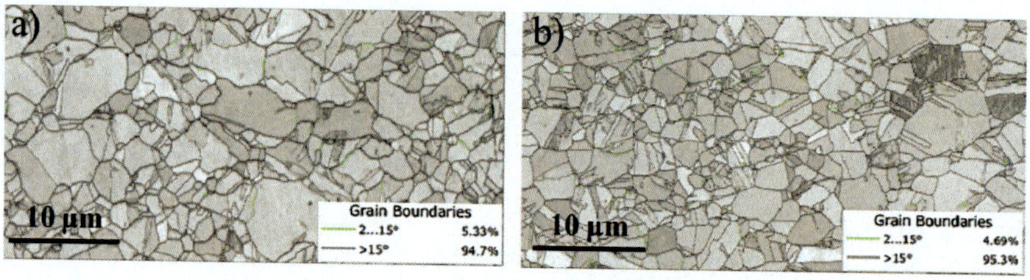

Fig. 8. Grain boundary maps for a) MA1 and b) MA3.

The forming depth (h) and periodical length (l_p) of formed microchannels are obtained, as shown in Fig. 10(a), and then the aspect ratio can be calculated according to Eq. (1).

$$\delta = h/(l_p/2) \tag{1}$$

The forming depth and periodical length of MA1, MA2 and MA3 are 0.237, 0.242 and 0.245 mm, and 0.819, 0.811 and 0.814 mm, respectively, which are summarized in Fig. 12. In addition, the calculated aspect ratio increases from 0.580 to 0.602 with the

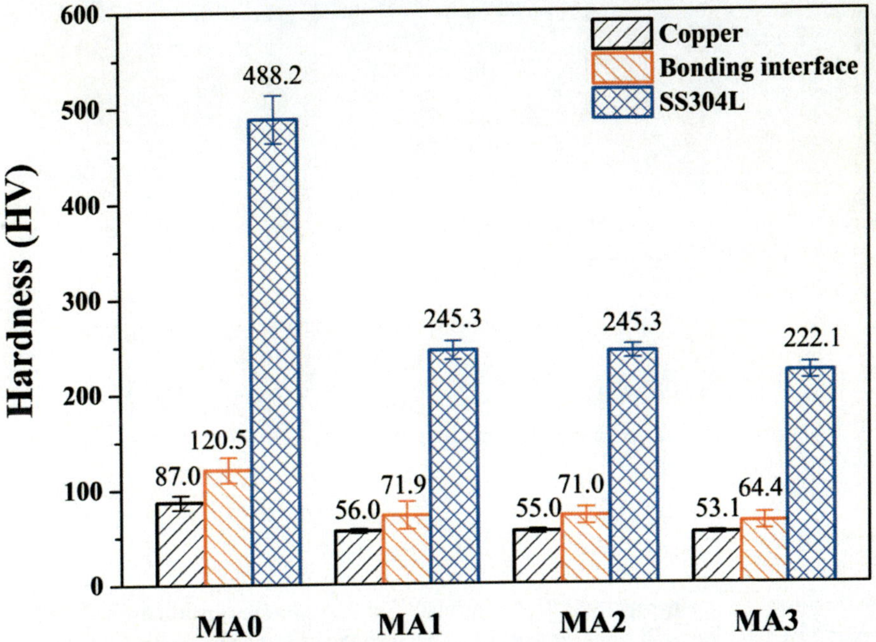

Fig. 9. Microhardness on the copper layer, bonding interface and SS304L layer for as-received (MA0) and annealed samples (MA1, MA2 and MA3)

increase of holding time. The bonding interface for MA1 before and after micro rolling is shown in Fig. 13. After micro rolling process, the stretched grain region is observed on the copper layer due to the severe plastic deformation, and the intimate bonding interface without defects can still be obtained.

Fig. 10. OM of the formed channels: a) MA1, b) MA2 and c) MA3

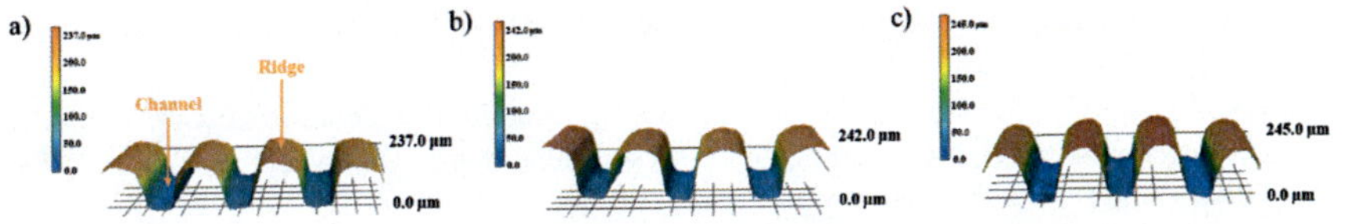

Fig. 11. 3D Geometrical morphologies of the formed channels: a) MA1, b) MA2 and c) MA3

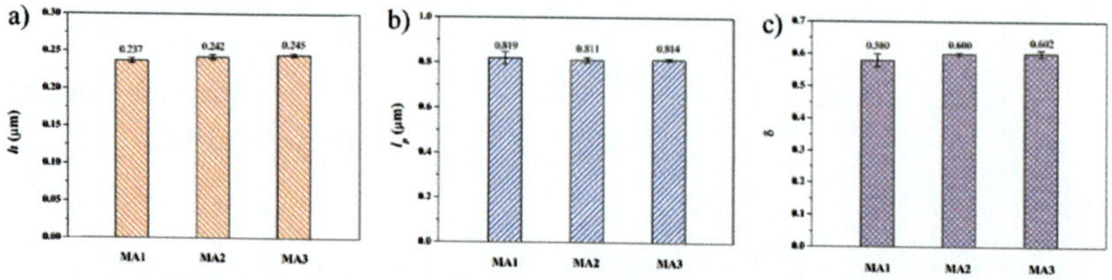

Fig. 12. Comparison of geometrical parameters: a) forming depth h, b) periodical length l_p and c) aspect ratio δ of MA1, MA2 and MA3

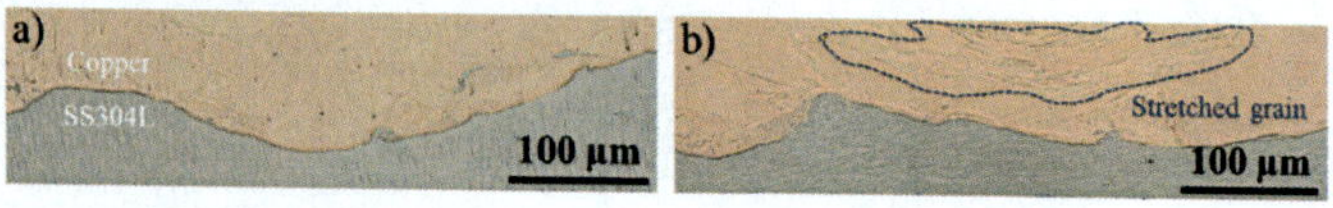

Fig. 13. Cross-sectional morphologies of the bonding interface for MA1 a) before and b) after micro rolling

4 Conclusions

After annealing treatment at 800 °C, the intimate bonding interface on the copper/SS304L composite foils can still be observed. The decreased hardness and improved formability after annealing are mainly attributed to the dislocation annihilation and the grain coarsening of the copper layer. Additionally, the formation of austenite on the SS304L layer also improves formability. Microchannels based on the annealed copper/SS304L composite foils with different holding times can be formed by the micro rolling process, and the stable and continuous microchannels can be observed.

Acknowledgements. The authors acknowledge the financial support from Australian Research Council (ARC, Grant Nos. DP190100408). The first author appreciates the University of Wollongong for the scholarship support.

References

1. Elyasi, M., Khatir, F.A., Hosseinzadeh, M.: Manufacturing metallic bipolar plate fuel cells through rubber pad forming process. Int. J. Adv. Manuf. Technol. **89**(9–12), 3257–3269 (2016). https://doi.org/10.1007/s00170-016-9297-6
2. Deng, D., Xie, Y., Chen, L., Chen, X.: Experimental investigation on laser micromilling of SiC microchannels. Int. J. Adv. Manuf. Technol. **101**(1–4), 9–21 (2018). https://doi.org/10.1007/s00170-018-2800-5
3. Zhang, X., Pan, T., Flood, A., Chen, Y., Zhang, Y., Liou, F.: Investigation of copper/stainless steel multi-metallic materials fabricated by laser metal deposition. Mater. Sci. Eng. A **811**, 141071 (2021)
4. Weiss, M., Zhang, P., Pereira, M.P., Rolfe, B.F., Wilkosz, D.E., Hodgson, P.D.: Understanding size effects and forming limits in the micro-stamping of industrial stainless steel foils. Metals **11**(1), 38 (2020)

5. Song, Y., et al.: Review on current research of materials, fabrication and application for bipolar plate in proton exchange membrane fuel cell. Int. J. Hydrog. Energy **45**(54), 29832–29847 (2020)
6. Abeyrathna, B., Zhang, P., Pereira, M.P., Wilkosz, D., Weiss, M.: Micro-roll forming of stainless steel bipolar plates for fuel cells. Int. J. Hydrog. Energy **44**(7), 3861–3875 (2019)
7. Huang, J., Deng, Y., Yi, P., Peng, L.: Experimental and numerical investigation on thin sheet metal roll forming process of micro channels with high aspect ratio. Int. J. Adv. Manuf. Technol. **100**(1–4), 117–129 (2018). https://doi.org/10.1007/s00170-018-2606-5
8. Ren, M., Lin, F., Jia, F., Xie, H., Yang, M., Jiang, Z.: Micro rolling fabrication of copper/SS304L micro composite channels. J. Manuf. Process. **90**, 1–13 (2023)
9. Xie, H.B., Manabe, K., Furushima, T., Tada, K., Jiang, Z.Y.: An experimental and numerical investigation on micro rolling for ultra-thin strip. Int. J. Mater. Form. **9**(3), 405–412 (2015). https://doi.org/10.1007/s12289-015-1229-8
10. Ren, M., et al.: Effect of heat treatment on the microstructure and mechanical properties of copper/SS304L composite sheets. Vacuum **204**, 111307 (2022)
11. Ren, M., et al.: Evolutions of microstructure and mechanical properties of copper/SS304L composite micro channels during micro rolling. Mater. Sci. Eng. A **871**, 144881 (2023)
12. Jiang, Z.: Ribbed strip rolling by three-dimensional finite element method combining extremely thin array of elements. Front. Mech. Eng. **5**(1), 52–60 (2009)

An Exploration of the Process Operating Window for Folding-Shearing in a Press-Tool

Rishabh Arora[1], Omer Music[1,2], Christopher J. Cleaver[1,3], and Julian M. Allwood[1](✉)

[1] Department of Engineering, University of Cambridge, Cambridge, UK
jma42@cam.ac.uk
[2] Mechanical Engineering, TED University, Ankara, Turkey
[3] DeepForm Limited, Cambridge, UK

Abstract. Folding-shearing, is a novel process, which in previous work has been used to create isolated shrink corners. The top surface of a blank was clamped, and the part formed by manual folding followed by a shearing process. For the first time, this paper demonstrates folding-shearing in a press-tool. A single set of tooling is used to crash form the upper surface of the part, followed by folding and in-plane shearing with all tools moving vertically, avoiding any requirements for cams or other mechanisms. The results of the paper demonstrate that folding-shearing can be implemented in a conventional press-tool. A process window is defined by process parameters and is limited by springback, thinning or thickening. These limits are influenced by process parameters including part radius, feed radius, fold geometry, part height and material parameters. An analytical description of the mechanics of the process is proposed and is validated though finite-element simulations and experimental trials.

Keywords: Sheet metal forming · folding-shearing · stamping

1 Introduction

Deep-drawing is typically used in automotive press lines, where a part is formed by pulling material radially into a forming die by the action of a punch, while gripping the perimeter of a blank in the blankholder region. Tearing and wrinkling are the two most common failure modes. The punch force is limited to the maximum tensile load that can be sustained by the part wall and hence limits the depth of the part that can be formed, before a fracture is seen. Wrinkling occurs once the compressive stresses are high enough to cause a flange instability [1]. While deep drawing is highly repeatable, at a high frequency, it suffers from a low material utilization. In the automotive industry, 44% of the sheet metal used in the production of passenger vehicles is scrapped [2]. This scrap occurs due to three reasons: firstly, when blanks are cutout from coils of sheet metal; secondly, when excess material is clamped in the blankholder region to help control the forming window between tearing and wrinkling and then cut off; thirdly, when the formed-part is trimmed to make holes and cutouts. Improving this material utilization by 70% could help reduce carbon emissions by 26% [2]. Efforts have been made to improve

K. Mocellin et al. (Eds.): ICTP 2023, LNME, pp. 70–77, 2024.
https://doi.org/10.1007/978-3-031-40920-2_8

this utilization; researchers have developed a stamping process without drawbeads, while optimizing the blank shape to reduce the material use [3]. In other cases, all the material from the blank is eventually fed into the die [4].

Today's low utilisation motivated the development of the folding-shearing technique in 2019 [5]. The process, which combines folding followed by in-plane shearing, aims to improve material utilization by removing the need for a blankholder. The workpiece is first folded into an intermediate geometry, with "beaks" formed in regions of incompatibility. These beaks are then "ironed-out" between conical dies, which work against the curvature of the beak to create a stress-state near to pure shear. Preliminary work investigated the shearing stage for a pre-folded sample and found that parts can be made with a thickness reduction of no more than 10–15%. Folding-shearing was first investigated to create an isolated shrink corner using a manual rig [6]. The process was limited mainly by thickening leading to a "locking-up" of the workpiece between the dies, and eventually tearing.

Previous trials required complex folding kinematics with two stages setup independently. For the first time, this paper explores folding-shearing in a press tool to predict the process operating window.

2 Process

One half of a U-channel part with a shrink corner is formed by folding-shearing. The target part geometry has a part height of 'h' with a part radius of 'R_p'.

The tooling consists of five parts (Fig. 1d): post, upper shear tool, lower shear tool and two upper clamps. The setup is assembled in a compression testing machine. In the first stage (Fig. 1b), a blank is folded along line AB. The blank is manually crash formed by the crash angle, δ, and is bolted to the post. At this point, the lower shear tool is in position. The upper shear tool is lowered at a constant speed and folds the workpiece along lines CE and DE over a radius. The geometry of the upper and lower shear tool defines the shape of the beak which is formed only in bending. During shearing, the upper die bears down on the beak, creating lateral compression and down-ward tension, which if balanced lead to pure shear along the beak-perimeter. At this perimeter, the workpiece also experiences bending and unbending as it passes through the feed radius, R_f, defined by the upper shear tool.

2.1 Fold Geometry

The angle at which the shearing tools are inclined downwards is defined as the fold angle, α, and the angle between the side faces of the beak is defined as the beak angle, β (Fig. 2). To determine the fold geometry, an origami model, assuming zero radius at folds, is developed to understand the relationship between process parameters. The beak, (β), and fold, (α), angles can be calculated as a function of the shear zone, (θ), and crash, (δ), angles as given in Eq. 1 and 2.

$$\beta = 2\sin^{-1}\left(\frac{\sin(\theta - \delta)}{\sin(\theta)}\right) \tag{1}$$

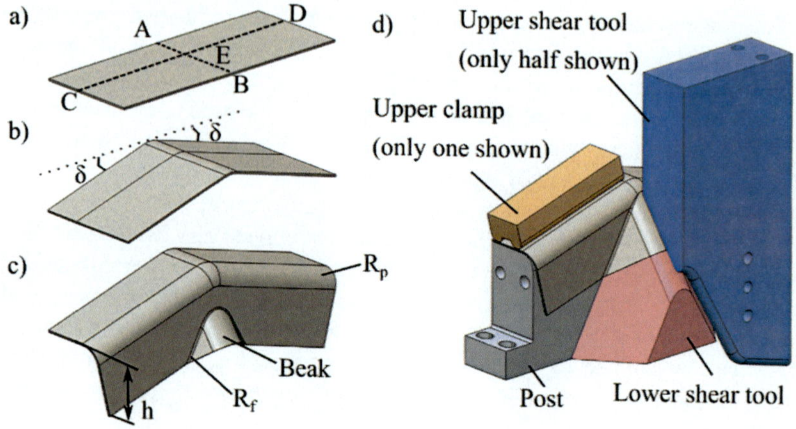

Fig. 1. (a) Initial blank (b) After crash forming (c) During shearing (d) Folded workpiece with tooling

$$\alpha = \sin^{-1}\left(\frac{\cos(\theta)}{\cos(\theta - \delta)}\right) \quad (2)$$

3 Methodology

In order to explain the process, a numerical simulation is created, validated against physical trials and an analytical model, and then used to explore the process operating window.

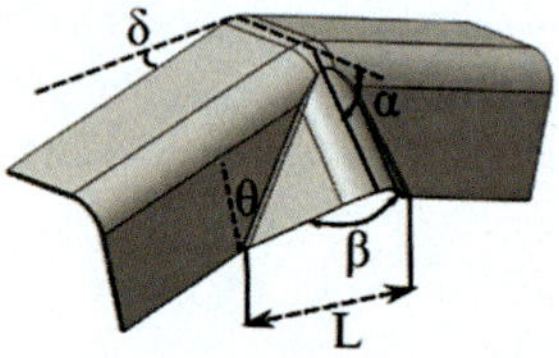

Fig. 2. Parameters of fold geometry

3.1 Physical Trials

The experimental apparatus (Fig. 3) is setup on a compression testing machine, where the upper steel is attached to the crosshead and the post is fixed to the base of the machine. In the first forming stage, the blank is crash formed by an angle of 15°. The upper steel is lowered at a speed of 1 mm/s and folds the part over the post radius of 10 mm to create the beak. The shearing dies are then bolted together and are moved vertically downwards until the process is completed. Force measurements were made using loadcells attached to the compression testing machine and thickness measurements were made using a micrometer and an ultrasonic precision thickness gauge.

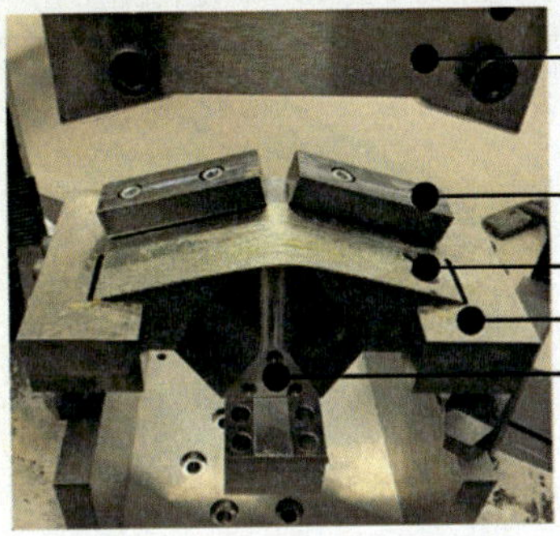

Fig. 3. Physical setup

3.2 Analytical Model

To estimate the force required to shear the beak, an energy approach has been used. Equation (3) considers the work done due to (i) shearing and (ii) bending and unbending of the sheet as it passes the feed radius. Work done due to friction was found to be negligible and is ignored.

$$WD_{total} = WD_{shear} + WD_{bending-unbending} \tag{3}$$

The force required to shear the beak is given by:

$$F_{US} = \frac{\delta W}{\delta x} = \frac{\gamma \tilde{\sigma} L t}{\sqrt{3}} + \frac{\sigma_y t^2 L}{2R_f} \tag{4}$$

where, γ is shear strain (Eq. 5), $\tilde{\sigma}$ is flow stress, L is projected length of contact between workpiece and shearing dies (Eq. 6), t is thickness, σ_y is yield stress and R_f is feed radius in the working direction. In Eq. 6, x is the distance travelled vertically, in the working direction.

$$\gamma = \frac{h}{2}\left(\frac{sin\theta}{\sin(\theta - \delta)} - sin\alpha - cos\alpha\right) \tag{5}$$

$$L = 2(h - x)\tan\left(sin^{-1}\left(\frac{sin(\theta - \delta)}{sin(\theta)}\right)\right) \tag{6}$$

3.3 Numerical Simulation

A finite element (FE) model was setup in commercial software Autoform R10, to explore the impact of process and part parameters on process window. A model was setup using rigid tools for the upper and lower shear dies. A deformable rectangular blank was used, and a mesh sensitivity study was conducted considering accuracy and computation time. A Coulomb-friction model was used with a coefficient of friction set at 0.1.

Fifty FE simulations were carried out to determine the process operating window. Table 1 lists the part and process variables that have been investigated via physical trials and/or by FE simulations. FE simulations were carried out with increasing part height, h, in increments of 5 mm, until a failure was identified. By controlling part radius, material, thickness and shear zone angle, a process window can be mapped to define failure at a certain height for a given crash angle. An important geometric criterion defines the crash angle limit; crash angle must always be less than the shear zone angle, else a beak cannot be formed. FE simulations were performed for all crash angles and a set of physical trials was conducted for a 15° crash angle.

Table 1. Part and process variables for physical and FE trials

	Experiment	FE
Part variables		
Part radius, R_p, [mm]	{10}	{10}
Height, h, [mm]	{25, 50}	{15, 20, 25 ... 80$^{\otimes}$}
Material	{1050-H111}	{1050-H111}
Thickness, t, [mm]	{1.2}	{1.2}
Process variables		
Shear zone angle, θ, [deg]	{45}	{45, 60}
Crash angle, δ, [deg]	{15}	{5, 10, 15 ... 45, 50*, 55*, 60*}

$^{\otimes}$Until failure is seen
* Only for 60° shear zone angle

4 Results and Discussion

4.1 Validation

To validate the numerical simulation and force model, one half of a U-channel part is made using Aluminum 1050-H111, with 50 mm part height and 1.2 mm part thickness. Figure 4a compares the force evolution by the upper shear tool against the ratio of sheared distance over the part height. Numerical simulations match the physical trials closely. Analytical model predicts the force only during the shearing stage and is marked on the graph. As the process continues, the length of contact between shearing dies and workpiece decreases, hence the force decreases as the process continues. Considering a line of symmetry, Fig. 4b marks six points where thickness was measured. Physical trials found a maximum thinning and thickening of 6.6% and 15%, respectively, whereas FE estimates are 7.5% and 13.5%.

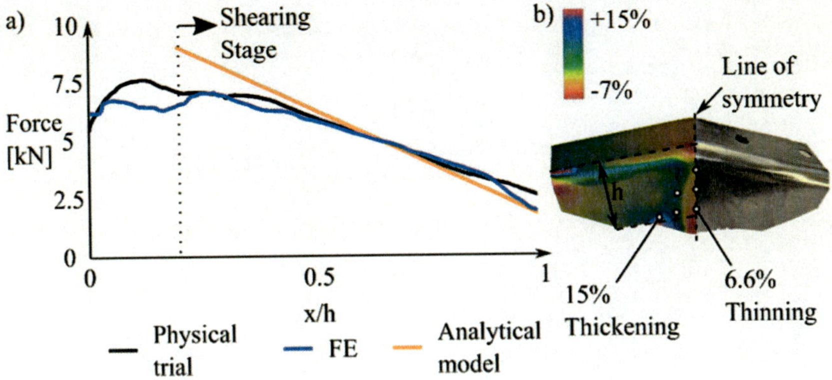

Fig. 4. (a) Force evolution (b) Thickness distribution

4.2 Process Operating Window

Figure 5 plots limiting height for a given crash angle for Al1050-H111, with 1.2 mm part thickness, 10 mm part radius and a shear zone angle of 45°. A viable process window has been defined considering three failure modes: springback, thinning and thickening. A trial is considered successful if thinning is less than 5%, thickening is less than 10% and springback is less than 0.5 mm. Part height is limited by springback for 0° – 4.5° crash angles, by thinning for 4.5°–16.5° crash angles and by thickening for 16.5° – 20° crash angle. The maximum feasible part height reaches a maximum of 65 mm, falling to 0 mm as the crash angle reaches 20°. There is no feasible part height for crash angles greater than 20°. In the figure, 'X_1, X_2, X_3' is labelled if a trial has failed due to springback, thinning or thickening, respectively, and 'O', if the trial is successful.

The failure modes are now explained; as crash angle tends to 0°, the process tends to pure bending and springback is the dominant failure mode, whereas thinning is negligible. As the crash angle increases, thinning replaces springback as the dominant failure mode. With increase in crash angle, more material is folded into the beak, resulting in a larger shearing force. Simultaneously, at a fixed crash angle, as the part height increases the line of contact between the beak and the shearing dies increases. Both increase in crash angle and height results in higher probability of thinning. As the crash angle tends to shear zone angle, the beak incompatibility increases, resulting in thickening of workpiece. Eventually, the thickness increases, and 'lock-up' of the workpiece between the two shearing dies occurs. Both physical trials and FE show that maximum thickening occurs at the bottom corners of the beak.

To verify the process window, two physical trials (* in Fig. 5) were conducted at part heights of 25 and 50 mm. Part height of 25 mm was a successful trial; however,

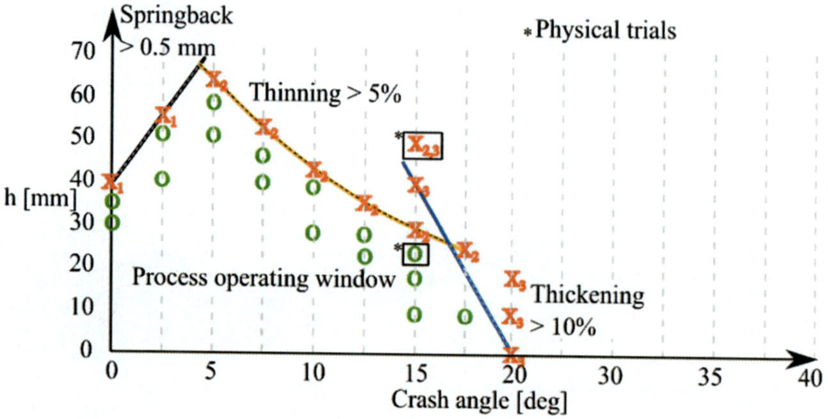

Fig. 5. Process window for folding-shearing in a press-tool for 45° shear zone angle

50 mm part height fails by exceeding both thinning and thickening criteria. From FE, it was found that the part first fails due to thinning at a part height of 30 mm, followed by a thickening failure at 40 mm. Both these trials agree with the process window, derived from FE simulations.

4.3 How Does the Shear Zone Angle Impact the Process Window?

The shear zone angle can be varied independent of the target part geometry. Figure 6 compares the process window for shear zone angles of 45° and 60°. The process window has increased with a larger shear zone angle. Crash angles greater than 20° are now feasible; start of the thickening limit, for 60° shear zone angle, occurs at a crash angle of 30°, compared to 17° for a 45° shear zone angle. With increase in shear zone

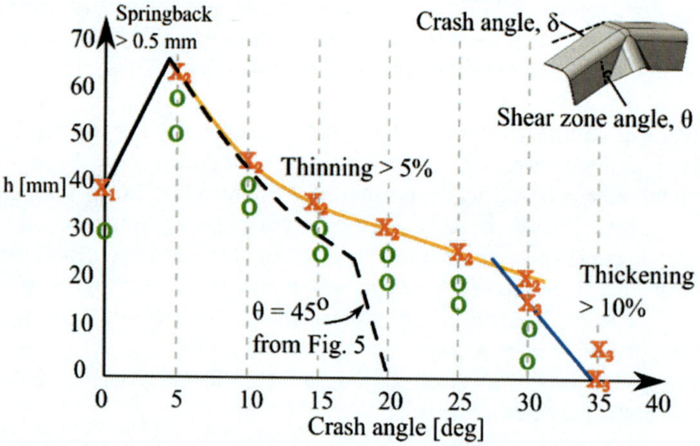

Fig. 6. Process window for shear zone angles 45°and60°

angle, for the same part height, more material is stretched within the beak. This helps reduce the incompatibility of the beak and pushes the thickening limit to larger crash angles. Depending on the target part geometry, a suitable shear zone angle can be chosen accordingly.

5 Conclusions

For the first time, this paper reports the implementation of folding-shearing in a press tool. A numerical solution has been implemented and validated and used to anticipate the process operating window for two part geometries. The three distinct lines of failure revealing the process window provide a clear direction for future analytical work, building on the model presented here, to predict thinning and thickening across all possible geometries and materials.

References

1. Colgan, M., Monaghan, J.: Deep drawing process: analysis and experiment. J. Mater Process. Technol.**132**, 35–41 (2003)
2. Horton, P., Allwood, J.: Yield improvement opportunities for manufacturing automotive sheet metal components, J. Mater. Process. Tech. **249**, 78–88 (2017)
3. Shim, H.: Digital tryout based on the optimal blank design toward realization of beadless stamping. In: Proceedings of the Institution of Mechanical Engineering, vol. 227, pp. 1514–1520 (2013)
4. Smartform. https://www.thyssenkrupp-steel.com/en/industries/automotivetrucks/smartform/smartform.html. Accessed 19 Feb 2022
5. Allwood, J.M., Cleaver, C.J., Loukaides, E, Music, O., Nagy-Sochacki, A.: Folding-shearing: shrinking and stretching sheet metal with no thickness change. CIRP Ann. Manuf. Technol. **60,** 285–288 (2019)
6. Cleaver, C.J., Arora, R., Loukaides, E., Allwood, J.M.: Producing isolated shrink corners by folding-shearing. CIRP Ann. Manuf. Technol. **71,** 217–220 (2022)

Plate Roll Embossing Process - The Efficient and Flexible Embossing of Sheet Metals

David Briesenick(✉), Mathias Liewald, and Pascal Heinzelmann

Institute for Metal Forming Technology (IFU), University of Stuttgart, Stuttgart, Germany
david.briesenick@ifu.uni-stuttgart.de

Abstract. Embossings on sheet metal surfaces allow a targeted enhancement of local part strength and affect thinning in deep drawing. However, conventional manufacturing of tailor embossed blanks on press machines with simultaneous intrusion of embossing punches results in high force demands and tool wear. Incremental forming of each indentation reduces the embossing forces required but increases cycle time. Embossing by rolling provides limited flexibility in embossing patterns. Therefore, presented paper deals with the development of a new approach for an efficient and flexible embossing of sheet metals – the Plate Roll Embossing Process (PRE Process). Here, a modular design of an upper and lower tool is used, into which various embossing, counter punch or blank positioning inserts can be integrated. The sheet metal is placed between the tools and the setup is moved through a defined roll gap of a mill stand. The line contact between rolls and tool defines a relatively small forming zone, which is moved translationally through the workpiece by the rotation of the rolls. Numerical and experimental investigations were performed with aluminium AA6014 and steel DC01, showing homogenous embossing results with reduced loads compared to conventional embossing. Furthermore, the paper demonstrates that the high flexibility of the tool and process design enables embossing of flat workpieces as well as the simultaneous embossing and forming of curved parts. Finally, tensile tests performed with embossed specimens show that the PRE Process represents a fast and efficient sheet metal forming method for the targeted modification of component properties.

Keywords: Embossing · Flexibility · Sheet Metal Forming · Force Reduction

1 Introduction

The mobility sector is currently facing a growing demand for sustainable manufacturing and use of products. For both environmental and financial benefits, manufacturers are therefore aiming to implement lightweight design and efficient ways of using material in production by establishing new engineering and manufacturing technologies [1]. Thin high-strength sheet metal parts play an essential role in this context, but their full potential can only be exploited through tailored functional properties. These so-called tailored blanks are manufactured by welding, heat treatment or rolling processes and differ in terms of their formability and processing capabilities [2].

K. Mocellin et al. (Eds.): ICTP 2023, LNME, pp. 78–89, 2024.
https://doi.org/10.1007/978-3-031-40920-2_9

Near-surface embossing of sheet metals allows a local modification of part properties and thus offers new possibilities for manufacturing and use of tailored blanks. Local work hardening caused by embossed surface indentations impacts the mechanical properties [3] as well as the stiffness and rigidity of the sheet metal [4]. Furthermore, Walzer and Liewald [5] found a change in plastic flow behavior and formability depending on embossing depth and pattern for DP600 sheet metal. Abe et al. [6] and Namoco et al. [7] provided corresponding results for embossed aluminium sheets and also showed an enhanced formability. Beyond that, embossing allows targeted introduction of residual stresses into sheet metals, which can be exploited for improved fatigue strength [8] or tailored magnetic properties [9] of respective components.

Presented ongoing research work on tailor embossed blanks offers promising opportunities for enhanced lightweight design and material utilization. However, simultaneous embossing of sheet metals with large tool contact areas significantly increases the forming force, exceeds the working capacities of presses and amplifies effects of tool deformation and wear [10]. Incremental sheet-bulk metal forming processes with point contact decreases the total load at a time, but causes non-economic cycle time [11]. In contrast, roller embossing enables efficient embossing of simple patterns under reduced loads due to a line contact, but is limited in flexibility [12].

Against this background, the presented paper deals with the development of a new forming process for an efficient and flexible embossing of various patterns and sizes under reduced loads.

2 New Forming and Tool Concept for Sheet Metal Embossing

New Plate Roll Embossing Process (PRE Process) originates as a combination of conventional and roller embossing. Furthermore, basic process kinematics can also be found in roller die cutting machines for the segmentation of different patterns from hard foils, cardboard and paper. The new PRE Process is performed with a straightforward and modular tool and machine setup which uses the relative motion and force application of rolls. This implies a line-shaped tool contact area with a relative movement to the blank, thus minimizing the simultaneously acting forming zone and leading to a significant reduction of embossing forces.

Figure 1(a) shows an exemplary tool setup, consisting of scalable tool base plates with a standardized mounting interface for the implementation and arrangement of arbitrary designable inserts. The blank is placed on top of the lower tool half and the whole setup is moved through a defined roll gap of a mill stand. Thereby, the sheet metal surface is incrementally formed, depending on the mounted insert geometries, the forced elastic tool deformation and tool base plate tipping. In doing so, a relatively small forming zone is moved translationally through the workpiece by the opposing rotation of the rolls.

Accordingly, different process variants for several use cases can be derived, as shown in Fig. 1(b) and (c). For instance, embossing inserts with positive indentation geometries are mounted at the interface of the upper and lower tool base plates (PRE 2×2). Alternatively, flat surface inserts on one side and embossing inserts on the opposite side towards the sheet metal surface represent a single-sided embossing process (PRE 2×1). Positioning of tool base plates with block guidance ensures proper alignment of forming inserts and allows a slight elastic tool deformation and tipping for a reduced forming contact zone. On contrary, a single tool base plate with inserts is used for a one-sided embossing of the blank, placed between the lower tool and the upper roll. As shown in Fig. 1 c), the translational movement of the tool and blank compound caused by the rotation of the rolls results in a simultaneous embossing and bending due to the absence of a supporting counter tool (PRE 1 × 1).

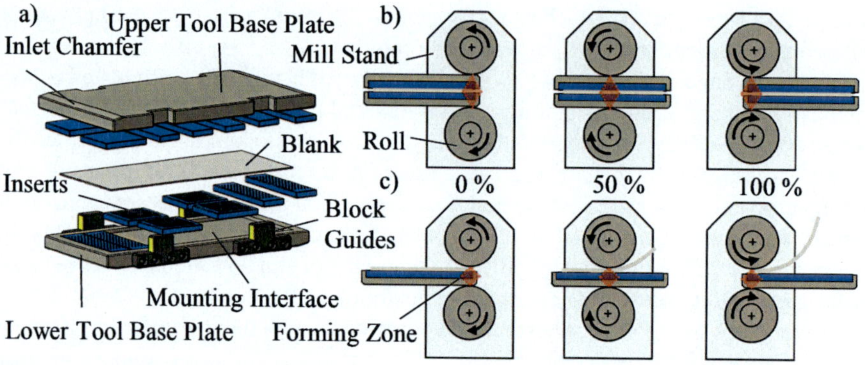

Fig. 1. a) Concept of a modular tool design for b) flat (PRE Process 2 × 1 and 2x2)) or c) curved embossed sheet metals (PRE Process 1 × 1).

Figure 2 presents an overview of a possible design of positioning, embossing and forming inserts for the single- and double-sided processing of sheet metals. The accurate positioning of blanks in combination with high flexibility on blank size is ensured by guiding inserts with positioning edges, as shown in Fig. 2(a). User-defined embossing geometries can be easily designed, manufactured and arranged within the tool base plate interfaces for the single- or double-sided PRE Process. Additionally, hollow embossing of form elements with small heights could be performed with the opposite arrangement of punch and die inserts, see Fig. 2(d).

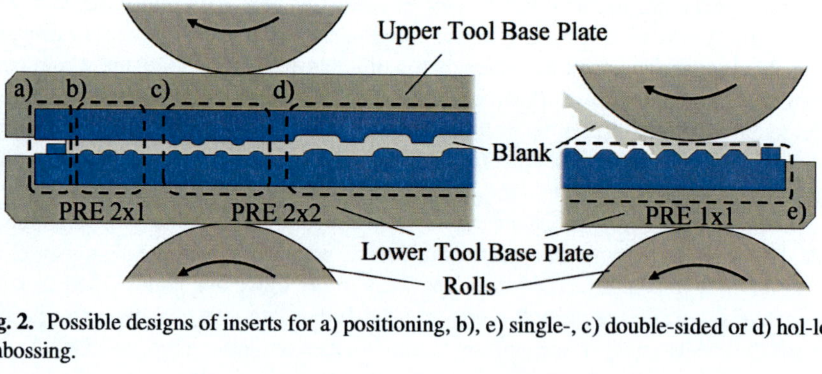

Fig. 2. Possible designs of inserts for a) positioning, b), e) single-, c) double-sided or d) hol-low embossing.

In the following, the feasibility and potentials of the new PRE Process is pointed out and the numerical and experimental work carried out in this respect is presented. Here, the single-side embossing with one (PRE 1 × 1) and two tool base plates (PRE 2 × 1) was selected and investigated.

3 Material

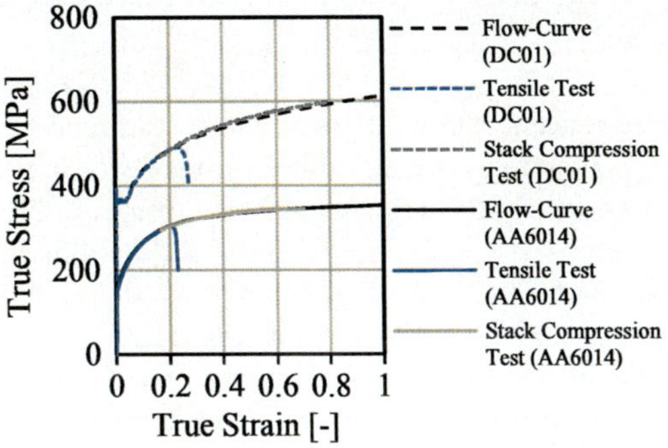

Fig. 3. Experimental stress-strain and extrapolated flow curves.

Table 1. Material properties.

	DC01	AA6014
t [mm]	2.00	2.50
E [GPa]	205.8	70.2
YS [MPa]	356.4	149.4
UTS [MPa]	396.9	248.0
UE [%]	17.0	19.2
n [–]	0.17	0.20
r0 [–]	0.79	0.86
r45 [–]	0.99	0.67
r90 [–]	0.89	0.54

Mild steel DC01 (t = 2.0 mm) and aged aluminium alloy AA6014, T4 condition (t = 2.5 mm) were selected for presented investigations. The characterization of elastic and plastic sheet metal material properties was performed by standardized tensile tests (ISO 6892-1) with ISO II specimen shape on the universal testing machine Roell + Korthaus RKM 100. For optical strain measurement, the GOM ARAMIS 12M system was used. Additional experimental data for higher strains was achieved by stack compression tests on the material testing machine Dynamic Systems Gleeble 3800c with a height-diameter ratio of 1.25 (AA6014) and 1.19 (DC01). The material properties identified and the flow curve derived according to the combined Swift/Hockett-Sherby law are shown in Table 1 and Fig. 3. Corresponding material parameters as well as the extrapolated flow curves were implemented in the elastic-plastic material model MAT_024 of LS-DYNA.

4 Numerical and Experimental Preliminary Study on the Feasibility of PRE-Process

Based on the elaborated concept of the new embossing method and the selected specimen material, an experimental and numerical feasibility study was performed preliminary. Initially and as a first approach, the experimental tool setup for the first investigations on PRE Process 1 × 1 and 2 × 1 were implemented in a simplified manner. Therefore, a hardened steel insert with frustums of pyramids on the top surface served as the lower tool base plate as well as the positive embossing form. The rectangular blank was manually placed in the center on top of it and the compound of tool and blank was moved through a defined roll gap smaller than its height (z-increment $\Delta z = 1.0$ mm) for the PRE Process 1 × 1, see Fig. 4(a). For a total of five repetitions, vertical reaction forces were measured by the strain transducer HBM SLB700A mounted on the roll housing of a Buehler EW150x80A duo mill stand. Embossed specimens were globally digitized by GOM ATOS 5M and specific local embossing depths were measured at the start (M_{Start}), center (M_{Center}) and end (M_{End}) of the specimen with the help of the confocal microscope nanofocus μsurf, see Fig. 4(b).

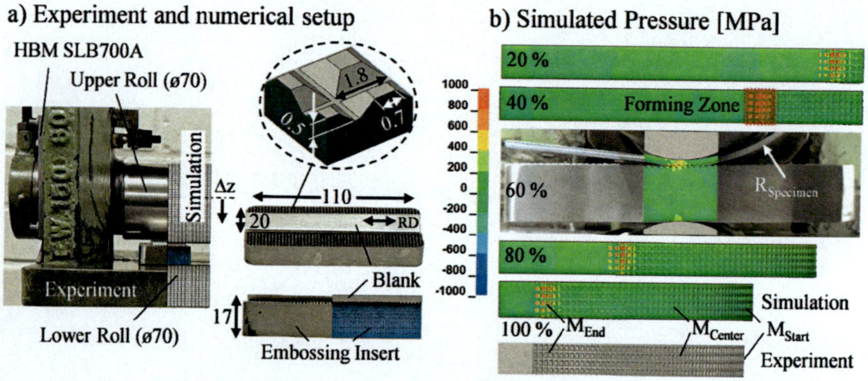

Fig. 4. a) Experimental and numerical investigations for the feasibility study of PRE Process 1 × 1 with b) exemplary illustrated results for DC01.

The experimental test described was afterwards numerically modeled within LS-DYNA FE solver using a half-symmetric model with elastic tool (MAT_001) and the elasto-plastic models MAT_024 of AA6014 and DC01. Sheet metal specimens were discretized by solid elements with a homogenous edge length of 0.2 mm, whereas the tool base plates were mapped by tetrahedrons with minimal edge length of 0.2 mm. The upper and lower roll of the mill stand with a diameter of 70 mm were described by rigid shell elements and a constant opposing rotation applied. In the simulation, the z-increment was adjusted step by step to achieve a comparable simulated embossing depth at the three evaluation areas and to compensate any elastic effects of the mill stand. Contact forces of the rolls were evaluated in z-direction and the embossing depth was measured within post-processing of simulation results in LS-PrePost.

As demonstrated by the numerical results for DC01, the sheet metal is incremental embossed and bent by the PRE Process 1 × 1. Evaluation of the calculated pressure at different time steps confirms, that the opposing rotation of the rolls and the resulting translational movement of the workpiece reduces the size of the forming zone and thus the required embossing force. Compared to a conventional embossing process with the simultaneous embossing of the specimen, the PRE Process 1 × 1 exhibits an average reduction of forming zone size of 92% for DC01 and 94% for AA6014. Furthermore, the pressure distribution indicates high compression inside the roll gap and thus demarcates the forming zone where a localized incremental forming of the blank occurs. Due to the locally increased pressure, an elastic deformation of the embossing insert was observed and resulted in a deflection of the tool up to 20 μm when embossing AA6014 and up to 37 μm for DC01. Therefore, a specific tool elasticity must be taken into account when developing new tool concepts for PRE Process in the future.

A comparison of the simulated and experimentally embossed specimens was performed by evaluating the embossing depth distribution and the global specimen curvature. Both, numerical predicted and experimental manufactured specimens show a scattered and inhomogeneous distribution of embossing depths. Simultaneous embossing and bending of AA6014 by the PRE Process 1 × 1 results in an experimental determined average indentation depth of 401 μm and a corresponding mean deviation of 57 μm. The experimentally determined specimen curvature radius amounted 181 mm, whereas the simulation underestimates the curvature by a radius of 167 mm. Experimental embossing and bending of mild steel DC01 revealed an average indentation depth of 367 μm and a mean variation of 14 μm. Maximum indentation depth was mostly located in the center area of specimen. The curvature radius of 76 mm was also underestimated by the simulation (R65 mm). In general, it can be stated that the formed blank curvature depends on material properties and embossing depth. Additionally, a measurement of experimental specimen dimensions before and after embossing showed a domination of the material flow in translational direction. The length of the specimens increased by 14.5% from an initial length of 110 mm for both materials, whereas the width increases slightly between 2% (DC01) and 3% (AA6014).

Figure 5 shows the setup for the numerical and experimental feasibility study of the PRE Process 2 × 1 for manufacturing flat embossed blanks. Here, a hardened and plane steel plate served as an upper tool base plate, numerically approximated as an elastic body in the simulation (LS-DYNA). All other dimensions, assumptions and boundary conditions for the experimental und numerical investigations remained as described for the PRE Process 1 × 1.

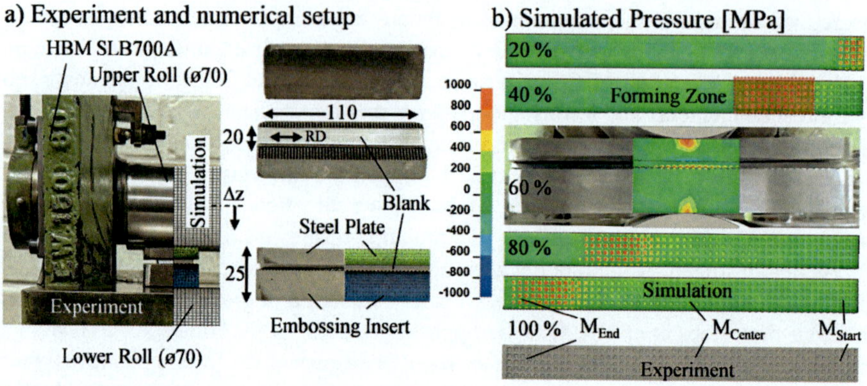

Fig. 5. a) Experimental and numerical investigations of the feasibility study of PRE Process 2 × 1 with b) exemplary illustrated results for DC01.

Adding a steel plate as a counter tool led to a fundamental change of the forming process and result. Due to the simplified tool setup for preliminary investigations, the absence of guiding the lower and upper tool base plates caused relative translational movements and resulted in a large number of failed attempts. A robust and precise guiding system is therefore essential for the PRE Process 2 × 1 and must be considered for the future design of the tool. However, post-processing of the pressure distribution at selected time steps indicated again an incremental embossing process. The supporting upper tool caused a slight increase of the forming zone, which is also moved translationally through the workpiece. In detail, embossing DC01 material by PRE Process 2 × 1 decreased the forming zone by 74% compared to a simultaneous embossing of the whole specimen (forming zone = 2,200 mm^2) and 75% for AA6014. High pressure in the area of the roll gap towards the contact zones between rolls and tools appeared remarkable and increased the elastic tool deformation. Processing the mild steel DC01 caused a maximum deflection of the upper steel plate by 28 μm (41 μm for AA6014) and 61 μm (46 μm for AA6014) for the lower embossing insert. The elastic tool deformation therefore depends on the material strength, but can be compensated by an adjustment of the roll gap (Δz).

For both the experimental and numerical experiments, the evaluation of the specimens embossed by using the PRE Process 2 × 1 provided a similar and homogeneous distribution with a slight decrease of indentation depth at the center area. Surface measurements revealed an experimentally embossed averaged indentation depth of 199 μm for DC01 (302 μm for AA6014) with a mean deviation of 29 μm for both materials. The specimen dimensions after embossing increased in length (DC01: + 1% & AA6014: + 1.4%) and width (DC01: + 1.6% & AA6014: + 5%). A tendency for width elongation can be stated for the investigated embossing process and specimen geometry. The shape of the forming zone, its moving direction and the plastic material properties therefore affect the material flow in the embossing process.

A comparison of experimentally measured and numerically calculated vertical forming forces for both processes and materials is shown in Fig. 6. The embossing of flat sheet metals by the PRE Process 2 × 1 resulted in increased vertical forces compared to the embossing and bending by PRE Process 1 × 1 for both materials investigated. The experimentally measured force values by the strain transducer revealed forces up to 210 kN for the PRE Process 2 × 1 with DC01 material and minor force demand for the PRE Process 1 × 1 and AA6014 material. On contrary, the manufactured indentation depth appeared higher for embossing the aluminum alloy without an upper tool plate. A reason can be found in the different localization of the forming zone and tool contact for both PRE Process variants, shown in Fig. 4 and 5. The more the forming zone is localized, the more the force is reduced and the embossing depth is increased. In general, the numerical process model tends to underestimate the vertical force demand by an average of 21% for similar embossing depths. Here, further model optimization is needed in the future in terms of enhanced material models for a proper consideration of tension and compression asymmetry of sheet metal material. Furthermore, the accuracy of the force calculation is reduced by the elasticities of the mill stand, which have not yet been taken into account.

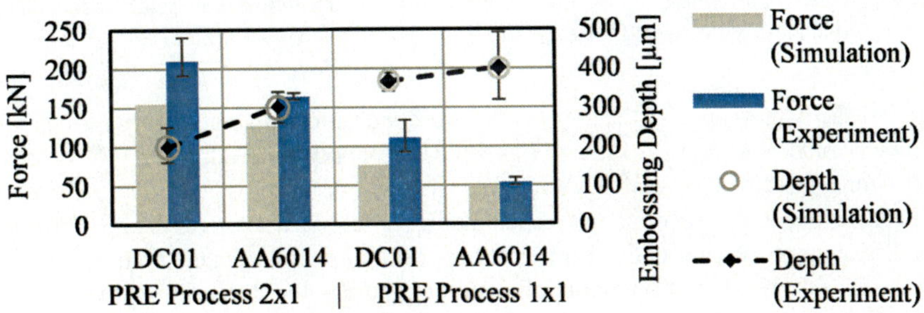

Fig. 6. Evaluation of resulting vertical forces for PRE Process 1 × 1 and 2 × 1 with comparable embossing depths – experiment and simulation.

5 Experimental Comparison of the Embossing Processes and Resulting Specimens

Based on the elaborated tool concept (Sect. 2) and the findings of the experimental and numerical feasibility study (Sect. 4), a new prototype tool for manufacturing flat and tailor embossed blanks with the PRE Process 2 × 1 was designed and machined. The tool design is shown in Fig. 7 a) and consists of a flat upper tool base plate positioned with block guides towards a positive embossing form on the lower base plate and positioning pins for the blank.

Hence, this tool setup can be used for the conventional embossing in a hydraulic press machine (Metallkraft WPP 160 HBK) as well as for the new embossing technique PRE Process 2 × 1 in the Buehler EW150 × 80A mill stand. The new embossing method was compared to the conventional embossing by a hydraulic press applying two different embossing depths for the both investigated materials. In order to ensure a comparable averaged embossing depth, same nine chosen indentations were selected on the respective embossed specimen for both embossing processes investigated and

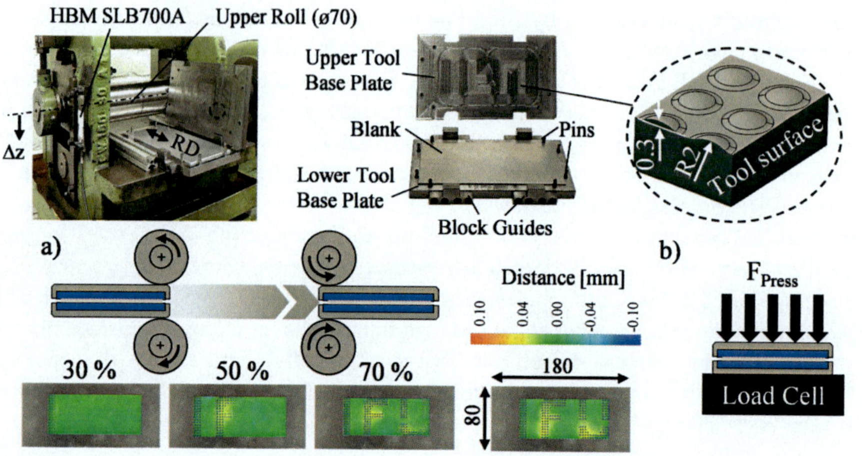

Fig. 7. Prototype tool for the a) PRE Process 2 × 1 and b) conventional embossing of exemplary tailor embossed blanks in AA6014 and DC01.

were precisely measured by the confocal microscope nanofocus μsurf. First, a medium averaged embossing depth of 115 μm was achieved for both materials in the PRE Process 2 × 1 for z-increments Δz of 0.5 mm (AA6014) and 0.7 mm (DC01). The maximum achievable indentation depth for each specimen material was limited for the PRE Process 2 × 1 by material strength, the chamfered edge of tool base plates and the torque of the mill stand. For the aluminium material, maximum embossing depth amounted 210 μm (Δz = 1.0 mm) and 190 μm (Δz = 1.25 mm) for steel material. Comparable indentation depths at the evaluated areas were achieved for conventional embossing by iteratively increasing the force of the hydraulic press.

After embossing a set of five specimens of each process and material, a global surface measurement of the embossed specimens was conducted using the GOM ATOS 5M system, as shown for maximum indentation depth in Fig. 8. Measured distances between the specimen surface and a fitted plane provide details about the flatness within the embossed area as well as a distribution of the embossing depth. For the PRE Process 2 × 1 (Fig. 8a) and the conventional embossing (Fig. 8b), this surface measurement revealed an enhanced flatness and homogenous distribution of embossing depths for the new embossing process by using the prototype tool with a guiding of tool halves and blank. On contrary, a decreasing indentation depth in the center area of the specimen was observed. Due to the in-plane material displacement while embossing, frictional and material effects might have limited the intrusion of the embossing tool in the central area. This effect appeared more pronounced for the PRE Process 2 × 1 and might be intensified by the elastic deflection of the rolls.

Furthermore, vertical process forces for the conventional embossing with a simultaneous intrusion of positive embossing form were captured by the load cell HBM C6A and compared to the estimated forces of corresponding PRE Processes by strain transducer HBM SLB700A. Figure 9 illustrates the maximum forces for embossing aluminium and steel blanks with medium and maximum indentation depth. Due to the increased yield

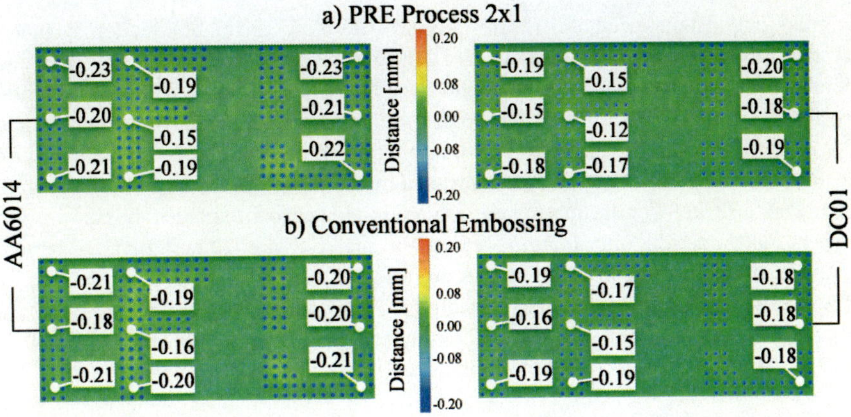

Fig. 8. Surface measurement of AA6014 and DC01 specimens embossed by a) PRE Process 2 × 1 and b) conventionally.

strength, high force demand (809 kN) was observed when conventionally embossing DC01 material. Embossing a similar indentation depth by using the PRE Process 2 × 1, maximum vertical forces were reduced by an average of 67% for DC01 and 62% for AA6014. This force reduction effect is caused by the new incremental forming kinematic of the PRE Process, where the line contact between rolls and tool defines a reduced forming zone, moving translationally through the workpiece.

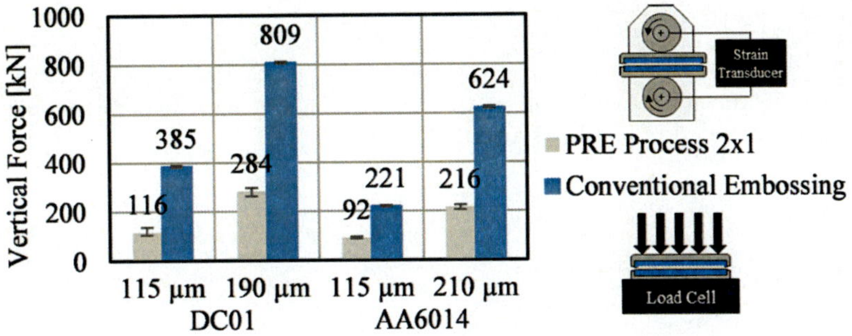

Fig. 9. Comparison of embossing force for conventional (blue) and PRE Process 2 × 1 (grey).

Conclusively, the impact of embossing on the tensile strength of the material was investigated. For this purpose, the area with straight-line indentations and the neighboring unembossed region were separated from three specimens of each parameter configuration, as shown in Fig. 10a). The obtained and non-standard miniature tensile specimens provided a range of embossed portion of sheet metal surface between 12% (115 µm embossing depth for both materials) and 23% for maximum embossing depth of 210 µm (AA6014). Tensile testing was done on the universal testing machine Roell + Korthaus RKM 100 and the maximum force versus displacement curves were recorded for constant speed of 0.1 mm/s. Figure 10c) shows the elaborated tensile material strengths represented by the maximum tensile force before failure for embossed and unembossed steel and aluminium specimens. The mild steel DC01 exhibits a maximum uniaxial force of 10.6 kN for the unembossed specimen and an increasing strength up to 10.9 kN due to the embossing. The maximum tensile force for unembossed aluminium specimens amounted 7.3 kN and was increased by the work hardening of the embossing process up to 7.6 kN. For the investigated materials and due to the small embossed portion of sheet metal surface, only marginal effects on material strengthening was observed. Nevertheless, plastic material properties are affected in terms of increased hardening rates and reduced pronounced yield strength for DC01 material, as shown in Fig. 10 b). Adjusting the material properties by embossing process is thereby not affected, whether the new PRE Process 2 × 1 or the conventional embossing process is used.

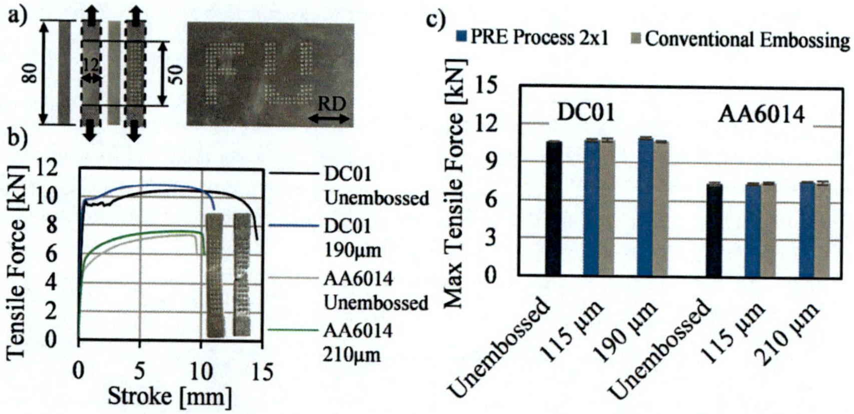

Fig. 10. a) Specimen separation and b) tensile testing for c) maximum tensile forces of unembossed and embossed DC01 and AA6014 sheet metal.

6 Conclusion and Outlook

The presented paper discusses a new approach for an efficient and flexible embossing of sheet metals, called the Plate Roll Embossing Process (PRE Process). A conceptual design followed by numerical and experimental studies with aluminium (AA6014) and steel (DC01) sheet metals revealed the general feasibility of this new embossing method as well as a pronounced reduction of embossing forces. In particular, new tool design and forming kinematics reduce the simulatenous acting contact areas between tool and blank. This leads to a downsizing of the forming zone, which is moved translationally through the workpiece by the rotation of the rolls. Process derivations for the one-sided embossing with (PRE Process 2 × 1) and without an upper tool (PRE Process 1 × 1) resulted in significantly different formed parts. The latter minimizes the line-shaped forming zone due to the direct contact with the upper roll and forms an embossed and bent sheet metal part under reduced loads. Experimental investigations on manufacturing tailor embossed blanks by the PRE Process 2 × 1 using a prototype tool remarkably reduces the force demand up to 69% compared to conventional embossing on a hydraulic press. An impact of the embossing processes on the final geometry and material property adjustment appeared to be neglectable. Therefore, new PRE Process provides a manufacturing alternative for embossing sheet metals under reduced force demand compared to existing embossing techniques.

Based on these promising feasibility study, future research work is needed to transfer the PRE Process into industrial practice, which will necessitate the development of new forming machines and tool designs. Focusing on the numerical process model to enhance accuracy and cost may support the process development by the identification of significant parameters and correlations. Furthermore, new applications such as hollow embossing of metallic bipolar plates by a PRE Process 2 × 2 might enable new manufacturing potentials with reduced forming loads and increased formability.

Acknowledgement. The authors would like to thank Novelis Switzerland SA for the donation of aluminium material.

References

1. Horton, P.M., Allwood, J.M.: Yield improvement opportunities for manufacturing automotive sheet metal components. J. Mater. Process. Technol. **249**, 78–88 (2017)
2. Merklein, M., Johannes, M., Lechner, M., et al.: A review on tailored blanks—production, applications and evaluation. J. Mater. Process. Technol. **214**(2), 151–164 (2014)
3. Walzer, S., Liewald, M.: Studies on the influence of embossing on the mechanical properties of high-strength sheet metal. In: ESAFORM Proceedings, Vitoria-Gasteiz, Spain (2019)
4. Briesenick, D., Walzer, S., Liewald, M.: Study on the Effect of embossing on the bending properties of high-strength sheet metals. In: Daehn, G., Cao, J., Kinsey, B., Tekkaya, E., Vivek, A., Yoshida, Y. (eds.) Forming the Future. TMMMS, pp. 2585–2595. Springer, Cham (2021). https://doi.org/10.1007/978-3-030-75381-8_213
5. Walzer, S., Liewald, M.: Novel approach to decrease sheet thinning during sheet metal forming by using embossing technique. Procedia Manuf. **50**, 795–799 (2020)
6. Abe, Y., Mori, K.-I., Maeno, T., Ishihara, S., Kato, Y.: Improvement of sheet metal formability by local work-hardening with punch indentation. Prod. Eng. Res. Devel. **13**(5), 589–597 (2019). https://doi.org/10.1007/s11740-019-00910-6
7. Namoco, C.S., Iizuka, T., Narita, K., et al.: Effects of embossing and restoration process on the deep drawability of aluminum alloy sheets. J. Mater. Process. Technol. **187–188**, 202–206 (2007)
8. Walzer, S., Liewald, M., Simon, N., et al.: Improvement of sheet metal properties by inducing residual stresses into sheet metal components by embossing and reforming. Appl. Sci. Eng. Progr. **15**(1) (2021)
9. Gilch, I., et al.: Impact of residual stress evoked by pyramidal embossing on the magnetic material properties of non-oriented electrical steel. Arch. Appl. Mech. **91**(8), 3513–3526 (2021). https://doi.org/10.1007/s00419-021-01912-6
10. Cotton, D., Maillard, A., Kaufmann, J.: Improved coining force calculations through incorporation of key process parameters. IOP Conf. Ser. Mater. Sci. Eng. **967**(1) (2020)
11. Sieczkarek, P., Kwiatkowski, L., Tekkaya, A.E., et al.: Improved tool surfaces for incremental bulk forming processes of sheet metals. Key Eng. Mater. **504–506**, 975–980 (2012)
12. Halmos, G.T.: Roll Forming Handbook. Taylor & Francis, Boca Raton (2006)

Tailoring the Hardness in Multi-stage Press Hardening of 22MnB5 Sheet Material in a Progressive Die

Juri Martschin[1](✉), Malte Wrobel[2], Joshua Grodotzki[1], and A. Erman Tekkaya[1]

[1] Institute of Forming Technology and Lightweight Components, TU Dortmund University, Dortmund, Germany
juri.martschin@iul.tu-dortmund.de
[2] Institute of Mechanical Process Engineering and Mechanics, Karlsruhe Institute of Technology, Karlsruhe, Germany

Abstract. Multi-stage press hardening enables a stepwise control of the thermo-mechanical history during press hardening, thereby facilitating a flexible hardness tailoring. Further degrees of freedom arise from utilizing rapid austenitization by induction heating within the first process stage. To assess the potential of rapid austenitization and a subsequent multi-stage thermo-mechanical history for tailoring the hardness, near-process characterizations are carried out with 22MnB5 steel. By a variation of the austenitizing temperature and dwell time, when cooling at a rate of 30 K/s, a hardness between 310 and 448 HV was set. By inducing a strain of 0.1 after rapid austenitization, the hardness was reduced by 15% due to an accelerated ferrite and bainite transformation. However, it was observed that a second, subsequent induced strain can potentially increase the hardness by work hardening. Using the example of multi-stage press hardening with rapid austenitization implemented in a progressive die, the hardness of 22MnB5 sheet material is tailored. A hardness between 350 and 445 HV was set by varying the austenitization parameters and the strain induced with a controllable blank holder.

Keywords: Multi-stage; Press hardening · Tailoring · Rapid heating

1 Introduction

Sheet metal components with a tailored hardness distribution are key to safety-relevant lightweight structures in many engineering contexts. Established manufacturing methods include the thermal process control via blank and tool tempering as well as the use of tailor welded blanks [1]. Such approaches exhibit limitations with regard to the components complexity. A new development is tailoring by multi-stage press hardening with a step by step control of the thermal history during quenching [2]. One related concept is press hardening in a progressive die with rapid induction heating, wherein the hardness is set by variation of the austenitization parameters [3]. Combining both approaches and utilizing the capabilities of multi-stage forming provides increased flexibility in terms

K. Mocellin et al. (Eds.): ICTP 2023, LNME, pp. 90–100, 2024.
https://doi.org/10.1007/978-3-031-40920-2_10

of the components complexity and for tailoring the hardness. However, the influence of a multi-stage thermo-mechanical history after a rapid, optionally non-homogeneous austenitization on the microstructure evolution and ultimately on the hardness is not fully understood. So far, the majority of the studies conducted on rapid heating in press hardening of 22MnB5 - as summarized in [4] - have focused on achieving a homogeneous austenitic state before quenching, either to replace conventional furnace heating or to improve product properties. The effects of the different strain levels induced after potentially inhomogeneous austenitization in conjunction with the influence of different forming temperatures and cooling rates have not been considered in depth. Hence, in context of tailoring the hardness of 22MnB5, this paper first examines the potential of rapid austenization in combination with different cooling rates, as well as the influence of a subsequent single- and multi-stage forming operation using near-process characterization methods. Finally, a multi-stage hardness tailoring by multi-stage press hardening with rapid austenization combined with an extended actuator setup implemented in a progressive die is demonstrated.

2 Methods

2.1 Setup for Press Hardening with Tailored Properties in a Progressive Die

To tailor the hardness by press hardening in a progressive die, a sequence of four die stages (Fig. 1) is utilized. These stages are arranged after the precutting stages of a progressive die, which is mounted in a servo press (MSD2–400, Schuler). The material investigated is the press hardening steel 22MnB5 (Table 1) with a sheet thickness t_0 of 2 mm. Initially, the microstructure has a ferrite grain size averaging 12 μm and consists of 70% ferrite, 22% sorbite, 5% pearlite, and 3% fine carbides.

Table 1. Chemical composition in wt. % of the used material batch of 22MnB5 slit strip according to the melt analysis (Anal.) and tolerance range (Min. – Max.) according to [5].

Element		Fe	Mn	Si	Cr	C	Al	Ni	Ti	Cu
wt.%	Anal	97.935	1.13	0.33	0.27	0.19	0.036	0.03	0.028	0.02
	Min. – Max.	–	1.10–1.40	0.00–0.40	0.15–0.35	0.19–0.25	0.020–0.060	–	0.02–0.05	–
Element		**P**	**Mo**	**N**	**As**	**S**	**B**	**V**	**Sn**	**Nb**
wt.%	Anal	0.01	0.004	0.0037	0.003	0.003	0.0023	0.002	0.002	0.001
	Min. – Max.	0.000–0.025	–	–	–	0.000–0.015	0.0008–0.0050	–	–	–

In this study, the cycle time per process stage t_{cs} is 10 s resulting from a stroke rate of the press f_{SR} of 6 min^{-1}. In the first stage, the pre-cut blank is inductively heated with a rate r_h of 300 K/s to the austenitizing temperature $T_{\gamma.}$, which is then held for the dwell time t_γ. By means of the control variables of austenitization, the progress of the homogenization of the microstructure can be varied [6] and hereby the achievable hardening [7]. In the second stage, a temperature profile can be set via air cooling (area 1) and resistance heating (area 2), which additionally enables an extension of the dwell time for area 2. In the third stage, a hat-shaped profile is stretch-formed. Here, a blank holder with controllable force F_{BH} enables the sheet draw-in E and thus the thinning $\Delta t(l_{\mathrm{h}})$ to be set. In the fourth stage, the frame located in area 2 is bent. Consequently, the strains resulting from the preceding stretch forming operation are superimposed by bending strains. In both the third and fourth stage of the presented progressive die, the blank is quenched by contact with the dies, which are water cooled.

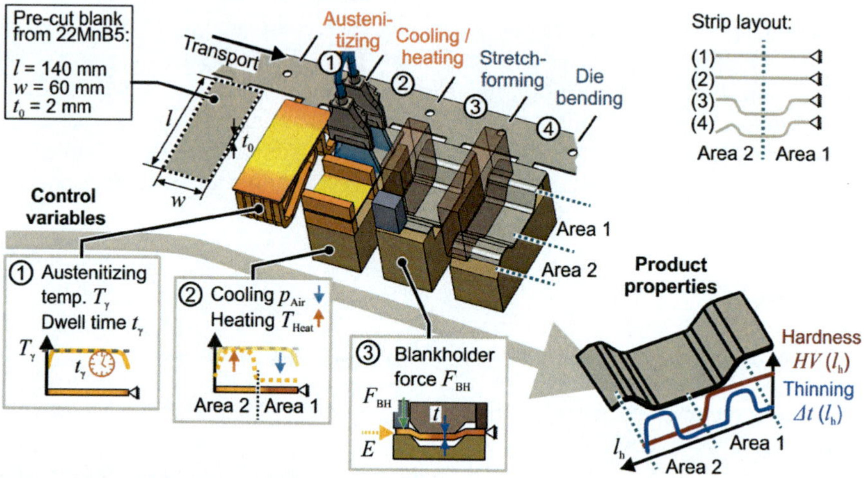

Fig. 1. Multi-stage press hardening of 22MnB5 in a progressive die with extended actuator arrangement for tailoring the hardness in conjunction with the thinning in area 1 and 2.

For supplying the power for inductive heating in the first stage, a 40 kW generator (TrueHeat MF 7040, Trumpf) is used. To ensure homogeneous heating during austenitization, a two-part coil arrangement is utilized consisting of a primary coil that generates a longitudinal field and a return conductor connected to the pre-cut blank [6]. Further details for the specific coil layout are given in [7]. The temperature during austenitizing is measured with a pyrometer (Metis M318, Sensortherm), which is focused on the center line of the pre-cut blank, and controlled with a PID-controller. The emissivity coefficient was calibrated for each used combination of austenitization temperature and dwell time in a separate heating setup by comparing pyrometer measurements with thermocouple measurements. After calibration, the average deviation between the pyrometer and the thermocouple is $\pm$ 5 K. In the second stage, the air pressure p_{Air} for cooling of area 1 is set with a controllable valve (VPPM, Festo) and area 2 is heated by a 44 kW DC generator (LAB/HP 4020, FE-Systems). In the third stage, the blank holder force F_{BH} is applied by a hydraulic cylinder, with the hydraulic pressure being set by a pressure-limiting valve (PMV 41–43, HAWE). The temperature distribution is recorded for all four process stages with a thermal imaging camera (TIM-M1, Micro Epsilon) during the transport phase.

2.2 Dilatometry and Multi-stage Hot Tensile Tests

A hot tensile test setup (Fig. 2-A) implemented in a universal testing machine (LFV-100-HH, Walter + Bai) is used to analyze the effect of rapid austenitization and of a multi-stage thermo-mechanical history in the context of tailoring the hardness. A near-process thermo-mechanical history corresponding to that of the multi-stage process, cf. Chapter 2.1, is imposed on the tensile test specimen (Fig. 2-B). First, the test sample with a parallel length of 55 mm and width of 6 mm is heated with $r_h = 300$ K/s by a 10 kW DC generator to the austenitizing temperature T_γ. The latter is maintained for the dwell time t_γ. Then the sample is cooled to the respective forming temperatures of the progressive die stages ($T_{\varepsilon\text{-S3}}$, $T_{\varepsilon\text{-S4}}$) with stage-dependent cooling rates ($r_{c\text{-S2}}$, $r_{c\text{-S3}}$, $r_{c\text{-S4}}$), which are averaged with the cooling rate $r_{T\gamma\text{-Ms}}$. The steps are timed according to the stage cycle time t_{cs}. Cooling of the specimen is achieved by compressed air. A pyrometer (Metis M318, Sensortherm) is used to measure the temperature, which is set by means of the DC generator's power output. The emissivity coefficients identified as described in Sect. 2.1 are used. At the start of the third and/or the fourth stage, the specimen is strained with the rate $\dot{\varepsilon}_\Phi$ to the strain of $\varepsilon_{\Phi\text{-S3}}$ and/or $\varepsilon_{\Phi\text{-S4}}$. For the dilatometry and strain measurements, an extensometer meeting the ISO 9513 class 0.5 accuracy requirements (Model 3542, Epsilon) is employed. In order to maintain a uniaxial tension stress state within the specimen during the tensile tests, and to hereby ensure comparability with other shapes of tensile test specimens, the tensile tests were stopped prior to the onset of necking.

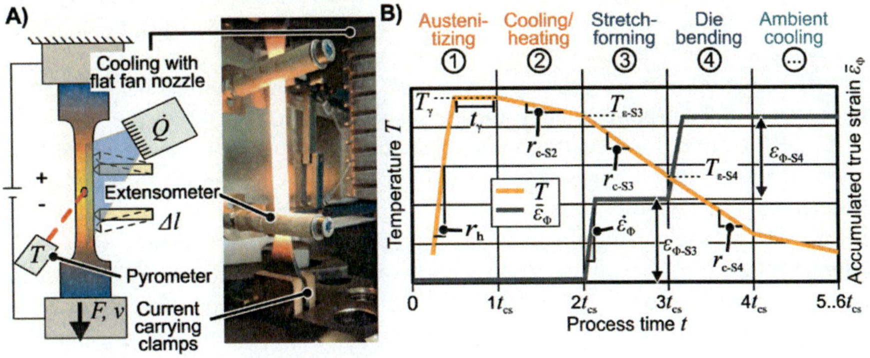

Fig. 2. **A)** Setup for the dilatometry and multi-stage tensile tests; **B)** Temperature and true strain over process time during the near-process tensile tests.

2.3 Measurement Methods

To create micrographs, a sample is cut out from the center of the tensile test specimens. The cuts for separating the sample were made along the width direction of the tensile test specimen. The cut surface perpendicular to the drawing direction of the tensile test is evaluated respectively. Other cutting directions, e.g. along the length of the tensile specimens, are not investigated under the assumption of isotropic material behavior at the elevated temperatures studied. The assumption needs to be verified in future work to make valid predictions about the microstructure evolution in all three dimensions of the samples. After separation, the sample is embedded in phenolic resin, then polished (last step - 1 μm suspension) and etched: first, for 15 s in a solution of 100 ml 4 % picric acid and 1 ml concentrated hydrochloric acid and then for 5 – 10 s in a 10 % sodium metabisulfite solution. Due to etching, martensite turns light brown, banite dark brown to black and ferrite white. The micrographs are acquired with a microscope (Axio Im-ager M1m, Zeiss) at a 1000x magnification. The hardness measurements of dilatometry and tensile tests specimens are conducted on the polished surface of embedded speci-mens with a hardness tester (Dia Testor 2 RC, Wolpert) according to HV10. To evaluate the hardness of the parts produced with multi-stage press hardening, those parts were cut along the centerline and the cut surface was polished (last polishing step – polish agent grain size of 3 μm). Then the hardness was measured as described above. The thinning $\Delta t(l\mathrm{h})$ of the parts produced with multi-stage press hardening is measured along the centerline of the parts with a coordinate measuring machine (PRISMO VAST 5 HTG, Zeiss).

3 Experimental Results

3.1 Influence of Austenitization Parameters on the Hardenability

To assess the potential for tailoring the hardness via the control variables austenitizing temperature T_γ and dwell time t_γ after rapid heating with r_h = 300 K/s, the hardness of heat-treated specimens and the relative elongation measured during quenching is compered (Fig. 3). The temperature profile over time for the heat treatment as well as the variation of the heat treatment parameters is specified in Fig. 3-A.

In the literature, for 22MnB5, the critical cooling rate for a complete martensitic transformation is in the range of 25 K/s [8] to 30 K/s [9]. According to the Maynier model [10], with the given chemical composition of the investigated material batch, the hardness after a complete martensitic transformation is approximately 455 HV. In contrast, after rapid austenitization and cooling with a cooling rate of 30 K/s, a hardness within the range of 310 HV (T_γ = 950 °C, t_γ = 1 s) to 448 HV (T_γ = 1050 °C, t_γ = 5 s) is obtained (Fig. 3-B), thus indicating, that before quenching, different homogenization states were present. In general, within the investigated parameter space of this study, an increase in hardness with higher austenitizing temperatures, longer dwell times as well as with faster cooling rates can be observed. With increasing austenitization parameters the hardness converges towards a maximum hardness value. The calculated theoretical maximum value of 455 HV (see above) was reached with T_γ = 1050 °C, t_γ = 5 s and r_c = 50 K/s. The increase in hardness with the austenitization parameters can be traced back to the progress of homogenization and in this context mainly the progress of carbide dissolution. The latter increases the carbon content available for a martensitic transformation while reducing the critical cooling rate until a threshold is reached when a homogeneous austenitic state is established. Furthermore, rising austenitization parameters yield larger austenite grains [11], delaying the transformation during cooling.

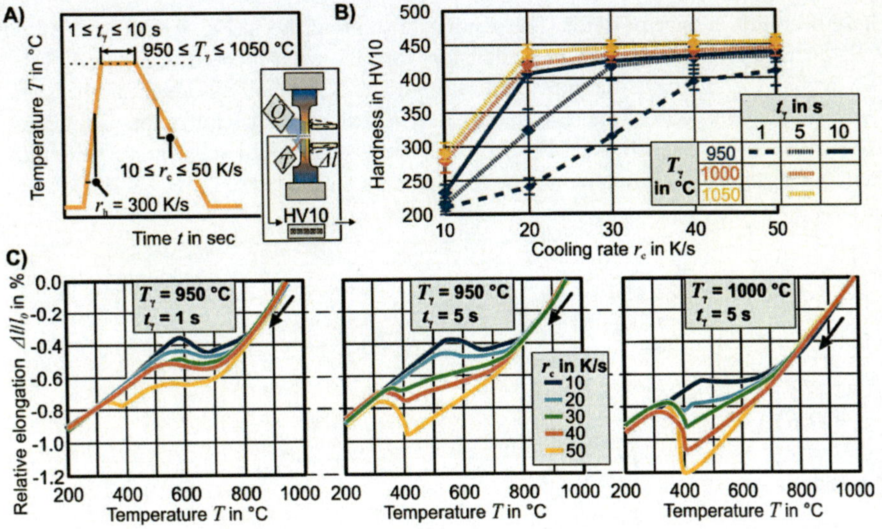

Fig. 3. **A)** Temperature time profile of the dilatometry; **B)** Hardness after different austenitizing conditions; **C)** Relative elongation during cooling after different austenitizing conditions.

The above described relations are supported by the relative elongation $\Delta l/l_0$ measured during quenching of the specimens (Fig. 3-C). With increasing austenitizing temperature and dwell time, on average the transformation points indicated by a change of the gradient of the relative elongation shift to lower temperatures. In addition, an increase in the amplitude of the jump in the relative elongation at the martensite start temperature (~ 408 C) can be observed. The $\Delta l/l_0$-jump increases from 0.2% (for $T_\gamma =$ 950 °C) to 0.4% ($T_\gamma = 1000$ °C) with the same cooling rate of 50 K/s and dwell time of $t_\gamma = 5$ s. Both above stated observations are correlating with the assumed progress of homogenization. For $T_\gamma = 950$ °C, $t_\gamma = 1$ s and $r_c = 50$ K/s, the progression of the relative elongation shows several changes in the gradient indicating a ferrit, bainite and martensite formation yielding a hardness of 410 HV. The range of phase transformations at this high cooling rate could be due to the non-uniform distribution of carbon in the initial microstructure. This entails that former pearlite and sorbite regions can transform to martensite, while in the former ferrite regions, the carbon from the carbides could not yet diffuse sufficiently into the austenite. When the austenitizing temperature is increased to $T_\gamma = 1000$ °C and the dwell time to $t_\gamma = 5$ s, a microstructure with a near maximum hardness of 445 HV is obtained correlating with the only discontinuity in the relative elongation around the martensite start temperature (~408 °C). In the case of T_γ $= 950$ °C and $t_\gamma = 5$ s, an intermediate state from the two discussed above is attained.

3.2 Influence of Multi-stage Forming

To estimate the influence of deformation induced acceleration of diffusion driven phase transformation [12] in the context of multi-stage forming, the hardness and micrographs of purely heat treated specimens are compared with hot tensile test specimens, which have been drawn once or twice. The results are summarized in Fig. 4. In Fig. 4-A and –B, the underlying thermo-mechanical history for each test is illustrated. Figure 4-B and -D display the measured hardness and induced strain for the individual micrographs, respectively. The color of the surrounding box indicates the austenitization parameters. The captions in Fig. 4-B and -D specify the respective multi-stage press hardening process history that is emulated by the near-process tensile tests.

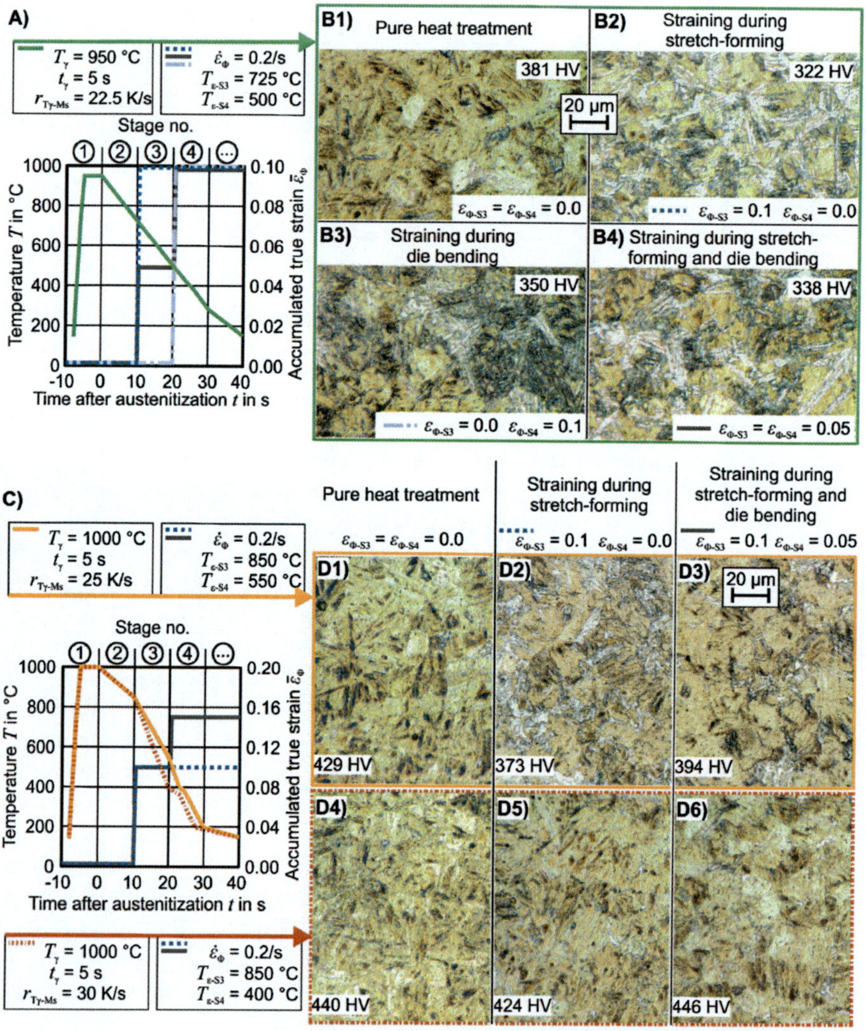

Fig. 4. **A + C)** Thermal and mechanical history of the near-process tensile test; **B + D)** Images of the formed and heat treated specimens under the light microscope and resulting hardness.

With an austenitizing temperature $T_\gamma = 950$ °C, a dwell time $t_\gamma = 5$ s and an average cooling rate $r_{T\gamma\text{-}Ms} = 22.5$ K/s, a martensitic-banitic microstructure with a resulting hardness of 381 HV is obtained (Fig. 4-B1). If additionally in the third stage at the temperature $T_{\varepsilon\text{-S3}} = 725$ °C a strain $\varepsilon_{\Phi\text{-S3}} = 0.1$ is applied, the fraction of upper bainite increases and ferrite islands appear, while the martensite fraction decreases and a hardness of 322 HV is achieved (Fig. 4-B2). The drop in hardness of 15% is associated with a strain-induced shift of the diffusion driven phase transformation to higher cooling rates. If the strain $\varepsilon_{\Phi\text{-S4}} = 0.1$ is applied at the temperature $T_{\varepsilon\text{-S4}} = 500$ °C, the proportion of upper bainite is more pronounced (Fig. 4-B3) and a hardness of 350 HV results. The influence of strain on the phase transformation appears to be less prominent at the lower forming temperature, which could be attributed to reduction of the remaining time for diffusion until the martensite start temperature is reached. If the strain of 0.1 is divided among the forming stages ($\varepsilon_{\Phi\text{-S3}} = \varepsilon_{\Phi\text{-S4}} = 0.05$), an intermediate state of the two previously discussed with a hardness of 338 HV is obtained (Fig. 4-B4). It should be noted that the hardness resulting from the phase fractions can be superimposed by work hardening due to forming at a lower temperature of $T_{\varepsilon\text{-S4}} = 500$ °C. This mechanism could not be separated in this study.

With an austenitizing temperature $T_\gamma = 1000$ °C and a dwell time $t_\gamma = 5$ s, non-continuous cooling with an average of $r_{T\gamma\text{-Ms}} = 25$ K/s, results in a hardness of 429 HV (Fig. 4-D1) and $r_{T\gamma\text{-Ms}} = 30$ K/s results in a hardness of 440 HV (Fig. 4-D4). The latter corresponds to the measured hardness values after continuous cooling (Sect. 3.1).

If additionally a strain $\varepsilon_{\Phi\text{-S3}} = 0.1$ is applied at a temperature $T_{\varepsilon\text{-S3}} = 850$ °C followed in the first case by cooling with $r_{c\text{-S3}} = 30$ K/s and $r_{c\text{-S4}} = 35$ K/s, the hardness is reduced by 13.05% to 373 HV due to the formation of ferrite and upper bainite (Fig. 4-D2).

In the second case, when cooling the sample with $r_{c\text{-S3}} = 45$ K/s and $r_{c\text{-S4}} = 35$ K/s, the hardness is reduced by 3.64% to 424 HV due to the formation of small ferrite grains (Fig. 4-D5). When an additional strain of $\varepsilon_{\Phi\text{-S4}} = 0.05$ at the temperature $T_{\varepsilon\text{-S4}} = 550$ °C (Fig. 4-D3) or $T_{\varepsilon\text{-S4}} = 400$ °C (Fig. 4-D6) is induced after straining with $\varepsilon_{\Phi\text{-S3}} = 0.1$, the hardness increases relative to the cases displayed in Fig. 4-D2 and Fig. 4-D5, respectively. Despite the comparable resulting microstructures (cf. Figure 4-D2 with -D3 and -D5 with -D6), the second induced strain at $T_{\varepsilon\text{-S4}} = 550$ °C and $T_{\varepsilon\text{-S4}} = 400$ °C yields an increase in hardness, indicating the presence of work hardening.

3.3 Hardness and Thinning in Multi-stage Press Hardening

To demonstrate the feasibility for tailoring the hardness in multi-stage press hardening with the actuator setup described in Sect. 2.1, the austenitizing temperature T_γ and the blank holder force F_{BH} during stretch forming are varied, while maintaining a dwell time of $t_\gamma = 5$ s (Fig. 5). In addition, in the case of parameter set 5, with the use of DC heating in the second process stage the dwell time for area 2 is extended to $t_\gamma = 10$ s. After the die try-outs, the hardness of the formed and hardened parts produced during the parameter variation is measured (cf. Chapter 2.3) along the center fiber at the points indicated in Fig. 5, and the thinning is measured along the entire length of the center fiber. The hardness distribution over the sheet thickness and over the width of the produces parts is not evaluated. Therefore, the following statements are only applicable to the center fiber.

When heating the blank to $T_\gamma = 1000$ °C and setting the blank holder force to $F_{BH} = 1$ kN, an average hardness of ~ 440 HV is obtained while the thickness strain ε_t, does not exceed 0.027 (set 3). This average hardness value is reduced to ~ 430 HV when setting a lower T_γ of 950 °C (set 1). If the blank holder force is increased to $F_{BH} = 10$ kN, the sheet draw-in E is restricted, resulting in increased material thinning of $\varepsilon_t \approx 0.062$ in the frame of area 1 and $\varepsilon_t \approx 0.048$ in the frame of area 2. This in turn triggers a reduction of the hardness of up 17.6% for $T_\gamma = 950$ °C (set 2) and 8.1% for $T_\gamma = 1000$ °C (set 4) due to the strain-induced acceleration of diffusion driven phase transformation. However, there is only a slight influence of the thickness strain on the hardness in the bending radii, in which the strain peaks are present. This can be related to high contact pressures in the radii during forming causing here higher cooling rates than in the other sheet segments. Furthermore, in contrast to the radii R_1, which are formed during stretch-forming, the radius R_2 is bend afterwards in the fourth stage at a temperature $T_{\varepsilon\text{-S4}}$ of ~ 550 °C. As revealed with the near-process tensile tests, forming at this temperature after a previous forming at a higher temperature can negate softening by ferrite and bainite formation through work hardening. If, with an austenitizing temperature T_γ of 950 °C, the dwell time t_γ of area 2 is extended to 10 s by the DC heating in the second process stage and the blank holder force F_{BH} is set to 5 kN (set 5), a thickness strain ε_t of 0.035 is obtained in area 2. In addition, a hardness profile is obtained in area 2, which is similar to that from parameter set 1, despite the larger induced strains. Therefore, by extending the dwell time for area 2, it is possible to avoid softening by strain induced accelerated phase transformation.

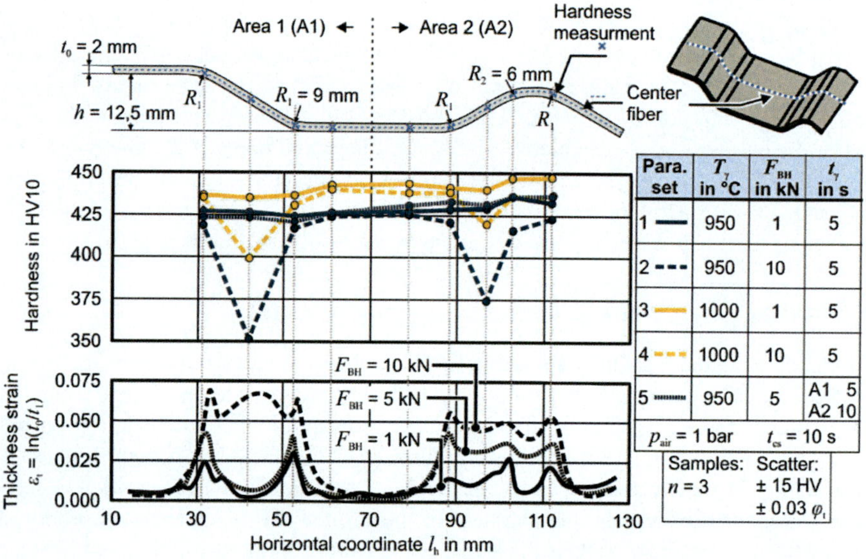

Fig. 5. Hardness and thickness strain of parts produced with multi-stage press hardening.

4 Conclusion

Using the example of multi-stage press hardening of 22MnB5 in a progressive die with rapid austenitization, the potential for tailoring the hardness by gradually adapting the thermo-mechanical history throughout the process was demonstrated. With a near-process heat treatment, the effect of the austenitizing temperature T_γ and the dwell time t_γ after rapid heating with $r_\text{h} = 300$ K/s on the hardenability was examined. When cooling at a rate of $r_\text{c} = 30$ K/s, a hardness between 310 HV ($T_\gamma = 950$ °C, $t_\gamma = 1$ s) and 448 HV ($T_\gamma = 1050$ °C, $t_\gamma = 5$ s) was achieved. The underlying mechanism, which can be utilized for a hardness tailoring, is the progress of the homogenization of the austenite which depends on T_γ and t_γ. With near-process tensile tests, the effect of single and multiple forming steps on the hardness after a rapid austenitization was investigated. Inducing a strain of $\varepsilon = 0.1$ at a temperature of 725 °C during cooling with a rate of 22.5 K/s reduced the hardness by 15% due to an accelerated ferrite and bainite transformation. On the other hand, a second induced strain can potentially increases the hardness due to work hardening. An analysis of multi-stage press-hardened components revealed that hardness distributions with a range of 350 to 445 HV could be tailored by controlling the thermo-mechanical history via inductive heating, a subsequent DC resistance heating as well as via a controllable blank holder.

Acknowledgement. The authors thank the German Research Foundation (DFG) for the financial support of project 424334660 within the SPP2183 "Property-controlled forming processes".

References

1. Karbasian, H., Tekkaya, A.E.: A review on hot stamping. J. Mat. Proc. Tech. **210**(15), 2103–2118 (2010)
2. Nakagawa, Y., Mori, K., Suzuki, Y., Shimizu, Y.: Tailored tempering without die heating in hot stamping of ultra-high strength steel parts. Mat. & Des. **192**, 108704 (2020)
3. Löbbe, C., Hering, O., Hiegemann, L., Tekkaya, A.E.: Setting mechanical properties of high strength steels for rapid hot forming processes. Materials **9**(4), 229 (2016)
4. Andreiev, A., Grydin, O., Schaper, M.: Evolution of microstructure and properties of steel 22mnb5 due to short austenitization with subsequent quenching. Steel Res. Int. **87**(12), 1733–1741 (2016)
5. Spittel, M., Spittel, T.: Steel symbol/number: 22MnB5/1.5528. In: Martienssen, W., Warlimont, H. (eds.) Metal Forming Data of Ferrous Alloys - Deformation Behaviour 2C1, pp. 930–935. Springer, Heidelberg (2009)
6. Tekkaya, A.E., Meya, R., Löbbe, C.: Heizvorrichtung zur Erzeugung von Joule'scher Wärme bei der Erwärmung von vorzugsweise dünnwandigen Blechteilen mittels mittelfrequentem Strom. Patent application. DE 10 2017 004 935 A1 (2017)
7. Löbbe, C.: Temperaturunterstütztes Biegen und Wärmebehandeln in mehrstufigen Werkzeugen. Dissertation. TU Dortmund University (2018)
8. Naderi, M.: Hot stamping of ultra high strength steels. Dissertation. RWTH Aachen University (2007)
9. Geiger, M., Merklein, M., Hoff, C.: Basic investigations on the hot stamping steel 22MnB5. AMR **6–8**, 795–804 (2005)

10. Maynier, P., Jungmann, B., Dollet, J.: Creusot-Loire system for the prediction of the mechanical properties of low alloy steel products. Symp. Hardenab. Concepts Appl. Steel **1977**, 518–545 (1978)
11. Sun, C.N., Zhang, H.H.: Microstructural evolution and quenching properties of 22MnB5 steel for hot stamping during resistance heating. AMR **849**, 75–80 (2013)
12. Naderi, M., Saeed-Akbari, A., Bleck, W.: The effects of non-isothermal deformation on martensitic transformation in 22MnB5 steel. Mat. Sci. Eng. **487**(1–2), 445–455 (2008)

Evaluation of Required Diameter Adjustment of a Novel Ironing Punch Concept for Reducing Wear During Retraction

Kaarel Siimut[1,2], Úlfar Arinbjarnar[1], Kasper M. Madsen[2], Ermanno Ceron[2], Thomas L. Madsen[2], and Chris V. Nielsen[1(✉)]

[1] Department of Civil and Mechanical Engineering, Technical University of Denmark, Kgs. Lyngby, Denmark
cvni@dtu.dk

[2] Department of Mechanics Technology and Innovation, Group Advanced Manufacturing Engineering, Grundfos A/S, Bjerringbro, Denmark

Abstract. Tools for high-volume sheet metal forming processes generally sacrifice flexibility for productivity. For a tribologically severe process like ironing, this lack of flexibility makes mitigating galling during internal ironing challenging. In a novel concept of an adjustable ironing punch, the diameter of the punch can be rapidly and accurately adjusted to control the part height and decrease contact pressure during punch retraction, thereby reducing tool wear and lubrication requirements, while also making the process more flexible. The present study shows numerical analysis of the process of ironing the interior of a deep drawn cup made from steel grade 1.4301 and puts forward a method for estimating the required punch contraction to substantially reduce wear during retraction. Different punch diameters, cup wall thicknesses and reduction ratios are considered in a simplified process setup, with the effect of simplifications being explained based on selected examples. Results show that the required absolute punch contraction is positively correlated, and the strain required to reach that punch contraction is negatively correlated to the punch diameter. Both response variables were found to be positively correlated to the thickness change performed in the ironing process. Consequently, the design requirements of an adjustable ironing punch are the loosest for large diameter cups and low thickness changes.

Keywords: Forming tools · ironing · numerical analysis · tribology · flexibility

1 Introduction

In metal forming processes, flexibility and productivity are often conflicting requirements as aptly stated by Yang *et al.* [1] in their comprehensive review on flexibility in metal forming. At one extreme of the flexibility-productivity scale lay incremental forming processes, which suffer from a slow production speed, but offer highly customizable geometries and are thus interesting from a mass customization perspective. At the other extreme are processes with high productivity and low flexibility, such as progressive

K. Mocellin et al. (Eds.): ICTP 2023, LNME, pp. 101–109, 2024.
https://doi.org/10.1007/978-3-031-40920-2_11

die forming and roll forming. The push for Industry 4.0 creates a demand for increased flexibility in mass manufacturing, which often implies increased complexity in process mechanics and control.

In sheet metal forming processes suitable for progressive dies, flexibility has been demonstrated in multiple niches. Löbbe *et al.* [2] applied closed-loop control to a bending operation assisted by induction heating in a progressive die, achieving bend angle adjustability within ±1.5°. The control of the blank holder force has been demonstrated in deep drawing [3] by using blank draw-in measurements in multiple locations as input for the control loop, and in stamping by controlling material flow using hydraulically actuated segmented dies [4]. In-line tuning of strip flatness in cold rolling is yet another example of flexibility in high-volume production processes [5]. In all the studies above, the increased flexibility was achieved at the cost of higher complexity in both hardware and software of the production setup.

Ironing is a forming process often integrated in progressive dies following one or more deep drawing steps. It is one of the most tribologically demanding processes due to sliding contact under high interfacial pressure. When ironing the interior of the workpiece using a punch, retraction of the punch is usually performed before cup ejection from the die due to geometrical restrictions, as opposed to some cases, e.g. beverage cans, where external ironing with a die is used. Galling is a common problem when forming high-strength materials due to high tool-workpiece contact pressure. It was shown by Dong *et al.* [6] that if the contact pressure exceeds a critical value, galling is initiated. Üstünyagiz [7] observed heavy galling during ironing punch retraction in internal ironing and explained it with the reduction of punch-workpiece contact area, and the formed material being strain-hardened, leading to increased contact pressure, and the lubricant having been wiped off during the forming stroke. This problem could be mitigated by reducing the diameter of the ironing punch before retraction as suggested by Nielsen *et al.* [8].

The required magnitude of punch diameter reduction is an important parameter for selecting the principle of diameter change. Ideally, the punch contracts enough before retraction to eliminate any punch-cup contact during the entire retraction stroke. According to Archard's wear model, the worn volume is proportional to the normal pressure and sliding length [9]. Thus, even if the punch diameter cannot be contracted sufficiently to eliminate contact, wear is still reduced by decreasing the interface pressure.

The aim of this paper is to investigate the relationship between the punch diameter reduction and the resulting punch-cup contact length during punch retraction. The effect of nominal punch diameter (D), nominal thickness reduction ratio (R), and initial blank thickness (T) are studied to provide tool designers practically relevant data and methods for evaluating the design requirements for an adjustable ironing punch.

2 Methods

2.1 Simplifications

In practice, ironing is often performed after deep drawing, implying uneven wall thickness and the presence of residual stresses in the cup wall before ironing.

The cup diameter (which is similar to the punch diameter D for thin-walled cups) and blank thickness influence the residual stresses in the cup wall. Some factor settings like large diameter thin-walled cups or small diameter thick-walled cups result in geometries that cannot be deep drawn at a high draw ratio in a single step. Additionally, a blank holder force must be applied in deep drawing to control material flow. The selection of the correct blank holder force in deep drawing is often driven by empirical knowledge, making the choice subjective. If a blank holder is used for clamping the cup during ironing, the blank holder force will influence the amount of material in the cup opening radius that is drawn in during the process.

Considering that the goal of the study was to gain qualitative knowledge about the effects of the selected variables, the following simplifications were made:

- Ironing was simulated on an ideal cup with a uniform initial wall thickness T. No initial strains or residual stresses were present at the start of the process.
- The blank holder for clamping the cup during ironing was not included. Instead, the radial translation of the peripheral nodes on the cup flange was constrained as shown in Fig. 1a.

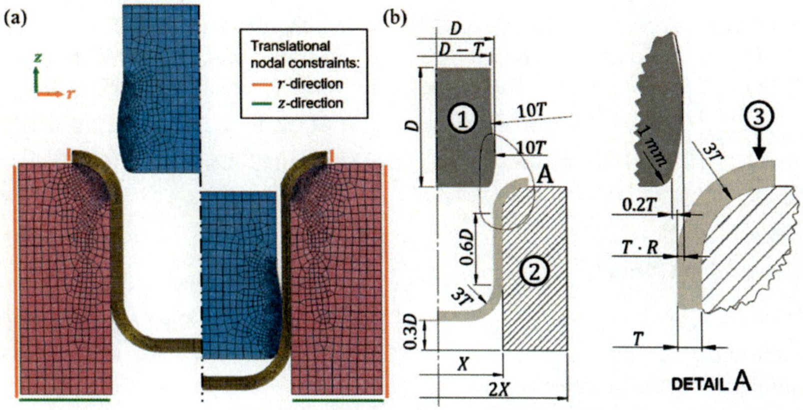

Fig. 1. (a) Simulation setup before (left) and after (right) ironing, (b) Parametric axisymmetric drawing views of (1) punch, (2) die and (3) cup used for numerical analysis. Dimensions scale with the three factors. Inner die diameter is $X = D + 2T(1 - R)$. Factor settings $D = 20\,\text{mm}$, $T = 1.5\,\text{mm}$ and $R = 30\%$ are used for the drawing.

2.2 Numerical Analysis

LS-DYNA finite element software was used for simulating the ironing process on geometries defined in Fig. 1b. Volume-weighted 2D axisymmetric solid elements were used for simulations. The mesh size was reduced near round edges and the element size for the cup was $0.1T$, giving 10 elements through the wall thickness. Both tools were modelled as elastic using the room temperature elastic modulus (205 GPa) and density (7730 kg/m^3)

of Sleipner® [10], whereas the EN 1.4301 austenitic stainless steel cup was modelled using a piecewise-linear plasticity material model. The stress-strain curve for this material model had been previously recorded in an in-plane torsion test [11] and was shown in [12]. The density of the cup material was set to 7900 kg/m^3, the elastic modulus to 200 GPa, and the initial yield strength to 300 MPa, all common values for austenitic stainless steel. A Poisson's ratio of 0.3 was assumed for all materials. Frictional contact was modelled by the Coulomb friction model with a constant friction coefficient of 0.1 in all contact pairs.

First, the ironing stroke was analysed without retraction. After that, springback of only the die and the formed cup was resolved, which enabled eliminating the effect of the punch during the springback. The ironing stroke was analysed using dynamic explicit simulation. The punch velocity was kept constant at 2000 mm/s and mass scaling was applied to increase the timestep size and reduce computation time. The end of the forming stroke was registered when any punch node outside of the ironing nose contacted the cup, indicating contact with the cup bottom radius. Springback was resolved in a single timestep using quasi-static implicit analysis.

The effect of the simplifications was investigated by substituting the ideal cup in the ironing simulation with a cup deep drawn from a flat blank with an initial thickness T, using a formulation similar to the ironing simulation. Before ironing, springback of the drawn cup was resolved. The used deep drawing die diameter was 0.2 mm smaller than the ironing die diameter, and the punch-die clearance was $u_D = 1.22T$. A draw ratio (DR) of 1.9 was selected. The blank holder force for deep drawing was $F_{BH} = A_{BH} \cdot 0.003S_u\left[(DR-1)^3 + \frac{0.005d_0}{T}\right]$, where $S_u = 700\,\text{MPa}$ was assumed, and where A_{BH} is the initial blank-die contact area and d_0 is the initial diameter of the blank [13, p. 111]. In ironing and subsequent springback, a blank holder force of $10F_{BH}$ was used.

2.3 Simulation Output

The simulation output was extracted from the cup geometry after final springback. The axial and radial coordinates of the nodes on the fully ironed surface were used for computing the punch diameter change required to reduce contact during retraction. As the ironed surface is not perfectly straight after ironing, the necessary punch contraction varies during the retraction stroke. Punch contraction (PC) was calculated using

$$PC(z) = D - 2x(z), \tag{1}$$

where x is the radius of the ironed surface at axial position z, which is always smaller than D/2 after springback. Note that the ironed surface comprises points that have contacted the widest portion of the punch.

A method was developed for relating the punch contraction to the change in punch-cup contact length during retraction. While iterating over the range of PC values, the ratio $\frac{\sum_i l_{xi}}{l_\Sigma}$ was computed. The length l_Σ measures the range of axial coordinates of the nodes that contacted the widest point on the punch, which is the total sliding length during ironing. It is graphically shown in Fig. 2b. The lengths l_{xi} measure the sections, where the punch would contact the cup during retraction if its diameter was reduced by x µm,

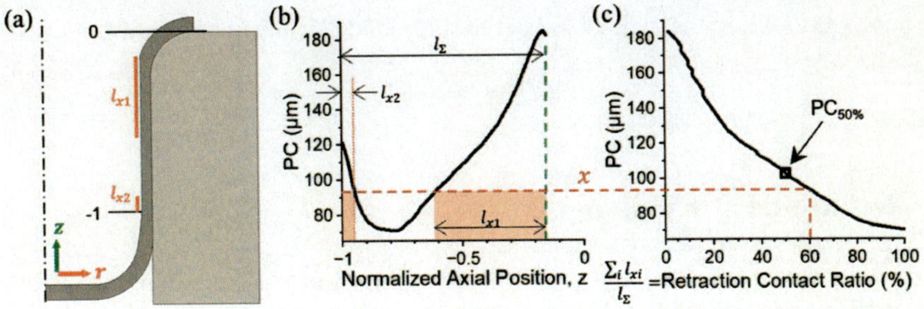

Fig. 2. Analysis of simulation output: (a) cup and die after springback with normalized notation of origin ($z = 0$) and bottom dead centre ($z = -1$), (b) cup wall profile, where PC is calculated according to Eq. (1), and (c) punch contraction curve compiled using the cup wall profile in (b).

and $\sum_i l_{xi}$ is the total contact length during retraction. Thus, $\frac{\sum_i l_{xi}}{l_\Sigma}$ is the contact ratio during punch retraction for a certain PC value if plastic deformation at the contacting areas during retraction is ignored.

2.4 Design of Experiments

The effect of the three factors was studied using a 2^3 full factorial design. Table 1 presents the selected factors and their levels which were selected to encompass most of the components produced by Grundfos.

Table 1. Factors and their levels selected for the 2^3 full factorial experiment.

Factor name	Symbol	Unit	Levels	
			Low	High
Punch diameter	D	mm	20	60
Blank thickness	T	mm	0.3	1.5
Reduction ratio	R	%	15	30

The responses were defined using the punch contraction curve shown in Fig. 2c. To eliminate punch-cup contact during retraction, the highest PC value shown in the curve should be chosen. However, the punch contraction values at both extremes of retraction contact ratio may vary differently when the factor settings are changed (as later shown in Fig. 4), negatively impacting predictability. The slope of the curves was relatively constant between about 25% and 75%, so the punch contraction at 50% retraction contact ratio was selected as one of the responses ($PC_{50\%}$ shown in Fig. 2c).

To introduce adjustability to the punch diameter, it is necessary to strain the punch material either mechanically, thermally, or otherwise, and the punch material must withstand many deformation cycles to be industrially feasible. Another response, relative punch contraction ($RPC_{50\%}$), which is the radial or circumferential nominal strain that the punch must withstand, was added to the study, and calculated as follows:

$$RPC_{50\%} = \frac{PC_{50\%}}{D} \tag{2}$$

3 Results and Discussion

The main effects of the factorial experiment are shown in Fig. 3a, where the effect of punch diameter is emphasized. Increasing the punch diameter leads to an increase in the absolute necessary punch contraction ($PC_{50\%}$), while the strain required in the punch to reach that contraction ($RPC_{50\%}$) decreases when T and R are constant. The negative correlation between D and $RPC_{50\%}$ is expected because the relative difference in punch diameter levels (3×) is higher than the relative difference in $PC_{50\%}$ in all cases. As both responses increase with an increase in T or R, the combined effect of these factors was measured by thickness change (ΔT), which is a product of these factors. Figure 3b reveals a positive correlation between both responses and thickness change. Thus, if the diameter change mechanism of an adjustable ironing punch is strain-limited, cases with lower thickness change and larger part diameter lead to less restrictive material selection criteria for the ironing punch.

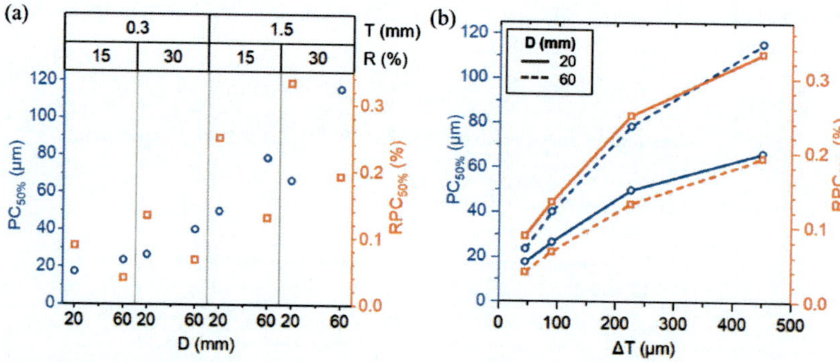

Fig. 3. Absolute (blue) and relative (red) punch contraction at 50% retraction contact ratio at (a) all factor settings and (b) as a function of thickness change, $\Delta T = T \cdot R$.

The effect of simplifications described in Sect. 2.1 can be seen in Fig. 4. Comparing the wall profiles (Fig. 4a) of simplified simulations (solid lines) to more realistic simulations (dashed lines), the combined effect of residual stresses, uneven wall thickness and blank holder force can be seen. In the simplified cases, the ironed surfaces are almost perfectly straight after ironing, only showing a minor draw-in peak near the opening. Realistic cups have an uneven wall thickness even after ironing, because elastic tool deflection is a function of the local thickness change, which itself is connected to the axial wall thickness profile after deep drawing.

The imperfect cup surfaces in Fig. 4a cause the slope of the corresponding punch contraction curves in Fig. 4b to decrease, confirming the need for such curves when the full process chain is simulated. $PC_{50\%}$ (marked in Fig. 4b) is consistently 31% to 40% higher in the realistic cases compared to corresponding simplified cases. As $RPC_{50\%}$ is proportional to $PC_{50\%}$, $RPC_{50\%}$ is larger by the same magnitude in the realistic case.

A pairwise comparison of the series in Fig. 4a ((1) to (2), (3) to (4), etc.), shows that the height of the draw-in peak close to the cup opening (near $z = 0$) is connected to the reduction ratio. Ironing with a higher reduction ratio causes higher axial stresses in the cup wall and hence radial stresses in the flange, making draw-in harder to avoid. The draw-in peaks are generally less pronounced in simplified cases compared to the realistic cases, where a blank holder was included. This is likely due to the radial boundary conditions applied to the cup flange in the simplified case.

In realistic cases, the slope of the middle portion of series (6) and (8) in Fig. 4a is lower than (5) and (7), indicating that higher reduction ratio leads to a more constant PC value. Therefore, even though low reduction ratio reduces the requirement for punch contraction, very low reduction ratios can lead to poorer cylindricity of the ironed surface when the ironing punch diameter is reduced before retraction. The positive slopes in series (5) and (7) stem from thickening of the flange in the prior deep drawing, while this effect is almost completely removed by the ironing with higher reduction ratio in series (6) and (8).

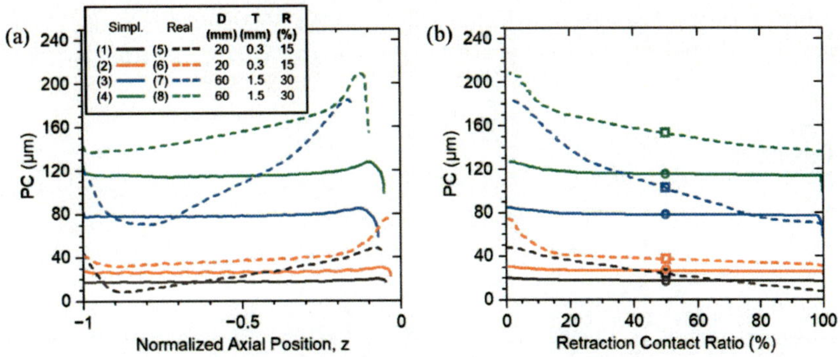

Fig. 4. (a) Final cup wall profiles and (b) PC curves with $PC_{50\%}$ marks of four factor combinations, including simplified cases (1)–(4), where only ironing was simulated as described in Sect. 2.2, and realistic cases (5)–(8), where one deep drawing step was simulated before ironing.

4 Conclusions and Outlook

The following conclusions can be drawn from the results of the study:

- The absolute punch contraction ($PC_{50\%}$) is positively correlated to the punch diameter, whereas correlation between D and the relative response ($RPC_{50\%}$) is negative if T and R are kept constant.

- Both responses are positively correlated to the thickness change, the product of initial wall thickness T and nominal reduction ratio R.
- In the studied range of process parameters, a maximum necessary $PC_{50\%}$ was 115 μm ($D = 60$ mm, $T = 1.5$ mm and $R = 30\%$), and the maximum $RPC_{50\%}$ was 0.33% ($D = 20$ mm, $T = 1.5$ mm and $R = 30\%$) for simplified cases.
- The simplifications made led to a more cylindrical ironed wall profile than cases, where the simplifications were not made, and deep drawing was simulated. Including the deep drawing step in the analysis increased the necessary $PC_{50\%}$ and $RPC_{50\%}$ by 31% to 40% in the four studied cases.
- If the deep drawing and ironing are simulated, lower reduction ratio leads to a poorer cylindricity of the ironed cup than a higher reduction ratio.

All in all, the diameter change requirement is lowest for large diameter cups, where the thickness change during ironing is low. While the results of the study provide some practically relevant knowledge, future design of adjustable ironing punches should include verification of the diameter reduction requirement for each specific case.

As suggested in [8], an adjustable ironing punch could also unlock the possibility of closed-loop control of product properties, e.g. compensating for variations in part length caused by fluctuating strip thickness. This function poses another requirement to the punch contraction magnitude, which is independent of the results of the present study and depends on the tolerances of the specific part and material to be ironed.

Acknowledgements. The authors would like to thank the Independent Research Fund Denmark (grant DFF – 0136-00159B) and Innovation Fund Denmark (grant 2052-00040B) for their support.

References

1. Yang, D.Y., et al.: Flexibility in metal forming. CIRP Ann. **67**, 743–765 (2018)
2. Löbbe, C., Hoppe, C., Becker, C., Tekkaya, A.E.: Closed loop springback control in progressive die bending by induction heating. Int. J. Precis. Eng. Manuf. **16**, 2441–2449 (2015)
3. Fischer, P., Harsch, D., Heingärtner, J., Renkci, Y., Hora, P.: Inline feedback control for deep drawing applications. IOP Conf. Ser.: Mater. Sci. Eng. **159**, 012006 (2016)
4. Cao, J., Kinsey, B.L., Yao, H., Viswanathan, V., Song, N.: Next generation stamping dies – controllability and flexibility. Robot. Comput. Integr. Manuf. **17**(1–2), 49–56 (2001)
5. Allwood, J.M., et al.: Closed-loop control of product properties in metal forming. CIRP Ann. **65**, 573–596 (2016)
6. Dong, W., Xu, L., Lin, Q., Wang, Z.: Experimental and numerical investigation on galling behavior in sheet metal forming process. The Int. J. Adv. Manuf. Technol. **88**(1–4), 1101–1109 (2016)
7. Üstünyagiz, E.: An Off-line Methodology to Determine Limits of Lubrication in Sheet Metal Forming. PhD Thesis, Technical University of Denmark (2018)
8. Nielsen, C.V., et al.: A novel ironing punch concept with adjustable tool diameter. CIRP Ann. Manuf. Technol. **72**, 213–216 (2023). https://doi.org/10.1016/j.cirp.2023.03.001
9. Archard, J.F.: Contact and rubbing of flat surfaces. J. Appl. Phys. **24**, 981–988 (1953)
10. Uddeholms, A.B.: Uddeholm Sleipner. https://www.uddeholm.com/app/uploads/sites/54/2018/11/Tech-Uddeholm-Sleipner-EN.pdf. Accessed: 23 May 2023

11. Traphöner, H., Clausmeyer, T., Tekkaya, A.E.: Material characterization for plane and curved sheets using the in-plane torsion test – An overview. J. Mater. Process. Technol. **257**, 278–287 (2018)
12. Siimut, K.: Development of a Novel Metal Forming Punch with Adjustable Diameter. MSc Thesis, Technical University of Denmark (2022)
13. Altan, T., Tekkaya, A.E. (eds.): Sheet Metal Forming: Fundamentals. ASM International (2012)

Investigation of Biaxial Prestrain Influence on Bending-Under-Tension Springback

Y. Yang[1,2], C. Baudouin[2], H. Chalal[5], G. Vincze[3,4], and T. Balan[2(✉)]

[1] State Key Lab of Solidification Processing, School of Materials Science and Engineering, Northwestern Polytechnical University, Xi'an 710072, China
[2] Arts et Metiers Institute of Technology, Université de Lorraine, LCFC, 57000 Metz, France
tudor.balan@ensam.eu
[3] Center for Mechanical Technology and Automation, Department of Mechanical Engineering, University of Aveiro, Campus Universitário de Santiago, 3810-193 Aveiro, Portugal
[4] LASI – Intelligent Systems Associate Laboratory, 4800-058 Guimarães, Portugal
[5] Arts et Metiers Institute of Technology, Université de Lorraine, LEM3, 57000 Metz, France

Abstract. Springback notoriously depends on the strain-path history of the sheet material during forming. The bending-under-tension (BUT) test is used in this work for a discriminant characterization of springback in various conditions. In order to generate non-linear strain-paths, tensile prestrain is usually applied in the literature. In this work, an original biaxial prestrain device was designed and build. It allowed for the uniform prestrain of large samples of 500 mm diameter in DP600, up to 9.5% of strain. Subsequent samples for bending-under-tension were cut and tested at different combinations of tool radius and backforce. Classical models were used to describe the material's behavior within finite element simulations of the BUT test. These experiments allowed for a discriminant comparison between the springback predictions of competing constitutive models.

Keywords: DP600 · Biaxial prestrain · Bending-Under-Tension · Springback

1 Introduction

Advanced high strength steel (AHSS) has been used to solve the challenge of light-weight vehicle manufacturing. However, compared with the aluminum alloy and classical mild steels, the AHSS has a higher yield strength which induces more severely unloading springback. The AHSS formed car components usually undergo complex multi-stage forming processes. This induces strain-path changes during car component forming, requiring careful modeling for their further use in industrial automotive applications.

The sheet metal properties are usually affected by loading path changes which generate pre-strain from the previous loading step. Moreover, the different prestrain types and the accumulated deformation levels may influence material properties and eventually generate forming defects [3, 4]. Numerous researches including constitutive modeling and experimental investigations have been devoted to strain-path dependency over the last two decades [5–8]. It is worth noting that most studies on prestrain effects in metal

K. Mocellin et al. (Eds.): ICTP 2023, LNME, pp. 110–115, 2024.
https://doi.org/10.1007/978-3-031-40920-2_12

forming are confined to uniaxial tensile or simple shear pre-strain modes. Thus, it would be useful and necessary to investigate other prestrain effects, such as biaxial state, on material forming process.

In this work, biaxial prestrain are pre-deformed by applying Marciniak-type experiments on a hydraulic press. Moreover, material characterization tests of biaxial pre-strained specimen including uniaxial tension (UT), simple shear (SS) and reverse shear (RS) are performed to model strain-path effect. Consequently, the developed strain-path dependent hardening models (isotropic hardening, isotropic-kinematic hardening) with rigorously identical predictions under monotonic loading has been employed in finite element simulation. Finally, the bending-under-tension springback investigation is applied to reveal the influence of biaxial prestrain effects.

2 Material Testing and Modeling

2.1 Biaxial Prestrain Experiments

In this paper, a commercial advanced high strength steel DP600 with 1 mm thickness was tested. First, biaxial prestrain were generated via Marciniak-type stamping tests illustrated in Fig. 1(a). The prestrain level was measured at the bottom of the deformed stamping part where the deformation state is considered quasi-biaxial. Finally, 5.5% biaxial prestrain level was obtained. The initial blank is a square of 1000 mm size, while the resulting, uniformly pre-strained material is a circular blank of approximately 500 mm in diameter. This very large sample size gives the opportunity for relatively large pre-strained samples to be cut out for subsequent tests, in view FLD or different other stamping tests.

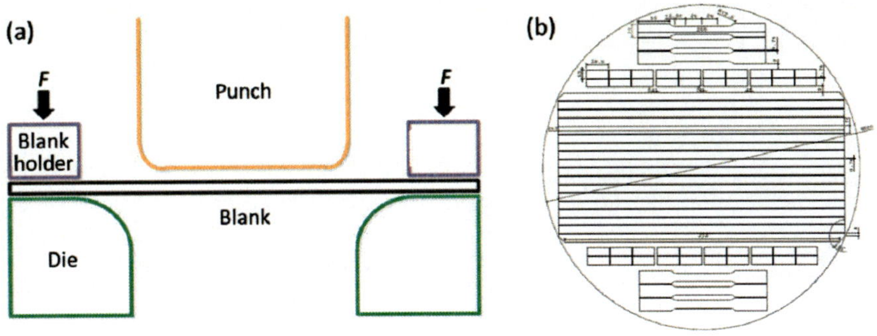

Fig. 1. Schematic of: (a) Biaxial pre-strain test; (b) example of subsequent samples cutting plan: uniaxial tension (UT), simple shear (SS), reverse shear (RS) and bending under tension (BUT) samples within the same pre-strained blank.

Subsequently, uniaxial tension (UT), simple shear (SS), reverse shear (RS) and bending under tension (BUT) samples were cut from the bottom area of prestrained parts, as illustrated in Fig. 1(b). The specimens with biaxial prestrain for UT, SS and RS were used in material characterization tests. In addition, the specimens with biaxial prestrain for BUT were applied in bending-under-tension test to investigate its effects on springback.

2.2 Material Characterization and Modeling

Two types of hardening models, Voce-Swift isotropic hardening (IH) and modified isotropic-kinematic hardening model (Mod IKH) were used. The detailed constitutive equations and modeling process were discussed in previous work [1, 2], including model parameters and model predictions under monotonic and reverse loading. The two models exhibit identical predictions (by construction) under any monotonic (proportional) loading paths, while the isotropic-kinematic hardening model predicts significantly better reverse shear strain paths.

Figure 2 illustrates predictions of IH (blue line) and Mod IKH (green line) models under simple shear following 5.5% biaxial prestrain, along with the corresponding experimental results (black line). In contrast with the results under reverse loading, this particular loading sequence is more poorly predicted by the isotropic-kinematic hardening model. It is noteworthy that the biaxial-shear sequence corresponds to an orthogonal strain-path change, which activates completely different deformation modes within sheet metals. Thus, this new pre-strain technique extends the possibilities for validation and parameter identification of advanced constitutive models.

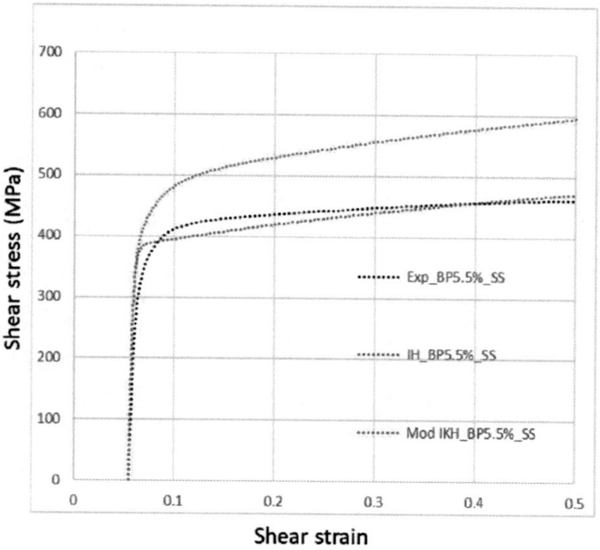

Fig. 2. Predictions of simple shear with 5.5% biaxial prestrain level via IH (blue lines) and Mod IKH (green lines), and their comparison with experimental results (black lines).

3 BPBUT Springback

The biaxial pre-strained specimen for bending-under-tension (BPBUT) test was performed in LCFC. For bending under tension springback investigation, backforce factor k and the ratio of bending radius vs. sheet thickness (R/t) were studied. The backforce factor can be defined by Eq. (1):

$$k = F_b/\sigma_0 \cdot w \cdot t \tag{1}$$

where F_b is the back force applied in BPBUT, σ_0 is flow stress corresponding to the strain at 0.02%, while w and t are width and thickness of the specimen. Table 1 gives backforce factors and tools radii investigated in BPBUT test.

Table 1. Influence factors of springback in BPBUT test

Biaxial pre-strain	Bending radius (R)	*R/t*	Backforce factors
5.5%	6 mm	4.92	0.29, 0.67, 1.04

After the BPBUT springback test, the deformed specimens were scanned using a GOM-ATOS measurement platform, and the springback angle θ was measured (see [1] for methodology details). Springback angle for 5.5% biaxial prestrain level and without biaxial prestrain are shown in Fig. 3. It appears that the springback angle θ is significantly larger for the pre-strained samples, although the pre-strain level is relatively small.

4 Confrontations and Discussion

In BPBUT springback simulation, Voce-Swift isotropic hardening (IH) and modified isotropic-kinematic hardening (mod IKH) models were used, as mentioned in Sect. 2.2. Combing the BPBUT experimental and simulation results, the confrontations of BPBUT springback angle between experiment and simulation for each k of DP600 with 5.5% biaxial prestrain level are shown in Fig. 4.

It appears that both IH model and mod IKH model underestimate springback angle for these pre-strained configurations. The IKH model shows better predictions, the gap to the experimental results being about twice smaller compared to that of the IH model. In addition, the simulation results from IH and mod IKH show quasi-linear decreasing trend, with the backforce factor k increasing. It is noteworthy that in the absence of pre-strain and for the same k and R/t factors, the two models would predict almost identical springback angles, which lay close to the experimental ones [1]. It is thus apparent that this new pre-strain test enlarges the possibilities for advanced constitutive model validation. Future work will be devoted to the validation of advanced constitutive models, which also include the description orthogonal strain paths.

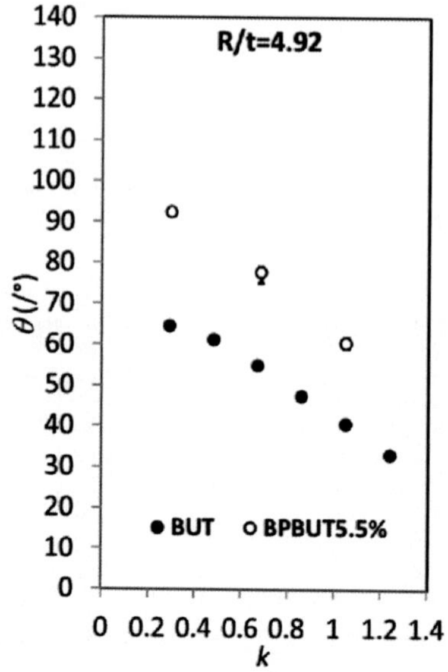

Fig. 3. BPBUT experimental mean value of springback angle with 5.5% prestrain level and without biaxial prestrain for R/t = 4.92.

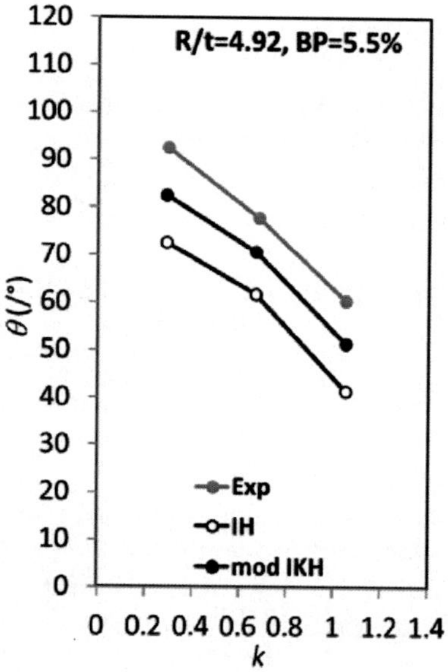

Fig. 4. Confrontations of BPBUT springback angle between experiment and simulation for each *k* of DP600 with 5.5% biaxial prestrain for R/t = 4.92.

Acknowledgements. Y. Yang acknowledges partial financial support from National Natural Science Foundation of China (No. 52205418). G. Vincze acknowledges support from the Portuguese

Foundation of Science and Technology (FCT), in its State Budget component (OE) through project 2022.05783.PTDC. All authors thank Dr Leopold Wagner in voestalpine Stahl GmbH, Austria for the fruitful discussions and for providing the DP600 material. The experiments were developed and performed on the Vulcain platform at Arts et Métiers Metz with the contributions of Ludovic Freund, Sébastien Burgun, Alexandre Fendler, Daniel Boehm, Sylvio De Paolis and Stéphane Mathieu, who are warmly acknowledged.

References

1. Yang, Y., Chalal, H., Baudouin, C., Vincze, G., Balan, T.: Prediction of springback after bending under tension. Int. J. Mater. Form. **15**, 34 (2022)
2. Yang, Y., Vincze, G., Baudouin, C., Chalal, G., Balan, T.: Strain-path dependent hardening models with rigorously identical predictions under monotonic loading. Mech. Res. Comm. **114**, 103615 (2021)
3. Lian, J., Liu, W., Gastañares, X., Juan, R., Mendiguren, J.: Plasticity evolution of an aluminum-magnesium alloy under abrupt strain path changes. Int. J. Mater. Form. **15**, 40 (2022)
4. Schlosser, F., Signorelli, J., Leonard, M., Roatta, A., Milesi, M., Bozzolo, N.: Influence of the strain path changes on the formability of a zinc sheet. J. Mater. Process. Tech. **271**, 101–110 (2019)
5. Wroński, M., Arul Kumar, M., McCabe, R.J., Wierzbanowski, K., Tomé, C.N.: Deformation behavior of CP-titanium under strain path changes: experiment and crystal plasticity modeling. Int. J. Plast. **148**, 103129 (2022)
6. Hama, T., Yagi, S., Tatsukawa, K., Maeda, Y., Takuda, H.: Evolution of plastic deformation behavior upon strain-path changes in an A6022–T4 Al alloy sheet. Int. J. Plast. **137**, 102913 (2021)
7. Yoshida, K., Okada, N.: Plastic flow behavior of fcc polycrystal subjected to nonlinear loadings over large strain range. Int. J. Plast. **127**, 102639 (2020)
8. Jinjin, H., Baral, M., Korkolis, Y.P.: Ductile fracture of an aluminum sheet under proportional loading. J. Mech. Phys. Solids **132**, 103685 (2019)

Simulation-Based Investigation of the Heat Exchange Within a Partial Hot Stamping Process

Alborz Reihani[1(✉)], Darwin Badroosian[1], Sebastian Heibel[1], Thomas Schweiker[1], and Marion Merklein[2]

[1] Mercedes-Benz AG, Béla-Barényi-Straße 1, 71059 Sindelfingen, Germany
alborz.reihani_masouleh@mercedes-benz.com
[2] Lehrstuhl Für Fertigungstechnologie, Friedrich-Alexander Universität Erlangen-Nürnberg, Egerlandstr. 13, 91058 Erlangen, Germany

Abstract. Studies on crash performance show that hot stamped components with homogeneous mechanical properties can reach their limits in highly stressed car body areas. To face this challenge, hot stamped components can be optimized by integrating ductile areas. A suitable technology for adjusting tailored properties is based on a special furnace chamber within a multi-chamber furnace in which a cooled aluminum mask protects the blank locally from incident heat radiation. Simultaneously, the mask absorbs the blank's own radiation. In addition to the temperature and radiation properties, this heat exchange depends in particular on the orientation and position of the radiating surfaces in the furnace chamber. For this reason, the mask distance to the blank can influence the width and characteristics of the transition zone after partial hot stamping. In order to predict the correlation between hard and ductile area, the process of partial hot stamping with varying mask distances is modelled in LS-DYNA® and validated experimentally. The results indicate that it is feasible to simulate the influence of the mask distance on the resulting hard, soft and transition zone after partial hot stamping. Furthermore, the results demonstrate that transition zones on a partially hot stamping part can be designed more flexibly, thus extending the limits in terms of part complexity.

Keywords: Hot Stamping · Heat Treatment · Tailored Properties · Finite-Element Simulation

1 Introduction

The reduction of fuel consumption and CO_2-emissions as well as improving the passenger's safety are legally demanded technical targets that have framed the development of future automotives [1]. To meet these requirements lightweight design by hot stamping has become an essential approach by reducing the vehicles total mass while providing high intrusion resistance [2]. The produced martensitic microstructure resulting from forming of fully austenitized boron manganese steels followed by rapid cooling results

K. Mocellin et al. (Eds.): ICTP 2023, LNME, pp. 116–123, 2024.
https://doi.org/10.1007/978-3-031-40920-2_13

in a hardness and tensile strength of up to 500 HV and 1600 MPa respectively [3]. However, the latest research has shown consensus that strategically integrated areas of lower strength and higher ductility can improve the crash performance and joinability of parts in downstream treatments [4]. Reason for that is a raised energy absorption potential in a crash-like scenario compared to a solely homogeneous martensitic microstructure [5]. The use of semi-finished products like tailor-welded blanks, tailor-rolled blanks or patchwork blanks as well as several variants of the conventional process like partial austenitization, differential cooling with tempered tools and tailored tempering have been introduced as possible methods to achieve ultra high-strength steels with tailored properties [6]. One of these is partial austenitization by partial pre-cooling using the TemperBox®. Instead of forming in a fully austenitized state, the blank is being transferred into the TemperBox® where it is positioned between water-cooled aluminum masks that shield the blank from radiation and simultaneously absorb its own emitted heat. The masked zones cool down to create soft microstructures while uncovered zones are reheated to maintain the austenite and produce martensite after forming and cooling. This process variant is characterized by several advantages compared to alternative tailoring approaches: It neither requires expensively designed tempered tools as used for differential cooling nor additional process step as in tailored tempering (e.g., laser annealing) and it further allows a flexible adjustment of hard and soft areas with complex workpiece geometries. This paper contains the experimental investigation of partial austenitization by partial pre-cooling and the simulative prediction of the hardness in LS-DYNA®.

2 Experimental Setup and Simulation Model

2.1 Material and Experimental Setup

The workpiece material of this study is the aluminum-silicon-coated manganese boron steel 22MnB5 AS150. The chemical composition of 22MnB5 in weight percent (wt. %) is summarised in Table 1. The blank thickness is 1.5 mm and the size of the blank is 500 mm × 500 mm.

Table 1. Chemical composition of 22MnB5 in wt. %

Alloying elements	C [%]	Mn [%]	B [%]	Cr [%]	Si [%]	Al [%]	Ti [%]	N [%]	Mo [%]
22MnB5	0.224	1.183	0.0026	0.192	0.254	0.034	0.039	0.005	0.003

In order to analyse the influence of the mask distance on the heat exchange and the resulting phase transformation after quenching, blanks are hardened with flat tools and full-surface contact pressure. For this purpose the blank is laser cut to size and heated for 400 s at 950 °C in the prototype chamber furnace of the AP&T Sweden AB. The chamber furnace consists of three chambers: Two chambers are used for heating and austenitising and an innovative chamber, the so-called TemperBox® (Fig. 1a), for targeted pre-cooling in covered areas and simultaneous reheating in uncovered areas. In the TemperBox® the blanks are partially pre-cooled with different mask distances d_{mask} of 2 mm and 8 mm and pre-cooling times $t_{pre\text{-}cooling}$ of 45 s and 90 s. For each parameter combination, three blanks are partially hot stamped. In the investigated mask recess, chamfers are milled along the thickness direction in order to specifically influence the radiant heat exchange in the transition area (Fig. 1b). To avoid contact between the mask and the heated blank and to ensure controlled cooling through the defined mask distances, spacer pins are integrated into the mask. Following partial pre-cooling, the blank is positioned in the tool after a transfer time of 11 s and quenched after 15 s by tool contact on both sides with a holding time of 25 s and a press force of 1250 kN. To ensure consistent thermal boundary conditions and to prevent premature cooling of the blank due to one-sided tool contact, four spring-mounted spacer pins are integrated into the lower tool half on which the blanks are positioned. The Imposite 2500 hydraulic press from Dieffenbacher GmbH is used for the quenching process.

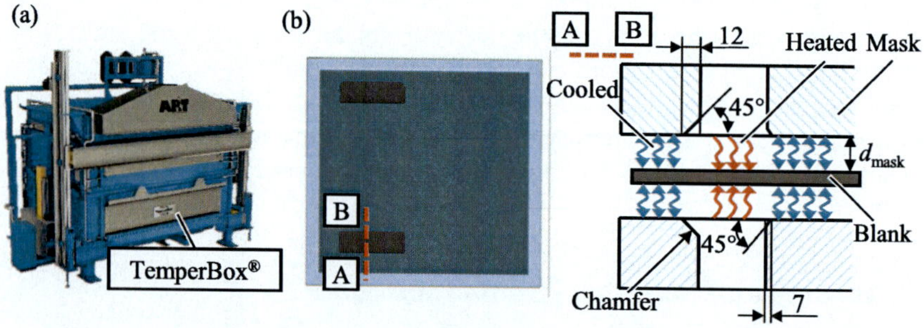

Fig. 1. (a) Chamber furnace and (b) examined mask area in the sectional view

To identify the influence of the parameter settings on the resulting mechanical component properties and microstructure, microhardness tests according to Vickers are carried out using the semi-automatic hardness tester DuraScan G7 from Struers GmbH and the associated software Ecos Workflow from EMCO-TEST GmbH. For the hardness measurements, the samples are cut with a laser from the partially hot stamped blanks and prepared metallographically in cross-section.

2.2 Simulation Model

LS-DYNA® (mpp_R12.1) is used to calculate the heat exchange in the TemperBox® and the results after partial hot stamping. Within the TemperBox®, the temperature change of the blank takes place mainly through the radiation exchange with a cooled aluminium mask. The calculation of the radiation exchange is carried out with the keyword "*BOUNDARY_ENCLOSURE_RADIATION". This involves calculating a view factor matrix between all segments in the enclosure and solving the radiation problem. The view factor results from the photometric fundamental law [7] and is given for two finite surfaces A_i and A_j by

$$F_{ij} = \frac{1}{A_i} \int_{A_i} \int_{A_i} \frac{\cos\theta_i \cos\theta_j}{\pi S^2} \delta_{ij} dA_i dA_j. \quad (1)$$

δ_{ij} describes the visibility of the finite surfaces, S the distance between the surfaces and θ_i as well as θ_j the angles between the common connecting line and the normal vector of the surfaces. In addition to the calculation of the radiative heat exchange, the numerical model considers the convective heat exchange with the environment and the latent heat that occurs during the decomposition of austenite into ferrite, pearlite, bainite and martensite. For the calculation of the heat exchange, the blank is modelled with "*MAT_244" and solid elements (elform = 2), as shell elements cannot represent the bidirectional radiation exchange due to the unidirectional element normals. The material model "*MAT_244" associates the thermal properties, the temperature and strain rate dependent flow behavior and occurring phase transformations which enables the prediction of the hot stamping process including the hardness calculation of 22MnB5 components. The model implemented in "*MAT_244" is based on the results of Åkerström [8]. The phase transformation starting temperatures are derived from the CCT diagram of Aranda et al. [9] and were manually implemented into the material model "*MAT_244". The heating coils, the enclosure and the cooled mask (Fig. 2a) are modelled as shell elements (elform = 9) using the material models "*MAT_RIGID" and "*MAT_THERMAL_ISOTROPIC", since the initial temperatures and the temperature profile are given as boundary conditions and are not calculated specifically.

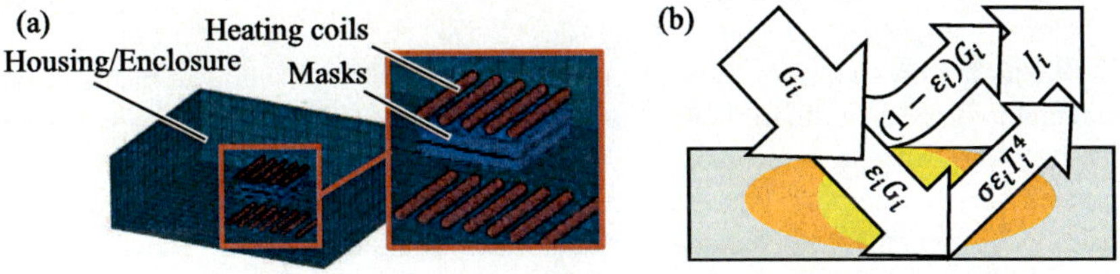

Fig. 2. (a) Simulation model of the TemperBox® and (b) radiation exchange at zone i

In the following, the radiation model for solving radiation exchange problems in LS-DYNA® is derived. Using the energy balance at a zone i (Fig. 2b), the heat flow $\dot{Q}_i$ can be described as the difference between emitted radiation J_i and irradiation G_i [7]. As the emitted radiation includes the reflected radiation of the irradiation G_i, the irradiation can be eliminated and the heat flow $\dot{Q}_i$ can be described with the following equation:

$$\dot{Q}_i = A_i(J_i - G_i) = \frac{A_i\varepsilon_i}{(1-\epsilon_i)}(\sigma T_i^4 - J_i), \quad (2)$$

where ε_i is the emissivity, σ the Stefan–Boltzmann constant and T_i the temperature of the zone i. By linking the incident radiant flux $A_i G_i$ with the sum of all other zones, a second relationship between the heat flow $\dot{Q}_i$ and emitted radiation J_i is obtained. In conjunction with the reciprocity law $A_jF_{ji} = A_iF_{ij}$, the total radiation incident on zone i can be described as followed [7]:

$$A_iG_i = \sum_{j=1}^{n} A_jF_{ji}J_j = A_i\sum_{j=1}^{n} F_{ij}J_j. \tag{3}$$

By substituting Eq. (3) into Eq. (2) and eliminating the unknown heat flow $\dot{Q}_i$, the radiation exchange can be solved by the following equation:

$$A_i(J_i - (1-\varepsilon_i)\sum_{j=1}^{n} F_{ij}J_i) = A_i\varepsilon_i\sigma T_i^4, i = 1, 2, \ldots m. \tag{4}$$

Equation (4) is used for zones 1,2... m, where the temperature T_i of the zone i is given and all coefficients are known. In matrix form, Eq. (5) is called the radiosity model or radiosity matrix equation and can be formulated as followed [7]:

$$\begin{bmatrix} A_1-(1-\varepsilon_1)A_1F_{11} & -(1-\varepsilon_1)A_1F_{12} & \cdots & -(1-\epsilon_1)A_1F_{1n} \\ -(1-\varepsilon_2)A_2F_{21} & A_2-(1-\varepsilon_2)A_2F_{22} & \cdots & -(1-\epsilon_2)A_2F_{2n} \\ \vdots & \vdots & \ddots & \vdots \\ -(1-\varepsilon_m)A_mF_{m1} & -(1-\varepsilon_m)A_mF_{m2} & \cdots & A_n-(1-\varepsilon_n)A_{mn}F_{mn} \end{bmatrix} \cdot \begin{bmatrix} J_1 \\ J_2 \\ \vdots \\ J_m \end{bmatrix} = \begin{bmatrix} \varepsilon_1A_1\sigma T_1^4 \\ \varepsilon_2A_2\sigma T_2^4 \\ \vdots \\ \varepsilon_mA_m\sigma T_m^4 \end{bmatrix}. \tag{5}$$

3 Results and Discussion

Surface temperature measurements are performed on the blank to validate the numerical model and to determine the influence of the mask distance on the heat exchange as well as the resulting phase transformations. The experimentally determined and numerically calculated time-temperature curves are compared in Fig. 3. Based on the experimentally determined results, the mask distance has an influence on the course of the blank temperature in the heated and in the cooled zones. The evaluation of the measuring point in the heated zone shows that with a mask distance d_{mask} of 2 mm (Fig. 3a) the blank reaches a temperature of 838 °C after 45 s, while with a mask distance d_{mask} of 8 mm (Fig. 3b) the temperature rises up to 827 °C. This can be explained by the fact that with a mask distance d_{mask} of 8 mm, the distance between the upper heating coils and the blank is 6 mm greater than with a mask distance d_{mask} of 2 mm. Furthermore, the comparison of the temperature curves in the cooled zone demonstrates that the blank temperature in the partial pre-cooling phase (>10 s) decreases faster with a mask distance d_{mask} of 2 mm compared to a mask distance d_{mask} of 8 mm. The higher cooling rate with the mask distance d_{mask} of 2 mm leads to the effect that the released thermal energy due to the phase transformation from austenite to ferrite is already recognisable after approx. 35 s, whereas with the mask distance of 8 mm the latent heat is recognisable after approx. 45 s. The correlation between mask distance and cooling rate in the covered area of the blank as well as the lower blank temperature due to the increased distance to the heating coils can be calculated in the simulation with a maximum deviation of 4.5%.

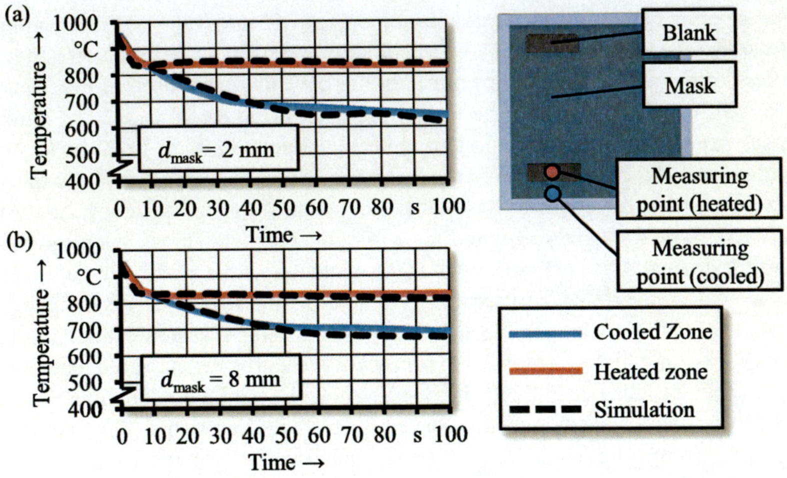

Fig. 3. Experimentally determined and numerically calculated temperature curves during transfer to the TemperBox® and in the partial pre-cooling phase

The hardness analyse is based on the measurements of all three samples for each parameter set respectively that are carried out in 1 mm point intervals and averaged. In Fig. 4, the experimentally and numerically determined hardness curves are compared for different mask distances d_{mask} and partial pre-cooling times $t_{\text{pre-cooling}}$.

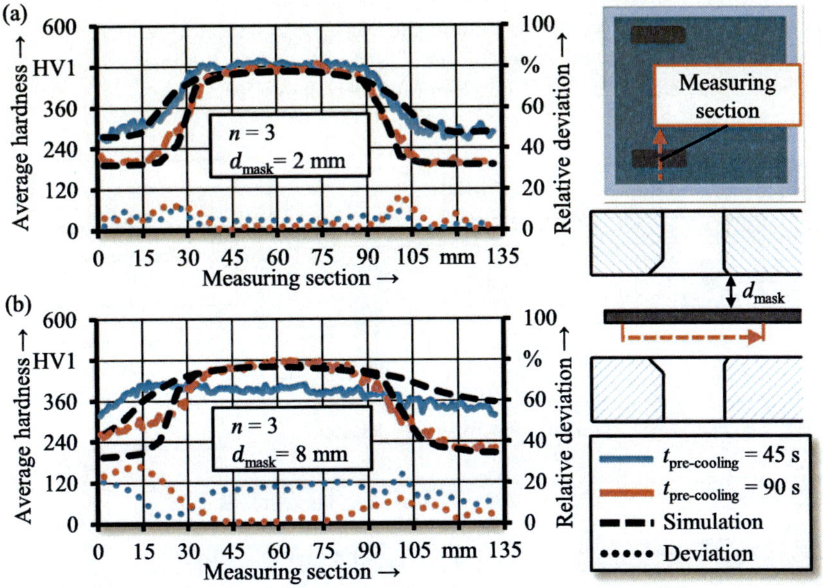

Fig. 4. Numerically and experimentally determined hardness curves at different mask distances

The experimental results for a mask distance d_{mask} of 2 mm (Fig. 4a) show a homogeneous hardness level in the heated area (35–90 mm) with values between 464–502 HV, regardless of the partial pre-cooling time. In the cooled areas a drop in hardness can be observed, which depends on the partial pre-cooling time $t_{pre\text{-}cooling}$. The hardness decreases due to the unidirectional internal heat conduction effects from the heated to the cooled area. Furthermore, with a mask distance d_{mask} of 2 mm, the phase transformation in the partial pre-cooling phase can be simulated with good accuracy. Analysis of the hardness curve for a mask distance d_{mask} of 8 mm (Fig. 4b) shows that a completely martensitic structure does not develop in the heated area with a partial pre-cooling time $t_{pre\text{-}cooling}$ of 45 s due to the hardness levels of approx. 380–405 HV. This indicates that this area does not have a homogeneously austenitized state, as austenitization is a time-dependent process and does not occur abruptly after reaching the eutectoid temperature. This assumption is confirmed by the hardness values at a partial cooling time $t_{pre\text{-}cooling}$ of 90 s, where a hard zone with values between 455–475 HV is formed due to the longer heating time. In the simulation a fully austentized blank is assumed for the decomposition of austenite into pearlite, ferrite and bainite, which results in deviations in the hardness calculation with a pre-cooling time of 45 s (Fig. 4b). Analysis of the hardness curves in the cooled area shows that the drop in hardness is not symmetrical. This can be explained by the fact that, depending on the chamfer width, more radiant heat reaches the covered surface and the phases are re-austenitized as a result. The consideration of the re-austenisation did not lead to any changes in the calculation of the phase composition and the resulting hardness, as the hardness calculation in "*MAT_244" is based on the cooling rate at 700 °C and the chemical composition, without taking re-austenisation into account.

To illustrate the correlation between the mask distance and the length of the soft zone, the transition zone and the hard zone, the averaged lengths are compared in Fig. 5. The distinction between soft zone and hard zone is based on the interpolation of the hardness values listed in the CCT diagram of Naderi et al. [10].

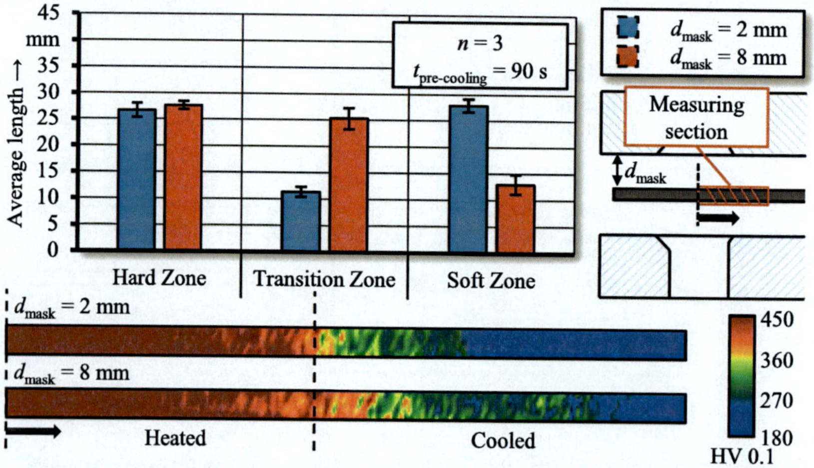

Fig. 5. Comparison of the average length of soft zone, transition zone and hard zone

The results show that the mask distance has a significant influence on the transition zone, while the width of the hard zones are at the same level regardless of the mask distance (see Figure 5). As the mask distance increases, more transverse radiation can be transmitted into the covered area. Due to the transverse radiation and the resulting re-austenitisation, the length of the transition zone is extended by an increased proportion of hard phases such as martensite and bainite in the microstructure. The hardness maps in the cross-section determined by HV0.1 measurements confirm this correlation between the mask distance and transition zone.

4 Summary

In this paper, the influence of the mask distance in the TemperBox® on the hardness of partially hot stamped blanks has been investigated numerically and experimentally. The results show that the soft zone and transition zone can be tailored more flexible by varying the mask distance and partial pre-cooling time, which enables a better adaption to the crash requirements. In order to set these tailored properties, different temperature and sheet thickness-dependent pre-cooling times are required. Furthermore, the simulation results show that the "*MAT_244" is not suitable to simulate the decomposition of austenite into ferrite, pearlite and bainite as well as the re-austenitization during partial hot stamping with TemperBox® starting from a not fully austenitized blank.

References

1. Mohrbacher, H.: High-performance steels for sustainable manufacturing of vehicles. In: Singh, M., Ohji, T., Asthana, R. (eds.) Green and Sustainable Manufacturing of Advanced Material, pp. 135–155. Elsevier, Amsterdam, Netherlands (2016)
2. Horn, A., Merklein, M.: Functional optimization of hot-stamped components by local carburization. Int. J. Lightweight Mater. Manuf. **3**, 43–54 (2020)
3. Karbasian, H., Tekkaya, A.E.: A review on hot stamping. J. Mater. Process. Technol. **210**(15), 2103–2118 (2010)
4. Bao, L., Wang, B., You, X., Li, H., Gu, Y., Liu, W.: Numerical and experimental research on localized induction heating process for hot stamping steel sheets. Int. J. Heat Mass Transf. **151**, 119422 (2020)
5. Merklein, M., Wieland, M., Lechner, M., Bruschi, S., Ghiotti, A.: Hot stamping of boron steel sheets with tailored properties: a review. J. Mater. Process. Technol. **228**(1), 11–24 (2016)
6. Billur, E. (ed.): Hot stamping of ultra high-strength steels. Springer, Cham (2019). https://doi.org/10.1007/978-3-319-98870-2
7. Baehr, H.D., Stephan, K.: Heat and Mass Transfer. Springer Berlin Heidelberg, Berlin, Heidelberg (2006). https://doi.org/10.1007/3-540-29527-5
8. Åkerström, P.: Modelling and simulation of hot stamping, Doctoral theses, Luleå University of Technology, Sweden (2006)
9. Aranda, L.G., Chastel, Y., Pascual, J.F., Negro, T.D.: Experiments and simulation of hot stamping of quenchable steels. In: Proceedings of the 7th International Conference on Technology of Plasticity, pp. 1135–1140. Yokohama, Japan (2002)
10. Naderi, M., Uthaisangsuk, V., Prahl, U., Bleck, W.: A numerical and experimental investigation into hot stamping of boron alloyed heat treated steels. Steel Re. Int. **79**(2), 77–84 (2008)

Check for updates

Influencing Parameters in the Deep Drawing of Fiber Metal Laminates with Low Viscous Matrix

Moritz Kruse[1](✉) and Noomane Ben Khalifa[1,2]

[1] Institute for Production Technology and Systems, Leuphana University of Lüneburg, Universitätsallee 1, 21335 Lüneburg, Germany
moritz.kruse@leuphana.de

[2] Institute of Materials and Process Design, Helmholtz-Zentrum Hereon, Max-Planck-Straße 1, 21502 Geesthacht, Germany

Abstract. Multi-materials, such as fiber metal laminates, offer several advantages over monolithic materials, i.e. the possibility of tailoring the material properties. The in-situ hybridization process combines deep drawing and thermoplastic resin transfer moulding (T-RTM) to manufacture fiber metal laminates with complex geometries in a single process step. However, due to strong fluid-structure interactions between fabric, metal and the low viscous matrix during forming and infiltration, other parameters besides those in pure metal deep drawing affect the process and part quality. Therefore, the influence of blank holder force, injection time, punch velocity and matrix viscosity on the process is experimentally investigated on part scale. The manufactured parts are subsequently assessed by measuring the metal blank surface strains and the thickness of the glass fiber reinforced layer to evaluate the forming behavior and matrix flow. Based on these findings, the most influential parameters and parameter interactions during the deep drawing of fiber metal laminates with a low viscous matrix are identified, and their influence on the process is discussed.

Keywords: Deep drawing · Fiber metal laminates · Multi-materials

1 Introduction

Lightweight design is increasingly important to reduce CO_2 emissions and increase range in the transportation sector. Fiber-reinforced polymers (FRP) are frequently utilized due to their high strength-to-weight ratio. However, FRPs are susceptible to impact damage. Fiber metal laminates (FMLs), consisting of stacked FRP and metal layers, offer improved impact and fatigue resistance while retaining a high strength-to-weight ratio [1]. However, manufacturing FMLs with complex geometries typically requires multiple forming and bonding steps, which is time-consuming [2]. To address this, investigations regarding forming processes, such as deep drawing, and bonding of the layers in one step were performed in the last years. Deep drawing of FMLs was performed with thermoplastic [3] and semi-cured thermoset [4] pre-impregnated fabric materials. Strong

K. Mocellin et al. (Eds.): ICTP 2023, LNME, pp. 124–134, 2024.
https://doi.org/10.1007/978-3-031-40920-2_14

interactions between fabric, metal and high-viscous matrix often cause delamination, matrix accumulations and wrinkling or tearing of metal sheet and fibers [5]. The blank holder force was identified as a major influencing parameter on the formability of FMLs [6, 7]. Blala et al. [8] furthermore found that the drawing depth could be significantly increased when using semi-cured compared to fully cured fabric materials.

To improve forming by reducing the fluid-structure interaction, Mennecart et al. [9] proposed a combined thermoplastic resin transfer moulding and deep drawing process with a low viscous matrix (Fig. 1). For improved adhesion, the inner side of the metal sheets is ground in 0° and 90° direction as well as in small circles. A bonding agent (Dynasalan® Glymo by Evonik) is then applied to improve adhesion between the matrix and the metal sheets. In the in-situ hybridization process, the reactive matrix is injected through the punch into the middle fabric layer during the deep drawing process. Bulging of the upper metal sheet can occur due to high injection pressures and high fabric compaction in the flange from the blank holder pressure. The resulting low permeability hinders the flow of the matrix to the flange and causes hydro stretch forming of the upper sheet of the FML during injection. During further deep drawing and compression between die and punch, the matrix is distributed to the flange. After forming, the FML part is compressed between punch, blank holder and die until the matrix is fully polymerized.

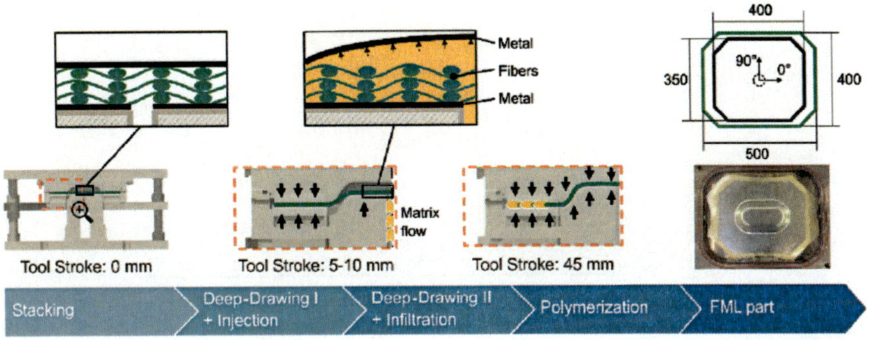

Fig. 1. Combined deep drawing and resin transfer moulding process (in-situ hybridization) [9].

The formability of metal sheets is reduced by the fabric, as Nakajima tests have shown [10]. This is due to high local normal loads at the roving's crossings, which restrict the relative movement of the layers under high contact pressures. Thus, high friction between dry fabric and metal occurs, which is reduced when applying the matrix that acts like a lubricant [11]. Previous investigations of the interaction between fabric and metal during the in-situ hybridization process were conducted in deep drawing experiments without matrix injection [12].

As previously described, strong fluid-structure interaction is present when introducing the matrix into the process. Thus, different parameters than in pure metal deep drawing influence the forming process and part quality. This work aims to investigate the influence of these parameters, such as matrix viscosity and punch velocity, on the fluid-structure interaction during the process. The part quality is evaluated by measuring metal surface strains and FRP layer thickness. Eventually, the most influential parameters are identified and implications for processing are derived.

2 Materials and Method

The metal-fiber-metal stack consists of DC04 (1.0338) steel sheets with 1 mm thickness as the top and bottom layer with six twill-woven E-glass fabric (280 g/m^2, Interglas 92125 FK800) plies in between (dimensions shown in Fig. 1). A 0.025 mm PTFE foil is placed between top blank and tool for lubrication. The bottom metal layer has a 19 mm hole in the center for matrix injection. 300 ml of the reactive thermoplastic matrix (Arkema Elium® 150) mixed with 2.5% hardener (dibenzoyl peroxide Perkadox® GB-50X) is used to ensure full infiltration of the part. The initial matrix viscosity is 100 mPas and the tool is heated to 70 °C for faster polymerization. The viscosity increase during the polymerization is shown in Fig. 2. After reaching the final drawing depth of 45 mm, the part is compressed for an additional 20 min for the matrix to fully polymerize before demoulding. A tool with double-dome geometry is used.

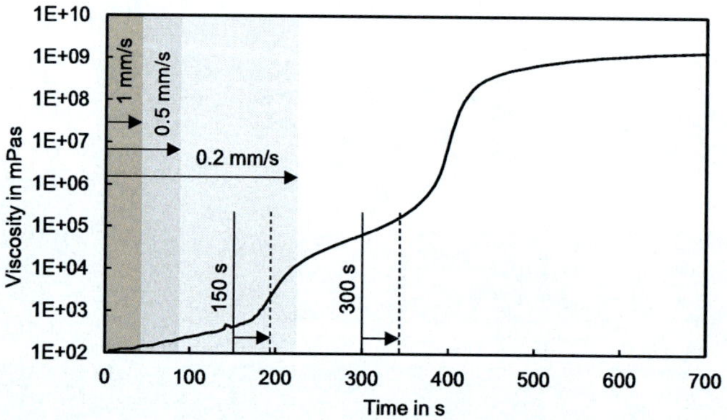

Fig. 2. Viscosity during polymerization of the reactive thermoplastic matrix at 70 °C. Increasing viscosities during deep drawing with different parameters indicated.

The FML stack is initially deep drawn to 15% of the total drawing depth, which ensures the contact pressure between the bottom metal sheet and punch is high enough to ensure sealing during infiltration. At 15%, the matrix injection is started. Three different deep drawing hold times after injection (0 s ≙ 100 mPas, 150 s ≙ 1,000 mPas, 300 s ≙ 100,000 mPas, see Fig. 2) are used to investigate the effect of various matrix viscosities during deep drawing. For comparison, one test with injection during continuous deep drawing is performed, as carried out by Mennecart et al. [9]. Additionally, three different punch velocities (0.2 mm/s, 0.5 mm/s, 1 mm/s) and blank holder forces (100 kN ≙ 1 MPa, 160 kN ≙ 1.6 MPa, 190 kN ≙ 1.9 MPa) are investigated. For the experiments in this investigation, the inner side of the metal sheet is not ground but only treated with a release agent (Henkel Loctite® Frekote HMT2) so that each layer can be investigated separately after manufacturing. A comparison of parts manufactured with both surface treatment methods showed no difference in observed metal strains or fiber draping, which demonstrates the validity of the experiments performed with the release agent. Ram and blank holder forces are measured by the hydraulic press (Röcher RZP 270.21 NZ), while the injection pressure is measured with a pressure transducer. A GOM ARGUS optical surface strain measurement system is used for the evaluation of the metal sheets after forming. In addition, the thickness of the glass fiber reinforced polymer (GFRP) layer is measured with a GOM ATOS laser scanning system.

3 Results and Discussion

3.1 Matrix Viscosity and Injection Time

The experiments were performed with a blank holder force of 190 kN, punch velocity of 1 mm/s, no vacuum during injection and a part target thickness of 3 mm. The outer metal sheets were previously found to exhibit higher in-plane strains than the inner metal sheet, with the most critical location being the double-curved area in the punch radius due to biaxial tensile strains [12]. Thus, only the strains in the outer metal sheet are used for comparison.

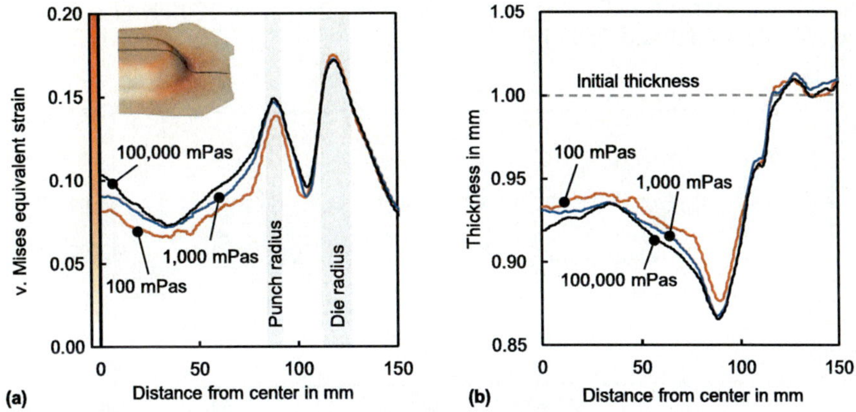

Fig. 3. Influence of different matrix viscosities on (a) strains in the outer metal sheet (part shown with heatmap: 100 mPas; black line in heatmap indicates the cross-section cut), (b) thickness of the outer metal sheet.

Contrary to pure metal deep drawing, high strains and thus, a thickness reduction of approximately 7–8% in the part bottom are present because of the initial bulging during matrix injection (Fig. 3a/b). Biaxial tensile strains in the part bottom and wall area persist in the final part due to bulging. A higher matrix viscosity results in higher strains in the part bottom, punch radius and wall area, but does not affect the strains in the flange (Fig. 3a). While the ultimate tensile strength of the FML is primarily affected by fiber draping and void content of the GFRP layer, the elastic properties of the FML mainly depend on the metal sheet [13]. With higher matrix viscosities, more strain hardening and possibly higher resulting yield strength in the bottom and wall area may be present. However, excessive thinning in the punch radius might lead to lower yield strength of the part manufactured with high matrix viscosity. Lower overall strains also indicate higher possible drawing depths when using a low viscous matrix.

According to Darcy's law, the flow rate Q is calculated by [14]

$$Q = -\frac{K * A * \Delta p}{\eta * I} \tag{1}$$

where K is the fabric's permeability, η is the viscosity of the injected silicone oil, A is the flow channel cross-sectional area, I is the specimen length, and Δp is the pressure difference between injection pressure and ambient pressure. Hence, the matrix flow rate from part bottom to the flange is proportional to the internal pressure and inversely proportional to the viscosity. Higher viscosities lead to reduced outflow to the flange during deep drawing and thus, the bulge being present for longer. The matrix is then squeezed out to the flange when the part is compressed between punch and die. The contact between the upper metal sheet and die is indicated by the sudden increase in punch force at 85–90 s (Fig. 4). A higher matrix viscosity leads to higher internal fluid pressures during compression of the FML and squeezing of the matrix to the flange (Eq. (1)). This results in higher punch forces during the first die contact as well as during the final compression at the end of deep drawing (100 mPas: 500 kN; 1,000 mPas: 530 kN; 100,000 mPas: 570 kN).

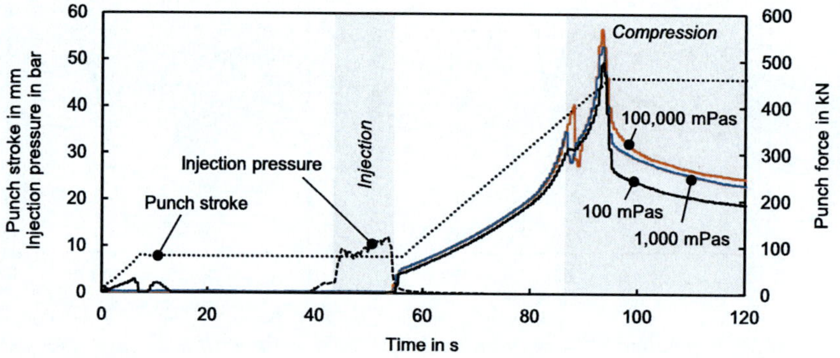

Fig. 4. Influence of different matrix viscosities on the punch forces during the process.

The influence of the injection time on the punch force is negligible, with 510 kN after injection during continuous deep drawing, compared to 500 kN after injecting during hold (Fig. 5a). No difference in metal strains was observed, either. However, the injection pressure increases during continuous deep drawing (up to 21 bar) compared to the injection during hold (11 bar). Two effects contribute to this: First, with continued deep drawing, the compaction of the fabric in the die radius increases, thus the permeability decreases which hinders flow to the flange. Second, the continued deep drawing results in strain hardening of the metal sheet. Consequently, higher injection pressures occur, when the metal sheet bulges during continued deep drawing.

More matrix accumulations and a generally thicker GFRP layer result from the later injection (Fig. 5b). No effect on the overall GFRP layer thickness can be observed with different matrix viscosities. Differences mainly occur in the double-curved punch radius, where higher matrix viscosities result in increased thinning of the GFRP layer. Higher biaxial tensile metal strains in the punch radius cause increased contact pressure on the fabric material during forming. As a result, higher strains due to higher matrix viscosities directly correlate with the thickness reduction in the punch radius.

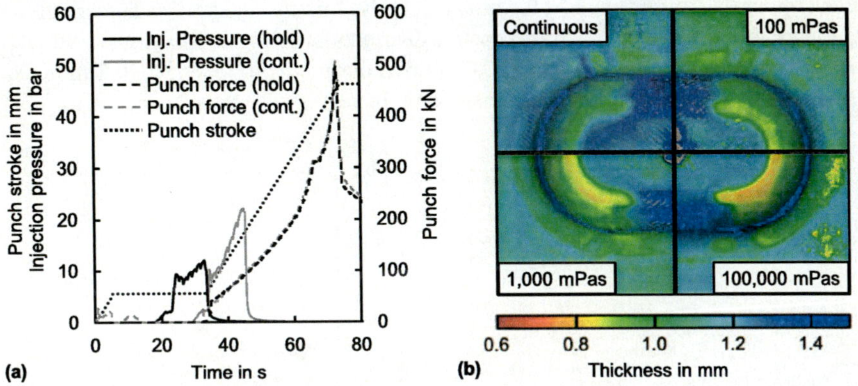

Fig. 5. (a) Comparison of punch forces and injection pressure with matrix injection during hold and during continuous deep drawing, (b) thickness of the GFRP layer.

In addition, high matrix viscosities lead to high fluid forces during flow to the flange, resulting in fiber washing in the flange area (Fig. 6). Only little fiber washing close to the outer edge occurs with a matrix viscosity of 1,000 mPas, but strong fiber washing in the whole flange area occurs with 100,000 mPas.

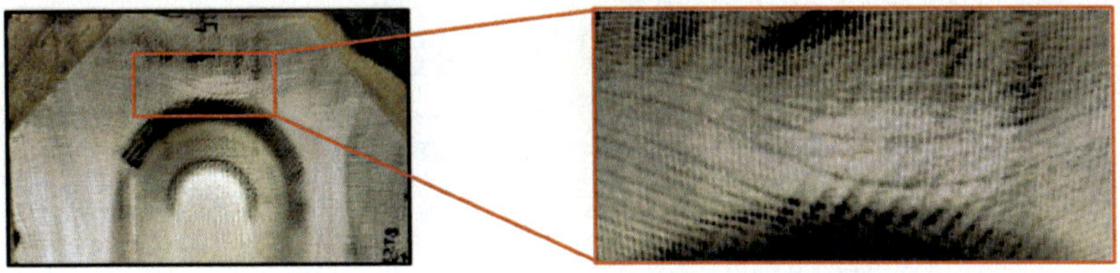

Fig. 6. Fiber washing in the GFRP layer (100,000 mPas).

3.2 Punch Velocity

The experiments were performed with a blank holder force of 190 kN, matrix viscosity of 100 mPas, vacuum in the tool until injection start and a part target thickness of 2.8 mm. When using a low punch velocity, the strains and thinning in the final part are reduced, in particular in the part bottom and critical punch radius (Fig. 7a). Thinning of the GFRP layer in the punch radius and matrix accumulations are reduced as well (Fig. 7b). This is because the matrix has more time to advance from the bulge to the flange during the deep drawing process at low velocities. As a result, the bulge is reduced towards the beginning of the drawing process, leading to lower strains in the final part and a more homogeneous metal and GFRP layer thickness. However, counteracting that is an increasing matrix viscosity over time, leading to higher viscosities towards the end of the drawing process when using low velocities (Fig. 2). This leads to higher maximum punch forces (0.2 mm/s: 660 kN; 0.5 mm/s: 560 kN; 1 mm/s: 560 kN), as higher internal fluid pressures are necessary for the matrix distribution during compression. Although the matrix viscosity increases up to 10,000 mPas during the process with 0.2 mm/s, no fiber washing occurs because the viscosity only increases after the majority of forming.

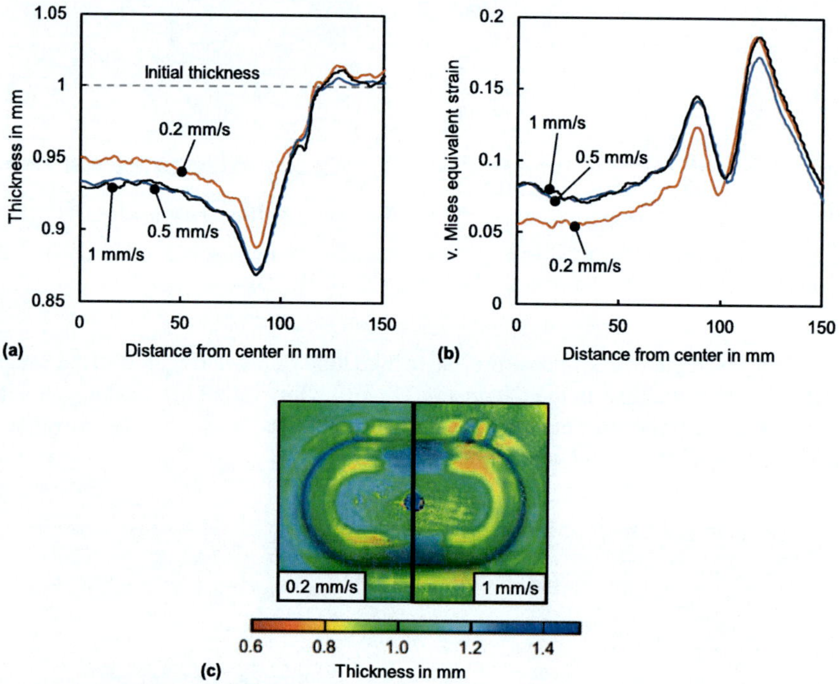

Fig. 7. Influence of different punch velocities on (a) thickness of the outer metal sheet, (b) strains in the outer metal sheet and (c) thickness of the GFRP layer.

3.3 Blank Holder Force

The experiments were performed with a matrix viscosity of 100 mPas, punch velocity of 0.2 mm/s, vacuum in the tool until the end of the injection and a part target thickness of 2.8 mm. As expected in deep drawing, decreasing the blank holder force results in reduced metal strains and thickness reductions in the part bottom, punch radius and wall

(Fig. 8a). The reduced friction in the flange allows the metal to flow more easily and causes fewer thickness reductions at lower blank holder forces. This results in lower punch forces during the deep drawing (Fig. 8b). Lower blank holder forces reduce the fabric compaction in the flange, allowing the matrix to flow more easily from the bulge to the flange, but also reduce the strains in the metal. Reduced strain hardening then possibly reduces the pressure on the fluid during deep drawing. Higher maximum punch forces at lower blank holder forces during the compression phase (100 kN: 670 kN; 160 kN: 640 kN; 190 kN: 620 kN) indicate that this effect results in higher flow rates to the flange during the deep drawing stage. As a result, less matrix remains in the bulge that has to be distributed during the compression phase, leading to higher punch forces for lower blank holder forces during deep drawing, but lower maximum punch forces during compression. Slight wrinkling did occur at 100 kN blank holder force, leading to matrix flowing between metal and tool at the outer edge of the flange.

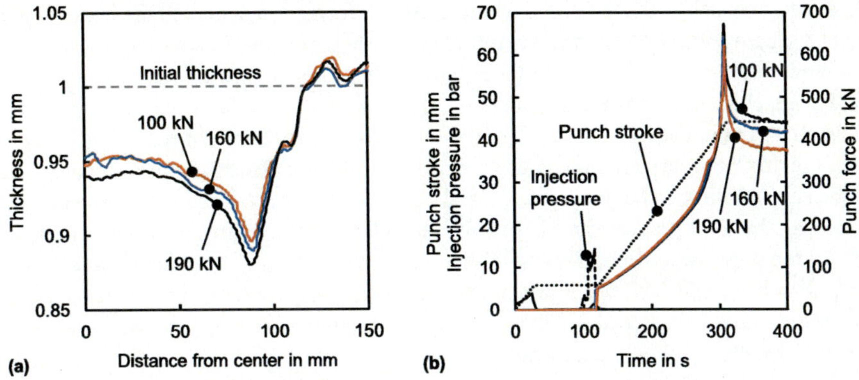

Fig. 8. Influence of different blank holder forces on (a) the thickness of the outer metal sheet and (b) punch forces during the process.

It has been shown that the deep drawing of fiber metal laminates differs from pure metal deep drawing, where the blank holder force (and part height) is the most significant parameter affecting metal strains and part quality. In contrast, in fiber metal laminate deep drawing, other parameters, such as punch velocities and matrix viscosity, strongly influence metal strains and thinning. Typically, punch velocities only significantly influence the metal strains when strain rates are higher than quasi-static, like in high-speed forming [15]. However, due to the strong fluid-structure interaction, low punch velocities and matrix viscosity had a similar or greater influence on the metal strains and thinning than the blank holder force. Metal strains were found to correlate with GFRP layer thinning in the punch radius due to higher compressions at higher strains caused by strain hardening. Matrix accumulations were reduced by injecting the matrix during hold and using a lower punch velocity. Although lower punch velocities resulted in higher viscosities towards the end of the deep drawing, it still proved favorable for metal strains and GFRP layer thickness, whereas a high matrix viscosity during forming is unfavorable. Fiber washing can occur due to high fluid forces at matrix viscosities of 1,000 mPas and higher, possibly even below. However, this not only depends on the matrix viscosity but also the fabric compaction and flow rate. Thus, fiber washing can possibly also be prevented by reducing the punch velocity, injected matrix amount and blank holder force. However, too low blank holder forces can cause wrinkling and matrix flow between tool and part.

The maximum punch force occurs towards the end of the process when the part is compressed between the punch and die. It is largely influenced by the amount of remaining matrix in the bulge and the matrix viscosity during the compression phase. Counterintuitively, lower blank holder forces resulted in higher punch forces because they allowed more matrix to be distributed during the deep drawing, thereby reducing the flow rate and internal fluid pressure during compression. In contrast, lower matrix viscosities and slower drawing speeds resulted in slightly lower punch forces by enabling the distribution of more matrix during the deep drawing stage.

The process can possibly be further improved by avoiding the initial bulging during the injection. This can be done by reducing the injection pressure to reduce the internal fluid pressures below the threshold where plastic deformation occurs in the upper metal sheet during injection. However, this would greatly lengthen the processing time, as only low flow rates can be achieved without bulging because of high fabric compaction in the flange. Hence, to increase the flow rate, the blank holder force could be temporarily lowered during injection when no deep drawing takes place. Furthermore, using an even lower matrix viscosity and just enough matrix to infiltrate the whole part might result in faster infiltration and overall lower metal strains and more homogeneous thicknesses. However, the total process time is mainly dependent on the matrix polymerization. To reduce the processing time, but avoid high matrix viscosities during deep drawing, the deep drawing could take place just below the matrix polymerization temperature. The tool is then quickly heated above the polymerization temperature during the compression phase to enable rapid polymerization.

4 Conclusion

The deep drawing of fiber metal laminates involves fluid-structure interaction, which leads to different parameters influencing the process than in pure metal deep drawing. It has been found that, in addition to the blank holder force, the matrix viscosity and punch velocity affect the punch forces as well as the metal strains and thicknesses of the final part. The following conclusions can be drawn:

- Low matrix viscosities, low blank holder forces, low punch velocities and an early matrix injection during deep drawing hold can reduce metal strains and lead to a more homogeneous GFRP layer thickness.
- Punch forces are mainly dependent on the FML part compression towards the end of the process. Thus, higher matrix viscosities and flow rates during compression cause higher punch forces.
- High matrix viscosities above 1,000 mPas can lead to fiber washing due to high fluid forces during deep drawing.
- Avoiding bulging during injection would be favorable. This could be achieved by reducing injection pressures and blank holder forces during injection, as well as using a low viscous matrix.

In future works, further interactions between these parameters will be investigated to identify the process window and limitations for the deep drawing of FML with a low viscous matrix. Moreover, other injection parameters, such as applying a vacuum inside the tool or changing the injection pressure and injected matrix amounts, are expected to influence the process due to the fluid-structure interaction and should be investigated further.

Acknowledgments. The authors would like to thank the German Research Foundation (DFG) for funding the projects BE 5196/4-1 and BE 5196/4-2. The matrix system and peroxide were kindly provided by the Arkema Group. The authors would like to thank Mr. Marvin Gerdes for the help in performing experiments and Mr. Henrik O. Werner for the help in planning experiments and discussing results.

References

1. Etri, H.E., Korkmaz, M.E., Gupta, M.K., Gunay, M., Xu, J.: A state-of-the-art review on mechanical characteristics of different fiber metal laminates for aerospace and structural applications. Int. J. Adv. Manuf. Technol. **123**(9–10), 2965–2991 (2022)
2. Min, J., et al.: Fabrication processes of metal-fiber reinforced polymer hybrid components: a review. Adv. Compos. Hybrid Mater. **5**(2), 651–678 (2022)
3. Wollmann, T., et al.: Thermoplastic fibre metal laminates: stiffness properties and forming behaviour by means of deep drawing. Arch. Civ. Mech. Eng. **18**(2), 442–450 (2018)
4. Heggemann, T., Homberg, W.: Deep drawing of fiber metal laminates for automotive lightweight structures. Compos. Struct. **216**, 53–57 (2019)
5. Ding, Z., Wang, H., Luo, J., Li, N.: A review on forming technologies of fibre metal laminates. Int. J. Lightweight Mater. Manuf. **4**(1), 110–126 (2021)
6. Hwang, S.-F., Li, Y.-R.: Deep drawing behavior of metal-composite sandwich plates. Materials **15**(19), 6612 (2022)
7. Rajabi, A., Kadkhodayan, M., Manoochehri, M., Farjadfar, R.: Deep-drawing of thermoplastic metal-composite structures: experimental investigations, statistical analyses and finite element modeling. J. Mater. Process. Technol. **215**, 159–170 (2015)
8. Blala, H., Lang, L., Li, L., Sijia, S., Alexandrov, S.: Mechanical properties and formability behavior of glass-reinforced aluminum laminate, developed by semi-cured process. J. Braz. Soc. Mech. Sci. Eng. **44**(12), 590 (2022)
9. Mennecart, T., Werner, H., Ben Khalifa, N., Weidenmann, K.A.: Developments and analyses of alternative processes for the manufacturing of fiber metal laminates. In: Volume 2: Materials; Joint MSEC-NAMRC-Manufacturing USA. American Society of Mechanical Engineers (2018)
10. Mennecart, T., Gies, S., Ben Khalifa, N., Tekkaya, A.E.: Analysis of the influence of fibers on the formability of metal blanks in manufacturing processes for fiber metal laminates. JMMP **3**(1), 2 (2019)
11. Kruse, M., et al.: Investigation of the friction behavior between dry/infiltrated glass fiber fabric and metal sheet during deep drawing of fiber metal laminates. Prod. Eng. Res. Devel. **17**, 37–46 (2022)
12. Kruse, M., Lehmann, J., Ben Khalifa, N.: Parameter investigation for the in-situ hybridization process by deep drawing of dry fiber-metal-laminates. In: Liewald, M., Verl, A., Bauernhansl, T., Möhring, H.-C. (eds.) Production at the Leading Edge of Technology. Lecture Notes in Production Engineering, pp. 122–130. Springer International Publishing, Cham (2023). https://doi.org/10.1007/978-3-031-18318-8_13

13. Werner, H.O., Liebig, W., Weidenmann, K.A.: Mechanical properties of in-situ polymerized fiber-metal-laminates. In: Hausmann, J.M., Siebert, M., Hehl, A. von, Weidenmann, K.A. (eds.) Proceedings 4th International Conference Hybrid 2020 Materials and Structures (2020)
14. Arbter, R., et al.: Experimental determination of the permeability of textiles: a benchmark exercise. Compos. Part A Appl. Sci. Manuf. **42**(9), 1157–1168 (2011)
15. Neugebauer, R., et al.: Velocity effects in metal forming and machining processes. CIRP Ann. **60**(2), 627–650 (2011)

Quantification of the Impact of Forming-Induced Residual Stresses on Subsequent Cutting Operations for Aluminum AA7075 Thick Sheets

Michael Ott[1(✉)], Moritz Mayer[2], Yiran Li[1], Florian Steinlehner[1], Michael F. Zaeh[2], and Wolfram Volk[1]

[1] Chair of Metal Forming and Casting, Technical University of Munich, Munich, Germany
michael.ott@utg.de

[2] Institute for Machine Tools and Industrial Management, Technical University of Munich, Munich, Germany
moritz.mayer@iwb.tum.de

Abstract. Residual stresses in metal components induced by forming operations significantly influence subsequent manufacturing steps. In particular, two aspects for the final processing of formed thick sheets by cutting and machining operations result: First, the achievable geometric accuracy is limited by distortion due to the modification of residual stresses across the process step sequence. In addition, the residual stresses present in the final component are determined by the process parameter settings of the individual substeps. In this work, the influence of the forming-induced residual stresses on a subsequent cutting operation was quantitatively investigated. For this purpose, the step sequence of forming and wet abrasive cutting for aluminum AA7075 thick sheets was implemented both numerically and experimentally. This allows the residual stress redistribution to be taken into account in future virtual setups of the production chain by specifically adjusting the process parameters.

Keywords: residual stress · thick sheet forming · milling · FEM

1 Introduction and State of the Art

The machining of complex geometries from bulk source material leads to a high material removal volume. According to [1] a machining volume of more than 90% is common for components in the aviation industry. As an alternative manufacturing approach, a combined step sequence of forming and subsequent machining can be employed. Examples of industrial applications based on this strategy include aerospace structural components such as a cargo door shown by [2] and a windshield frame part shown by [3]. The resulting reduction in machining volume leads to reduced wear on the cutting tools and less raw material is required for production. Furthermore, this step sequence offers the technical potential for improving component properties due to the possibility

K. Mocellin et al. (Eds.): ICTP 2023, LNME, pp. 135–145, 2024.
https://doi.org/10.1007/978-3-031-40920-2_15

of influencing the residual stresses in the part. The residual stresses in turn affect a wide range of application-relevant component properties such as fatigue strength in welding [4], forging [5] and shear cutting applications [6], stress corrosion cracking [7] and magnetic material properties [8]. However, for the reliable production of components with tight geometric tolerances, the distortion due to the residual stresses in the formed geometry must be taken into account in the subsequent cutting and machining processes. For the design of the process sequence, this raises the question of the achievable accuracy of a numerical process model that can predict the residual stresses and the component geometry along the two manufacturing steps.

1.1 Adjusting Residual Stresses in Metal Forming

The adjustment of residual stresses in formed geometries by adapting process parameters and selecting specific process variants was investigated for multiple forming processes. For single point incremental forming, [9] show that it is possible to influence the residual stresses by adjusting process parameters such as the tool radius. In contrast, the influence of the tool path strategy is described as negligible [10]. For the freeform bending of tubes, [11] investigated the manufacturing of identical specimen geometries with different residual stresses by using adjusted bending strategies. As a method to generate a uniform residual stress field in sheet metal components [12] investigated the application of multiple one-sided embossing points. The effect of the fluid pressure in hydroforming of sheets was quantified by [13], who describe a correlation between an increasing pressure and lower residual stresses.

1.2 Residual Stress-Induced Distortion During Cutting and Machining Operations

In a similar manner to the forming processes, approaches for influencing the development of residual stresses in the final geometry exist for various cutting and machining processes. [6] show an influence on residual stresses by the choice of the blanking process, such as fineblanking and precision blanking. The effect on near-surface residual stresses during milling was studied by [14] and [15] for titanium. [16] and [17] investigated the predictability of milling related redistribution of pre-existing residual stresses and its influence on geometric distortion. Compensation methods can be applied to achieve small geometric tolerance fields. For this purpose, [18] describe laser treatment and deep rolling methods for profile grinding. In the case of sequential manufacturing processes, the influence on the follow-up steps can be taken into account during process development. For the step sequence of forming and machining, [19] show the influence of lubricant and tool geometry parameters of the milling step. [3] and [20] investigated the part distortions for machining operations on forged semi-finished products by experiment and simulation.

2 Methods

Experimental Approach. The main research objective was to investigate the predictability of the residual stresses over the process chain of thick sheet forming and a subsequent cutting process. The procedure includes the experimental implementation of the process step sequence for a demonstrator geometry. For this purpose, an experimental forming tool was developed for the first process stage and wet abrasive cutting was used as an exemplary subsequent cutting process. By setting a very slow cutting speed in combination with the use of cooling lubricant, mechanical and thermal sources for the induction of additional residual stresses by the second process stage are to be reduced. This is thus chosen as a model process to investigate the formation of geometric distortions predominantly due to the release and redistribution of the pre-existing forming-induced residual stresses. In parallel, a numerical process model was developed to determine the achievable accuracy of prediction of the residual stresses across the process stages. To compare the predictions of the simulation model with the experimental results, measurement data of the component geometry, the strain at the component surface and the residual stresses at two locations were compared. These investigations are intended to provide a basis for investigating more complex and industrially relevant machining finishing processes such as milling in future work.

Material and Specimen Preparation. AA7075-T651 sheets with a thickness of 11 mm were used for the experimental investigation of the process chain. According to DIN EN 515, T651 denotes the material state as solution annealed, stress relieved by controlled stretching and artificially aged [21]. Waterjet cutting was selected for generating the specimen outer contours.

Forming Stage. The setup of the forming stage is shown in Fig. 1. A set of active elements was designed for an existing forming tool, which contains a top plate (I), a blankholder plate (II) and a base plate (III). The set consists of the punch ①, the corresponding die ② and the contact surfaces of the blank holder ③. The vertical die movement is guided by four column guides ④ with bushings. All of the forming experiments were performed on a hydraulic press DXU 320 B, Dieffenbacher, Eppingen, Germany, with a maximum press force of 3500 kN. A blankholder force of up to 1250 kN is transmitted via cylindrical rods by a bottom-mounted hydraulic drawing pad through the circular openings in the base plate visible in Fig. 1. Platinol B804/3 COW-1, Oest, Freudenstadt, Germany was used as the forming lubricant. The forming tool was designed to be used with several specimen geometries to generate various forming-induced residual stress states. The approach is thus comparable to setting different strain states by adjusting the blank geometries used by the Nakajima test, by the Marziniak test and by the forming tool for generating non-linear strain paths in [22]. For the surfaces of the punch an uniaxially curved area was combined with a hemispherical region. This geometry causes local varying uni- and biaxial bending loads on the specimen. For the investigations carried out in this work, the specimen geometry shown in Fig. 2b was developed. The shape with a central recess was chosen to generate a measurable distortion after the cutting step during the subsequent experiments.

Cutting Stage. The specimen geometry in Fig. 2b is shown including the cutting position and cutting direction. Wet abrasive cutting was performed on a precision cut-off machine Brillant 220, ATM, Mammelzen, Germany. The test setup with lateral clamping of the specimen geometry is shown in Fig. 2a. The rotational frequency was set to 4000 rpm, the feed rate to 0.04 mm/s. A corundum cut-off wheel with a diameter of

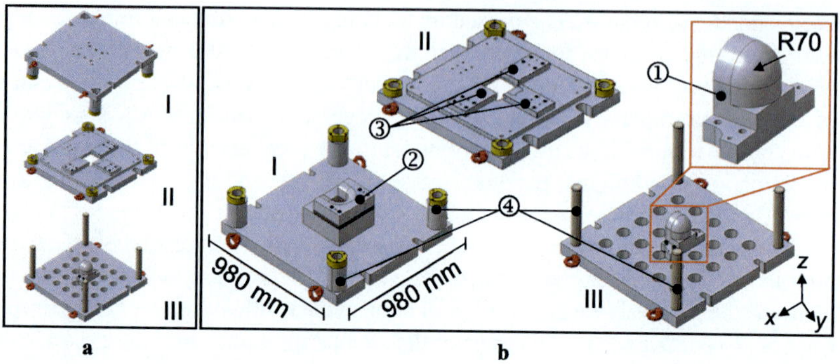

Fig. 1. Experimental tool for thick sheet forming: Mounting position (a) and overview of the tool components (b).

150 mm and a thickness of 0.45 mm was used. The red laser displayed in Fig. 2a helped to align the specimen.

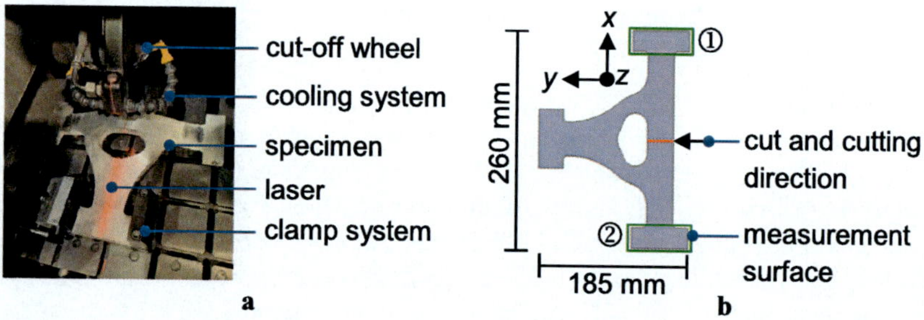

Fig. 2. Test setup on the ATM Brillant 220 precision cut-off machine (a); specimen with position of the cut marked in red and indication of the cutting direction and measurement surfaces marked in green (b).

Measurement Methods. The optical 3D digitizer ATOS Compact Scan, GOM, Braunschweig, Germany, was used for full-surface measurement of the component geometry after forming. The system uses a fringe projection method with a stereo camera system. To measure the geometry changes caused by the cutting process a coordinate measuring machine LH87, Wenzel, Wiesthal, Germany, was employed with a touch-trigger probe using a measuring sphere diameter of 4 mm. Moreover, to measure the strains on the top and bottom surface of the component after forming, the ARGUS system, GOM, Braunschweig, Germany, was used.

The residual stresses after forming were determined using the hole drilling method on a MTS3000-Restan, SINT Technology, Calenzano, Italy [23]. The milling tool used for the measurements has a diameter of 1.6 mm and the maximum achievable evaluation depth is 1 mm. Strain was measured with the strain gauge shown in Fig. 3a at the two positions displayed in Fig. 3b. The residual stresses were calculated with a model-based approach according to Schwarz and Kockelmann, which is implemented in the evaluation software of the MTS3000-Restan.

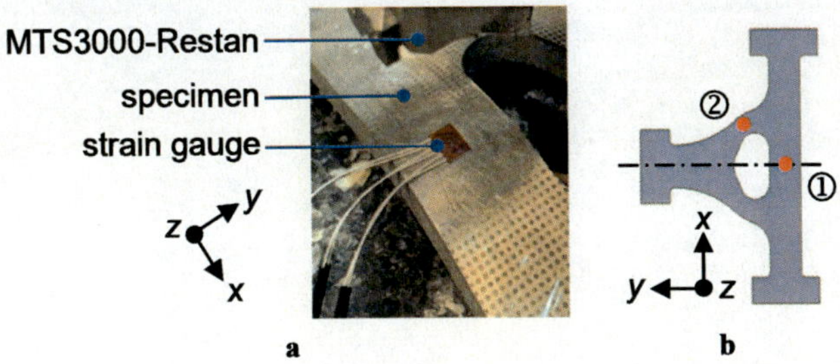

Fig. 3. Set-up of the hole drilling method (a) and position of the measurement spots marked with red (b).

Numerical Process Model. Dassault Abaqus 2022 and LS-Dyna R12 as market available software were chosen to set up the numerical process model in order to be able to realistically assess the transferability to industrial processes. The model comprises an initial explicit forming step and an implicit springback step. This is followed by the removal of the elements located on the cutting trajectory. In Abaqus, a predefined element set can be used to identify these elements and in LS-Dyna arbitrary trim curves can be defined. An implicit integration procedure is likewise employed for this purpose. The goal of this step is to obtain the updated residual stress state and the associated geometric distortion. Spatial discretization of the thick sheet was executed by hexahedral continuum elements with reduced integration and the plastic material behavior was modeled based on the data from tensile tests with samples taken an angles of 0°, 45° and 90° to the rolling direction. Plots were generated using the tool developed by [24].

3 Results

Prediction of the Component Geometry Along the Process Chain. Figure 4 provides an overview over the predicted geometries after the three subsequent stages of the numerical process model. The development of the stress component along the *x*-axis, which is orthogonal to the cutting plane, is visualized.

Geometry after the Forming Stage. Specimens with a fixed blankholder force of 100 kN and drawing depths of 23.5 mm and 27.5 mm were evaluated. The results of the GOM ATOS measurements of the components show a high repetition accuracy of the geometry after the first process stage. For the occurring springback, a height difference of the specimens of 1.9 mm was determined from the simulation model at a drawing depth of 27.5 mm. For the design of the process chain for industrial components, existing algorithms such as the stress-based method in [25] or the displacement-based method in [26] can be applied for compensation. In Fig. 5 a comparison of the strains on the top and bottom surface with the numerical model is shown. The resulting similarity is a necessary condition for a realistic prediction of the residual stresses.

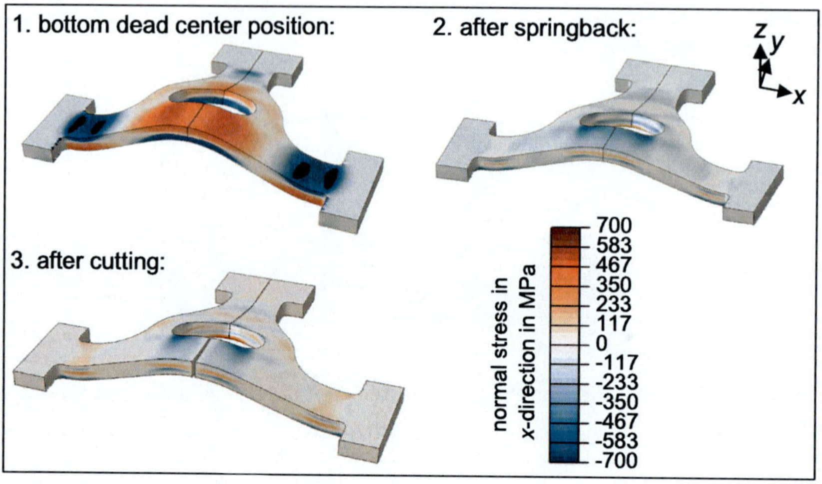

Fig. 4. Predicted geometry and stress distribution of the subsequent stages of the numerical process model.

Geometry after the Cutting Stage. In Fig. 6, the local area around the cut is compared between the numerical model created in Abaqus and the experimental results. Qualitatively, the angle formed in the experiment is reproduced in the simulation.

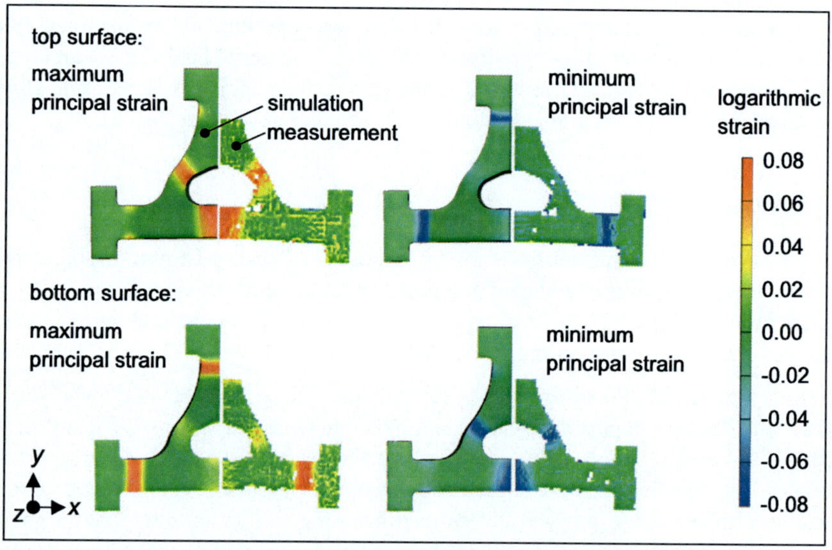

Fig. 5. Maximum and minimum principal strain: Comparison between the predicted numerical model (left half section) and the measured values (right half section) on the top and bottom surfaces.

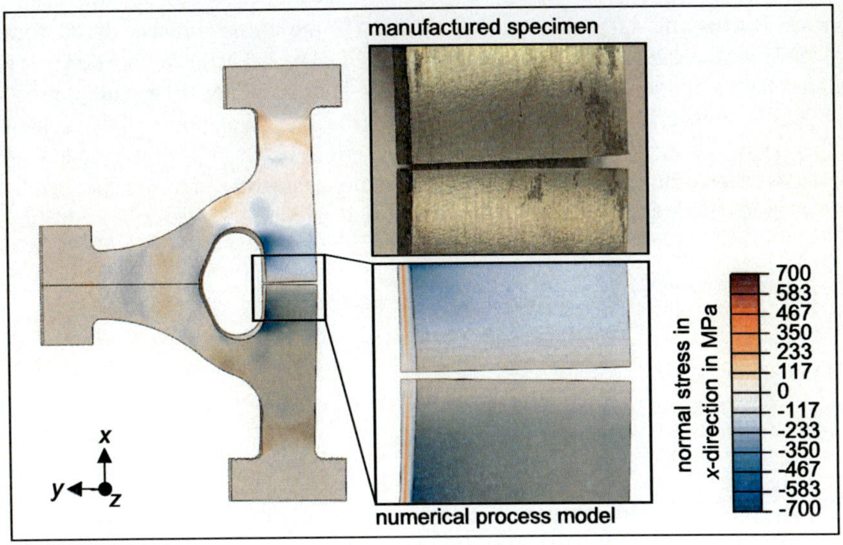

Fig. 6. Comparison of the local springback in the vicinity of the cut between the numerical process model and the manufactured specimen.

As a global measure of the geometric distortion caused by the cut, the angles of the two lateral contact surfaces under the blank holder with respect to each other are used. For this purpose, the two surfaces were gauged with the coordinate measuring machine before and after the cut. Eight measurement points per surface were probed and a plane was derived using a least square fit. The measured values are shown in Table 1 following the numbering from Fig. 2b. Qualitatively the deflection is predicted correctly, quantitatively there are deviations to the measured values.

Table 1. Angular changes of the measurement planes.

planes	source	angle after forming stage	angle after cutting stage	angular difference
1 vs. 2	measurement	1.504°	0.667°	0.837°
	simulation	1.418°	1.057°	0.361°

Stresses Before the Forming Stage. A hole drilling measurement of the AA7075 raw material revealed values of the von Mises stress ranging from 19.2 MPa at the surface to 13.1 MPa in the interior of the component as deep as the measurable depth of 1 mm. They can be regarded as minor compared to the forming-induced residual stresses. Possible stresses previously induced during rolling or heat treatment were thus greatly relieved by the controlled stretching during the T651 sequence.

Stresses During and After the Forming Stage. For the stresses induced during forming, in Fig. 7a the values calculated using the numerical model at the bottom dead center position for measurement point 1 are shown. The corresponding strains are plotted in Fig. 7b. The stress and strain data were read out at the integration points of the elements and an interpolation of the progression of the quantities between them was added. The point at which the cut is placed is thus predominantly characterized by uniaxial bending with symmetric plastically deformed edge areas and an elastically deformed core area.

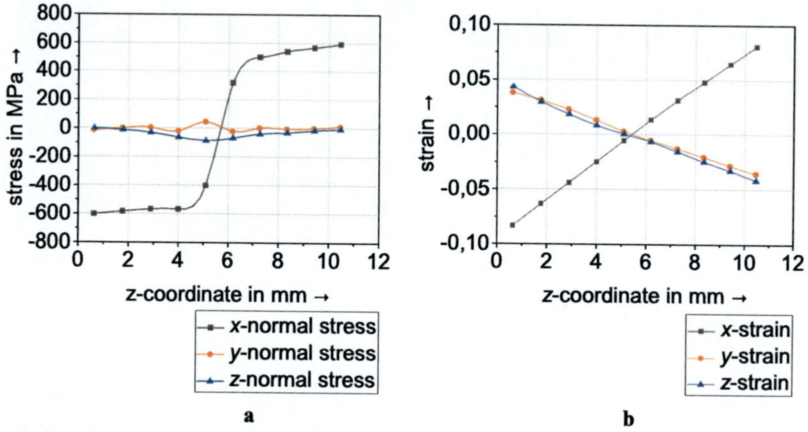

Fig. 7. Calculated normal stress (a) and strain (b) components in the bottom dead center position at measurement point 1 as a function of the sheet thickness (z-coordinate).

Concerning the residual stresses, the measurements were carried out on a specimen after forming (drawing depth of 23.5 mm, blankholder force of 100 kN) at the two measurement points 1 and 2 from Fig. 3b. The measured values are compared to the results from the numerical process model in Fig. 8. An average value of nine adjacent elements is displayed for each data point and the associated bar is calculated from the maximum and minimum value of these elements. Therefore, the length of the bars can be seen as an indicator of the local homogeneity of the stress values. The hole drilling method measures minimum and maximum principal stresses in the planes orthogonal to the drilling axis. In addition, the angle between the principal axis system and the coordinate system of the strain gage is calculated. At measurement point 1, the principal axis system matches with the *xy*-plane, at measurement point 2 the results are shown in the principal axis system.

For measurement point 1, the result is a residual stress profile that matches the expected profile after springback following a uniaxial bending load. At measurement point 1, the numerical process model reproduces the gradient of the residual stress measured during the hole drilling measurements in the depth direction along the *x*-axis and the *y*-axis. For the absolute values of the residual stresses, there are deviations especially in the y-direction.

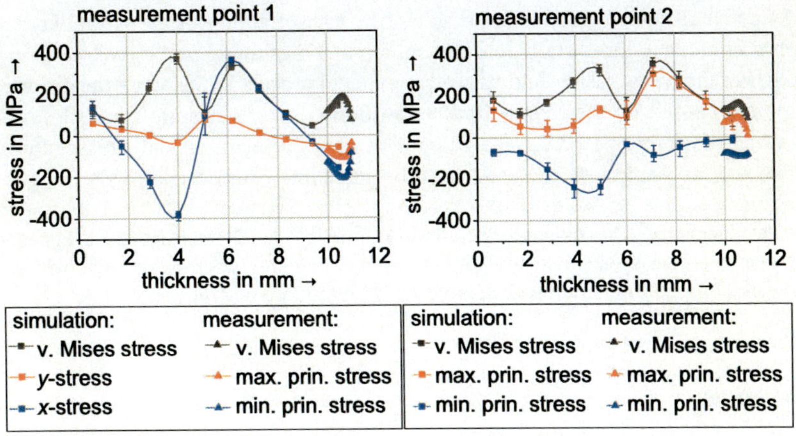

a b

Fig. 8. Residual stress measurement using the hole drilling method on the formed specimen geometry compared to the calculated stress profile in the numerical process model at measurement point 1 (a) and measurement point 2 (b).

4 Summary and Outlook

The process step sequence of forming and an exemplary cutting process for aluminum AA7075 thick sheets was performed both virtually and experimentally. For the specimen geometry, a good qualitative correlation with the simulation results was found for the local cut-related distortion. For the residual stresses after the forming step, a quantitative agreement with the predicted values at the position of the cut was obtained. Numerical prediction of resulting component distortion due to the release and relocation of residual stresses over the entire process chain can thus be considered as a feasible approach applicable for complex geometries. This represents a necessary step towards a completely virtual design of the process chain layout. In contrast to existing works for the process development by [3] and [20], the residual stresses for the simulation of a cutting or machining stage were not applied as initial conditions based on residual stress measurements. Instead, the residual stresses were calculated based on the modeling of their physical origin mechanism. Initially, this resulted in a higher effort for the parameterization of the plastic material model for the forming simulation. In return, the number of complex, expensive and, for the methods by [3] and [20], also non-destructive measurements of the residual stresses can be reduced. In particular, the observed inhomogeneity of the residual stresses over the thick sheet geometry must be taken into account. For the hole drilling method used in this work, which generates measurement data only on a small spot, this results in a high number of necessary measurements. Another advantage of a purely virtual layout is the ability to take into account the effects of a wide range of process parameter settings on residual stresses and geometric distortion. In addition, unlike for most measurement methods, all components of the stress tensor are accessible in the numerical model.

As potential future research approaches, the transferability of the virtual layout of the process chain to industrially relevant cutting and machining processes can be investigated. For the wet abrasive cutting used as a model process in this work the distortions are mainly caused by the release and redistribution of pre-existing residual stresses. In contrast, for other processes the generation of additional residual stresses through thermal and mechanical mechanisms must be taken into consideration.

Acknowledgements. The authors gratefully acknowledge the German Research Foundation (Deutsche Forschungsgemeinschaft, DFG) for supporting this work carried out within the framework of the collaborative research project 457270653 (VO 1487/63-1).

References

1. Wang, Z.-J., Chen, W.-Y., Zhang, Y.-D., et al.: Study on the machining distortion of thin-walled part caused by redistribution of residual stress. Chinese J. Aeronaut. **18**, 175–179 (2005). https://doi.org/10.1016/S1000-9361(11)60325-7
2. Groeneveld, H., Mordechai, C.: Airplane Cockpit Reduced from Two Hundred Parts to One: Sheet Metal Simulation Tested for Explosive Forming in Aerospace. https://formingworld.com/aerospace-autoform-3d-metal-forming/ (2021). Last accessed 19 Apr 2023
3. Liu, L., Sun, J., Chen, W., et al.: Study on the machining distortion of aluminum alloy parts induced by forging residual stresses. Proc. Inst. Mech. Eng., Part B **231**, 618–627 (2017). https://doi.org/10.1177/0954405415583805
4. van den Berg, N., Xin, H., Veljkovic, M.: Effects of residual stresses on fatigue crack propagation of an orthotropic steel bridge deck. Mater. Des. **198**, 109294 (2021). https://doi.org/10.1016/j.matdes.2020.109294
5. Ball, D.L., James, M., Bucci, R.J., et al.: The impact of forging residual stress on fatigue in aluminum. In: 56th AIAA/ASCE/AHS/ASC Structural Structural Dynamics Material Conference (2015). https://doi.org/10.2514/6.2015-0386
6. Stahl, J., Müller, D., Pätzold, I., et al.: The influence of residual stresses induced by near-net-shape blanking processes on the fatigue behavior under bending loads. IOP Conf.: Ser. Mater. Sci. Eng. **651**, 12086 (2019). https://doi.org/10.1088/1757-899X/651/1/012086
7. Bai, L., Jiang, K., Gao, L.: The influence and mechanism of residual stress on the corrosion behavior of welded structures. Mat. Res. **21**, 162 (2018). https://doi.org/10.1590/1980-5373-MR-2018-01660
8. Gilch, I., et al.: Impact of residual stress evoked by pyramidal embossing on the magnetic material properties of non-oriented electrical steel. Arch. Appl. Mech. **91**(8), 3513–3526 (2021). https://doi.org/10.1007/s00419-021-01912-6
9. Maaß, F., et al.: Forming mechanisms-related residual stress development in single point incremental forming. Prod. Eng. Res. Devel. **13**(2), 149–156 (2018). https://doi.org/10.1007/s11740-018-0867-3
10. Maaß, F., Hahn, M., Dobecki, M., et al.: Influence of tool path strategies on the residual stress development in single point incremental forming. Procedia Manuf. **29**, 53–58 (2019). https://doi.org/10.1016/j.promfg.2019.02.105
11. Maier, D., Stebner, S., Ismail, A., et al.: The influence of freeform bending process parameters on residual stresses for steel tubes. Adv. Manuf. **2**, 100047 (2021). https://doi.org/10.1016/j.aime.2021.100047

12. Walzer, S., Simon, N., Liewald, M., Gibmeier, J.: Local creation of uniform residual stresses into sheet metal components by means of multiple one-side embossing. In: Daehn, G., Cao, J., Kinsey, B., Tekkaya, E., Vivek, A., Yoshida, Y. (eds.) Forming the Future. TMMMS, pp. 2357–2368. Springer, Cham (2021). https://doi.org/10.1007/978-3-030-75381-8_197
13. Kleiner, M., Krux, R., Homberg, W.: Analysis of residual stresses in high-pressure sheet metal forming. CIRP Ann. **53**, 211–214 (2004). https://doi.org/10.1016/S0007-8506(07)60681-7
14. Wimmer, M., Rinck, P., Kleinwort, R., Zah, M.F.: Methodology for a model-based control of the boundary zone properties during milling of TI-6AL-4V. MM Sci. J. **2019**(04), 3214–3219 (2019). https://doi.org/10.17973/MMSJ.2019_11_2019073
15. Abboud, E.: Characterization of Machining-Induced Residual Stresses in Titanium-Based Alloys. PhD Thesis, McGill University, Montreal, Canada (2015)
16. Li, B., Jiang, X., Yang, J., et al.: Effects of depth of cut on the redistribution of residual stress and distortion during the milling of thin-walled part. J. Mater. Process. Technol. **216**, 223–233 (2015). https://doi.org/10.1016/j.jmatprotec.2014.09.016
17. Robinson, J.S., Tanner, D.A., Truman, C.E., et al.: Measurement and prediction of machining induced redistribution of residual stress in the aluminium alloy 7449. Exp. Mech. **51**, 981–993 (2011). https://doi.org/10.1007/s11340-010-9389-4
18. Schieber, C., Hettig, M., Zaeh, M.F., et al.: Evaluation of approaches to compensate the thermo-mechanical distortion effects during profile grinding. Procedia CIRP **102**, 331–336 (2021). https://doi.org/10.1016/j.procir.2021.09.057
19. Denkena, B., Leon, L.D.: Milling induced residual stresses in structural parts out of forged aluminium alloys. Int. J. Mach. Mach. Mater. **4**, 335 (2008). https://doi.org/10.1504/IJMMM.2008.023717
20. Werke, M., Hossain, M., Semere, D. et al.: Machining Distortion Analysis of Aerospace Components using the Contour Method, pp. 216–222. https://doi.org/10.3384/ecp19162025
21. Deutsches Institut für Normung: DIN EN 515 – Aluminium and aluminium alloys – Wrought products – Temper designations (2017)
22. Weinschenk, A., Volk, W.: FEA-based development of a new tool for systematic experimental validation of nonlinear strain paths and design of test specimens. In: AIP Conference Proceedings, vol. 1896, pp. 020009-1–020009-6 (2017). https://doi.org/10.1063/1.5007966
23. SINT Technology MTS3000-Restan product page. https://www.sintechnology.com/de/produkte/mts3000-restan/. Last accessed 19 Apr 2023
24. Baeker, M.: SpectrumBaker – better contour plots for Abaqus. Python plugin. https://doi.org/10.13140/RG.2.2.10738.91845 (2016). Last accessed 19 Apr 2023
25. Scandola, L., et al.: Development of a numerical compensation framework for geometrical deviations in bulk metal forming exploiting a surrogate model and computed compatible stresses. Int.J. Mater. Form. **14**(5), 901–916 (2021). https://doi.org/10.1007/s12289-020-01603-7
26. Gan, W., Wagoner, R.H.: Die design method for sheet springback. Int. J. Mech. Sci. **46**, 1097–1113 (2004). https://doi.org/10.1016/j.ijmecsci.2004.06.006

Approximation of Pressure Fields Generated by High-Voltage Discharges in Liquid on a Flat Wall

M. Knyazyev, M. Holzmüller(✉), and W. Homberg

Forming and Machining Technology, Paderborn University, Warburger Str. 100, Paderborn, Germany
mh@luf.uni-paderborn.de

Abstract. Pressure fields in electrohydraulic forming often have a very complex character due to the combination of the many physical phenomena involved. Among several energy-force factors generated by underwater discharge, direct shock waves play a leading role in the deformation of the sheet blank. The measuring method based on a multipoint membrane pressure gauge makes it possible to obtain detailed information about shock-wave pressure distribution over a large area (field). The analysis of pressure fields shows three types of loaded segments: zones of direct shock wave loading, zones of shock waves' interaction, and zones of concentration along lateral rigid walls. In their profile, the pressure distributions in the above-mentioned zones are very similar to the Gaussian distribution curve. The Laplace exponential normalised function was thus selected as a basis for the test approximation dependencies. In the proposed approximation method, the loaded zones are assumed to be independent pressure sources and the principle of superposition is valid. The resulting pressure value at each point is determined as the sum of pressure values from all the loaded zones. The approximation results correlate well with the test data for two simultaneous discharges in a rectangular chamber. The approximations of pressure fields give more exact data on the configuration of each loading zone and its share in the total field formation and can serve as a technological characteristic of a discharge chamber.

Keywords: impulse · electrohydraulic · pressure · field · approximation

1 Introduction

The impulse release of energy in a closed chamber generates the following main energy-force factors: direct shock wave, hydraulic flow, and quasi-static pressure over the entire volume. Other factors appear at different stages of impulse loading and sheet blank deformation: reflected and secondary shock waves, cavitation, interference and interaction of shock waves, rarefaction waves, etc. [1, 2]. Direct shock waves play a leading role in the impulse loading of flexible sheet blanks through the creation of cavitation in a liquid layer attached to the moving blank that further accelerates a hydraulic flow [2–4]. Also, the formation of small high-stiffness elements and the final shape of the sheet part (calibration) are determined by the peak pressure of the shock wave [4].

K. Mocellin et al. (Eds.): ICTP 2023, LNME, pp. 146–157, 2024.
https://doi.org/10.1007/978-3-031-40920-2_16

For planning electrohydraulic impulse sheet forming (EHF) operations, it is important to know the technological characteristics of a specific design of discharge chamber. The distribution of the shock-wave pressure (pressure field) on a flat wall (simulating a flat sheet blank) located under the exhaust window of a discharge chamber can characterise the technological capabilities in terms of qualitative and quantitative indexes.

Characteristic pressure fields of this type for each discharge chamber equipped with a single electrode pair can be obtained from computer simulations [1, 5]. The highly complex nature of the actions and interactions of the different energy-force factors, the intricate geometric shapes of the chamber, electrode units and auxiliary elements, the deviations in the discharge channel position and the energy release all mean that the computer codes include many assumptions and require test verification and corrections.

A measuring method based on the application of multi-point membrane pressure gauges has the capability to register detailed pressure maps over vast areas [1, 6–8]. This method is also characterised by its simplicity of use, its capability for the adaptation (selection) of a metallic membrane for the pressure field parameters intended for measurements, its simple and inexpensive design, and the easy procedure for processing the test results. A membrane pressure gauge works on the principle of the dependency between the deformation value of a thin membrane (metallic sheet) deflected into the hole of a gauge body and the intensity of the pressure pulse at the point under consideration. Vagin et al. [8] showed that membrane deflection is proportional to the pressure amplitude p_{max} if the duration of the membrane's plastic deformation is much less than the characteristic duration of the pressure pulse. The duration of the membrane's deformation can be calculated from the Cole formula [9] and, for small hole diameters d, is measured in the range of several microseconds; the characteristic duration of shock waves for EHF is usually tens of microseconds [8].

An analysis of the test pressure maps revealed a non-uniform pressure distribution on a flat wall facing the exhaust window of the discharge chamber [6]. A single discharge channel generates two types of loaded zone on a flat wall: a zone of loading with a direct shock wave under the discharge channel, and zones with a concentration of energy-force factors located along the rigid side walls of the chamber.

The application of two or more simultaneous discharges in the same work volume greatly changes the pressure field [6]. The interaction of two or more shock waves in a liquid is highly non-linear in nature. This phenomenon results in the creation of a further type of loaded zone – a zone of shock-waves interaction. This appears along the distance between the sources of the pressure pulses in the form of an increased-pressure spot on a flat wall. Its characteristics (peak pressure, sizes) depend on the intensity of the direct shock waves, the distance between the discharge channels and the flat wall [6].

In this work, it is proposed to apply the experimental dependencies that relate to the loaded zones and the superposition principle to construct the approximation method. Employing this approach, the loaded zones are conventionally considered as independent sources of pressure. It allows to compile the total pressure distribution as a collection of several pressure (loaded) zones with the summation of a number of particular values at each point of the pressure field.

This work includes the consideration of test pressure fields generated by the simultaneous explosion of two wires in a rectangular discharge chamber [10]. The measured pressure fields [10] are used as an example for the development of a general approximation method and specific dependencies in this research.

The materials are submitted in the following order: experimental set-up and measurement results, development of the approximation method and dependencies, computational results and discussion, conclusions.

2 Experimental Setup and Measurement Results

The experimental tooling has the simple design of a box-type discharge chamber with a flat top plate and straight side walls (Fig. 1). All four electrodes were of the insulated type in order to realise the sequent connection of spark gaps. The electrode insulator material is polyoxymethylene and the electric conductive rods are carbon-steel studs with a M12 thread. This feature allows external cable connections (Fig. 1a) and the reliable fixation of wires inside the chamber (Fig. 1b). Such a connection ensures a high synchronicity of two underwater discharges.

The transmitting work medium (tap water) was supplied through the nipple in the lower point of the discharge chamber, and the air evacuation took place through the nipple in the top plate of the chamber (see Fig. 1a).

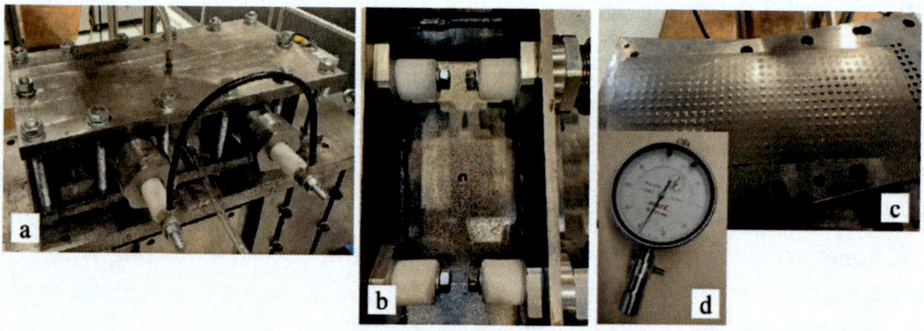

Fig. 1. General view of the experimental tooling (a), discharge chamber with electrodes and fixed wires (b), multipoint pressure gauge body and metallic membrane after impulse loading (c), dial indicator with adaptor (d)

Electrodes are energised by an SSG-0620FS pulse current generator (Poynting GmbH, Germany) [11]. The selected charge parameters are voltage $U_0 = 20$ kV, capacitance $C = 15$ μF, and charge energy $W_0 = 3$ kJ.

Discharges were initiated by wires of electrotechnical copper with a diameter of 0.4 mm. The length of the straight wires is 20 mm and is determined by the geometrical configuration of the discharge system (electrodes). Their purpose is to ensure easy breakdown and to stabilise the shape and length of the discharge channels and hence to ensure their equal cross-sections, lengths, resistances and generation of shock waves.

The measurements of the pressure fields were performed with a multipoint membrane pressure gauge. The bottom plate attached to the discharge chamber was made as a body gauge plate with 396 holes of 6 mm in diameter (see Fig. 1c).

The measurement area corresponds to the size of the exhaust window of a discharge chamber of 400 × 150 mm (Fig. 2). The 0.5-mm sheets of steel DC01 (1.0330, standard EN 10130) were used for the function of metallic membrane. The deflections h_i of the membrane (heights of dimples) at each point (hole) of the grid were measured with the aid of a dial indicator (0.01 mm per division) equipped with a special fixture (see Fig. 1d).

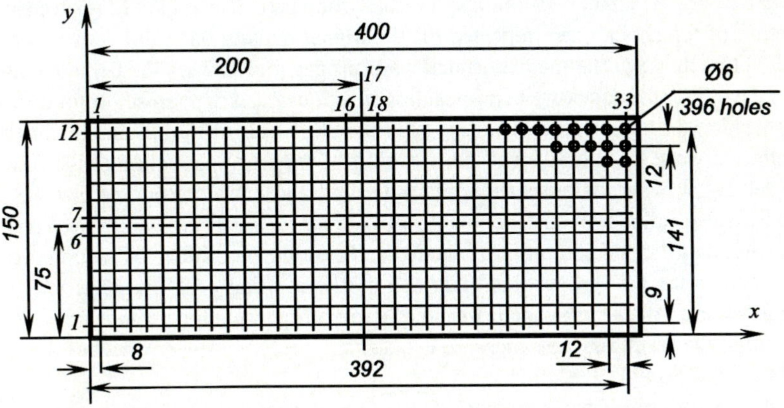

Fig. 2. Loading and measurements area of the membrane pressure gauge body and the locations of the measurement points (holes)

Each i-point (hole axis in the gauge body) has coordinates (x_i, y_i) (see Fig. 2). Each line of holes along coordinate x starts from the point at 8 mm and has 33 points. The column of holes along coordinate y starts from the point at 9 mm (12 points in each column). The pressure gauge can then be presented as a coordinate matrix (12 × 33) with 12 lines and 33 columns. The 17th column is on the middle line of the loaded zone (coordinate $x = 200$ mm) between the axes of the electrode pairs. The middle line that goes crosswise the axes of the electrode pairs has coordinate $y = 75$ mm and is located between the 6th and 7th lines of holes (see Fig. 2).

The peak pressure value p_i for each i-point (hole) is calculated from the famous Laplace formula for a spherical shell segment as a function of the radius r_i of a formed dimple

$$p_i = \frac{2\sigma_{sdi} t_i}{r_i}, \tag{1}$$

where t_i – thickness of membrane after plastic deformation; σ_{sdi} – flow stress at which plastic deformation occurs at impulse load p_i.

The initial measuring data (x_i, y_i, h_i) were processed to calculate the p_i values and obtain the tables (x_i, y_i, p_i) for plotting the pressure maps (for a detailed description of the processing procedure see the earlier publication [10]).

3 Development of Approximation Method and Dependencies

The approximation method is based on general laws relating to the physics of shock waves, including their propagation in a liquid medium, their reflection from the solid wall with a higher density, their multiple reflections in corner elements and the non-linear interaction of two or more shock waves in a liquid volume, etc. Thus, the shock wave front propagates in a liquid along direct lines originating from the middle point of the discharge channel (assuming spherical symmetry). The angle of the propagation direction of the shock wave reflected off the higher-density flat solid wall is equal to the direction angle of the incident shock wave propagation toward the perpendicular to the surface. When a shock wave propagating in a liquid strikes against a solid wall with a higher density and a large thickness (like the steel wall of the discharge chamber), its reflected wave loses only small portion of its energy. An interaction of incident and reflected shock waves takes place near to the rigid wall. It is considered that the peak pressure at the solid wall surface doubles. In corner elements created by two or more walls, the phenomenon of multiple reflection and multiple non-linear interactions occurs, which causes the non-proportional increase in the resulting peak pressure. When two or more shock waves strike against each other in a liquid medium, a non-proportional increase in the resulting peak pressure is observed – the non-linear effect of the shock waves' interaction [3, 9, 12, 13].

The analyses of the pressure maps (Figs. 3 and 4) revealed three types of zones with increased pressure: 1) zones with main loading by direct shock waves (two areas); 2) zones with non-linear interaction of direct shock waves (single zone); 3) several zones with a concentration of direct and reflected shock waves along the lateral walls of the discharge chamber (Fig. 5).

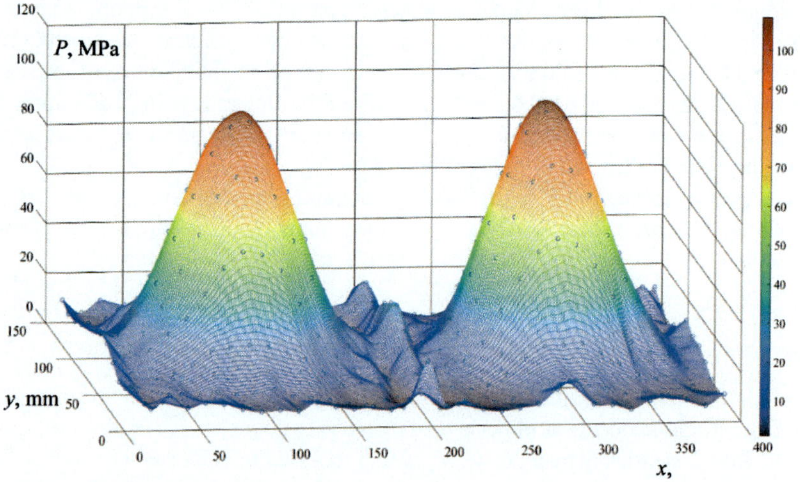

Fig. 3. Three-dimensional representation of the pressure field generated by the simultaneous explosion of two wires plotted from the mean test values

In the proposed approximation method, these loading zones are conventionally considered as independent sources of pressure located on the loaded flat wall (not discharge channels between electrodes). Therefore, the principle of superposition can be applied, i.e. the pressure values from all the zones can be summarised at each *i*-point. Thus, the influence of each revealed zone on the total pressure distribution is taken into account.

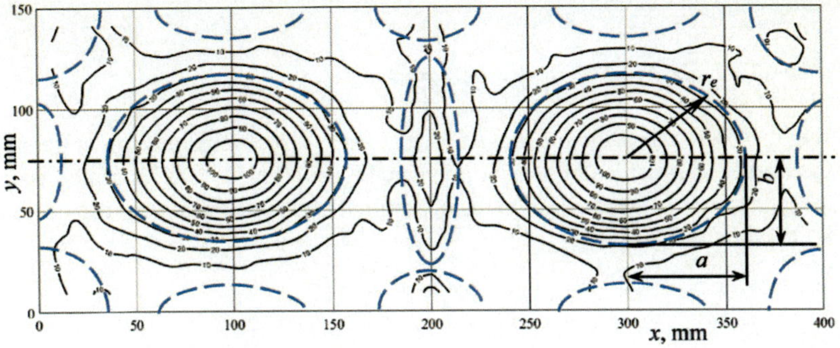

Fig. 4. Contour representation of the pressure field generated by simultaneous explosion of two wires, plotted from the mean test values: digits on isobars are in MPa; curved dotted lines are sample approximation lines for isobars

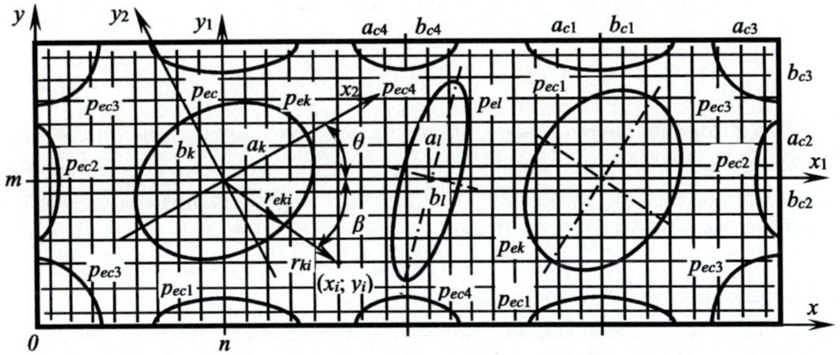

Fig. 5. Schematic diagram for calculations with approximation dependencies for arbitrary orientations of elliptical isobars in zones of direct shock wave loading and interaction

The first-type zones are mainly created by direct shock waves and located directly under two discharge channels with the centre coordinates (100; 75) and (300; 75) respectively (see Figs. 4 and 5). These two zones are characterised by the largest pressure values on the test pressure map. The pressure decreases rapidly along the specific curve from the central point to the periphery of each loading zone (see Fig. 3).

The isobars of direct-loading zones are evidently oval. It is assumed that isobars can be approximated by elliptical lines (see the dotted-line ellipses in Fig. 4) and that all ellipses (isobars) have the same central point and angular orientation. Each ellipse has the radius r_e varying from the largest a to the smallest b values depending on direction. The pressure in the central point equals the maximum pressure value p_{max}. For the considered test example, the mean maximum pressure is 104.748 MPa. In Fig. 5 the elliptical isobars of direct-loading zones are shown rotated at an angle θ as a general approach. In the considered test pressure map, $\theta = 0$ (see Fig. 4).

It was noticed [6] that, in the vertical section of the three-dimensional test map, the pressure curve (see Fig. 3) is very similar to the Gaussian normal distribution curve.

The Laplace distribution function (in a normalised form) is more suitable for an approximation than the Gauss law

$$t = \frac{1}{\sqrt{2\pi}} e^{-\frac{z^2}{2}}. \tag{2}$$

Applying the similarity principle, t can be assigned as a pressure value p_{ki} at the i-point from the k-zone of direct loading, and dimensionless coefficient $1/\sqrt{2\pi}$ can be assigned as the p_{kmax} maximum pressure value. In the original Laplace formula, the dimensionless argument is a ratio

$$z = \frac{\Delta x}{\sigma} = \frac{x - x_m}{\sigma}. \tag{3}$$

The $(x - x_m)$ value (x is an argument of the Gaussian distribution function $y(x)$) here should be considered as a distance r_{ki} from the centre of the k-zone to the i-point. The r_{ki} value is calculated from the coordinates $(x_i; y_i)$ of the i-point and the coordinates $(x_{k0}; y_{k0})$ of the centre of the k-zone respectively in the selected system of coordinates (see Fig. 5). Thus, the approximation formula takes the form

$$p_{ki} = p_{kmax} e^{-\frac{r_{ki}^2}{2\sigma^2}} = p_{kmax} e^{-\frac{z_{ki}^2}{2}}. \tag{4}$$

With this approach, the σ value should be the parameter responsible for the elliptical form of the isobars (see Fig. 4) and the sharpness of the distribution curve (see Fig. 3). Parameter σ is a varying parameter similar to the varying radius r_{eki} of the ellipse (distance from the ellipse centre to the ellipse curve) at which the t_{ek} and z_{ek} values are constant. These latter parameters can be calculated with the aid of the Laplace normal distribution function in the form of a table.

The Laplace distribution function shows that the t value is reduced by coefficient $1/\sqrt{2\pi}$ and it should therefore be considered that

$$t_{ek} = \frac{p_e}{p_{kmax}\sqrt{2\pi}}, \tag{5}$$

where p_e – pressure value on the elliptical approximation line from the k-zone of direct loading (equals a pressure value on the selected isobar).

Using the Laplace normal distribution function, the z_{ek} value is determined from the calculated t_{ek} value. Then σ_{eki} value then follows as

$$\sigma_{eki} = \frac{r_{eki}}{z_{ek}}. \tag{6}$$

Hence, parameters t_{ek} and z_{ek} are constant along the ellipse line, and parameters r_{eki} and σ_{eki} vary as a function of the direction to the selected i-points from the centre of the ellipse (see Fig. 5).

The r_{eki} value is determined from the common solution of the ellipse formula and the formula for the straight line that goes through the centre of the k-zone

$$\frac{x^2}{a_k^2} + \frac{y^2}{b_k^2} = 1;\ y = qx. \tag{7}$$

Here, the values of the largest a_k and smallest b_k ellipse radii are determined from the graph (see Fig. 4). The direction of radius r_{eki} is determined by the direction of segment r_{ki} (see Fig. 5) from the centre of the k-zone, with coordinates (x_{kO}; y_{kO}), to the i-point, with coordinates (x_i; y_i).

In general, both possible linear displacements (m and n) of the ellipse and its angular position θ in the selected system of coordinates should be considered (see Fig. 5).

The middle zone of increased pressure ($x = 200 \pm 25$ mm) is explained by the interactions of two direct shock waves [6]. Pressure distribution in this zone is also similar to the Gaussian distribution curve (see Fig. 3). It can therefore be approximated with a similar function and hence the pressure value at each i-point of the calculation field from the l-zone of interaction is

$$p_{\mathrm{li}} = p_{\mathrm{lmax}}\, e^{-\frac{r_{\mathrm{1i}}^2}{2\sigma_{\mathrm{li}}^2}} = p_{\mathrm{lmax}}\, e^{-\frac{z_{\mathrm{li}}^2}{2}}. \tag{8}$$

Earlier, in terms of application to EHF it was confirmed [6, 14] that the physical phenomenon of the interaction of two shock waves is non-linear in nature. In the considered case, the interaction effects give $p_{lmax} = 22.16$ MPa.

The calculation diagram in Fig. 5 shows an arbitrary angular orientation of the interaction zone for a general approach. In the considered pressure map $\theta = \pi/2$ (see Fig. 4), and this value is used in the approximation example.

The analysis of the experimental pressure maps revealed four subtypes of concentration zones (see Fig. 5): 1) four zones along the longer sides of the chamber with the largest a_{c1} and smallest b_{c1} radii; 2) two zones along the shorter sides with the radii a_{c2} and b_{c2}; 3) four corner zones with the radii a_{c3} and b_{c3}, which are approximately equal; 4) two zones along the longer sides of the chamber with the radii a_{c4} and b_{c4}, which are created by the interaction of direct shock waves on lateral rigid walls. The isobars in these zones are approximated by half-ellipses and quarter-ellipses, i.e. defined by the configuration of rigid walls.

The pressure distribution in these zones is also approximated with a similar exponential function for each c-zone of concentration.

The approximation dependencies above thus make it possible to calculate the pressure values at each i-point of the calculated area from all the zones. With this approach, the loading zones are considered as independent sources of pressure on the loaded flat wall; therefore, the principle of superposition can be applied, i.e. the pressure values from all the zones can be summarised at each i-point

$$p_i = \sum\nolimits_{\mathrm{k=1}}^{\mathrm{DZ}} p_{\mathrm{ki}} + \sum\nolimits_{\mathrm{l=1}}^{\mathrm{IZ}} p_{\mathrm{li}} + \sum\nolimits_{c=1}^{\mathrm{CZ}} p_{ci} \tag{9}$$

where DZ is quantity of zones with loading through a direct shock wave; IZ is the quantity of interaction zones and CZ is the quantity of concentration zones.

4 Computational Results and Discussion

The programming of the approximation algorithm was performed in Microsoft Excel and the map plotting in MATLAB (Fig. 6), with the cubic polynomial for surface interpolation. In a general view, the approximation map tightly corresponds to the experimental map in terms of the principal details (see Fig. 3 for comparison).

In the submitted approximation example, the same coordinates (x_i; y_i) and the same number (396) of points as in the measurement membrane gauge are used. This approach allows to exclude the influence of interpolation on a comparative analysis of the approximation and test pressure maps and make it more exact and convenient.

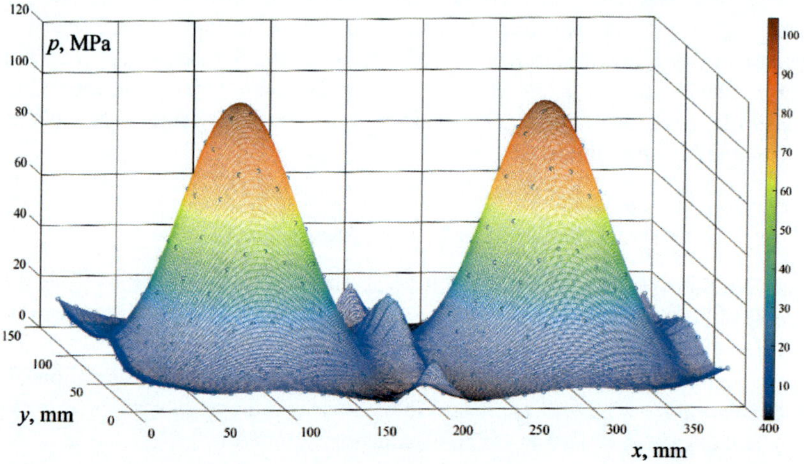

Fig. 6. Three-dimensional approximation pressure map based on exponential functions

Approximation exponential functions generate smooth surfaces. But the experimental map has many steps, with irregular lines especially along peripheral zones located along the lateral rigid walls (see Fig. 3). This feature indicates a very complex interaction of direct shock waves and a large number of waves reflected from the lateral, upper and bottom walls at very different angles as well as from elements of the electrode units. Here, the absolute pressure values are rather small, but their deviations are comparable with absolute values.

During construction of the approximation, zones of lower pressure and indefinite geometric parameters (mainly the concentration zones) required several iterations in order to satisfy the experimental map through the selection of proper values a, b, z_e, p_{max}. Parameters of other zones were also tuned slightly for better accuracy (Table 1).

The criteria for the tuning (accuracy optimisation) of the approximation map are selected as follows: general configuration of the approximation map and its correspondence to the experimental map; minimum difference between the mean pressure value of the approximation map p_m and the experimental value p_{exm}; minimum value of the mean relative deviation (with sign); minimum value of the mean absolute relative deviation; minimum value of the largest absolute relative deviation $\overline{\Delta p}_{max}$.

Table 1. Main approximation parameters of the loading zones

Parameters	Zones of loading with direct shock wave		Interaction zone (single)	Zones with a concentration of direct and reflected shock waves along the lateral rigid walls			
	Left	Right		1) 4 zones	2) 2 zones	3) 4 zones	4) 2 zones
a, mm	60.2	60.2	50.0	90.0	50.0	40.0	17.0
b, mm	42.4	42.4	12.0	12.0	12.0	25.0	15.0
z_e	1.65	1.65	1.8	2.0	1.5	1.4	1.8
p_{max}, MPa	104.748	104.748	22.16	7.0	21.0	13.0	27.0

The tuning of the total approximation pressure surface with regard to the experimental pressure map was first conducted in relative deviations $\overline{\Delta p_i}$ with consideration to the sign at each i-point (Fig. 7). The aim is for the total approximation surface to coincide with the experimental one with a total (summarised) relative deviation equal to zero. The submitted approximation map (see Fig. 6) is performed with a mean relative deviation of $\overline{\Delta p_m} = -5.02\%$, i.e., the negative deviations (lower pressure values) prevail. The mean pressure value of the test pressure map is thus $p_{exm} = 24.87$ MPa, while the approximation map gives $p_m = 23.36$ MPa, which is 6.1% less, mainly at the expense of the peripheral loaded areas.

For the submitted approximation, the mean absolute relative deviation is $\overline{\Delta p_{ma}} = 21.46\%$ with the largest deviation being $\overline{\Delta p_{max}} = 140.47\%$. The largest deviations are attributed to the zones of low pressure (see Fig. 7), where even a small absolute deviation creates a very large relative error, while zones of high pressure have absolute and relative deviations near to zero.

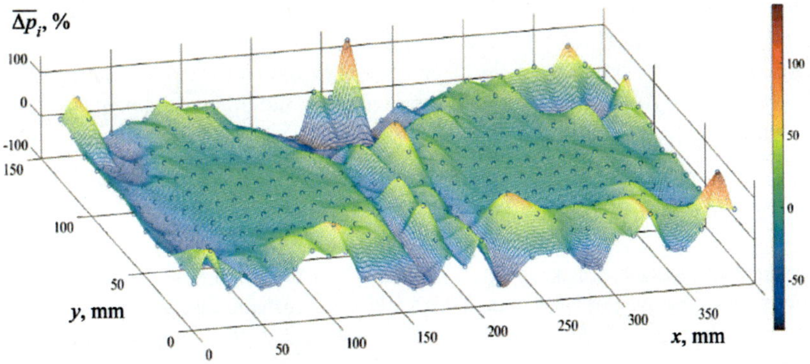

Fig. 7. Three-dimensional representation of the relative errors of the approximation pressure map (see Fig. 6) in relation to the test map (see Fig. 3)

5 Conclusions

The approximation results confirmed the selected general strategy based on fundamental physical laws for shock waves, the use of an exponential function based on the *e* number (Laplace distribution function), and the use of an ellipse formula for the approximation of the isobars. The tuning of the approximation pressure map can be conveniently performed with the Laplace function dimensionless argument z, the maximum pressure value p_{max}, the ellipse radii a and b, and the angle for the ellipse orientation for a corresponding loading zone.

The method for the approximation of test pressure fields is constructed with the measurement results from a multipoint membrane pressure gauge that allows to obtain the detailed pressure maps (large number of measuring points with small distances between them) at a relatively low cost. Other pressure measuring methods can, however, be used for improving the accuracy of measured pressure values.

The approximation of test results made it possible to determine the parameters of three types of zones with increased pressure: zones with main loading by direct shock waves, zones of non-linear interaction of direct shock waves, and zones with a concentration of direct and reflected shock waves along the lateral rigid walls of the discharge chamber. The approximation procedure provides an opportunity to describe their physical and geometrical parameters with a good level of accuracy.

Test and approximation pressure maps present the qualitative and quantitative characteristics of a given design of discharge chamber and can be used for the development and optimisation of EHF processes.

The logical next step in the development of this strategy can be the creation of a simulation method based on pressure measurement results, developed approximation functions and new experimental dependencies for different distances from the discharge channel to the measured object and energies of discharges. Principally, the types of loading zones will be the same, but the maximum pressure values p_{max} and the geometric parameters a and b of the elliptical isobars will be the changing values. The varying energy of discharges will influence the absolute pressure values at the points, but the geometric configurations of the loading zones will be the same. An experimental simulation computer program of this type can constitute an extended characteristic of the technological capabilities of a discharge chamber and be useful for planning EHF operations.

References

1. Mamutov, A.V., Golovashchenko, S.F., Mamutov, V.S., Bonnen, J.J.: Modeling of electrohydraulic forming of sheet metal parts. J. Mater. Process. Technol. **219**, 84–100 (2015)
2. Anuchin, M.A. (ed.) Shtampovka vzrivom. Osnovi teorii. Pod redakczieiy. Moskva Mashinostroenie (Stamping explosion. Fundamentals of the theory. Moscow Engineering) (1972). https://lib-bkm.ru/load/85-1-0-1821
3. Mazurovsky, B.Y., Siziov, A.N.: Elektrogidravlicheskiy effect v listovoi shtampovke (Electrohydraulic effect in sheet forming), 192 p. Naukova Dumka, Kiev (1983)
4. Knyazyev, M.K., Chebanov, Y.I.: Multi-electrode discharge blocks for electrohydraulic forming and their capabilities. Eastern-Eur. J. Enterp. Technol. **3**(9), 13–20 (2004)

5. Avrillaud, G., Mazars, G., Cantergiani, E., Beguet, F., Cuq-Lelandais, J.-P., Deroy, J.: Examples of how increased formability through high strain rates can be used in electro-hydraulic forming and electromagnetic forming industrial applications. J. Manuf. Mater. Process. **5**, 96 (2021)
6. Knyazyev, M.K.: Investigations, Development and Implementation of Manufacturing Processes of Electrohydraulic Forming with Application of Multi-Electrode Discharge Chambers/Dissertation for Candidate of Engineering Science, 223 p. KhAI, Kharkov (1995)
7. Homberg, W., Djakow, E., Damerow, O.: Process reliability and reproducibility of pneumo-mechanical and electrohydraulic forming processes. In: Proceedings ICHSF 2014, Daejeon, Korea, 26–30 May 2014
8. Vagin, V.A., Zdor, G.N., Mamutov, V.S.: Metody issledovaniya vysokoskorostnogo deformirovaniya metallov. [Methods of research of high-speed deformation of metals.], 208 p. Navuka i tekhnika, Minsk (1990)
9. Cole, R.H.: Underwater Explosions. Princeton Univ. Press, Princeton (1948)
10. Knyazyev, M., Holzmüller, M., Homberg, W.: Investigation of pressure fields generated by two simultaneous discharges in liquid initiated by wires. J. Manuf. Mater. Process. **7**, 40 (2023). https://doi.org/10.3390/jmmp7010040
11. Bedienungsanleitung SSG-0620FS, Version 1.02, letzte Änderung: 04/2012, Poynting GmbH, Dortmund, Germany. – 33 s
12. Ghoshal, R., Mitra, N.: Underwater oblique shock wave reflection from submerged hydraulic structures. Ocean Eng. **209**, 107324 (2020). https://doi.org/10.1016/j.oceaneng.2020.107324
13. Ridah, S.: Shock waves in water. J. Appl. Phys. **64**, 152 (1988). https://doi.org/10.1063/1.341448
14. Surkaev, A.L., Kul'kov, V.G., Talyzov, G.N.: Experimental investigation of the interaction of two acoustic shock waves. Tech. Phys. Lett. **27**(6), 487–488 (2001). https://doi.org/10.1134/1.1383833

Investigation of Necking and Fracture Strain Limits of Medium-Mn Steel Sheet Under Hot Stamping Conditions

Chenpeng Tong[1], Ruiqiang Zhang[1], Dechao Xu[2], and Zhusheng Shi[1](✉)

[1] Department of Mechanical Engineering, Imperial College London, London SW7 2AZ, UK
zhusheng.shi@imperial.ac.uk

[2] Shougang Research Institute of Technology, Shougang Group Co., Ltd., Beijing 100041, China

Abstract. Medium-Mn (MMn) steels with low austenitisation temperatures have recently attracted much research attention because of their suitability for hot stamping to save cost and improve productivity. However, research on their formability under conditions that mimic industrial hot stamping is still lacking. In this study, an innovative method for biaxial testing, which includes a cruciform specimen design, is used on a Gleeble 3800 simulator to obtain the forming limit curves (FLCs) and fracture forming limit curves (FFLCs) of a representative MMn (8Mn) steel. The localised necking strain and fracture strain are measured using a newly developed spatio-temporal method. The FLCs and FFLCs are constructed using strain limits in different loading states, e.g. uniaxial, plane strain and equi-biaxial stretching. The results show that the loading state conditions are acceptably simulated using the biaxial test system, and the FLCs and FFLCs are well-constructed, displaying a V shape. The data will contribute to the design of forming process. In addition, the forming strain limits of the MMn steel are much higher than those of a conventional boron steel (22MnB5) under the same hot stamping conditions, suggesting that the MMn steel has great potential and benefit as an alternative in hot stamping.

Keywords: Medium Manganese Steel · Forming Limit Curve · Hot Stamping · Biaxial Tensile Test

1 Introduction

Hot stamping is an advanced sheet metal forming technology developed for manufacturing ultra-high-strength steels (UHSS) in the automotive industry, giving light weight, good formability and negligible springback [1]. The process involves a full austenitisation heat treatment, followed by forming at elevated temperatures and in-die quenching to room temperature simultaneously. Currently, boron steels are the most widely used materials in hot stamping, but their high working temperature (over 900 °C) leads to high cost and poor productivity, and their post-formed ductility is low, restricting a wider application of this technology [2]. Using medium-Mn (MMn) steels in hot stamping has recently attracted much research attention as one of the most promising approaches to

K. Mocellin et al. (Eds.): ICTP 2023, LNME, pp. 158–166, 2024.
https://doi.org/10.1007/978-3-031-40920-2_17

overcoming these limitations [3]. This enables the working temperature of the process to be greatly reduced, thus giving cost savings and improving productivity; at the same time, the mechanical properties after hot stamping are superior to those of conventional boron steels [4].

An evaluation of the formability of metal sheet is often done by using forming limit curves (FLCs) and fracture forming limit curves (FFLCs) to determine its ability to form complex-shaped panel components in hot stamping [5]. These curves show the critical combination of major (ε_1) and minor (ε_2) strains at the onset of localised necking and fracture in a metal sheet, respectively. With the help of the curves, the allowable maximum strain level during stamping can be determined and the stamping process can be optimised [6]. However, obtaining these forming limit curves under hot stamping conditions is quite challenging and the research on formability of hot-stamping MMn steels is lacking.

In this study, the formability of a representative medium-Mn (8Mn) steel under hot stamping conditions is evaluated. The FLCs at the onset of localised necking and FFLCs at fracture are obtained through a novel biaxial tensile test developed in-house. The results are discussed in comparison with those of a boron steel.

2 Experimental Programme

A medium-Mn (MMn) steel investigated in this study contains alloying elements of 0.1 C, 8 Mn and 0.8 Al (wt%). The equilibrium transformation temperatures A_{e1} and A_{e3}, calculated using ThermoCalc, are 580 °C and 730 °C, respectively. In order to perform the formability test, cruciform specimens were machined from the received cold-rolled material with their horizontal arms parallel to the rolling direction. A diagram illustrating the specimen geometry is shown in Fig. 1a. Several notches have been incorporated into the turnings of the arms to reduce the stress concentration at these locations. The thickness of the specimen was decreased in a circular central area, called the gauge area, to localise the deformations. The slits are confined to the horizontal direction to ensure a good uniformity of the temperature distribution in the gauge area during electrical resistance heating. A biaxial testing system developed in-house was used on a Gleeble 3800 simulator, and the setup of the test is shown in Fig. 1b. Input uniaxial force from the Gleeble was converted to output biaxial forces using the test apparatus; details can be found in [7]. As shown in Fig. 1b, the central area of the specimen was prepared with speckle patterns by spraying with heat resistant paint, and this was used in conjunction with a digital image correlation (DIC) system for measuring the strain [8]. A pair of K-type thermocouples was welded onto the top edge of the circular gauge area and connected to the Gleeble for temperature control.

To construct FLCs, at least three loading states are required, namely uniaxial stretching, plane-strain stretching and equi-biaxial stretching. Schematic diagrams illustrating these different stretching modes are shown in Fig. 2a. The production of the different loading states is controlled by the ratio of the vertical displacement Δy to the horizontal displacement Δx. A displacement ratio $\Delta y/\Delta x$ of −0.5 was used to give a uniaxial loading state and a value of 0 corresponds to plane-strain loading. However, a ratio of 1 (corresponding to an equi-biaxial loading state) is challenging in the present work;

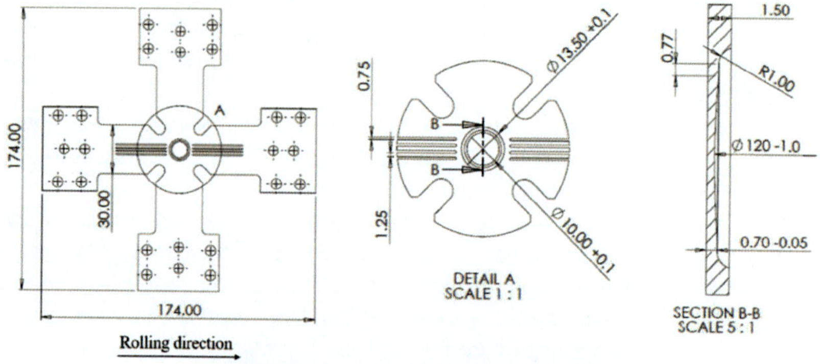

(a) Specimen design for formability test

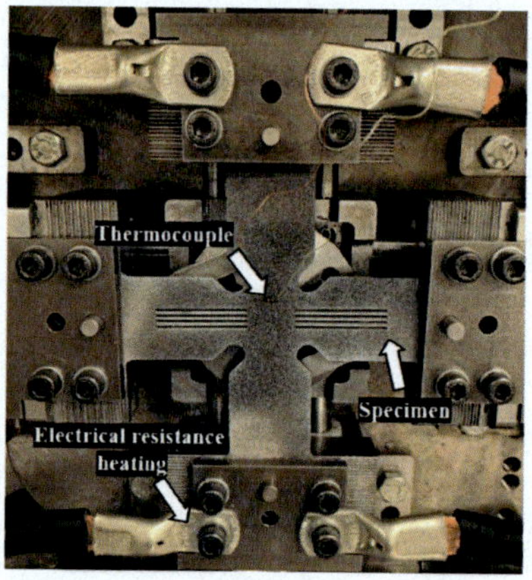

(b) Setup of the formability test

Fig. 1. The geometry of the cruciform specimen and setup of the formability test in Gleeble.

instead, a ratio of 0.8, corresponding to the closest configuration to equi-biaxial loading, and an intermediate ratio of 0.4, was used to give a total of four points on the forming limit curves, enabling the shape of the curves to be determined more precisely. The specimens were heated to 780 °C at the rate of 25 °C s^{-1} and held for 300 s for a complete austenitisation based on a previous study [9]. Following this, they were quenched to the deformation temperature (750 °C) then stretched until fracture at different strain rates (0.01 and 0.1 s^{-1}). At least two tests were conducted for each condition and the results of FLCs and FFLCs are shown as average values.

The innovative spatio-temporal method developed in-house [10] is employed in this study to detect the onset of localised necking; a corresponding schematic diagram of this procedure is shown in Fig. 2b. The main concept of this method is to identify the point of the onset of a discrepancy between the thickness strain on average within a narrower square zone immediately surrounding the eventual crack (base zone BZ with width W_{BZ}) and within a wider refence zone (RZ with width W_{RZ}). The dimension W_L of both zones

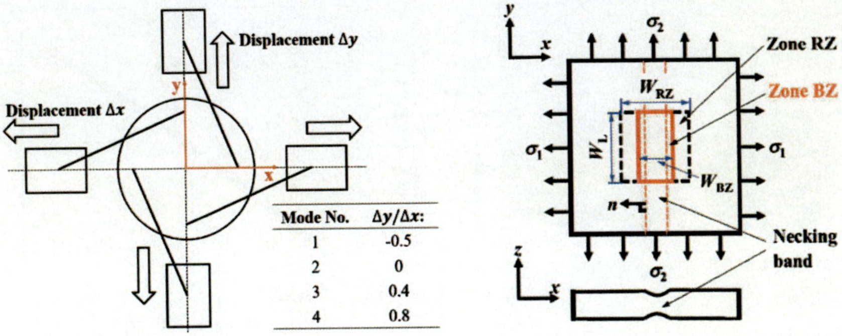

(a) Stretching modes (b) Selection of zones BZ and RZ [10]

Fig. 2. Schematic diagrams illustrating different stretching modes to deform the specimens and the base zone and reference zone for determination of initiation of localised necking.

is the same in the direction parallel to the crack length and the same central point where fracture occurs. It is assumed that localised necking begins when the difference in the two average thickness strains appears: this is detected as a change in slope on a plot of the average thickness strains in RZ against that in BZ. The slope is ideally unity in the early stage of deformation but becomes smaller once necking occurs, at which zone BZ has greater average thickness strain than that of zone RZ. More details are presented in the results part (Sect. 3) with an illustrative plot of this type. After identifying the initiation of localised necking, the strain limits can then be obtained from zone BZ. In addition, this method can be used to determine the fracture strain within BZ immediately preceding the fracture. As recommended in [10], the dimensions of these two zones are $W_{RZ} = 2W_{BZ} = W_L = 2h$, where h is the initial thickness of the gauge area near centre ($h = 0.55$ mm in this study).

3 Results and Discussion

3.1 Onset of Localised Necking and Strain Path

(a) Uniaxial tension

Figure 3 displays the determination of the initiation of localised necking for the sample tested at 750 °C and 0.1 s^{-1} under uniaxial tension. Figure 3a displays a map of the thickness strain $|\varepsilon_3|$ in the central area at the beginning of deformation and at the time immediately preceding fracture (t_F). As can be observed, deformation appears primarily to have taken place near the specimen centre. According to the data processing method introduced in Sect. 2, two zones BZ and RZ were drawn, both with their widths parallel to the length of the crack. In Fig. 3b, the thickness strain on average in the RZ is plotted in comparison to that in the BZ during the whole deformation process. Straight lines fitted to uniform deformation in the early part of the test (blue dashed line) and very late deformation near fracture (red dashed line) are indicated in Fig. 3b; localised necking

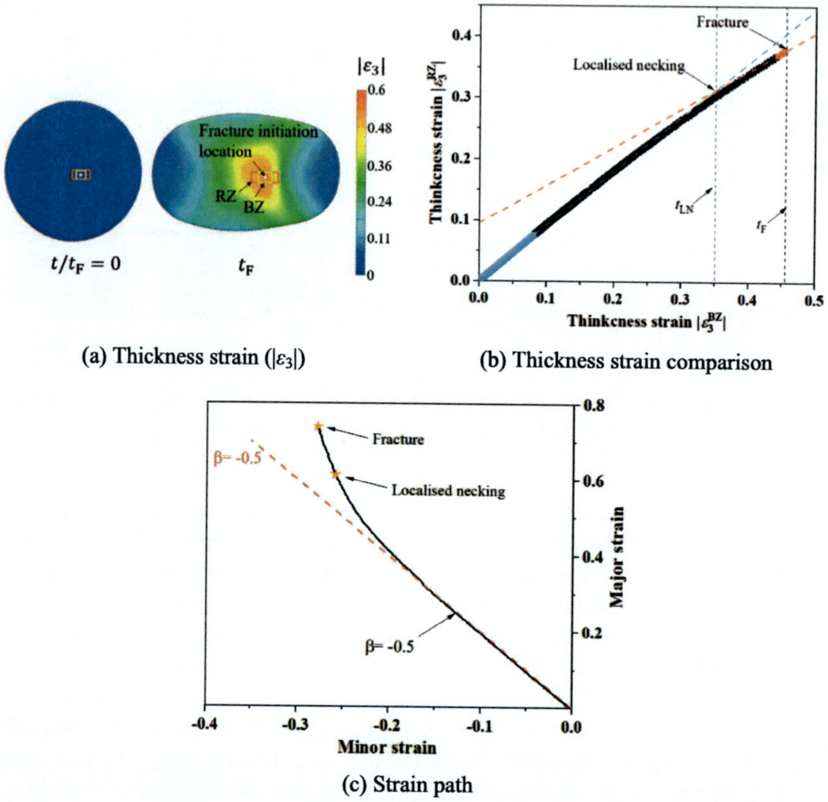

Fig. 3. Method for determination of the initiation of localised necking, and strain path for the sample deformed at 750 °C and 0.1 s^{-1} under uniaxial tension.

begins at the intersection between the two lines with a time point marked as t_{LN}. The determination of these two straight fitting lines is based on the method in [10].

In Fig. 3c, the major strain ε_1 is plotted against the minor strain ε_2 during the whole deformation process based on zone BZ. The strain path is nearly linear during the first-half of the deformation with a strain ratio $\beta = d\varepsilon_2/d\varepsilon_1 = -0.5$, showing that a uniaxial tension state is achieved; the strain path deviates from a straight line in the later stages of deformation, which is consistent with findings from other formability assessments performed at room temperature [11, 12]. Several factors may contribute to the deviation from linearity, such as the geometry of the specimen and the experiment conditions. The onsets of both localised necking and fracture are marked (orange stars) in the strain path and correspond to the times t_{LN} and t_F, as determined in Fig. 3b.

(b) Plane-strain and equal-biaxial tension

The strain paths with marked limit strain pairs (strains at the onset of localised necking and fracture; orange stars) for the sample deformed at 750 °C and 0.1 s^{-1} under plane-strain and equal biaxial tension are displayed in Fig. 4a and Fig. 4b, respectively.

As can be observed, the strain paths are nearly linear during the entire deformation process, without an obvious deviation at the later stages of deformation, and the average strain ratios β are -0.19 in plane-strain tension and 0.72 in equal biaxial tension, not far from the set values (0 and 0.8, respectively); these are beneficial for obtaining reliable forming limits data for sheet metals. Although the data are valid, further improvement is required for the current design to achieve pre-set strain paths. The deformation for both loading conditions occurs around the gauge area centre according to the colour plots of the major strain ε_1 in the gauge area at the onset of localised necking and fracture (Fig. 4c and d).

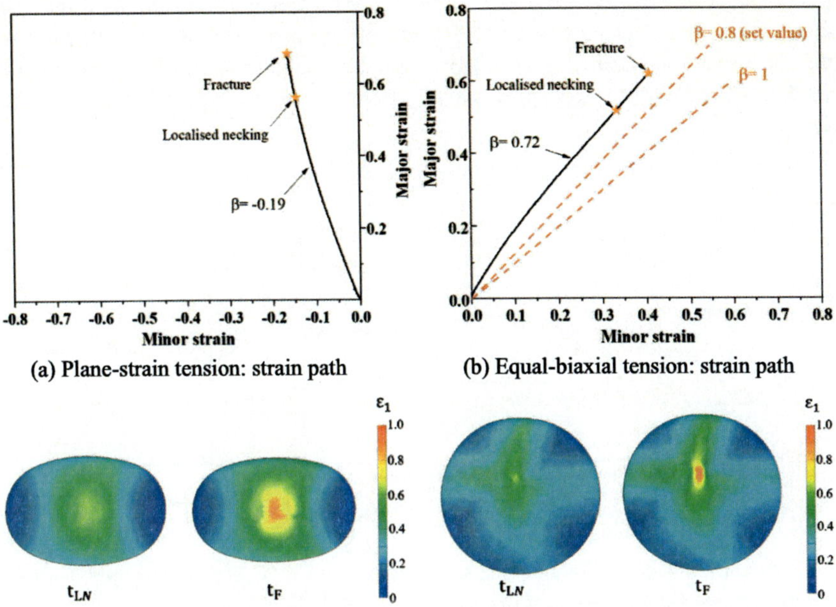

(a) Plane-strain tension: strain path (b) Equal-biaxial tension: strain path

(c) Plane-strain tension: major strain (d) Equal-biaxial tension: major strain

Fig. 4. Strain path and map of major strain in the gauge area for the sample deformed at 750 °C and 0.1 s^{-1} under plane-strain and equal-biaxial tension.

3.2 Forming Limit Curves

Figure 5a illustrates the collected limit strain pairs (strains at the onset of localised necking and fracture) plotted on axes of ε_1 versus ε_2 for sample deformed at 750 °C and strain rates of 0.1 s^{-1} and 0.01 s^{-1}. The FLCs and FFLCs are generated by connecting the limit strain pairs for the same deformation conditions using smooth curves. It can be seen that the shape of these curves is consistent with that of the conventional forming limit curves at room temperature, displaying a V shape, i.e. there is a lower limit ε_1 in the near plane-strain tension compared with other tension states [13, 14]. Moreover, it can be seen that the forming limit strains of the MMn steel with regard to localised necking and fracture slightly decrease at a lower strain rate (less than 2% in the limit major stains), while the shape of the curves remains similar; this is consistent with the experimental results from uniaxial hot tensile testing of the same material in a previous study [9], suggesting that the ductility properties of the MMn steel studied here are not sensitive to the strain rates investigated. Although further work is required to cover wider temperature and strain rate ranges, the FLCs and FFLCs determined in the present study can provide experimental basis for developing and calibrating modelling to predict forming limits of the MMn steel for hot stamping applications and optimise the hot stamping process. Figure 5b shows a comparison of FLCs an FFLCs at 750 °C and 0.1 s^{-1} for the MMn steel and a conventional boron steel (22MnB5); the data for the latter is from [15] and was obtained using the same testing method. It can be seen that the forming limit strain values of the MMn steel are much higher than those of boron steel in the conditions used with the lowest localised necking major strain of around 0.44 and 0.33 respectively, suggesting that the MMn steel appears to be a promising alternative to boron steel in hot stamping applications.

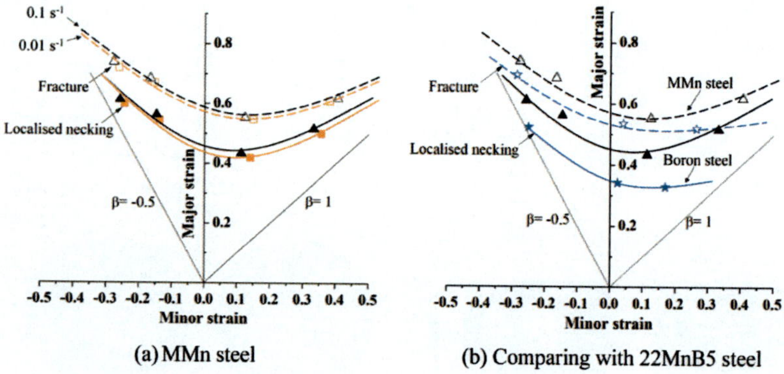

Fig. 5. Forming limit curves for the MMn steel deformed at 750 °C, and at strain rates of 0.1 s^{-1} and 0.01 s^{-1}, and a comparison to conventional boron steel (22MnB5).

4 Conclusions

In this study, both forming limit curves (FLCs) at the onset of localised necking and fracture forming limit curves (FFLCs) at the fracture for a representative MMn (8Mn) steel are investigated under hot stamping conditions using a biaxial testing. The following conclusions are drawn:

- It has been demonstrated that the recently developed method of biaxial testing can be used to construct FLCs and FFLCs for the MMn steel under hot stamping conditions.
- The FLCs and FFLCs obtained at a deformation temperature of 750 °C are relatively insensitive to strain rates in the range investigated (0.01–0.1 s^{-1}), with the limit major strain varying less than 2%. All curves have a V shape, i.e. the lowest limit major strain is located near the plane strain state, which are similar to those determined at room temperature using the conventional standard method.

- The formability of the MMn steel is much greater than that of a conventional boron steel (22MnB5) under the same deformation conditions (750 °C, 0.1 s^{-1}), with the lowest limit necking major strain of around 0.44 and 0.33 respectively, suggesting that the MMn steel has great potential and benefit as an alternative for conventional hot-stamping boron steel.

Acknowledgement. Financial support from Shougang Research Institute of Technology, Shougang Group, China for this research is gratefully acknowledged. The research was conducted at the Shougang-Imperial Lab for Lightweight Steel-Based Systems for Impact Resistant Automotive Applications at Imperial College London.

References

1. Karbasian, H., Tekkaya, A.E.: A review on hot stamping. J. Mater. Process. Technol. **210**(15), 2103–2118 (2010)
2. Ta, T., Clough, A.: Critical review of automotive hot-stamped sheet steel from an industrial perspective. Mater. Sci. Technol. **34**(7), 809–861 (2018)
3. Tong, C., et al.: New developments and future trends in low-temperature hot stamping technologies: a review. Metals **10**(12), 1652 (2020)
4. Li, X., Chang, Y., Wang, C., Hu, P., Dong, H.: Comparison of the hot-stamped boron-alloyed steel and the warm-stamped medium-Mn steel on microstructure and mechanical properties. Mater. Sci. Eng., A **679**, 240–248 (2017)
5. Ma, B., Wan, M., Li, X., Wu, X., Diao, S.: Evaluation of limit strain and temperature history in hot stamping of advanced high strength steels (AHSS). Int. J. Mech. Sci. **128**, 607–613 (2017)
6. Keeler, S.P.: Determination of forming limits in automotive stampings. SAE Trans. **74**, 650535–650936 (1966)
7. Shao, Z., Li, N., Lin, J., Dean, T.: Formability evaluation for sheet metals under hot stamping conditions by a novel biaxial testing system and a new materials model. Int. J. Mech. Sci. **120**, 149–158 (2017)
8. Khoo, S.-W., Karuppanan, S., Tan, C.-S.: A review of surface deformation and strain measurement using two-dimensional digital image correlation. Metrol. Meas. Syst. **23**(3), 461–480 (2016)
9. Tong, C., et al.: Investigation of deformation behaviour with yield point phenomenon in cold-rolled medium-Mn steel under hot stamping conditions. J. Mater. Process. Technol. **306**, 117623 (2022)
10. Zhang, R., Shi, Z., Shao, Z., Dean, T.A., Lin, J.: A novel spatio-temporal method for determining necking and fracture strains of sheet metals. Int. J. Mech. Sci. **189**, 105977 (2021)
11. Mu, L., Jia, Z., Ma, Z., Shen, F., Sun, Y., Zang, Y.: A theoretical prediction framework for the construction of a fracture forming limit curve accounting for fracture pattern transition. Int. J. Plast. **129**, 102706 (2020)
12. Noder, J., Butcher, C.: A comparative investigation into the influence of the constitutive model on the prediction of in-plane formability for Nakazima and Marciniak tests. Int. J. Mech. Sci. **163**, 105138 (2019)
13. Hu, Q., Zhang, L., Ouyang, Q., Li, X., Zhu, X., Chen, J.: Prediction of forming limits for anisotropic materials with nonlinear strain paths by an instability approach. Int. J. Plast. **103**, 143–167 (2018)

14. Basak, S., Panda, S.K.: Failure strains of anisotropic thin sheet metals: experimental evaluation and theoretical prediction. Int. J. Mech. Sci. **151**, 356–374 (2019)
15. Zhang, R., Shi, Z., Yardley, V.A., Lin, J.: Experimental studies of necking and fracture limits of boron steel sheet under hot stamping conditions. J. Mater. Process. Technol. **302**, 117481 (2022)

A Novel Insight into the Mechanism of Plastic Instability Behavior from the Mesoscopic-Scale Strain Analysis in Medium Mn Steel

Shuai Zhao, Renbo Song(✉), Yingchao Zhang, Weifeng Huo, YongJin Wang, and Changhong Cai

School of Materials Science and Engineering, University of Science and Technology Beijing, Beijing 100083, China
songrb@mater.ustb.edu.cn

Abstract. The spatiotemporal coupling relationship of strain in different local deformation regions was characterized by digital image correlation (DIC). The Lüders band nucleates at the deviation point of the elastic stress-strain line and begins to propagate at the lower yield point. Propagation of the Portevin-Le-Chatelier (PLC) band leads to the formation of the Band Propagation Region and No-band Region in the entire deformation region. The occurrence of stress serrations is attributed to the periodic strain transferring between the Band Propagation Region and the No-band Region at the mesoscopic scale.

Keywords: Strain Transferring · Deformation Band · Plastic Instability Behavior · Medium Mn steel

1 Introduction

It is well known that the plastic instability behavior represented by the Lüders effect and Portevin-Le-Chatelier (PLC) effect is common to occur in medium Mn steel (containing 3–12 wt.% Mn) [1, 2]. The yield plateau and stress serrations are complex responses of typical Lüders and PLC bands on the mechanical curves [3–5]. Although many scholars believe that strain transferring and strain partitioning occur in this process and have given more micro-scale explanations such as dislocation depinning effect for the plastic instability behavior from the microscopic scale, few direct and corresponding characterizations between strain behavior and stress serrations were reflected in the current studies [6–8]. In this study, digital imaging correlation (DIC) technology was used to characterize dynamic strain distribution in space and time at the characteristic local deformation regions during tensile test deformation, which provides a novel insight into the interpretation of plastic instability behavior in 3 Mn steel at the mesoscopic scale.

K. Mocellin et al. (Eds.): ICTP 2023, LNME, pp. 167–173, 2024.
https://doi.org/10.1007/978-3-031-40920-2_18

2 Materials and Methods

The chemical composition of the experimental steel is 0.37C-3.25Mn-0.43Al-1.53Si-0.048Nb-0.26Cr-0.15Mo-Fe (wt%). After annealing at 725 °C for 30 min, the hot-rolled samples were air-cooled and water-quenched, respectively, followed by tempering at 300 °C for 30 min. The obtained tempering samples are called IAT (Intercritical annealing-Air cooling-Tempering) and IWT (Intercritical annealing-Water quenching-Tempering), respectively. The microstructure was observed through the scanning electron microscope (ZEISS GeminiSEM 500). The tensile tests were conducted on sub-sized tensile specimens (3 mm in thickness, 6 mm in width, 25 mm in gauge length, and 80 mm in total length) at an initial strain rate of 0.001 s^{-1} at room temperature through the universal testing machine (MTS Exceed E45) and the kinematic field measurement was carried out with a visible-light camera (Macro 100F2,8D) and software (Lavision, Davis 10). The sampling frequency of the camera is 10 frames per second.

3 Results and Discussion

Figure 1 shows the curves of mechanical properties and the images of microstructures of IAT and IWT. It is evident that the IWT sample exhibits obvious plastic instability behavior, characterized by a yield plateau and stress serrations, whereas the IAT sample maintains a continuous yield state. As shown in Fig. 1b and c, more coarsening lath grains and more precipitations are found in the IAT sample compared with the IWT sample, which should be the reason for their different plastic behavior in curves of mechanical properties. But at present, it is failed to characterize the correspondence between the variation of microstructure and the appearance of plastic instability behavior (yield plateau and stress serrations) in real-time. In order to gain a more comprehensive understanding of the mesoscopic structural evolution in the deformation region during the emergence of the yield plateau and stress serrations, we employ the DIC technique to analyze the in-situ dynamic tensile behavior of the mesostructured [9, 10].

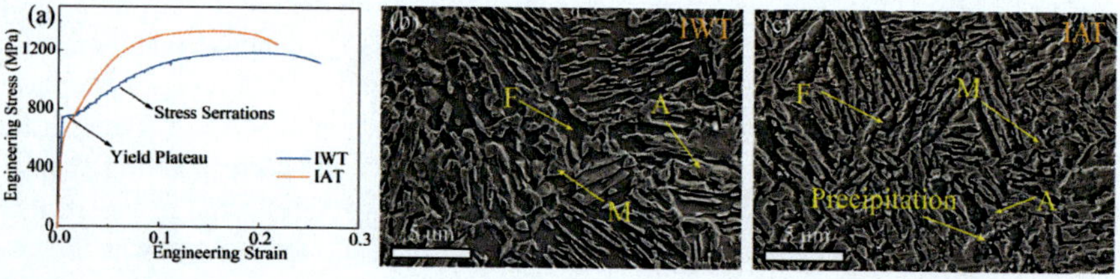

Fig. 1. (a) Engineering stress-strain curves of IAT and IWT. (b) and (c) show the microstructure of IWT and IAT, respectively.

According to Fig. 2, IWT exhibits conspicuous plastic instability behavior during tensile deformation, while IAT shows continuous strain hardening. DIC images of marked points on the stress curves of IAT and IWT in Fig. 2a are shown in Fig. 2b and c, respectively. Deformation band A and B of IAT samples are initiated at one end and the middle of the gauge section, respectively. The bands begin nucleating at point 2 on the stress curve, which marks the deviation point of the elastic stress-strain line. Once formed, Bands A and B propagate toward each other. In addition to the band propagation region, there exists a non-propagating region that is not affected by the band, referred to as the No-Band Region, located below Band B.

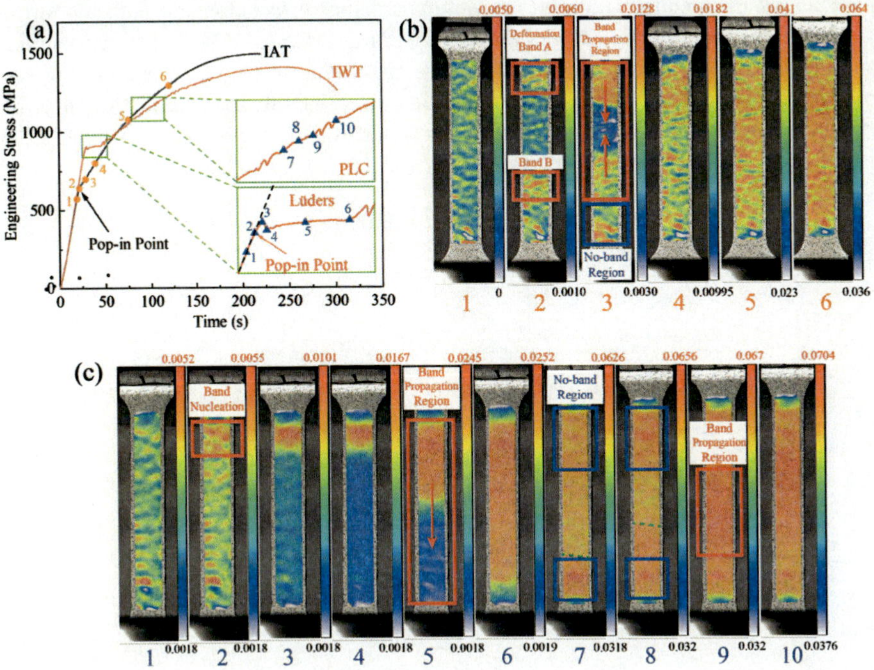

Fig. 2. (a) Stress-time relationship of IAT and IWT. (b) and (c) show the strain distribution of IWT and IAT at certain time points, respectively. The color map is distribution of engineering strain.

The obvious Lüders and PLC effect was observed in IWT. In many studies, the Lüders band was believed to nucleate at the moment of the stress drop or the macroscopic yield point [11, 12]. However, this conclusion was built on assumptions of a lack of experimental evidence and support [13]. As shown in Fig. 2b, the same as the nucleation time of the local deformation band in IAT, the nucleation of the Lüders band was also observed at the deviation point of the elastic stress-strain line before the upper yield point. The Lüders band begins to propagate from the low yield point (point 4). After the Lüders band extends to the whole deformation region, the PLC band propagates periodically but doesn't extend to the whole gage region, which forms the No-band Region. The range of point 7 to point 10 in Fig. 2c is a cycle of PLC band propagation.

After the Lüders band propagated in IWT, the whole gage region is mainly divided into three parts, Region A, Region C, and the band-propagation region between Region A and C. Region B is the part of the Band Propagation Region. Figure 3e and f show the strain-time curves of Regions A, B and C. During the propagation of the Lüders band, the strain of Region A becomes gentle from a sharp increase, which means that the Lüders band begins to spread outward from Region A. At this time, the strain of Region A is the critical strain of Lüders band migration. With the extension of the Lüders band, the critical strain increases slowly and linearly until the end of propagation. The higher value of the strain means a stronger strain-hardening effect. Thus, the propagation of the Lüders band gradually increases the strain hardening effect along the path and forms a strain hardening gradient, which plays an important role in strain transferring in the subsequent process. Line 1 in Fig. 3e shows the variation of critical strain with the propagation time. The relationship of the critical strain ε_1 with the propagation time t is shown in Eq. (1) and this relation also corresponds to the relationship between critical strain ε1 and propagation position from Region A to C.

$$\varepsilon_1 = 0.0004t + 0.0065 \quad (1)$$

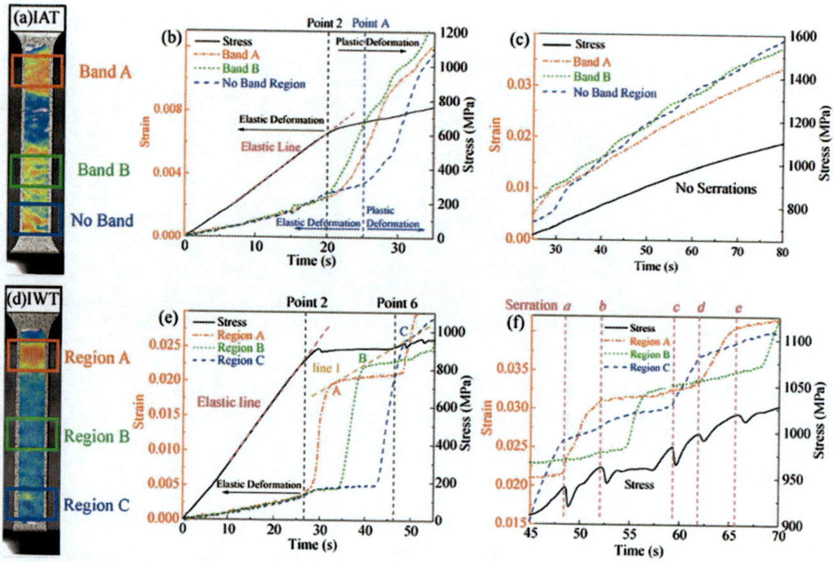

Fig. 3. The variation of the strains in the local region: images (a) and (d) show the position of the selected special regions in IAT and IWT, respectively. Graphs (b) and (c) show the strain variation of regions marked in (a). Graphs (d) and (e) shows the strain variation of regions marked in (b).

Figure 3f shows that the stress serrations always appear at the time point when the strain state of Region A and C become abrupt. The abrupt variation of the strain state is induced by the strain transferring, accompanied by the occurrence of stress serrations. The specific details of the dynamic strain distribution in space are shown in Fig. 4.

As shown in Fig. 4a, The linear distribution of strain along the propagation path of the Lüders band. Figure 4c shows the strain distribution at the formation time of serration ***a***. At this time, the propagation of the Lüders band has just finished and Region C is

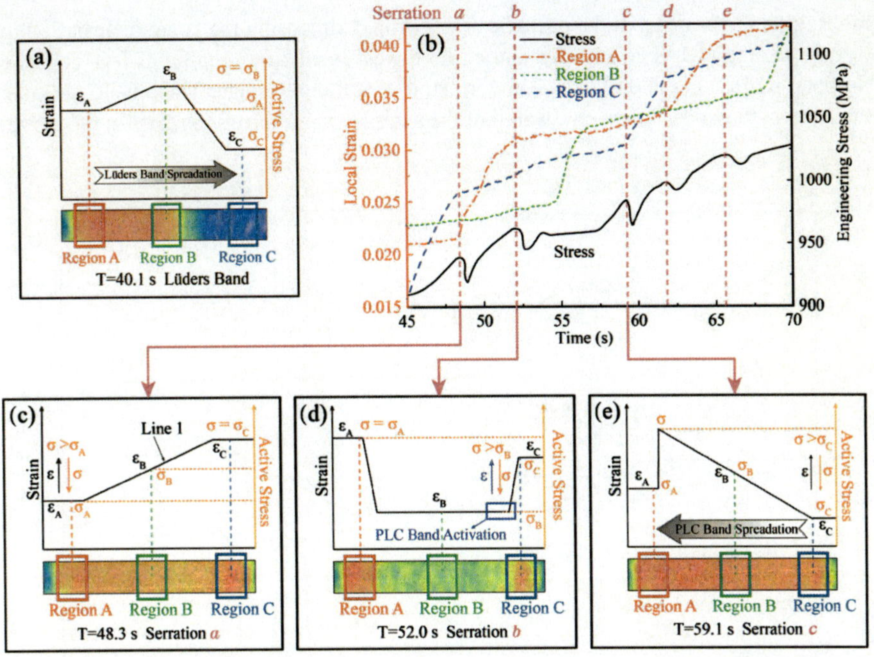

Fig. 4. Lüders band propagation: (a) the strain-space distribution at the yield plateau. (b) the serrations in Fig. 3f. Graphs (c), (d) and (e) show the formation of serrations *a, b* and *c* in (b), respectively. The right axis in graphs (a), (c), (d) and (e) is the active stress satisfied with the increase of strain at the corresponding position. The loading displacement end is located near Region A.

at the local deformation position currently. The strain ε_C and stress σ_C of Region C is highest under the role of strain increment along the band propagation path and the stress σ_C works as the working stress σ of the sample. At this time, the plastic deformation is transferred to Region A with the lowest strain, and the strain ε_A would be greatly improved [14]. However, due to the stress σ_A satisfying the growth of current strain ε_A in Region A is lower than the current working stress $\sigma = \sigma_C$. Then the value of the working stress will be greatly reduced to form the serration ***a*** in Fig. 4b. When Region A is under a high strain, the higher deformation resistance causes the strain to transfer to the region with the lowest strain. The stress distribution diagram is shown in Fig. 4d, and the current working stress $\sigma = \sigma_A$. As strain transfers to the propagation region of the PLC band with the lowest strain σ_B. The working stress drops and the stress serration ***b*** is formed, which results in the switching on of the PLC band. The propagation direction of the PLC band is in contradiction to the Lüders band. The critical strain of PLC band propagation continues to increase along the propagation path, which plays an important role in the next transfer of strain. After the propagation of the PLC band, the stress drops again under the condition of the sharp increase of the Region C strain, and serration ***c*** is formed. Therefore, the reason for the strain transferring should be the periodic variation of the lowest local strain region among Region A, the propagation region of the PLC band

and Region C, resulting in the periodic stress drop. Comparing the strain distribution in each local area of IAT in Fig. 2.b and c, the lowest local strain region of IAT is in the No-band Region at the early stage and in band A at the late stage. There is no periodic variation in the lowest local strain area of the sample, so no serrations exist in the curve.

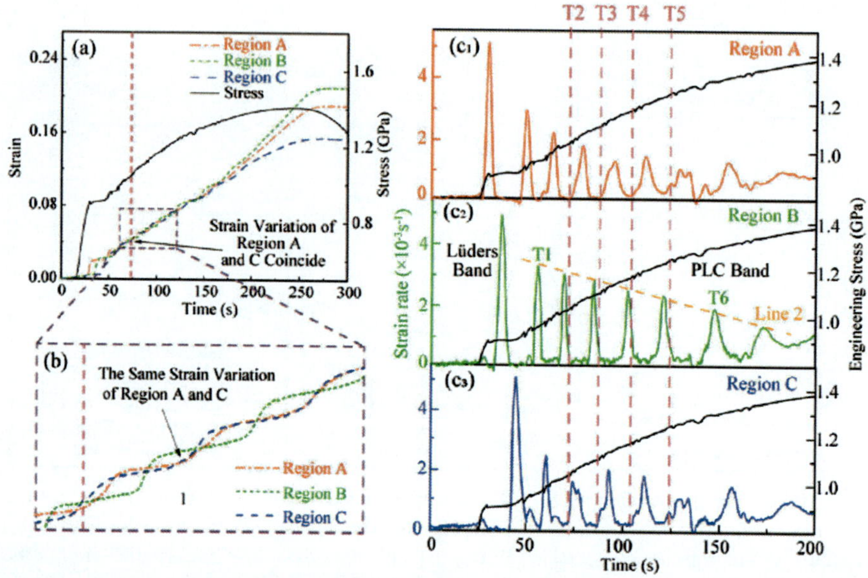

Fig. 5. The relationship between the periodic appearance of serrations and local strain rate: (a)The strain of Region A, B and C. (b)At the late stage, the strain variation of Region A and C begins to coincide. (c) the strain rate of Region A, B and C.

Figure 5 shows the subsequent variation of strain in different regions. It can be seen that after the synchronized variation of the strain in Regions A and C, the periodicity of serrations becomes more obvious. When the strain rate of Region A and C increases suddenly, which indicated the strain transferring to Region A and C, the serrations appear in time. Thus, the serrations are still related to the abrupt variation of strain in Region A and C. Basic on the apparent periodic appearance of serrations corresponding to the synchronized variation of strain in Region A and C, the strain transferring between the band propagation region and No-band regions is the main reason for serrations. Figure 5b shows that the maximum strain rate during the propagation of the lüders band is higher than that of the PLC band. The maximum strain rate of Region B decreases gradually with the propagation of the PLC band. The maximum strain rate is linear with time. Equation (2) describes the relationship between the maximum strain rate ε_{max} of line 2 and the propagation times T of the PLC band.

$$\varepsilon_{max} = -0.0003T + 0.0037 \tag{2}$$

When the strain rate of the band propagation region decreases to a certain extent, the PLC band would be not initiated again.

4 Conclusion

In summary, the Lüders band nucleates at the deviation point of the elastic line and starts to propagate at the lower yield point. The periodic strain transferring between the band propagation region and No-band regions is the main reason for serrations. This offers a novel insight into the appearance of plastic instability behavior in medium Mn steel from the mesoscopic scale. Although the relationship between the microstructure and mesoscopic region cannot be given directly at present, it is clear that the spatiotemporal coupling rules of the deformation band on the mesoscopic scale have a direct corresponding relationship with the occurrence of plastic instability behavior in medium Mn steel.

References

1. Wang, Y., Ma, Z., Song, R., et al.: Correlation between cementite precipitation and Portevin-Le Chatelier effect in a hot-rolled medium Mn steel. Mater. Lett. **258**, 126796 (2020)
2. Borisova, Y., Yuzbekova, D., Mogucheva, A., et al.: Characterization of the spatiotemporal behavior of the PLC bands in austenitic steel during deformation at elevated temperature. Mater. Lett. **304**, 130584 (2021)
3. Yang, F., Luo, H., Pu, E., et al.: On the characteristics of Portevin-Le Chatelier bands in cold-rolled 7Mn steel showing transformation-induced plasticity. Int. J. Plast. **103**, 188–202 (2018)
4. Lebyodkin, M., Dunin-Barkowskii, L., Bréchet, Y., et al.: Spatio-temporal dynamics of the Portevin–Le Chatelier effect: experiment and modelling. Acta Mater. **48**, 2529–2541 (2000)
5. Li, X., Song, R., Zhou, N., et al.: An ultrahigh strength and enhanced ductility cold-rolled medium-Mn steel treated by intercritical annealing. Scr. Mater. **154**, 30–33 (2018)
6. Hojo, T., Koyama, M., Kumai, B., et al.: Comparative study of stress and strain partitioning behaviors in medium manganese and transformation-induced plasticity-aided bainitic ferrite steels. Scr. Mater. **210**, 114463 (2022)
7. Hu, J., Du, L., Xu, W., et al.: Ensuring combination of strength, ductility and toughness in medium-manganese steel through optimization of nano-scale metastable austenite. Mater. Charact. **136**, 20–28 (2018)
8. Danilov, V.I., Orlova, D.V., Gorbatenko, V.V., et al.: Kinetics of strain-induced phase transformation in TRIP steel. Mater. Lett. **301**, 130289 (2021)
9. Suthar, H., Bhattacharya, A., Paul, S.K.: Determination of local constitutive properties in similar and dissimilar friction stir welded joints from DIC based surface strain measurement in two mutually perpendicular surfaces. Mech. Mater. **160**, 103930 (2021)
10. Min, J., Lin, J., Sun, B.: Effect of strain rate on spatio-temporal behavior of Portevin–Le Châtelier bands in a twinning induced plasticity steel. Mech. Mater. **68**, 164–175 (2014)
11. Wang, X.G., Wang, L., Huang, M.X.: Kinematic and thermal characteristics of Lüders and Portevin-Le Châtelier bands in a medium Mn transformation-induced plasticity steel. Acta Mater. **124**, 17–29 (2017)
12. Gao, S., Bai, Y., Zheng, R., et al.: Mechanism of huge Lüders-type deformation in ultrafine grained austenitic stainless steel. Scr. Mater. **159**, 28–32 (2019)
13. Qiu, H., Inoue, T., Ueji, R.: In-situ observation of lüders band formation in hot-rolled steel via digital image correlation. Metals **10**, 530 (2020)
14. Zhang, Y., Ding, H.: Ultrafine also can be ductile: on the essence of Lüders band elongation in ultrafine-grained medium manganese steel. Mater. Sci. Eng. A **733**, 220–223 (2018)

Investigation of Hot Stamping with Cooled and Textured Tools Manufactured by Directed Energy Deposition

Anna Komodromos(✉), Gabriel Marín, Joshua Grodotzki, and A. Erman Tekkaya

Institute of Forming Technology and Lightweight Components, TU Dortmund University, Dortmund, Germany

anna.komodromos@iul.tu-dortmund.de

Abstract. Additive manufacturing of hot forming tools provides increased flexibility regarding geometry and proximity of the cooling channels to the tools' surface. In this field, the application of Directed Energy Deposition has not been investigated extensively. The conventional manufacturing route of cooling channels in hot stamping tools comprises the segmentation of the tool and hole drilling. This leads to the disadvantages of possible leakages and low flexibility regarding the channels' size and location. Thus, a novel combination of Directed Energy Deposition and ball burnishing for leveling the additively manufactured surfaces is established to manufacture hot stamping tools for the forming of hat profiles. In the additively manufactured punch, near-surface cooling channels are integrated after a load-adapted definition of channel geometry and path, both based on numerical and experimental analysis. On the blank holder side, an additively manufactured texture, targeting the control of the heat balance and material flow, is applied. With these tools, hot stamping tests with a hat profile (22MnB5) are conducted and compared with conventionally manufactured tools with drilled cooling channels. It is demonstrated that the near-surface cooling channels are able to decrease the punch temperature in the hot stamping process by up to 50% in the investigated configuration. The textured blank holder creates a more homogeneous temperature distribution over the hat profile during forming. As a result, part properties can be enhanced by a lowered sheet thickness reduction and a higher and evenly distributed hardness. Numerical simulations can be used to depict those effects during process planning.

Keywords: Hot stamping · Directed Energy Deposition (DED) · near-surface cooling channels · temperature control

1 Introduction

Hot stamping of automobile parts is frequently used when high-strength steel parts with low specific weights are required. Due to the high process temperatures, the advantages of high formability during forming and strengthening of the part by subsequent hardening are combined. The most common materials used in hot stamping are manganese-boron

K. Mocellin et al. (Eds.): ICTP 2023, LNME, pp. 174–183, 2024.
https://doi.org/10.1007/978-3-031-40920-2_19

steels, such as 22MnB5, due to their ability to transform to a martensitic microstructure in the process of combined forming and quenching. This results in high-strength parts of up to 1500 MPa [1]. Stamping tools with integrated cooling channels perform the simultaneous forming and quenching processes. The tools must fulfill the requirement of a minimum cooling rate of 27 K/s for the martensite transformation on the 22MnB5. Additive manufacturing (AM) of hot forming tools provides increased flexibility regarding geometry and proximity of the cooling channels to the tool surface, e.g. to avoid hot spots [2]. Regarding the powder-based Directed Energy Deposition (DED) method, numerical simulations of cooling channels close to the surface contour shows that it achieves a reduction in tool temperature of up to 40 °C compared to machined cooling channels, which do not follow the contour [3]. Using a drop-shaped cross-section, the fabrication of the cooling channel by AM is shown to be feasible [4]. In that analysis, a suitable channel path was milled into the base material. No experimental testing in a forming tool took place. Regarding the closing of cooling channels by DED for use in hot stamping tools, Cortina [3] and Chantzis [4] show that the closing strategy strongly depends on the channel width. The exact formation of the seam geometries requires the consideration of, among other things, the laser power, powder feed rate, nozzle angle, and feed speed. Using a model tool, in which the tool surface is coated with the material Cu85, it is shown that the cooling channels close to the surface enable a homogeneous cooling process [5]. A method for influencing the material flow in a hot stamping process has been presented using a hat profile piece made from aluminum (AA6082-T6, $t = 1.5$ mm). By machining a macro-texture onto the blank holder surface, the drawability can be improved [6]. It has been shown that a texture parallel to the material flow has a more significant influence on the material flow than a texture perpendicular to it. In this way, sheet thinning in the component is prevented. In summary, there are preliminary investigations regarding the additive manufacturing of cooled hot stamping tools but none with actual use in the forming process or with a combination of cooling channels and textures manufactured by DED. Hence, a novel DED and ball burnishing combined process is presented for the use in cooled and textured hot stamping tools for forming 22MnB5 hat profiles. By the use of near-surface cooling channels, a low tool temperature is to be maintained in order to keep the cooling rate at the required level through several forming cycles. The aim of using additively manufactured textures consists in the control of the heat balance and material flow during hot stamping.

2 Methodology

2.1 Directed Energy Deposition and Ball Burnishing

The metal powders (Ferro 55® by voestalpine) are processed in a hybrid machine Lasertec 65 3D (DMG MORI). The machine includes a DED laser unit and a five-axis milling unit. A nozzle with a diameter of 3 mm is used for the deposition of the metal powder. Further parameters are defined as follows: in-plane nozzle feed of 1000 mm/min, 14 g/min powder rate, and an initial laser power of 2000 W for one layer. The laser power is decreased by 200 W per layer until 800 W to avoid an extensive reheating of the previous layer (Fig. 1a). The subsequent ball burnishing of the additively manufactured surfaces is carried out on a five-axis milling machine DMU 50 (DMG MORI). A hydrostatic ball burnishing tool HG13 (ECOROLL) with a ball diameter of 12.7 mm is used. Using an external high-pressure pump, the burnishing pressure is adjusted to the targeted value at the burnishing ball. The ball burnishing process is carried out with a constant tool feed of 33 mm/s.

A workpiece with a hat profile (Fig. 1b) is chosen to be formed by the stamping tools including additively manufactured cooling channels. In the setup, the upper part of the punch and its cooling channels are fabricated by DED (via Lasertec 65 3D) using the tool steel powder. The cooled upper part of the punch is deposited on a pre-machined base material (tool steel 1.2367/X38CrMoV5-3) to save time and costs. The proximity of the cooling channels to the tool's surface as well as its continuity are the two prime advantages of using AM as the manufacturing method. The channel requires only one inlet and one outlet for the water coolant.

2.2 Hot Stamping Experiments

During the hot stamping tests, the temperature of the punch was measured by thermocouples located 2 mm beneath the surface. The following cases are compared:

- Conventionally manufactured punch (tool steel 1.2367) with two drilled cooling channels with a diameter of 8 mm and a distance to the tool surface of approx. 15 mm (due to restrictions of the deep-hole drilling process).
- Additively manufactured punch with approx. 4.5 mm distance between the tip of the drop-shaped cooling channels and the top surface (Fig. 1b). The width of the drop-shaped channels is 6 mm, and the distance between each is 8 mm.

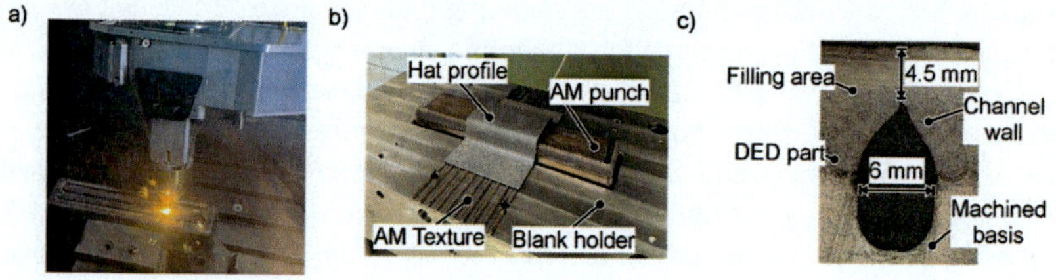

Fig. 1. a) DED process of the punch, b) hot stamping setup, c) micrograph of cooling channel.

2.3 Hot Stamping Simulations

During hot stamping, the material exhibits elastic visco-plastic characteristics and phase transition [7]. For this investigation, the temperature, and strain rate-dependent behaviour are described by the constitutive equations developed by Åkerström [7] and Olsson [8]; and executed according to the description of 22MnB5 done by Lechler [9] and Hochholdinger [10].

The simulations are performed using the commercial Finite Element Program LS-DYNA, following the methodology proposed by Hochholdinger [10]. The model used to describe the stamping process is shown Fig. 2a.

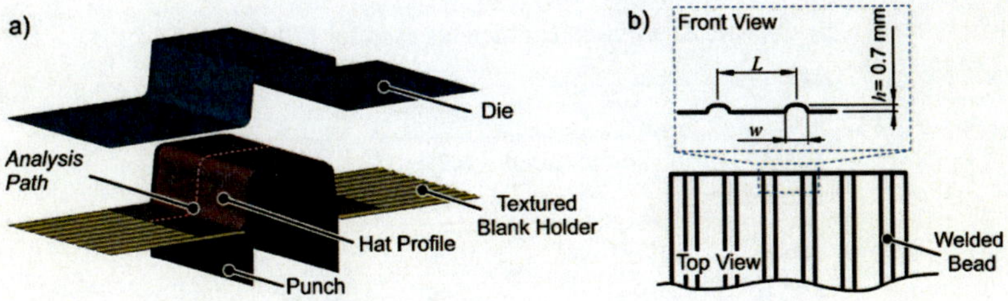

Fig. 2. a) FE model, b) texture definition on the blank holder.

The stamping tools and the sheet are discretized and represented by square Belytschko-Tsay shell elements (shell element type 16). The tools are treated as rigid bodies (MAT020) with properties of the tool steel 1.2367, while the 22MnB5 sheet is treated as a thermo-elastoplastic body (MAT244). The hot stamping process is defined by six steps:

1. Transfer: transport of the austenitized sheet at $T_\gamma = 950$ °C from the furnace to the stamping tools. Heat losses are estimated by a radiation and convection model [10].
2. Positioning: the heated sheet is positioned on the blank holder, which has an initial temperature of 25 °C. Due to the gravity effect on the sheet dropping, the thermo-mechanical effects are estimated with the dynamic implicit method.
3. Clamping: closing the die on the sheet, with a gap of 0.5 mm. In this stage, the punch and blank holder are spatially fixed with an initial temperature of 25 °C.
4. Drawing: the sheet is formed in a hat shape. The friction conditions between the tool and sheet, which change according to the surface texture (Fig. 2b), are experimentally determined as a function of the temperature, drawing velocity, and contact pressure (Table 1). Due to the high strain rates over short process times, both stages 3 and 4 thermo-mechanical effects are estimated with an explicit calculation method.
5. Holding: cooling of the hat profile while the tools are closed. The interfacial heat transfer coefficients used to estimate the heat transfer between tool and blank were experimentally determined as a function of the contact pressure (Fig. 3).
6. Opening: die and punch are retracted from the stamped piece, letting it cool down under combined radiation, convection, and conduction conditions on the blank holder. Stages 1, 5 and 6 thermic effects are estimated with an implicit convergence method.

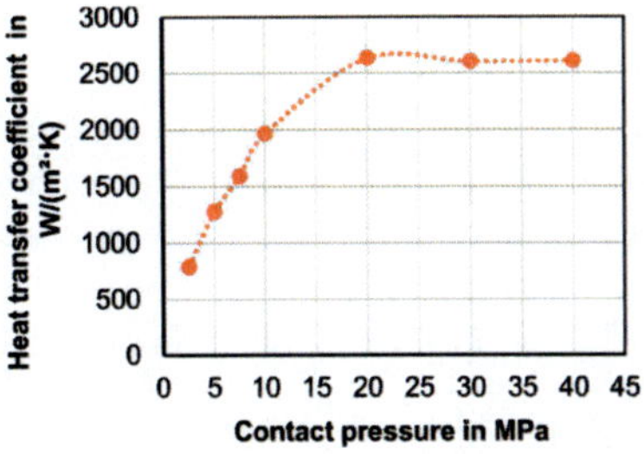

Fig. 3. Interface heat transfer coefficient as a function of the contact pressure.

Table 1. Friction coefficients used for FEM.

		Textured	
Blank temperature	Non textured	$w/L = 0.3$	$w/L = 0.2$
500 °C	0.41	0.57	0.49
650 °C	0.40	0.59	0.54
800 °C	0.39	0.65	0.57

Drawing velocity: 60 mm/s, contact pressure: 5 MPa

3 Numerical Investigations

Comparing the blank temperatures at the analysis path (Fig. 2a) obtained using the conventional (non-textured) and the textured blank holder reveals significant differences. At the clamping stage, an important aspect of the textured blank holder becomes evident: the temperature of the blank stays more homogeneous and at a higher level than the conventional blank holder (Fig. 4a). This effect is continued during the forming stage (Fig. 4b).

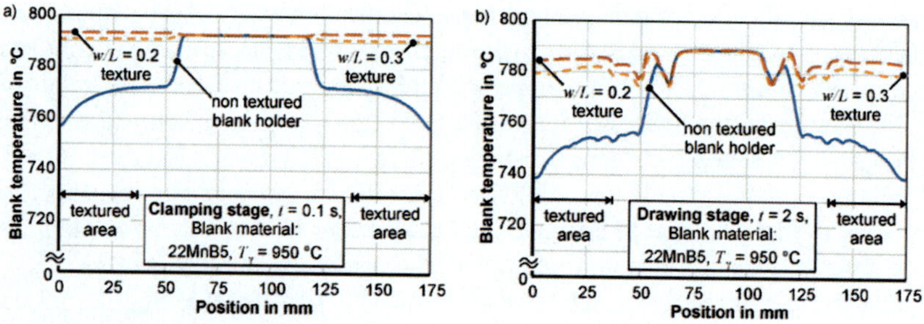

Fig. 4. Numerical analysis of temperature distribution in the blank during hot stamping at the end of the a) clamping stage and the b) forming stage.

At the end of the holding stage (Fig. 5a), the blank is cooled down in the top area of the hat profile while the flange area remains at a higher temperature level. After the opening of the tools, the textured area influences the blank temperature in such a way that the flanges remain at a higher temperature than with the conventional, i.e. non textured, blank holder. That means the cooling is slowed down by the texture, with a more considerable effect by a coarser texture (Fig. 5b). This bears the potential that specific areas of the formed part (e.g. the flange) can produce more ductile properties than other areas.

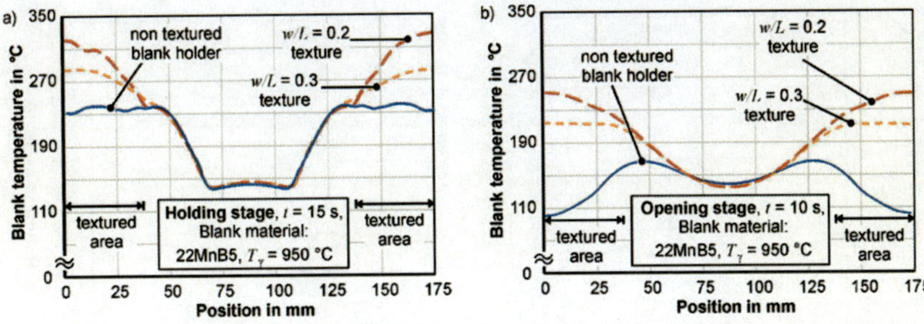

Fig. 5. Numerical analysis of temperature distribution in the blank during hot stamping at the end of the a) holding stage and the b) opening stage.

4 Experimental Investigations

The comparison of the punch temperatures measured by thermocouples located 2 mm under the tool surface shows that the additively manufactured cooling channels not only reduce the temperature in the hot stamping process by up to 50%. It also keeps the temperature constant during consecutive forming cycles, whereas the temperature of the conventionally manufactured punch increases (Fig. 6).

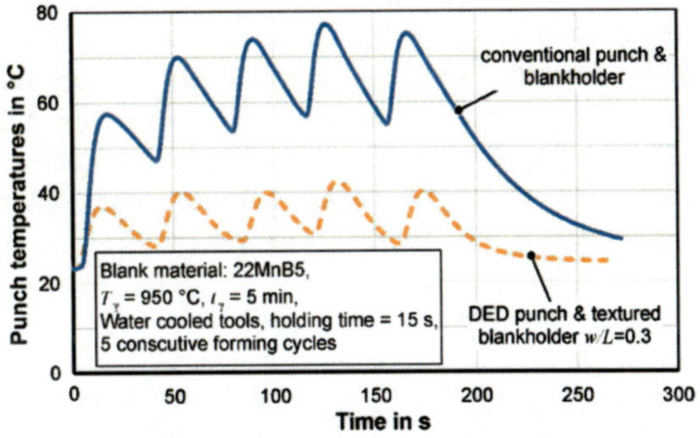

Fig. 6. Punch temperatures during five hot stamping cycles with conventionally and AM tools.

After the hot stamping tests, hardness measurements are performed on the hat profiles. The comparison of the experimentally with the numerically obtained values exhibits a deviation of around 6%. The results indicate that the top region and the wall region of the profile are harder when using near-surface cooling channels manufactured by AM (Fig. 7).

By contrast, the texture on the blank holder reduces the hardness in the flange area and the adjacent areas of the profile as a result of the altered thermal history during the process. The hardness distribution across all areas is more homogeneous compared to conventionally manufactured parts. Those parts exhibit a sharp drop in the hardness from the top to the wall area. Similar observations can be made regarding the sheet thickness and its distribution (with a deviation of 4% to the numerical results). Using

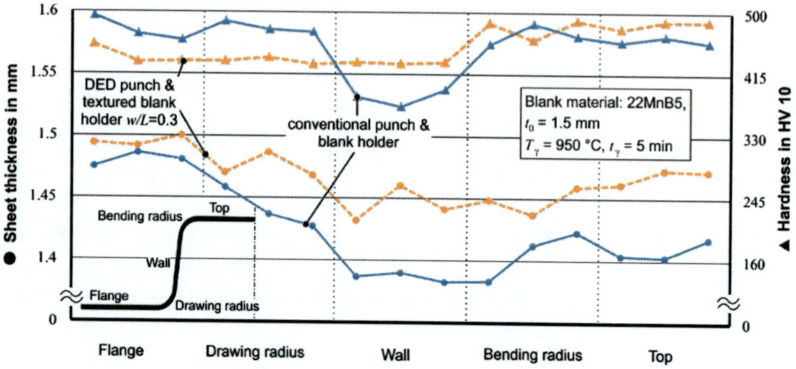

Fig. 7. Sheet thicknesses and hardness measurements on the hot stamped hat profile after hot stamping.

additively manufactured textures, the part has a higher thickness across all areas and the distribution is significantly more homogeneous (Fig. 7). For the conventional parts, the drop in hardness in the wall correlates with the lowest sheet thickness observed in the same region. Based on these findings, the effects of the additively manufactured near-surface cooling channels and surface textures on the hot-stamped hat profile are now qualitatively known. Through the textured tool surface, the heat is kept longer in the blank, and an early cooling of specific areas of the part is prevented. In the case of the conventional blank holder, the blank in the flange area is cooling down faster. Consequently, the material flow is impeded in this case, which causes the thinning in the wall of the hat profile. Keeping the heat for a longer time in certain areas can be useful for several applications. In the case of the flange area, it can be used for the facilitation of post-processes like cutting or welding.

The temperatures of the formed hat profile are evaluated by thermal imaging right after the opening of the tools. Regarding the flange temperature of the hat profile formed with the use of the conventional blank holder compared with the one formed with the textured blank holder, the temperature is kept up to 20% higher with the additively manufactured texture (Fig. 8a). This effect is sustained through the adjacent parts of the hat profile. Hence, the material flows throughout the entire process benefit from the higher and more homogeneous temperature distribution in all sections of the hat profile, promoting a uniform deformation through the hat profile piece [11]. Thus, a more homogeneous and fine microstructure grain size is obtained, and undesired effects, such as orange peel (Fig. 8b), which indicates a coarse grain size and excessive thinning in the wall section, can be avoided. Therefore, the intended outcome of using cooled and textured tools for hot stamping can be regarded as successfully achieved. The comparison of the temperature obtained experimentally and numerically exhibits a deviation of around 10%, indicating an overall good agreement.

Force-displacement measurements from standard tensile tests compare the mechanical performance in T-bone specimens taken from the flange, walls, and top areas of the hot stamped hat profile (Fig. 9). Normalized stress-strain results are not used since the thinning of the sheets behaves differently in the investigated hot stamping methods.

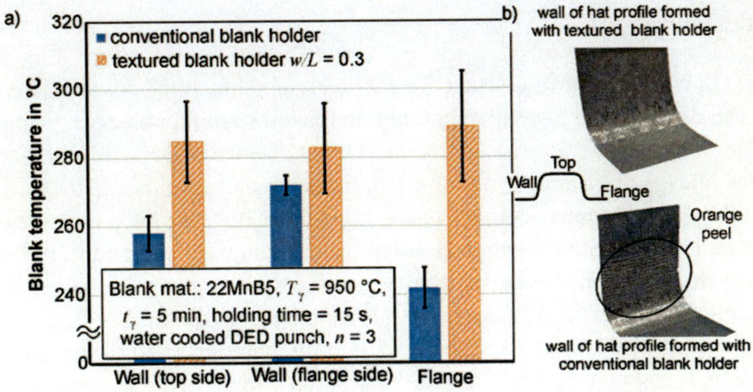

Fig. 8. a) Temperature evaluation of the 22MnB5 hat profile after hot stamping, b) surface defect in form of orange peel caused by premature cooling with conventional blank holder.

Regarding the maximum tensile force obtained by the specimens taken from the hat profiles, those formed with the conventional blank holder exhibit a force drop of 41% at the wall area compared to the flange and the top areas, while those formed with the textured blank-holder exhibit a drop of only 12%. This evidence the more homogeneous and finer grain size [12] achieved by the more homogeneous temperature distribution generated by the use of the textured blank holder. Similarly, the more significant displacement achieved by the specimen from the wall of the hat profile formed with the textured blank holder indicates a higher toughness of the microstructure, which is a sign of a finer grain size as well [13].

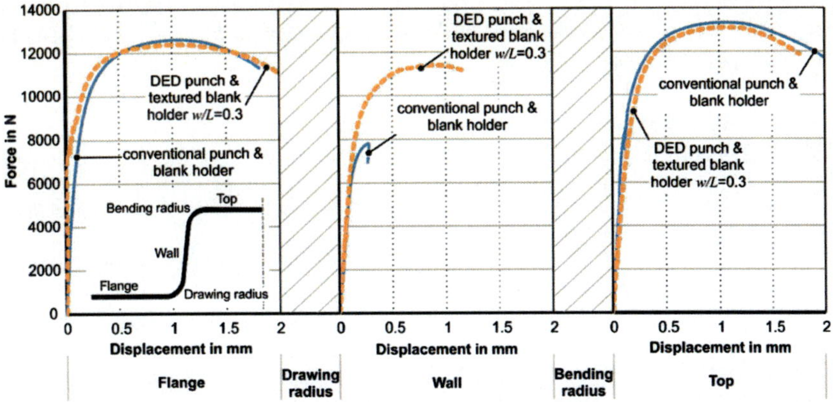

Fig. 9. Force-displacement measurements generated with standard tensile tests on the hot stamped 22MnB5 hat profile.

5 Conclusions

The presented work demonstrates that the additively manufactured near-surface cooling channels can decrease the punch temperature in the hot stamping process. Compared to the conventionally manufactured punch, the Directed Energy Deposition process provides the advantage of locating the cooling channels three times closer to the tool surface and generating a continuous coolant cycle in the punch. Because the punch temperature is reduced and kept constant through consecutive hot stamping operations, the potential to reduce cycle times and prevent overheating of the tools is given. Moreover, the near-surface cooling channels can increase the hardening effect on the hot stamped part in the cooled contact area. In addition to the successful manufacturing of the near-surface cooled punch, it is shown that the DED process is suitable for manufacturing textures on the tool surfaces as well. Employing a textured blank holder, the thermal history of the part is altered. The additively manufactured texture retains the heat inside the blank for a longer time, allowing a higher ductility of the part during the hot stamping and potential reduction of forming forces. Further, the textured blank holder provides the advantage of a homogenized temperature distribution over the hat profile during forming, which enables fine and well-distributed grain size at the microstructure. This directly influences the formed part's sheet thickness, strength, and hardness. Properties that show higher and more homogeneously distributed values compared to the conventional tool route. It is then temperature control is an enabler for tailored properties on the hot-stamped components. Numerical simulations of the process can also predict these effects accurately. In conclusion, the feasibility and characteristics of manufacturing hot stamping tools by DED are proven such that the findings can be used for further developments in this field, like the application of both-sided textures on the tools.

Acknowledgements. The authors thank the German Research Foundation (DFG) for the financial support of the project "Functionalization of additively manufactured hot stamping tools by ball burnishing" (project number 417202720).

References

1. Karbasian, H., Tekkaya, A.E.: A review on hot stamping. J. Mater. Process. Technol. **2**, 2103–2118 (2010)
2. Hölker, R., Jäger, A., Ben Khalifa, N., Erman Tekkaya, A.: Controlling heat balance in hot aluminum extrusion by additive manufactured extrusion dies with conformal cooling channels. Int. J. Precis. Eng. Manuf. **14**(8), 1487–1493 (2013). https://doi.org/10.1007/s12541-013-0200-1
3. Cortina, M., Arrizubieta, J.I., Calleja, A., Ukar, E., Alberdi, A.: Case study to illustrate the potential of conformal cooling channels for hot stamping dies manufactured using hybrid process of laser metal deposition (LMD) and milling. Metals **8**, 102 (2018)
4. Chantzis, D., et al.: Review on additive manufacturing of tooling for hot stamping. Int. J. Adv. Manuf. Technol. **109**(1–2), 87–107 (2020)
5. Vollmer, R., Kolleck, R., Schwemberger, P.: Herstellung oberflächennaher Kühlkanalstrukturen für das Presshärten mittels Laserauftragsschweißen. In: Merklein, M. (Hrsg.), Tagungsband zum 9. Erlanger Workshop Warmblechumformung, pp. 61–73. Meisenbach Verlag, Bamberg (2014)

6. Zheng, K., Politis, D.J., Lin, J., Dean, T.A.: An experimental and numerical in-vestigation of the effect of macro-textured tool surfaces in hot stamping. Int. J. Mater. Form. **10**, 241–254 (2017)
7. Åkerström, P.: Modelling and Simulation of Hot Stamping. Doctoral dissertation of Luleå University of Technology, Luleå, Sweden (2006)
8. Olsson, T.: An LS-DYNA material model for simulations of hot stamping processes of ultra-high strength steels. In: 7th European LS-DYNA Conference (2009)
9. Lechler, J.: Beschreibung und Modellierung des Werkstoffverhaltens von presshärtbaren Bor-Manganstählen. Doctoral dissertation of der Friedrich-Alexander-Universität, Erlangen-Nürnberg, Germany (2009)
10. Hochholdinger, B.: Simulation des Presshärteprozesses und Vorhersage der mechanischen Bauteileigenschaften nach dem Härten. Doctoral dissertation of ETH Zürich, Switzerland (2012)
11. Verlinden, B., Driver, J., Samajdar, I., Doherty, R.: Thermo-Mechanical Processing of Metallic Materials, 1st edn. Elsevier (2007)
12. Morris, J.W.: The Influence of Grain Size on the Mechanical Properties of Steel. Lawrence Berkeley National Laboratory (2001)
13. Li, W.-L., Li, J.C.M.: The effect of grain size on fracture toughness. Philos. Mag. A **59**(6), 1245–1261 (1989)

Forming Limit of Dual Phase Steel: An Experimental and Numerical Investigation

Martina Müller[1(✉)], Niklas Fehlemann[3], Tim Herrig[1], David Lenz[3], Markus Könemann[3], Thomas Bergs[1,2], and Sebastian Münstermann[3]

[1] Laboratory for Machine Tools and Production Engineering, WZL of RWTH Aachen University, Campus-Boulevard 30, 52074 Aachen, Germany
martina.mueller@wzl.rwth-aachen.de

[2] Fraunhofer Institute for Production Technology IPT, Steinbachstrasse 17, 52074 Aachen, Germany

[3] Institute of Metal Forming IBF, RWTH Aachen University, Intzestrasse 10, 52074 Aachen, Germany

Abstract. Forming limit diagrams and forming limit curves are used to determine the failure limits of sheet metal materials by conducting a forming limit analysis. The aim is to enable the selection of suitable materials, geometries and process parameters at an early stage in the component development. In this way, engineering time and costs can be reduced. In addition to experimental forming limit analyses, different models and model approaches exist to numerically predict damage and failure. All have in common that calibration is necessary based on experimental investigations. The correspondingly determined parameters specify when ductile damage initiation begins, how it evolves, and how the material finally fails. Thus, the calibration and the parameters determined have a significant impact on the models' accuracy. The present work focuses on the numerical prediction of DP 800 dual phase steel in a Nakajima test using a modified Bai-Wierzbicki model. Besides the calibration influence of the damage model coming from a non-unique parameter selection is investigated. A sensitivity analysis of the experimentally determined parameters of the damage model is carried out with the aim of predicting the crack initiation with high accuracy and to determine the influence of the calibrated parameters, especially the range between the damage initiation locus, the failure locus and the damage evolution rate. The objective is to determine which damage model parameters possess a high influence on the failure prediction, in order to be able to make a risk assessment with regard to the prediction accuracy for later applications.

Keywords: Deep drawing · dual phase steel · damage modeling · forming limit curve

1 Introduction

The automotive industry is one of the main contributors to CO_2 emissions. With the European Union's goal of becoming a net-zero greenhouse gas emitting economy by 2050 [1], it is vital to find ways to significantly reduce emissions in this sector. One

K. Mocellin et al. (Eds.): ICTP 2023, LNME, pp. 184–193, 2024.
https://doi.org/10.1007/978-3-031-40920-2_20

effective approach is through the implementation of lightweight design. By using less material and being more resource-efficient, lightweight design can help to reduce CO_2 emissions throughout the entire life cycle of a vehicle. In this context, high-strength materials such as dual phase steel are frequently used. Dual phase steel combines high strength with good formability, thus providing a suitable choice for weight optimized components [2]. Deep drawing is widely applied in the manufacturing of vehicle bodies in the automotive industry. Since 30% of the total mass of a vehicle is accounted for the vehicle body, deep drawing offers a lever for saving CO_2 through the implementation of lightweight design measures [3]. However, the components used still must meet strict requirements in order to ensure the safety of passengers. This means that the methodologies utilized for product and process design must be reliable and safe. Numerical process analysis has become an indispensable part of product and process design, as it allows for the estimation of critical component areas and the optimization of the process to achieve greater stability and resilience. In order to identify critical areas of a deep drawn component, forming limit analyses are performed using forming limit curves (FLC) from forming limit diagrams (FLD). The procedure is nowadays standardized by DIN EN ISO 12004–2 [4] with the different approaches according to Nakajima [5] and Marciniak [6]. In terms of precise process modeling, however, there is a multitude of challenges to overcome in sheet metal forming. The models employed in this field are becoming increasingly intricate and sophisticated, with a need to accurately capture the interactions between the simultaneously changing geometry and material properties, as well as the location- and time-dependent conditions that exist within the effective area between the workpiece and the tool [7]. In the field of damage prediction and thus also for the identification of critical areas in the component, damage mechanics models are frequently used. Nowadays two types of models exist: coupled and uncoupled models [8]. Coupled models use a damage variable to reduce the yield potential based on the material's ductile damage during deformation. These models are often micromechanical in nature, such as the Gurson-Tvergaard-Needleman (GTN) model [9–11], or phenomenological, such as the Lemaitre model [12, 13]. Micromechanical models depict physical phenomena like void nucleation using a set of parameters requiring an extensive iteration process for parameter calibration. Phenomenological models, on the other hand, treat damage evolution in a macroscopic manner, utilizing mathematical formulation to describe the occurring effects. In contrast, uncoupled models describe the material behavior including fracture without taking damage into account. Notable examples of these include Johnson-Cook [14] and the Bai-Wierzbicki (BW) [15] model. Considering the BW model, further extensions to the model were realized. For instance, Lian et al. combined the strengths of uncoupled and coupled models into a modified Bai-Wierzbicki (MBW) model. The model now takes into account a locus for the damage initation point, which depends the stress triaxiality and the lode angle, and, in addition, a set of critical values for the damage variable, at which material fracture was postulated in the numerical simulation [16]. Further developments consider a locus for the fracture and non-proportional loading path [17] as well as anisotropic damage and fracture [18]. The MBW damage model is widely used these days because it is convenient to apply and calibrate, while accurately representing damage behaviour [19]. However, no investigations have yet focused on the explicit influence of the calibration parameters on the prediction of damage, especially the range between the damage initiation locus, the failure locus and the damage evolution rate. Therefore, the objective of the present work is to determine which damage model parameters possess a high influence on the failure prediction, in order to be able to make a risk assessment regarding the prediction accuracy for later applications. Overall, the goal is to provide a better understanding of the factors that influence the prediction accuracy of the damage model. This will enable to make more informed decisions when designing and manufacturing components by means of deep drawing.

2 Materials and Methods

2.1 Set up for Nakajima Tests

The Nakajima tests were carried out in accordance with DIN EN ISO 12004 [4]. Different geometries (Fig. 1a) of dual phase steel DP800 with galvanized surface were used to account for different strain paths. As lubrication, different layers of PTFE and PVC were used (Fig. 1b)).

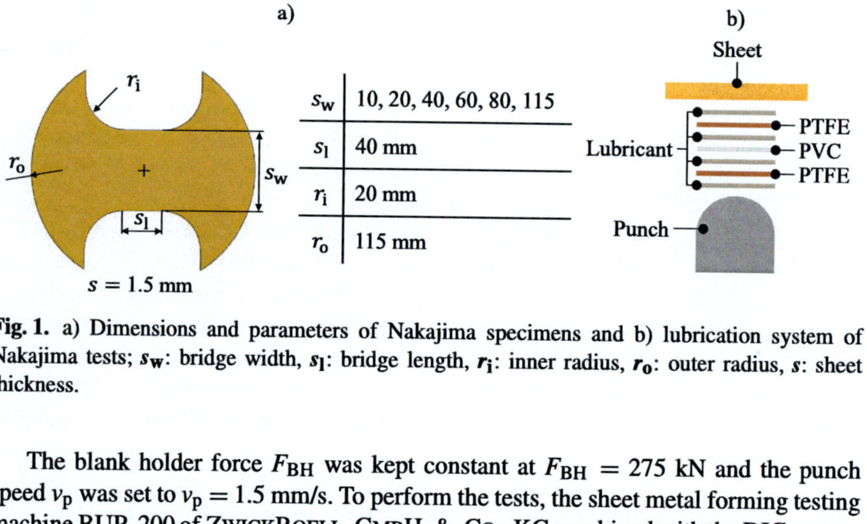

Fig. 1. a) Dimensions and parameters of Nakajima specimens and b) lubrication system of Nakajima tests; s_w: bridge width, s_l: bridge length, r_i: inner radius, r_o: outer radius, s: sheet thickness.

The blank holder force F_{BH} was kept constant at $F_{BH} = 275$ kN and the punch speed v_p was set to $v_p = 1.5$ mm/s. To perform the tests, the sheet metal forming testing machine BUP 200 of ZwickRoell GmbH & Co. KG combined with the DIC system Aramis SRX of Carl Zeiss GOM Metrology GmbH were utilized.

2.2 Process Model

The process was modeled using Abaqus/explicit 2021. A general contact with a uniform Coulomb friction coefficient of $\mu = 0.05$ was applied to model the specimen's interactions with the tools. The mesh parameters are listed in Table 1. A MBW damage model with an extension for non-proportional loading paths was used to estimate the damage behaviour. Details on the calibration for DP800 and the model itself can be found in the

Table 1. Mesh parameters and types of bodies for process model

Name	Mesh	Type of body
Sheet	C3D8R	Solid deformable body
Punch, blank holder, die	R3D4	Discrete rigid

publication work of Pütz et al. [19]. The implementation of the model was realized as a VUMAT subroutine. The mesh size of all models is 0.3 mm.

The MBW model considers the influence of stress triaxiality and lode angle parameter for different damage states, utilizing different loci to calculate the plastic equivalent stress at failure and the damage variable. The equations characterizing the loci are as follows:

$$f_{\text{di}}\left(\eta_{\text{avg}}, \overline{\theta}_{\text{avg}}\right) = \left[\frac{1}{2}(D_1 e^{-D_2\eta_{avg}} + D_5 e^{-D_6\eta_{avg}}) - D_3 e^{-D_4\eta_{avg}}\right]\overline{\theta}^2_{\text{avg}} + \frac{1}{2}(D_1 e^{-D_2\eta_{avg}} - D_5 e^{-D_6\eta_{avg}})\overline{\theta}_{\text{avg}} + D_3 e^{-D_4\eta_{avg}} \tag{1}$$

$$f_{\text{df}}\left(\eta_{\text{avg}}, \overline{\theta}_{\text{avg}}\right) = \left[\frac{1}{2}(F_1 e^{-F_2\eta_{avg}} + F_5 e^{-F_6\eta_{avg}}) - F_3 e^{-F_4\eta_{avg}}\right]\overline{\theta}^2_{\text{avg}} + \frac{1}{2}(F_1 e^{-F_2\eta_{avg}} - F_5 e^{-F_6\eta_{avg}})\overline{\theta}_{\text{avg}} + F_3 e^{-D_4\eta_{avg}} \tag{2}$$

In this case, D_1-D_6 and F_1-F_6 denote the damage initiation locus (DIL) and the ductile fracture locus (DFL), accordingly. If $D_1 = D_5$ and $D_2 = D_6$ respectively $F_1 = F_5$ and $F_2 = F_6$, both loci obtain a symmetry regarding the lode angle paramete $r\overline{\theta}$. As a result, only four parameters are necessary for the corresponding loci which was initially exploited in this investigation. Regarding the triaxiality η, according to previous investigations, a cut-off value is fixed at $\eta = -\frac{1}{3}$, assuming ductile damage is not triggered under pressure effects [17]. The simulations were performed with parameters taken from Pütz et al. [19].

2.3 Approach

In a first step, Nakajima tests are performed using different specimen geometries to generate a variance in strain and stress states. A digital image correlation (DIC) system is used to record the strain state over time and to evaluate the crack initiation point. In a second step, the forming limit curve is derived. After this, a calibrated MBW model is used to represent the Nakajima tests numerically. Subsequently, the numerical and experimental FLCs are compared and analyzed with respect to their deviations as a function of minor strain $\varepsilon_{\text{minor}}$ and major strain $\varepsilon_{\text{major}}$ as well as the corresponding stress states. Following this a sensitivity analysis of selected damage model parameters is carried out with the aim to determine the influence on the failure prediction. Based on the results, the damage model parameters with high influence are identified and a strategy is discussed and derived to improve the prediction quality of the failure prediction of the damage model for DP 800 dual phase steel in Nakajima tests and for later applications.

3 Results

Figure 2 shows the comparison between the experimentally determined and the simulated FLC. The shape of the two curves does not match well. Although the curves are in a similar order of magnitude in the range of a minor strain ε_{minor} of $\varepsilon_{minor} < -0.1$. Yet, the simulated FLC deviates strongly in the other ranges. In particular, there are deviations in the range of positive minor strains ε_{minor}, which leads to a clear underestimation of the time of failure in the area of biaxial stress by the simulation.

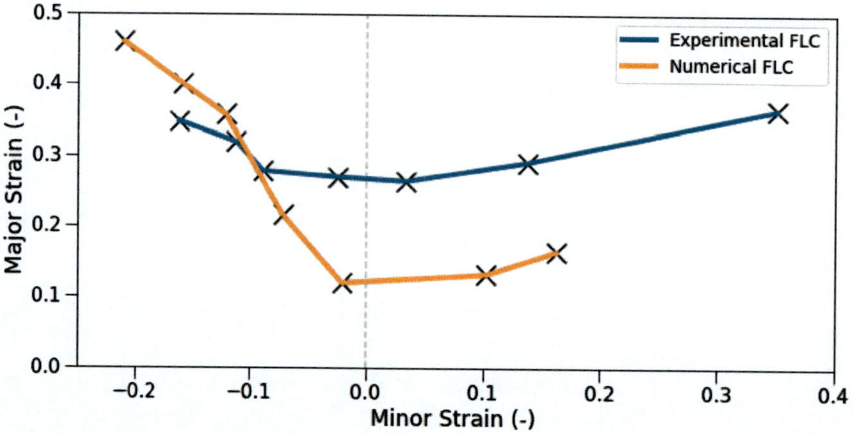

Fig. 2. Comparison between experimentally determined FLC and the simulated FLC using the parameters from [19]

Since in the MBW model the failure locus is spanned by stress triaxiality η and lode angle parameter $\overline{\theta}$, a sensitivity study was conducted with respect to the failure locus to systematically analyze the effect of damage model parameters on the numerically determined failure locus. The variation of the loci is listed in Table 2. It should be noted that the damage initiation locus was not changed. However, this does not affect the failure point noticeably.

The results of the change in the model parameter F_1, which controls the model's sensitivity with respect to lode angle parameter $\overline{\theta}$ (regarding failure) are depicted in Fig. 3. It becomes evident that these change of the failure locus leads in particular to an influence of the FLC in the range of a minor strain ε_{minor} of $\varepsilon_{minor} < -0.1$. By reducing the parameter F_1, an improved prediction quality is achieved in this area and the deviation from the experimentally determined FLC is slightly reduced. In the range from a minor strain ε_{minor} of $\varepsilon_{minor} > -0.1$, on the other hand, these changes show hardly any influence on accuracy of the failure prediction.

Table 2. Loci for the parameter study

	Original	Higher $\overline{\theta}$ dependence	Lower $\overline{\theta}$ dependence	Higher η dependence	Lower η dependence
F1	0.7	1.05	0.35	0.7	0.7
F2	1	1	1	0.5	1.5
F3	0.366	0.366	0.366	0.366	0.366
F4	2	2	2	1	3

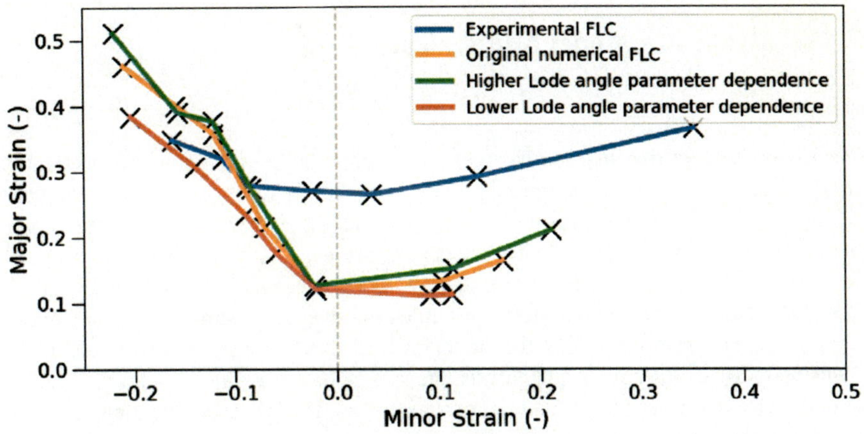

Fig. 3. Results of the parameter study with respect to the lode angle parameter

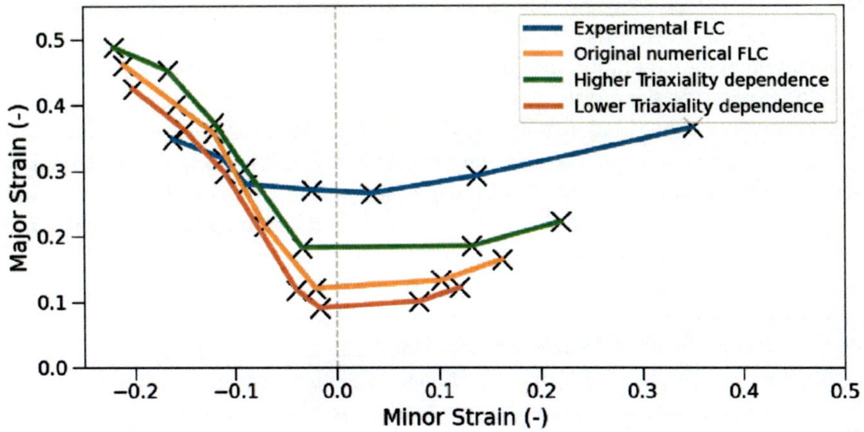

Fig. 4. Results of the parameter study with respect to the stress triaxiality

Figure 4 shows the results of the changes in the parameters F_2 and F_4 which control the model's sensitivity with respect to the stress triaxiality η. As can be seen from the courses, a higher dependency on the stress triaxiality leads to a deterioration of the failure prediction in ranges for a minor strain $\varepsilon_{\text{minor}}$ of $\varepsilon_{\text{minor}} < -0.1$. In ranges above a minor strain $\varepsilon_{\text{minor}}$ of $\varepsilon_{\text{minor}} > -0.1$, on the contrary, the deviation could be reduced. For a lower dependency, an improvement is shown in the range of a minor strain $\varepsilon_{\text{minor}}$ of $\varepsilon_{\text{minor}} < -0.1$. In contrast, the values show larger deviations for failure prediction in the range a minor strain $\varepsilon_{\text{minor}}$ of $\varepsilon_{\text{minor}} > -0.1$. In general, the variation in stress triaxiality η dependence leads to stronger variations in the FLC, which in turn has a stronger influence on the simulations.

4 Discussion and Model Adjustment

4.1 Discussion

The results indicate that making changes to the lode angle parameter $\overline{\theta}$ or triaxiality η dependence in the original calibrated model, whether increasing or reducing it, does not lead to an accurate prediction of the FLC. In fact, the deviations from the model compared to the experimentally generated FLC are especially noticeable in areas of increasing triaxiality η to biaxial stress states with a decreasing lode angle parameter $\overline{\theta}$. This leads to an underestimation of the possible tolerable major strains $\varepsilon_{\text{major}}$ and minor strains $\varepsilon_{\text{minor}}$ before failure. Thus, the model gives an overly conservative estimate in the downstream use in forming limit analysis, meaning that it may predict a lower limit for the maximum deformation the material can withstand. It is believed that the specimens used for calibration in Pütz et al. [19] are insufficient in representing all the stress states observed in the Nakajima experiments. Due to the selection of notched tensile test specimens, the stress states characterized by negative lode angle parameters $\overline{\theta}$ are not fully covered. Accordingly, the deviation in biaxial stress states is clearly visible in the FLC. In addition, stress triaxialities occur during biaxial tensile loading that are outside the spectrum of the calibration specimens. A resulting insufficient extrapolation can also have an influence on the results.

Furthermore, a threshold method was used for the model to determine the numerical damage onset, which has an influence on the plastic material behaviour, and the failure locus. For this purpose, the numerical damage onset was identified as the point at which the deviation between simulated and experimental force and displacement curves was obvious. An analogous approach was applied for the failure prediction [19]. Therefore, the numerical simulations were based on tensile tests from which the yield curve was derived and modified. However, it is assumed that not sufficiently consideration was given to the occurrence of damage development during the tensile tests.

After all, as is known from studies, damage in the form of voids occurs in particular under tensile load [20]. This damage, which must be explicitly distinguished from failure [21], already has an influence on the plastic material behavior. A corresponding derivation of the yield curve from the tensile test therefore means that the values are already susceptible to damage. So that the beginning of the damage initiation locus and the failure locus may not have been determined with sufficient accuracy leading to a deviation of the numerical and experimental FLC. In addition, the assumption of symmetry of the lode angle parameter $\overline{\theta}$ as the basis of the model to simplify the calibration is suspected to be insufficiently accurate and remains questionable.

4.2 Model Adjustment

To check if a lode angle parameter $\overline{\theta}$ asymmetry of the failure locus will improve the results, the VUMAT subroutine was modified to incorporate asymmetric behavior. For this, the original formulation of Eq. (1) was used, adding F_5 and F_6 parameter with individually assigned values without assumption of symmetry defining the locus. The results are depicted in Fig. 5. It becomes clear that the results show a smaller deviation between the numerical and experimental FLC. Furthermore, there is much less underestimation in regimes of biaxial tension. These first results sustain the presumption that incorporating asymmetry into the failure locus can lead to more accurate results in predicting the FLC. However, further investigations must show whether the asymmetry improves the quality of the model per se or whether a symmetrical locus can also lead to accurate numerical predicted results with a more elaborate calibration. For this purpose, mathematical optimization and a set of calibration samples covering a broader stress state can be used.

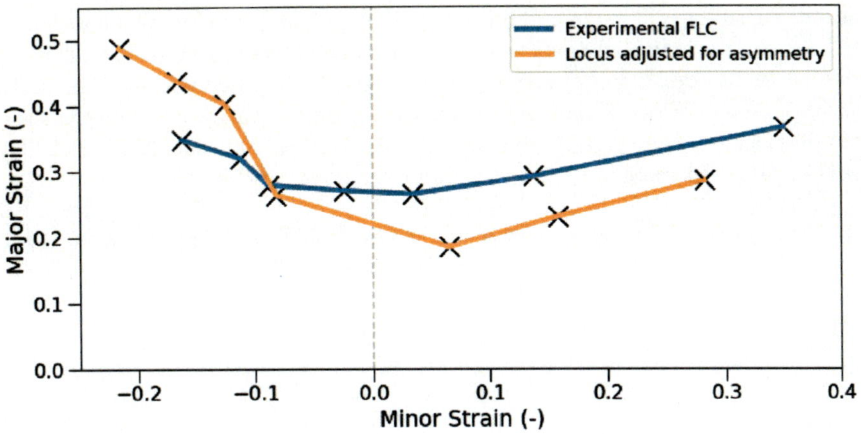

Fig. 5. Simulations with asymmetric locus; $F_1 = 0.7, F_2 = 0.5, F_3 = 0.366, F_4 = 1, F_5 = 1.4, F_6 = 1.0$

In case that one of the two possibilities enables the accurate prediction of the FLC, and subsequently the MBW model becomes available to accurately represent the common stress-strain state in sheet metal forming, it will be a significant improvement as a tool for both damage and failure prediction. Previous studies have primarily focused on failure prediction using uncoupled models or other failure criteria that neglected the influence of damage on the plastic behavior of the material, e.g. in [22–24], which limited their effectiveness. In contrast, the MBW model not only enables the prediction of failure but also the prediction of damage development, thereby allowing for the consideration of damage in product and process design. Incorporating damage prediction in the design process can help identify critical damage accumulation areas, which can lead to the development of more robust and reliable products. As a result, this method can unlock potential opportunities for product and process optimization.

5 Summary and Outlook

In summary, a model calibrated solely on different notched tensile tests with a symmetry assumption in lode angle parameter $\overline{\theta}$ is not able to simulate a FLC determined with Nakajima tests. Thereby, the model massively overestimates the sensitivity of the tests with respect to the stress state: the maximum difference in the major strain $\varepsilon_{\text{major}}$ in the test is about $\varepsilon_{\text{major}} = 0.08$, while in the simulation the largest difference is about $\varepsilon_{\text{major}} = 0.28$. Even a change in the damage model parameters in the form of triaxiality η and lode angle parameter $\overline{\theta}$ resulted in hardly any changes in the prediction quality of the FLC. Simultaneously, the assumption of symmetry of the lode angle parameter $\overline{\theta}$ as the basis of the model appears to be not accurate enough and remains questionable. In contrast, the assumption of a non-symmetric failure locus was able to improve the simulative prediction of the FLC.

For further investigation of the accuracy of the model, the scope of geometries of the specimen used for calibration will be extended. Therefore, biaxial tension tests which include negative lode angle parameters $\overline{\theta}$ will be considered. It is assumed that the coverage of the stress states can be improved in this way. Furthermore, the yield curve for the threshold method to determine the beginning of the damage initiation locus and the failure locus will be derived from upsetting tests instead of tensile tests which may lead to a further increase in the accuracy of the numerically predicted FLC.

Acknowledgements. This research was funded by Deutsche Forschungsgemeinschaft (DFG, German Research Foundation; Project number 278868966 – TRR 188, projects A06 and B05; Damage Controlled Forming Processes). Simulations were performed with computing resources granted by RWTH Aachen University under project rwth0744.

References

1. European Commision. https://climate.ec.europa.eu. Accessed 20 Feb 2023
2. Horvath, C.: Materials, Design and Manufacturing for Lightweight Vehicle. Woodhead Publishing, Oxford (2010)
3. Tisza, M., Czinege, I.: Comparative study of the application of steels and aluminium in lightweight production of automotive parts. Int. J. Lightweight Mater. Manuf. **1**(4), 229–238 (2018)
4. ISO 12004 Metallic materials - Sheet and strip - Determination of forming-limit curves (2021)
5. Nakajima, K., Kikuma, T., Hasuka, K.: Study on the formability of steel sheets. Yamata Technical report, pp. 8517–8530 (1968)
6. Marciniak, Z.: Stability of composite shell–type structures. Arciwum mechaniki sto-sowanej, pp. 253–428 (1965)

7. Liewald, M., et al.: Perspectives on data-driven models and its potentials in metal forming and blanking technologies. Prod. Eng. Res. Devel. **16**, 1–19 (2022). https://doi.org/10.1007/s11740-022-01115-0
8. Besson, J.: Continuum models of ductile fracture: a review. Int. J. Damage Mech **19**(1), 3–52 (2010)
9. Gurson, A.L.: Continuum theory of ductile rupture by void nucleation and growth. J. Eng. Mater. Technol. **99**(1), 2–15 (1977)
10. Tvergaard, V., Needleman, A.: Analysis of the cup-cone fracture in a round tensile bar. Acta Metall. **32**(1), 157–169 (1984)
11. Tvergaard, V.: Influence of voids on shear band instabilities under plane strain conditions. Int. J. Fract. **17**(4), 389–407 (1981)
12. Lemaitre, J.: A continuous damage mechanics model for ductile fracture. J. Eng. Mater. Technol. **107**(1), 83–89 (1985)
13. Lemaitre, J.: A Course on Damage Mechanics. Springer, Berlin, Heidelberg (1996). https://doi.org/10.1007/978-3-642-18255-6
14. Johnson, G.R., Cook, W.H.: Fracture characteristics of three metals subjected to various strains, strain rates, temperatures and pressures. Eng. Fract. Mech. **21**(1), 31–48 (1985)
15. Bai, Y., Wierzbicki, T.: A new model of metal plasticity and fracture with pressure and Lode dependence. Int. J. Plast **24**(6), 1071–1096 (2008)
16. Lian, J., Sharaf, M., Archie, F., Münstermann, S.: A hybrid approach for modelling of plasticity and failure behaviour of advanced high-strength steel sheets. Int. J. Damage Mech. **22**(2), 188–218 (2013)
17. Wu, B., Li, X., Di, Y., Brinnel, V., Lian, J., Münstermann, S.: Extension of the modified Bai-Wierzbicki model for predicting ductile fracture under complex loading conditions. Fatigue Fract. Eng. Mater. Struct. **40**(12), 2152–2168 (2017)
18. Shen, F., Münstermann, S., Lian, J.: Investigation on the ductile fracture of high-strength pipeline steels using a partial anisotropic damage mechanics model. Eng. Fract. Mech. **227**, 106900 (2020). https://doi.org/10.1016/j.engfracmech.2020.106900
19. Pütz, F., Shen, F., Könemann, M., Münstermann, S.: The differences of damage initiation and accumulation of DP steels: a numerical and experimental analysis. Int. J. Fract. **226**(1), 1–15 (2020). https://doi.org/10.1007/s10704-020-00457-z
20. Voyiadjis, G.: Handbook of Damage Mechanics: Nano to Macro Scale for Materials and Structures. Springer, New York (2015)
21. Müller, M., Weiser, F., Herrig, T., Bergs, T.: Numerical prediction of the influence of process parameters and process set-up on damage evolution during deep drawing of rectangular cups. Eng. Proc. **26**(6), 1–10 (2022)
22. Lumelskyj, D, Rojek, J, Tkocz, M.: Numerical simulations of Nakazima formability tests with prediction of failure T (2022). J. Techn. Sci. **60**(3), Eng. Proc. **26**(6), 184–194 (2015)
23. Karajibani, E., Hashemi, R., Sedighi, M.: Forming limit diagram of aluminum-copper two-layer sheets: numerical simulations and experimental verifications. Int. J. Adv. Manuf. Technol. **90**(9–12), 2713–2722 (2016). https://doi.org/10.1007/s00170-016-9585-1
24. Lou, Y., Huh, H.: New ductile fracture criterion for prediction of fracture forming limit diagrams of sheet metals. Int. J. Solids Struct. **49**(25), 3605–3615 (2012)

Development of the One-Step Hybrid Forming Process to Produce an Al-GMT-Hybrid Crash Management System

Amir Hajdarevic(✉), Saarvesh Jayakumar, Lorenz Stolz, and Xiangfan Fang

Institute of Automotive Lightweight Design, University of Siegen, Siegen, Germany
amir.hajdarevic@uni-siegen.de

Abstract. The newly developed hybrid forming process allows producing steel-long fiber reinforced plastic hybrid components in a single process step. The hybrid components produced in this way exhibit good failsafe behavior and enormous energy absorption potential combined with high structural integrity. They are also cost-efficient and environmentally friendly. To explain the field of application, this process was further investigated to produce hybrid components with aluminum-LFT and –PP-GMT (glass mat thermoplastic). Therefore, it is suitable for developing automotive structural components, e.g., crash management systems (CMS). To develop a lightweight hybrid-CMS with approx. 15% weight reduction compared to a full aluminum (Al) reference, different requirements must be fulfilled by Al and GMT. To achieve this, firstly, tool concept for axial crash and 3-point bending components were developed. Corresponding components were produced and tested for their performance.

Keywords: Hybrid forming · Multi-material design · Crash management system · Lightweight design

1 Introduction

In recent years, the global goal to reduce CO2 emissions has become a severe concern for almost every industrial sector. For example, in the European Union, road transport is responsible for 21% of the total carbon dioxide (CO2) emissions, the main greenhouse gas [2]. Consequently, the lightweight design of car bodies is a crucial goal in the automotive industry to reduce gasoline consumption or extend the driving range of pure battery electric vehicle. For ICE vehicles (internal combustion engines), a weight reduction of 100 kg leads to a gasoline consumption reduction of 0,3–0,5 l per 100 km [3, 4]. These findings hold roughly also for BEV (battery electric vehicles), where weight reduction of 100 kg means a saving in energy consumption of 0,6 kWh [5]. Therefore, besides lightweight metals like magnesium or aluminum (Al), using fiber-reinforced plastics (FRP) becomes increasingly important to realize a lightweight design.

To produce hybrid components, different materials are combined using different production methods, such as sheet forming, die casting, and extrusion. According to the

K. Mocellin et al. (Eds.): ICTP 2023, LNME, pp. 194–205, 2024.
https://doi.org/10.1007/978-3-031-40920-2_21

current state of the art, to produce hybrid components made of short FRP and sheet metals, either post-molding assembly (PMA) or in-mold assembly (IMA) can be applied [6–8]. In a PMA process, plastic and metal components are manufactured separately and joined together in a third operation. However, these three separate process steps increase the manufacturing cost and cycle time. To avoid these drawbacks, metal-FRP hybrid components are commonly manufactured using the IMA process. In this process, preformed sheet metal components are inserted into an injection mold and over-molded by FRPs. Another two-step hybrid process was developed based on a similar principle, which can be found in [9]. Firstly, a hot-formed B-Pillar was produced by deep drawing and then reinforced with the organic sheet and long fiber thermoplastics (LFT) by impact extrusion process. Additionally, the metal part used for this process can be pre-coated with the bonding agent to create an adhesive bonding. However, these multistage processes are always associated with additional costs for tools and processes, as well as longer cycle times. Therefore, such techniques are scarcely used in large-scale production.

Furthermore, in injection molding, the long fibers of thermoplastic materials are cut or broken into short fibers with fiber lengths much less than 2 mm [17]. Compared to injection molding, compression molding can realize much longer fiber length in the final component (up to 20 mm) [18]. The fiber length of FRP has a significant influence on the strength and impact resistance of the final parts. To better utilize the ability of FRP in crash load cases, a fiber length of more than 50 mm is required [10]. For these reasons, glass mat thermoplastics (GMT) materials with longer fiber lengths were investigated in this work.

Although FRPs have the advantages of low density and high specific weight properties, the mechanical properties of metals, like stiffness and energy absorption capabilities, can not be overlooked. The ductile failure behavior of metals is also of great importance in automotive applications like chassis development. Metals show crack formations and stabile crack growth during cyclic loading, so a fail-safe behavior is guaranteed.

In this work, two basic Al-GMT profiles were produced through a newly developed tool concept with one-step hybrid forming process. The developed profiles were tested for their corresponding load cases. These results served as a base for the development of the final hybrid crash management system.

2 Principle of the One-Step Hybrid Forming Process

2.1 Process Description

To overcome all these above mentioned issues, a novel single-stage process for manufacturing metal-FRP components, "hybrid forming," was developed in a previous study [1]. Compared to multi-stage processes, the processing and forming of two different materials (for example, sheet metal and LFT/GMT) can be done simultaneously in one step using a single forming die in hybrid forming. Further, hybrid forming uses compression molding as the forming process for the long fiber-reinforced thermoplastics in combination with the deep drawing of sheet metal in the same die and process step. Thereby, the molten LFT/GMT material is used as a forming medium for the final form of the metal sheet, which is similar to the methodology of a hydro-mechanical forming technique [11]. Finally, bonding agents are used to join the sheet metal and FRPs. In this way, hybrid components with optimized mechanical properties and optimal lightweight potential can be manufactured.

In contrast to injection molding, in hybrid forming the fiber lengths remain much longer (up to 50 mm) in the final part. Furthermore, due to the processing temperature of the polymer, 200 °C PP GMT or 280 °C PA LFT, the formability of the metal sheet is also influenced.

The manufacturing steps of the hybrid forming process are illustrated in Fig. 1. First, Steel or Al metal sheets used for this process are pre-coated with a bonding agent (e.g. VESTAMELT by Evonik) [12]. The thermoplastics materials (usually polyamide PA6 or polypropylene PP) with long glass fiber are prepared in a twin screw compounder. In addition to this direct processing, pre-fabricated products like GMT can be reheated in an IR heater to a semi-molten state before transferring them to the press tool. The pre-coated metal sheets are preheated to 200 °C to activate the bonding agent. Eventually, the steel or aluminum sheet, together with the molten extrudate is transferred into the press tool, which is controlled by a hydraulic press.

Subsequently, the press tool closes with the pressing force, whereby the punch presses and forms the molted LFT/GMT, which simultaneously forms the sheet metal with hydrostatic pressure. Depending on the punch design, ribs for reinforcement are created through the cavities at the same time. In the last step, after approx. 20 s die closing time, which makes a major part of the cycle time, LFT/GMT is firmly bonded to the steel or Al, and the part can be removed from the press.

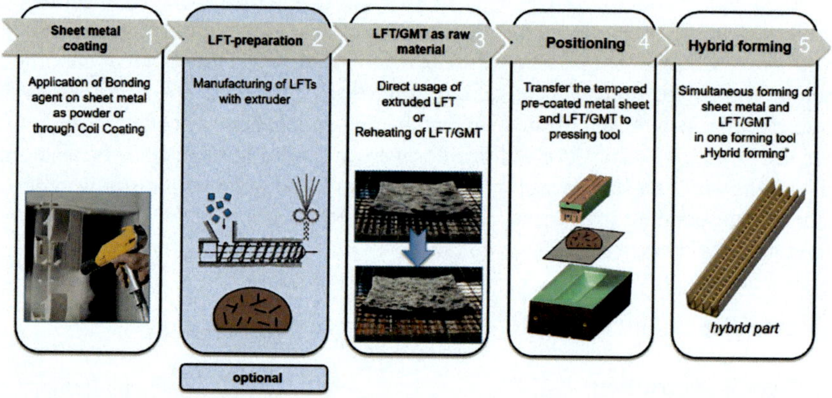

Fig. 1. Process steps of novel manufacturing technique "hybrid forming" combining metal-forming and LFT/GMT compression molding

In addition to the manufacturing of hybrid components with very long fiber lengths (GMT), another advantage of this process is that LFT/GMTs are fully covered by sheet metal. This ensures that crash elements of a vehicle, for example, a crash box absorb the energy without allowing fractured LFT/GMT fragments to pollute the environment. Therefore, the good energy absorption of LFT/GMT can be utilized effectively.

2.2 Tool Concept for Hybrid Forming Process

Various tool technologies are available for manufacturing FRP components. A distinction can be made between open and closed tools [13]. One major issue in the process development of hybrid forming is to work out tool concepts. A challenge in hybrid forming tool development is to realize the simultaneous metal forming and compression molding by avoiding thermoplastic melt leakage or rupture in the metal sheet. Various tool and sealing concepts for an open (U-profile), as well as a closed profile (tube geometry) were developed and investigated in previous studies [1, 14]. The hybrid forming tool consists basically of a pure sheet metal forming tool and were developed for use in a hydraulic metal forming press.

New sealing strategies must be developed for a crossmember or crash box that is completely open. On this basis, tool and sealing concepts for the hybrid forming of aluminum and GMT for a homogeneous U-profile (see Fig. 2d) have been developed in this work. The tool consists of 3 main parts: die, punch, and blank holder. In contrast to [1, 14], no high level of blank holder force for the sealing element is needed by using the new tool concept. Furthermore, the sealing elements do not move separately from the punch and are thus a part of the punch. The die is located in the movable upper tool half, and the punch is in the fixed lower tool half. The blank holder plate creates a free volume for the GMT before the forming tool completely closes and thereby builds an enclosed space. Two lateral ledges realize the lateral sealing against GMT melt leakage in the mold in punch. The sealing is achieved by specifically adjusting the drawing gap between the punch and die (drawing gap $\approx$ sheet thickness, see Fig. 2c). These lateral sealing elements are replaceable and can be easily disassembled. In addition, the concept can be used to achieve different sheet thicknesses and makes it possible to obtain plastic-free areas at the edge of the metal sheets, which can be used for joining to other components.

Furthermore, this sealing concept consists at the end of the punch of a 2 mm step (thickness of GMT) as a barrier for the plastic melt (see Fig. 2b). A gap of 1,5 mm between the punch and die, which equals exactly the metal sheet thickness, makes the metal sheet itself the sealing element. For different sheet metal thicknesses, a gap (the width of the sealing element) must be changed accordingly.

3 Materials and Sheet Metal Formability

In this work, the Al-alloys AA5182 were chosen for the crash box and AA6451 for the cross member due to their properties. AA5182 has high ductility and thus energy absorption during axial crashing, while AA6451 shows high strength for bending loads. As fiber-reinforced thermoplastic investigated in this work, a glass mat thermoplastic material PP-GMT (40 vol % of glass fibers) was chosen. This material has a complex architecture comprising both continuous fiber (glass mat layers and plain woven layers) and discontinuous fibers in a polypropylene matrix. This material has superior mechanical properties and a very homogeneous glass fiber distribution which is relevant for crash applications. For bonding, the bonding agent VESTAMELT (Evonik) was applied to aluminum sheets before manufacturing.

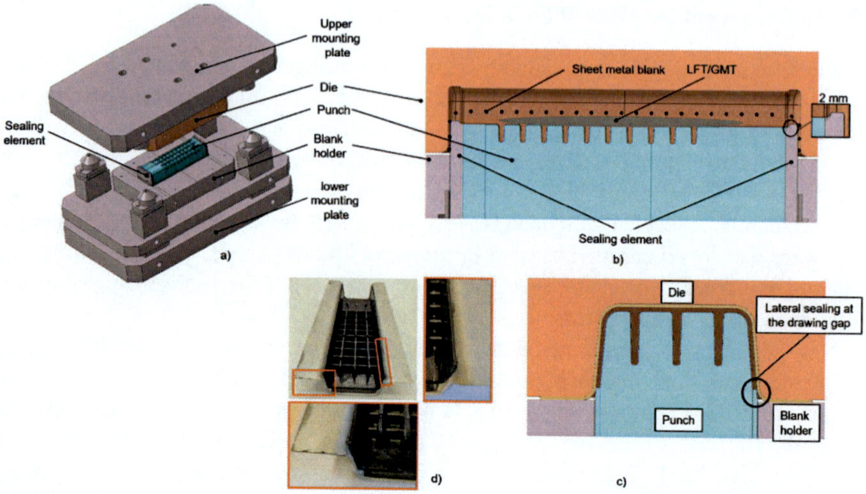

Fig. 2. a) Tool concept for homogeneous open-ended U-profile, b) Hybrid forming sealing concept at the end of the punch, c) Lateral sealing edge at the drawing gap, d) Sealing concepts "Hybrid crash box".

Since the aluminum sheets have to be preheated to the activation temperature of the bonding agent VESTAMELT (190–200 °C) and formed together with 200 °C PP-GMT, the formability of aluminum alloys at different temperatures was determined. The Nakajima tests were carried out according to DIN EN ISO 12004-2 to create the forming limit diagram [15].

Figure 3 shows the forming limit diagrams (FLDs) of AA5182 and AA6451 determined at RT, 100 and 200 °C. It was found that the temperature dependencies of these two aluminum alloys are very different. For AA5182 alloy, no aging effect is observed, as temperature increased, limit strains increased in all the modes of deformation. The formability in the warm working temperature range was found to be much higher (up to 100% at 200 °C) than at room temperature due to thermal softening and enhanced ductility of AA5182 (see Fig. 3b). Compared to AA5182 alloy, the formability of AA6451 T4 alloy exhibited irregular changes at higher temperatures due to the aging effect. Furthermore, with the help of forming limit diagrams (FLD), the formability and manufacturing of real hybrid components can be predicted.

4 Experimental Drop Tower Testing of Al-GMT Hybrid Parts

4.1 Experiment Setup and Procedure

Regarding to the development of the hybrid CMS, first, the process reproducibility must be validated for simpler components. Using the sealing concept (see Fig. 2d), two different components consisting of the AA6451/AA5182 with 1,5 mm sheet thickness and PP-GMT were developed and hybrid formed. These generic hybrid components can be considered as the state-of-the-art design for body-in-white structures exposed to high

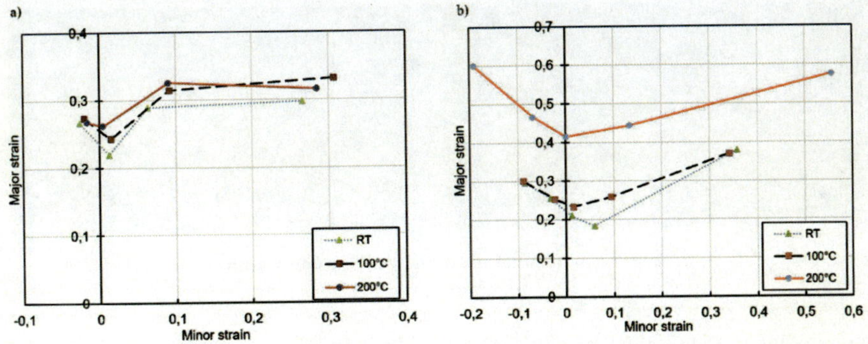

Fig. 3. Forming limit diagrams determined at three different temperatures for a) AA6451 and b) AA5182

bending loads and are the basis for CMS development in this work. To investigate the failure behavior and mechanical properties of hybrid structures, they were tested under crash loading in a drop tower.

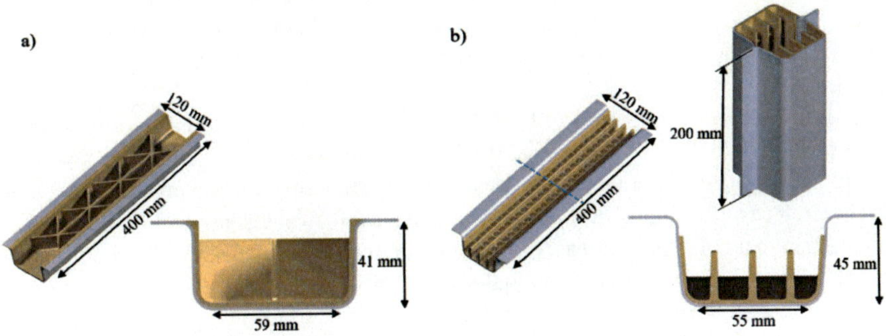

Fig. 4. Component dimension: a) 3-point bending, b) axial crash

Figure 4 shows the detailed geometrical dimensions and optimized rib design of the Al-GMT hybrid profile for both 3-point bending and axial crash profiles. The internal GMT-rib design and dimension of the hybrid components were obtained after the optimization, which showed superior specific energy absorption for each load case. Therefore, the rib patterns are different. The geometry of the generic crossmember is optimized for the 3-point bending load cases and consists of an X cross ribs pattern with cylinders connecting each pair of X cross ribs (see Fig. 4a). This type of ribs exhibits high stiffness against bending and torsion loading. To ensure that no failure of the bonding agent occurs, the hybrid-formed bending profile was closed with a 1,5 mm Al sheet metal using spot welding.

The test setup of the 3-point bending drop tower test and sequential illustration of failure are shown in Fig. 5. Compared to the axial crash, the impactor used for the 3-point bending drop tower test was a hemispherical head with a radius of 25 mm. The

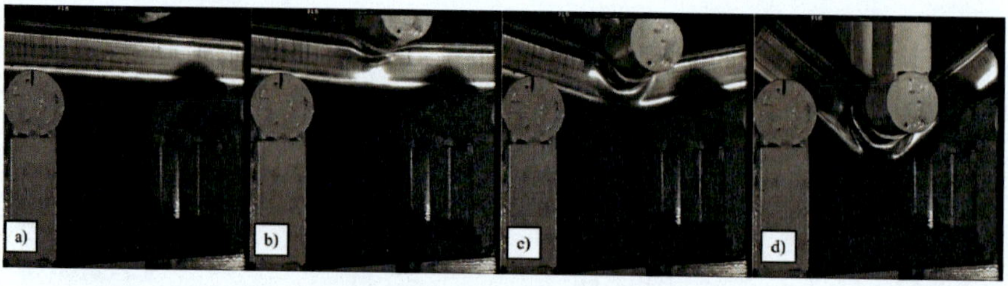

Fig. 5. Sequential illustration of 3-point bending

components were not fixed, which allowed sliding along the cylindrical supporting pins. The velocity used for dynamic 3-point bending is similar to the low-speed Research Council for Automobile Repairs (RCAR) crash test. With a falling mass of 200 kg and an impact velocity of 4,4 m/s, a kinetic energy of 2 kJ was used in the tests. In the experiments, the impact with the falling mass led to a failure of the adhesive between Al and GMT only in the contact region where the impactor meets the hybrid profile.

These hybrid profiles were also manufactured for the axial crash load cases to determine the deformation and failure behaviors of Al-GMT hybrid profiles during high-speed loading. The axial crash profiles consist of an orthogonal ribs pattern, which enables high energy absorption at a nearly constant force level. In addition, the length of the profile for axial crushing is shorter than the bending profile due to profile stability issues during crushing.

The axial crash components of Al-GMT were joined by riveting with a resulting crash box length of 200 mm (see Fig. 4b). To position the samples in the drop tower, the hybrid parts described above were centrally fixed. The impactor is a massive steel plate that is screwed to the falling drop tower head. The drop tower tests were conducted with a total falling mass of 375 kg and a striking velocity of 6,2 m/s, resulting in a kinetic energy of 7kJ. Figure 6 shows the experimental setup, consisting of a falling mass, hybrid part, and fixture. Furthermore, to initiate progressive failure of GMT and reduce the initial peak force, a local weakening in the form of a 45° chamfer trigger was employed in the hybrid components as shown in Fig. 6B. The reaction force is measured by a load cell below the base plate. Figure 7 shows a sequential illustration of axial crash component failure, and the drop tower test results are presented in the force-displacement diagram (see Fig. 8).

4.2 Experiment Results

The reinforcement of the aluminum components with GMT by hybrid forming significantly increases the absorbed energy under dynamic loading.

In order to analyze the exact course of the crash test and thus to be able to capture and interpret the individual failure mechanisms, the impact process of the Al-GMT profile is shown in Fig. 7 using a series of images. The first image shows the profile when the impactor makes first contact time t = 0 ms. Initial buckling of the profile occurs at a force of approx.110 kN (t = 1 ms). Due to the reduced cross-section area (trigger), the force was transmitted first in the outermost layer. The 45° chamfer trigger ensures

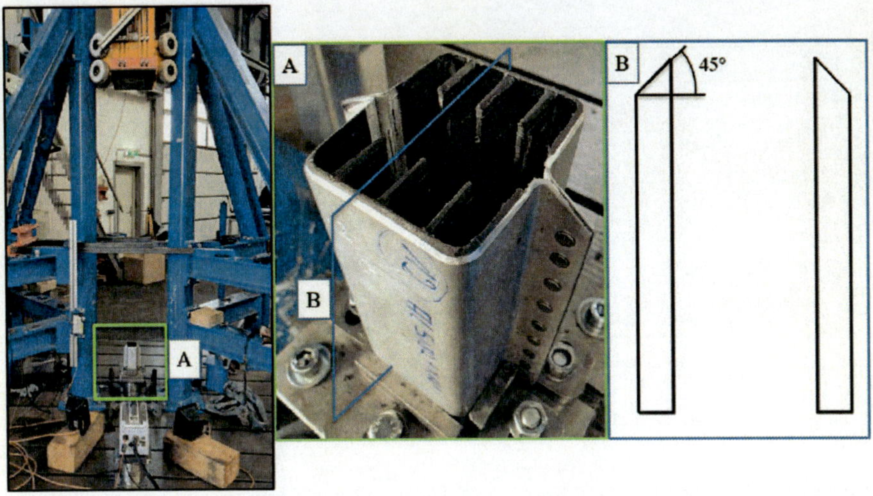

Fig. 6. Setup for axial drop tower (left), axial crash component (A) and trigger form (B)

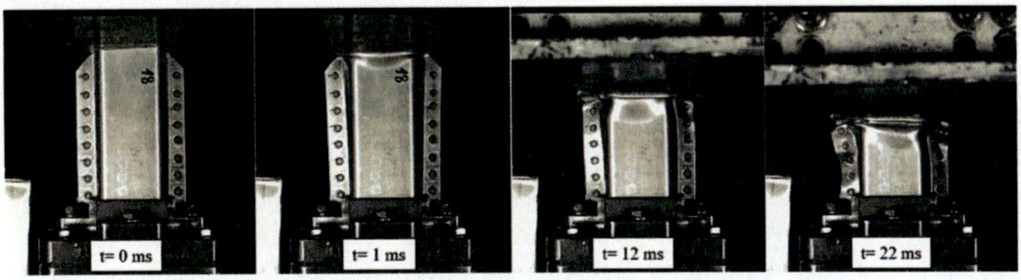

Fig. 7. Sequential illustration of component failure

that the entire component surface is in contact after approx. 23 mm of crosshead travel, thus reaching the maximum force of 130 kN. The sharp drop in force to approx. 70 kN results from the initial buckling of the aluminum. At this time no crack of the structure can be detected. Furthermore, the force increases to a level of 120 kN. At this time ($t = 12$ ms), the second fold of the profile is formed. After $t = 22$ ms the energy is completely absorbed and the impactor is at standstill at this time.

Figure 8 shows the force-displacement curves and deformation patterns of 3-point bending and axial crash tests used to validate the process reproducibility. As predicted, the crash box of a vehicle absorbs energy efficiently during impact by deforming the aluminum and crushing the GMT without the sharp-edge GMT fragments reaching the outside environment. An aluminum crash box produces folds through progressive failure during axial crushing. The presence of GMT reinforcement combines with aluminum folding during testing to provide a superior mean force level providing high energy absorption and hence reduced intrusion. The detailed numerical modelling and comparisons of axial crash box with experiments are explained in a separate paper [19].

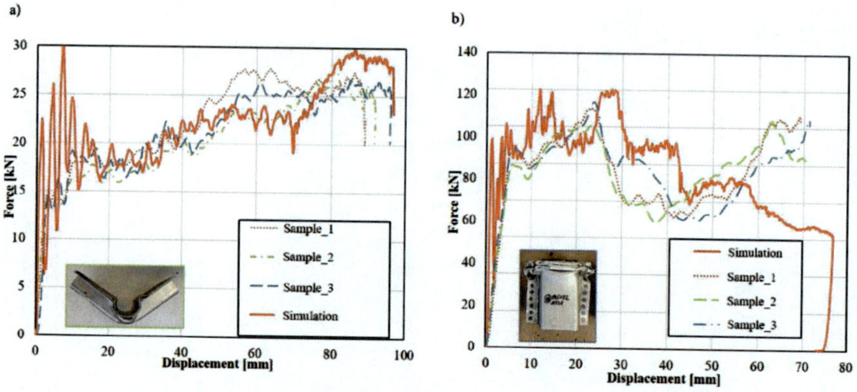

Fig. 8. Drop tower test results: a) 3-point bending, b) axial crash

In this work, the numerical results are only compared with the experiments for validation. Moreover, a positive fail-safe behavior of the hybrid components is achieved both in experiments and simulations, which serves the major requirements for automotive structural parts in drivetrain and body-in-white applications.

From the 3-point bending test results, it can be concluded that structures manufactured with hybrid forming show good energy absorption. The reinforcements of the metal structures led to a steeper increase of the force at the beginning of the experiment until the first wrinkling or buckling of the flanks occurs. The initial peak level reaches 15 kN at a displacement of 4 mm. The post-peak behavior of this 3-point bending results occurs after 4 mm displacement. Subsequently, the sidewalls of the component start to buckle at about 43 mm, due which force level increases to approx. 25 kN. The energy is completely absorbed at a displacement of approx. 90 mm. In the middle area of the 3-point bending component, a slight detachment of the GMT could be seen in the bottom area, but no catastrophic failure behavior of Al-GMT hybrid profile could be detected. Due to the ductile support of the aluminum hat profile, complete breakdown is prevented.

These test results were later used for the validation of simulation results [16]. This workflow laid the foundation for designing and developing complex automotive structures such as CMS.

5 Application of Hybrid Forming Process for Al-GMT-CMS

After studies on the basic profiles concerning simulation and process reproducibility, the component design methodology described above was implemented to design the Al-GMT hybrid CMS. This hybrid CMS consisting of a crossmember and crash box was developed for the design space of a compact class vehicle. Different rib thicknesses and designs of the crash box and the crossmember were investigated.

The following load case simulations were carried out for the development of hybrid CMS: RCAR low-speed structural test, RCAR bumper test, and towing simulation. Since the simulation work is beyond the scope of this topic, the numerical investigations carried out with the LS-DYNA solver will be reported in a separate paper. The hybrid

CMS optimized and developed from FE simulations for different above mentioned load cases are shown below in Fig. 9.

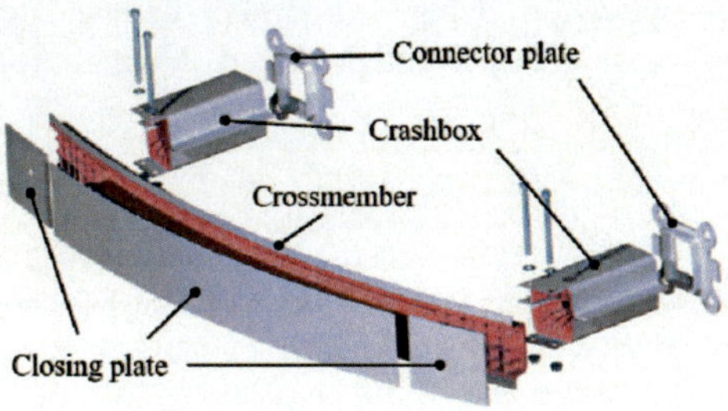

Fig. 9. CAD assembly of Al-GMT hybrid CMS

Based on the simulation results and the mentioned hybrid forming process, crossmember and crash box were manufactured and shown below in Fig. 10. Compared to reference aluminum CMS, the CMS developed through hybrid forming could achieve a significant weight reduction of 15% with comparable energy absorption. Further weight reduction is possible through employing e.g. laser or friction stir welding process. These welding techniques enable the connection of the two crash halves and attachement of the crash box with the crossmember by eliminating the flange regions and additional components like screws.

The future steps of this work are the assembly of hybrid CMS and carrying out the necessary crash tests, RCAR low-speed structure crash test, RCAR bumper test and finally, verifying the simulation results.

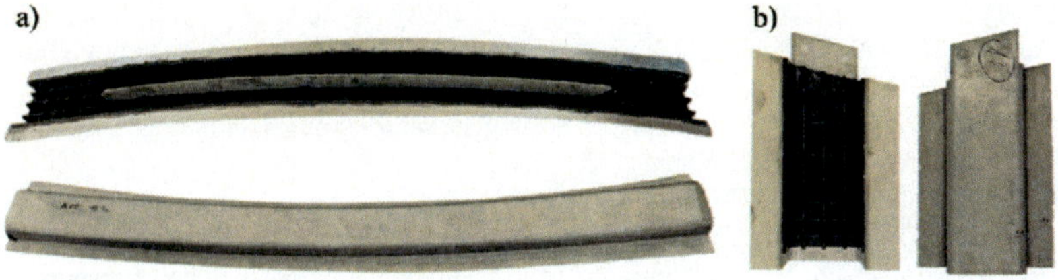

Fig. 10. a) Crossmember and b) Crashbox, produced by hybrid forming of Al-GMT.

6 Summary and Conclusions

The increasing use of multi-material structures in automotive lightweight design for vehicle body and chassis structures requires efficient processes. Most of the established processes for hybrid parts incorporate multi-stage manufacturing steps, in which metal and FRP parts are produced separately and joined together in a subsequent process step. In this work, the new manufacturing technique, hybrid forming for economical manufacturing of metal-FRP hybrid structural parts, was successfully realized. Hybrid Al-GMT hat profiles were produced using this single-stage hybrid forming process. These profiles were subjected to bending and axial crash loads. They were successfully deformed with high energy absorption and good fail-safe behavior, which is mandatory for automotive structure parts in body application. Using the hybrid process and developed die concept, a hybrid Al-GMT CMS was developed and produced. A weight reduction of up to 15% was achieved.

References

1. Fang, X., Kloska, T.: Hybrid forming of sheet metals with long fiber-reinforced thermoplastics (LFT) by a combined deep drawing and compression molding process. Int.J. Mater. Form. **13**(4), 561–575 (2019). https://doi.org/10.1007/s12289-019-01493-4
2. European Commission: Road transport: Reducing CO2 emissions from vehicles, http://ec.europa.eu/clima/policies/transport/vehicles_en. Accessed 08 Feb 2023
3. Klein, B.: Leichtbau-Konstruktion, Berechnungsgrundlagen und Gestaltung, 9th edn. Vieweg-Verlag, Wiesbaden (2007)
4. Mende, O.: Entwicklung des Fahrzeugunterbodens vor dem Hintergrund CO2-Effizienz. In: Proceedings, pp. 67–81. Springer, Wiesbaden (2016).https://doi.org/10.1007/978-3-658-14144-8_5
5. Weiss, M., Cloos, K.C., Helmers, E.: Energy efficiency trade-offs in small to large electric vehicles. Environ. Sci. Eur. **32**(1), 1–17 (2020). https://doi.org/10.1186/s12302-020-00307-8
6. Modler, N., Adam, F., Maaß, J.: Intrinsic lightweight steel-composite hybrids for structural components. In: Material Science Forum, pp. 401–408. Trans Tech Publication Ltd., Switzerland (2015)
7. Ehrenstein, G.W., Amesöder, S., FernandezDiaz, L., Niemann, H., Deventer, R.: Werkstoff- und prozessoptimierte Herstellung flächiger Kunststoff-Kunststoff und Kunststoff-Metall-Verbundbauteile. In: Geiger, M., Ehrenstein, G.W. (eds.) Robuste, verkürzte Prozessketten für flächige Leichtbauteile, pp. 150–177. Meisenbach Verlag, Bamberg (2003)
8. Ehrenstein, G.: Handbuch Kunstsoff-Verbindungstechnik. Carl Hanser Verlag, München (2004)
9. Maaß, J.: Höchstbelastete PKW-Karosseriestrukturen in Multi-Material-Design, Dresdener Transferbrief 3 (2015)
10. Bürkel, E., Sieverding, M., Mitzler, J.: Spritzgiessverarbeitung von langfaser-verstärktem. Carl Hanser Verlag, München, pp. 47–50 (2003)
11. Siegert, K., Häussermann, K., Lösch, B., Rieger, R.: Recent developments in hydroforming technology. J. Mater. Process. Technol. **98**(2000), 251–258 (2000)
12. EVONIK Homepage. https://corporate.evonik.com/en/media/press-releases/resource-efficiency-archive/production-of-hybrid-components-in-an-automated-process-with-vestamelt-hylink-100247.html. Accessed 10 Feb 2023
13. Lauter, C.: Entwicklung und Herstellung von Hybridbauteilen aus Metall und Faserverbundkunststoffen für den Leichtbau im Automobil. Ph.D. thesis. Automotive Lightweight Design (LiA), University of Paderborn (2014)
14. Fan,g X.F., Kloska, T.: Hybridforming: a combined deep drawing and compression molding process for forming metal sheets with thermoplastics (LFT) to hybrid parts. In: 3rd International Conference on Hybrid Materials and Structures, pp. 18–19, Bremen (2018)
15. DIN EN ISO 12004, Metallic materials Sheet and strip - Determination of forming-limit curves – Part 2: Determination of forming-limit curves in the laboratory
16. Jayakumar, S., Stolz, L., Anand, S., Hajdarevic, A., Fang, X.F.: FE crash modeling of aluminum-FRP hybrid components manufactured by a hybrid forming process. Key Eng. Mater. **926**, 2050–2059 (2022). https://doi.org/10.4028/p-w0hue6

17. Schemme, M.: LFT – development status and perspectives. Plast. Addit. Compound. **10**(02), 38–43 (2008). https://doi.org/10.1016/S1464-391X(08)70060-6
18. Tatara, R.: 14 – compression molding. In: Applied Plastics Engineering Handbook, 2nd edn., pp. 291–320. Elsevier, Amsterdam (2017)
19. Jayakumar, S., Hajdarevic, A., Anand, S., Fang, X.F.: Experimental and FE analyses of the crushing and bending behaviors of GMT and hybrid-formed Al-GMT structures. Thin-Walled Struct. **186**, 110648 (2023)

Simplified Calibration of an Academic Refined Model for Industrial Use

W. Liegard[1,2(✉)], L. Charleux[1], E. Roux[1], P. Balland[1], and L. Tabourot[1]

[1] Univ. Savoie Mont Blanc, SYMME, FR-74 000 Annecy, France
wilfried.liegard@bystronic.com
[2] Bystronic Laser AG, Niederönz, Switzerland

Abstract. Material heterogeneity has a considerable effect on spring back. Integrating heterogeneity in a simulation to obtain a more accurate spring back prediction is complex and often requires a significant number of tests and parameters. Whilst in the industry, the necessity of predicting forming process outcomes, e.g. the spring back of the material, is becoming essential, it does inevitably require a performant and easy to calibrate model. At academic level, a compartmentalized model [1] considers the heterogeneity of the material as one of the parameters driving its plastic behaviour. Its ability to describe load-unload cycles with a small number of adjustable parameters has demonstrated [2]. In this paper, we propose to optimize the identification process on an industrial material.

The chosen material, a DP 600 steel, already presents high heterogeneity level due to its initial structure. The main objective is to calibrate the compartmentalized model by using only a monotonous tensile test. The identified model is integrated in a finite element code and simulations of a loading/unloading path are then carried out to evaluate the prediction accuracy of the material behaviour at small strain.. This possible new method could help reduce the number of tests required to obtain the required prediction accuracy, especially when a large spring back effect is present. This would allow to achieve more accurate optimization of the process parameters in the bending operation.

Keywords: Heterogeneity · Simulation · Bending · Sheet Metal · Industrialisation

1 Introduction

Sheet metal forming is frequently used for producing parts. It covers many processes such as deep drawing, punching, hydroforming or air bending. Most of those processes are used for serial production with a large number of parts. The air bending process on the other hand, can be used for single part production or serial production.

In each process, the goal is to not only reduce the set-up time and the overall production cost but also the number of parts needed to obtain the first good part. It is therefore necessary to simulate the process. Nowadays, simulations are becoming more and more accurate in predicting failure modus. They are nevertheless still not accurate enough to predict the final shape of the part especially when large spring back effects are present.

K. Mocellin et al. (Eds.): ICTP 2023, LNME, pp. 206–212, 2024.
https://doi.org/10.1007/978-3-031-40920-2_22

The key to getting a good part at the first bend is to be able to accurately estimate the spring back of the material and obtain a realistic prediction of industrial operations.

Models, such as the QPE model [3] describe this effect; unfortunately, most require many tests to identify their numerous parameters. These tests generally require special equipment and cannot be conducted during the production. Even if the simulation is lead once to set the production, it is necessary to consider batch variations for fine tuning or adaptation.

Therefore, a model that is simpler to calibrate and does not require extensive testing could change the way productions is executed. New machine set up and calibration, due to a new batch of the same material, could be supported by simulation and not rely on manual correction. This is a possibility that the compartmented model could offer. Based on the study by Chen et al. [4], not only can this model simulate with a certain accuracy the behaviour of a DP 980 but it also appears that only one load/unload/load tensile test was necessary to calibrate it [1].

As the compartmentalised model proposes a new way of simulating material behaviour, it is necessary to fully understand its limits. One of the assumptions that could be made in this paper, is that to calibrate the model, only a single tensile test could be necessary. This was tested with new calibration procedure based on data from new experiments performed on a DP 600 sheet metal.

In this paper, the compartmentalised model and its calibration for the simulation of a cyclic loading path is briefly summarized. Conclusions about the interest of the identification process based on the presented results are subsequently drawn.

2 Presentation of Compartmentalised Model and Straightforward Parameters Identification

Making use of a compartmentalized model requires the incorporation of material heterogeneity into the simulation to match the experimental evidence, particularly with regard to spring back. This is obtained by applying different properties on each compartment of the simulated geometry. Practically, each element has a different material property whose overall distribution is ruled by adapted statistical functions. It has been shown by Charleux et al. [2] that only 5 parameters are required to achieve a correct result and most importantly to simulate the natural behavior of heterogeneous material such DP steels. The result can be seen in the Fig. 1 which was extract from Charleux et al. [2].

It must be noted that as the distribution is randomized, for each group of parameters 10 distributions are tested. This is allowed to evaluate the stability of the optimized solution and to identify the mean, minimum and a maximum values for the optimisation results.

In the same study by Charleux et al. [1] it is shown that the compartmentalised model can inherently simulate the non-linear effect of the material unloading/loading, as it can also be seen in the Fig. 1. The study [1] further supports the simulation's results with a comparison to the measured Young's modulus issued from Chen et al. [3] and the one resulting from the simulation. It was found that the simulation and practical measurement are correlated and within a very small margin of each other, suggesting that the model is effective in simulating this type of loads.

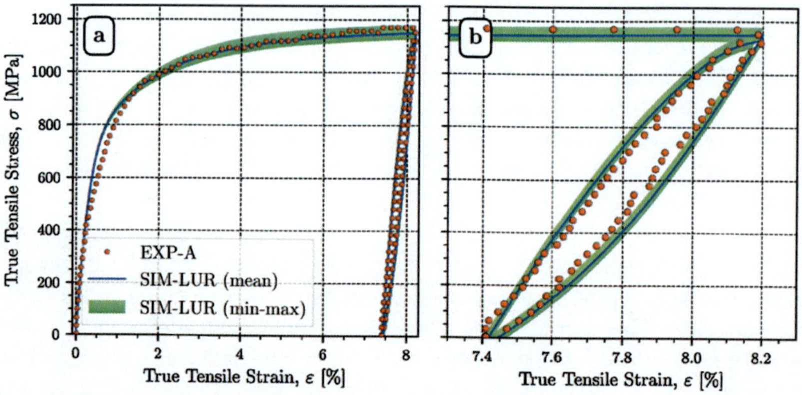

Fig. 1. Result of the optimization of the compartmented model based on the experiment issue from the literature on a DP980 [1]. The points EXP-A are the points extracted from Chen et al. [3], the SIM-LUR represent the mean value of the optimisation results and the SIM-LUR represent the mean and max value of this optimisation.

This first test with a more complex behaviour, multiple loading/unloading/reloading, was calibrated on one of the loading/unloading/reloading instances and the coefficients were identified thereafter. Another simulation was then run with the remaining cycles at tensile strain of 0.02, 0.04, 0.06. The results suggest that the model achieves its purpose, however the minimum required test to calibrate it was not established. Therefor This paper aims to describe the possibility of calibrating the model with only tensile test.

3 Experimental Database for DP600 Steel

The cyclic test method described and used in the first test in Chen et al. [4] for DP 980 [1] was reproduced for DP 600. In addition, a tensile test was done for the straightforward identification of the parameters. Each test was repeated 3 times to obtain a mean value.

The machine used for the testing was a 5569 Instron and the strain was measured with a 12.5 mm width extensometer. A pre-stress of 3 MPa was applied, the tests were run with a control speed of 8 mm/min, the unload stopped when the 0 MPa was reached and the load was interrupted at various strain increments ($\Delta\varepsilon$) of 0.02, 0.04, 0.06, 0.08 and 0.1. The same displacement speed was imposed for each tensile test.

This experimental database allows for the adoption of different schemes to identify the parameters of the model. It can then be possible to choose if the model should be with experimental data involving all the measurement until 30% strain or only up to 10% strain for example. It is of interest to determine the minimum amount of measurement necessary to obtain similar simulated results as the measured ones.

For each test, the Young's modulus is evaluated following the method explained in Chen et al. [4]. The Young's modulus measured from the first loading was used as a fixed value for the identification, here 203 GPa.

Figure 2 shows the simulation result of a tensile test up to 25% strain and the result after running the cycle simulation at 2% of strain. The left side represents the superposition of tensile test and the optimised simulation. The right side represents the simulation results of loading/unloading/reloading at 0.02 strain compared with the measured data.

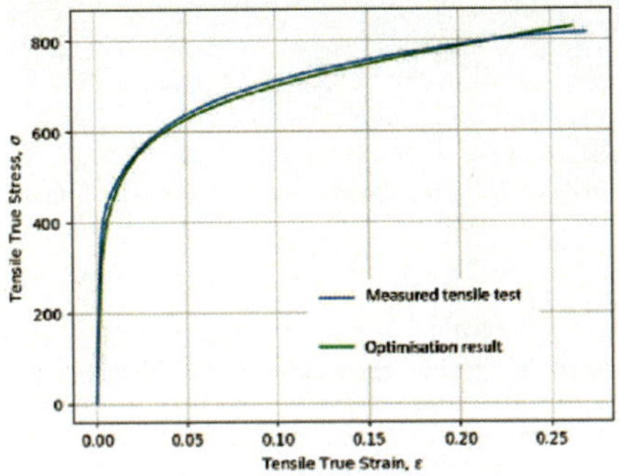

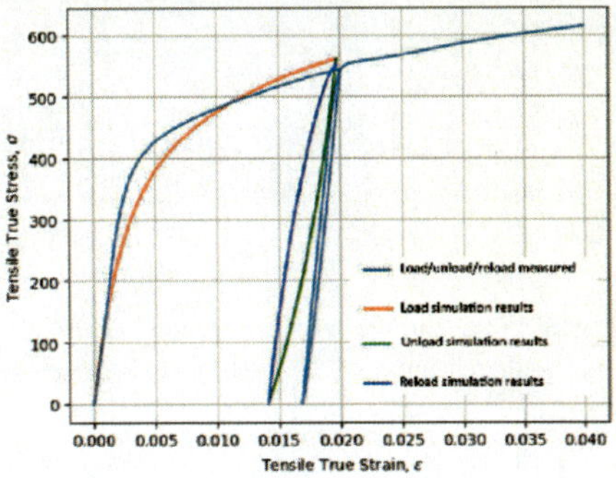

Fig. 2. Simulation result of a tensile test until 25% and the result of the simulation with an unload/load at 2% strain. In the left figure the green corresponds to the simulation and the blue to the tensile test. In the right figure red colour corresponds to the simulation of the first loading, the green to the simulation of the first unloading and the blue corresponds to the simulation of the reloading.

It can be observed that even if the calibration was run on only a tensile test, the model was able to naturally model the non-linearity of the elasticity, unfortunately, it was not closely correlated to the measured data.

Other optimisations were run with a reduced portion of the tensile test, where only the first 10% or only the first 5% strain were taken in account. The objective was to calibrate the model with more points closer to the transition between the plastic and elasto-plastic behaviour.

There again a new set of parameters was determined and applied to the exact same load path as the previous one and compared to the experimental data. The results of load/unload/reload at 0.1 strain after an optimisation on a tensile test reduce to the 10% can be seen in Fig. 3.

By calibrating the model with the curve reduced at 10% of strain, a new set of parameters was found. These parameters were different as the one found with the calibration with 20% strain. Even if better results could be achieved with this calibration, the Young's moduli were still underestimated. The spring back was not properly simulated.

Similar results were achieved with a calibration with only 5% strain. The model had the tendency to be stiffer and increase the Young's moduli values but they were still underestimated. Some parameters examples can be found in the following Table 1.

The compartmented model uses a double Weibull distribution to distribute the elements. It can be seen one distribution for "hard" elements and one for "soft" elements. Hard elements will enter plasticity at higher constrain than the soft ones.

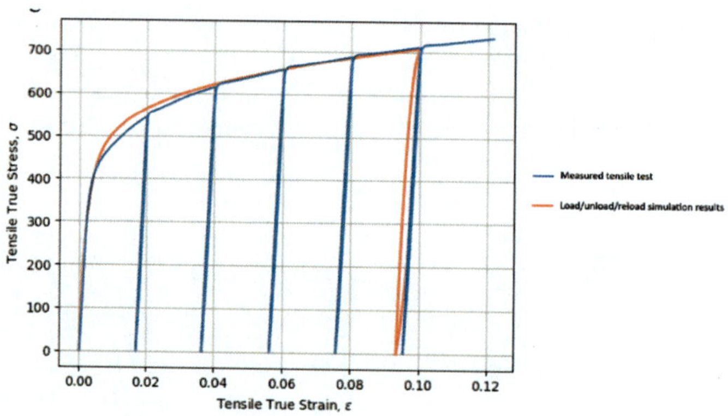

Fig. 3. Result of the simulation with a unload/load at10% strain after calibrating the model with the tensile test cut at 10%. The blue curve corresponds to the test and the red curve to the simulation results.

Table 1. Summary of parameters after the calibrations

Parameters	Calibration with 25% strain	Calibration with 10% strain	Calibration with 5% strain
k1	1.12878	1.36266	1.94715
k2	3.14022	2.55392	1.31353
l1	0.00239	0.00208	0.00215
l2	0.01605	.021518	0.03581
w1	0.77552	0.76541	0.82543

The parameter k influences the shape of the repartition of the plastic value and the parameter l influences the value of the elastic value of the elements. The w parameter influences the proportion of "hard" and "soft" elements. Parameters denoted with 1, influence the "hard" elements and those denoted with 2 influence the "soft" elements.

In Table 1, only the parameter w1 is noted as, even if each Weibull distribution is defined by the group of 3 parameters (k, l, w), the proportion of the distribution is always equal to 1. This is why it not necessary to show the parameter w2, even if it used in the distribution for the modelling.

Despite a slight decrease of the proportion of the "hard" elements noticed for the calibration at 10%, it was compensated for by the decrease of the value of the "soft" elements which makes the overall material behaviour more elastic and explains why the simulated results are approaching the measured ones.

4 Conclusion and Perspectives

One of the most important observations is that each time a new calibration was done on the same dataset, a different set of parameters was found. This suggests that there is no single solution to describe the loading path. This further implies that more information is required to achieve results suitable for production optimisation.

This study also confirmed the natural ability of the compartmentalised model to simulate the spring back. This is still an advantage for a possible industrialisation as when looking at the minimum known test used to calibrate this model (loading/unloading/reloading), it is becoming easier to calibrate a simulation and implement it prior to a new start of the production.

Based on this conclusion further work and testing is required to validate the hypothesis of using one loading test to calibrate the compartmented model and then applying the extracted parameters to run more complex simulations such a loading following by a loading.

One test to consider is the initial Young's modulus as a variable and not as a fix value. In fact, this E coefficient cannot usually be accurately measured nor is it provided in the datasheet provided by material suppliers which then may constrain the model with some values that do not have a physical meaning.

Changing the optimization process is also a solution that will be evaluated. A split in the dataset with a point weighting based on the influence of each portion of the measured curve can be tested in order to model the leading path with higher accuracy.

Another evaluation that will be conducted is the calibration with different loading system at different deformation value. This will then bring insight into the stability of the model strain and its calibration.

As aforementioned, the model is believed to be optimal, seeing that it naturally simulates the non-linearity of the elastic behaviour. The need for a simple loading/unloading/reloading to calibrate it is already a major advantage compared to other models for which more tests with more expensive machinery are required.

It is essential to define the most accurate calibration to be able to simulate more complex loads. As having a simulation with a heterogenous distribution of the material properties on each compartment does not seem to increase the computing of the model (simulation times for these loads were similar to a standard simulation), it possible to suppose that the calibration step could be in the future the bottle neck of the full simulation. So by reducing the process to a simple tensile test and followed by an optimization of a few parameters should be still one of the major goals for a inline process optimization in the industry.

References

1. Bizet, L., Charleux, L., Balland, P., Tabourot, L.: Influence of heterogeneities introduced into the modelling of a ring compression test. Arch. Civil Mechanical Engineering **17**(2), 365–374 (2016). https://doi.org/10.1016/j.acme.2016.11.002
2. Charleux, L., Tabourot, L., Roux, E., Farah, I., Bizet, L.: Dependency of the Young's modulus to plastic strain in DP steels: a consequence of heterogeneity? Mater. Today Commun. **24**, 1000972 (2020)

3. Sun, L., Wagoner, R.H.: Complex unloading behavior: nature of the deformation and its consistent constitutive representation. Int. J. Plast. **27**(7), 1126–1144 (2011). https://doi.org/10.1016/j.ijplas.2010.12.003
4. Chen, Z., Gandhi, U., Lee, J., Wagoner, R.H.: Variation and consistency of Young's modulus in steel. J. Mater. Process. Technol. **227**, 227–243 (2016). https://doi.org/10.1016/j.jmatprotec.2015.08.024

Development of Warm and Cold Tube Forming of Ti-6Al-4V Alloy by Press Forming

Y. Okude(✉), T. Iwaoka, I. Nakamura, T. Muraoka, and T. Katagiri

Mechanical Technology Group, Tokyo Metropolitan Industrial Technology Research Institute, Aomi, Koto-Ku, Tokyo 135-0064, Japan
okude.yusuke@iri-tokyo.jp

Abstract. The low ductility of Ti–6Al–4V alloy at forming temperatures below 780 °C is still a problem. Ti–6Al–4V is usually hot press formed at 780 to 900 °C. Hence, it is necessary to develop a new press forming for Ti–6Al–4V alloy sheets at forming temperatures below 780 °C to improve productivity. We developed a method to achieve of warm and cold press forming of Ti–6Al–4V alloy sheets by combining press motion control and clearance between a die and a blank. In this study, we tried tube forming by developed press forming method. In this method, the following two processes were separated: punch motion process without BHF and BHF motion process without punch motion. These two processes were carried out alternately during deep drawing or drawing and ironing. Three forming parameters, namely, the punch speed S_{pi}, punch stroke Psi, and blank holding force BHF_i, were controlled separately and the effect of forming conditions on deformation behavior was investigated by using experiment and finite element analysis (FEA). As a result, warm and cold tube forming of Ti–6Al–4V alloy was achieved without any undesirable deformation by applying the developed press forming method.

Keywords: Tube forming · Warm forming · Cold forming · Press motion

1 Introduction

Ti-6Al-4V titanium alloy has been actively used in medical equipment and aero-space due to its characteristics of light weight, corrosion resistance, and high specific strength [1–3]. A hot press formed Ti–6Al–4V alloy has been utilized in various devices such as medical and aerospace equipment [4]. However, hot press forming increases cost, especially because it uses a furnace. To reduce cost, hot stamping by electric heating has been applied to Ti sheets or continuous resistance heating has been applied to the bar forming of Ti–6Al–4V alloy. Both electric heating and continuous resistance heating reduce the heating time in the forming temperature range from 780 to 900 °C [5, 6]. Hence, it is necessary to develop a new press forming for Ti–6Al–4V alloy sheets at forming temperatures below 780 °C to improve productivity.

We developed a method to achieve of warm and cold press forming of Ti–6Al–4V alloy sheets by combining press motion control and clearance between a die and a blank

K. Mocellin et al. (Eds.): ICTP 2023, LNME, pp. 213–222, 2024.
https://doi.org/10.1007/978-3-031-40920-2_23

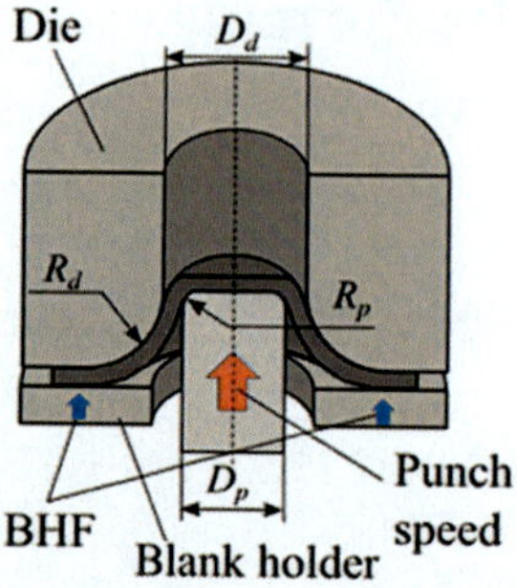

Fig. 1. Schematic of deep drawing process

Table 1. Dimension of dies of deep drawing process

Die	Diameter: D_d /mm	35.0
	Shoulder radius: R_d /mm	3.0
Punch	Diameter: D_d /mm	32.5
	Shoulder radius: R_d /mm	5.0

[7, 8]. In this study, we tried tube forming by developed press forming method. In this method, the following two processes were separated: punch motion process without BHF and BHF motion process without punch motion. These two processes were carried out alternately during deep drawing or drawing and ironing. Three forming parameters, namely, the punch speed S_{pi}, punch stroke P_{si}, and blank holding force BHF_i, were controlled separately and the effect of forming conditions on deformation behavior was investigated by using experiment and finite element analysis(FEA). As a result, warm and cold tube forming of Ti–6Al–4V alloy was achieved without any undesirable deformation by applying the developed press forming method.

2 Experimental Apparatus and Formability of Ti-6Al-4V Alloy Sheet at Warm Temperature of Deep Drawing Test

In order to understand the basic formability of cylindrical deep drawing of Ti-6Al-4V alloy sheet, cylindrical deep drawing with a general blank holding force (BHF) was first performed. Figure 1 shows a schematic of deep drawing process, and Table 1. the dimensions of the dies of deep drawing process. The die material used was SKH51. A Ti-6Al-4V alloy plate and a blank with a thickness of $t_0 = 0.5$ mm were used as test materials. The test was performed with a blank diameter of $\phi = 60$ mm (Drawing ratio, $DR = 1.85$). Cylindrical deep drawing was carried out at the condition of 300 °C forming temperature. Test was performed at the same temperature as the blank for the drawing die and the wrinkle control plate. The punch was kept at room temperature under all conditions. The punch speed S_p was set to three conditions of 9, 90 and 900 mm/min, and the constant wrinkle suppressing force BHF_{const} was applied at 4 kN. Figure 2 shows the punch load-punch stroke curve and photographs of cups drawn under each set of conditions. In each condition, fracture occurred at punch shoulder. Therefore, the low formability of Ti-6Al-4V alloy sheet was confirmed.

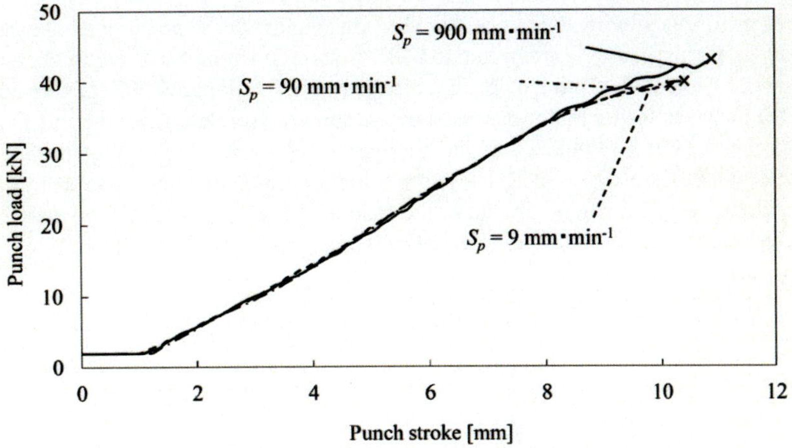

(a) Punch load—punch stroke curve

Conditions	Punch speed S_p [mm/min]		
	9	90	900
Photo. of drawn cups			
	Occurrence of fracture at punch shoulder	Occurrence of fracture at punch shoulder	Occurrence of fracture at punch shoulder

(b) Photographs of cups drawn under each set of conditions

Fig. 2. Effects of punch speed on deformation behaviour of Ti-6Al-4V alloy sheet at 300 °C

3 Incremental Press Forming Process Method and Formability of Ti-6Al-4V Alloy Sheet

In warm deep drawing with a constant blank holding force, cracks occurred at the punch shoulder at any punching speed at a forming temperature of 300 °C. In this study, we focused on the fact that in normal cylindrical drawing, wall thinning progressed due to tension due to wrinkle suppressing force from the start of the punch until the maximum punch load was reached. When the material is pressed into the mold with a punch under the condition of applying wrinkle suppressing force, the thickness of the punch shoulder is significantly reduced. Therefore, in order to perform press forming in the sequential forming developed in this study, two processes were separated: a punch motion process and BHF without punch motion wrinkle force loading process. Figure 3 shows the schematic of incremental press forming. FORMKOTE T-50 was applied as a lubricant between the blank and the mold. In the sequential forming, forming was performed under the conditions of S_{pi} = 9 mm/min, P_{si} = 0.2 mm, and BHF_i = 20 kN. For comparison, a cylindrical drawing was performed under the same lubricating conditions with a punch speed of 9 mm/min and blank holding force of 4 kN. Figure 4 shows the effect of sequential forming on formability. However, the longitudinal fracture occurred at the flange as shown in figure caused by the compressive deformation of the flange. In addition, the effect of punch feed on the formability was investigated by forming under three conditions of $P_{si} = 0.2$, 0.4, and 0.8 mm, with the punch speed fixed at a fixed condition of $S_{pi} = 9$ mm/min. As in the case of the punch speed, the formability was different depending on the feed amount of the punch.

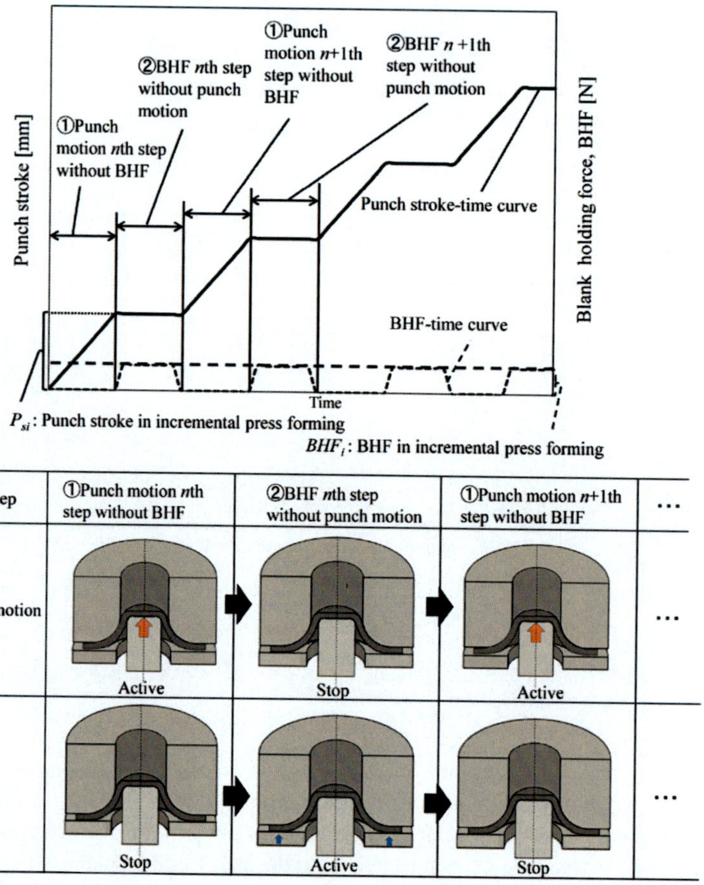

Fig. 3. Schematic of incremental press forming

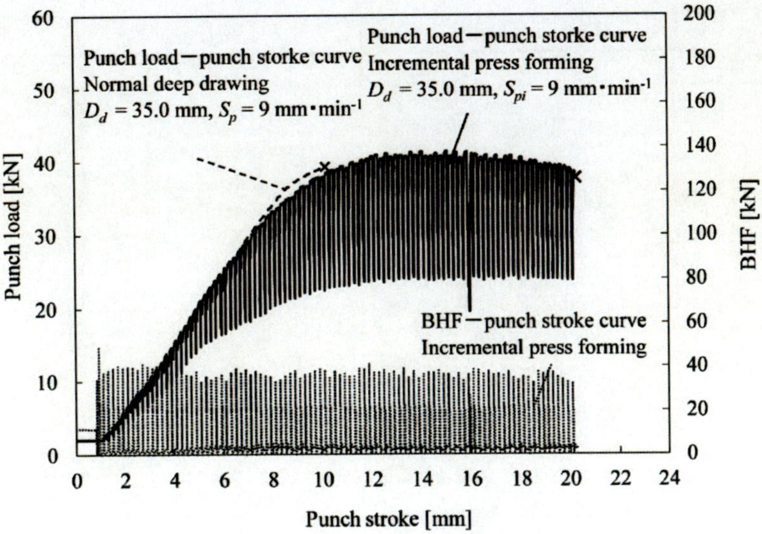

(a) Punch load—punch stroke curve

Method	Incremental press forming	Normal deep drawing with lubrication
Photo. of drawn cups	 Occurrence of fracture at flange end	 Occurrence of fracture at punch shoulder

(b) Photographs of cups drawn under each set of conditions

Fig. 4. Effect of incremental forming on deformation behavior at 300 °C

4 Drawing and Ironing by Incremental Press Forming and Formability of Ti–6Al–4V Alloy Sheets

It was difficult to prevent undesirable deformations in the deep drawing of Ti–6Al–4V alloy sheets, so metal flow control by the combining press motion control and clearance between a die and a blank was focused on. Figure 5 shows the schematic of the deformation behaviour observed during drawing and ironing [7, 8]. The deformation behaviour, in metal flow was controlled by press motion and clearance.

4.1 FEA Model and Conditions

FEA of drawing and ironing of Ti–6Al–4V alloy sheets was calculated. The finite element analysis code LS-DYNA3D was used for drawing and ironing simulation. Representative

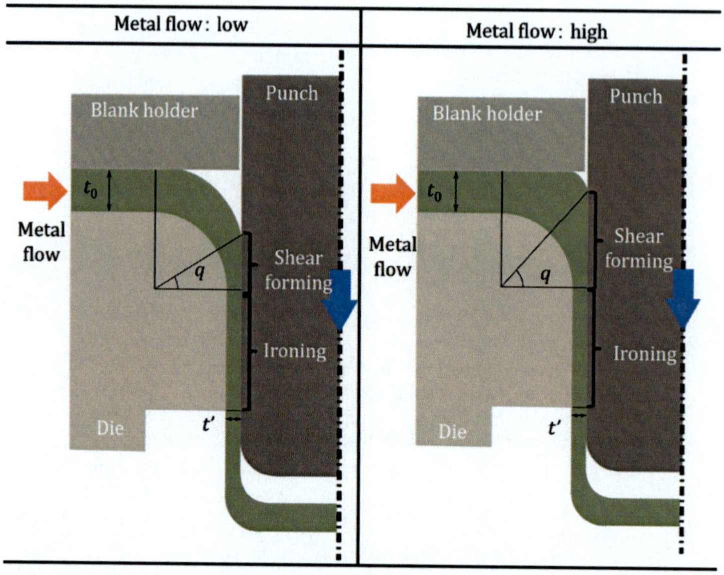

Fig. 5. Schematic of deformation behaviour during drawing and ironing [7, 8]

finite element models of workpieces (Ti–6Al–4V, $t_0 = 0.5$ mm, blank diameter = 60 mm) were constructed. Table 2. indicates the properties of the sheet materials of Ti–6Al–4V alloys (workpiece, at room temperature). An elastoplastic material was assumed, satisfying an n power strain hardening law with the constitutive equation as follows,

$$\sigma = C(\varepsilon_n + \overline{\varepsilon_p})^n \tag{1}$$

where, C is plastic modulus, ε_n elastic strain, $\overline{\varepsilon_p}$equivalent plastic strain, n work hardening exponential.

Figure 6 shows the FEA model of deep drawing process. The dimensions of dies were obtained as same as experiment. In FEA, the friction coefficients between dies and blank are set to 1.0 without lubrication(punch and blank, blank holder and blank), and 0.5 with lubrication (die and blank). In the incremental forming, press forming was performed under the conditions of $P_{si} = 0.4$ mm, and $BHF_i = 20$ kN.

Table 2. Material property of workpiece at room temperature

Material	Ti-6Al-4V
Density, ρ[kg/m^3]	4.51×10^3
Modulus of elasticity E [GPa]	106
Poisson's ratio	0.3
C value [MPa] *	1228
n value*	0.04

$^*\sigma = C(\varepsilon_n + \overline{\varepsilon_p})^n$

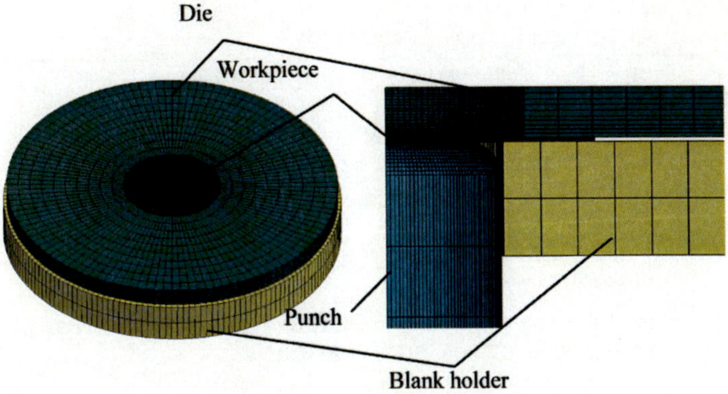

Fig. 6. FEA model of deep drawing process

4.2 FEA Model and Conditions

Figure 7 shows results of drawing and ironing process of Ti–6Al–4V alloy sheets (shear strain distribution, $P_{si} = 0.4$ mm, $BHF_i = 20$ kN). At the point A, the workpiece was formed for the bottom of drawn cup by drawing process only. Hence, the punch load gradually increased until starting the drawing and ironing (point B). When the starting point of drawing and ironing is reached, the punch load increased sharply. And shear strain caused by ironing process occurred at the thickness of workpiece near the die radius end. Then, shear strain increased with the progress of ironing process as shown in Fig. 7 (point C). And Fig. 8 shows the formed cup of equivalent plastic strain and equivalent stress distributions in FEA. As the results, the drawing and ironing process was calculated by FEA, and ironing due to shear strain could be estimated by simulation. The FEA results indicate the tube forming of Ti–6Al–4V alloy was achieved by the drawing and ironing with developed incremental press forming.

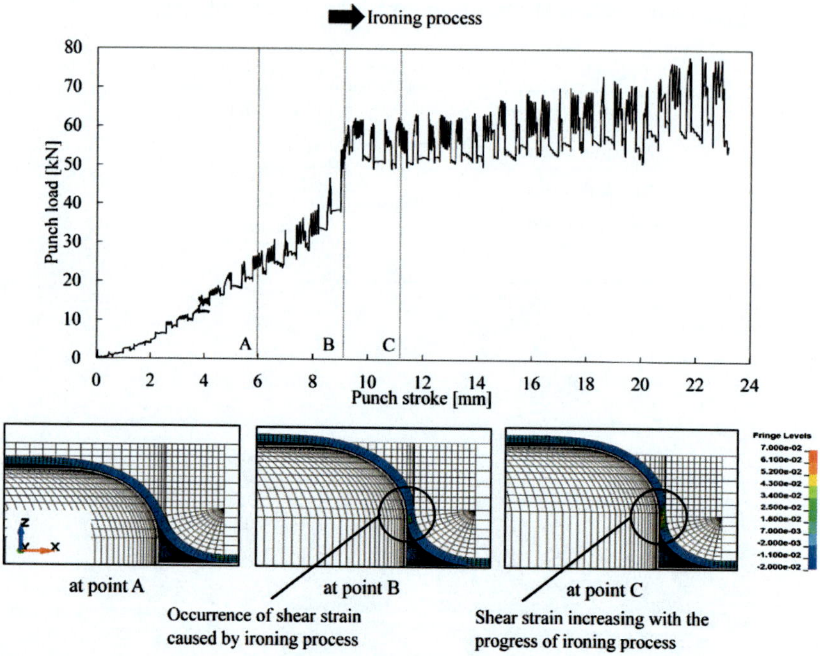

Fig. 7. FEA results of drawing and ironing process of Ti-6Al-4V alloy sheets (Shear strain distribution, $t_0 = 0.5$ mm, $P_{si} = 0.4$ mm, $BHF_i = 20$ kN)

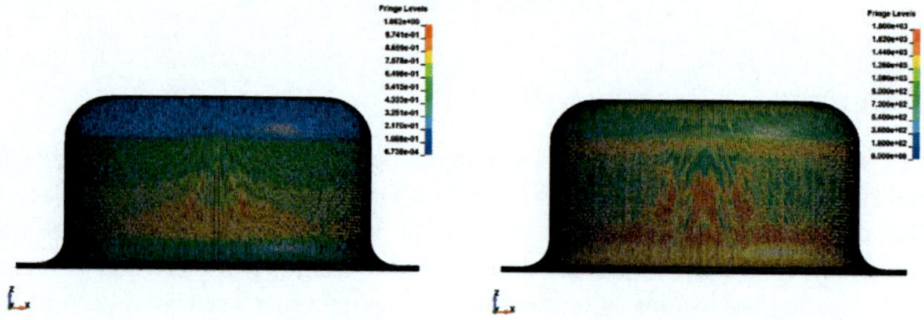

(a) Equivalent stress distribution (b) Equivalent plastic strain

Fig. 8. Formed cup of Ti-6Al-4V alloy in FEA ($t_0 = 0.5$ mm, $P_{si} = 0.4$ mm, $BHF_i = 20$ kN)

4.3 Experimental Results

To confirm the effect of metal flow controlling on deformation behaviour, drawing and ironing of Ti-6Al-4V alloy sheets (blank diameter $\phi = 60$ and 65 mm) were carried out using two dies ($D_d = 33.8$ and 31.3 mm). For, the conditions of $D_d = 33.8$ mm, cold drawing and ironing processes of $\phi = 60$ mm were carried out at the conditions of $S_{pi} =$ 90 mm/min, $P_{si} = 0.4$ mm, and $BHF_i = 20$ kN. On the other hand, and cold drawing and ironing processes of $\phi = 65$ mm were carried out at the conditions of $S_{pi} = 90$ mm/min,

$P_{si} = 0.5$ mm, and $BHF_i = 50$ kN for the tube forming with $D_d = 31.3$ mm condition. Figures 9 and 10 show the punch load – punch stroke curves for the cold drawing and ironing processes respectively. And, Fig. 11 shows the photograph of the formed cups. By applying the developed method and an 80% clearance of the initial blank thickness, the cold tube forming by drawing and ironing of Ti–6Al–4V alloy sheets were achieved.

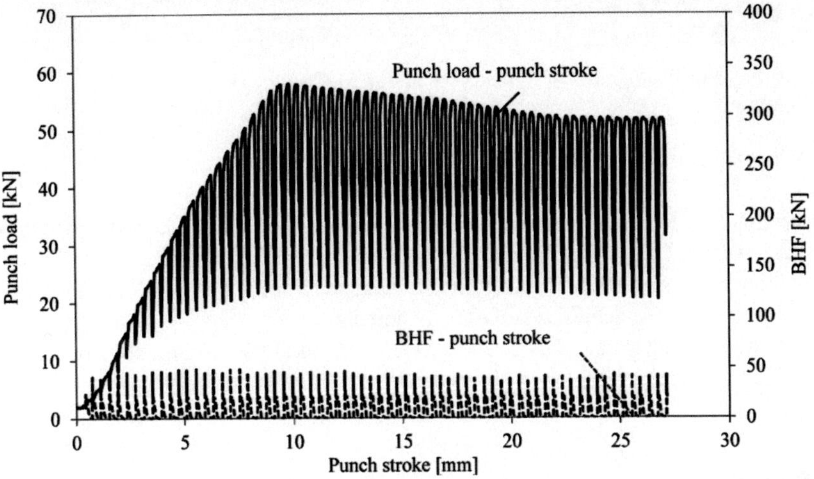

Fig. 9. Punch load – punch stroke curves for the cold drawing and ironing processes ($\phi = 60$ mm, $D_d = 33.8$ mm)

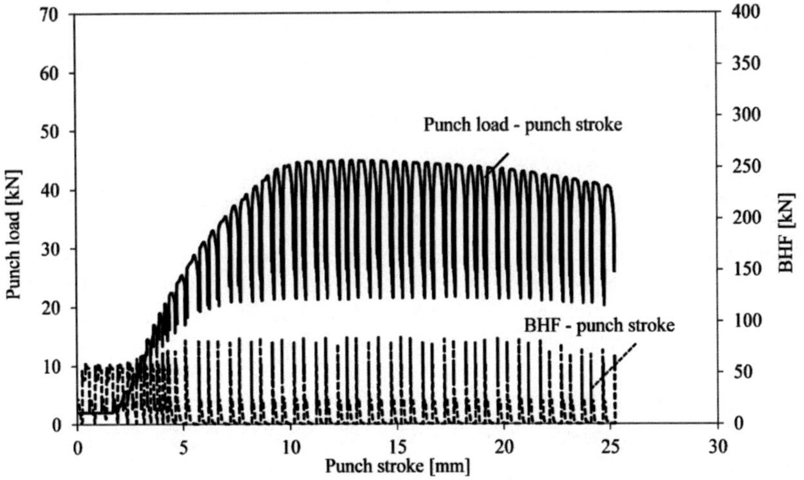

Fig. 10. Punch load – punch stroke curves for the cold drawing and ironing processes ($\phi = 65$ mm, $D_d = 31.3$ mm).

Fig. 11. Photographs of formed cups of Ti-6Al-4V alloy sheets in experiments

5 Conclusions

1) By applying incremental forming, it was possible to form the Ti-6Al-4V titanium alloy without cracking after reaching the maximum punch load at 300 °C.
2) By applying incremental forming, the drawing and ironing of Ti-6Al-4V alloy sheets of $DR = 2.11$ was achieved without any defects.
3) This study indicates the possibility of Ti-6Al-4V alloy tube forming by drawing and ironing with incremental forming method.

References

1. Maeno, T., Mori, K., Hamedon, Z.: Proceedings of Manufacture, 14th International Conference on Metal Forming, pp. 287–290 (2012)
2. Maeno, T., Mori, K., Yamashita, Y., Nakayama, M., Hamedon, Z.: Proceedings of 2011 Japanese Spring Conference for the Technology of Plasticity, pp. 345–346 (2011)
3. Maeno, T., Mori, K., Yamashita, Y.: Proceedings of 2012 Japanese Spring Conference for the Technology of Plasticity, pp. 67–68 (2012)
4. Beal, D.J., Boyer, R., Sanders, D.: ASM Handbook, 14B, Metalworking: Sheet Forming, pp. 656–669 (2006)
5. Yamazaki, K., Yamaguchi, T., Murayama, Y., Mori, K., Maeno, T.: Development of hot stamping process for aircraft titanium alloy parts using resistance heating. J. Jpn. Soc. Technol. Plast. **55**(647), 1097–1101 (2014)
6. Yanagimoto, J., Izumi, R.: Continuous electric resistance heating—Hot forming system for high-alloy metals with poor workability. J. Mater. Process. Technol. **209**, 3060–3068 (2009)
7. Okude, Y., Iwaoka, T., Nakamura, I., Katagiri, T.: Development of warm forming method for Ti-6Al-4V alloy sheets with press motion control. J. Jpn. Soc. Technol. Plast. **61**(714), 159–164 (2020)
8. Okude, Y., Iwaoka, T., Nakamura, I., Katagiri, T., Muraoka, T.: Development of method of cold incremental drawing and ironing Ti-6Al-4V alloy sheets TI-6Al-4V. J. Jpn. Soc. Technol. Plast. **62**(729), 133–138 (2021)

Real Time Monitoring of Friction Variation in Stamping Process Using Die-Embedded Sensing System

Ming Yang[1(✉)] and Takunori Kyuno[2]

[1] Tokyo Metropolitan University, Tokyo 191-0065, Japan
yang@tmu.ac.jp
[2] ADD.Q Co., Sagamihara City, Japan

Abstract. A die-embedded sensing system by embedding a 3-axis piezo sensor in the die was developed to measure the dynamic characteristics of sliding friction between materials and dies during stamping process. A hat-bending of steel sheets of high-strength steels was performed using the developed monitoring system. The influence of material strength, slide speed, cushion pressure, etc. on the dynamic characteristics of friction was evaluated. It is found that the friction coefficient changes dynamically depending on the processing conditions and material strength and the mixed lubricating state lasted for a certain duration time, and the duration time was theoretically calculated based on the squeeze effect theory. A pulse motion was adapted to the hat-bending process based on the calculation and the effect of the friction reduction was verified.

Keywords: die-embedded sensing system · dynamic friction characteristics · sheet forming

1 Introduction

Metal forming is one of the most fundamental processes for making metallic parts and components in the industry. With increase of requirement for light weight and high strength of parts, process of difficult-to-work materials, such as high strength steel, Titanium alloy, Aluminum alloy, becomes more significant. The process with these materials faced on some issues, such as, lower formability or higher flow resistance, and design of the dies and process conditions depends on the knowledge of expertise. On the other hand, the servo press technology has been commercialized more than 20 years and paid attentions to solve these problems. The servo press has controllability as a digital processing technology and has high potential capability to improve formability of working materials and accuracy of products by utilizing various slide motions [1–3]. A deep drawing process of sheet material using servo press, named JIM-FORM (JFE Intelligent Multistage Forming), was proposed [4]. In the process, the die and the workpiece are positively separated during the formation and as a result, lubricating oil is re-inflowed between the die and the surface of the material at that moment. There is an effect of preventing lubricating oil film break from the high contact pressure between the workpiece

K. Mocellin et al. (Eds.): ICTP 2023, LNME, pp. 223–232, 2024.
https://doi.org/10.1007/978-3-031-40920-2_24

and die. By using slide position control and servo controllable die cushion with a servo press, the slide is moved by a pulse motion, and the die cushion position is also adjusted accordingly. As a result, the punch load was reduced by 20%. It is seen that re-inflow of lubricating oil by pulse motion promotes refill of the lubricating oil to the interface and keeps the frictional force low. Against the same drawing height, the process can withstand a larger blank holding force (BHF) and has a larger margin depending on process conditions. However, in the case of deep drawing process using pulse motion, the process time becomes longer, and the reduction in productivity becomes a problem. It is important to raise the re-lubrication effect while suppressing the lowering of the production efficiency by appropriate motion setting according to the change of the lubrication state. Monitoring of the lubrication state during forming and design of the proper motions for the process becomes a key issue. Developments of die-embedded sensors for monitoring contact force or friction coefficient were studied previously. Embedding pins with strain gauges into die was developed to measuring sliding friction [6–8]. A structured device for measuring multi-axis forces was developed for measuring normal and shear forces and then calculating friction coefficient [9]. Monitoring the contact gap between blank holder and workpiece and feeding back to control the blank holding force was conducted in deep-drawing process [10–12]. However, these studies did not focus on the friction characteristics in dynamic motions. It is important to understand the dynamic friction characteristics during the various motions.

In this study, a sensor system with sensor-embedded die has been developed to monitor the contact surface pressure and various sliding friction between the die and the material. 3-axis piezo force sensor was utilized to measure the force in the normal direction and sliding direction for monitoring the sliding friction. A hat-bending of steel sheets of high-strengths steels was performed using the developed monitoring system. The influence of material strength, slide speed, cushion pressure, etc. on the dynamic characteristics of friction could be evaluated. The dynamic characteristics of the friction was calculated based on the squeeze effect theory by using the experimental results. A pulse motion was adapted to the hat-bending process based on the model for verifying the effect of the friction reduction.

2 Experiments and Conditions

2.1 A Die-Embedded Sensing System

A die-embedded sensing system was developed for sensing properties of the various friction between the die and workpiece in stamping process. The schematic configuration is shown in Fig. 1. The dies are designed for deep drawing of a cylindrical cup in diameter of 100 mm and a load cell (Kyowa Corporation, LCW-C-300KN60SA3) is placed under the punch so that the punch load can be measured. An enlarged view of the blank holder is shown in the lower left. The blank holder has a two-piece structure, an upper plate and a lower plate. 3-axis piezo sensors (Kistler, model 9017c) were embedded in the upper plate with a cover plate that come into contact with the blank and tightened from the lower plate with bolts to provide preload in the vertical direction of the piezo sensor. The piezo sensor has high response speed and is suitable for dynamic force measurement. It is utilized to measure the load in the vertical direction and the shear direction during the process, and then, the sliding friction can be calculated by dividing the shear force by the vertical force. In order to accurately measure the shear force, a groove was provided between the cover plate on the top of the sensor and the blank holding plate to prevent contact between the cover and the blank holding plates. An O-ring was added to prevent oil from entering the groove. In addition, a radius of 0.5 mm was provided on the shoulder of the contact plate to prevent rocking the sliding material. After assembling the blank holder, the cover plate was ground simultaneously with the upper plate to align the height. This die was settled in a servo press machine (AMADA's SDE-2025) for performing deep drawing and bending processes.

2.2 Materials and Process Conditions

High-strength steels (JSC590Y and JSC980Y, T = 1.2 mm) and Mild steel (JSC270F, T = 1.2 mm) are used for the experiment. A strip blank (300 mm × 38 mm) was prepared. A hat bending using the deep drawing dies was implemented in this study since the friction characteristics are more stable and easier to evaluate during hat bending process than deep drawing process. Process conditions with blank holding pressure from 3MPa to 10 MPa and slide speed from 3 to 40 spm are utilized to measure the sliding friction. In this case, the slide speed of the strip is not constant while using the crank motion; Speed of the strip starting slide is about 35 mm/s for 3 spm and about 512 mm/s for 40 spm. Three kinds of lubricating oil in various viscosities of low viscosity (1 mm^2/s), medium viscosity (10 mm^2/s), and high viscosity (31.4 mm^2/s) were used for the experiment.

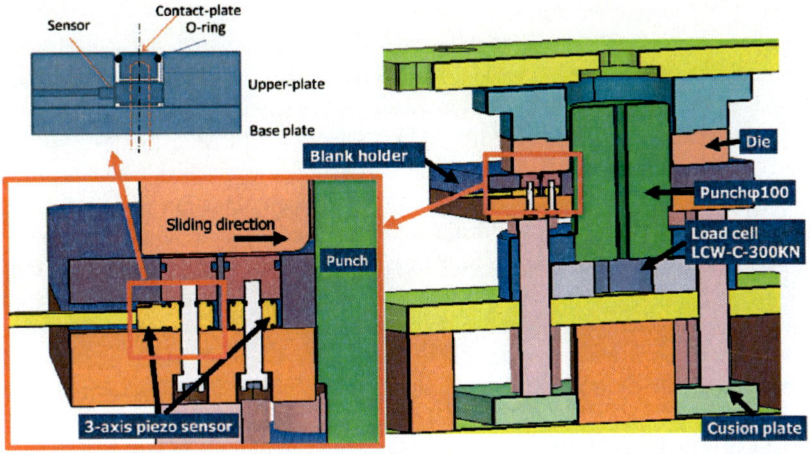

Fig. 1. Schematic configuration of the sensor-embedded die and sensing system.

3 Results and Discussions

3.1 Influence of Process Conditions on Friction Variation

Figure 2 shows the sensing data monitored in the hat bending with the cushioning force of 20 to 300 kN and slide speed of 3 spm. The blank holding pressure becomes 0.7 to 11.2 MPa. In this case, mild steel (JSC270F) and the lubricant oil in medium viscosity (10 mm^2/s) were used. Figure 2 (a) shows the vertical force of the outer piezo sensor, (b) the shearing force of the outer piezo sensor, (c) the vertical load of the inner sensor, (d) the vertical load of the inner sensor, (e) the punch load, and (f) the displacement of the slide. When the cushioning force is increased, the vertical load and shearing load increase both the outer and inner sides, and the punch load also increases. In this case, the workpiece in rolling direction was used as a representative. Figure 3 shows a typical curve of the friction coefficient μ calculated by dividing the shear load by the vertical load for the outer sensor with a cushion force of 200 kN. The punch load and slide displacement were also plotted for reference. The slide moves downward and the upper-die contacts on lower die by sandwiching the blank between the die and the blank holder. The punch touches the blank and the punch load begins to increase from point B and the bending starts and workpiece starts to slide toward the punch. μ increases steadily in range C to D and settles almost constantly in range D to E. We think that in range C to D a squeeze effect of the lubricant occurs between workpiece and die that prevents oil discharging from the gap and the metal contact area between the die and the blank gradually widens, and as a result, it supports a mixed friction regime. When point D is reached, excess oil is removed and the metal contact area between the die and the blank becomes stable, and so the μ also becomes stable.

The bending test with various cushion pressures and slide speeds were performed for high-strength steels JSC590Y and JSC980Y with a lubricant of viscosity 31.4 mm^2/s. Variations in the friction coefficient were measured as shown in Fig. 4. As the cushion pressure is increased (3,5,7.5.10 MPa for JSC980, 5,7.5.10 MPa for JSC590Y), the friction coefficient increases slightly when a certain value is reached, which is different from that for the mild steel. This is because as the strength of the material increases, the yield stress also increases, and the plastic deformation of the material surface asperity is less likely to occur, and the pressure-bearing effect of the material surface lubricating oil caused by high pressure is less likely to occur. Regarding the dynamic characteristics of the friction coefficient, the friction coefficient at beginning of the hat bending gradually increases from a low number, and finally approaches a certain value and stabilizes. The duration for the friction coefficient transiently increasing decreases with increasing surface pressure. This is because it takes time for the lubricant film thicker than the material asperity at the interface to be removed from the interface due to the squeeze effect. The time for the lubricating oil to be removed is inversely proportional to the surface pressure. By comparison of the results for various slide speeds, we can see that the changing in the duration for the friction coefficient transiently increasing is minor, but the increase in the sliding speed shortens duration of the hat bending significantly. As a result, the average friction coefficient reduced significantly for higher speed. The flow characteristics of the lubricant film at the interface can be evaluated by using Reynolds' equation with the influential factors of the surface pressure, velocity of the slide motion and viscosity of lubricant.

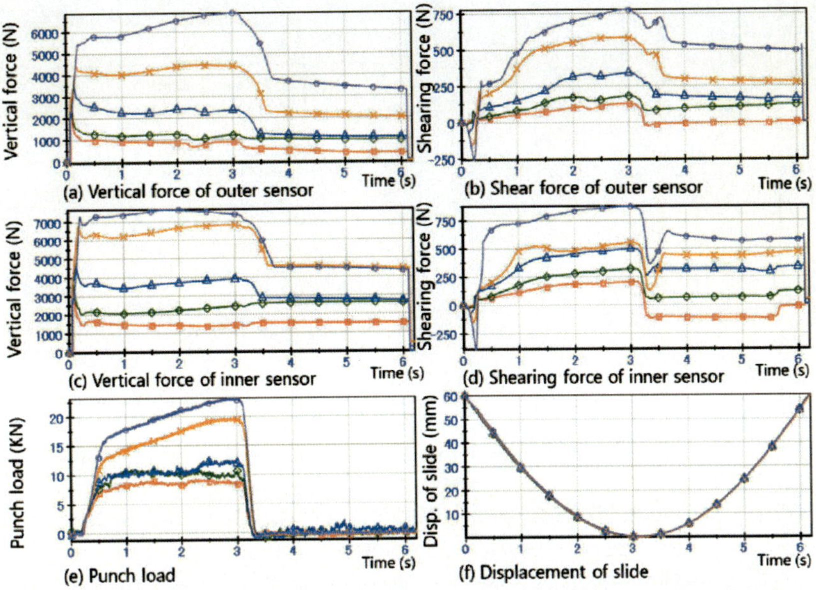

Fig. 2. Typical data monitored in the hat bending process in a condition.

3.2 A Model for Friction Variation with Dynamic Motions

A schematic model for the squeeze effect can be expressed as Fig. 5. While the blank holder following motion of the slide moves in vertical direction, the lubricant between the die and workpiece flows through the narrow gap and pressure arises due to viscous resistance. By assuming that the gap between two flat surfaces is narrow enough and the descending speed of the upper surface is slow, the lubricant flow may be laminar and the lubricant inertia effect can be neglected [5]. Reynolds' equation can be introduced for the rectangular surfaces as follow:

$$\frac{\partial^2 p}{\partial x^2} + \frac{\partial^2 p}{\partial y^2} = \frac{Uh}{2} - \frac{12\eta}{h^3} \cdot \frac{dh}{dt} \tag{1}$$

Coordinate system determined on rectangular surface is shown in Fig. 5. *p* is the pressure generated in the lubricant, U the sliding velocity of the workpiece. η, *h* is the viscosity of lubricant and oil film thickness, respectively and *t* is the elapsed time of the oil film. d*h*/d*t* is the flow rate of oil from the gap. We can see that flow rate is a function of Laplacian of pressure, sliding speed and viscosity of the oil and these parameters are related to the process conditions and lubricant used. We can evaluate the squeeze effect empirically from the experimental results shown in Fig. 4. Then, we can design

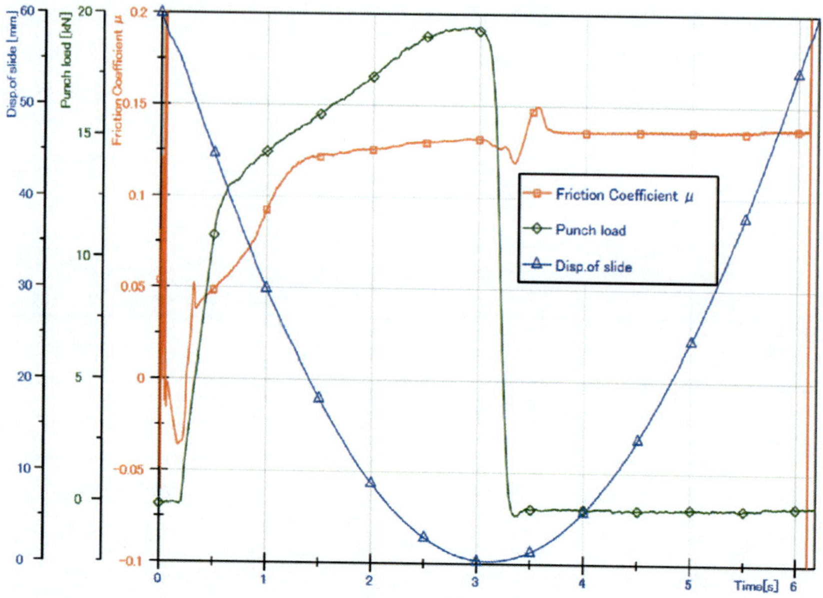

Fig. 3. Friction coefficient calculated from the monitoring data for the outer sensor.

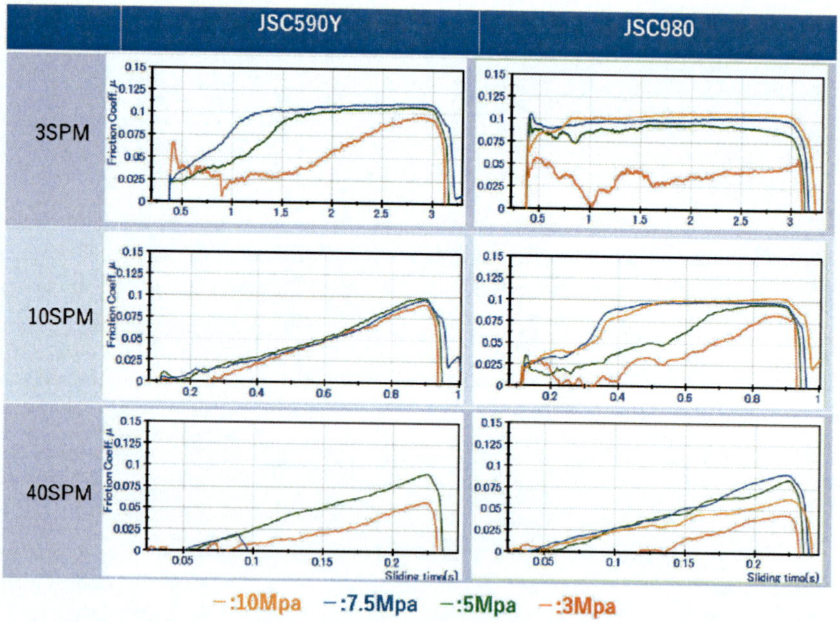

Fig. 4. Data monitored in the hat bending process in different conditions.

motion of the slide for reduction of the friction coefficient, especially for design of pulse motions.

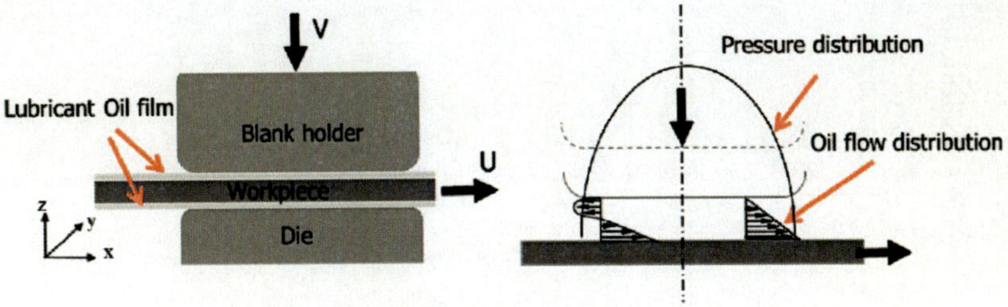

Fig. 5. A schematic model for the squeeze effect.

3.3 Friction Variation in a Pulse Motion

Bending test with a pulse motion (4 mm down and 2 mm up in a motion) was performed. The material used was 590 materials, the cushioning force was 120 kN (surface pressure 5.1 MPa). Lubricants in viscosity of 1 mm^2/s and 31.4 mm^2/s were used for comparison. The results are shown in Fig. 6. In the pulse motion, the cushion is set not to follow the slide displacement when the slide rises, the punch could depart from the workpiece. As s result, the punch load and blank holding force become to zero during every upward motion as shown in Fig. 6(a). The vertical loads are almost the same for outer and inner sides of sensors, but the shearing loads are reduced for the high viscosity oil (Fig. 6(b) and (c)). As a result, the duration of the mixed lubrication state for high viscosity is longer than that for low viscosity (Fig. 6(d)) and therefore, the forming load decreased due to reduction of average friction force. It can be confirmed that the punch load for the high-viscosity lubricant is about 30% lower than that of the low viscosity. In this case, the pulse motion was designed to adopt the lubricant in high viscosity. By increasing each motion in a pulse following the squeeze effect theory, same effect can be achieved with the lubricant in lower viscosity. It is effective to design pulse motion with lubricant in proper viscosity for stamping process in the mixed lubrication region in order to reduce forming load and increase forming limit. The biggest issue when using pulsed motion is that the process time is longer than the crank motion. The increase in process time can be minimized by reducing the duration time of each pulse by understanding the dynamic friction characteristics.

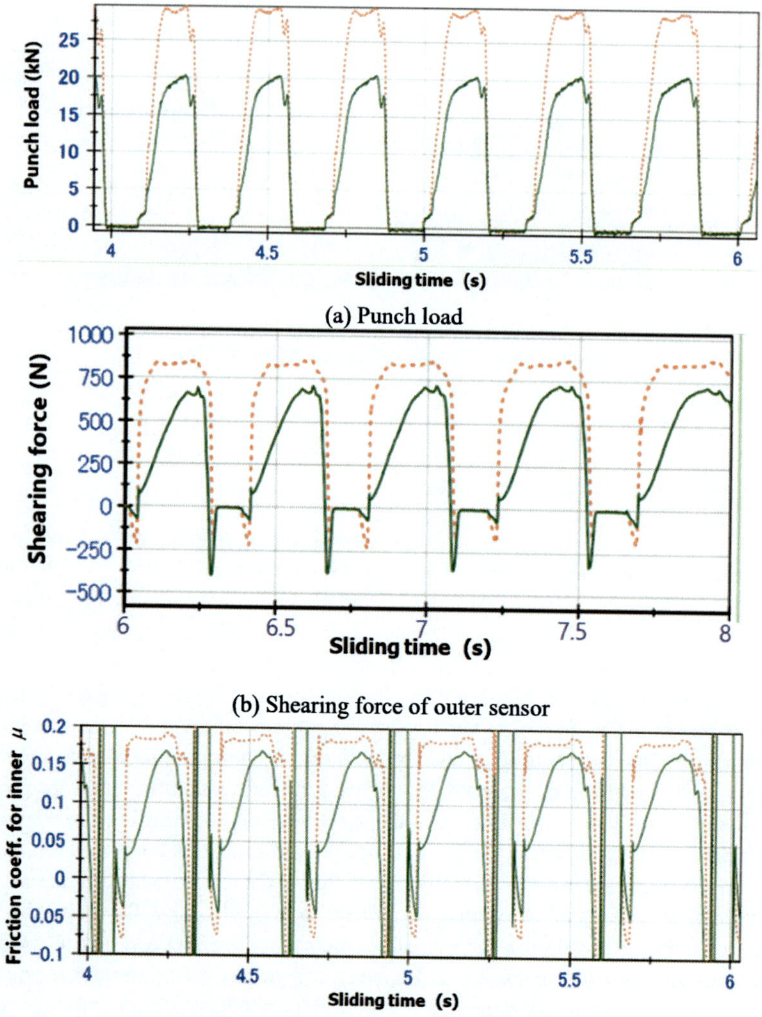

(a) Punch load

(b) Shearing force of outer sensor

(c) Friction coeff. for inner sensor

---- : Low viscosity, ---- : High viscosity

Fig. 6. Comparison of the dynamic change in the friction coefficient between the low-viscosity and the high-viscosity lubricants with pulse motion.

4 Conclusions

A sensor system with sensor-embedded die has been developed to monitor the contact surface pressure and various sliding friction between the die and the material. A hat-bending of steel sheets of high-strengths steels was performed with various process conditions using the developed monitoring system. The influence of material strength, slide speed, cushion pressure, etc. on the dynamic characteristics of friction was evaluated. The dynamic characteristics of the friction was calculated based on the squeeze effect theory by using the experimental results. It can obtain not only the average friction coefficient but also the dynamic friction behavior, which can be used for motion design with selected lubricating oil. A pulse motion was adapted to the hat-bending process. The effect of the friction reduction was verified and as a result, the punch load for the process was reduced by 30% for the case using higher viscosity lubricant with the squeeze effect.

Funding. This work is supported by Tokyo Government and Japan Forming Machinery Association.

References

1. Osakada, K., Mori, K., Altan, T., Groche, P.: Mechanical servo press technology for metal forming. CIRP Ann. Manuf. Technol. **60**, 651–672 (2011)
2. Nishimura, N.: Advancement of press processing using servo press. Press Working **48**(11), 18–23 (2010)
3. Ming, Y.A.N.G.: Smart metal forming with digital process and IoT. Int. J. Lightweight Mater. Manuf. **1**, 207–214 (2018)
4. Tamai, Y., Yamazaki, Y., Yoshitake, A., Imura, T.: Improving deep drawability of steel sheets by motion control of servo press. J. Japan Soc. Technol. Plast. **51**(592), 68–72 (2010). (in Japanese)
5. Turns, S.R.: Annular squeeze film with inertia effects. ASME J. Lubric. Technol. **105**(3), 361 (1983)
6. Andersen, C.B., Ravn, B.G., Wanheim, T.: Development of a commercial transducer for measuring pressure and friction on a model die surface. J. Mater. Process. Technol. **115**, 205–211 (2001)
7. Fratini, L., Lo Casto, S., Lo Valvo, E.: A technical note on an experimental device to measure friction coefficient in sheet metal forming. J. Mater. Process. Technol. **172**, 16–21 (2006)
8. Wang, S.L., Ramaekers, J.A.H.: Measurement of friction and material flow-stress by a plane-strain compression tribometer. J. Mater. Process. Technol. **57**(3–4), 345–350 (1996). https://doi.org/10.1016/0924-0136(95)02082-9
9. Groche, P., Brenneis, M.: Manufacturing and use of novel sensoric fasteners for monitoring forming processes. Measurement **53**, 136–144 (2014)
10. Gong, H., Lou, Z., Zhang, Z.: Studies on the friction and lubrication characteristics in the sheet steel drawing process. J. Mater. Process. Technol. **151**(1–3), 328–333 (2004). https://doi.org/10.1016/j.jmatprotec.2004.04.082
11. Zhang, Z.L., Chen, L.: Study of the friction characteristics of lubricants and their affecting factors during drawing. J. Mater. Process. Technol. **63**, 144–147 (1997)
12. Rosochowsk, A.: Technical feasibility of a three-axis force transducer for measuring pressure and friction on the model die surface - prototype development. J. Mater. Process. Technol. **115**, 192–204 (2001)
13. Doege, E., Seidel, H.-J., Griesbach, B., Yun, J.-W.: Contactless on-line measurement of material flow for closed loop control of deep drawing. J. Mater. Process. Technol. **130–131**, 95–99 (2002). https://doi.org/10.1016/S0924-0136(02)00763-X
14. Siegert, K., Ziegler, M., Wagner, S.: Closed loop control of the friction force. Deep drawing process. J. Mater. Process. Technol. **71**(1), 126–133 (1997). https://doi.org/10.1016/S0924-0136(97)00158-1

15. Wagner, S., Ziegler, M.: Tendencies in presses and dies for sheet metal forming processes. J. Mater. Process. Technol. **98**, 259–264 (2000)
16. Ben, N., Zhang, Q., Meng, D., Lee, M.-G.: Analysis of real contact area and reubrication in oscillating bulk forming process by corrosion method. J. Mater. Process. Tech. **253**, 178–194 (2018)

Contact Conditions and Temperature Distribution During Cryogenic Deep Drawing with Macro-structured Tools

Marc Tulke(✉), Rémi Lafarge, Alexander Wolf, and Alexander Brosius

Chair of Forming and Machining Processes, Technische Universität Dresden, 01062 Dresden, Germany
marc.tulke@tu-dresden.de

Abstract. In this paper, the process model for cryogenic forming with macro-structured tools is analyzed regarding the continuously changing thermal conditions during the deep drawing process. For this purpose, the transient contact conditions within the macro-structure are investigated for different tool designs of the macro-structure as a function of the immersion depth and resulting contact pressure in a numerical analysis. The determination of the heat transfer coefficient between the aluminum sheet metal and the tools as a function of the contact pressure is carried out experimentally. Numerical investigations are used to determine an improved macro-structure in terms of low-heat flow and reliable suppression of wrinkling. The influence of the tool design on the temperature distribution is presented and compared with the conventional deep drawing process.

Keywords: Deep drawing · Aluminum · Cryogenic forming

1 Introduction

Demands to increase fuel efficiency and reduce CO_2 emissions are leading to higher material efficiency in production processes and the substitution of steel by lightweight materials. The AA5xxx, AA6xxx and AA7xxx aluminum alloys are often used as lightweight materials due to material properties such as high strength and corrosion resistance. Unfortunately, their mechanical properties pose major challenges due to a relatively low formability at room temperature. However, their ductility increases significantly at elevated part temperatures that can lead to reduction in strength [1]. Alternatively, the formability of aluminum alloys increases at cryogenic temperatures. Schneider et al. demonstrate in [2] a general increase in strength and fracture strain with decreasing temperature for different aluminum alloys and Sotirov et al. show improved formability in a forming limit diagram at low temperatures down to $\vartheta = -100$ °C [3]. Tulke et al. present in [4] a reduced springback behavior for deep drawing with macro-structured tools at cryogenic blank temperature compared to room temperature. Grabner et al. show the positive effect of cryogenic deep drawing of a downscaled B-pillar that was only successfully deep drawn at a tool temperature of $\vartheta = -150$ °C. Yuan et al. achieved an

K. Mocellin et al. (Eds.): ICTP 2023, LNME, pp. 233–243, 2024.
https://doi.org/10.1007/978-3-031-40920-2_25

increased drawing ratio from $\beta = 1.8$ at room temperature up to $\beta = 2.08$ with nitrogen cooled tools at $\vartheta = -60$ °C [5]. Contrary, Tulke et al. demonstrated a method for improving the forming limits in deep drawing at cryogenic blank temperatures without tool cooling, where macro-structured tools are used to reduce the contact area between the tool and blank to reduce the heat flow [6]. The advantage of the presented tool design is the lower complexity since there is no cooling system required.

The correlations of thermal and mechanical conditions require a FE-model for the complex process optimization. The accuracy of a thermomechanical model depends on the accuracy of the thermal parameters. For that reason, the heat transfer coefficient at the contact area between blank and tool need to be precisely determined. However, the heat transfer coefficient depends on the temperature and contact pressure [7].

2 Approach

The objective of the present work is the analysis of the contact conditions while deep drawing with macro-structure tools in the cryogenic range. Indeed, cryogenic deep drawing is expected to be achieved for higher drawing ratio by leveraging the improved formability of aluminum at low temperature [5]. In this context, the influence of the different designs of the macro-structure on the contact area is analyzed. Because the heat transfer is dependent on the contact pressure, the heat transfer coefficient is determined experimentally for an appropriate process design by a finite element analysis.

2.1 Macro-Structured Deep Drawing at Cryogenic Temperatures

As described in [8] macro-structured tools reduce the contact area between the tools and the blank which results in a reduced friction force. The wavy structure induces an alternate bending depending on the immersion depth δ and wavelength λ (see Fig. 1), which prevents wrinkling and enables deep drawing without lubricants. The aluminum blanks are cooled in a liquid nitrogen bath before deep drawing, while the tools remain at room temperature. Here, the reduced contact area is mainly used to obtain the blank temperature in a cryogenic range as long as possible. The punch is designed hollow at the end surface to reduce the contact area to the blank. The tool design for cryogenic deep drawing without tool cooling is presented in Fig. 1.

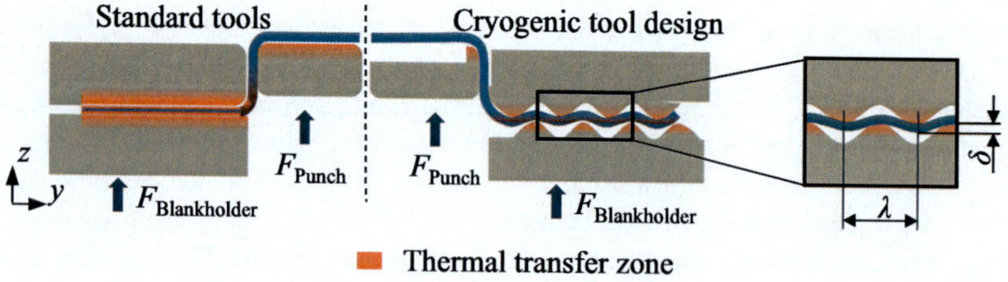

Fig. 1. Cryogenic deep drawing with macro-structured tools according to [6]

2.2 Determination of Heat Transfer Coefficient

Experimental Setup for Thermal Characteristics. For determining the heat transfer coefficient, an experimental setup similar to [7] is used. The experimental setup, shown in Fig. 2, consists of two punches made of steel (1.7225) and an aluminum plate (AA6014) with a thickness $t_{Al} = 10$ mm. The diameter of the aluminum plate and the punches is $D_{Al} = D_{punch} = 70$ mm. Each punch is equipped with one thermocouple (TC) at the contact surface between the punch and the aluminum plate. Three TC are located in the punches with a distance of 4 mm, 11 mm, 30 mm and 55 mm from the contact surface. The positions of the thermocouples result from the assumption that the temperature profile is approximately exponential while the temperature change between the measuring points decreases with increasing distance to the contact surface. The thermocouples are located in the center of the punches. The holes for the thermocouples are arranged horizontally to suppress the thermal influence of the holes. Finally, one thermocouple is positioned in the center of the aluminum plate. The blank is cooled in liquid nitrogen prior the tests whereby the tool temperature remains at room temperature. The tests are performed on the material testing machine Inspekt 250, HEGEWALD&PESCHKE. The contact pressure varies from 0.1 MPa $< \sigma_p <$ 50 MPa due to the relative low contact pressure at the macro-structured tools in deep drawing.

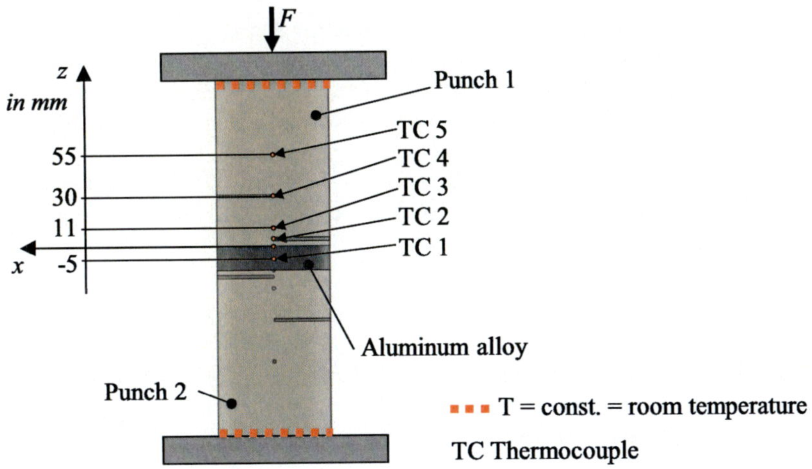

Fig. 2. Experimental setup for heat transfer coefficient in principle

Parameter Identification. From the measurement of temperature, the value of the heat transfer coefficient can be derived. Due to the two different materials and the associated complex system of equations, the problem is analyzed numerically. A first model was implemented to solve the forward problem – determination of the temperature change knowing the value of h_c – and used with an optimizer to solve the inverse problem and identify the real value of h_c. The complete experimental setup can be simplified and be modeled as one dimensional. The hypothesis is that the heat flow in the x and y directions are negligible compared to the z direction. Within the bulk of the blank and the tool, the heat transfer equation simplifies to:

$$\frac{\partial T}{\partial t} - \frac{\lambda_{th}}{\rho \cdot C_p} \frac{\partial^2 T}{\partial z^2} = 0 \tag{1}$$

with λ_{th} the thermal conductivity, ρ the density and C_{p} the thermal capacity.

The heat transfer between the tool and the blank can be described by the following:

$$\rho_{\mathrm{Tool}} C_{\mathrm{p,\,Tool}} \frac{\partial T_{\mathrm{Tool}}}{\partial t} - \rho_{\mathrm{Blank}} C_{\mathrm{p,\,Blank}} \frac{\partial T_{\mathrm{Blank}}}{\partial t} = h_c \cdot (T_{\mathrm{Tool}} - T_{\mathrm{Blank}}) \tag{2}$$

with h_{c} the contact heat transfer coefficient.

The temperature at the end of the tool is taken as constant. Because the system is symmetric, the second order derivative of the temperature is zero in the middle of the blank. This gives a fully determined, well-posed, problem to solve. This can be done using the finite difference method, which is easy to employ with the heat transfer equation. An explicit scheme for time derivation and central differentiation for spatial derivation was used, which enable to rewrite the system as:

$$\left[T^{n+1}\right] = A \cdot \left[T^n\right] + B \tag{3}$$

with $\left[T^{n+1}\right]$ and $\left[T^n\right]$ the vector of the temperature value at the node values at time step $n+1$ and n respectively, A a square tridiagonal matrix of real coefficients and B a vector of coefficient. Because the scheme is explicit, the value of the time step and spatial resolution must be carefully selected so that convergence is achieved [9]. The temperature vector at an arbitrary time step can be computed as:

$$\left[T^n\right] = A^n \cdot \left(\left[T^0\right] - C\right) + C \tag{4}$$

with:

$$C = (A - I)^{-1} B \tag{5}$$

The hypothesis is made that the thermal properties of the material are constant, so that the value of A is constant during the simulation. It follows that computation of the power of A can be significantly fastens by using a diagonalization so that the equation becomes:

$$\left[T^n\right] = P^{-1} A_d^n P \cdot \left(\left[T^0\right] - C\right) + C \tag{6}$$

with A_d diagonal. The complete model is implemented in Python using Numpy as SciPy. The values of λ_{th}, ρ and C_p are given in Table 1.

Using the described model, the value of h_{c} is determined by solving the inverse problem. For that purpose, the least square fitting is realized between the model output and the measured temperature curve, with h_{c} as parameter. To that effect, the minimize_scalar function of SciPy using the bounded Brent method, was employed [12].

An implementation of the described algorithm is published in [13].

Table 1. Material characteristics used for the optimization

	Blank (AA 6014) [10]	Tool (1.7225) [11]
Density ρ in $kg \cdot m^{-3}$	$2.7 \cdot 10^3$	$7.72 \cdot 10^3$
Thermal conductivity λ_{th} in $W \cdot m^{-1} K^{-1}$	120	45
Thermal capacity C_p in $J \cdot kg^{-1} K^{-1}$	600	465

2.3 Numerical Setup for Characterizing the Tool Design

Contact Length. A numerical analysis is used for determining the contact length of the tools, depending on the geometry of the macro-structure and the immersion depth δ. In order to save the calculation resources associated with running a full forming simulation, the following simplified FE-model was used. The influencing geometrical parameters of the macro-structure are the wavelength λ and the wave radius R_{wave}, see Fig. 3. The 2D-model is built in LS-Dyna 12.0.0 with a plane-strain element type (ELFORM 13). The blank is meshed with an element size of 0.05 mm in general. In the range of contact between the tools and the blank, the mesh size for the blank and tool is 0.0125 mm for higher accuracy. The material behavior of the blank is described by material type MAT133 with temperature dependent flow curves. The flow curves are determined by tensile tests according to ISO 6892 at different cryogenic temperatures and a subsequent extrapolation using Voce law [14]. The contact conditions are modeled by the Coulomb friction model with a friction coefficient $\mu = 0.15$. The model is reduced to one wave of the die and the blank holder due to symmetry.

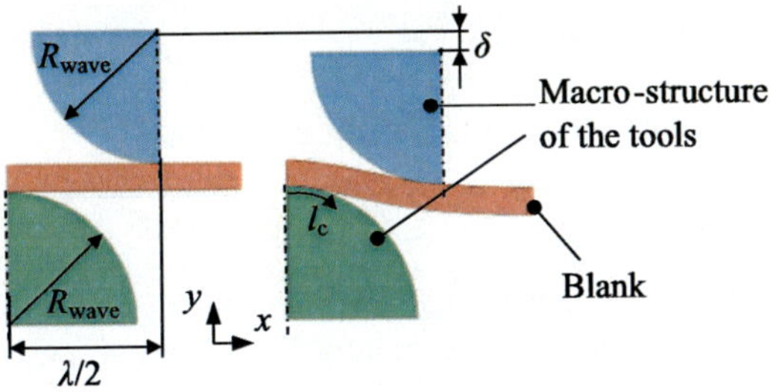

R_{wave} radius of macro-structure

λ wavelength

δ immersion depth

l_c contact length

Fig. 3. FE-model for characterizing the contact length in the macro structure

Sixteen different tool designs are examined regarding the contact length due to the combination of each wavelength with every single wave radius, summarized in Table 2.

Deep Drawing. A second numerical setup describes the deep drawing process of a rotational symmetrical cup with macro-structured tools at cryogenic temperatures with the same material model as described above. That FE-model is used to validate selected geometrical variants of the macro-structure regarding wrinkling and bottom cracks. Here, the blank is meshed with an element size of 1.0 mm with a fully integrated 3D-shell

Table 2. Parameters of different tool designs

Wavelength λ in mm	8	10	12	14
Wave radius R_{wave} in mm	2	3	5	8

element (ELFORM 16) with nine integration points over the shell thickness. The FE-model of the rotational symmetrical cup is reduced to a quarter to reduce calculation time. The tool setup in principle and the dimensions of the die are presented in Fig. 4. The initial temperature of the blank is $T = 123$ K while the temperature of the tools is $T = 293$ K. Due to the results reached in [6] the investigated initial blank diameter is $D_0 = 200$ mm which results in a drawing ratio of $\beta = 2.0$. The punch diameter is $d_{punch} = 100$ mm with a die edge radius $r_{die} = 10$ mm. The cups are deep drawn with a punch velocity $v_{punch} = 20$ mm/s.

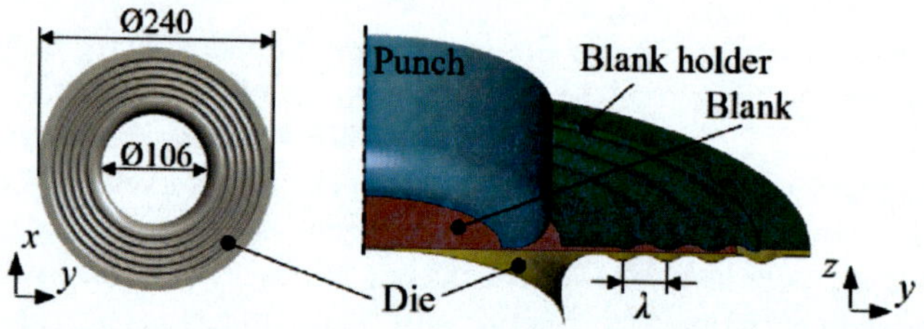

Fig. 4. Tools setup in principle for macro-structured deep drawing a rotational symmetrical cup

3 Influence of the Tool Design on the Deep Drawing Process

3.1 Influence on the Contact Length

The contact length l_c increases with the immersion depth, but is also strongly dependent on the geometry of the macro-structure. The varied geometrical parameters are the wavelength λ and the wave radius R_{wave} at the contact. The result of the investigation shows that the contact length l_c is negatively correlated to the wavelength λ on the one hand and positively correlated to the wave radius R_{wave}. It is recognizable that the impact of the wave radius on contact length increases with decreasing wavelength. It is illustrated for different immersion depths in Fig. 5. Regarding the contact length, a large wavelength with a small wave radius is preferable to minimize the heat flux. However, the transferability to the deep drawing process needs to be examined due to the dependency on the immersion depth.

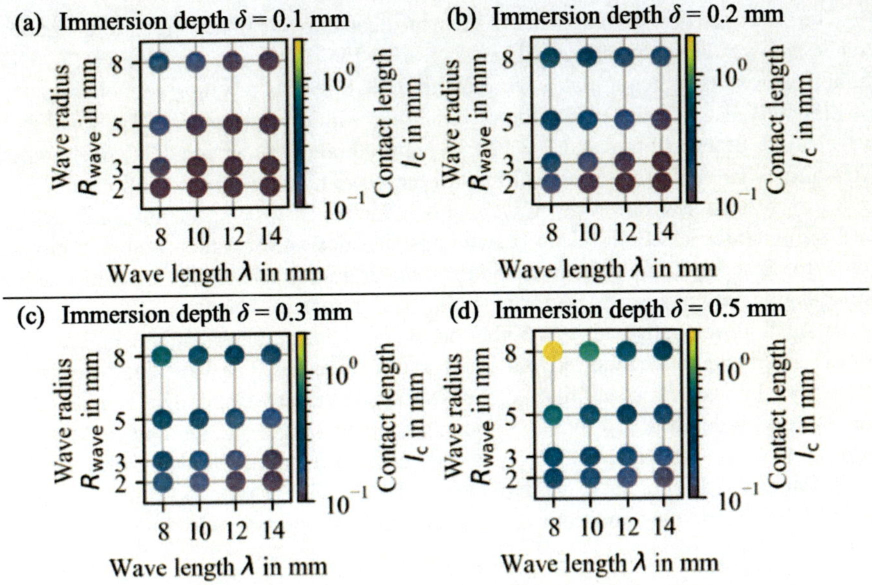

Fig. 5. Contact length l_c depending on the tool design and immersion depth δ

3.2 Influence on the Cryogenic Deep Drawing Process

Based on the results regarding a reduced contact length l_c, selected combinations of wavelength λ and wave radius R_{wave} are used for validating their usability for cryogenic deep drawing. The challenge here is to find the right immersion depth for each combination. If δ is too large, bottom cracks occur and if it is too low, the blank wrinkles. Increasing the immersion depth increases the contact length proportionally. That results in higher heat flux, friction force and the induced bending increases. As a result, the blank temperature rises faster and the strength of the aluminum alloy decreases.

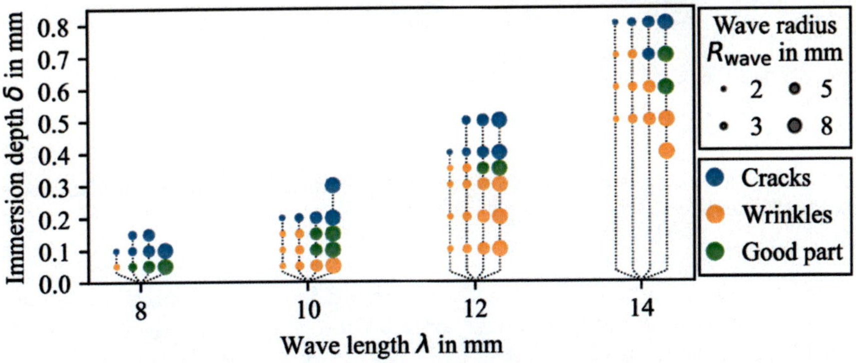

Fig. 6. Influence of the tool design on the deep drawing result

The examined designs and their effect on the deep drawing process regarding bottom cracks and wrinkling are presented as function of the immersion depth in Fig. 6. The numerical results indicate that a wavelength $\lambda = 8$ mm is advantageous, because it enables fully deep drawn cups without wrinkling with a wide range of R_{wave}. Higher wavelengths of $\lambda = 10$ mm and $\lambda = 12$ mm require higher immersion depth to suppress wrinkling where only the combination with the wave radius $R_{wave} = 5$ mm and $R_{wave} = 8$ mm was successful. If the wavelength is increased to $\lambda = 14$ mm, only $R_{wave} = 8$ mm enables successful deep drawn cups. In contrast, no valid combination was found for $R_{wave} = 2$ mm in the examined parameter field. These results show that larger wavelengths require an increased wave radius and immersion depth.

A small wavelength leads to a higher number of contact zones, but the total contact area is dependent on the required immersion and wave radius. The wavelength primarily influences the geometric stability, i.e. the resistance against wrinkling, as a function of the immersion depth. R_{wave} mainly affects the contact length and thereby contact area depending on δ. Consequently, R_{wave} is responsible for the development of the blank temperature and thus for the strength of the blank material. The process limits are defined by the thermodynamic conditions on one hand and the onset of wrinkling on the other hand.

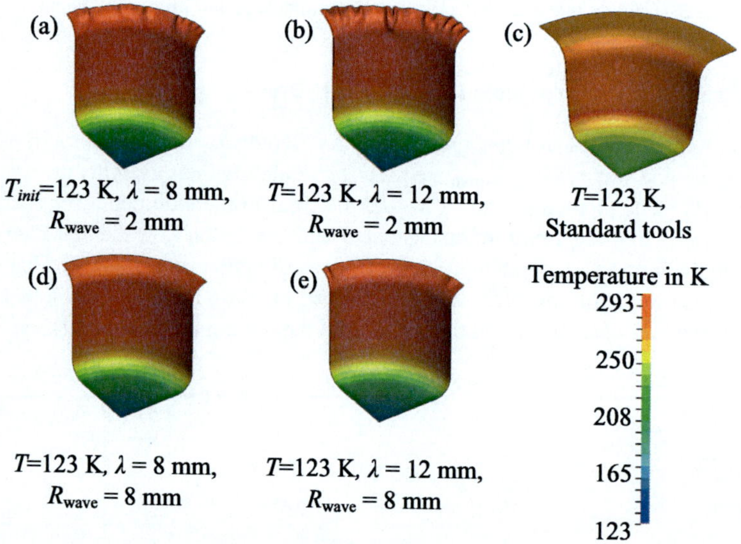

Fig. 7. Numerical deep drawn cups for different designs of the macro-structure and comparative with standard tools

Figure 7 shows exemplarily the numerical results for $\lambda = 8$ mm and $\lambda = 12$ mm in case of a successfully deep drawn cup or wrinkling. In contrast, deep drawing with a drawing ratio $\beta = 2.0$ with standard tools leads to bottom cracks at lower drawing depth. The high contact area of the standard tools results in a high heat flux and the blank temperature rises faster. Consequently, the strength of the blank decreases and bottom crack occurs even with a low drawing depth. The cup deep drawn with standard tools is depicted in Fig. 7(c), at the time step before the bottom crack occurs.

4 Results of the Contact Heat Transfer Coefficient

Figure 8 presents the results of the parameter identification. Overall, a good fit between the experimental values and the model are observed. The dependency of the h_c values on the contact pressure, depicted in Fig. 8 is quite clear, whereby a saturation of h_c is recognizable with increasing contact pressure. Furthermore, the values are consistent with other sources, e.g. [15]. However, the fit in case of a contact pressure $p_c = 0$ MPa shows a higher difference that is explained by the inhomogeneous contact conditions at low contact pressure. Nevertheless, this is negligible since the contact pressure in the presented deep drawing process is greater than 10 MPa. The temperature dependency of λ_{th} and C_p are neglected here. An evolution of the model where these thermal properties are dependent upon the temperature may provide results that are more accurate.

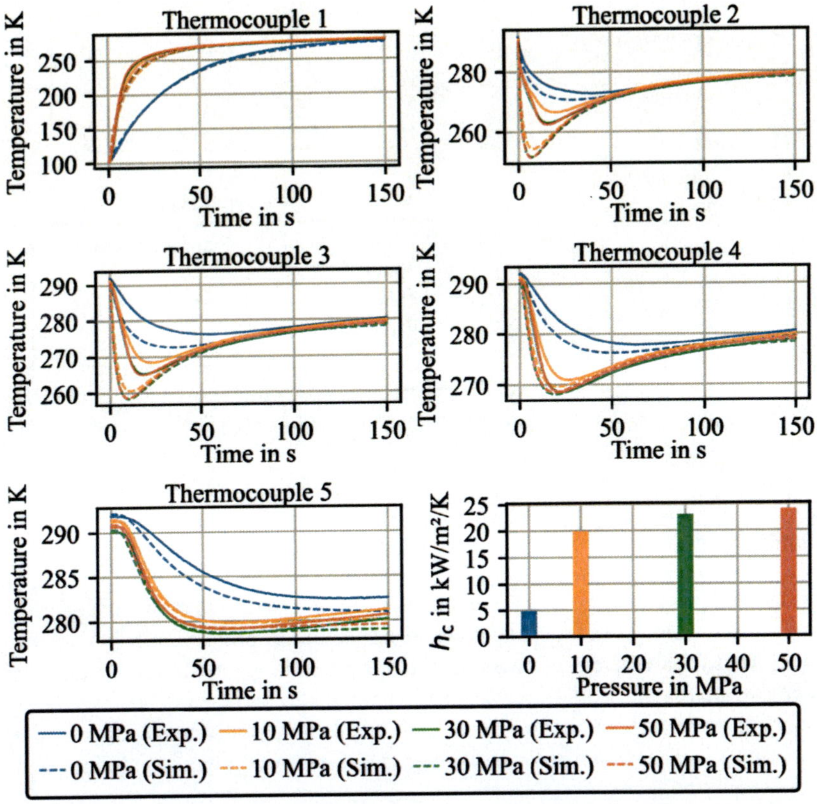

Fig. 8. Experimental and numerical temperature development of the different thermocouple locations and the resulting specific heat transfer coefficient h_c as function of the contact pressure

The influence of a contact pressure dependent heat transfer coefficient on the temperature development in the numerical simulation is shown in Fig. 9. It is clearly visible that the temperature development differs from that of a static h_c, especially at the beginning of the process. At the beginning, the flange is still large so that the contact pressure remains low. Accordingly, the heat transfer coefficient is small, and the predicted blank temperature is significantly lower compared to the calculation with a constant h_c. The predicted blank temperatures of both simulations are almost the same at the end of the process, but the temporal course of the blank temperature as function of the drawing path is significantly influenced. This is essential for a successful deep drawing process. Besides the flange, the temperature in the critical area for bottom cracks near the punch edge is predicted lower as illustrated in Fig. 9. The lower temperature correlates with a higher strength of the aluminum sheet metal. Consequently, the resistant against bottom crack rises that leads to higher drawing depth.

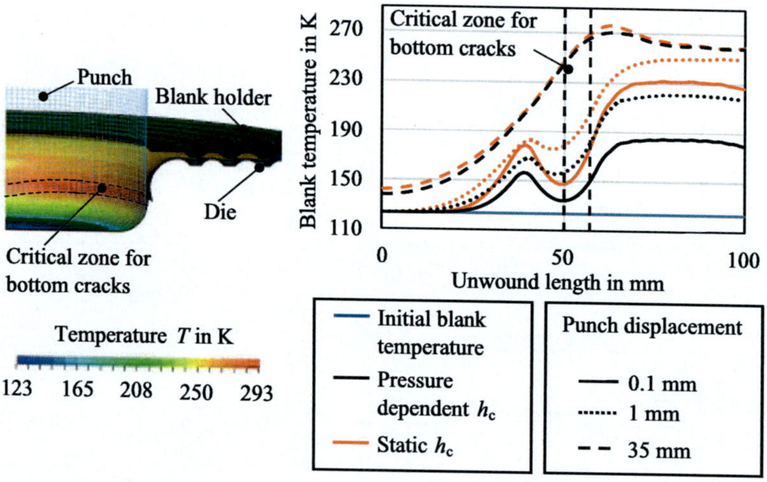

Fig. 9. Temperature development for static and pressure dependent h_c at different forming stages

5 Conclusion

The presented work demonstrates the influence of the tool design on the contact length between macro-structured tools and blank as a function of immersion depth. However, the geometrical parameters in order to obtain both a reduced surface contact, while avoiding wrinkling and bottom cracks need to be chosen carefully. The usage of small wavelength combined with a low immersion depth turned advantageous regarding a small contact length, which reduces the heat transfer between blank and tools, and for suppressing wrinkling.

The heat transfer coefficient shows a significant dependency on the contact pressure which needs to be considered in simulation. Otherwise, the blank temperature will be predicted incorrectly that leads to errors in process calculation.

References

1. Sotirov, N., Grabner, F., Schneider, R., Mozdzen, G.: Cryogenic forming behaviour of AW-6016-T4 sheet. Trans. Nonferrous Metals Soc. China **27**, 1257–1263 (2017). https://doi.org/10.1016/S1003-6326(17)60146-8
2. Schneider, R., et al.: Constitutive flow curve approximation of commercial aluminium alloys at low temperatures. Mater. Des. **88**, 659–666 (2015). https://doi.org/10.1016/j.matdes.2015.09.034
3. Sotirov, N., et al.: Improved formability of AA5182 aluminium alloy sheet at cryogenic temperatures. Mater. Today: Proc. **2**, S113–S118 (2015). https://doi.org/10.1016/j.matpr.2015.05.027
4. Tulke, M., Wolf, A., Brosius, A.: Cryogenic deep drawing of aluminum alloy AA6014 using macro-structured tools. IOP Conf. Ser.: Mater. Sci. Eng. **1157**(1), 012053 (2021). https://doi.org/10.1088/1757-899X/1157/1/012053
5. Yuan, S., Cheng, W., Liu, W., Xu, Y.: A novel deep drawing process for aluminum alloy sheets at cryogenic temperatures. J. Mater. Process. Technol. **284**, 116743 (2020). https://doi.org/10.1016/j.jmatprotec.2020.116743
6. Tulke, M., Wolf, A., Lafarge, R., Grunow, F., Brosius, A.: Influence of macro-structured tools on the formability of aluminum alloys in the cryogenic temperature range. Key Eng. Mater. **926**, 760–767 (2022). https://doi.org/10.4028/p-nhu9o5
7. Malinowski, Z., Lenard, J.G., Davies, M.E.: A study of the heat-transfer coefficient as a function of temperature and pressure. J. Mater. Process. Technol. **41**(2), 125–142 (1994). https://doi.org/10.1016/0924-0136(94)90057-4
8. Brosius, A., Mousavi, A.: Lubricant free deep drawing process by macro structured tools. CIRP Ann. **65**(1), 253–256 (2016). https://doi.org/10.1016/j.cirp.2016.04.060
9. Allaire, G.: Analyse numérique et optimisation: Une introduction à la modélisation mathématique et à la simulation numérique. Éd. de l'École polytechnique, Palaiseau (2005)
10. NIST: https://www.nist.gov/mml/acmd/aluminum-6061-t6-uns-aa96061. Last accessed 28 Feb 2023
11. Suzhou Waldun Steel: https://waldunsteel.com/products/din-1-7225-steel. Last accessed 28 Feb 2023
12. Brent, R.P.: Algorithms for Minimization Without Derivatives. Dover Publications Inc, Mineola, New York (2013)
13. Lafarge, R., Tulke, M., Brosius, A.: tud-if-ff/CryoHTC: CryoHTC_Contribution_ICTP2023 (2023). https://doi.org/10.5281/zenodo.7845166
14. Voce, E.: The relationship between stress and strain for homogeneous deformation. J. Inst. Met. **74**, 537–562 (1948)
15. Engineers edge: https://www.engineersedge.com/heat_transfer/thermal_contact_conductance_13849.htm. Last accessed 28 Feb 2023

Numerical and Experimental Forming of a Cylindrical Cup in 7075-T6 Under Warm Temperature Conditions

S. Royne[1(✉)], H. Laurent[1], and A. Maillard[2]

[1] Univ Bretagne Sud, UMR CNRS 6027, IRDL, 56100 Lorient, France
sylvain.royne@univ-ubs.fr

[2] CETIM, Technical Center for Mechanical Industries, 60300 Senlis, France

Abstract. The use of aluminium alloys with high mechanical properties is very interesting to reduce the weight of parts, especially in the automotive industry. However, forming of these alloys at room temperature (RT) is difficult, especially for complex parts where fracture often occurs before the end of the forming process. The warm forming process, which consists of heating aluminum alloys to a temperature between 150 and 250 °C, increases formability and thus allows complex shapes to be formed. In this study, forming of a cylindrical cup using a specific two-step drawing device by two reductions of the blank diameter is performed at RT and 200 °C for a 7075-T6 alloy. These two forming steps are also numerically simulated using Abaqus. Forming at RT leads to the failure of the cup in the first stage, which is also obtained by numerical simulation with plastic deformations higher than what the material can support. At 200 °C, the cylindrical cup can be formed in two steps, which results experimentally and numerically in a significant reduction of the forming forces.

Keywords: Warm forming · Aluminium Alloy · 7075-T6 · Two-step forming

1 Introduction

High strength aluminum alloys, such as the 7xxx series, are an attractive alternative to conventional steels in the automotive and aerospace industries. However, the poor formability of these aluminum alloy sheets is a serious obstacle to their use in the manufacture of complex lightweight components. This disadvantage can be reduced with the increase of the temperature which leads for temperatures higher than 150 °C to a strong increase of the formability. Nevertheless, most of the studies carried out concern a part made in a single step and use heating methods which are not necessarily compatible with the current industrial environment [2,3,7]. Only a few studies focus on multi-stage warm forming of sheets to analyze the improvement of formability of high strength aluminum alloys [1,5]. It is also important to keep the best performing T6 state of these

K. Mocellin et al. (Eds.): ICTP 2023, LNME, pp. 244–251, 2024.
https://doi.org/10.1007/978-3-031-40920-2_26

alloys and in our previous study [6], we showed that temperature forming should be performed for times between 10 and 20 s at 200 °C. A longer duration leads to the formation of precipitates that no longer maintain the T6 state and then requires additional heat treatment costs to recover this state. The objective of this study is to investigate the warm formability of a 7075-T6 alloy using a new drawing and redrawing test configuration of a cylindrical cup. The simulation of these two-step forming at RT and 200 °C is also numerically simulated using Abaqus and the forming forces are compared to the experimental during these two steps.

2 Material and Methods

2.1 Material

AA7075-T6 sheet aluminum alloy with 0.8 mm thickness is used in this work. Commercially purchased aluminum was directly received in T6 state. The chemical composition of this alloy is presented in Table 1. Its mechanical properties measured in the T6 receiving conditions are given in Table 2.

Table 1. Chemical composition of AA7075-T6 sheet in %wt.

Zn	Mg	Cu	Si	Cr	Fe	Mn	Ti	Al
5.1-6.1	2.1-2.9	1.2-2.0	0.4	0.18-0.28	0.5	0.3	0.2	Balance

Table 2. Mechanical properties at RT of the AA7075-T6 measured at receiving condition.

Conditions	$Rp_{0.2\%}$ [MPa]	R_m [MPa]	$A\%$
Receiving	500	557	14.6

2.2 Warm Deep Drawing and Redrawing of a Cylindrical Cup

The warm forming and redrawing experiments were carried out separately in two stages on a press equipped with several heating and forming stages with a specific device manufactured by CETIM. Figure 1 presents the schematic of the arrangement of the heating station and the two forming stages stations in the press.

The blank initial diameter is 157 mm. The device consists of a first stage of heating the blank by clamping contact to reach a uniform temperature of 200 °C in 0.45 s. In the next stage, the blank is manually transferred to the first forming

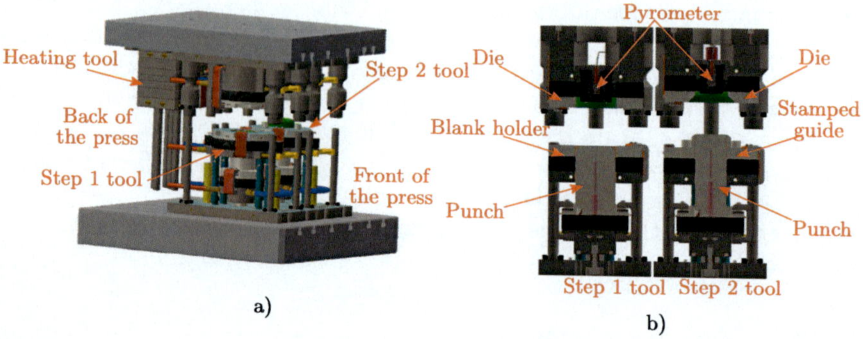

Fig. 1. a) Schematic of the press setup with the heating station and the two forming stages stations, b) Cutting view of the two forming tools.

station as presented in Fig. 2-a and once the stamping is done, the blank is manually transferred to the second drawing station presented in Fig. 2-b. These two forming stations are placed next to each other to limit the transfer time. All tools are heated with heating cartridges and their temperature is controlled by K-type thermocouples. A clamping force of 66 kN is applied on the first stamping step to the blank holder and no clamping force is applied on the second forming step. The reaction force during both drawing processes is measured by a force sensor under the punch and the temperature of the blank is monitored by a pyrometer located at 20 mm from the blank.

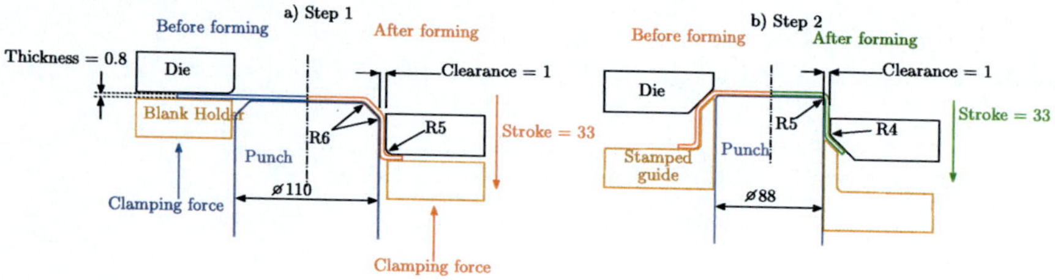

Fig. 2. Dimensions and movements of tools (in mm) during a) the first step of forming and b) the second step of forming

It is assumed that all the stamping and transfer stages are carried out under isothermal conditions. The three steps (heating, first and second forming) are carried out in less than 20 s to maintain the mechanical characteristics of the T6 state [6].

Forming test are performed at RT and 200 °C with a JELT 5411 lubricant. At RT, forming is impossible and causes the blank to break during the first drawing. At 200 °C, four tests are performed to check the reproducibility but the average result will be used as a representative. Figure 3 presents the theoretical dimensions of the part after the two stamping operations and a picture of the parts obtained at the end of these two operations at 200 °C.

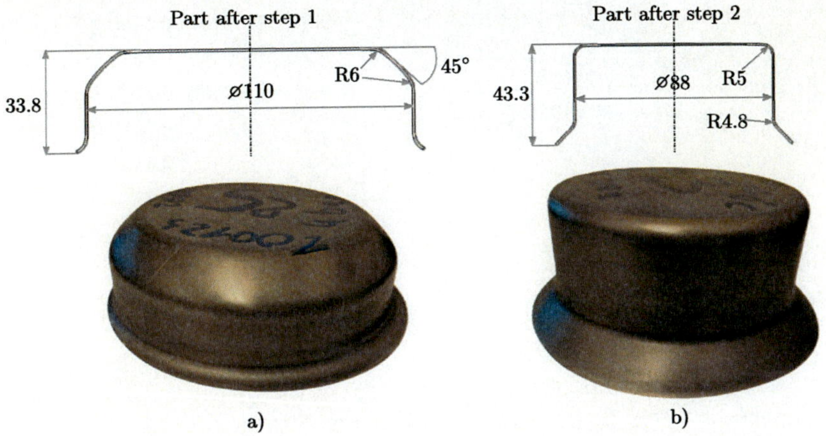

Fig. 3. a) Dimensions (in mm) of the part after step 1 a) and b) after step 2.

2.3 Numerical Simulations of the Warm Forming Process

Isothermal numerical simulations of the forming process are conducted with Abaqus/Standard. A 2D axisymetric model is used with tools modeled as rigid surfaces. A section of the blank is meshed with 2D quadratic quadrilateral elements CAX4. A mesh size of approximately 3.5 mm is used in the punch zone before the punch radius whereas the size decrease to approximately 0.5 mm after this radius. Moreover, 11 elements are used through the thickness. The tensile tests show little striction and therefore a Swift law has been chosen to identify the behavioral laws. Several Swift law (1) are identified by tensile tests at room temperature and 200 °C for strain rates of 2×10^{-4}, 2×10^{-3}, 2×10^{-2} and 0.2s^{-1}. The blank has therefore a strain rate dependent elastic-plastic behavior. The parameters of the hardening law at RT and 200 °C are summarized in Table 3.

$$\bar{\sigma} = K(\varepsilon_0 + \bar{\varepsilon})^n \tag{1}$$

K, ε_0, n are material parameters and $\bar{\varepsilon}$ is the equivalent plastic strain

Table 3. Swift's law parameters at RT and 200 °C identified through tensile tests at different strain rates.

Temperature [°C]	Strain rate [s^{-1}]	K [MPa]	ε_0 [-]	n [-]
20	2×10^{-4}	841.7	0.0405	0.1410
	2×10^{-3}	869.4	0.0457	0.1590
	2×10^{-2}	859.2	0.0468	0.1530
	2×10^{-1}	859.4	0.0458	0.1523
200	2×10^{-4}	300.3	0.0114	0.0077
	2×10^{-3}	431.1	0.1695	0.1497
	2×10^{-2}	479.9	0.2378	0.1713
	2×10^{-1}	500.5	0.0401	0.0680

3 Results and Discussion

The experimental results of the reaction force of the punch at RT and 200 °C during the first step forming are shown in Fig. 4 as well as the temperature measured by the pyrometer during the forming at 200 °C.

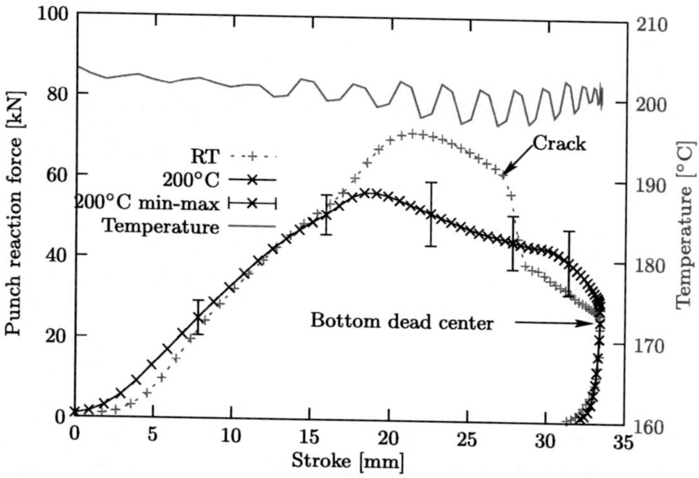

Fig. 4. Experimental punch force as a function of punch displacement at RT and 200 °C with a clamping force of 66 kN during the first stage of drawing and temperature of the blank during forming measured by pyrometer for the target temperature of 200 °C.

The temperature measured during forming is 201 ± 1.67 °C and thus confirms that the drawing is indeed performed at the target temperature. Oscillations in the temperature measurement may be due to slight vibrations during the drawing process and the fact that the lubricant generates some bubbles during heating, which may slightly disturb the measurement. The error bar on the force curve at 200 °C represents the variation of the force over the four tests performed, while at 20 °C the force is similar for all tests.

The punch force decreases with the temperature as classically observed by several authors [4,7]. At RT, a fracture of the cup during this first stage of drawing appears at about 26 mm of the punch displacement whereas the cup is correctly formed at 200 °C. The force is not zero when the bottom dead center of the press is attained, force that is gradually cancelled out when the tools start to move up. No wrinkles appear on the cup at the end of this first drawing with the clamping force of 66 kN at 200 °C (see Fig. 3-a).

Figure 5 presents the results of numerical simulations of the first drawing stage at RT and 200 °C for a friction coefficient of 0.05, 0.08 and 0.1. The small jump in force observed numerically at about 16mm displacement is due to the contact of the blank on the second punch radius. At RT, a coefficient of friction of 0.05 is the closest to the experimental curve until the cup breaks, while at 200 °C, a coefficient of friction of 0.08 gives the best result especially after the maximum force. Numerically, the failure of the cup during the first stage is detected with plastic deformations beyond what the material can withstand.

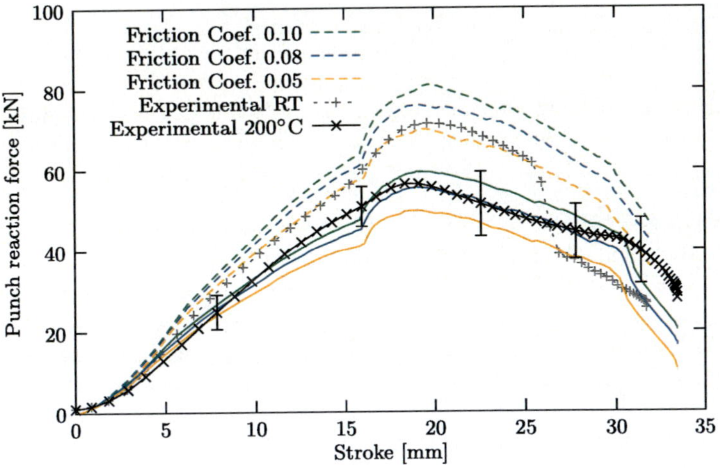

Fig. 5. Experimental and numerical punch force as a function of punch displacement for the first step of stamping at RT and 200 °C and influence of friction coefficient.

Figure 6 presents the comparison between experimental and numerical results of the punch force in function of the punch displacement during the second drawing stage at 200 °C. The punch force is much weaker during this second stage than during the first. The stabilization of punch force at about 6 mm of displacement is due to the blank wrapping around the punch radius. The increase observed at about 23 mm is due to the contact with the bottom radius of the blank obtained during the first drawing stage which comes in contact with the die of the second stage. This evolution of the contact force is more marked numerically. The influence of the friction coefficient is much smaller during this step and the choice of an effective friction coefficient is much more difficult to give.

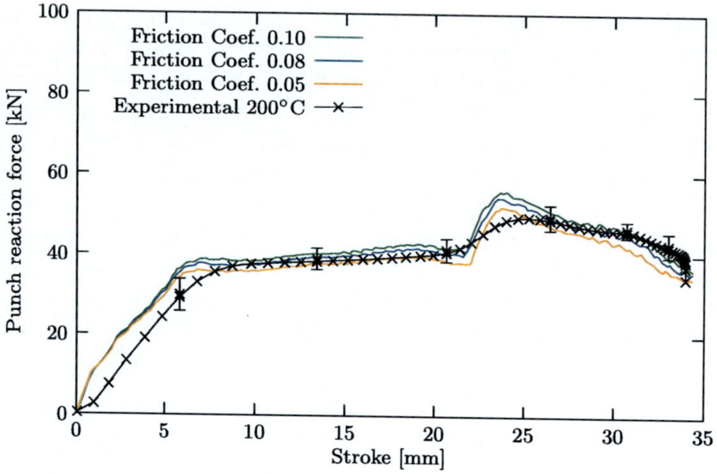

Fig. 6. Experimental and numerical punch force as a function of punch displacement for the second step of stamping at 200 °C and influence of friction coefficient.

The friction coefficients found numerically are low compared to the standard value in stamping. Additional simulations are being carried out because it is possible that the clamping force applied is lower than that currently envisaged in the device. This lower clamping force could be explained by the presence of a constant clearance between the die and the blank holder and the total force envisaged of 66 kN would not fully apply. Numerical simulations are still in progress to verify this hypothesis both numerically and experimentally. Finally, the chosen behavior law can also influence the numerical results of force-displacement. Thus, a study of the influence of the choice of the constitutive law using a saturating Voce law on the results in force and on the friction coefficients is currently in progress.

4 Conclusion

Warm forming in two successive steps of a cylindrical cup part of 7075-T6 alloy was performed at RT and 200 °C with a dedicated process. Numerical force displacements were compared with the experimental results and the influence of friction coefficient was tested. It is observed that the cylindrical cup breaks at RT during the first drawing step, whereas at 200 °C, it forms correctly in two steps. In addition, the different changes in force direction observed on the force displacement curves at RT and 200 °C appear to be much more pronounced numerically. The numerical results obtained are in the same order of magnitude as the experimental results but not for the same friction coefficient in function of temperature. Finally, the various friction coefficients found numerically seem low compared to the standard values. A lower clamping force could allow to find a higher friction coefficient and numerical simulations are still in progress to confirm this hypothesis.

Acknowledgments. The authors gratefully acknowledge the financial support of the Brittany Region (France) via the program RB EMBHOTAL.

References

1. Günzel, J., Hauß, J., Gaedigk, C., Bergmann, J., Groche, P.: Development of a process chain for multi-stage sheet metal forming of high-strength aluminium alloys. IOP Conf. Seri. Mater. Sci. Eng. **1238**(1), 012014 (2022). https://doi.org/10.1088/1757-899X/1238/1/012014. https://iopscience.iop.org/article/10.1088/1757-899X/1238/1/012014
2. Huo, W., Hou, L., Zhang, Y., Zhang, J.: Warm formability and post-forming microstructure/property of high-strength AA 7075-T6 Al alloy. Mater. Sci. Eng. A **675**, 44–54 (2016). https://doi.org/10.1016/j.msea.2016.08.054. http://www.sciencedirect.com/science/article/pii/S0921509316309698
3. Kumar, M., Sotirov, N., Chimani, C.M.: Investigations on warm forming of AW-7020-T6 alloy sheet. J. Mater. Process. Technol. **214**(8), 1769–1776 (2014). https://doi.org/10.1016/j.jmatprotec.2014.03.024. https://www.sciencedirect.com/science/article/pii/S0924013614001150
4. Laurent, H., Coër, J., Manach, P.Y., Oliveira, M.C., Menezes, L.F.: Experimental and numerical studies on the warm deep drawing of an Al-Mg alloy. Int. J. Mech. Sci. **93**, 59–72 (2015). https://doi.org/10.1016/j.ijmecsci.2015.01.009. https://www.sciencedirect.com/science/article/pii/S0020740315000132
5. Panicker, S.S., Prasad, K.S., Sawale, G., Hazra, S., Shollock, B., Panda, S.K.: Warm redrawing of AA6082 sheets and investigations into the effect of aging heat treatment on cup wall strength. Mater. Sci. Eng. A **768**, 138445 (2019). https://doi.org/10.1016/j.msea.2019.138445. https://www.sciencedirect.com/science/article/pii/S0921509319312316
6. Royne, S., Laurent, H., Maillard, A.: U-channel forming of an 7075–t6 in warm conditions. IOP Conf. Ser. Mater. Sci. Eng. **1238**(1), 012011 (2022). https://doi.org/10.1088/1757-899X/1238/1/012011. https://doi.org/10.1088/1757-899X/1238/1/012011
7. Wang, H., Luo, Y.b., Friedman, P., Chen, M.h., Gao, L.: Warm forming behavior of high strength aluminum alloy AA7075. Trans. Nonferrous Metals Soc. China **22**(1), 1–7 (2012). https://doi.org/10.1016/S1003-6326(11)61131-X. https://www.sciencedirect.com/science/article/pii/S100363261161131X

The Bending Method for Sheet Metal Having Widely Thickness Change in More Than Three Times Thickness Range by Press Brake with Variable Punches

Hideki Okada[1(✉)], Naoya Hirano[2], Tsuyoshi Kimura[2], Hiroki Oribe[1], Takumi Wada[1], Shunya Azami[2], and Takashi Kitahata[2]

[1] Kawasaki Heavy Industries Ltd. Aerospace System Company, 1. Kawasaki-Cho, Gifu-Pref, Kakamigahara City 504-8710, Japan
okada_hide@khi.co.jp

[2] Kawasaki Heavy Industries Ltd. Corporate Technology Division, 1-1. Kawasaki-Cho, Hyogo-Pref, Akashi City 673-8666, Japan

Abstract. Skin sheets of metal aircrafts have complex thickness change patterns. Thus, their forming with press brake is quite difficult.

To automate and ensure high quality, a cutting-edge press brake equipment with synchronized material handling system, position variable punches, die, curvature measurement devices, and bending mechanism with curvature feedback/feedforward technology has been developed. In order to determine specifications of many innovative devices such as material handling systems, variable punches, die, etc., and to realize new bending process, a huge number of studies and development have been conducted.

Here in this paper, a bending characteristics of Aluminum Alloy 2524-T351 clad and an optimization of each variable punch strokes are described. The bending characteristics are affected by the phase in which only clad aluminum plasticity, the phase in which the clad layer re-yields upon unloading, and the phase in which the base material plasticizes, resulting in a complex curve. It depends on the thickness, die width and previous bending in sequential bending.

Keywords: Aircraft skin · Aluminum Alloy 2524-T3 clad · Sequential bending · Press brake · Optimization of punch strokes · Die

1 Introduction

A fuselage of a metal aircraft consists of aluminum alloy called skin sheets which have wide range of thickness changes in order to reduce weight. Generally, in single contour skin, the forming process known as sequential bending with press brake is used to improve yield rate. The stretch-forming is also known as similar purpose. However, it requires material excess during forming process which makes the part size smaller, and it increases aircraft weight due to overlapping areas. In case if skins are roll-formed or sequentially bent before thickness milling process, forming itself is stable but achieving

K. Mocellin et al. (Eds.): ICTP 2023, LNME, pp. 252–260, 2024.
https://doi.org/10.1007/978-3-031-40920-2_27

desired shape is more costly and difficult due to changing in internal stress balance during thickness milling process after bending. Therefore, bending a thickness-changed skin is preferable for high precision and efficiency. Moreover, the sequential bending with press brake after thickness milling process has many benefits in the aspect of the material cost, an aircraft weight and forming accuracy.

Nonetheless it is difficult to bend the thickness-changed skins to the specified curvature with press brake. In aircraft skins, the thickness changes as multiple steps, tapers and cutouts, resulting in complex thickness change patterns. Before the new press brake with new forming process is developed, workers had to bend these skins with great difficulties with those process following; fill the thin area with cardboards to make its virtual thickness flat, form few press lines, measure the formed curvature, adjust forming parameters such as press strokes or thickness of cardboards, form the same area again, and those processes are repeated with their experience and intuition, which takes costs and time.

To automate and ensure high quality, the cutting-edge press brake with the synchronized material handling system, the variable punches (see Fig. 1 as conceptual), the die to be suitable for the variable punches, the curvature measurement devices and bending mechanism with curvature feedback/feedforward were developed.

Fig. 1. Press brake with variable punches to bend thickness changed materials

The variable punches can change the stroke of punches depending on the local thicknesses of the materials at every location. This is to say, the longitudinal shape of the whole punch device can be varied.

To bend the material into the exact shape of a aircraft fuselage's large curvature radius without any kinks or appearing bend line on inner/outer surface, obviously the strain to be introduced is very small and the amount of bending angle must be less than 0.5 degrees per one bend. Therefore, ultra-precise punch stroke accuracy is required. Even if the exact same strokes of the punches are set, slight variations in milled thickness or differences in material properties within tolerances cause differences in bending results. In addition, an edge of the skin with complex shape must be precisely bent for overlapping skins nearby.

A complete understanding of the material properties and the bending characterization were imperative for this slight and precise bending. The first step in achieving automation is to determine precise relationship between the punch strokes and the bending angles. Internally it is called as Master Curve(s), MC for short in the following.

In this study, Aluminum Alloy 2524-T351 clad (AA2524C-T3) is used and has a pure aluminum layer only on the outside surface (shown as Fig. 2). For this reason, it is not simple relationship between the punch strokes and the bending angles or curvature.

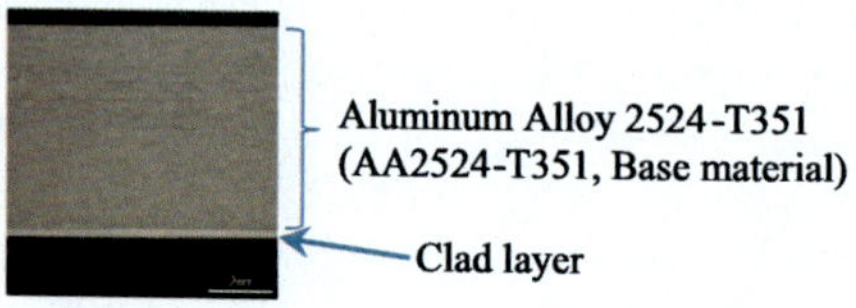

Fig. 2. Cross section of AA 2524C-T3

2 Punch Strokes and Bending Angles

A classical bending theory, which contains practical and sufficient knowledge could lead the punch strokes from the desired bending angles. However, it is not suitable for clad materials or the sequential bending from the aspects of a material properties including residual stress. This section describes the theoretical examination of MC. The effects of various factors on MC were examined under varying conditions. Those factors attributed to the materials affecting MC are represented by whole thickness to be bent, clad layer thickness, base material properties, clad material properties, initial residual stress, residual stress due to previous bending, etc.

The material should be bent by the parameters with taking into consideration of variability of the material properties, or without any affect by the variability, however, it is not realistic to measure and identify all of those factors for every material just before bending. Errors also exist within the tolerance of material thickness due to thickness milling process. The determination of MC itself, either experimentally or by FEM, also requires a great deal of time and effort. Therefore, baseline MC should be logically identified based on the material properties obtained from average, nominal or examination such as stress-strain curve.

2.1 Master Curve Calculation for AA2524C-T3 in Sequential Bending

The classical bending theory which contains practical and sufficient knowledge needs to be expanded for the sequential bending of thickness milled clad material in the following points.

1 Clad material; The material properties are different depending on the location in the thickness direction.
2 Clad material; The neutral surface can be set at any position due to the asymmetry of material properties in the thickness direction.
3 Clad material; The material re-yields at the specific area in spring-back.

4 Thickness milled material; Residual stress in the thickness direction in thickness milled material need to be considered.

Considering the sequential bending gives only a small amount of deformation and material size would be as large as 10 m, Bernoulli-Navier hypothesis is reasonable. Even great material models that can adequately simulate a cyclic tensile testing, reflecting the residual stress imparted by material manufacturing process and thickness machine-milling is difficult. Therefore, $\Delta\sigma$ was used to directly taken into account the effects of residual stress as shown in Fig. 3. In essence, assuming that the stress-strain curve is offset by residual stress, the stress is expressed as the one subtracted the residual stress.

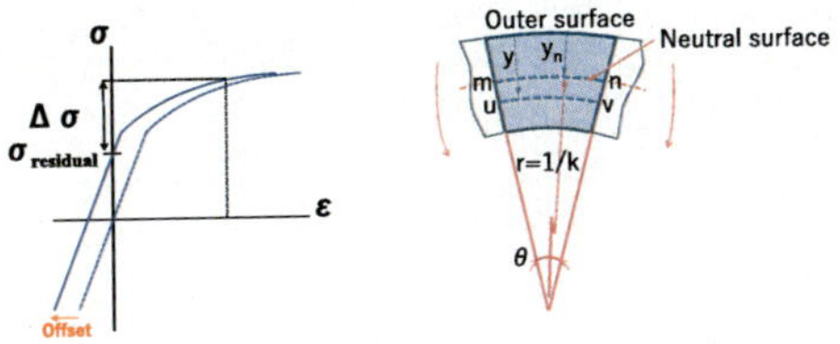

Fig. 3. Stress delta considered residual stress, and bending diagram

Where κ is the curvature of the material of neutral surface, t is the thickness of the material, y is the position in the thickness direction from the outer surface of the material, and y_n is the position of the neutral surface from the outer surface of the material.

$\Delta\varepsilon$ is described in the Eq. (1) from the definition of strain [1];

$$\Delta\varepsilon = \frac{(\widehat{uv} - \widehat{mn})}{\widehat{mn}} = -\kappa\left(\mathrm{y} - \mathrm{y_n}\right) \tag{1}$$

At the neutral surface, summation of $\Delta\sigma$ with respect to $\Delta\varepsilon$ is zero, which can be expressed in the Eq. (2);

$$\int_0^t \Delta\sigma(\Delta\varepsilon)dy = \int_0^t \Delta\sigma(y, \kappa, y_n)dy = 0 \tag{2}$$

It is also known that bending behavior of the clad material is not exactly same as it of single material model [2, 3]. Accordingly, the material properties such as $\Delta\sigma$ were considered as different in the thickness direction when integrating in the thickness direction. Thus, in the base material area, the $\Delta\sigma$ of AA2524-T351 is used; in the clad area, the $\Delta\sigma$ of pure aluminum is used. With these assumptions, the bending moment during bending in the curvature of κ is expressed in the Eq. (3);

$$\begin{aligned} \mathrm{M} &= \int_0^t \Delta\sigma(\Delta\varepsilon) * (y - y_n)dy = \int_0^t \Delta\sigma(y, \kappa, y_n) * (y - y_n)dy \\ &= \int_0^t \Delta\sigma(y, \kappa, y_n) * ydy \end{aligned} \tag{3}$$

The stress-strain of AA 2524-T351, and the stress-strain of pure aluminum could be identified by tensile test, etc. Solving for κ from the Eq. (3) gives the relationship between κ and M. Thereby the curvature distribution can be obtained. By integrating the curvature distribution twice, the bent shape can be obtained, which can determine the punch strokes during bending.

Where $\Delta\kappa$ is the curvature delta from bending, and κ^* is curvature after spring-back. The moment after spring-back is as shown as the Eq. (4)

$$\mathrm{M} = \int_0^t \Delta\sigma\left(y, \Delta\kappa, y_n'\right) * y dy \quad (4)$$

$$\kappa^* = \kappa + \Delta\kappa \quad (5)$$

where y_n' is the new neutral surface.

$$\Delta\varepsilon = -\Delta\kappa(\mathrm{y} - \mathrm{y_n'}) \quad (6)$$

$$\int_0^t \Delta\sigma(\Delta\varepsilon)dy = \int_0^t \Delta\sigma\left(y, \Delta\kappa, y_n'\right)dy = 0 \quad (7)$$

Re-yield of materials during the spring-back stage is a well-known phenomenon [4], and in the sequential bending which is small amount of bending, re-yield is particularly observed in the clad material. Only the outer surface of aircraft skins is pure aluminum with low proof strength, while the majority of the thickness is AA 2524-T351 with high proof strength. As a result, the outer surface where the largest strain is observed easily re-yields in the spring-back stage. Due to the above, the neutral surface in the spring-back stage is not exactly the same as the bending stage. Therefore, the strain may increase rather than spring-back at the specific local area. By considering these points, MC can be determined. Then κ^* and the punch strokes can be theoretically calculated by MC. Figure 4 shows the typical MC of 3mm for whole thickness, 0.2 mm for clad layer thickness with no residual stress.

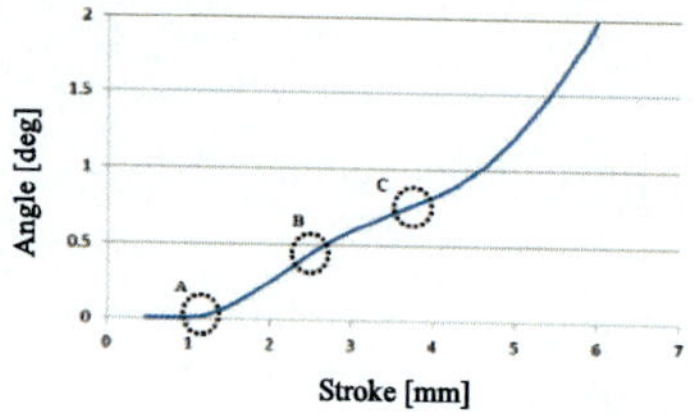

Fig. 4. MC of 3 mm thickness with 0.2 mm clad layer, no residual stress

A; First inflection point, the yield point of the outermost surface of the clad which the highest bending moment. In the A-B section, the bending angle increases as the plastic region gradually increases in the clad region. At this point, even the outside surface of the base material has not yet reached plasticity.

B; Second inflection point. This occurs when the clad layer re-yields during spring-back. Compared to the case of elastic change, the stress delta corresponding to the strain delta becomes smaller, and the amount of spring-back becomes larger than the case of no re-yielding. In B-C section, the base material is still in the elastic region.

C; Third inflection point. The base material reaches plasticity.

2.2 Evaluation of Factors Affecting Master Curve

Figures 5 and 6 show the difference of MC according to the load distribution given by the punches and the width of V-shape die. Compared to the punch load distribution, the die width has a larger impact on MC. Once determined, the die width may not be changed. The study about dies is described in another paper [5].

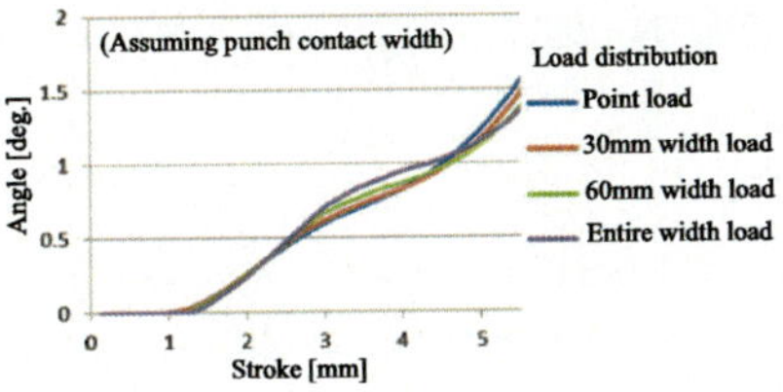

Fig. 5. Influence of punch load distribution

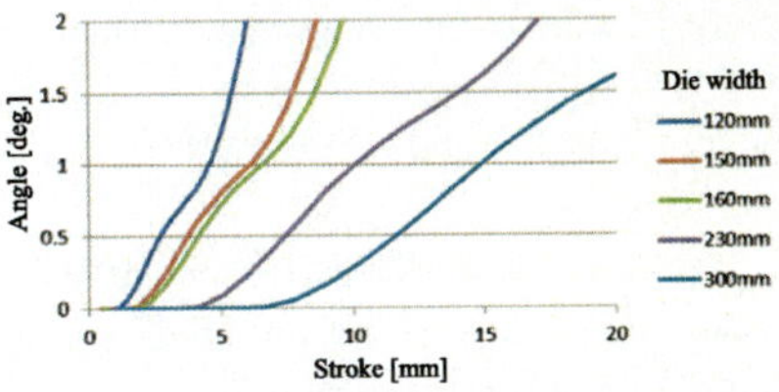

Fig. 6. Influence of die width

Figures 7, 8 and 9 show the difference of MC of the base material with different thicknesses, of with 0.20 mm clad materials on outer side, and of clad materials with different clad layer thickness. Those results suggest that the influence of clad material cannot be ignored.

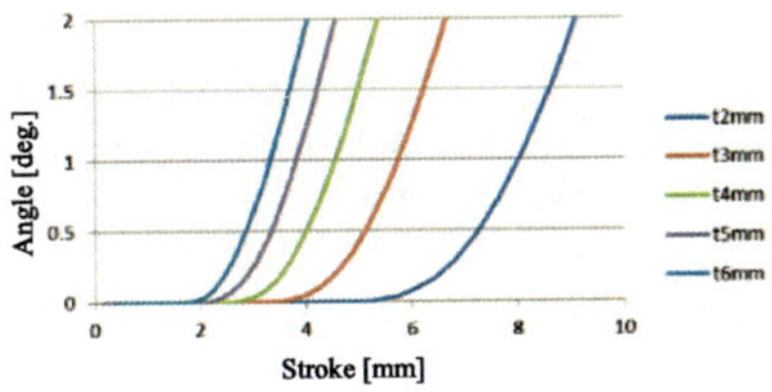

Fig. 7. MC of base material (AA2524-T351)

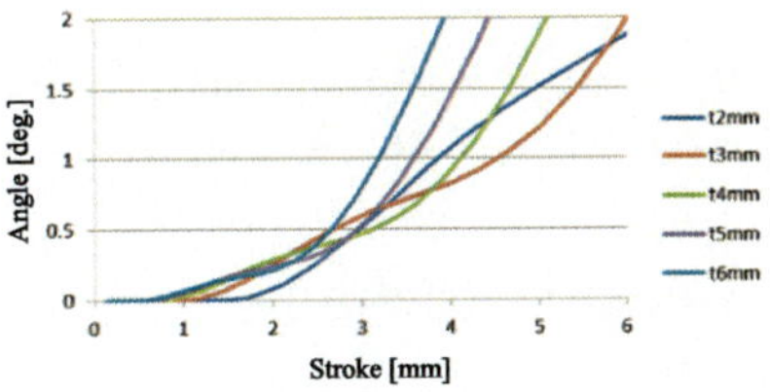

Fig. 8. MC of AA2524C-T3 with 0.20 mm clad layer on outer surface

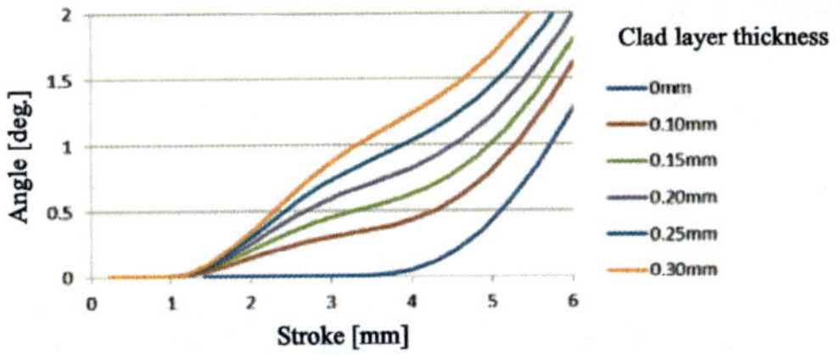

Fig. 9. Influence of clad layer thickness (Total 3 mm)

Figures 10 and 11 show the difference of MC according to residual stresses on the outer surface. If the outside is in compression, the large punch strokes are needed due to the increase in apparent proof stress. This is a good example for the sequential bending where the effect of the previous bending remains.

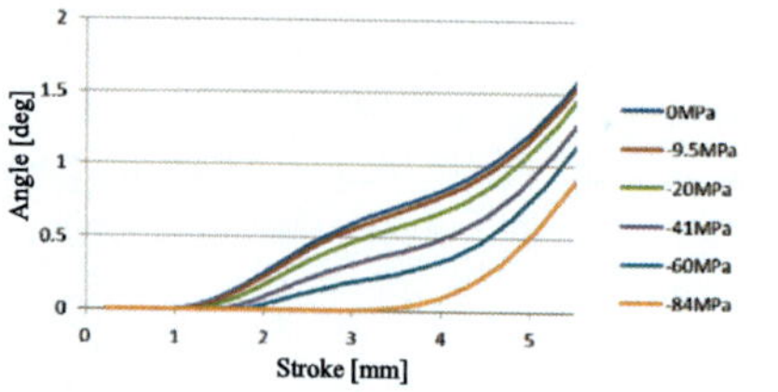

Fig. 10. Influence of compressive residual stress

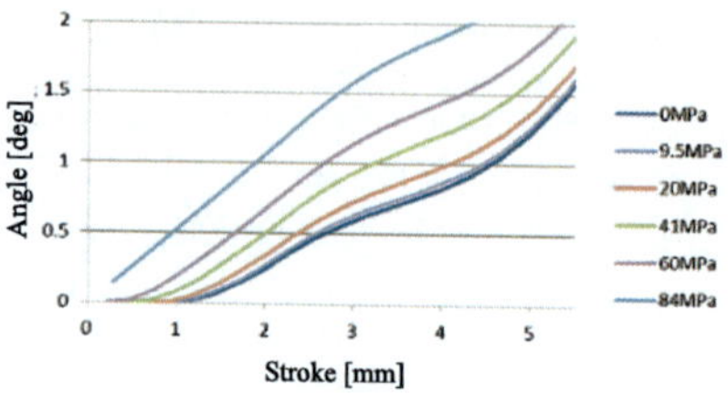

Fig. 11. Influence of tensile residual stress

Figure 12 shows the difference of MC between 1st bending and 2nd bending. Figure 13 shows MC when varying the bending interval in the multiple sequential bending (here, 4 points) to achieve the same curvature radius.

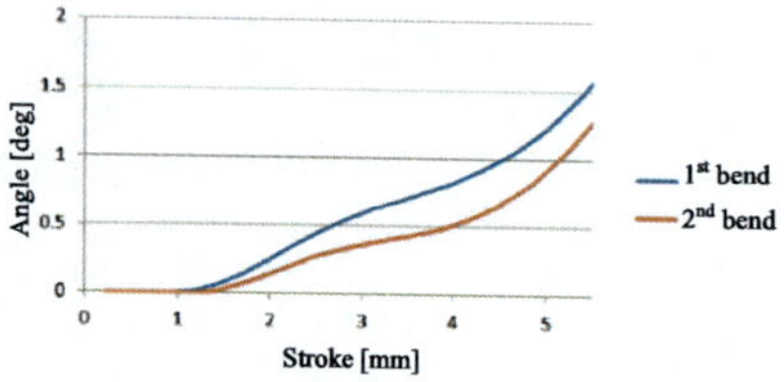

Fig. 12. MC of 1st bend and 2nd bend Note that 1st bend is 0.46 deg, 25 mm interval

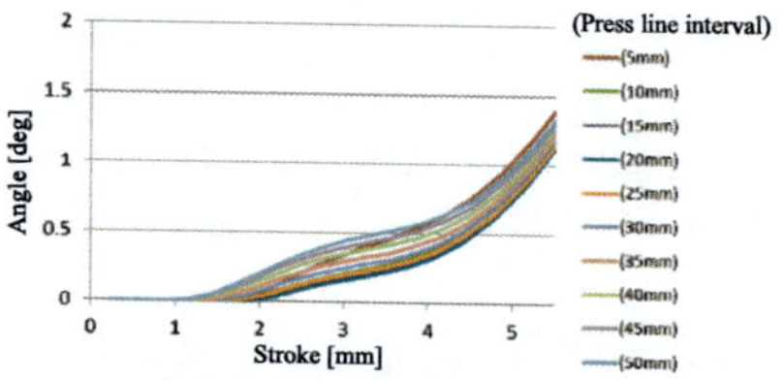

Fig. 13. MC of 4th Bend Note that bend angle depends on the interval to achieve target product curvature radius

Once the bend line interval is set, the target angle to be bent at one bend is determined from required design. All of MC are calculated, the punch strokes of each positions for the variable punches can be determined from the thickness of local area of the skin. Some skins need to be bent with more than 100 bend lines, requiring a large amount of MC calculations. Calculating MC from the above equations by iterative method is not so difficult but the number of calculations increases, resulting in high calculating cost. For the purpose of reducing calculating time, one of good practice is to approximate with a cubic function based on representative MC. This can help to reduce time to determine the punch strokes and in the actual bending process, this can also make it easy to compensate MC as the result of the feedback after bending [6].

3 Determination of Variable Punch Shape

It can be explained from Figs. 4, 5, 6, 7, 8, 9, 10, 11, 12 and 13 that the thickness difference and the punch strokes are not the same, e.g., thickness difference of 5.7 mm leads a punch stroke difference of 14 mm. This causes problems especially in the areas with steps or large tapers. The material is continuous and is affected by neighboring. Too much plastic strain is introduced on the punch contacted region where a punch stroke is much larger compared to its neighbors. Therefore, the punch strokes must be moderated even with the ideal punch stroke calculated from MC, considering a continuity of the material.

The calculated results of the moderately modified punch strokes for a steeply tapered part are shown in Fig. 14. This modification is calculated taking into account the stiffness and continuity of the material. For instance, since the stiffness is related to the cube of the thickness, the cube of the thickness could be used to moderate variable punch strokes. Additionally, a press brake equipped with automated measuring device would help with those optimization [7].

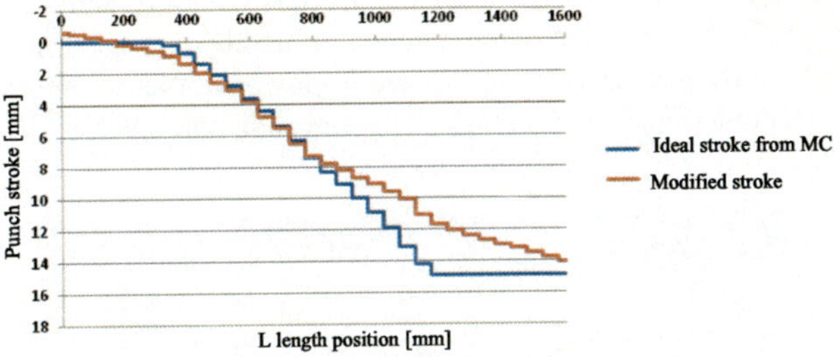

Fig. 14. Modified punch stroke to adapt tapers

As a moderate-size-testing, the bending test using coupon material are conducted. This coupon material is 1600 mm * 1200 mm size including constant thickness areas, tapered areas where a thickness gradually change from 6.35 mm to 2.54 mm, several thickness steps and a cutout. As shown in Fig. 15, the pressure-sensitive film was placed on the punches contact surface of the test coupon to check the contact condition. It was confirmed that all of punches contacted with same range of contact force by moderating punch strokes and no kink where the excessive plasticity was loaded was observed by checking its curvature. Although only one case is shown in this paper, these results from several coupons with representative taper or step suggested that the punch strokes were moderated correctly, stepped or tapered material could be evenly bent.

Fig. 15. Examination using coupon material

4 Conclusions

The bending characteristic of AA2524C-T3 are studied. The relationships between the punch strokes and the bending angles are affected by the phase in which only clad aluminum plasticity, the phase in which clad layer re-yields upon unloading, and the phase in which the base material plasticizes, resulting in the complex curve. It needs to be optimized according to material thickness variations in a bend line direction i.e., tapers and steps. With the moderately modified punch strokes, the variable punches make it possible to bend skin sheets with thickness variations automatically.

With those technologies described above, Kawasaki Heavy Industries, Ltd. produces over 10 m-size aircraft skin using newly developed equipment, referred Fig. 16. This automated equipment provides significant improvement in an accuracy and great reduction in a manufacturing time compared to conventional methods.

Fig. 16. New equipment in operation

References

1. Masujiro, M.: Soseigaku to soseikakou [in Japanese]. Ohmsha (1969)
2. Fusahito, Y., Noriaki, U., Masanobu, O.: The effect of stretching on the spring-back of clad sheet metals. J. Jpn. Soc. Technol. Plast. **30**(338), 358–364 (1989)
3. Fusahito, Y.: Sekisoufukugouban no dansoseihenkei no rikigaku [in Japanese]. J. Jpn. Soc. Technol. Plast. **32**(360), 20–26 (1991)
4. Fusahito, Y., et al.: Elastic–plastic behavior of steel sheets under in-plane cyclic tension–compression at large strain. Int. J. Plast. **18**(5–6), 633–659 (2002)
5. Hideki, O., et al.: Intelligence press brake for Aircraft skin bending, THERMEC2023, Vienna (2023)
6. Hideki, O., Takuya, S., Kristof, T., Alexander, C.: WO2020262678A1 (2020)
7. Hideki, O., Wataru, H., Kristof, T., Alexander, C.: JP.Patent 7228041 (2023)

Development of a Demonstator Plant for Hot Stamping of Metal Sheets with a Machine Learning Assisted Anomaly Detection Control System

Felix Neubürger(✉), Joachim Arens, Thomas Kopinski, and Matthias Hermes

University of Applied Sciences South Westphalia, Lindenstr. 53, 59872 Meschede, Germany

neubuerger.felix@fh-swf.de

Abstract. Due to the demographic developments of an increasingly aging society, the number of employees retiring is also rising. Likewise, the fluctuation within the individual companies has increased drastically within the last few years. As a result, experience gained over many years is being lost. On the other hand, there is a strong trend towards an ever-increasing number of very complex manufacturing processes with a large number of process influencing variables, where the human monitoring capabilities are limited. In this work we study the complex process of hot stamping of metal blanks for car body parts with a demonstrator plant. The demonstrator plant consists of five main components, comprising a magazine for material feed, an industrial robot for transferring the raw material and the finished part, an annealing furnace, and a hydraulic press with the temperature controlled hot forming tool inside. All machines have open communication interfaces with which all sensor data can be accessed. The demonstrator plant is additionally outfitted with different sensors measuring crucial process parameters. We use temperature sensors in the punch and die and make use of capturing the structure-borne sound and mechanical vibrations during the process to closely monitor the process. This sensor data is analyzed and methods of machine learning are used to detect anomalies in the process data. The goal of this work is to develop a methodology for the systematic detection of anomalies in the hot forming process using machine learning methods within a demonstrator plant that are suitable for real time anomaly detection for a high level control system. We use unsupervised machine learning techniques and neural networks to detect anomalies with great accuracy on our test data.

Keywords: Hot stamping · industrial control systems · industrial AI · smart manufacturing

K. Mocellin et al. (Eds.): ICTP 2023, LNME, pp. 261–273, 2024.
https://doi.org/10.1007/978-3-031-40920-2_28

1 Introduction and Related Work

Due to the demographic development and the resulting aging society, the number of expert employees retiring is also increasing [1,2]. As a result, experiential knowledge acquired over many years is lost and a lack of experience in the maintenance, set-up and monitoring of highly complex manufacturing processes occurs [3]. Likewise, the fluctuation rate within for example german companies is high [4]. In contrast, there is an ever-increasing level of automation and the associated substitution of comparatively simple work, for example by camera systems for control tasks in quality assurance. However, the high degree of mechanization and automation leads to more complex faults, so that fault analysis and the elimination of faults is increasingly becoming an engineering task [5]. If this expert knowledge is lacking, for example due to fluctuation or the establishment of a new production site abroad, faulty production and more rejected goods can quickly result in the event of a plant malfunction. In this work and the overarching project we want to address this problem. The research project WiTraPres [14] aims to counteract these effects. WiTraPres stands for the conception and development of a knowledge transfer system (KTS, or german WTS) for complex manufacturing processes. This intelligent system is to be set up using the example of press hardening of profiles made of the lightweight manganese boron steel 22MnB5. This process involves a large number of process influencing factors and a concrete problem environment at the industrial cooperation partners manufacturing site. The constantly increasing demands with regard to quality requirements and the large number of influencing factors make a broad process knowledge of the processes during forming indispensable [6]. The development of this system will focus on automated process control, guided troubleshooting, maintenance support and the development of training formats. In this particular work we focus on the analysis of data taken during experimental runs. In this data we collect only using temperature and already built in sensors that do not influence the process itself. From this sensor data we detect anomalies with unsupervised machine learning techniques, including artificial neural networks. The main technique used to analyze the acquired data is a Long Short Term Memory Autoencoder in combination with thresholding on the reconstruction error of the autoencoder for the binary classification of normal/anormal. This setup gives us the ability to detect anomalies in real time and warn the operators to shut down the plant, thus possibly preventing critical damage to machine parts. The process we developed can be adapted and run in a production process without larger interventions or additional cost. Anomaly detection is well studied in computer science due to the enormous impact it has on industrial applications [7]. In contrast to normal conditions, anomalies are infrequent in real-world applications. Furthermore, annotated data might not always be available for such applications. Hence, research in the literature primarily emphasizes unsupervised anomaly detection. A comprehensive review on used methods has been done by Nassif et al. [8] Sect. 2 describes the manufacturing process and experimental setup in more detail. In Sect. 3 we describe the measurement process and machine learning methods used for the analysis. The results of the data analysis are presented and discussed in Sects. 4 and 5.

2 Experimental Setup

The following section will provide a brief overview of the press hardening process. This will be followed by a closer look at the material fundamentals required for a better understanding of high-strength and ultra-high-strength steel grades and the manganese-boron steels to be classified under them, such as the substrate material 22MnB5 that is used in this project. In addition, the fundamentals of press hardening and its process variants, direct and indirect press hardening, are briefly explained.

The manufacturing steel is cut from blanks, then heated in an oven and subsequently transferred to a press where the sheet metal is reshaped while rapidly cooling off. A schematic illustration of the process steps is displayed in Fig. 1.

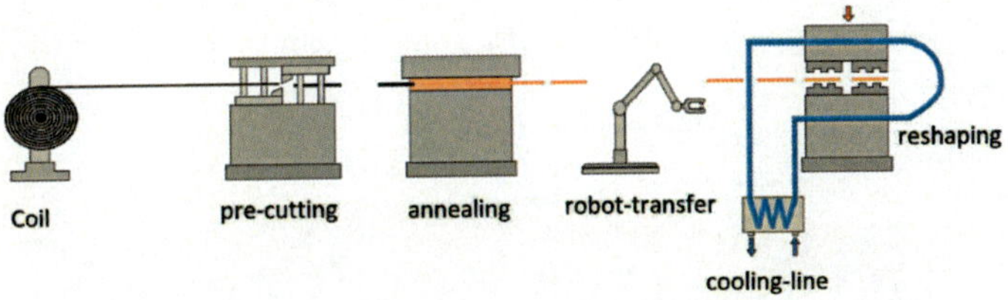

Fig. 1. Schematic overview of the direct press hardening process used in this work.

This process is especially used for automotive steels which, due to their chemical composition and thus very good hardening properties. They usually contain alloying elements such as Mn, Cr, Mo and Ni [9]. Due to their high yield strengths of well above 560 MPa and tensile strength above 700 MPa [10] they are particularly suitable for components subjected to high dynamic loads. The tensile strength is more than tripled by quenching and tempering. In the quenched and tempered condition, these steels subsequently exhibit very high tensile strength with comparatively good toughness. Quenched and tempered steels are generally hypoeutectoid steels with carbon contents of 0.25–0.8 %. The microstructure consists of ferrite and pearlite [11]. Low and medium carbon contents cause a strong decrease of the critical cooling rate to −27 K/s during the quenching and tempering process [10]. This is of crucial importance for the hardening process included in the quenching and tempering. These currently represent the highest strength class used in automotive construction [10]. In the unhardened state, they usually have a fine-grained, ferritic-pearlitic microstructure with a hardness of about 170 HV10 and yield strengths in the range of 350 to 600 MPa. After the hardening process, they have a fully martensitic microstructure with hardness values between 450 and 480 HV10 and a tensile strength of up to 1500 MPa

[12]. The most commonly used sheet material in press hardening is the quenched and tempered steel 22MnB5. Press hardening is a thermomechanical manufacturing process increasingly used in auto body construction to produce complex geometries. It refers to the forming of manganese-boron-alloyed carbon steels in a red-hot, austenitized condition of approximately 900–950 °C with subsequent quenching in the forming tool. For hot stamping, mainly steel sheets made of quenched and tempered steel with a thickness of 0.3 to 3 mm are used.

The data for this study was obtained from a demonstrator plant consisting of five primary components. These include a material feed magazine, an industrial robot for transferring the raw material and finished products, an annealing furnace, and a hydraulic press containing a temperature-controlled hot forming tool. These components form a single unit within a safety fence and are controlled through a central Programmable Logic Controller (PLC) running on an industrial PC (IPC). The system is also monitored through a human-machine interface (HMI) located outside the unit. All machines have open communication interfaces with which sensor data can be accessed using a standardized and secure communication protocol. A CAD rendering of the demonstrator plant is shown in Fig. 2. The finished products are a simple U-shaped profile to study the characteristics of the process. The forming die and instrumentation can be seen in Fig. 3.

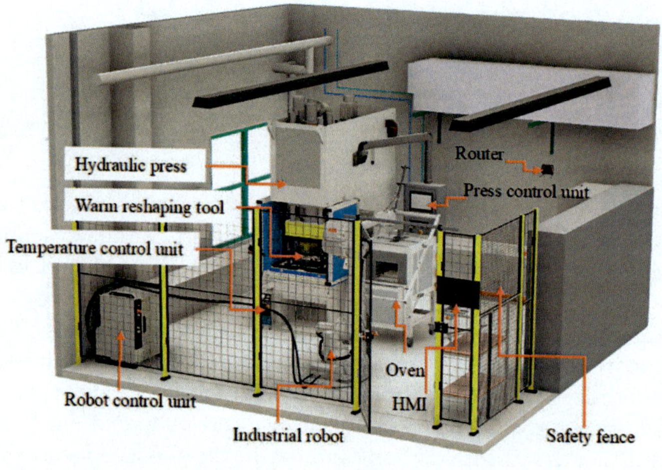

Fig. 2. CAD rendering of the demonstrator plant. The automated cell can be controlled with the human machine interface (HMI) outside of the safety fence.

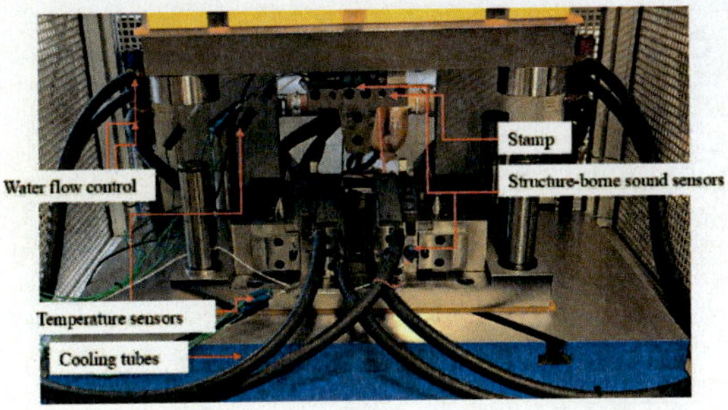

Fig. 3. Annotated image of the warm reshaping tool with added instrumentation.

3 Data and Machine Learning Methodology

Data acquisition was done by running the laboratory process normally while later introducing various mechanical faults that may also occur during a realistic process scenario. The data captured in this experiment consists of temperature sensors that are embedded in the forming die and punch close to the surface. This instrumentation is used to monitor the heat transfer from metal sheet to the die. In a previous work we showed that this data can also be used to estimate the cooling curves of the sheet metal while being reshaped [13]. In addition to monitoring the temperatures we acquire data from the press directly. This data includes the pressing force, and positions of the stamp. Another external sensor is the QASS structure-borne sound sensor that includes a PC for its own data acquisition. [15] The structure-borne sound data is Fourier-transformed in real-time with specialized FPGA hardware to ensure a high sampling rate 12207 Hz for a high temporal resolution. The Fourier-transformed data is the processed and analyzed to find anomalies in the process (Fig. 4). In this work we try to find the introduced anomalies purely from the gathered data. This technique can be used in production to find problems in the production in real time. The machine learning techniques we used can be classified as unsupervised anomaly detection. We use a Long Short Term Memory (LSTM) autoencoder architecture to estimate if the current state is normal or not. The LSTM Autoencoder is built with the TensorFlow Framework [16] for the python programming language [17] Unsupervised anomaly detection with LSTM autoencoders is a technique used to identify anomalies or outliers in a dataset without the need for labeled data. LSTM stands for Long Short-Term Memory, which is a type of neural network architecture that is commonly used for sequence modeling and time-series analysis. An autoencoder is a neural network that is trained to reconstruct its input data. It consists of two main components: an encoder and a decoder network. The encoder takes the input data and maps it into a lower-dimensional representation, that is called latent space, while the decoder takes this latent space representation and reconstructs the original input. The autoencoder is trained on the normal or expected behavior of the time series data, with the objective of minimizing the reconstruction error between the input and output of the autoen-

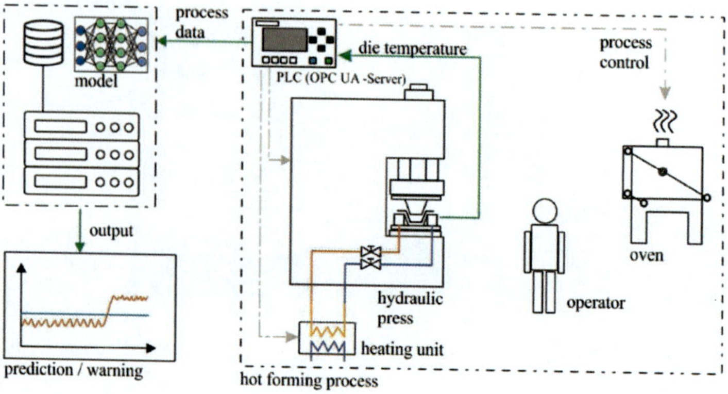

Fig. 4. Schematic overview of the implemented control system. The process data is gathered and analyzed using machine learning models to predict anomalies.

coder. Once the autoencoder is trained, it can be used to detect anomalies or outliers in new time series data. This is done by calculating the reconstruction error for each time step in the new data and comparing it to a threshold. If the reconstruction error exceeds the threshold the time step is flagged as an anomaly. LSTM autoencoders are particularly effective for detecting anomalies in time series data because they are able to capture temporal dependencies and patterns in the data. They can also handle variable-length sequences, which is important for many real-world applications. To evaluate the performance of the different anomaly detection models we use the confusion matrix and accuracy scores. The confusion matrix is a table that summarizes the performance of a binary classification model on a set of test data, by comparing the predicted class labels to the true class labels. The confusion matrix consists of four elements: true positives (TP), false positives (FP), true negatives (TN), and false negatives (FN).

- True positives (TP) are the cases where the model correctly predicted the positive class. In this case anomaly.
- False positives (FP) are the cases where the model incorrectly predicted the positive class, when it was actually negative (no anomaly).
- True negatives (TN) are the cases where the model correctly predicted the negative class.
- False negatives (FN) are the cases where the model incorrectly predicted the negative class, when it was actually positive.

The accuracy metric is one way to measure the performance of a binary classification model, and is calculated as the ratio of the number of correctly classified samples to the total number of samples in the dataset. It is expressed as a percentage, and ranges from 0 to 100 %. The formula for accuracy is:

$$\text{Accuracy} = \frac{TP + TN}{TP + FP + TN + FN}. \tag{1}$$

In addition to the accuracy we use the recall, precision, and F1-score which are evaluation metrics used to assess the performance of a binary classification model. The recall is the ratio of true positive predictions to the total number of actual positive samples. It measures the proportion of actual positives that are correctly identified by the model. The formula for recall is:

$$\text{Recall} = \frac{TP}{TP + FN}. \tag{2}$$

The precision is the ratio of true positive predictions to the total number of positive predictions made by the model. It measures the proportion of positive predictions that are correct. The formula for precision is:

$$\text{Precision} = \frac{TP}{TP + FP}. \tag{3}$$

The F1-score is a measure of a model's accuracy that combines recall and precision. It is the harmonic mean of the two metrics and provides a balance between them. The F1 score is calculated as follows:

$$\text{F1score} = \frac{2 \cdot \text{Precision} \cdot \text{Recall}}{\text{Precision} + \text{Recall}}. \tag{4}$$

The F1-score ranges from 0 to 1, with 1 indicating perfect precision and recall, and 0 indicating poor performance. In summary, recall measures how well a model identifies all relevant instances, precision measures how well a model identifies only relevant instances, and the F1 score is a balance between the two.

4 Experimental Results

We compare the performances of the machine learning models for different data selections. First we benchmark the simplest possible experimental setup omitting the QASS structure-borne sound measurements. Several different standard machine learning algorithms are tested to predict if a given sequence is an anomaly or a normal machine state. This supervised learning benchmark yields the results presented in Table 1.

Table 1. The scores of the supervised learning benchmark. The anomaly detection carried out on the press and temperature sensor data and derived features. The supervised models are validated in a 3-fold cross-validation. The supervised models are trained on a subset of data where normal and anormal states occur. We find that a Logistic regression classifier works best on the given data with an accuracy of 0.92 ± 0.05.

Classifier	accuracy	standard deviation
Logistic Regression	**0.92**	**0.05**
Naive Bayes	0.85	0.06
Random Forest	0.86	0.08

We expect a worse performance for the unsupervised methods because of the lack of information given to the unsupervised models. In a real-world scenario however the abundance of normal state data and the lack of correctly labelled anormal data makes this method of supervised learning unfeasible. For the unsupervised methods we used a LSTM autoencoder that uses the preprocessed data from the structure-borne sound sensor as well as the other process data. The training procedure is carried out on normal state data and tested on a dataset where anomalies and normal state sequences are almost equally present. After training the distribution of the mean absolute error (MAE) is calculated and displayed in Fig. 5. From this MAE distribution we set the anomaly threshold to 0.375. This threshold is chosen qualitatively by setting it above the maximum MAE found in the training data. With that threshold we calculated the MAEs for the test set and flagged the anomalies when the threshold is exceeded. The MAE for each time step in the whole dataset and the threshold can be seen in Fig. 6. In this figure it becomes clear that there are sequences where the LSTM Autoencoder cannot reconstruct the input data accurately, thus hinting at an anomaly. The time points in which the MAE is below the threshold can be interpreted as seemingly normal states of the system. The anomaly appears to have an impact later in the process and is then captured by the autoencoder. When translated into the time domain with the annotation of anomaly we can see qualitatively that the model does not classify normal states as anomalies while correctly having a high reconstruction error on anormal states.

Table 2. The classification report from the unsupervised anomaly detection carried out on the press and temperature sensor data. The total accuracy of the classification is 0.70. The precision, recall and F1-score are calculated for each class separately. The scores are then averaged with their respective sample weights to get the final scores.

Class	precision	recall	F1-score	samples
no anomaly	0.66	0.91	0.76	23
anomaly	0.83	0.48	0.61	21
weighted average	**0.74**	**0.70**	**0.69**	44

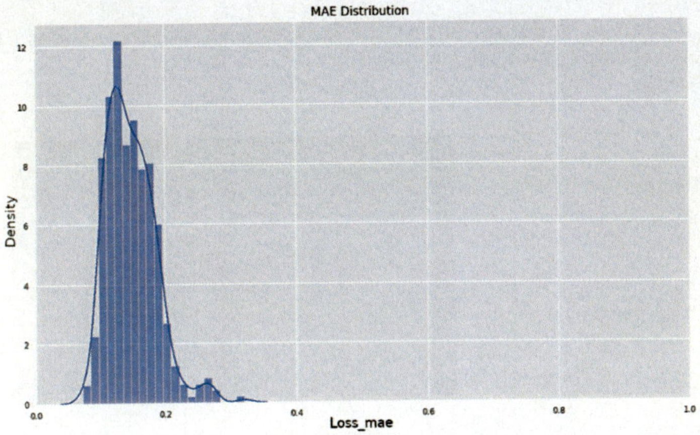

Fig. 5. MAE distribution for the training data set. The threshold for an anomaly detection is set at 0.375 for the following classification task. This threshold is chosen due to the MAE for the reconstruction task not exceeding it. This ensures that known normal state data is not misclassified and flagged as an anomaly.

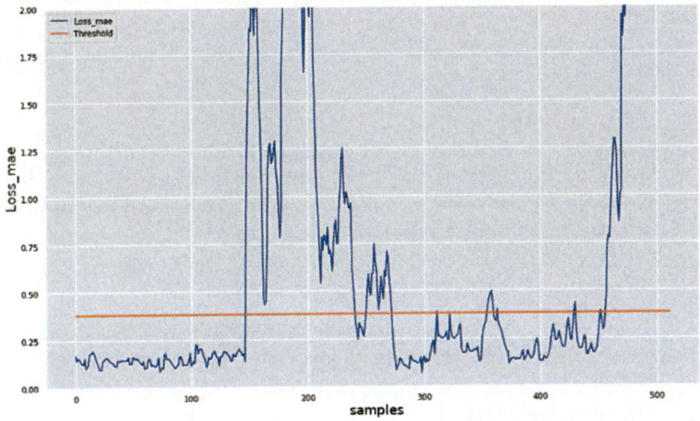

Fig. 6. Temporal distribution of the MAE for the structure-borne sound data set. The threshold for an anomaly detection is set at 0.375 for the following classification task. The temporal axis was transformed to fit all data into the plot and thus has arbitrary time units. The data containing the anomaly states start in the last half of the plot.

We quantitatively evaluate the test set according to a binary classification task. The evaluation metrics can be found in Table 2.

In a second step we only used the structure-borne sound sensors to detect the anomalies. A sample of the Fourier transformed data can be seen in Fig. 7. When compared to a Fourier spectrum of a process where an anomaly was introduced a difference in the summed Fourier amplitudes can be observed. Data from an anomalous process is displayed in Fig. 8.

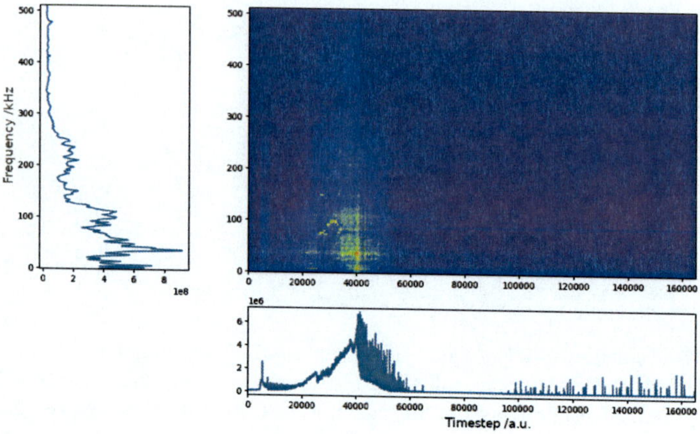

Fig. 7. Fourier-spectrum of the structure-borne sound measurements of a normal process. On the left and bottom the marginal sums are displayed. In this work we focus on the bottom data, which is the sum of Fourier spectrum amplitudes for each time step.

Analogous to the previous analysis we calculate the metrics of the classification performance. The results are presented in Table 3. The total accuracy, precision, recall and f1-score are lower than from the other sensor data. This can be explained by the aggregation of the data down to the fourier amplitudes. This extracted feature does not contain enough information alone to yield the performance we can reach by specifically instrumenting the demonstrator plant and handcrafting features from that process. By reducing the full fourier spectrum to one less dimension we do not take the spectral shape across the frequencies into account. In addition to not considering the frequency domains shape we also lose the two dimensional information and possible correlations in the full spectrum. In a future work we want to expand our analysis methods on that higher dimensional data space. For that we want to use more complex machine learning algorithms like convolutional neural networks in combination with LSTMs.

The code, supplementary figures, analysis and data to reproduce the results can be found under:

https://github.com/DataScienceLabFHSWF/WiTraPresAnomaly.

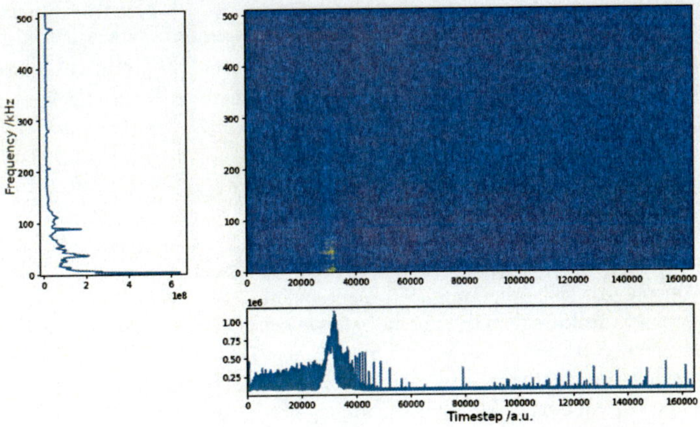

Fig. 8. Fourier-spectrum of the structure-borne sound measurements of an anormal process. On the left and bottom the marginal sums are displayed. In this work we focus on the bottom data, which is the sum of Fourier spectrum amplitudes for each time step. In the amplitude one can observe the different shapes compared to normal state data.

Table 3. The classification report from the unsupervised anomaly detection carried out on the structure-borne sound data. The total accuracy of the classification is 0.52. The precision, recall and F1-score are calculated for each class separately. The scores are then averaged with their respective sample weights to get the final scores.

Class	precision	recall	F1-score	samples
no anomaly	0.60	0.65	0.63	46
anomaly	0.36	0.31	0.33	29
weighted average	**0.48**	**0.52**	**0.51**	85

5 Discussion and Outlook

We developed a demonstrator plant for the hot stamping of metal and outfitted it with instrumentation that acquires data for a cyber physical system. We then benchmarked several machine learning techniques to detect the anomalies introduced in the experiments. We find that the unsupervised detection of anomalies is accurate with a precision of 74 % without adding any expert knowledge to the model. The usage of structure-borne sound sensors proved promising but needs further investigation and more complex analysis. The advantages of using this system is the easy fitting of the sensors on existing hardware. All of the methods developed in this work can be used in real-time applications when deployed on an edge server. When deployed in production these methods can add guidance to employees that are not very familiar with the process. The results can be aggregated into a simple binary classification that is easy to understand even for non-machine learning experts. In conjunction with a secondary fault classification that is based on the FMEA and more experimental data a maintenance suggestion can be suggested to the machine operator. The control system as a whole can thus be used to help new operators to benefit from the knowledge of retired experts.

References

1. Statistisches Bundesamt, Statistischer Bericht - 15. koordinierte Bevölkerungsvorausberechnung - Deutschland - 2021 BIS 2070 (2021). https://www.destatis.de/DE/Themen/Gesellschaft-Umwelt/Bevoelkerung/Bevoelkerungsvorausberechnung/Publikationen/Downloads-Vorausberechnung/statistischer-bericht-bvb-deutschland-2070-5124202219005.html
2. Denkena, B., Merwart, M.: Wissensmanagement in der Zerspanung: Potenziale der Digitalisierung im Sinne der ISO 9001: 2015. Zeitschrift für wirtschaftlichen Fabrikbetrieb **110**(7–8), 444–447 (2015). https://doi.org/10.3139/104.111362
3. Freith, S., Gondecki, L., Kreimeier, D., Kuhlenkötter, B.: Wissenserhalt durch den Einsatz digitaler Lerninhalte: Digitale Lerninhalte sichern Wettbewerbsvorteile durch Verringerung des Wissensverlusts hervorgerufen durch den demographischen Wandel. Zeitschrift für wirtschaftlichen Fabrikbetrieb **110**(6), 340–343 (2015). https://doi.org/10.3139/104.111344
4. Smettan, J., Kleineidam, T.: Hohe Mitarbeiterfluktuation - Fluch oder Segen? Wie sich die Ursachen der Fluktuation veraendern. Berufsakademie Sachsen - Wissen im Markt 2018, pp. 55–61 (2018)
5. Ahrens, D.: Neue Anforderungen im Zuge der Automatisierung von Produktionsprozessen: Expertenwissen und operative Zuverlässigkeit. AIS-Studien **9**(1), 43–56 (2016). https://doi.org/10.21241/ssoar.64825
6. Karbasian, H., Tekkaya, A.E.: A review on hot stamping. J. Mater. Process. Technol. **210**(15), 2103–2118 (2010). https://doi.org/10.1016/j.jmatprotec.2010.07.019
7. Chalapathy, R., Chawla, S.: Deep Learning for Anomaly Detection: A Survey, CoRR (2019). https://doi.org/10.48550/arXiv.1901.03407
8. Nassif, A.B., Talib, M.A., Nasir, Q., Dakalbab, F.M.: Machine learning for anomaly detection: a systematic review. IEEE Access **9**, 78658–78700 (2021). https://doi.org/10.1109/ACCESS.2021.3083060
9. Jacobs, O.: Werkstoffkunde (3. Auflage). Vogel Communications Group, Würzburg (2015). ISBN 978-3-8343-3350-6
10. Hyunwoo, S.: Scherschneiden von Mangan-Bor-Stählen unter Temperatureinfluss, München: Universitätsbibliothek - Technische Universität München (2015). https://nbn-resolving.de/urn/resolver.pl?urn:nbn:de:bvb:91-diss-20150921-1229369-1-7
11. Seidel, W.W., Hahn, F.: Werkstofftechnik: Werkstoffe - Eigenschaften - Prüfung - Anwendung, p. 473. Hanser, München (2014). ISBN 9783446456884
12. Lechler, J.: Beschreibung und Modellierung des Werkstoffverhaltens von presshärtbaren Bor-Manganstählen. Meisenbach, Bamberg (2009). https://doi.org/10.25593/978-3-87525-286-6
13. Neubürger, F., Arens, J., Vollmer, M., Kopinski, T., Hermes, M.: Coupled finite-element-method-simulations for real-time-process monitoring in metal forming digital-twins. In: 2022 10th International Conference on Control, Mechatronics and Automation (ICCMA), Belval, Luxembourg, pp. 260–265 (2022). https://doi.org/10.1109/ICCMA56665.2022.10011608

14. Kopinski, T., Hermes, M., Schwung, A.: WiTraPres Project (2019). https://www.forschung-fachhochschulen.de/fachhochschulen/shareddocs/projekte/de/fhprofunt/witrapres.html
15. QASS GmbH, Optimizer 4D. https://www.qass.net/wp/optimizer4d/
16. Abadi, M., et al.: TensorFlow: large-scale machine learning on heterogeneous systems (2015). https://tensorflow.org/
17. Van Rossum, G., Drake, F.L.: Python 3 Reference Manual. CreateSpace, Scotts Valley (2009)

On Forming Sheet Metal Parts in Single Curvature with English Wheel

Kuang-Jau Fann(✉) and Yi-Hsien Wu

Department of Mechanical Engineering, National Chung Hsing University, Taichung, Taiwan
kjfann@nchu.edu.tw

Abstract. As trade patterns have shifted to high variety and small batches, metalworking techniques used in metalcraft have been reconsidered. Therefore, the purpose of this article is to study one of these techniques, the English Wheel, which can produce a curved sheet metal part by pulling and pushing the sheet, clamped under a pressing force, back and forth along the working path between the roller and the anvil wheel inside a giant C-clamp. In this study, SPCC sheets were used to form parts with a single curvature. As a result, the curvature of the parts formed by the English Wheel was measured and determined by a 3D optical digitizer and approximated by its circumscribed circular cylinder. The curvature of the part formed by a sawtooth path is the same as the superimposed curvature of each part formed by individual parallel paths decomposed from the sawtooth path. Based on these findings, this study then provides a processing scheme for forming with English Wheel, which is validated by the finite element method on an example with a given curvature.

Keywords: English Wheel · Sheet Metal Forming · Finite Element Analysis

1 Introduction

In today's world, we are surrounded by sleek and beautiful design. Whether it's in architecture or artwork [1], the smooth and flowing curves of the design are crucial to its appeal. As a result, many industries have turned to designing objects with curved surfaces to emphasize their distinctive features. This has also become a popular trend in the automotive and motorcycle industries [2], where customers demand customized designs of specialty exterior sheet metal panels that reflect their individuality.

When it comes to creating these unique and special shapes on sheet metal, traditional stamping techniques are not always the best option. They can be expensive due to the high cost of the die set, and they are only capable of producing a single shape [3–5]. Instead, a flexible forming technique is needed to meet to the demand for custom shapes [6–8]. Of the various flexible forming technologies available for forming free-form sheet surfaces, English Wheel technology is best suited for forming complex curved surfaces in sheet metal. It not only enables the production of special curved surface shapes with simple tools, but also saves processing costs, making it suitable for use in the customization industry.

K. Mocellin et al. (Eds.): ICTP 2023, LNME, pp. 274–284, 2024.
https://doi.org/10.1007/978-3-031-40920-2_29

As shown in Fig. 1, English Wheel processing mainly involves manual manipulation of the sheet metal while using two separate wheels, the roller and the anvil wheel, to press it between the wheels and form it into the desired curve by pushing and pulling it back and forth or toward and away from the wheels as forming paths [8]. Various complex shapes can be achieved on to the sheet metal by using different types of paths and anvil wheels. However, due to the scarcity of literature on the English Wheel process, craftsmen must rely solely on years of experience to produce the desired curved surfaces, which makes it difficult for those who are new to the technique to master the English Wheel. Therefore, the purpose of this study is to experimentally investigate how the English Wheel process works by studying various parameters such as rolling path, path spacing, and number of passes on the results of cold-rolled steel sheet, while using finite element analysis to evaluate the effects of these parameters. The results will help improve the current understanding of the English Wheel process and can serve as a reference for those who want to use it. By making the process more accessible, it is expected that industries will develop faster and become more efficient in producing custom shapes.

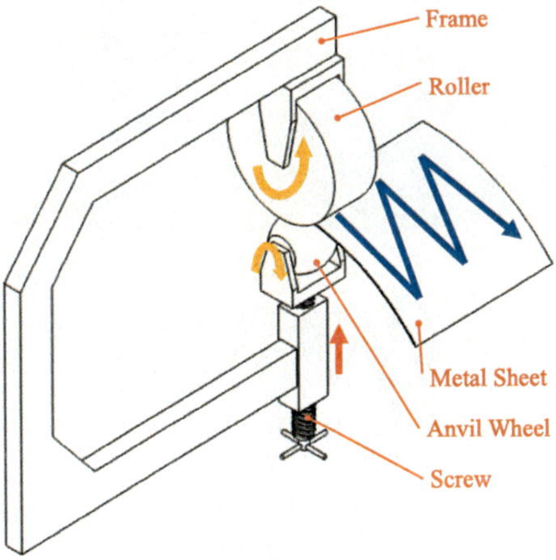

Fig. 1. The English Wheel process.

2 Experiment Setups

As shown in Fig. 2, a computer numerical control universal testing machine (model HT-2402, Hongta Instruments, Taiwan) is used in this study to apply a controlled downward pressure in the Z direction locally to an experimental blank sheet. A roller wheel of circular cylinder shape was fixed to the crosshead at the top of the testing machine, while the anvil wheel of ellipsoid shape was fixed to the base of the testing machine, which was arranged differently without a clamping screw as shown in Fig. 1.

The sheet metal was clamped to a fixture attached to a jig mounted on a linear module using a ball screw driven by an AC motor to allow the sheet metal to move linearly in the X direction away from or toward the testing machine. To ensure accurate forming of the sheet metal with the assigned forming path, a matrix of positioning holes was designed on the jig according to the rolling path mode and spacing, while the fixture was used to fix the sheet at the correct locations according to the rolling path requirements, preventing it from slipping or shifting. The sheet metal used in the experiment was the commercially acquired SPCC cold-rolled steel sheet with a size of 200 mm × 200 mm × 1 mm. The chemical composition of the SPCC sheet is listed in Table 1, which was obtained using a spectrometer (model PDA-MF, Shimadzu, Japan).

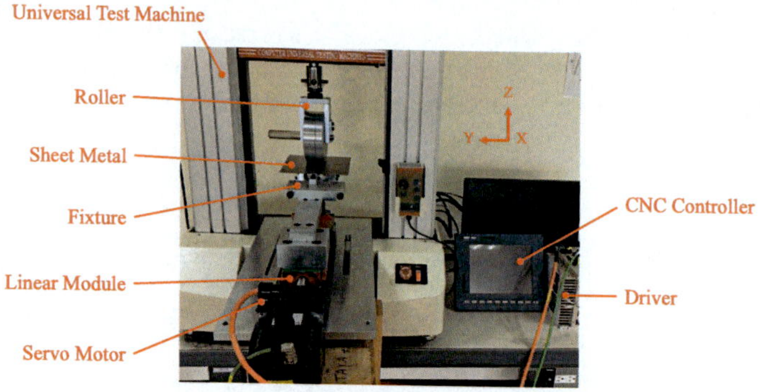

(a)

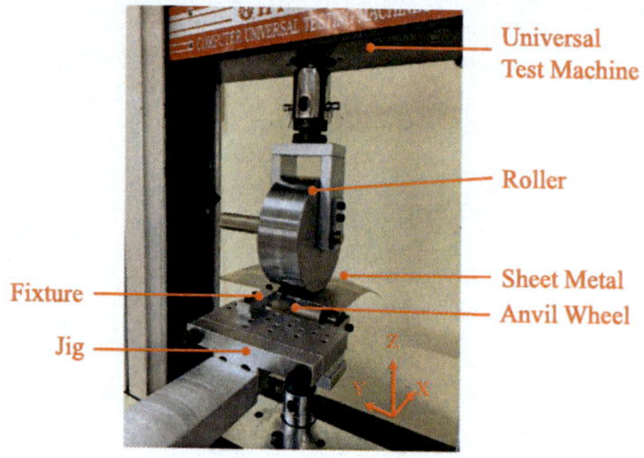

(b)

Fig. 2. The experimental setup in this study (a) overall setting (b) detailed setting of the fixture.

As mentioned above, although there have been some studies involving in finite element analysis on the relevant parameters of the English Wheel in the literature [9, 10], a similar finite element analysis is still required for the experimental setup in this study. Figure 3 shows the finite element analysis model used in this study, in which the roller and the anvil wheel are simplified using shell elements as rigid bodies, while the sheet metal is formulated using shell elements with elastic-plastic behavior. The corner point A of the sheet metal is set as the origin, while the center positions of the roller and

Table 1. Chemical composition of SPCC sheet in weight percentage (wt%) obtained from spectrometer analysis.

Element (wt%)						
C	Mn	P	S	Si	Al	Fe
0.1	1.3	0.16	0.3	1.0	0.31	Bal

anvil wheel are located at (40 mm, 40 mm) in the X and Y directions with their axes both parallel to the Y axis. Square meshes are used for the roller, anvil wheel, and sheet metal, with a mesh size of 2 mm for the sheet metal and a mesh size of less than 4.5° for the roller and anvil wheel on their respective circumferences to maintain their curvature characteristics. During the forming process, the roller presses down on the sheet metal from top to bottom, and then the sheet metal moves along a path while the roller and anvil wheel rotate accordingly.

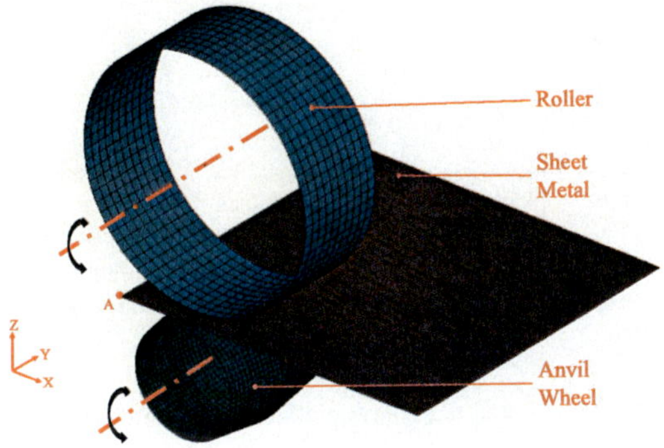

Fig. 3. A typical finite element analysis model in this study.

Since this study used the explicit solver of the commercial finite element analysis software LS-DYNA, the size of the time step in the simulation calculation must be proportional to the mesh size to obtain correct results. Therefore, in this study, the step size for a wheel to pass through a single sheet element was set to at least 1,000 steps to ensure the accuracy of the calculations.

The elastic-plastic material properties of the sheet metal used for finite element analysis were obtained by cutting the sheet into 5 specimens each at 0°, 45°, and 90° to the rolling direction using a CNC milling machine according to ASTM-E8 specifications. The specimens were then subjected to tensile testing at a constant rate of 0.1 mm/s using the universal testing machine. The test results were averaged and converted into true stress-strain curves to describe the material properties in LS-DYNA, as shown in Fig. 4.

Table 2 lists the process parameters, each with three parameters, including the pressing force (5 kN, 10 kN, and 20 kN), the rolling path (straight lines parallel and oblique to the side edge as well as sawtooth), the path spacing (10 mm, 20 mm, and 30 mm), and the number of rolling passes (3, 5, and 7). The individual straight line pathes parallel and oblique to the side edge can be considered as a decomposition of the continuous sawtooth rolling path. In addition, the geometry of the anvil wheel was considered as the geometry parameter for the English Wheel in this study, as shown in Table 3, and the difference lies in the geometric features. The geometric feature of W_A is a surface of revolution of a circular segment, while W_B and W_C are surfaces of revolution of straight line segments with the same conical ends, but different widths of the cylindrical part in the middle (5 mm and 10 mm). The diameter of the three anvil wheels is 70 mm, and the width is 75 mm. Anvil wheels W_B and W_C can be considered modifications of anvil wheel W_A, because they are easily made on conventional lathes. The diameter and width of the roller are 149.5 mm and 57 mm, respectively.

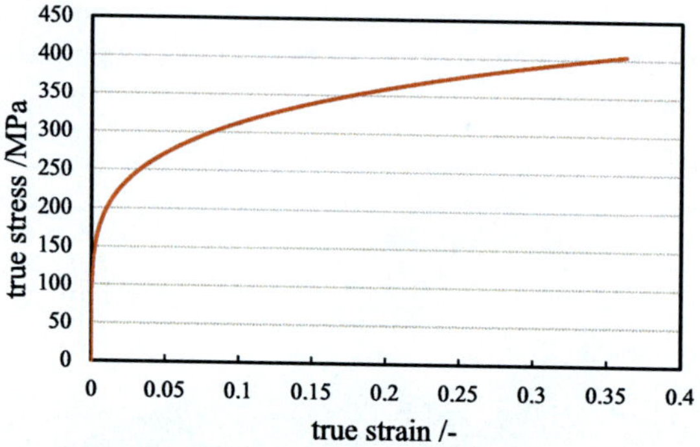

Fig. 4. The flow stress of the sheet for finite element analysis in this study.

Table 2. Parameters related to experiment and finite element analysis model.

Parameters				
Pressing force	Path type	Path spacing	Rolling passes	Anvil wheel
5 kN	Parallel to edge	10 mm	3	W_A
10 kN	Oblique to edge	20 mm	5	W_B
20 kN	Sawtooth	30 mm	7	W_C

3 Results and Discussion

Different rolling passes were used to create indentations in the SPCC sheet. The paths consisted of straight lines parallel to the side edge and spaced 20 mm apart using anvil wheel W_B with a pressing force of 20 kN, which made the indentations more noticeable. In order to observe the relationship between the indentations and the forming geometry, the curved sheet steel part was scanned using a 3D optical digitizing system (ATOS 1

Table 3. Anvil wheel features for experiments and finite element analysis models.

Anvil wheel features			
Anvil wheel	Radius of the generatix	Taper of the conical ends	Width of the middle cylinder
W_A	50.8 mm	-	-
W_B	-	0.29°	5 mm
W_C	-	0.29°	10 mm

Rev. 02, GOM, Germany). The surface of the sheet was first sprayed with a mixture of titanium dioxide powder and alcohol to reduce reflectivity, and then placed on the 3D optical digitizer for scanning, resulting in a point cloud of approximately 20,000 points.

In order to clearly observe and compare the forming geometry obtained by different rolling passes (1, 3, 5, and 7), a YZ plane at the side edge of the formed parts in Fig. 5, which shows the point cloud of the scanned points, was selected to slice the cross sections, which are staked together by aligning their highest points as shown in Fig. 6 and observed from the X direction. It can be seen that each curve has a wavy pattern. When only one pass is rolled, the resulting indentation causes the sheet metal to bend, while the rest of the sheet remains straight due to the lack of rolling. When 3 passes are used, the straight sections with no bending are relatively reduced. Similarly, when 5 or 7 passes are used, the surface features of the sheet become more pronounced.

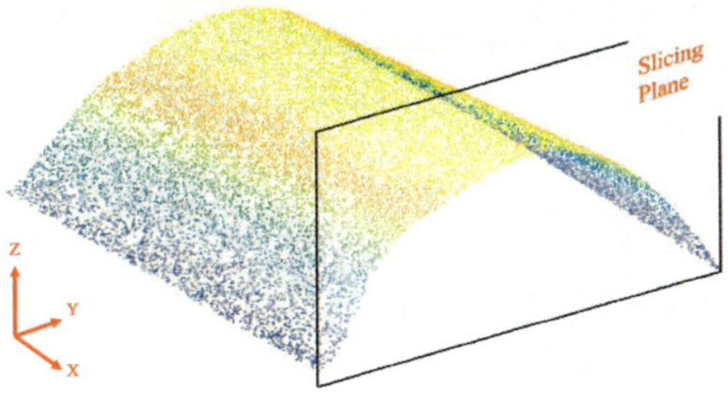

Fig. 5. The slicing plane to select the cross-section from the point cloud of the scanned points as the formed sheet part.

In the study, the geometric results of the sheet part formed with 7 rolling passes (solid line shown in Fig. 7a) were used as an example to demonstrate the potential geometric forming mechanism of the English Wheel. As mentioned above, each indentation impressed by a single rolling pass causes the sheet to bend at an angle θ to each unrolled straight wing, as shown in Fig. 7b, where one of the indentations was locally enlarged. With the average indentation angle θ_{avg} calculated from the all 7 indentations, here 176.6°, a circumscribed circle and an inscribed circle can be determined using line segments made by the indentation and the path spacing, which have the radii of

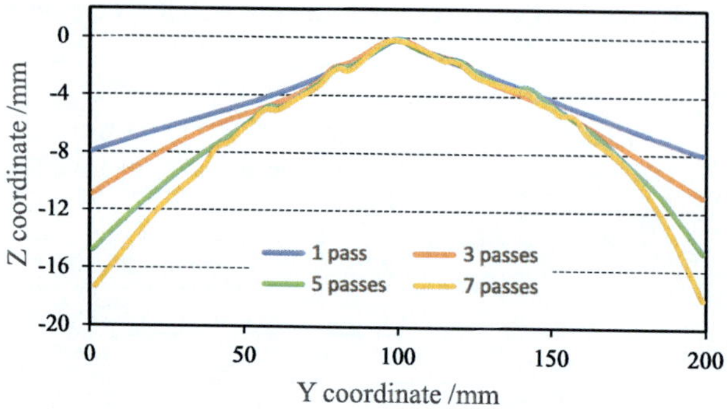

Fig. 6. Cross-section of the sheet metal parts formed with different rolling passes.

488.2 mm and 470.8 mm, respectively. By determining the maximum principal curvature k_1, 2.08 m^{-1}, of the formed sheet part through the Gaussian curvature K and the mean curvature H from the point cloud of scanned points [11], its curvature radius can be calculated as the reciprocal of k_1, 480.5 mm. This means that the error between the circumscribed circle and the experimental value was smaller than that of the inscribed circle. Therefore, as shown in Fig. 7a, the circumscribed circle segment with the radius ρ drawn in the dashed line using CAD can be considered to accurately predict the actual forming results, which demonstrates the enveloping forming characteristics of the English Wheel on rolled sheet parts.

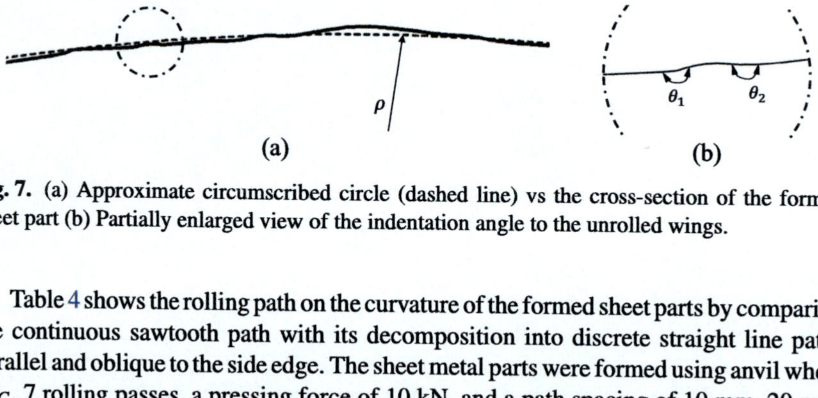

Fig. 7. (a) Approximate circumscribed circle (dashed line) vs the cross-section of the formed sheet part (b) Partially enlarged view of the indentation angle to the unrolled wings.

Table 4 shows the rolling path on the curvature of the formed sheet parts by comparing the continuous sawtooth path with its decomposition into discrete straight line paths parallel and oblique to the side edge. The sheet metal parts were formed using anvil wheel W_C, 7 rolling passes, a pressing force of 10 kN, and a path spacing of 10 mm, 20 mm, and 30 mm, respectively. The results show that the curvature value of the sawtooth path can be obtained by adding the results formed by the discrete parallel and oblique straight line paths separately. Therefore, based on these results, it was confirmed that the English wheel manufacturing process has the superposition characteristic of forming, that is, the curvature value of the continuous sawtooth path can be obtained by superimposing the results, which can be used as a reference for selecting the subsequent process parameters.

To verify the feasibility of the finite element model, simulations were performed with the same parameter setup for paths of parallel straight lines as above for Table 4, using the

Table 4. Curvature values on the sheet metal parts formed using straight line paths parallel, oblique to the side edge and sawtooth path with a path spacing of 10 mm, 20 mm, and 30 mm, respectively.

Path spacing (mm)	Curvature k_1 (m^{-1})		
	Parallel lines	Oblique lines	Sawtooth
10	1.76	1.85	3.56
20	1.04	1.14	2.15
30	0.34	0.48	0.78

experimental curvature values as the target values for comparison. As shown in Table 5, the root mean square error between the two results analyzed using the GOM software was be less than 1.5%. This confirms the feasibility of the finite element simulation.

Table 5. Comparison of the curvature values obtained from experiments and finite element analysis on the sheet metal parts formed using paths of straight lines parallel to the side edge with a path spacing of 10 mm, 20 mm, and 30 mm, respectively.

Path spacing (mm)	Curvature k_1 (m^{-1})		Error (%)
	Experiment	Finite element anaysis	
10	1.76	1.76	– 0.12
20	1.04	1.06	1.32
30	0.34	0.34	– 0.76

Table 6 shows the trend of the formed curvature of the sheet metal with respect to the pressing force and the comparison between the experimental and finite element simulation results, using the same parameter setup as in Table 4, but with a path spacing of 20 mm and a pressing force of 5 kN, 10 kN, and 20 kN, respectively. The results show that the larger the pressing force applied by the roller to the sheet, the larger the formed curvature value. In addition, the error percentage between the experiment and the finite element results was maintained below 1.5%.

Figure 8 shows the influence of different anvil wheel geometries on the sheet metal forming effects using 7 straight line paths parallel to the side edge with a pressing force of 10 kN and a path spacing of 10 mm, 20 mm, and 30 mm, respectively. As described above, they are anvil wheel W_A with a revolutionary surface of circular segment and its modifications anvil wheel W_B and W_C of conical shape with widths of 5 mm and 10 mm in their cylindrical central part. The results show that the smaller the path spacing, the larger the curvature value of the formed sheet part. At the same time, a linear relationship between the curvature and the path spacing can be found for all anvil wheels. It can also be observed that anvil wheel W_B with a simple geometry can achieve a similar result as anvil wheel W_A, which must be machined on an expensive CNC lathe. Furthermore, when comparing the results formed by anvil wheels W_B and W_C, a better forming effect

Table 6. Comparison of the curvature values obtained from experiments and finite element analysis on the sheet metal parts formed using straight line paths parallel to the side edge with a path spacing of 20 mm and a pressing force of 5 kN, 10 kN, and 20 kN, respectively.

Pressing force (kN)	Curvature k_1 (m^{-1})		Error (%)
	Experiment	Finite element anaysis	
5	0.68	0.68	0.08
10	1.04	1.06	1.32
20	1.64	1.63	– 0.57

can be achieved by anvil wheel W_B than by W_C. This indicates that the smaller the width at the cylindrical center of the anvil wheel, the larger the curvature value obtained on the part.

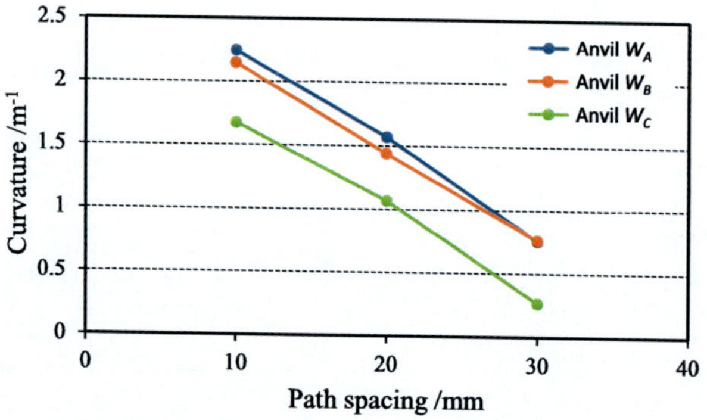

Fig. 8. Curvature values of the sheet metal parts formed by different anvil wheels using 7 straight line paths parallel to the side edge with a pressing force of 10 kN and a path spacing 10 mm, 20 mm, and 30 mm, respectively.

In summary, it can then be proposed as a process scheme for forming a sheet metal curved in one direction by English Wheel that one can first conduct a test by some straight line paths on the sheet metal using roller and anvil wheel having a specific geometry with a certain pressing force, measure the side angles of the indentation to the unrolled segments, and then draw an approximate circle of the given curvature based on the enveloping geometry made by indentation and unrolled segments. The path spacing can thus be determined by superimposing the paths, and then the operation can be performed based on the obtained values. In this study, a sheet with a given curvature value of 1.03 m^{-1} was taken as an example. According to the above proposed method, finite element analysis was used to validate the process. Specifically, the SPCC sheet, anvil wheels W_B and W_C, and a pressing force of 10 kN were used to obtain the indentation characteristics after each rolling pass. The path spacing for forming with anvil wheels W_B and W_C was then calculated to be 23 mm and 19 mm, respectively. The curvature value obtained on the formed parts was then calculated and listed in Table 7. The results showed an error falling about the range of 2%, indicating that the above proposed process design scheme is reliable for reference.

Table 7. Results obtained by performing finite element analysis models created according to the suggestion of the process design scheme proposed in this study to produce an SPCC sheet part with a given curvature of 1.03 m^{-1}.

Anvil wheel	Path spacing (mm)	Pressing force (kN)	Curvature k_1 (m^{-1})	Error (%)
W_B	23	10	1.05	1.86
W_C	19	10	1.05	2.07

4 Conclusion

This study is aimed to conduct a preliminary analysis of the English Wheel forming process using experiments and finite element analysis. The following conclusions were drawn: The English Wheel process produces indentations on the surface of the sheet metal by a force pressing the sheet metal clamped by the roller and anvil wheel, which causes the junction to the unrolled segment of the sheet metal to bend, thus forming the sheet metal. As the pressing force or indentation depth increases and the path spacing decreases, the curvature value of the formed sheet metal increases. Moreover, the curvature value increases as the width of the anvil wheel decreases. Experimental results showed that the curvature values obtained from the straight line paths parallel and oblique to the side edge added together were the same as those obtained from the sawtooth path, indicating the superimposed forming characteristic of the process. In addition, it was found that the error between the circumscribed circle and the experimental results is smaller. Therefore, it is recommended to use the circumscribed circle to accurately predict the actual forming results and prove the enveloping forming characteristic of the English Wheel process. Furthermore, this study proposed a process design scheme for forming a sheet metal part with a given curvature using English Wheel. The required path spacing for forming with a given anvil wheel under a specific pressing force can be determined by comparing its enveloping geometry composed of indented and unrolled segments with the desired geometry by a single test of a straight line path. The feasibility of the design scheme proposed in this study was validated through an example using finite element simulation. These conclusions can serve as a reference for industrial practitioners in the free-form manufacturing with English Wheel.

Acknowledgements. The authors would like to thank the National Science and Technology Council of the Republic of China with the grant coded NSTC 111-2221-E-005-053 for the financial support of this research.

References

1. Shisode, M.P., van den Boogaard, T., Hazrati, J.: Surface texture design for sheet metal forming applications. In: Proceedings of the 12th International Conference and Workshop on Numerical Simulation of 3D Sheet Metal Forming Processes, pp. 703–711. Springer, Cham (2022)
2. Barr, E.: Professional Sheet Metal Fabrication. Motorbooks International, Osceola (2013)
3. Quadrini, F., Santo, L., Squeo, E.A.: Flexible forming of thin aluminum alloy sheets. Int. J. Mod. Manuf. Technol. **2**(1), 79–84 (2010)
4. Schuh, G., Bergweiler, G., Fiedler, F., Bickendorf, P., Colag, C.: A review on flexible forming of sheet metal parts. In: Proceedings of the IEEE International Conference on Industrial Engineering and Engineering Management (IEEM), pp. 1221–1225. IEEE, Piscataway (2019)
5. Heo, S.-C., Seo, Y.-H., Ku, T.-W., Kang, B.-S.: A study on thick plate forming using flexible forming process and its application to a simply curved plate. Int. J. Adv. Manuf. Technol. **51**(1–4), 103–115 (2010)
6. Music, O., Allwood, J.M.: The use of spatial impulse responses to characterize flexible forming processes with mobile tools. J. Mater. Process. Technol. **212**(5), 1139–1156 (2012)
7. Downie, J.: Power Hammer Techniques and Applications for Creating Compound Curves in Sheet Metal. International Specialised Skills Institute, Melbourne (2010)
8. Rossi, G., Nicholas, P.: Re/Learning the wheel: methods to utilize neural networks as design tools for doubly curved metal surfaces. In: Proceedings of the 38th Annual Conference of the Association for Computer Aided Design in Architecture, pp. 146–155. Acadia, Dover (2019)
9. Bowen, D., Music, O., Erdinc, A.A., Shokrani, A., Loukaides, E.: Numerical modelling and deformation mechanics of the English wheel process. In: Daehn, G., Cao, J., Kinsey, B., Tekkaya, E., Vivek, A., Yoshida, Y. (eds.) Forming the Future - Proceedings of the 13th International Conference on the Technology of Plasticity, pp. 457–466. Springer, Cham (2021)
10. Fann, K.J.: Finite element study on forming metal sheets with an English wheel. IOP Conf. Ser.: Mater. Sci. Eng. **1222**, 012006 (2022)
11. Douros, I., Buxton, B.: Three-dimensional surface curvature estimation using quadric surface patches. In: Scanning 2002 Proceedings, Paris (2002)

Prediction of Wear in Roll Forming Using Data-Based Analysis and Modeling

Marco Becker(✉), Patrick Schuster, and Peter Groche

Institute for Production Engineering and Forming Machines, Technische Universität Darmstadt, Otto-Berndt-Straße 2, 64287 Darmstadt, Germany
marco.becker@ptu.tu-darmstadt.de

Abstract. Knowing exactly when a tool needs to be replaced helps to avoid unscheduled downtime due to failure. In roll forming, as in many other forming processes, especially abrasive wear is critical for tool life. Since numerical modeling of wear phenomena is complex and time-consuming due to required microscopic accuracy, a practical data-based approach is investigated to predict wear states. In roll forming, driven tool rolls are responsible for both the incremental profile bending at each forming step and the overall profile transport induced by all forming steps. The driving diameter of a tool roll is defined as being located at the position with rolling friction, i.e., without relative sliding velocity between the tool and the workpiece. The effective direction of the tool torque corresponds to the transport direction of the frictional force between tool and workpiece. Sensor data shows, that the torque of a driven tool roll is sensitive to changes in the tool gap between the rolls, to changes in transport velocity and to changes in geometrical bending parameters. Therefore, it is considered to be an important inline measured process variable. In this approach, idealized wear states in terms of enlarged tool radii are experimentally investigated using four discrete levels. As a result, the average torque level show correlations to the wear states for all the four investigated forming steps individually and simultaneously allowing to determine data trends. The trends are explained by a geometric displacement of the driving diameter at larger tool radii. Furthermore, stationary and non-stationary forming conditions are identified from data, allowing the forming process to be analyzed over time. Using the presented approach, four discrete wear states are classified for one of the forming stages by an artificial neural network. Using these results, torque monitoring can be applied to predict wear and estimate the end of the tool life.

Keywords: roll forming · wear prediction · inline sensory · data analysis

1 Introduction and State of the Art

Roll forming is a continuous production process where a sheet metal is formed between the rolls of sequential forming stands. Although roll forming is an established forming process, industry 4.0 technologies are rarely used in industrial application yet. Experience-based setup and adjustment processes are time-consuming and there is no

K. Mocellin et al. (Eds.): ICTP 2023, LNME, pp. 285–296, 2024.
https://doi.org/10.1007/978-3-031-40920-2_30

statistical data collected about tool life [1]. The circumferential driven tool rolls in roll forming are responsible for both the profile transport and the incremental profile bending. In contrast to the circumferential velocity of cylindrical rolls, which is the same along the width, the determination of the feed rate in roll forming is not trivial. The driving diameter of a tool roll is defined as being located at the position with rolling friction, i.e., without relative sliding velocity between the tool and the workpiece [2]. As in many other forming processes, in roll forming abrasive wear is critical due to high contact pressure and slip between tools and workpiece [3]. Tool wear is therefore mainly caused in areas of tool contours with diameter changes where slip with high sliding friction occurs [4]. To consider this and to understand the friction influences on roll forming, different friction values over the contact width were investigated in studies with test rigs [5] and finite element modeling [6]. However, the impact of tool wear on the roll forming process and the respective final profile geometry is mostly unexplored so far. In industrial roll forming lines, tools are replaced or reworked in predefined time intervals to prevent wear caused defects like surface scratches, geometry deviations and, in the worst case, load capacity reduction of the final product due to applied notches.

The reason why wear is not considered by several industrial roll formers is, that on the one hand, most profile products with low load capacity requirements are made of comparatively thin-walled metal sheets and low-strength materials. For these products, wear occurs comparatively rare after years of production and a lot of profiles. From these producers' point of view, it is not worth the effort for tracking wear on tool rolls in terms of scheduled or even predictive maintenance. On the other hand, there are several products from construction and automotive industries where profiles with thick-walled cross sections and high-strength materials are required due to high static or dynamic load capacities. In addition, some products in these sectors are made from hot-rolled steel with scale where the solid particles lead to excessive abrasive wear [7]. The trend towards higher-strength materials to save weight while maintaining or even improving the load capacity of the profiles has been increased in the past few years, especially in the mobility sector. For this reason, wear detection and wear prediction become more important as worn rolls lead to scrap production and unscheduled downtimes. A study of Galakhar et al. shows that the most significant wear in forming a U-profile is located in the bending zones of the convex rolls where the largest contact normal stresses occur [8]. In further research, they assume that not only high strength workpiece materials promote to abrasive tool wear, but also soft coated materials. Their theory is, that the high contact pressure in roll forming creates a hydrostatic fluid like behavior of the soft coating causing an opening of incipient cracks in the tool roll [9]. The objective of the present approach is investigating correlations between wear on roll forming tools and process measuring data to show the feasibility of predicting wear states for process observation. The aim is to build a foundation for implementation methods of predictive maintenance in roll forming.

2 Experiments

2.1 Experimental Setup

For this data-based approach, a roll forming line with four forming stages is used (Fig. 1). Each forming stand consists of one top and one bottom tool roll. Sheet metal strips with a length of 2000 mm, a width of 100 mm and a thickness of 1.0 mm are roll formed into a simple U-channel profile with a resulting angle of 56° (Fig. 2a). The sheet material is dual phase steel DP600 with a yield strength of 370 MPa and tensile strength of 590 MPa. The forming gap between the top and bottom rolls is set to 0.9 mm for reasons of deflection due to forming loads. This ensures a constant force transmission between rolls and sheet. The roll forming line operates with a feed velocity of 4.5 m/min. To guarantee comparability between all tests of the experimental series, no lubrication is used.

Each of the eight roll-shafts is equipped with a torque sensor. Furthermore, force and position sensors are integrated in each forming stand on both sides of the frame. Velocity sensors between the forming steps ensure a homogeneous feed rate. The overall data acquisition is performed by the PLC at a synchronized sampling rate of 50 Hz. After the last forming stand, a stand-alone optical sensor system for cross section measurement by laser triangulation is used to quantify variations in profile geometry.

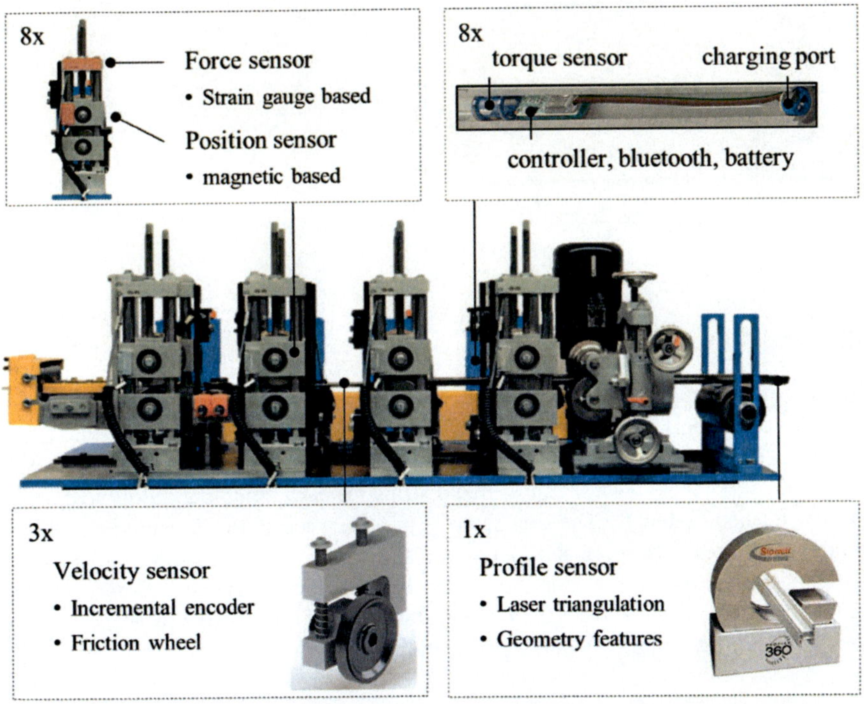

Fig. 1. Roll forming line with sensory equipment for experimental setup.

For each forming stage four convex top tool rolls with different forming radii are produced and systematically replaced during the test series. The different forming radii represent four idealized abrasive wear states. The bending radius of the first convex roll set is 1.0 mm (R1) which conforms to the initial roll design of brand new tools. Therefore, the setup with 1.0 mm on each of the four forming stands represents the tool condition without any wear and is defined as reference state. The other three convex roll sets are used to represent the appearance of abrasive wear in the bending radii. The respective radii in the bending areas are 2.0 mm (R2) for the second convex roll set, 3.0 mm (R3) for the third and 4.0 mm (R4) for the last as shown in Fig. 2b.

It should be noted that these are idealized wear states in which only the geometric contour of the roll forming radii is intentionally modified. The surfaces of all rolls are polished under same conditions. Therefore, this approach does not aim for a real industrial situation of wear occurring progressively during roll forming. Due to the lack of available data and information on wear in roll forming, these laboratory test conditions are meant to establish a starting point for systematic investigations with only one changing influence factor, the macroscopic geometry affected by wear. This facilitates the interpretability of the data and can subsequently be used to follow up with further investigations involving microscopic surface changes or even real stochastic wear phenomena in long-term investigations.

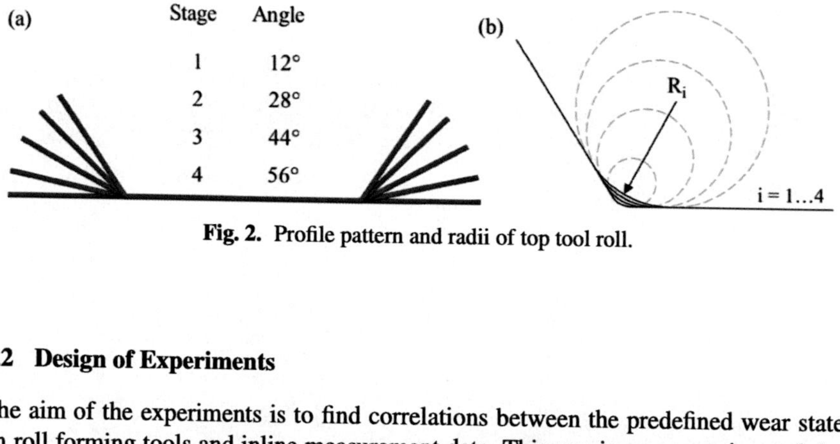

Fig. 2. Profile pattern and radii of top tool roll.

2.2 Design of Experiments

The aim of the experiments is to find correlations between the predefined wear states on roll forming tools and inline measurement data. This requires an experimental data set with a sufficient amount of data to statistically validate all states. A full factorial design of experiments with six sheet metal strips for each combination of wear states was established (Table 1). The first setup is the reference state with unworn 1.0 mm radius rolls for each forming step. Afterwards, the convex top rolls on each stage are systematically replaced one after another by the wear state rolls with 2.0 mm, 3.0 mm and 4.0 mm radii. Finally, the 4.0 mm wear state was set up for more than one forming step simultaneously to investigate superposition effects.

Table 1. Full factorial design of experiments.

Profile Number	Radius Stage 1	Radius Stage 2	Radius Stage 3	Radius Stage 4
01–06	1 mm	1 mm	1 mm	1 mm
07–12	1 mm	1 mm	1 mm	2 mm
13–18	1 mm	1 mm	1 mm	3 mm
19–24	1 mm	1 mm	1 mm	4 mm
25–30	1 mm	1 mm	2 mm	1 mm
31–36	1 mm	1 mm	3 mm	1 mm
37–42	1 mm	1 mm	4 mm	1 mm
43–48	1 mm	2 mm	1 mm	1 mm
49–54	1 mm	3 mm	1 mm	1 mm
55–60	1 mm	4 mm	1 mm	1 mm
61–66	2 mm	1 mm	1 mm	1 mm
67–72	3 mm	1 mm	1 mm	1 mm
73–78	4 mm	1 mm	1 mm	1 mm
79–84	1 mm	1 mm	4 mm	4 mm
85–90	1 mm	4 mm	4 mm	4 mm
91–96	4 mm	4 mm	4 mm	4 mm

2.3 Data Acquisition

The acquired data sets of the sensors are used for analyzing correlations to changing process variables. Typical time series data is shown in Fig. 3. The data is synchronous and the time characteristics of the roll forming process become visible. At $t = 7$ s, the sheet is moved into the first forming stage starting the run-in phase. It is then transported to the next forming stages in uniform time intervals due to constant feed velocity. After the first dashed line, where the sheet is simultaneously placed in all forming stages at $t = 18$ s, the quasi-stationary roll forming phase starts. Cyclic oscillations are present especially in forces, which result from the constant circulating movement of the tool rolls. However, at the top roll of forming stage 4 T_{4t}, a transient torque characteristic towards negative values is visible, so that no strictly stationary behavior can be assumed within the dashed lines. Behind the second dashed line from $t = 32$ s, the run-out phase starts in the same order.

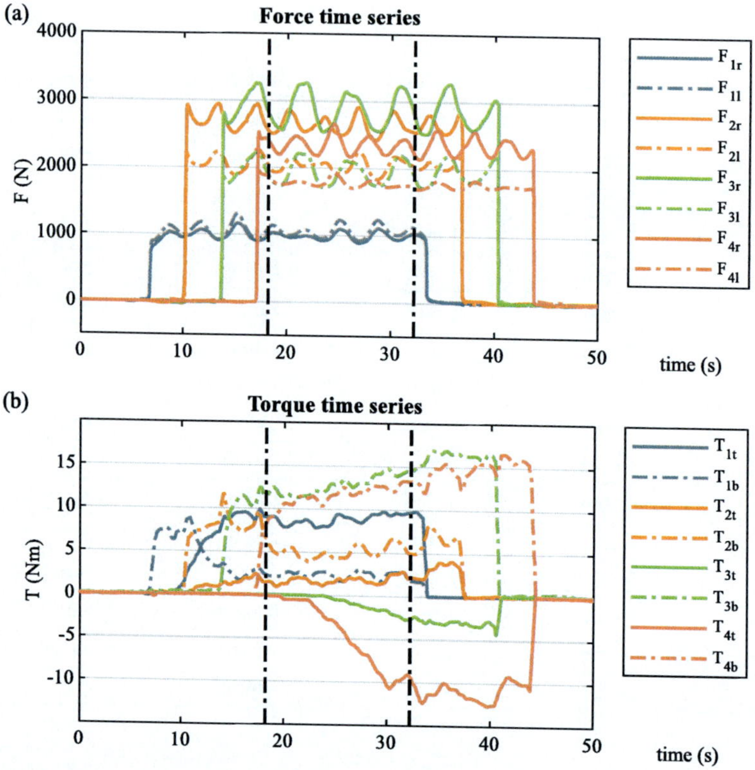

Fig. 3. Typical time series data of all four forming steps with (a) force on the left F_{il} and right F_{ir} side and (b) torque on the top T_{it} and bottom T_{ib} shafts.

3 Data Analysis Results

This chapter describes the data analysis of the experimental data using a data-based engineered approach to find some qualitative patterns first and using an Advanced Neural Network to validate and quantify these findings afterwards. Data from all sensors were recorded and analyzed during the experimental study. Velocity and tool position data are almost not affected by the investigated wear conditions and will not be discussed any further. Correlations between collected data and the investigated wear states were found for torque, force and geometry. These data will be compared and discussed within the next sections.

3.1 Data-Based Engineering Analysis

The engineering analysis aims to confirm the existence of correlations and the feasibility of finding some qualitative patterns to be applied as trend analysis for long-term data observations. Calculating the mean values from the quasi-stationary phase for all torque time series (see Fig. 3) leads to comparative average diagrams as shown in Fig. 4. The four wear classes at forming step 3 are compared in separate plots. The average torque is influenced by R_i on both forming step 3 and forming step 4. Whereas forming step 3 with the wear affected tool radius shows decreasing torque, forming step 4 shows the opposite behavior with increasing torque meanwhile (red arrows). This leads to the assumption that interactions between the forming steps occur due to wear affected geometrical changes of certain tool radii. Similar effects are obtained for the test results with wear on forming step 1, 2 and 4.

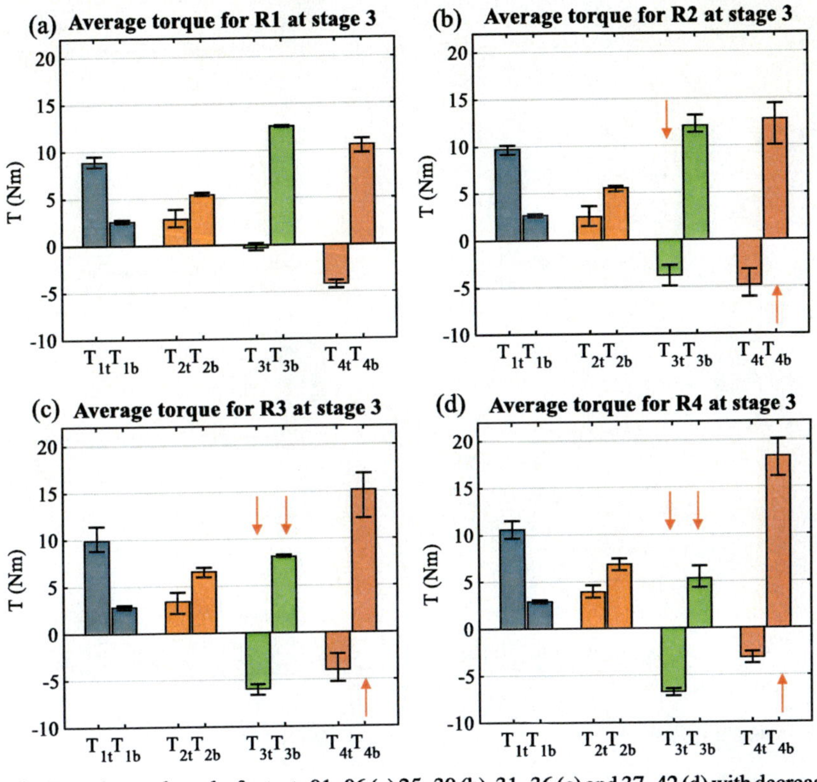

Fig. 4. Experimental results for tests 01–06 (a) 25–30 (b), 31–36 (c) and 37–42 (d) with decreasing torque at T_{3t} and T_{3b} and increasing torque at T_{4t} and T_{4b} for increasing wear states R1 to R4.

Evaluating these results, it can be stated that especially the drive torques of the wear affected forming stages as well as the next following forming stage are sensitive to different geometrical wear radii. Torque data of the top rolls are found to be the most sensitive indicating data. Therefore, Fig. 5 summarizes the average torque data with wear state R4 for each forming step separately (a–d) for comparison with the unworn reference state R1 on all stages (e) and wear state R4 at all forming stages simultaneously (f). Based on this data, clear patterns for the individual wear states can be identified and the superposition of wear at several stages simultaneously becomes traceable. Confirming the previous observations at forming stage 3, each wear affected forming roll shows changes in drive torque compared to the reference state. Solely the torque for wear at stage 1 is very similar to the unworn reference state.

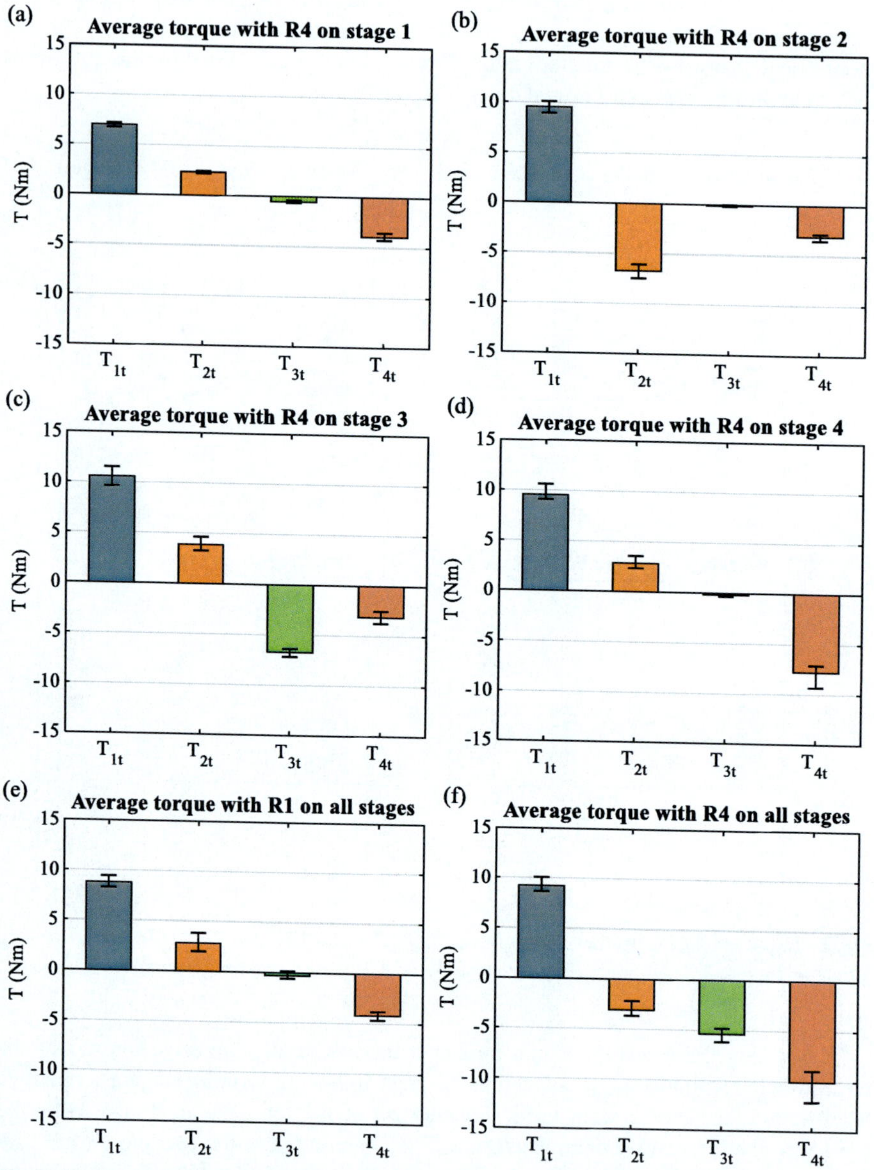

Fig. 5. Torque data of top rolls with R4 on all forming steps separately (a–d) and simultaneously (f) compared to the unworn reference state R1 at all forming stages (e).

To evaluate whether these data are suitable for detecting and predicting progressive wear, Fig. 6 shows interpolated trends for each top roll torque T_{it} across all four wear classes. For all stages despite stage 1 a clear trend can be observed depending on the wear class.

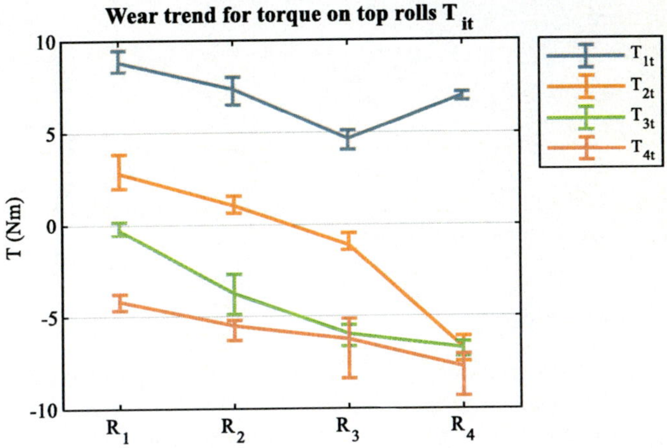

Fig. 6. Torque average trends for the wear classes of each top forming roll T_{it}.

Similar analyses were carried out for the force data. Correlations to the torque data and consequently also to the wear conditions are given, but significantly weaker. Physically, this can be explained as driving torque and contact force are directly related via the frictional force. With the direction of the frictional force acting in the same direction as the driving torque, torque data tend to react more sensitively to changes in the tribological system.

Finally, a graphical comparison is made to determine whether the geometry data of the final profile after the last forming stage can provide information on the wear state of the rolls. In contrast to inline torque acquisition, this measuring method is already being used in industry to guarantee geometric tolerance specifications. However, in practical applications, geometric characteristics are solely used to classify between good and worse geometry, and conclusions about the causes of produced scrap are not available. The geometry data shows that the wear radius affects the geometrical accuracy of the profile in terms of the inner and outer radius (Fig. 7). The geometry data shows that the wear radius affects the geometrical accuracy of the profile in terms of the inner and outer radius (Fig. 7). Wear at the last step (S4) has apparently a great influence on the final part geometry. For the data with wear in previous forming steps (S1, S2, S3), no clear correlations can be found since subsequent steps superimpose the effects on the formed profile. This observation is consistent with the findings of Groche et al. [10], who studied the influences on the final bending angle during roll forming by overbending intentionally at the last forming step.

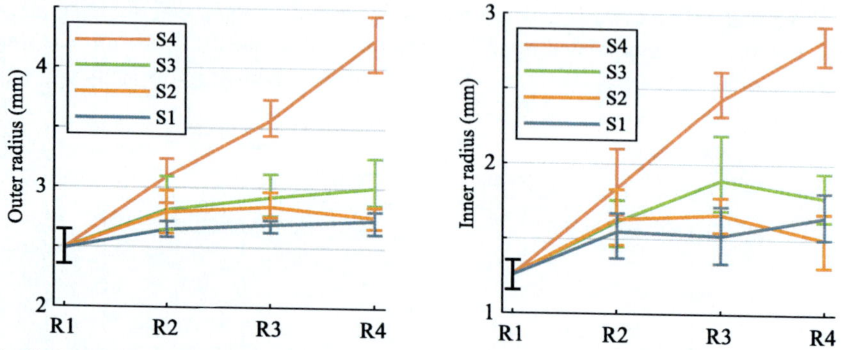

Fig. 7. Influence of wear states on the geometrical accuracy of the profile at the inner and outer radius with significant impact only at stage 4.

3.2 Data-Driven Analysis

For the data-driven approach, a single forming step from the whole data set was selected to test a lean model with a single sensor input. The four wear classes serve as outputs. Engineered features such as mean value and slope as well as features from Principal Component Analysis (PCA) are used to train an Artificial Neural Network (ANN). The results of the model show that the four discrete wear classes can be classified with an root-mean-square error (RMSE) of 22.214 ± 1.765 μm. The model and this result are presented in detail in a previous study [11]. Besides the roll forming process, the model has been used for wear prediction in a stamping process successfully. The transferability of the model approach to different forming processes is important, as the functionality for different data characteristics makes an extension to multisensory inputs promising in future applications. One modification of the previous model [11] shows that reducing the data to the cyclostationary phase, as in the average analysis of Sect. 3.1, leads to a worse model performance. It is suggested that significant information about wear, particularly in the transient phase of the process, can be found as prominent features by PCA. This is consistent with the time series observation showing an exponential decrease in torque during the run-in phase and the first part of the cyclostationary phase. The reason for this is the incremental forming behavior which leads to a shift in contact conditions even after the sheet metal strip run-in at the last forming step. This phenomenon is also seen in Fig. 3 for the reference process. A decrease in torque T_{4t} begins and ends within the cyclostationary phase. The duration respectively the distance of this decreasing torque corresponds exactly to the total length of the roll forming distance from stage 1 to stage 4. This proves that a stationary behavior of the drive torque during roll forming occurs only after twice the run-in distance. For verification, experiments with 4 m long sheet metal strips were carried out and verified this observation. The ANN results were improved adding the stationary phases of the 4 m sheet tests but remain below the model performance using both run-in and stationary data. Therefore, the data-driven analysis shows that the run-in phase contains important information for the regression of the wear state and should be collected for further improved model trainings.

4 Conclusion and Outlook

In the present approach, the assumption is made that enlarged bending radii can be considered as idealized abrasive wear classes in roll forming. This assumption bases on the limited literature available on this topic as well as non-scientific observations from industrial application. The four different tool radii serve as snapshots at discrete time intervals for the prediction of continuous wear progress. The abrasive wear is fundamentally considered on a macroscopic geometric level, whereas the microscopic surface is the same for each class. Consequently, influences from the adhesion and friction changes due to surface roughening can be excluded and only the macroscopic geometric differences of the tool rolls serve as an influencing factor.

As shown with the data-based analysis, the drive torques react sensitively to abrasive tool radius states. Since the driving diameter as a geometrical parameter depends on the tool contour, it shifts for different diameters and causes different slip behavior between the tool roll and the profile. This assumption agrees with the findings of Traub et al. who caused a shift of the driving diameter and therefore changes in driving torque by changing the rotational tool velocities in their investigations [12]. As seen in Fig. 8, the local spot of the circumferential speed of the top forming roll leads to increased slip between the tool and the sheet metal. This is due to the remaining sheet feed from the bottom forming roll and the neighboring forming steps. Based on the average torque of each top roll clear trends were found based on the investigated wear classes, with exception of stage 1. A possible reason for this is that comparatively low forces occur at stage 1 in combination with a larger contact area due to the small bending radius increment. This leads to a comparatively small shift of the driving diameter and inconsistent torque variance.

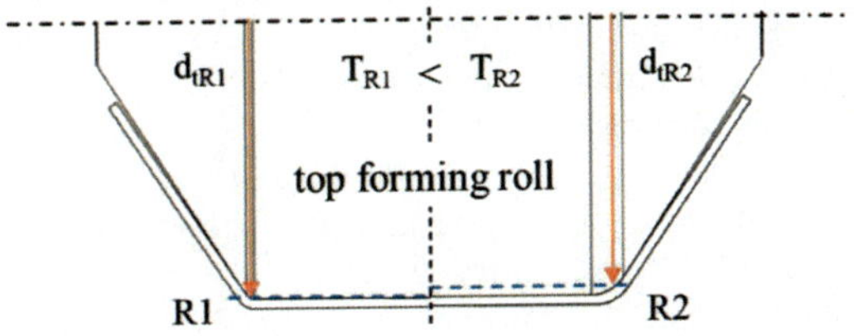

Fig. 8. Moving driving diameter depending on the tool contour at worn radius [11].

The data-driven results show that not only the average torque is suitable as a correlating feature. Several well performing PCA features were found for a ANN model of forming stage 3. The results show that run-in data is important for the regression as the transient torque time series contains significant features for the four different wear states.

As a conclusion, torque data from the individual forming steps is suitable for localization and prediction of wear states compared to optically measured geometry data of the final product. In future studies with the data-driven model approach and larger data sets, it remains to be proven that the present wear trends evaluated from discrete states can be observed and classified in industrial long-term application with progressive wear.

Acknowledgements. The results of this paper are achieved within the project Mittelstand-Digital Zentrum Darmstadt funded by the German Federal Ministry for Economic Affairs and Climate Action (BMWK). The authors wish to thank for funding and supporting this project. Furthermore, the authors would like to thank Dreistern GmbH & Co. KG for providing the sensory equipped roll forming line.

References

1. Christiany, M., Groche, P.: Reproducibility of wear tests and the effect of load on tool life in sheet metal forming. Adv. Mater. Res. **1018**, 293–300 (2014)
2. Eichler, U.: Walzprofilieren von Standardquerschnitten auf einer mehrgerüstigen Maschine mit einzeln angetriebenen Werkzeugwellen. Dissertation, Shaker-Verlag, Aachen (1987)
3. Galakhar, A., Meehan, P., Daniel, W., Ding, S.: A method of approximate tool wear analysis in cold roll forming. In: 5th Australasian Congress on Applied Mechanics (2007)
4. Filzek, J.: Tribologische Optimierung beim Walzprofilieren mit Hilfe von Laborversuchen. 6. Fachtagung Walzprofilieren, 2. Zwischenkolloquium SFB, Darmstadt (2008)
5. Galdos, L., Ulibarri, U., Gil, I., Ortubay, R., Sáenz de Argandoña, E.: Friction coefficient identification in roll forming processes. Key Eng. Mater. **611–612**, 425–435 (2014). https://doi.org/10.4028/www.scientific.net/kem.611-612.425
6. Mueller, C., Gu, X., Baeumer, L., Groche, P.: Influence of friction on the loads in a roll forming simulation with compliant rolls. Key Eng. Mater. **611–612**, 436–443 (2014). https://doi.org/10.4028/www.scientific.net/kem.611-612.436
7. Halmos, G.: Roll forming Handbook. CRC Press Tailor & Francis Group, Boca Raton (2006)
8. Galakhar, A., Daniel, W., Meehan, P.: Prediction of roll profile wear in the cold roll forming process. Key Eng. Mater. **410–411**, S. 643–660 (2009)
9. Galakhar, A., Gates, J., Daniel, W., Meehan, P.: Adhesive tool wear in the cold roll forming process. Wear **271**(11–12), 2728–2745 (2011). https://doi.org/10.1016/j.wear.2011.05.047
10. Groche, P., Beiter, P., Henkelmann, M.: Prediction and inline compensation of springback in roll forming of high and ultra-high strength steels. Prod. Eng. **2**(4), 401–407 (2008)
11. Kubik, C., Becker, M., Molitor, D., Groche, P.: Towards a systematical approach for wear detection in sheet metal forming using machine learning. Prod. Eng. Res. Devel. **17**, 21–36 (2023). https://doi.org/10.1007/s11740-022-01150-x
12. Traub, T., Güngör, B., Groche, P.: Measures towards roll forming at the physical limit of energy consumption. Int. J. Adv. Manuf. Technol. **104**(5–8), 2233–2245 (2019). https://doi.org/10.1007/s00170-019-03992-9

Estimation of Kinematic Hardening of Sheet Metals Based on Stress-Relaxation Behavior

Kouki Matsugi[1(✉)], Kazuhiro Ikeda[1], Takumi Araki[1], and Ryutaro Hino[2]

[1] Graduate School of Hiroshima University, 1-4-1, Kagamiyama, Higashi-Hiroshima 739-8527, Japan
d211336@hiroshima-u.ac.jp

[2] Graduate School of Advanced Science and Engineering, Hiroshima University, 1-4-1, Kagamiyama, Higashi-Hiroshima 739-8527, Japan
rhino@hiroshima-u.ac.jp

Abstract. This study investigates the relationship between stress relaxation behavior and kinematic hardening of sheet metals. Stress relaxation tests and tension-compression tests for a high-strength steel sheet and an aluminum sheet were carried out to discuss whether it is possible to estimate the back stress as the amount of kinematic hardening based on the results of the stress-relaxation tests. In the stress relaxation after uniaxial tension, stress decreases with time. On the other hand, in the stress relaxation after uniaxial tension followed by full unloading, stress increases. It was found that in the stress relaxation after uniaxial tension followed by unloading to a certain stress level, stress does not change during the relaxation test. As for the high-strength steel sheet, the unloading stress level where stress does not change in the relaxation test was found to be relatively high and its Bauschinger effect is significant. On the contrary, as for the aluminum sheet, the unloading stress level is found to be relatively low and its Bauschinger effect is weak. These results suggest that the externally applied stress and the internal back stress is balanced at a certain unloading stress level where stress does not change in the relaxation test, and the unloading stress level is correlated with the amount of kinematic hardening.

Keywords: Stress Relaxation · Kinematic Hardening · Bauschinger Effect of Sheet Metal

1 Introduction

In numerical analysis of sheet metal forming, it is of vital importance to consider the Bauschinger effect for accurate springback prediction. Tension-compression test is usually carried out to observe the Bauschinger effect of sheet metals, but the experiment cannot be performed using a standard testing machine and specimens. It requires an adhesively laminated specimen [1] or a specially designed apparatus [2] to prevent buckling in compressive loading.

One of the major mechanisms of the Bauschinger effect is explained as follows. As a result of plastic deformation, the dislocations accumulate at obstacles and produce

K. Mocellin et al. (Eds.): ICTP 2023, LNME, pp. 297–304, 2024.
https://doi.org/10.1007/978-3-031-40920-2_31

dislocation pileups. This dislocation structure generates internal stress i.e., local back stress opposing to the dislocation motion. The local back stresses assist the movement of dislocations in the reverse direction, and consequently the re-yield stress in the reversal compression becomes lower. The back stress in the macroscopic kinematic hardening model is regarded as an averaged local back stress.

On the other hand, the internal stress has a relation with the stress relaxation phenomenon as well. The stress relaxation is related to thermally activated motion of the dislocations under effective stress that is commonly described as the difference between the externally applied stress and the internal stress [3].

From the above, it is considered that the Bauschinger effect and the stress relaxation behavior are related through the back stress as the internal stress. In this study, two kinds of sheet metals (an advanced high-strength steel sheet and a pure aluminum sheet) were subjected to tension-compression tests and stress relaxation tests under various conditions to investigate the relationship between the stress relaxation behavior and the Bauschinger effect. The possibility of estimating the back stress level (that is, the amount of kinematic hardening) or the degree of the Bauschinger effect from the stress relaxation test results was investigated.

2 Experimental Method

2.1 Test Materials

Advanced high-strength steel sheet SPCN118Y, which is a 1180-MPa class dual-phase steel sheet whose thickness is 1.2 mm, and 2-mm thick pure aluminum sheet A1050-O were used. These two materials were selected since their Bauschinger effect and the elastic-plastic transient behavior were expected to be largely different from each other.

SPCN118Y specimen was cut out in the transverse direction (T.D.) and A1050-O specimen in the rolling direction (R.D.). The reason why SPCN118Y specimen is parallel to T.D. is that in some applications of this material major deformation orientation was expected to be T.D.

2.2 Tension-Compression Test

A tension-compression testing apparatus, which is equipped with an anti-buckling plate to prevent buckling during compressive loading, was attached to a testing machine (SHIMADZU AG-IS). A specimen shown in Fig. 1 was attached to the apparatus to carry out the tension-compression test.

Uniaxial tension was applied to the specimen up to target prestrain of 5%, followed by unloading and reversal compression until strain reached 0%. The strain was measured using a strain gauge attached to the center of the specimen. Crosshead speed in the tension-compression test was 0.35 mm·min^{-1}.

2.3 Stress Relaxation Test

A specimen shown in Fig. 2 and the same testing machine (SHIMADZU AG-IS) as above were used for the stress relaxation test. The strain was measured using a strain gauge attached to the center of the specimen.

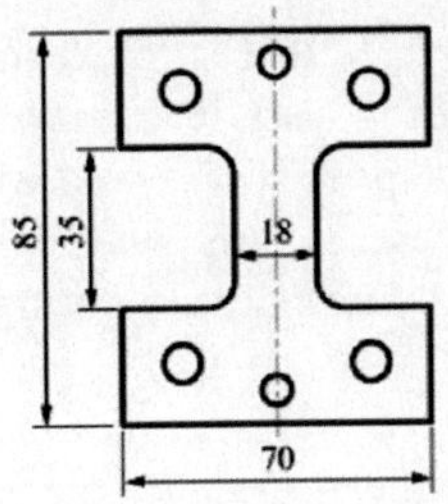

Fig. 1. Specimen for tension-compression test (in mm).

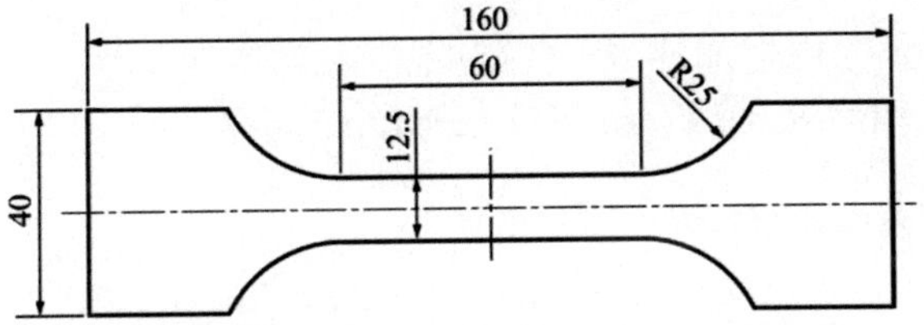

Fig. 2. Specimen for stress relaxation test (in mm).

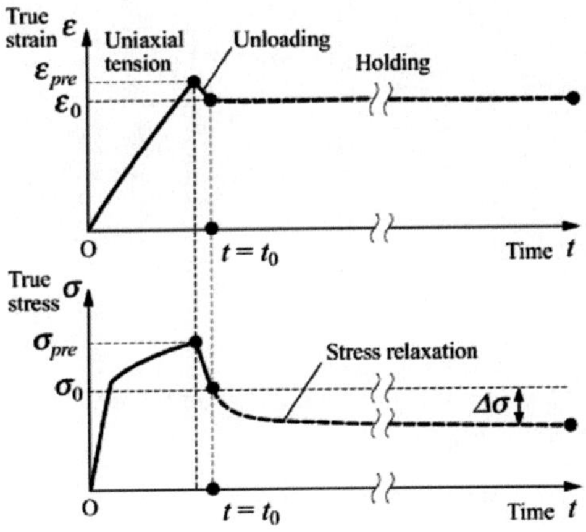

Fig. 3. Schematic diagrams of stress and strain histories in stress relaxation test.

Figure 3 shows schematics of stress and strain histories in the stress relaxation test. After applying uniaxial tension up to prestrain ε_{pre}, the specimen was immediately unloaded to a certain stress σ_0 and then the subsequent stress relaxation was measured holding the crosshead position of the testing machine. Assuming that σ_{pre} is stress value when the prestrain ε_{pre} is applied, stress level just after the unloading, i.e., stress level at starting point of the relaxation can be expressed as $\eta = \sigma_0/\sigma_{pre}$. When $\eta = 1$, that means ordinary stress relaxation test without unloading (tension-holding test). When $0 < \eta < 1$, that is relaxation test after incomplete unloading (tension-incomplete unloading-holding test). When $\eta = 0$, that is relaxation test after complete unloading (tension-complete unloading-holding test). It is known that ordinary stress relaxation where the stress decreases as time goes is observed when $\eta = 1$ while negative stress relaxation where the stress increases occurs when $\eta = 0$. There would be a certain condition of η in which the stress does not vary during the relaxation test in the range of $0 < \eta < 1$ [4].

Target value of the prestrain ε_{pre} was set to 5%. Crosshead speed for loading and unloading was 1 mm·min^{-1}. Crosshead speed for the tension-compression test and that for the stress relaxation test were determined so that strain rates in both experiments would be the same. Holding time of the stress relaxation was 5400 s at a minimum and 2 days at a maximum.

3 Experimental Results and Discussion

3.1 Results of Tension-Compression Test

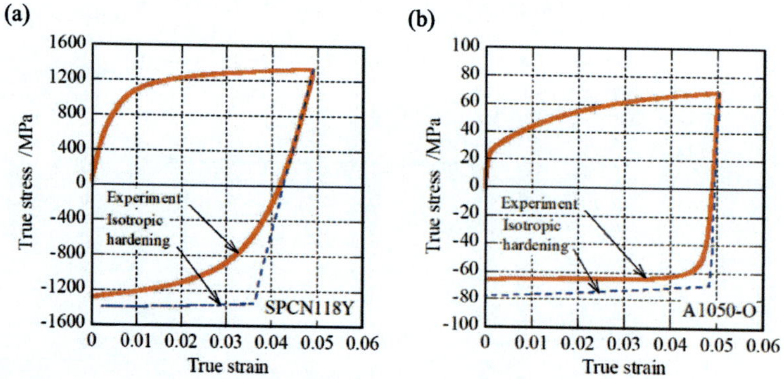

Fig. 4. Experimental results of tension-compression for (a) SPCN118Y and (b) A1050-O.

Stress-strain curves obtained from the tension-compression tests (solid curves) are shown in Fig. 4 in comparison with stress-strain behaviors when isotropic hardening is assumed (dashed curves). In the reversed loading, SPCN118Y exhibits stronger Bauschinger effect, i.e., the early re-yielding and the transient softening behavior compared to A1050-O.

3.2 Results of Stress Relaxation Test

Experimental results of the stress relaxation up to the holding time of 5400 s and 2 days are shown in Fig. 5. The vertical axis of the figure is the stress change $\Delta\sigma$ from the stress σ_0 at the starting point of the relaxation (see Fig. 3) and the horizontal axis shows the holding time. Logarithmic time axis is used for Figs. 5c and d.

In the ordinary stress relaxation without unloading ($\eta = 1$), the stress decreased with increasing holding time. SPCN118Y exhibited stress decrease $\Delta\sigma = -81.3$ MPa in 5400 s from the stress $\sigma_0 = 1307$ MPa at the relaxation starting point. A1050-O also showed stress decrease $\Delta\sigma = -12.6$ MPa from the stress $\sigma_0 = 66.2$ MPa.

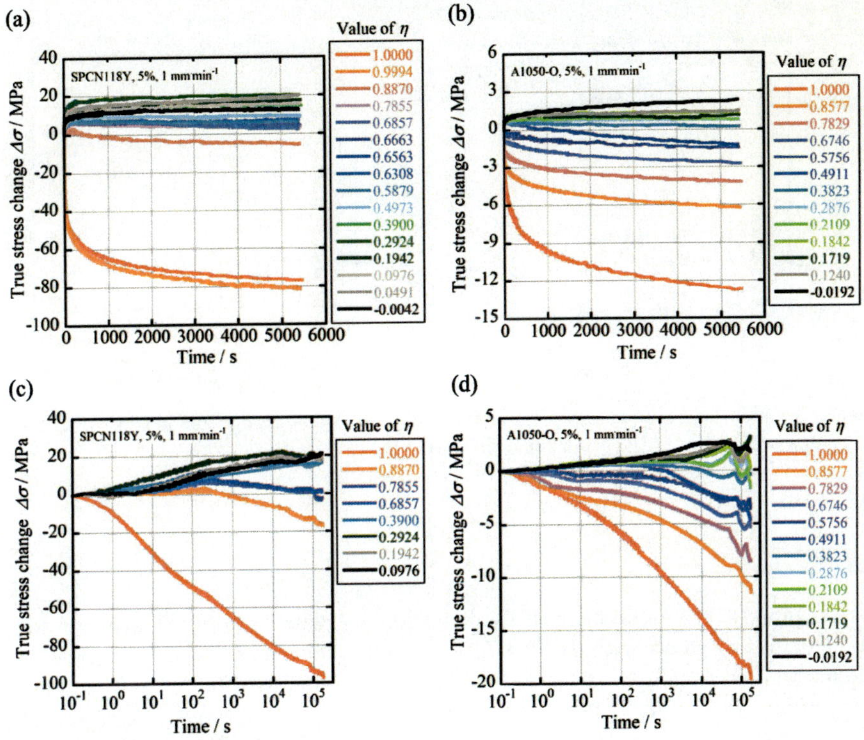

Fig. 5. Experimental results of stress relaxation ($\varepsilon_{pre} = 5\%$) for (a) SPCN118Y holding 5400 s, (b) A1050-O holding 5400 s, (c) SPCN118Y holding 2 days, (d) A1050-O holding 2 days.

On the contrary, under the conditions which were very close to the tension-complete unloading-holding test (to be specific $\eta = 0.0491$ for SPCN118Y and $\eta = -0.0192$ for A1050-O), the negative stress relaxation was observed in which the stress increased with increasing the holding time ($\Delta\sigma > 0$).

In the tension-incomplete unloading-holding test where $\eta = 0.6308$ for SPCN118Y and $\eta = 0.3823$ for A1050-O, slight increase in stress was observed in the relaxation. The stress change $\Delta\sigma$ in 5400 s was less than 3 MPa for SPCN118Y and 0.5 MPa for A1050-O.

All values of the stress change $\Delta\sigma$ at holding time of 5400 s and 2 days are extracted from all the obtained stress relaxation results and relationship between the stress change $\Delta\sigma$ and the corresponding unloading stress level η is summarized in Fig. 6. As shown in Fig. 6, the stress change $\Delta\sigma$ is negative when $\eta = 1$, but it increases with decreasing η and switches from negative to positive at a certain η. Assuming that η_0 is the partial unloading stress level at which the stress does not change during the relaxation test and $\Delta\sigma$ becomes almost zero, $\eta_0 \approx 0.7$ for SPCN118Y and $\eta_0 \approx 0.3$ for A1050-O. In other words, η_0 is relatively large for materials with strong Bauschinger effect, while η_0 is relatively small for materials with weak Bauschinger effect.

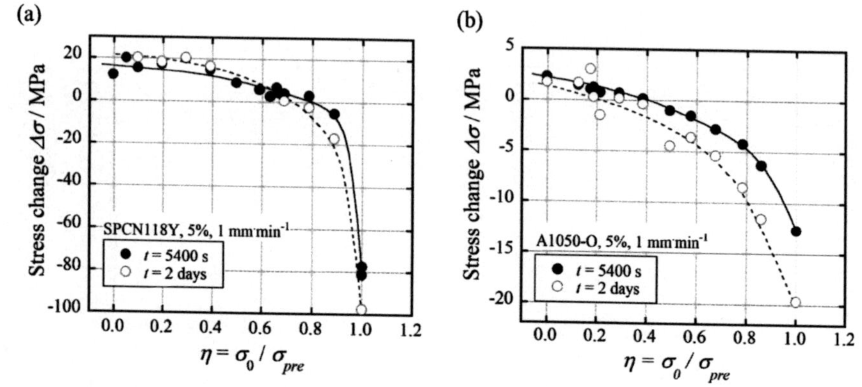

Fig. 6. Relationship between unloading stress level η and stress change $\Delta\sigma$ in stress relaxation tests ($\varepsilon_{pre} = 5\%$) for (a) SPCN118Y and (b) A1050-O.

3.3 Discussion

As shown above, the stress change in the relaxation $\Delta\sigma$ switches from negative to positive at the unloading stress level η_0. This fact shows that the externally applied stress and the internal stress balance at η_0, and the internal stress assists the reversal motion of dislocations. In addition, SPCN118Y, which has strong Bauschinger effect, has high internal stress, while A1050-O, which has weak Bauschinger effect, has low internal stress. Therefore, at least the material with high back stress also has high internal stress, and vice versa.

When assuming that the internal stress corresponds to the back stress and adopting combined (kinematic and isotropic) hardening model to describe the tension-compression behavior, tensile stress σ_{pre} just before unloading is composed of the initial yield stress Y, back stress α (kinematic hardening) and drag stress R (isotropic hardening) as schematically shown in Fig. 7. If it is assumed that the unloading stress level η_0 at which the stress change in the relaxation $\Delta\sigma$ is zero corresponds to the back stress level, then the following equations hold.

$$\sigma_{pre} = Y + \alpha + R = Y + \eta_0\sigma_{pre} + R \tag{1}$$

$$R = (1 - \eta_0)\sigma_{pre} - Y \tag{2}$$

Then the re-yielding stress Y' in the reverse compression is given by

$$Y' = \alpha - (Y + R) = (2\eta_0 - 1)\sigma_{pre} \tag{3}$$

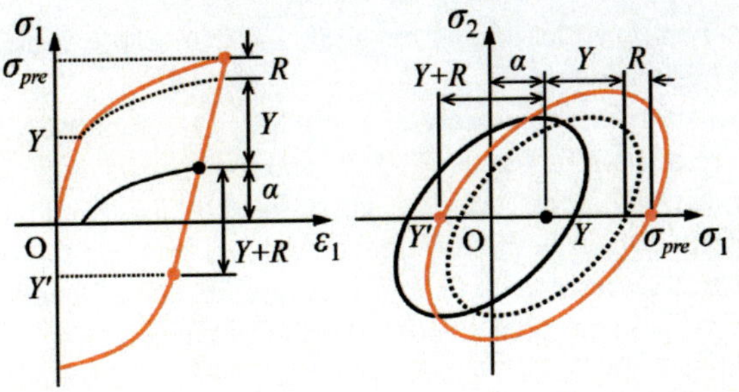

Fig. 7. Schematic illustration of stress-strain curve and yield surface evolution in combined (kinematic and isotropic) hardening model.

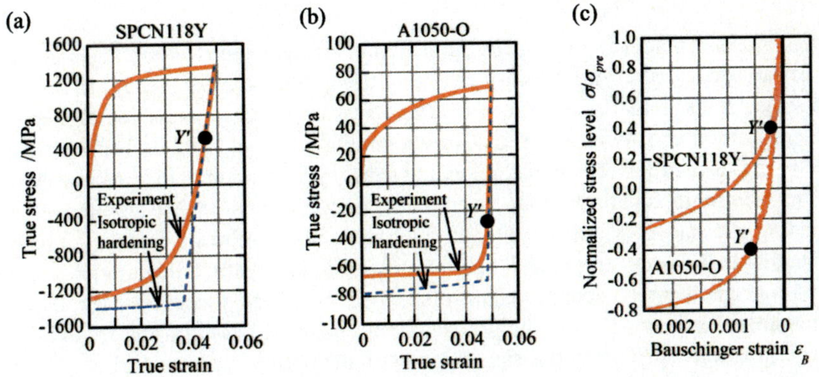

Fig. 8. Estimated re-yielding points in comparison with tension-compression test results of (a) SPCN118Y and (b) A1050-O. Estimated re-yielding points are also shown on (c) normalized Bauschinger curves.

The re-yielding point *Y'* of SPCN118Y calculated from Eq. (3) is shown on the stress-strain curve in Fig. 8a. The estimated re-yielding point is located in the unloading process rather than the reverse compression process, and this re-yield stress ($Y' = 536$ MPa) looks too high. Another problem is the contradiction that the drag stress *R* by Eq. (2) becomes negative. Therefore, either the assumption that η_0 corresponds to the back stress level or the conventional proof stress ($Y = 952$ MPa) as the starting point of plastic deformation might not be adequate for SPCN118Y. In addition, the influence of machine rigidity on the stress relaxation might be involved in the contradiction as well. However, the important point is that from this discussion it could be concluded that SPCN118Y is a fully kinematic hardening material, and that isotropic hardening does not need to be considered.

The re-yielding point of A1050-O estimated by Eq. (3) is shown in Fig. 8b in a similar way. The estimated re-yielding point is located in the reverse compression process, and this re-yield stress ($Y' = -27.7$ MPa) seems to be adequate. The amount of work hardening just before unloading in the tension-compression test is 39.5 MPa, of which 18.7 MPa would be the drag stress *R* and 20.8 MPa the back stress α. It could be concluded that A1050-O is a combined hardening material, and that both isotropic and kinematic hardening should be considered.

Figure 8c shows the estimated re-yielding points in comparison with normalized Bauschinger curves of both materials. The vertical axis shows stress in the unloading and reverse compression process, which is normalized by σ_{pre}, and the horizontal axis is Bauschinger strain, which is the difference in strain between the elastic unloading line and the measured Bauschinger curve. The Bauschinger strain corresponding to the estimated re-yielding point is about 0.0002 to 0.0006, which is extremely small. It seems that the re-yielding point by Eq. (3) gives a point at which extremely small compressive plastic strain has occurred.

4 Concluding Remarks

It was confirmed that in the stress relaxation after uniaxial tension followed by unloading to a certain stress level, stress hardly changes during the relaxation. As for sheet metal with strong Bauschinger effect, the unloading stress level where stress does not change in the relaxation was found to be relatively high. On the other hand, as for sheet metal with weak Bauschinger effect, that unloading stress level was found to be relatively low.

There is room for discussions in quantitative point of view, but qualitatively the unloading stress level where stress does not change in the relaxation seems to be corresponding to the back stress level, i.e., the amount of kinematic hardening. According to this concept, it can be determined to some extent whether a sheet metal is fully kinematic hardening material or combined hardening material based on the results of the stress relaxation test. This will assist selection of an appropriate material model and its parameter identification.

Acknowledgment. The authors are grateful to Professor Emeritus Fusahito Yoshida of Hiroshima University for his suggestion and help in the tension-compression testing. This work is supported by the JSPS KAKENHI Grant Number JP22K04760.

References

1. Yoshida, F., Uemori, T., Fujiwara, K.: Elastic-plastic behavior of steel sheets under in-plane cyclic tension-compression at large strain. Int. J. Plast **18**(5–6), 633–659 (2002)
2. Kuwabara, T., Kumano, Y., Ziegelheim, J., Kurosaki, I.: Tension–compression asymmetry of phosphor bronze for electronic parts and its effect on bending behavior. Int. J. Plast **25**(9), 1759–1776 (2009)
3. Shinoda, T.: Analysis of stress relaxation process using internal and effective stress as parameters. J Soc. Mater. Sci. Japan **20**(214), 872–878 (1971). (in Japanese)
4. MacEwen, S.R., Kupcis, O.A., Ramaswami, B.: An investigation of an incremental unloading technique for estimating internal stresses. Scr. Metall. **3**(7), 441–448 (1969)

Superplastic Rectangular Bulging of AA8090 Numerical Modeling: Validation and Development

Abdellah Lahbari(✉), Kenza Bouchaala, Mustapha Faqir, and Elhachmi Essadiqi

School of Aerospace and Automotive Engineering, International University of Rabat, LERMA Lab, Sala El Jadida, Morocco
abdellah.lahbari@uir.ac.ma

Abstract. Superplastic bulging forming is a hot manufacturing process widely used in the aerospace industry for its capability to produce complex shapes and decrease the joining operations but it has some drawbacks that make its monitoring quite difficult in terms of pressure regulation and strain rate control. A finite elements model of superplastic rectangular bulging of Al-Li alloy AA8090 was built and developed using Abaqus commercial code. The FE model was validated based on the thickness ratio distribution along the length and the width of the rectangular cavity of the die, besides the ratio of max creep strain rate over the target creep strain rate profile. The development of the model was proposed to enhance the accuracy of thickness distribution results by changing the friction coefficient of the contact between the die and the sheet. The obtained results were analyzed and explained and the new model gave promising results with good accuracy with respect to the experimental findings.

Keywords: Superplastic forming · Finite elements · Al-Li-Cu alloys

1 Introduction

Al-Li alloys are lightweight materials that have been used over the last century till now mainly in the aerospace industry, for instance: in upper wings, lower wings, and the fuselage of aircraft. The 3rd and the last generation is used for its good mechanical properties: Strength, toughness balance, fatigue crack growth resistance, thermal stability, and corrosion resistance [1, 2] However, The AA8090 alloy makes the exception and it's practically the only 2nd generation alloy still used and studied in some aerospace applications [3–5]. Hajjioui et al. [6] Explored the Al-Li alloys family by reviewing its characteristics, mechanical properties, and the forming process. For thin parts application, the Al-Li alloys are hot rolled in several passes distanced with annealing heat treatment and followed by solution heat treatment and aging [2]. The sheets of this family of alloys can be shaped using different manufacturing processes to obtain the final product. Bouchaâla et al. [7, 8] studied the deep drawing of different Al-Li alloys and investigated the optimum conditions for better formability. Wang et al. [9] studied the enhancement of warm sheet hydroforming formability of an Al-Li alloy.

K. Mocellin et al. (Eds.): ICTP 2023, LNME, pp. 305–313, 2024.
https://doi.org/10.1007/978-3-031-40920-2_32

For complex shapes manufacturing, high process temperature and low strain rate are necessary which consist of the process of low strain rate superplastic forming. Superplastic bulging or superplastic blow forming is one of the most used processes to obtain complex shapes. Lee et al. [10] conducted superplastic-free bulging on AA2195 to characterize the superplastic behavior of the alloy. Superplastic blow-forming of AA8090 Al-Li sheets in a hemispherical die was investigated [11]. Jia et al. [12] determined the superplastic forming limits of 2A97 Al-Li alloy by bulge test. Pancholi and Kashyap [13] conducted experiments on a three-layer microstructure AA8090 alloy by superplastic blow forming and deduced the optimum values of the governing parameters.

Meanwhile, the experiments of the superplastic forming are costly and time-consuming, that's why numerical models are needed before starting the experiments. Hwang et al. [14] developed a numerical model of superplastic bulging of AA8090 alloy in an elliptic-cylindrical die using DEFORM finite elements code. Balasubramanian et al. [15] developed a numerical model of superplastic blow forming of AA8090 alloy in the rectangular die using standard finite elements ABAQUS code.

In this study, validation and the development of a finite elements model of AA8090 Al-Li alloy using ABAQUS commercial code was conducted based on the experimental work of superplastic bulging in a rectangular die [16].

2 Theoretical Approaches

The material behavior in the viscous regime can be described by means of Rosserd's constitutive model [17]. It is used to identify the relation between flow stress (σ), deformation (ε), and strain rate ($\dot{\varepsilon}$) as follows:

$$\sigma = K \cdot \varepsilon^{n} \cdot \dot{\varepsilon}^{m} \tag{1}$$

where K is the strength coefficient, n is strain hardening and m is strain rate sensitivity.

At high temperatures, the strain hardening doesn't contribute much to the material behavior ($n = 0$), it gives Backofen's equation [18]:

$$\sigma = K \cdot \dot{\varepsilon}^{m} \tag{2}$$

We assume that K and m are temperature dependents only and since the process is isothermal, they are constant during the deformation.

In ABAQUS, the creep law can be expressed by the strain hardening form:

$$\dot{\varepsilon} = [A \cdot \sigma^{n'} \cdot \left((m+1) \cdot \dot{\varepsilon})^{m'}\right]^{\frac{1}{1+m'}} \tag{3}$$

where A, n', and m' are variables depending on temperature. When $m' = 0$:

$$\dot{\varepsilon} = A \cdot \sigma^{n'} \tag{4}$$

The die has a rectangular geometry identified by entry radius *R*, bottom radius *r*, cavity height *H*, length *a*, and width *b* as illustrated in Fig. 1:

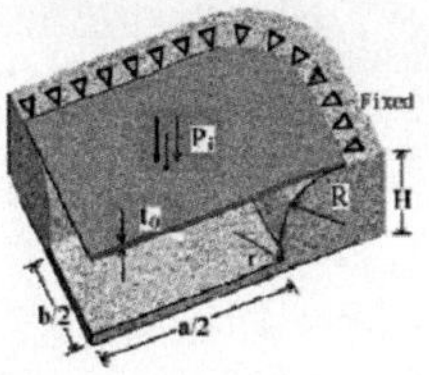

Fig. 1. Superplastic bulging in rectangular closed-die illustration [16]

3 FE Model Description

3.1 Material Properties

The chemical composition of the AA8090 Al-Li alloy is detailed in the Table 1 [3].

Table 1. Chemical composition of the AA8090 alloy

Chemical element	Li	Cu	Mg	Si	Fe	Zr
Wt. %	2.1–2.7	1.0–1.6	0.6–1.3	0.2	0.3	0.04–0.16

The material behavior is described in the elastic regime by Young's modulus $E = 77$ GPa and Poisson's ratio $v = 0.33$. Whereas, the creep behavior is defined by the means of a strain hardening form creep power law. The coefficients of the function are calculated based on the strain hardening sensitivity ($m = 0.48$) and the coefficient of strength ($K = 268$ MPa) [16].

3.2 Three-Dimensional Model

Taking advantage of the two planes of symmetry, the FE model is only a quarter of the initial model to decrease calculation time. The calculations were conducted in the implicit steps: Static General and VISCO. The first step was adopted to give stability to the model. No thermal effects were taken into consideration in this model.

The sheet is a membrane part with a circular shape with a diameter of 95 mm and a thickness of 1.65 mm. The die, as district rigid part, is a rectangular cavity with the following dimensions $a = 60$ mm and $b = 30$ mm length and width of the rectangle respectively, $R = 8$ mm die entry radius, die bottom radius $r = 1.5$ mm, and height of the die $H = 20$ mm. Figure 2 illustrates the assembly of the sheet and the die undeformed with and without meshing, and after deformation with meshing.

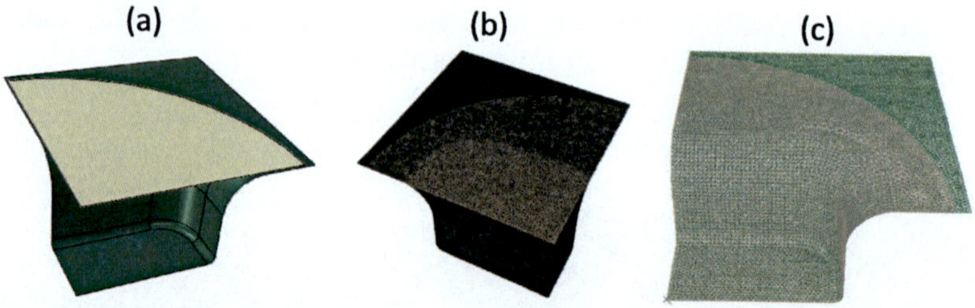

Fig. 2. Sheet and die assembly: (a) Undeformed without meshing, (b) undeformed with meshing (c) deformed with meshing

The sheet is pinned at the outer edge and experiences a nominal pressure *P* of 0.8 MPa in a VISCO step. The pressure is regulated so the maximum strain rate is close to the target strain rate which value is 8.10^{-4} s^{-1}.

The creep strain rate control is defined by the ABAQUS standard solution-dependent algorithm based on the creep strain rate ratio r_{max} which is the ratio of the maximum strain rate over the target strain rate to regulate the gas pressure *P*.

The friction in the model is handled as described in the experiments [16], the die and the sheet in two faces are brushed nitrides boron which is estimated to give a constant coefficient of friction *COF* $\mu = 0.3$ in every contact between the sheet and the die. It is adopted in the validation of the model. The numerical model is validated by the comparison between the results of the finite elements-built model, and the experimental work of Hwang et al. [16]. The compared results are:

o The distribution of sheet thickness ratio
o Ratio of maximum creep strain rate to target creep strain rate r_{max}

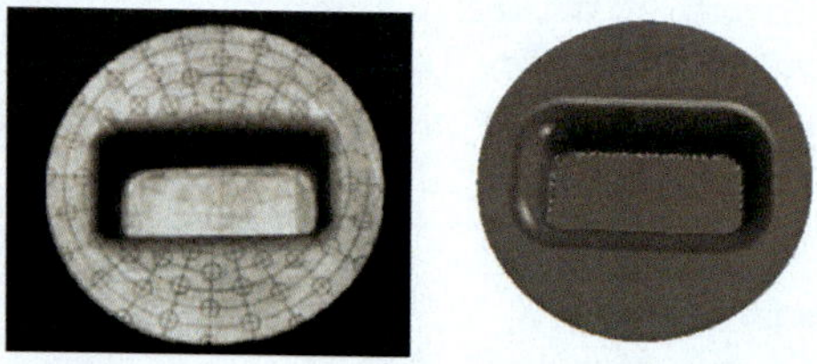

Fig. 3. Rectangular bulging results, experimental (left) & present numerical model (right)

3.3 Model Validation

Figure 4 illustrates the distribution of the sheet thickness ratio t/t_0 in the function of the length and width direction of the rectangle.

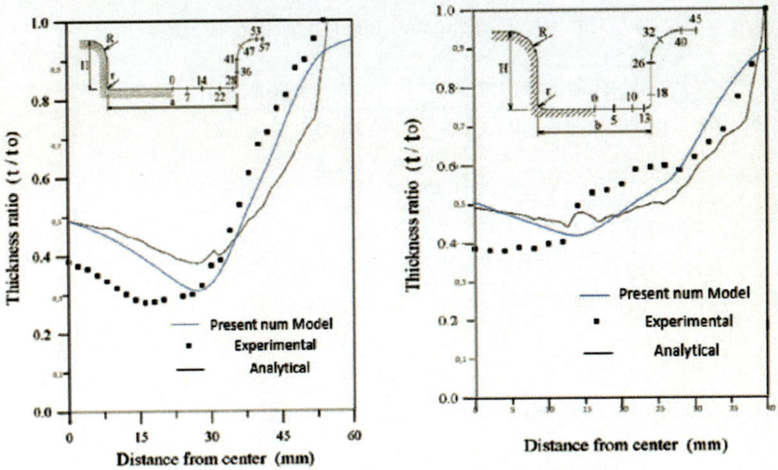

Fig. 4. Sheet thickness distribution in length and width

Table 2 presents the relative error between the experiment results and the analytical and the present numerical results.

Table 2. Experimental, analytical, and numerical results comparison

Thickness distribution along the length *a*						
Point	Distance from center	Exp [16]	Analytic [16]	Error	Present Numerical	Error
1	0	0.38	0.48	0.1	0.48	0.1
2	14	0.28	0.45	0.17	0.42	0.14
3	28	0.33	0.4	0.07	0.31	0.02
4	46	0.78	0.6	0.18	0.71	0.07
5	53	1	0.94	0.06	0.91	0.09
Along width *b*						
6	0	0.38	0.49	0.11	0.51	0.13
7	10	0.4	0.47	0.03	0.44	0.04
8	24	0.59	0.49	0.1	0.52	0.07
9	32	0.62	0.58	0.04	0.64	0.02
10	40	1	1	0	0.9	0.1

Table 3. Forming time for each friction value

Coefficient of friction	Forming time (s)
frictionless	4100
0.01	4000
0.1	2900
0.3	2068
1	2871
4	3245

The r_{max} of the model in function of time is shown in Fig. 5.

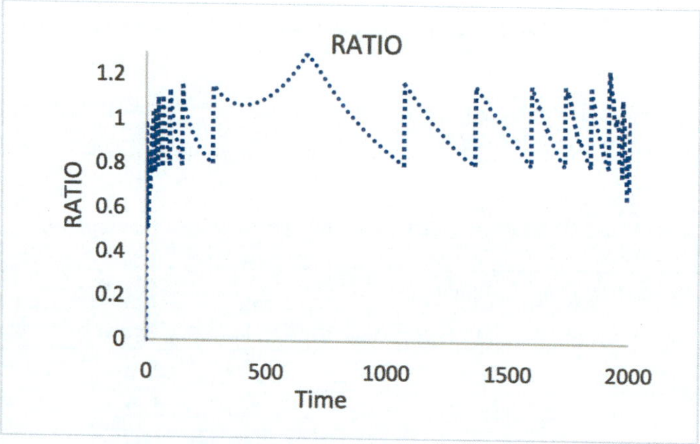

Fig. 5. Creep strain rate ratio in function of time

As shown in Table 2 the results of the comparison between the experimental results and the present numerical model fit better than those between the experimental results and the analytical results. With a max absolute error of 0.18 in length a and 0.14 in width *b*. The shape of the curve is quite similar with differences in some extreme points. The next section will discuss further those differences.

The r_{max} is oscillating at the value of 1 (Fig. 5), hence the model has good accuracy with respect to the target strain rate. Good agreement is achieved between the simulated formed cup profile and the existing experimental data.

4 Analysis and Model Development

For better model accuracy in terms of thickness distribution, a series of numerical simulations of the same validated model in the previous section but with different coefficients of friction. The pressure or the strain rate was not considered for better results because it was already optimized and the $P = 0.8$ MPa and the strain rate $= 8.10^{-4}$ gave promotive experimental outcomes. Table 2 presents the corresponding time for each simulation with different *COF* values.

Figure 6 illustrates the thickness distribution in function of true distance from the center along length *a* and length *b* respectively besides the cross sections of the rectangular die to display the different parts of the die: bottom, wall, and the entry.

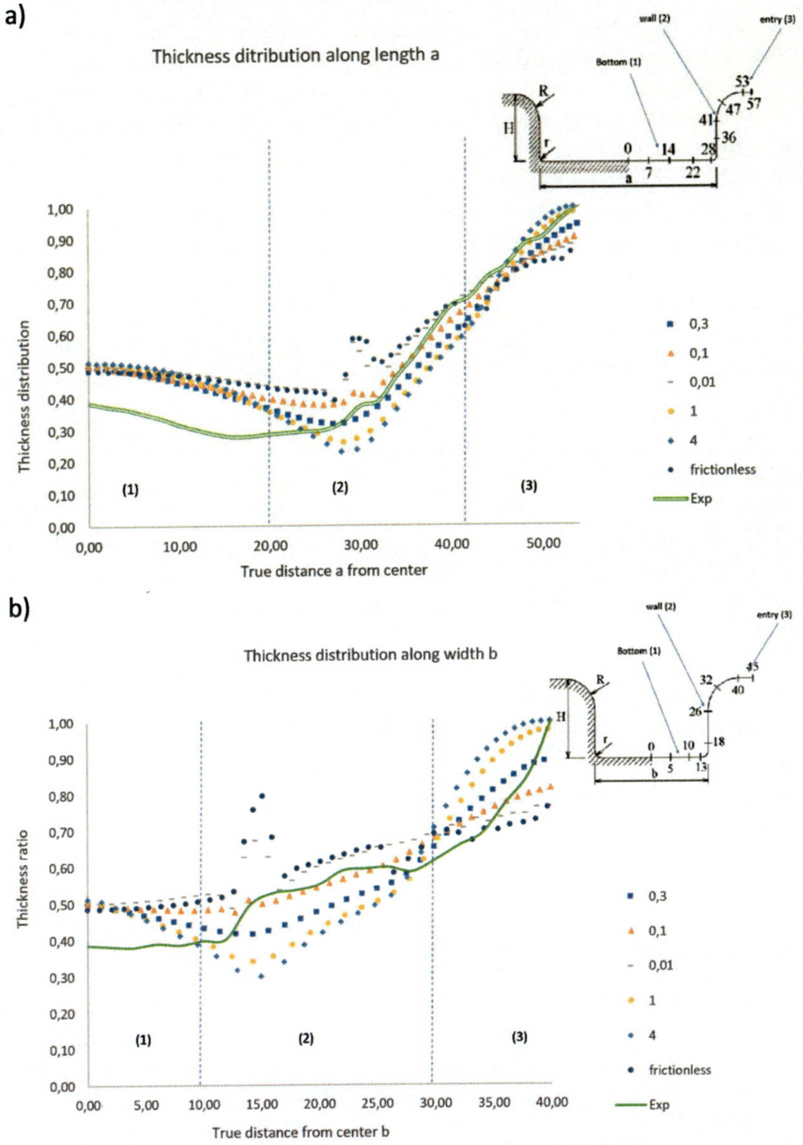

Fig. 6. Thickness distribution along a) length a b) length b

The forming time of each simulation indicates a minimum forming time in the case of the nominal friction and an increase if the coefficient of friction is either superior or inferior of 0.3 which indicates the difficulty of forming either by a slip or by gluey contact. In terms of thickness distribution, none of the simulations could match exactly the experimental result but they fit locally in the area. Along a length rectangle, the bottom of the cup is from 0 to 28 mm (1), from 28 to 47 mm is the wall (2), and from 47 to 53 mm is the entry of the die (3) (Fig. 6a). Concerning the direction of the width b, the wall is from 0 to 13 mm (1), the wall from 13 to 32 mm (2), and the entry from 32 to 40mm (3) (Fig. 6b). The frictionless model is the closest one to the experimental results at the bottom, 0.1 and 0.3 are the closest at the wall and 4 for the entry.

These findings are approved by the results of Chen et al. [19]; They estimated that the *COF* value is not constant everywhere in the contact between the sheet and the die and the more suitable model is a combination of friction coefficient as follows, entry: infinite, wall: 0.3 and bottom: 0.001. It was explained that in the superplastic forming of AA8090 alloy with boron nitride as lubricant- which is the case of the experiments- is long enough in time that allows the lubricant powder to slip from the entry to the bottom, hence is no lubricant is found at the entry and over lubrication is at the bottom and some trapped at the wall.

The pressure profile was not considered here, only the target strain rate. This parameter can explain the difference in thickness distribution, especially at the bottom of the sheet.

5 Conclusion

A model of the rectangular superplastic bulging is built using finite elements Analysis in the ABAQUS software based on the experimental work. The validation of the model was in terms of thickness distribution and ratio of maximum creep strain rate over target strain rate $r_{max.}$

The analysis and the development of the model in terms of coefficient of friction and the adopted approach found that no friction should be in the bottom, for the wall 0.1 & 0.3 gave the best results while for the entry 4 is the best.

References

1. Rioja, R.J., Liu, J.: The evolution of Al-Li base products for aerospace and space applications. Metall. Mater. Trans. A **43A**, 3325–3337 (2012)
2. Rioja, R.J.: Fabrication methods to manufacture isotropic Al-Li alloys and products for space and aerospace applications. Mater. Sci. Eng., A **257**(1), 100–107 (1998)
3. Panthri, S., Joshi, A., Pathak, M.K.: Effect of room temperature rolling followed by annealing on tensile and fracture behaviour of bulk ultrafine-grained Al–Li 8090 alloy. Metallogr. Microstruct. Anal. **8**, 581–590 (2019)
4. Abass, M.H., Mahmud, F., Alali, M.S., Abood, A.N.: The effect of annealing and anisotropic behaviour on tensile and fatigue properties of AA 8090. Arch. Mater. Sci. Eng. **99**(1/2), 5–12 (2019)
5. Shyamlal, C., et al.: Corrosion behavior of friction stir welded AA8090-T87 aluminum alloy. Materials **15**(15), 51–65 (2022)

6. Hajjioui, E.A., Bouchaâla K., Faqir, M., Essadiqi E. : A review of manufacturing processes, mechanical properties and precipitations for aluminum lithium alloys used in aeronautic applications. Heliyon **9**, e12565 (2022)
7. Bouchaâla, K., Ghanameh, M.F., Essadiqi, E.H., Faqir, M., Meziane, M., Mada, M.: Evaluation of the effect of contact and friction on deep drawing formability analysis for lightweight aluminum lithium alloy using cylindrical cup. Procedia Manuf. **46**, 623–629 (2020)
8. Bouchaâla, K., Ghanameh, M.F., Faqir, M., Mada, M., Essadiqi, E.: Investigation of contact impact in deep drawing for AA2198 Al-Li Sheet using ABAQUS/Explicit. In: Ezziyyani, M. (ed.) AI2SD 2019. AISC, vol. 1104, pp. 411–420. Springer, Cham (2020). https://doi.org/10.1007/978-3-030-36671-1_36
9. Wang, H., Cheng, S., Ye, Z., Wu, T., Jin, K., Guo, X.: Evaluation of the formability of AA2198-T3 Al-Li alloy in warm sheet hydroforming. Int. J. Adv. Manuf. Technol. **118**(1–2), 405–416 (2021). https://doi.org/10.1007/s00170-021-07961-z
10. Lee, H.-S., Yoon, J.-H., Yoo, J.-H., Min, K.-J.: Characterization of superplastic behavior of AA2195 by bulging test. In : International Symposium on Mechanical Engineering and Material Science (ISMEMS-16). Atlantis Press, vol. 93 (2016)
11. Chen, T.-R., Huang, J.C., Hwang, Y.M.: Fundamental characterisation of hemisphere free bulging using superplastic 8090 Al–Li sheets. Mater. Sci. Technol. **12**, 495–503 (1996)
12. Jia, L., Ren, X., Zhang, Y., Hou, H.: Materials science and technology Al-Li ALLOY. Mater. Technol. **54**(3), 397–405 (2020)
13. Pancholi, V., Kashyap, B.P.: Effect of layered microstructure on superplastic forming property of AA8090 Al–Li alloy. J. Mater. Process. Technol. **186**(1–3), 214–220 (2007)
14. Hwang, Y.M., Lay, H.S., Huang, J.C.: Study on superplastic blow-forming of 8090 Al–Li sheets in an ellip-cylindrical closed-die. Int. J. Mach. Tools Manuf. **42**(12), 1363–1372 (2002)
15. Balasubramanian, M., Ramanathan, K., Zhu, K.: Finite element modeling and numerical simulation of superplastic forming of 8090 Al-Li alloy in a rectangular die. Adv. Mater, Res. Trans Tech **487**, 116–121 (2012)
16. Hwang, Y., Lay, H.: Study on superplastic blow-forming in a rectangular closed-die. J. Mater. Process. Technol. **140**(1–3), 426–431 (2003)
17. Rosserd, C.: Rev. Metall. (Paris), **63**(3) (1966)
18. Backoften, W., Turner, J., Avery, D. : Superplasticity in an aluminum-Zinc alloy. Trans. ASM **57** (1964)
19. Chen, T.R., Huang, J.C., Hwang, Y.M., Wang, T.D.: Different friction coefficients at different positions of superplastically formed parts. Mater. Sci. Forum **304**, 741–746 (1999)

An Innovative Deep Drawing Process of Thin Curved Shells with Positive Local Bulging

Wen Sun[1], Wei Liu[1,2](✉), Yonggang Hao[1], Yongchao Xu[1,2], and Shijian Yuan[1,2]

[1] School of Materials Science and Engineering, Harbin Institute of Technology, Harbin 150001, China
liuw@hit.edu.cn

[2] National Key Laboratory for Precision Hot Processing of Metal, Harbin Institute of Technology, Harbin 150001, China

Abstract. Thin-walled shells with curved surfaces are widely used in the aerospace field. As the thickness to diameter ratio decreases, it becomes increasingly challenging to control inner wrinkling during the conventional deep drawing process. In this paper, a novel deep drawing process with positive local bulging was proposed to suppress inner wrinkles in the deep drawing of a semi-ellipsoidal part. The effects of loading paths, rigid ring numbers, and thickness to diameter ratio of blank on forming defects were analyzed by experiments. The results indicated that the defects-free parts with a small ratio of thickness to diameter were formed by the novel process, which cannot be formed by the conventional deep drawing process. Both inner wrinkling and splitting were suppressed by the optimized loading path of the reverse bulging force. Although the local thickness of the curved surface part was decreased by the reverse bulging tools, the minimum thickness was improved by the increased rigid rings. In addition, a larger reverse bulging force was required to prevent the inner wrinkles for a smaller thickness to diameter ratio. This research provides valuable experimental guidance for the development of a novel deep drawing process for curved surface parts.

Keywords: Curved surface parts · Deep drawing · Reverse bulging effect · Wrinkling

1 Introduction

The tank dome is a crucial load-bearing structure in pressure vessels, which is typically made of a thin-walled curved shell [1]. For a 2-m diameter dome, the thickness to diameter ratio is less than 3‰ [2]. At present, the conventional deep drawing of tank dome parts faces many challenges, such as wrinkling, local thinning, and splitting [3]. Especially, the parts with both small thickness to diameter ratio and poor plasticity material [4]. With the growing demand for lightweight components and materials with high strength-to-weight ratios [5], the thickness to diameter ratio of components will further decrease, and it is difficult to fabricate parts meeting the specified requirements by conventional deep drawing (CDD).

K. Mocellin et al. (Eds.): ICTP 2023, LNME, pp. 314–321, 2024.
https://doi.org/10.1007/978-3-031-40920-2_33

Sheet hydroforming process was developed in order to address the challenges of wrinkling and splitting. For the process, the stress state of the sheet in the unsupported area is regulated to biaxial tensile stress by the reverse bulging effect. Wrinkling was eliminated when the circumferential compressive stress was less than the critical wrinkling stress [6]. However, the forming force sharply increased due to the liquid pressure on the punch, especially for the large-size parts [7].

In this study, the deep drawing process with positive local bulging was proposed to overcome the above problems. In the forming process, the rigid rings were used to apply the reverse bulging force in the unsupported area of the sheet, which can alter the circumferential compressive stress and suppress wrinkles effectively. Based on the deep drawing setup, experiments were conducted to analyze the effects of the rigid ring numbers, loading path of reverse bulging force, and thickness to diameter ratio on the forming defects.

2 Mechanics Analysis of Reverse Bulging Area

The mechanical model of the reverse bulging area is shown in Fig. 1. The left boundary, top point, and right boundary of the reverse bulging area is point A, B, and C, respectively. The center of the rigid rings is point O. Where L_h is the section length of the reverse bulging area, the degree of the tangent line at point A is defined as α. The section length from points O and A to axis of symmetry is O_x and R_A, respectively. ρ is the radius of rigid tools. The reverse bulging stress is σ_c. The bending and unbending effects, and friction between blank and reverse bulging rings are neglected during the calculation process.

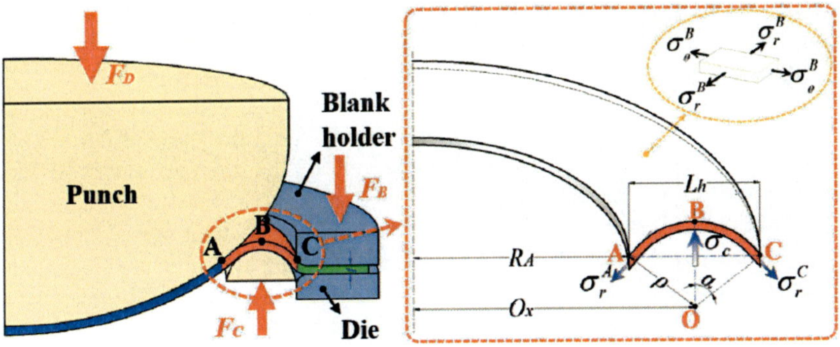

Fig. 1. The geometrical model of the blank in reverse bulging area

According to the equilibrium equation in the horizontal direction of the reverse bulging area, the σ_r^A, σ_r^C and σ_c satisfies:

$$\sigma_r^A \cdot 2\pi R_A \cdot t \cdot \sin\alpha + \sigma_r^C \cdot 2\pi \cdot (R_A + L_h) \cdot t \cdot \sin\alpha = \sigma_c \cdot 2\pi O_x \cdot L_h \tag{1}$$

Meanwhile, σ_c can be expressed by F_C as follows.

$$\sigma_c = \frac{F_C}{2\pi O_x \cdot L_h} \tag{2}$$

Combining Eqs. (1) and (2), σ_r^C can be expressed as follows:

$$\sigma_r^C = \frac{F_C}{4\pi \cdot (R_A + L_h) \cdot t \cdot \sin\alpha} \tag{3}$$

The radial stress σ_r^B at point B can be obtained through the equilibrium equation in the horizontal direction of the reverse bulging area.

$$\sigma_r^B \cdot 2\pi O_x \cdot t = \sigma_r^C \cdot 2\pi \cdot (R_A + L_h) \cdot t \tag{4}$$

Then σ_r^B can be expressed by σ_r^C as follows:

$$\sigma_r^B = \sigma_r^C \cdot \frac{(R_A + L_h)}{O_x} \tag{5}$$

When ρ is much smaller than O_x, $\varepsilon_\theta^B = 0$, the circumferential stress σ_θ^B can be expressed as

$$\sigma_\theta^B = \frac{F_C}{8\pi \cdot O_x \cdot t \cdot \sin\alpha} \tag{6}$$

According to the Eqs. (5) and (6), $\sigma_r^B > 0$ and $\sigma_\theta^B > 0$. The sheet in the unsupported area is bulging under the action of reverse bulging force. The stress state of the top point in the bulging zone is biaxial tensile stress, which effectively solves the wrinkle problem of the sheet in the unsupported area.

3 Process of Deep Drawing with Positive Local Bulging

3.1 Sheet Blank

The experiments material is 2024 aluminum sheets with a thickness of 0.8 mm and 1.2 mm. The designed thickness to diameter ratio is 2.35‰ and 3.53‰ with the same blank diameter of 340 mm. The mechanical properties of 2024 aluminum alloy were tested by Instron-1186 testing machine with the same initial strain rate of 1×10^{-3}/s. The yield stress σ_s is 101 MPa. The ultimate tensile strength σ_b, fracture elongation δ and anisotropy coefficient R are 254 MPa, 23%, and 0.67, respectively. The stress-strain relationship of 2024 aluminum alloy can be described with the Hollomon hardening law, and the flow stress curve was $\sigma = 706 \cdot \bar{\varepsilon}^{0.38}$ MPa.

3.2 Experimental Setup and Simulation Model

The deep drawing with positive local bulging experiments was conducted by a 3150 kN hydraulic press. To meet the experimental requirements, both the blank holder force and deep drawing force were precisely controlled.

As shown in the schematic diagram of Fig. 2a, there are four main components in the experimental setup: punch, blank holder, die, and reverse bulging tools. The reverse bulging tools consist of five parts: the rigid rings for reverse bulging, the ring bracket, the springs, and a displacement sensor. The displacement sensor was used to measure the real-time compression value of the springs, which were placed between the ring bracket and the spring fixing plate. The experimental dies are shown in Fig. 2b. The detailed tools parameters are shown in Table 1. In Fig. 2c, four loading paths were designed to analyze the effects of reverse bulging force on forming defects during the deep drawing process. The reverse bulging force of Paths 2, 3, and 4 was regulated by changing the compression of the springs or adjusting the number of springs, and the maximum force of each spring is 11800N. During the deep drawing process, variable blank holder force was used to ensure that the sheet can flow into the die with the increase of punch stroke. The unit blank holder force was 1.5 MPa, 3 MPa, and 3.5 MPa when the punch stroke was 0–30 mm, 30–60 mm, and 60–80 mm, respectively. The target part is a semi-ellipsoidal shell with a long axial diameter of 240 mm and an 80 mm drawing stroke. Three rigid rings with different radii were selected for the experiments. The radius of three rigid rings are 11 mm, 5 mm, and 3 mm, and the application range of drawing depth is 0–40 mm, 40–50 mm, and 50–56 mm, respectively.

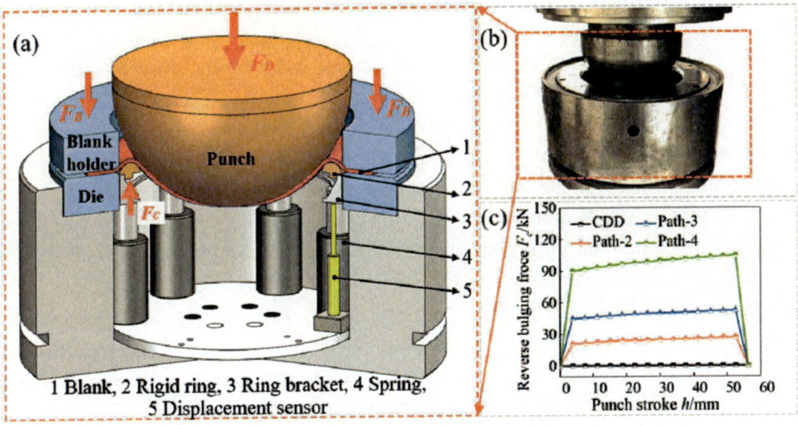

Fig. 2. The experimental setup with positive local bulging, (a) schematic diagram of deep drawing setup, (b) experimental dies, (c) loading paths of reverse bulging force.

The numerical simulations were performed using the Abaqus/Explicit software. The forming molds were built as rigid bodies. Meanwhile, the S4R elements with a size of 1 mm were used for the blank, and the total number of the elements is 186924. The friction coefficient of the punch and blank was 0.15, and the friction coefficient of the rigid tools, blank holder, die, and blank was 0.12. Von Mises was used in this simulation model.

Table 1. Basic parameters of molds.

Part	Geometrical size/mm
Punch	$R = 120, H = 75$
Gap between punch and die	3
Blank diameter	340
The radii of rigid rings	11, 5, 3
Diameter of rigid ring	244

4 Results and Discussion

4.1 Strain Path with Different Loading Paths

The strain path of the sheet at the corner of the die was analyzed to better understand the forming process, as shown in Fig. 3. In the deep drawing with local positive bulging, the sheet passes through the reverse bulging area and the attached area in turn with the increase of the punch stroke.

As shown in Fig. 3a, the radial strain increased with the increase of punch stroke before the sheet flowed to the peak of the reverse bulging area. The radial strain reached the maximum value of 0.18 at the peak of the reverse bulging area. The radial strain decreased with the increase of punch stroke after the sheet passes through the peak of the reverse bulging area. In the attached area, the radial strain was stable at 0.12. It is easy to understand the fluctuation of the radial strain in the reverse bulging area. The bend of the sheet in the reverse bulging area leads to the increase of radial tensile strain, while the unbend lead to the decrease of radial tensile strain. The trend of the circumferential strain was similar to that of the radial strain. Besides, compared with the results of CDD, the radial strain increased from 0.08 to about 0.18, and the circumferential strain also increased from 0.08 to 0.09, as shown in Fig. 3b. The reverse bulging effect enhanced the plastic deformation of the sheet in the reverse bulging area.

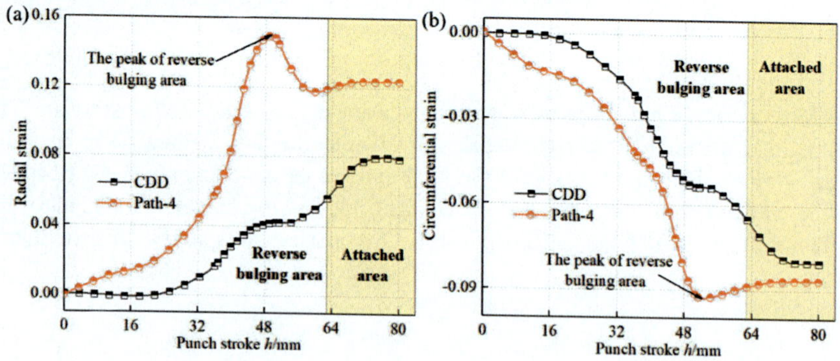

Fig. 3. Strain path under CDD and Path 4 (a) radial strain, (b) circumferential strain.

4.2 Effect of Loading Path on Forming Defects

The experimental results of the deep drawing with positive local bulging under four loading paths are shown in Fig. 4.

The formed dome parts by CDD, loading Path 2 and 3 both were wrinkled on the curved surface, and the number of wrinkles decreased with the increasing reverse bulging force (Figs. 4a, b, and c). The wrinkle-free dome part was obtained with the loading Path-4, as shown in Fig. 4d. It should be pointed out that the part was splitting at the corners of the die during the CDD. The splitting was caused by the sharply increased radial tensile stress. The thickness increased due to the wrinkling of the curved surface, and the material flow in the flange area was hindered. As a result, the radial tensile stress exceeded its ultimate tensile strength, which resulted that the dome ruptured. Compared with existing boundary control methods, the co-existence of wrinkling and cracking makes it very difficult to fabricate such parts with a small thickness to diameter ratio by conventional deep drawing. However, the above experiment results show that deep drawing with local positive bulging under appropriate loading paths can effectively eliminate wrinkling and improves formability.

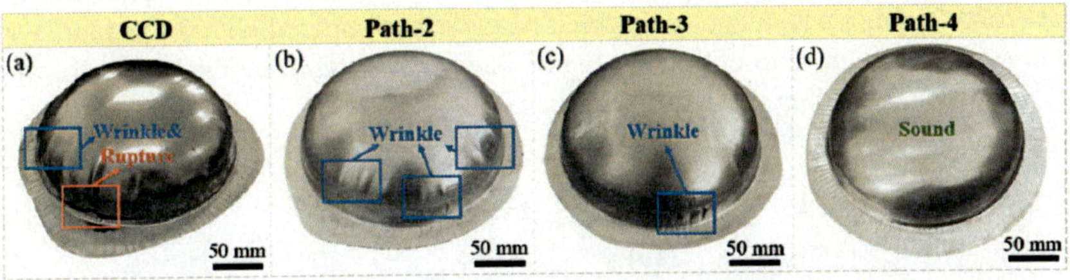

Fig. 4. Formed dome parts with four loading paths, (a) CDD, (b) Path-2, (c) Path-3, (d) Path-4.

4.3 Effect of Rigid Ring Numbers on Thickness Distribution

The cross-section thickness distribution of formed parts was selected to further analyze the influence of the rigid ring numbers, as shown in Fig. 5. Figure 5a shows the gauge points and the area with the minimum thickness. The minimum thickness point was located at point 27, which was the tangential point among rigid rings, punch, and blank during the forming process. The minimum thickness was 0.702 mm for two rigid rings, and 0.711 mm for three rigid rings (Fig. 5b). The maximum thinning ratio were 12.25% and 11.13%, respectively. Meanwhile, point -26 and -27 also showed a serious thinning.

The wrinkles were suppressed by the reverse bulging with rigid rings. However, more severe surface damage in the reverse bulging area was observed for the formed dome part with two rigid rings, compared with that of the part formed by three rigid rings. According to the thickness distribution along the central section, there is significant thinning in the reverse bulging area. The utilization of three rigid rings leads to a decreased thinning rate and an improvement in the thickness uniformity.

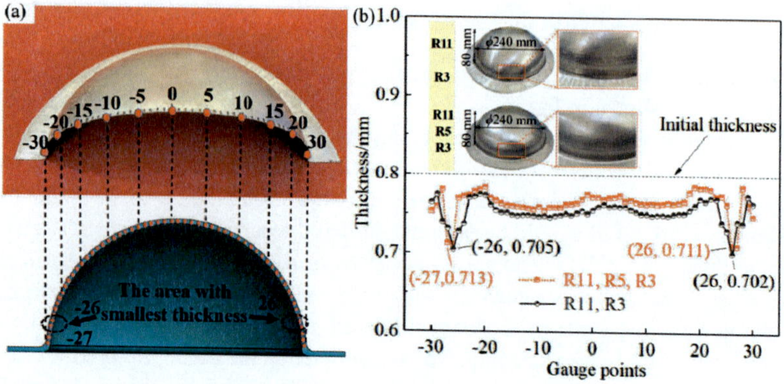

Fig. 5. Thickness distribution along the cross section for formed parts by two and three rigid reverse bulging rings, (a) gauge points, (b) the thickness distribution along cross section.

4.4 Effect of Thickness to Diameter Ratio on Winkling Behavior

The experiments of the sheet with thickness to diameter ratio of 3.53‰ was carried out under three loading paths, and the formed parts were compared with dome parts with thickness to diameter ratio of 2.35‰. Figure 6a shows the conventional deep drawing, and Fig. 6b-c are loading paths 2–3 respectively.

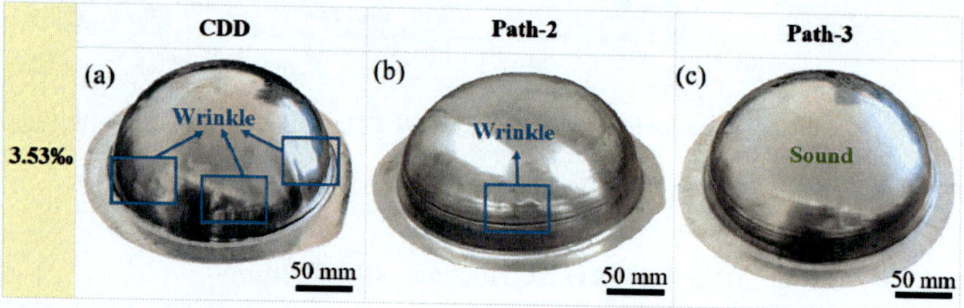

Fig. 6. Formed dome parts with thickness to diameter ratio of 3.53‰ under (a) CDD, (b) Path-2, (c) Path-3.

With the increase of reverse bulging force, the height of wrinkles decreases. The wrinkle-free dome part was obtained with loading Path-3. Figure 7 shows the wrinkle height and the number of wrinkles of the two dome parts with different thickness to diameter ratios. The wrinkle height was reduced by 88.6% when the thickness to diameter ratio was increased from 2.35‰ to 3.53‰, under the loading Path-2. Besides, the wrinkle height and the number of wrinkles of the dome pats with a thickness to diameter ratio of 3.53‰ are smaller than those with a thickness to diameter ratio of 2.35‰, under each loading path. The results show that the increased thickness to diameter ratio has a positive effect on preventing the formation of inner wrinkles. A larger reverse bulging force will be required to prevent the formation of wrinkles for a blank with a smaller thickness to diameter ratio.

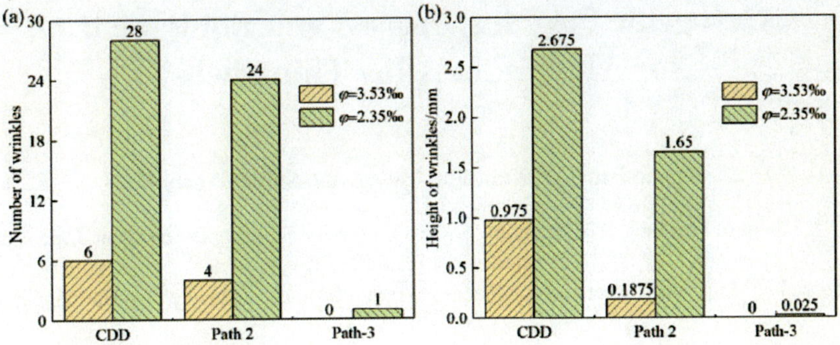

Fig. 7. Influence of thickness to diameter ratio on wrinkling behavior (a) the number of wrinkles, (b) the height of wrinkles.

5 Conclusion

In this paper, a novel deep drawing process with positive local bulging was developed, and the process experiments were carried out. The results show that the inner wrinkles and splitting both were suppressed by optimizing the loading path of the reverse bulging force with a smaller thickness to diameter ratio. The number of wrinkles decreased with the increased reverse bulging force. The maximum thickness thinning rate of the semi-ellipsoidal part decreased by using three rigid rings, compared with that by using two rigid rings. Moreover, a larger reverse bulging force was required to prevent the formation of wrinkles for a smaller thickness to diameter ratio.

This new process with positive local bulging provided the technical reserves for manufacturing ultra-large and thin-walled curved surface parts. In addition, the appropriate loading path and rigid ring numbers were crucial technical parameters, and further theoretical, simulation, and experimental research are needed in the future.

References

1. Salgado, M.C.V., Belderrain, M.C.N., Devezas, T.C.: Space propulsion: a survey study about current and future technologies. J. Aerosp. Technol. Manag. **10**, 23 (2018)
2. Fan, X.B., Yuan, S.J.: Innovation for forming aluminum alloy thin shells at ultra-low temperature by the dual enhancement effect. Int. J. Extreme Manuf, **4**(3), 5 (2022)
3. Atul, S.T., Babu, M.C.L.: A review on effect of thinning, wrinkling and spring-back on deep drawing process. Proc. Inst. Mech. Eng. Part B-J. Eng. Manuf. **233**(4), 1011–1036 (2019)
4. Zhang, H., Huang, G.S., Fan, J.F., Roven, H.J., Xu, B.S., Dong, H.B.: Deep drawability and drawing behaviour of AZ31 alloy sheets with different initial texture. J. Alloy Compd. **615**, 302–310 (2014)
5. Dursun, T., Soutis, C.: Recent developments in advanced aircraft aluminium alloys. Mater Des **56**, 862–871 (2014)
6. Liu, W., Chen, Y.Z., Yuan, S.J.: Mechanism analysis on thickness distribution of aluminum alloy hemispherical shells in double-sided sheet hydroforming. Int. J. Adv. Manuf. Technol. **89**(5–8), 2011–2020 (2016). https://doi.org/10.1007/s00170-016-9248-2
7. Yuan, S.J.: Fundamentals and processes of fluid pressure forming technology for complex thin-walled components. Engineering **7**(3), 358–366 (2021)

Cryogenic Forming Process and Equipment for Aluminum Alloy Thin Shells

Xiaobo Fan[1](✉), Xianshuo Chen[1], and Shijian Yuan[1,2]

[1] School of Mechanical Engineering, Dalian University of Technology, Dalian, China
xbfan@dlut.edu.cn
[2] School of Materials Science and Engineering, Harbin Institute of Technology, Harbin, China

Abstract. Aluminum alloy thin components are widely applied as structures in aerospace, aircraft, and electric vehicle. It is very challengeable to overcome the coexistence of wrinkling, splitting, and destroyed microstructure by the cold forming or hot forming processes when the components' size is larger and thickness is thinner. A phenomenon was found by authors that the ductility and hardening ability can be simultaneously enhanced at cryogenic temperatures, which can be used to develop a novel cryogenic forming process for overcoming those problems. In this paper, the bulging and drawing abilities were studied at cryogenic temperatures, which demonstrates that aluminum alloys also have excellent cryogenic formability at complex stress states. Thus, a novel forming process at ultra-low temperature gradient has been developed, and the related development and prospect were introduced. The world's first cryogenic forming device was developed with a drawing force of 22 MN and a platform size of 3 m. An integral 2219 aluminum alloy rocket tank dome with a diameter of 2.25 m was directly formed from a blank by the novel process at an ultra-low temperature gradient. The cryogenic forming process has considerable potential for large-sized and thin-walled components made of high-strength aluminum alloys, such as 2219 and 2195 aluminum alloys.

Keywords: Cryogenic forming · Aluminum alloy · Thin-shell components · Formability · Cryogenic hydro-bulging · Cryogenic deep drawing

1 Introduction

Aluminum alloy integral thin-walled components are widely applied as structures in aircraft, aerospace, and electric vehicle due to further requirements of the lightweight and reliability [1]. Large size, thin walls and deep cavities are the main characteristics of such integral components. For instance, the thickness of the integral rocket tank dome is less than 3 mm, and its diameter is more than 2 m. The coexistence of wrinkling and splitting is prone to occur in the integral forming of such components. The poor formability at room temperature of aluminum alloys is not enough to form such components [2]. Some hot forming processes have been developed with improved ductility by heating [3]. However, the weakening of strain hardening ability at elevated temperatures may

K. Mocellin et al. (Eds.): ICTP 2023, LNME, pp. 322–331, 2024.
https://doi.org/10.1007/978-3-031-40920-2_34

lead to concentrate deformation or even splitting [4]. In addition, complex heat treatment and truing processes were required to improve its service performance and attain shape and dimensional accuracy after hot forming.

During sheet metal forming, sufficient ductility and strain hardening ability are necessary to sustain and transfer deformation. In recent years, a phenomenon of dual enhancement effect (DEE) in aluminum alloys has been found [5], in which the ductility and strain hardening ability can be simultaneously enhanced at cryogenic temperatures. This phenomenon contributes to solving the coexistence of wrinkling and splitting in the integral forming of large-sized thin-walled components.

To utilize the DEE, some forming processes, such as cryogenic deep drawing [6], and cryogenic stamping [7], have been developed. Some aluminum alloy components with complex shapes and thin walls can be formed through cryogenic forming techniques. However, these methods are mostly conducted under uniform cryogenic temperatures. Large-sized blanks and forming tools are difficult to cool to ultra-low temperatures, and it is difficult to form large-sized and thin-walled components by these methods.

In this paper, cryogenic bulging tests and deep drawing tests were conducted to evaluate the cryogenic formabilities at complex stress states. a novel forming process at ultra-low temperature gradient for large-sized and thin-walled aluminum alloy components was developed by the DEE of ductility and hardening. And an integral 2219 aluminum alloy rocket tank dome with a diameter of 2.25 m was directly formed using the developed process.

2 Methodology of Cryogenic Forming at Ultra-low Temperature Gradient

2.1 DEE of Aluminum Alloys

Figure 1 shows the dual enhancement effect of ductility and hardening at cryogenic temperature. The fracture elongation δ at −196 °C is 42.6%, being 48.4% higher than that at 25 °C, indicating the significantly enhanced ductility at cryogenic temperature. The hardening index n at -196 °C is 0.38, being 52.0% higher than that at 25 °C, indicating the significantly enhanced hardening ability at cryogenic temperature. The phenomenon that the ductility and strain hardening ability are enhanced simultaneously at cryogenic temperature is called DEE.

DEE is affected by temperature, which is not significant until the temperature is decreased to a critical transition temperature. Figure 2 shows the mechanical properties of the 2219-O aluminum alloy sheet at different temperatures. Generally, the total elongation and strain hardening index are used to evaluate the ductility and strain hardening ability. Evidently, both ductility and strain hardening ability are enhanced as the temperatures decreased, especially when the temperature is less than −120 °C. Therefore, the temperature should be controlled below the critical transition temperature during cryogenic forming to obtain significantly improved formability.

Moreover, the degree of DEE is greatly affected by the heat treatment states. Figure 3 shows the mechanical properties of the 2219 aluminum alloy sheet at different heat treatments. Apparently, the most significant DEE is obtained in the annealed (O) and

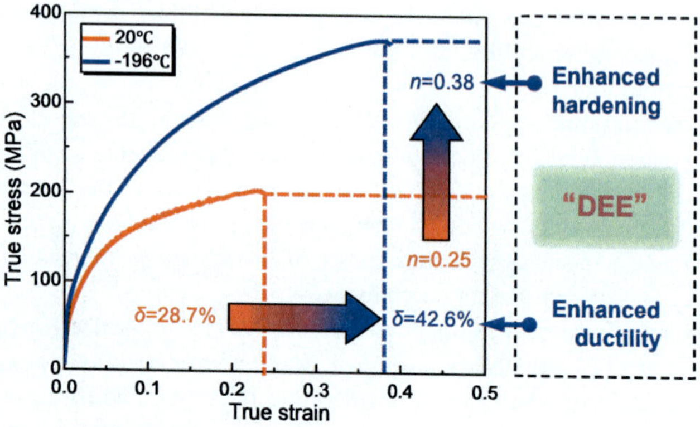

Fig. 1. Dual enhancement effect of ductility and hardening at cryogenic temperature. The material was a 2219-O aluminum alloy sheet.

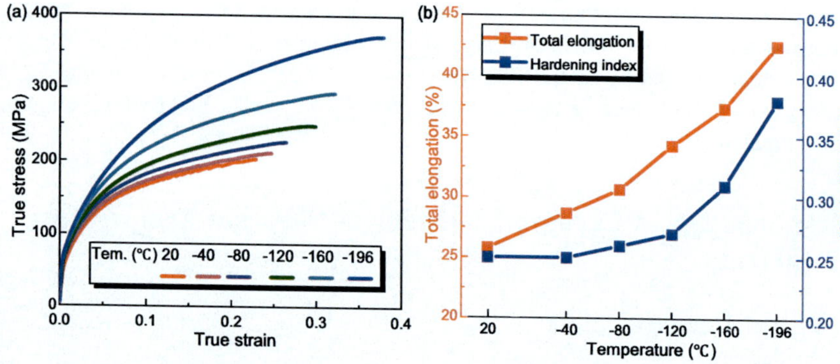

Fig. 2. Mechanical properties of the 2219-O aluminum alloy sheet at different temperatures: (a) true strain-stress curve, (b) total elongation and hardening index [8].

quenched (W) states. The total elongation at −196 °C of the annealed sheet is 42.7%, increasing by 48.4% compared with that at 20 °C. The DEE of quenched sheets is more significant than that of annealed sheets. The total elongation of quenched sheets reaches 32.5% at RT and further increases to 44.3% at −196 °C, which exhibited the best formability. In contrast, the DEE is not obvious in the artificially aged state (T4, T6 and T8). It is very difficult to form in aging states, not only because of the non-significant ductility but also because of the high strength. That is to say, in order to obtain the best formability during cryogenic forming, the sheet should be treated as annealed or quenched states.

In addition, in order to evaluate the applicability of DEE, the cryogenic mechanical properties of several typical aluminum alloys were tested, as shown in Fig. 4. Obviously, the ductility and strain hardening ability at −196 °C of the four types of aluminum alloy sheets are significantly improved compared with that at 20 °C. The fracture elongation at −196 °C of the four types of aluminum alloy sheets is more than 50% higher than

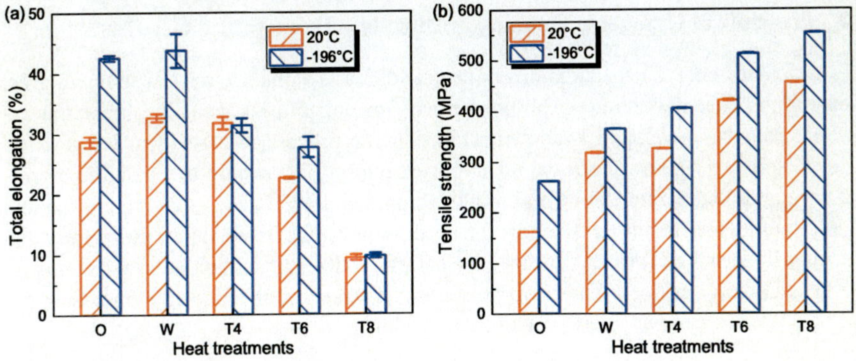

Fig. 3. Mechanical properties of the 2219 aluminum alloy sheet at different heat treatments: (a) total elongation, (b) tensile strength.

that at 20 °C. These results show that the DEE is a universal and stable phenomenon, which has great potential in sheet metal forming.

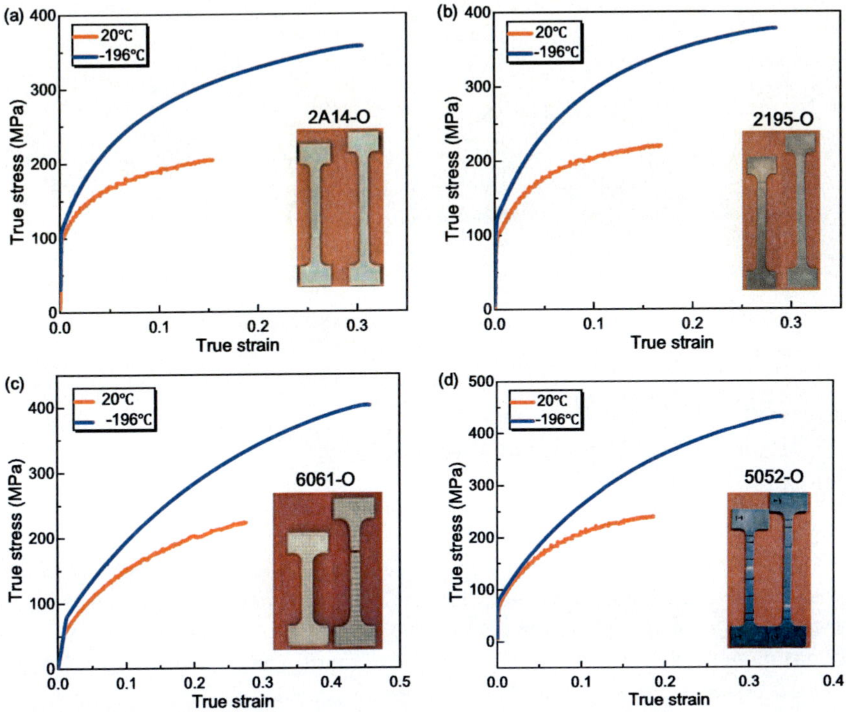

Fig. 4. Mechanical properties of the different types of aluminum alloys: (a) 2A14-O, (b) 2195-O, (c) 6061-O and (d) 5052-O.

2.2 Principle of Cryogenic Forming at Ultra-low Temperature Gradient

It is critical to cool fully the large-sized forming dies and blank to ultra-low temperatures. Considering the engineering feasibility, a novel forming process at an ultra-low temperature gradient was developed, as shown in Fig. 5. In this process, to avoid local deformation or even splitting, the most critical unsupported region (*TA*) should be cooled to a critical cryogenic temperature for obtaining the significant DEE. The gradient cooling and blank holding are then used to adjust the deformation. The deformation uniformity and forming limit can be further increased by enlarging the stress gradient between unsupported and flange regions. The hoop compressive stress can be reduced by increasing of blank-holder force (*Fb*), thereby avoiding wrinkling defects.

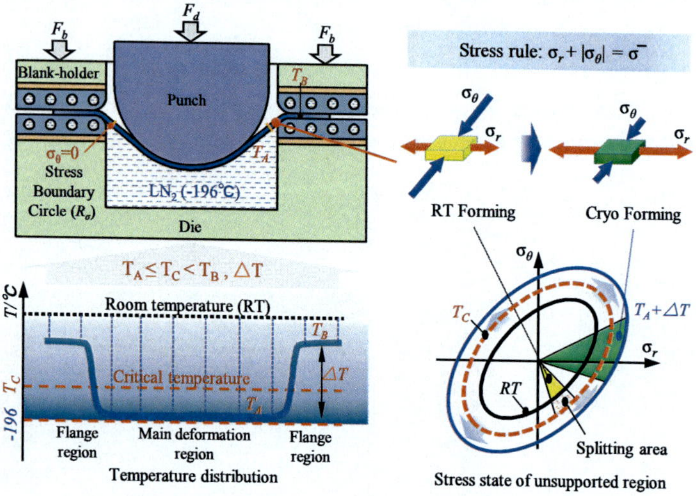

Fig. 5. Principle of cryogenic temperature gradient forming process [8].

3 Formability at Cryogenic Temperatures

3.1 Bulging Ability

To evaluate the cryogenic bulging formability, a cryogenic biaxial measuring system was created for aluminum alloy tubes and sheets (Fig. 6). The device used liquid nitrogen for cryogenic cooling and pressurization. The lowest temperature is −196 °C. And the maximum loading pressure of cryogenic temperature mediums can reach 100 MPa. During the bulging process, a laser distance sensor was used to collect the bulging height and a pressure sensor was used to measure the internal pressure, in real-time. The strain distributions of the bulging specimens were measured to evaluate the formability.

Figure 7 shows the bursting specimens and effective strain of the 2195 Al-Li alloy quenched sheets at 20 °C and −196 °C. The bursting height at −196 °C considerably

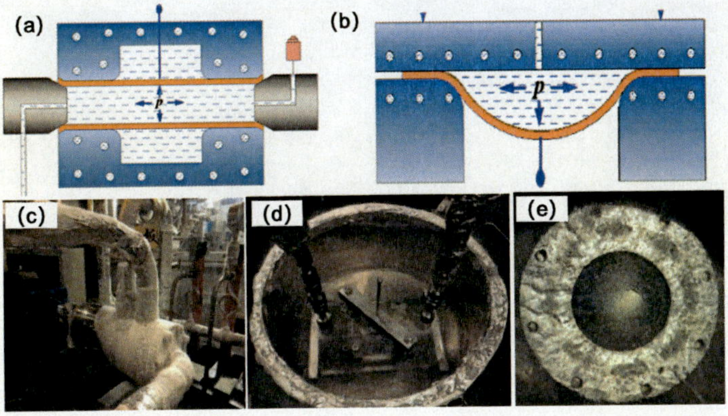

Fig. 6. Cryogenic bulging device: (a) and (d) are tube bulging dies, (b) and (e) are sheet bulging dies, (c) cryogenic pressurizing pump [9, 10] .

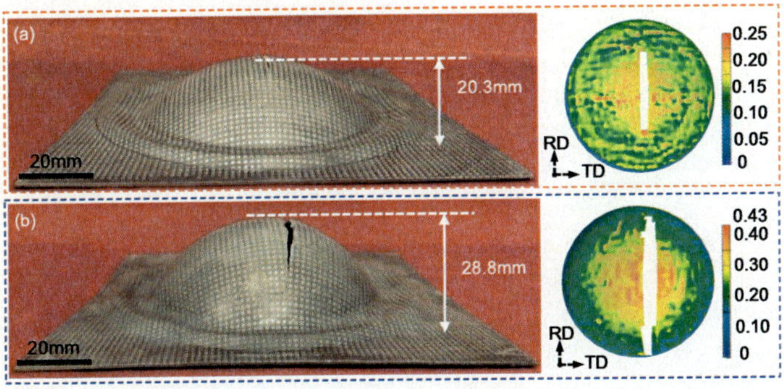

Fig. 7. Bursting specimens and effective strain of the 2195 Al-Li alloy quenched sheets at (a) 20 °C and (b) −196 °C [9].

increases to 28.8 mm, being 41.9% more than that at 20 °C. The maximum strain at −196 °C increases to 0.43, which is 72.0% more than that at 20 °C.

Figure 8 shows the bursting specimens and effective strain of the 6061 alloy quenched tubes at 20 °C and −196 °C. The bulging height is increased from 3.91 mm at 20 °C to 7.82 mm at −196 °C, being twice that at 20 °C. The maximum strain at −196 °C increases to 0.40, which is 66.7% higher than that at 20 °C. In addition, a more uniform strain distribution can be clearly observed at −196 ◦C. These results illustrate that the cryogenic biaxial formability was significantly improved compared to that at 20 °C.

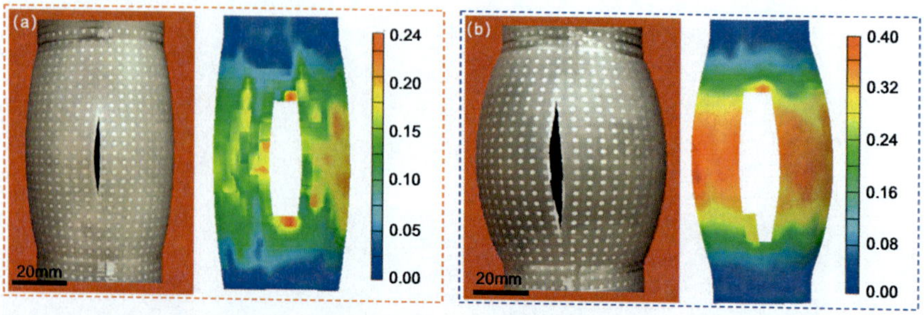

Fig. 8. Bursting specimens and effective strain of the 6061 alloy quenched tubes at (a) 20 °C and (b) -196 °C [10].

3.2 Drawing Ability

Figure 9 shows the cylindrical cups formed at 20 °C and cryogenic temperature gradient. The inner diameter and thickness of the cups are 100 mm and 1.0 mm, respectively. The temperature gradient corresponded to the temperature distribution of the unsupported region at −196 °C and the flange area at 20 °C. The LDR of the specimen formed at cryogenic temperature gradient increases to 2.2, being 22.2% higher than that at 20 °C. Generally, a larger LDR value illustrates the better ability to fabricate components with deep drawn and complex shapes.

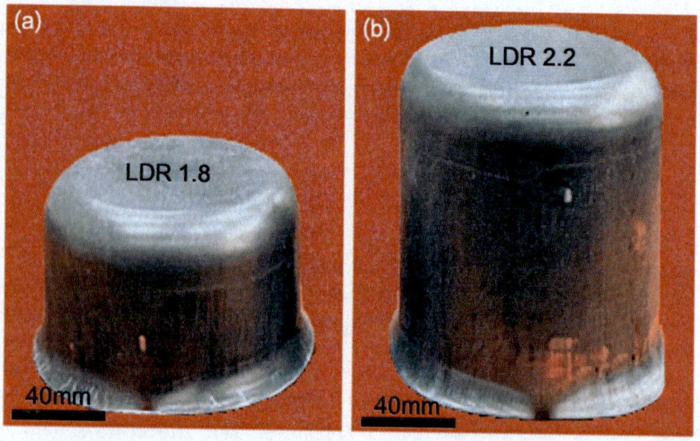

Fig. 9. The cylindrical cups formed at (a) 20 °C and (b) cryogenic temperature gradient.

Figure 10 shows the hemispherical specimens with a diameter of 150 mm and a thickness of 1.0 mm formed at 20 °C and cryogenic temperature gradient. A splitting defect appeared when drawing to 55.0mm at RT. By contrast, the intact hemispherical specimen can be formed at a cryogenic gradient temperature. And the depth reaches 85 mm, being 54.5% more than that at RT. These results illustrate that the cryogenic drawing ability is significantly improved compared to that at 20 °C, which has great potential for fabricating aluminum thin-walled deep cavity components.

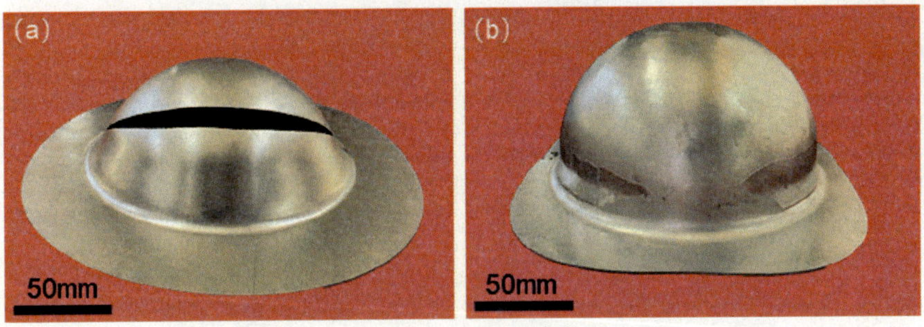

Fig. 10. The hemispherical specimens formed at (a) 20 °C and (b) cryogenic temperature gradient [8].

4 Application of Cryogenic Forming at Ultra-low Temperature Gradient

The above results demonstrate that aluminum alloys also have excellent cryogenic formability at complex stress states. Furthermore, to realize the engineering application of cryogenic forming at ultra-low temperature gradient, the world's first cryogenic forming device was developed with a drawing force of 22 MN and a platform size of 3 m. An integral 2219 aluminum alloy rocket tank dome with a diameter of 2.25 m was directly formed from a blank in the initial thickness of 4.5 mm by the novel cryogenic temperature gradient forming process (Fig. 11). The diameter-thickness ratio of the integral dome is only 1.6‰, further breaking the previous limiting value of 3.0‰. In fact, the integral rocket tank dome with a diameter of 3.35 m has also been formed by cryogenic forming at an ultra-low temperature gradient. The cryogenic forming process has considerable potential for large-sized thin-walled components made of high-strength aluminum alloys, such as 2219 and 2195 aluminum alloys.

Fig. 11. The integral dome with a diameter of 2.25 m and the cryogenic temperature forming press with 22 MN drawing force [8].

5 Conclusion

In this study, cryogenic bulging tests and deep drawing tests were conducted to evaluate the cryogenic formabilities at complex stress states. A novel forming process at ultra-low temperature gradient was proposed based on the dual enhancement effect at cryogenic temperature. The main conclusions are as follows:

(1) Dual enhancement effect of ductility and hardening at ultra-low temperatures is a universal and stable phenomenon. Many aluminum alloys have a stable and obvious dual enhancement effect after cooling to a critical temperature.
(2) The bulging and drawing properties of aluminum alloys at cryogenic temperature are significantly improved compared with room temperature, with an increase of more than 50%. These results demonstrate that aluminum alloy has excellent cryogenic formability at complex stress states, which has great potential for fabricating aluminum thin-walled deep cavity components.
(3) A novel forming process at ultra-low temperature gradient was proposed based on the DEE at ultra-low temperatures. The world's first cryogenic forming device was developed with a drawing force of 22 MN and a platform size of 3 m. An integral 2219 aluminum alloy rocket tank dome with a diameter of 2.25 m was directly formed from a blank by the novel cryogenic temperature gradient forming process.

Acknowledgements. This work was supported by the National Key Research and Development Program of China (No. 2019YFA0708804) and the Fundamental Research Funds for the Central Universities (No. DUT20ZD101).

References

1. Dursun, T., Soutis, C.: Recent developments in advanced aircraft aluminum alloys. Mater. Des. **56**, 862–871 (2014)
2. She, X., Jiang, X., Zhang, R., Wang, P., Tang, B., Du, W.: Study on microstructure and fracture characteristics of 5083 aluminum alloy thick plate. J. Alloy. Compd. **825**, 153960 (2020)
3. Zheng, K., Politis, D., Wang, L., Lin, J.: A review on forming techniques for manufacturing lightweight complex-shaped aluminum panel components. Int. J. Lightweight Mater. Manuf. **1**(2), 55–80 (2018)
4. Fakir, O., Wang, L., Balint, D., Dear, J., Lin, J., Dean, T.: Numerical study of the solution heat treatment, forming, and in-die quenching (HFQ) process on AA5754. Int. J. Mach. Tools Manuf. **87**, 39–48 (2014)
5. Fan, X., Yuan, S.: Innovation for forming aluminum alloy thin shells at ultra-low temperature by the dual enhancement effect. Int. J. Extrem. Manuf. **4**(3), 033001 (2022)
6. Yuan, S., Cheng, W., Liu, W., Xu, Y.: A novel deep drawing process for aluminum alloy sheets at cryogenic temperatures. J. Mater. Process Technol. **284**, 116743 (2020)
7. Grabner, F., Sterreicher, J., Gruber, B., Papenberg, N., Schlgl, C.: Cryogenic forming of Al-Mg alloy sheet for car outer body applications. Adv. Eng. Mater. **21**, 1900089 (2019)
8. Fan, X., Chen, X., Yuan, S.: Novel forming process for aluminum alloy thin shells at ultra-low temperature gradient. Int. J. Mach. Tool Manufact. **185**, 103992 (2023)
9. Chen, X., Fan, X., Xu, A., Yi, Z., Yuan, S.: Formability enhancement and deformation mechanism of 2195 Al-Li alloy sheet in cryogenic forming. Mater. Charact. **196**, 112577 (2023)
10. Wang, X., Fan, X., Chen, X., Yuan, S.: Cryogenic deformation behavior of 6061 aluminum alloy tube under biaxial tension condition. J. Mater. Process. Technol. **303**, 117532 (2022)

Comparison of Critical Shear Strain Evaluations for Simple Shear Test

Guofeng Han[1,2], Enzhen Ren[1,2], Ji He[1,2], and Shuhui Li[1,2](✉)

[1] State Key Laboratory of Mechanical System and Vibration, Shanghai Jiao Tong University, Shanghai 200240, China
lishuhui@sjtu.edu.cn

[2] Shanghai Key Laboratory of Digital Manufacture for Thin-Walled Structures, Shanghai Jiao Tong University, Shanghai 200240, China

Abstract. The simple shear test for sheet metals has become a topic of growing interest in recent years, whose significance could be reflected in two aspects: obtaining the mechanical behaviors under large strain and investigating the ductile fracture under low-stress triaxiality. In the current stage, the community lacks a widely-accepted and standardized procedure to conduct simple shear tests, and it makes the interpretation and comparison of experimental results to be indirect and difficult. Shear strain is commonly used to characterize the degree of deformation, but the method to estimate it in experiments is often different from one to another, and uncertainties will appear due to the inconsistency. In this work, some representative methods to obtain shear strain will be evaluated analytically and experimentally. The influence of the perturbation phenomenon (that is, the emergence of axial deformation and the tilt of shear direction with respect to the coordinate axis of strain measurement during simple shear) will be investigated in detail, and it seemed that there were no universal methods to estimate shear strain in experiments (to the best of the authors' knowledge). Some recommendations on this topic would be provided and the results were expected to enhance the understanding of the experimental mechanics of simple shear tests.

Keywords: Simple shear test · Shear strain evaluation · Effects of perturbation

1 Introduction

In the region of sheet metal forming, the uniaxial tension test has been widely adopted to characterize the mechanical properties (such as elastic modulus, strength, ductility, and anisotropy) of sheet metals, and the high degree of standardization makes it to be a mature and stable method. Whereas, its ability is restricted by the appearance of necking, and only the material property in small or moderate strain regions (generally <0.3) [16, 19] could be accessed. As a potential alternative to uniaxial tension, the simple shear test has become a topic of growing interest for its capacity to achieve large strain (sometimes >1.0) without plastic instability, and the reverse/cyclic loading could also be achieved easily [7–9]. Moreover, ductile fracture under low-stress triaxiality has received

K. Mocellin et al. (Eds.): ICTP 2023, LNME, pp. 332–340, 2024.
https://doi.org/10.1007/978-3-031-40920-2_35

great attention and simple shear is an indispensable way to study it [5, 12]. Up to now, the community lacks a widely-accepted and standardized procedure to conduct simple shear tests. The employed specimen geometry, the experimental apparatus, the strain measurement methods, and the shear strain evaluations often vary from one experiment to another [1, 2, 11, 15, 18, 20, 21], and it makes the interpretation and comparison on experimental results to be indirect and difficult. This work will summarize and evaluate the frequently adopted methods to estimate shear strain in experiments. It has been revealed that the strain perturbation phenomenon (the emergence of axial deformation during simple shear) will influence the results of some specific shear strain estimations at large deformation levels [6]. Another type of perturbation, the tilt of shear direction with respect to the coordinate axis of strain measurement will be introduced in this work, and the effects of which will be investigated analytically (in Sect. 2) and experimentally (in Sect. 3).

2 Analytic Comparison of Shear Strain Evaluations

2.1 Critical Shear Strain Evaluations

Two types of approaches are frequently applied to evaluate shear strain in simple shear tests: the direct and indirect methods [6, 19]. In the direct method, shear strain is obtained through the tangent of the shear angle [10, 13, 17], where the deformed configuration need not be pre-assumed. With the indirect method, the shear strain is often evaluated based on characteristics of deformation gradient or some strain tensors [1, 2, 18, 21], and pre-assumptions to the deformed configuration are generally required (such as the theoretical simple shear described by Eq. (1)). Table 1 shows some critical shear strain evaluations in literature, and in the following, they will be analyzed and compared under the perturbation of axial deformation and the tilt of shear direction.

$$\boldsymbol{F} = \begin{pmatrix} 1 & \gamma & 0 \\ 0 & 1 & 0 \\ 0 & 0 & 1 \end{pmatrix} \tag{1}$$

2.2 Perturbed Shear Kinematics

Due to the complexity of boundary conditions and material properties, the rigorous simple shear deformation could not be reproduced in experiments. The perturbed shear kinematics has been proposed to consider the effects of axial deformation (see Fig. 1b and 1c). This situation was defined as case I and the corresponding deformation gradient was expressed as:

Table 1. Summary of critical shear strain evaluations for the simple shear test.

No	Formulations	Meanings of parameters	Type	Ref
1	$\gamma_1 = \tan\theta,\ \theta = \arctan\left(\frac{U_{12}}{U_{11}}\right) + \arctan\left(\frac{U_{21}}{U_{22}}\right)$ or $\theta = \frac{\pi}{2} - \theta', \cos\theta' = \frac{g'_1 \cdot g'_2}{\lvert g'_1\rvert \lvert g'_2\rvert} = \frac{(Fg_1)\cdot(Fg_2)}{\lvert Fg_1\rvert \lvert Fg_2\rvert}$	θ: Shear angle U_{ij}: Components of right stretch tensor $\boldsymbol{F}$: Deformation gradient $\boldsymbol{g}_i, \boldsymbol{g}'_i$: Two adjacent fibers before and after deformation	Direct	[10, 11, 13, 14, 17]
2	$\gamma_2 = \sum 2d\varepsilon_{12}, d\boldsymbol{\varepsilon} = \frac{d\boldsymbol{F}\boldsymbol{F}^{-1} + (d\boldsymbol{F}\boldsymbol{F}^{-1})^T}{2}$	$d\boldsymbol{\varepsilon}$: Instantaneous strain tensor	Indirect	[2]
3	$\lvert\gamma_3\rvert = \frac{\lambda_1 - \lambda_2}{\sqrt{1-(\lambda_1+\lambda_2)/2}} \cdot \boldsymbol{E}_A = \frac{1}{2}\left(\boldsymbol{I} - (\boldsymbol{F}^{\mathrm{T}}\boldsymbol{F})^{-1}\right)$	E_A: Almansi strain tensor λ_i: Eigenvalues of E_A	Indirect	[18]
4	$\gamma_4 = \gamma_{opt.}\ s.t.\ \min \sum_{i=1,2} \left(\frac{\lambda_i^p(\gamma_{opt.}, b_{opt.}) - \lambda_i^{exp}}{\lambda_i^{exp}}\right)^2$	$\gamma_{opt.}$: Optimized shear strain $b_{opt.}$: Optimized perturbed parameter b $\lambda_i^p, \lambda_i^{exp}$: Predicted and experimental eigenvalues of E_A	Indirect	[18]
5	$\gamma_5 = 2\sinh\varepsilon_1, \boldsymbol{\varepsilon} = \ln \boldsymbol{V}$	$\boldsymbol{V}$: Left stretch tensor $\boldsymbol{\varepsilon}$: The Eulerian logarithmic (Hencky) strain tensor ε_1: Eigenvalue of $\boldsymbol{\varepsilon}$	Indirect	[1]
6	$\gamma_6 = \frac{F_{11}F_{12} + F_{21}F_{22}}{F_{12}^2 + F_{22}^2}$	F_{ij}: Components of deformation gradient	Direct	[15]
7	$\gamma_7 = F_{12}$	F_{ij}: Components of deformation gradient	Indirect	[21]

$$\boldsymbol{F} = \begin{pmatrix} 1+\delta_1 & \gamma(1+\delta_2) & 0 \\ 0 & 1+\delta_2 & 0 \\ 0 & 0 & 1+\delta_3 \end{pmatrix} \tag{2}$$

where δ_1 and δ_2 represented the degree of axial deformation along and perpendicular to the shear direction, respectively.

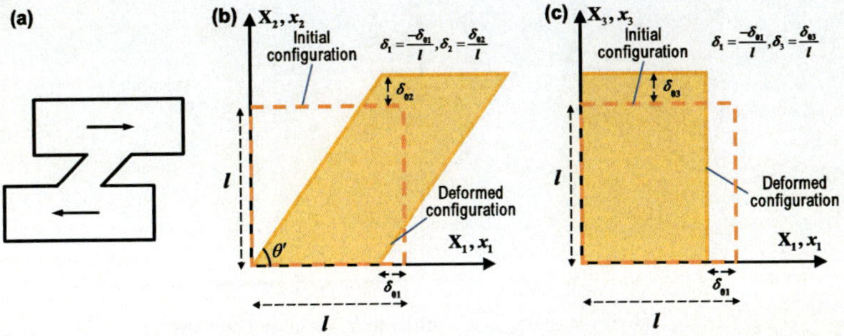

Fig. 1. A material element of the shear zone subjecting to the perturbation of axial deformation during simple shear (case I): (a) schematic of the specimen's configuration; (b) schematic of the perturbation in $X_1(x_1)$-$X_2(x_2)$ coordinate; (c) schematic of the perturbation in $X_1(x_1)$-$X_3(x_3)$ coordinate.

Moreover, in consideration of the non-idealities of experimental operations (such as the tilt of specimen or DIC camera) or the characteristics of specific experiments (such as multiaxial tests [3] and in-plane torsion test [4, 19]), the shear direction could not always coincide with the direction of the coordinate axis of strain measurement (DIC system), that is, there would exist a tilt angle α between them (see Fig. 2). It was defined as case II and the deformation gradient under this condition ($X'_1(x'_1)$-$X'_2(x'_2)$-$X'_3(x'_3)$ coordinate) was:

$$\boldsymbol{F}' = \boldsymbol{R}_\alpha \boldsymbol{F} \boldsymbol{R}_\alpha^T \tag{3}$$

where the transformation matrix $\boldsymbol{R}_\alpha$ was:

$$\boldsymbol{R}_\alpha = \begin{pmatrix} \cos\alpha & -\sin\alpha & 0 \\ \sin\alpha & \cos\alpha & 0 \\ 0 & 0 & 1 \end{pmatrix} \tag{4}$$

And it satisfied the condition:

$$\boldsymbol{R}_\alpha \boldsymbol{R}_\alpha^T = \boldsymbol{I} \tag{5}$$

Taking the derivative of both sides of Eq. (5), the result was expressed as:

$$\dot{\boldsymbol{R}}_\alpha \boldsymbol{R}_\alpha^T + \boldsymbol{R}_\alpha \dot{\boldsymbol{R}}_\alpha^T = \boldsymbol{0} \tag{6}$$

In the coordinate of $X'_1(x'_1)$-$X'_2(x'_2)$-$X'_3(x'_3)$, the left stretch tensor, the right stretch tensor, and the Hencky strain tensor were given as:

$$\boldsymbol{V}' = \boldsymbol{R}_\alpha \boldsymbol{V} \boldsymbol{R}_\alpha^T,\ \boldsymbol{U}' = \boldsymbol{R}_\alpha \boldsymbol{U} \boldsymbol{R}_\alpha^T,\ \boldsymbol{\varepsilon}' = \boldsymbol{R}_\alpha \boldsymbol{\varepsilon} \boldsymbol{R}_\alpha^T \tag{7}$$

Equation (2) and (3) were separately substituted into the shear strain evaluations of Table 1 and the derived results were summarized in Table 2 (γ_4 and γ'_4 could only be computed numerically and there were no analytical formulations for them. The numerical results of them would be discussed in Sect. 3). In case I, it was indicated that the

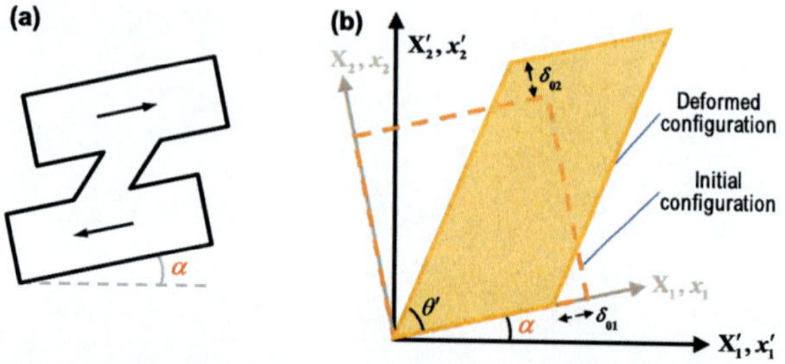

Fig. 2. A material element of the shear zone subjecting to the perturbation of axial deformation and the tilt of shear direction (case II): (a) schematic of the specimen's configuration; (b) schematic of the perturbation in $X_1(x_1)$-$X_2(x_2)$ coordinate and $X'_1(x'_1)$-$X'_2(x'_2)$ coordinate, respectively. The axis of $X'_3(x'_3)$ was supposed to be parallel to $X_3(x_3)$. In the figure, α represented the tilt angle.

emergence of axial deformation (δ_1 and δ_2) would influence to results of γ_2, γ_3, γ_5, γ_6 and γ_7, and the γ_1 was seemed to be a reasonable consequence. In case II, it was shown that the value of γ'_1, γ'_2, γ'_6 and γ'_7 were influenced by the tilt angle and thus they seemed not to be "objective". As an exception, if only the absolute value of shear strain was concerned, the γ'_1 could also lead to a rational result when $\alpha = 90°$. Although the results of γ'_3 and γ'_5 were not sensitive to tilt angle, they were also influenced by δ_1 and δ_2.

Table 2. Summary of shear strain evaluations under the perturbation of axial deformation (Case I) and the tilt of shear direction (Case II).

No	Case I (axial deformation)	Case II (axial deformation + tilt of shear direction)
1	$\gamma_1 = \frac{2\gamma(1+\delta_1)(1+\delta_2)}{\sqrt{4(1+\delta_1)^2(1+\delta_2)}} = \gamma$	$\gamma'_1 = \gamma\cos 2\alpha - \sin 2\alpha\frac{\left(\gamma^2+1\right)(1+\delta_2)^2-(1+\delta_1)^2}{21+2(1+\delta_1)(1+\delta_2)}$ $(\gamma'_1 = -\gamma,\ \text{when}\ \alpha = 90°)$
2	$\gamma_2 = \int d\gamma - \int \gamma\left(\frac{d\delta_1}{1+\delta_1} - \frac{d\delta_2}{1+\delta_2}\right)$	$\gamma'_2 = \int d\gamma\cos 2\alpha + \int(\sin 2\alpha - \gamma\cos 2\alpha)\left(\frac{d\delta_1}{1+\delta_1} - \frac{d\delta_2}{1+\delta_2}\right)$ $+\int d\alpha\frac{\cos 2\alpha(1+\delta_1)^2 + \left(\cos 2\alpha\left(\gamma^2-1\right) - 2\gamma\sin 2\alpha\right)(1+\delta_2)^2}{(1+\delta_1)(1+\delta_2)}$
3	$\lvert\gamma_3\rvert \approx \frac{2C_1\sqrt{C_2}}{\sqrt{C_2-A_2}}$	$\gamma'_3 = \gamma_3$
5	$\gamma_5 = \frac{B_1+C_1-2}{\sqrt{2(B_1+C_1)}}$	$\gamma'_5 = \gamma_5$
6	$\gamma_6 = \frac{\gamma(1+\delta_1)}{\left(\gamma^2+1\right)(1+\delta_2)}$	$\gamma'_6 = \frac{\gamma\cos 2\alpha(1+\delta_1)(1+\delta_2) - \sin\alpha\cos\alpha\left[\left(\gamma^2+1\right)(1+\delta_2)^2-(1+\delta_1)^2\right]}{\sin^2\alpha(1+\delta_1)^2+\cos^2\alpha\left(\gamma^2+1\right)(1+\delta_2)^2+\gamma\sin 2\alpha(1+\delta_1)(1+\delta_2)}$ $\left(\gamma'_6 = -\gamma\frac{1+\delta_2}{1+\delta_1},\ \text{when}\ \alpha = 90°\right)$
7	$\gamma_7 = \gamma(1+\delta_2)$	$\gamma'_7 = \gamma\cos^2\alpha(1+\delta_2) + \sin\alpha\cos\alpha(\delta_1-\delta_2),\ (\gamma'_7 = 0,\ \text{when}\ \alpha = 90°)$

where

$$\begin{cases} B_1 = (\gamma^2 + 1)(1 + \delta_2)^2 + (1 + \delta_1)^2 \\ C_1 = \sqrt{[\gamma^2(1 + \delta_2)^2 + (\delta_1 - \delta_2)^2][\gamma^2((1 + \delta_2))^2 + (\delta_1 + \delta_2 + 2)^2]} \\ A_2 = 2\delta_1 + 2\delta_2 + 8\delta_1\delta_2 + 4\delta_1\delta_2^2 + 4\delta_1^2\delta_2 - 2\delta_2\gamma^2 + \delta_1^2 + \delta_2^2 - \gamma^2 + 2\delta_1^2\delta_2^2 - \delta_2^2\gamma^2 \\ C_2 = 4(1 + \delta_1)^2(1 + \delta_2)^2 \end{cases} \tag{8}$$

To validate the analytic comparisons above, the experiment results of DP980-S1-in-plane-0°[6] would be reanalyzed, where S1 represented the specimen geometry 1. The proposed method could also be applied to other specimens without losing generality. The experimental procedure and the post-process of experimental data have been described in detail in [6] and would not be presented here again. A series of coordinates with different angles to shear direction were constructed in DIC software to investigate the effects of tilt angle.

3 Experimental Comparison of Shear Strain Evaluations

Figure 3 shows the evolution of shear strains with the variation of tile angle of shear direction for DP980-S1-in-plane-0°. In case I (see Table 2 and Fig. 3a), it was indicated that different shear strain evaluations would lead to different results, especially when the deformation was sufficiently large. The γ_1 was seemed to be a reasonable estimation in this condition [6] and it achieved the highest value, while the γ_6 would lead to the lowest value. The maximum difference between γ_1 and γ_6 was about 0.61, which was nearly 49% of γ_1. In case II (Fig. 3b–i), it was illustrated the values of γ_1', γ_2', γ_6' and γ_7' were considerably influenced by the tilt angle while the results of γ_3',γ_4' and γ_5' were not. These observations were consistent with the analytic results of Table 2. In addition, the value of γ_1' was more sensitive to the change of tilt angle than that of other evaluations. If only the absolute value of shear strain was concerned, the γ_1' could also be accepted under the condition of $\alpha = 90°$.

In summary, the shear strain evaluations listed in Table 1 would lead to different results in simple shear experiments. The γ_1 or γ_1' was seemed to be a reasonable choice if the shear direction is coincided with the coordinate axis of strain measurement because they would not be influenced by the axial deformation. Otherwise, the γ_3,γ_4 or γ_5 could be a potential alternative due to their insensitivity to the tilt of shear direction.

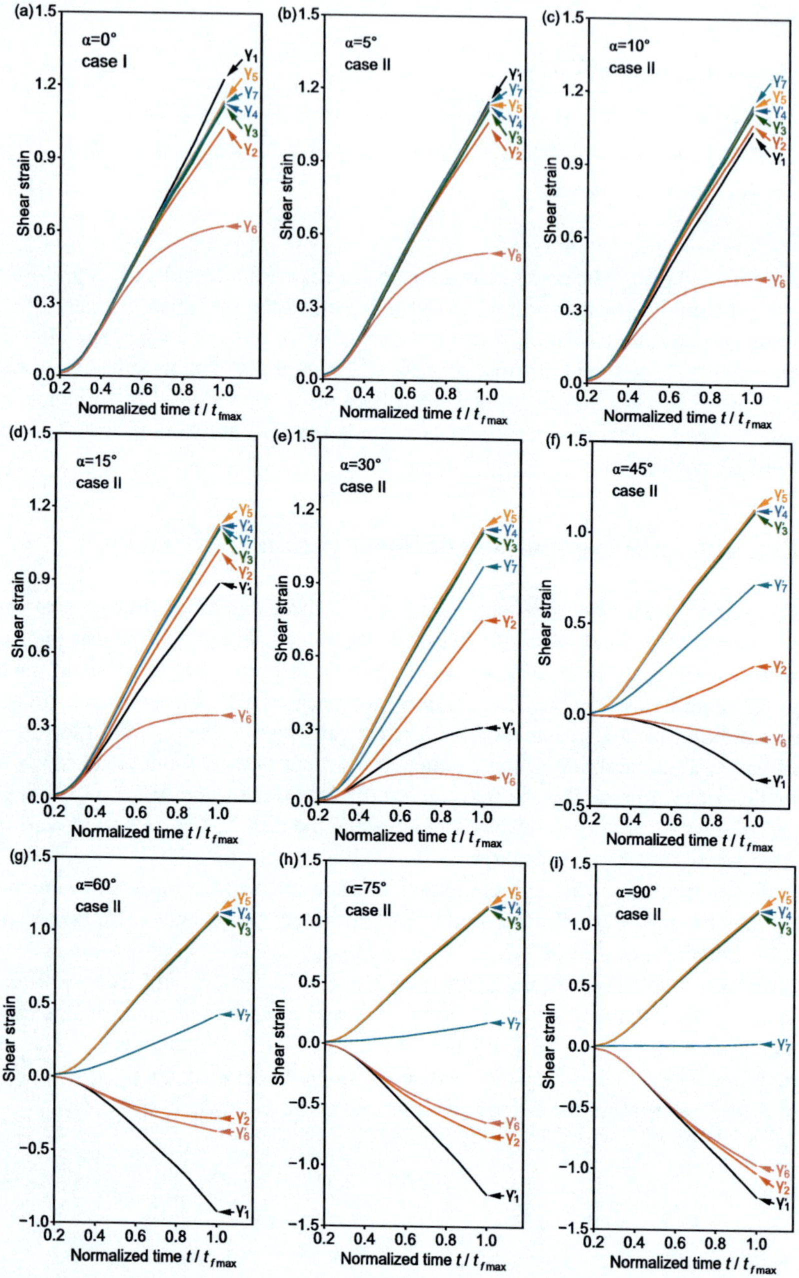

Fig. 3. Shear strains derived from the evaluations of Table 1 with the variation of tilt angle α for DP980-S1-in-plane-0° [6]: (a) $\alpha = 0°$; (b) $\alpha = 5°$; (c) $\alpha = 10°$; (d) $\alpha = 15°$; (e) $\alpha = 30°$; (f) $\alpha = 45°$; (g) $\alpha = 60°$; (h) $\alpha = 75°$; (i) $\alpha = 90°$. In the figure, t represented the current test time and t_{fmax} represented the time at which the specimen experiences the maximum force.

4 Conclusions

Shear strain is commonly used to characterize the degree of deformation in simple shear tests for sheet metals, but to the best of the authors' knowledge, there is no widely-accepted or standardized procedure to evaluate it in experiments up to now. The present work compared some critical shear strain evaluations and provided theoretical and experimental evidence to show that they would lead to different results in simple shear tests. The perturbation of axial deformation and the tilt of shear direction with respect to the coordinate axis of strain measurement were identified as two critical factors to influence the obtained shear strain. The γ_1 or γ_1' seemed to be a reasonable choice if there exists no tilt of shear direction. Otherwise, the γ_3,γ_4 or γ_5 was regarded as a potential alternative.

The presented result was also expected to provide some insights into the identification of shear strain in the multiaxial (combined shear tension) [3] and in-plane torsion test [4, 19], where the shear direction could not always coincide with the coordinate axis of strain measurement. The shear stress which conjugates with each shear strain mentioned above should be discussed in the future and the work conducted by B Reyne et al. [15] has provided a good guideline on this topic.

Acknowledgments. This work was supported by the National Natural Science Foundation of China (Grant No. 51975366) and the National Key Research and Development Program of China (Grant No. 2019YFA0708803).

References

1. Abedini, A., Butcher, C., Nemcko, M.J., et al.: Constitutive characterization of a rare-earth magnesium alloy sheet (ZEK100-O) in shear loading: STUDIES of anisotropy and rate sensitivity. Int. J. Mech. Sci. **128**, 54–69 (2017)
2. Chen, K.L., Kyriakides, S., Scales, M.: Effect of material frame rotation on the hardening of an anisotropic material in simple shear tests. J. Appl. Mech-T Asme **85** (2018)
3. Ferreira, I., Jesus, A., Machado, M., et al.: Experimental characterisation of fused filament fabrication printed parts under tension, shear, and combined shear–tension loads via Arcan test. Proc. Inst. Mech. Eng. Part L J. Mater. Des. Appl. **234**, 835–850 (2020)
4. Grolleau, V., Roth, C.C., Mohr, D.: Design of in-plane torsion experiment to characterize anisotropic plasticity and fracture under simple shear. Int. J. Solids Struct. **236–237**, 111341 (2022)
5. Gu, B., He, J., Li, S.H., et al.: Anisotropic fracture modeling of sheet metals: From in-plane to out-of-plane. Int. J. Solids. Struct. **182**, 112–140 (2020)
6. Han, G., He, J., Li, S.: Simple shear deformation of sheet metals: finite strain perturbation analysis and high-resolution quasi-in-situ strain measurement. Int. J. Plasticity **150**, 103194 (2022)
7. He, J., Han, G.F., Feng, Y.S.: Phase transformation and plastic behavior of QP steel sheets: TRansformation kinetics-informed modeling and forming limit prediction. Thin-Walled Struct. **173** (2022)
8. He J, Han GF, Guo C (2021) Non-associated anisotropic plasticity of metal sheets based on the distortional concept. Thin-Walled Structures 161
9. He, J., Han, G.F., Li, S.H., et al.: To correlate the phase transformation and mechanical behavior of QP steel sheets. Int. J. Mech. Sci. **152**, 198–210 (2019)

10. ASTM International: B831-14 Standard test method for shear testing of thin aluminum alloy products. In: ASTM International, West Conshohocken, PA (2014)
11. Kang, J.D., Wilkinson, D.S., Wu, P.D. et al.: (2008) Constitutive behavior of AA5754 sheet materials at large strains. J. Eng. Mater-T Asme **130** (2008)
12. Li, S.H., He, J., Gu, B., et al.: Anisotropic fracture of advanced high strength steel sheets: experiment and theory. Int. J .Plasticity **103**, 95–118 (2018)
13. GOM mbH (2013) ARAMIS User Manual-Software
14. Merklein, M., Biasutti, M.: Forward and reverse simple shear test experiments for material modeling in forming simulations. In: 10th International Conference on Technology of Plasticity, ICTP 2011. Pp. 702–707 (2011)
15. Reyne, B., Herault, D., Thuillier, S., et al.: Quality of the strain state in simple shear testing using field measurement techniques. Int. J. Mech. Sci. **208** (2021)
16. Stiebert, F., Traphöner, H., Meya, R., et al.: Characterization of flow curves for ultra-thin steel sheets with the in-plane torsion test. J. Manuf. Sci. Eng. **144**, 031011 (2022)
17. Tarigopula, V., Hopperstad, O.S., Langseth, M., et al.: A study of large plastic deformations in dual phase steel using digital image correlation and FE analysis. Exp. Mech. **48**, 181–196 (2008)
18. Thuillier, S., Manach, P.Y.: Comparison of the work-hardening of metallic sheets using tensile and shear strain paths. Int. J. Plasticity **25**, 733–751 (2009)
19. Traphöner, H., Clausmeyer, T,. Tekkaya, A.E.: Methods for measuring large shear strains in in-plane torsion tests. J. Mater. Process. Technol. **287,** 116516 (2021)
20. Yin, Q., Zillmann, B., Suttner, S., et al.: An experimental and numerical investigation of different shear test configurations for sheet metal characterization. Int. J. Solids Struct. **51**, 1066–1074 (2014)
21. Zang, S.L., Sun, L., Niu, C.: Measurements of Bauschinger effect and transient behavior of a quenched and partitioned advanced high strength steel. Mat. Sci. Eng. A-Struct. **586**, 31–37 (2013)

Mechanics and Materials for Sheet Metal Forming" In Honor of Prof. Fred Barlat

Affordable Multi-scale Numerical Simulation of Structures in Anisotropic Plasticity and Damage

Gilles Rousselier(✉)

Mines Paris, PSL Research University, MAT-Centre des Matériaux, CNRS UMR 7633, BP 87, 91003 Evry Cedex, France
gilles.rousselier@minesparis.psl.eu

Abstract. The "reduced texture methodology" enables the numerical modeling of laboratory specimens with mean field polycrystalline plasticity at an affordable cost. It makes it possible to model not only plastic anisotropy, anisotropic hardening and non-proportional loadings but also the various ductile fracture mechanisms observed in metallic alloys: initiation, growth, rotation, flattening and coalescence of microscopic voids, transgranular fracture, secondary voids nucleation at the grain scale and coalescence at the slip system scale. The extension to numerical modeling of processes and structural behavior can be achieved in the longer term.

Keywords: Polycrystalline Plasticity · Reduced Texture · Ductile Fracture · Porous Plasticity · Lode Variable · Non-proportional Loadings

1 Introduction

The numerical modeling of processes such as rolling, deep drawing, extrusion and of structural behavior is improved with advanced macroscopic models, but these models may fail to predict the complex distortion of the yield locus. The mean field polycrystalline plasticity framework has intrinsic advantages in describing plastic anisotropy, anisotropic hardening and non-proportional loadings, but the computational cost is prohibitive. The numerical calibration of a **reduced texture** (e.g. only 3 orthotropic texture components and 12 crystallographic orientations), initiated by Rousselier and Barlat [1], enables a drastic reduction of the computational cost. In this framework, various phenomena at multiple scales can be modeled: plastic anisotropy and texture evolution [2], distortion of the macroscopic yield surface and strength differential effect [3], Coulomb **ductile** fracture model with energy dissipation at the slip system scale [4], dynamic strain aging at the slip system scale [5, 6], secondary voids nucleation at the grain scale and coalescence at the slip system scale [7], Lode-dependent damage by void rotation and flattening for shear-dominated loadings [8], non-proportional loadings [9]. The objective of the paper is to highlight the capabilities and benefits of the reduced texture methodology with a few examples.

K. Mocellin et al. (Eds.): ICTP 2023, LNME, pp. 343–350, 2024.
https://doi.org/10.1007/978-3-031-40920-2_36

2 Applications of the Reduced Texture Methodology

The reduced texture parameters (Euler angles, volume fractions) are calibrated with mechanical tests only: tensile tests in various directions, biaxial tests, shear tests, shear-tension and tension-shear tests (as far as necessary), simultaneously with the plasticity model parameters. In Fig. 1, an example of comparison between experiments and the calibrated model is shown.

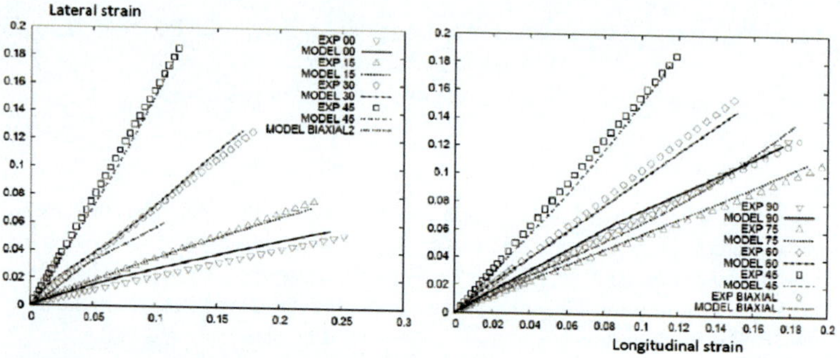

Fig. 1. Tensile tests. Result of parameter calibration with lateral strain (y axis) - longitudinal strain (x axis) curves (Lankford ratio), 2090-T3 aluminum alloy sheet.

An example of reduced texture is given in Fig. 2. The experimental texture is not used for calibration; nevertheless the two textures are in good agreement. In Fig. 3, the calibrated reduced texture polycrystalline model is used in deep drawing simulations [2]. The CPU time (2 h, single dual-core processor) is only twice that of the macroscopic Yld2004-18p model. Simulations are performed with ABAQUS.

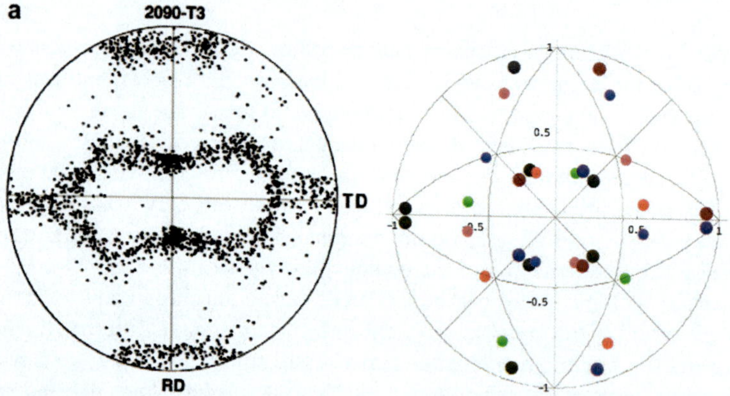

Fig. 2. {111} experimental (EBSD) texture and **calibrated reduced texture** of 2090-T3 aluminum alloy. Distinct color and size are given to each of the 8 crystal orientations. The 2 sets of 4 orientations are close to the S and Brass texture components

In forming operations of aluminum alloys, dynamic strain aging (DSA) generates Portevin-Le Chatelier (PLC) bands that can alter the sheet surface and interact with strain localization and ductile failure [5]. Synchrotron laminography (Fig. 4) shows the PLC incremental strain bands inside a CT specimen, compared with simulations in Fig. 5 [6].

Both DSA model at the slip system scale and porous plasticity model are implemented in the Finite Element software Z-set [10] for simulations with a reduced texture. The specimen failure is well predicted. The incremental strain field on the right of Fig. 5 shows a single band before the slant fracture.

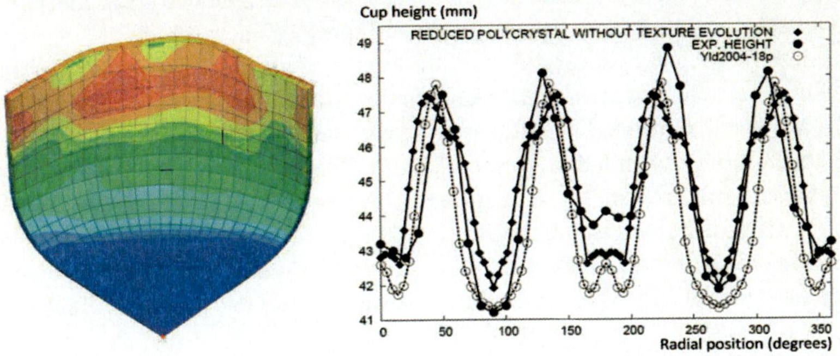

Fig. 3. Cup deep drawing of 2090-T3 aluminum alloy, von Mises stress, scale 4 MPa (blue) to 617 MPa (red), deformed mesh. Cup height (mm) vs. radial position 0–360°.

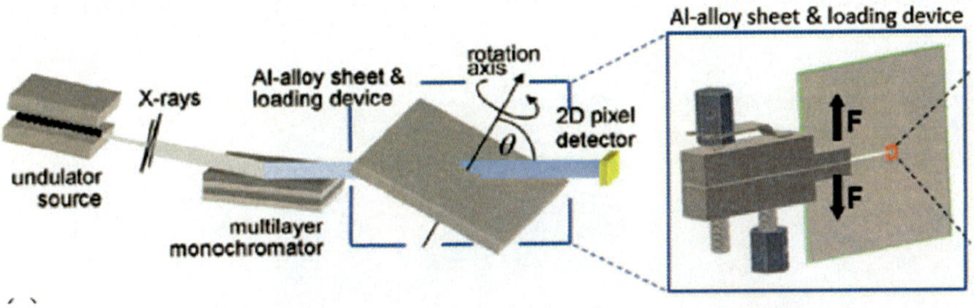

Fig. 4. Synchrotron laminography, thin CT specimen t = 1 mm.

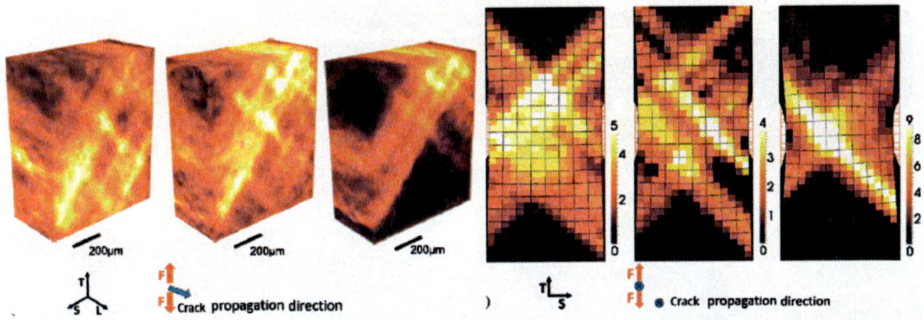

Fig. 5. Thin CT specimen t = 1 mm, 2198-T8R Al-Cu-Li alloy. Incremental strain fields (scales in %) at 1 mm ahead of the notch (radius 170 μm), for increasing loadings from left to right. Laminography (left) and simulation (right, deformed mesh).

In 2198-T8R aluminum alloy, several ductile damage mechanisms are involved: microscopic voids modeled with porous plasticity, submicroscopic voids initiated on dispersoids and modeled at the grain size, transgranular shear fracture modeled by the

Coulomb **ductile** fracture model at the slip system scale. At the notch of a thin KAHN specimen (t = 2 mm), the experimental crack, initiated at mid-thickness, is flat in a small triangle approximately 1.5 mm wide along the notch tip and 1.5 mm long in the notch plane. Two shear lips appear at the notch tip on both free surfaces of the specimen, one in the upper half of the specimen and one in the lower half. The two shear lips join after approximately 1.5 mm of crack propagation and form a slant crack.

One simulation needs several days with 8 cores of a single processor. The simulations would be impractical with the hundreds of orientations required to model plastic behavior with experimental textures. The experimental crack geometry is well modeled in the numerical simulation, as shown by the "broken" integration points in Fig. 6. In both the experiment and the simulation, the microscopic voids are primarily present in the flat triangular crack where the stress triaxiality is large. Transgranular fracture is observed in the slant crack (maximum Coulomb slip). Submicrometric voids contribute to the coalescence of large voids in the flat crack and they are also present in the slant crack [7].

In the slant crack of the KAHN specimen, the stress triaxiality is smaller than 0.5 and the Lode variable ξ is smaller than 0.3. These conditions of shear-dominated loading are known to favor failure by void rotation and flattening until coalescence. This damage mechanism can be modeled with an additional term g in the porous plastic potential and

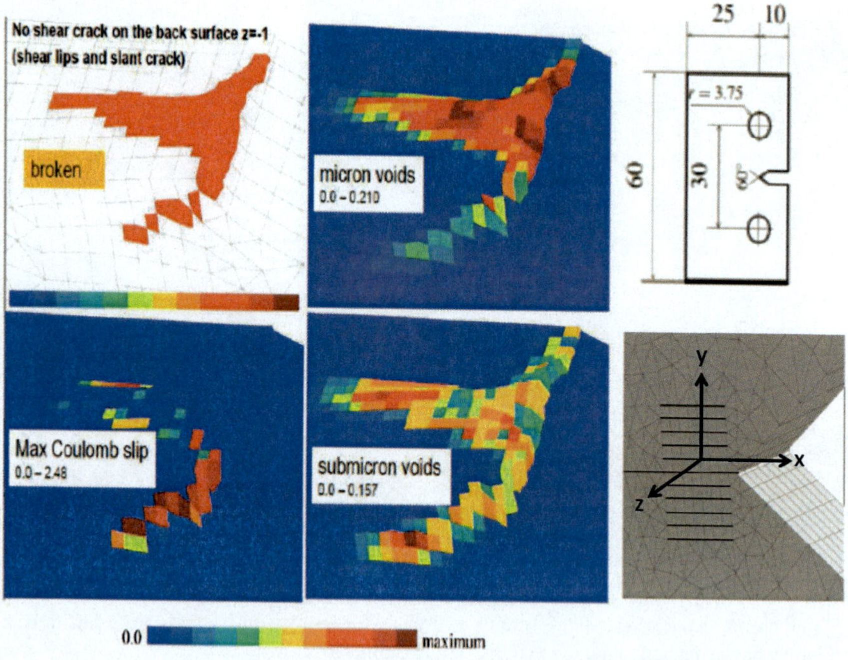

Fig. 6. KAHN specimen t = 2 mm, notch radius 60 μm, 12,976 elements and 117,453 dof, 8 elements in the thickness direction. Crack propagation in a Cartesian mesh (element initial size 0.2–0.25 mm) from the right (notch tip) to the left, shown in the lower half specimen y < 0: flat triangular crack in the plane y = 0 and slant crack intersection with the specimen front surface z = +1. Deformed mesh.

yield criterion F of Eq. (1), where σ_{eq} and σ_m are the von Mises stress and the mean stress, f and g the porosities, σ_1 and D_1 material parameters:

$$F(\underline{\sigma}, f, g) = \sigma_{eq} + \sigma_1 D_1 (f + g) \exp(\sigma_m/\sigma_1) - \sigma_p = 0 \tag{1}$$

An additional void rate for shear loadings was already used by Nahshon and Hutchinson [11] but with a single porosity f. The addition of a second porosity g is supported by the theoretical work of Madou and Leblond [12]. A simple evolution equation $\dot{g} = \kappa(1 - |\xi|^{\lambda})\dot{\xi}^{p}_{eq}$ is used here, with $k = \lambda = 2$ [8].

Although this rotation and flattening damage mechanism is not observed in the 2198-T8R alloy, it was added in the simulation in order to validate the model implementation in Z-set. It substitutes the submicrometric voids of Fig. 6.

In Figs. 7a and 7b, the same results as in Fig. 6 are obtained. The second porosity g-fields of Figs. 7c and 7d show the slant crack intersections with both free surfaces of

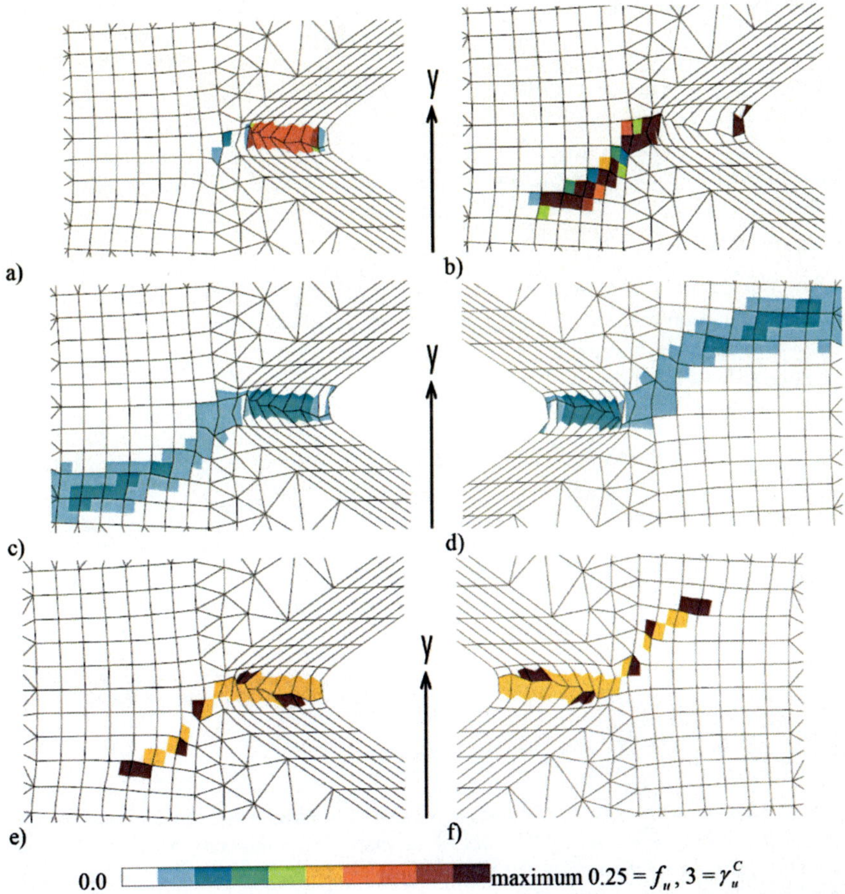

Fig. 7. KAHN specimen. a) porosity f, b) maximum Coulomb slip γ^C, c) second porosity g (in the lower half specimen), d) second porosity g (in the upper half specimen), e)f) broken integration points (in yellow for deleted elements). Deformed mesh.

the specimen. The notch is also a free surface where ε is small, so g also has significant values in the notch. The crack geometry is shown in Figs. 7e and 7f.

Non-proportional load paths are common in forming operations and structural behavior. In order to investigate the material behavior in plasticity and damage, cruciform specimens (Fig. 8) were designed, tested and simulated [9]. A first loading is applied in one direction (tension T or shear S of the ligament between the two notches), the specimen is unloaded and a second loading is applied in the perpendicular direction (TS and ST loadings).

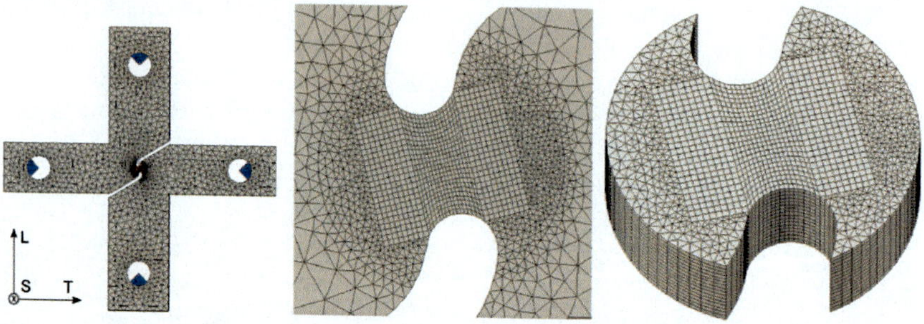

Fig. 8. Cruciform specimen (t = 1 mm) for tension-shear and shear-tension non-proportional loadings, ligament 1.4 mm. 70,816 elements, 16 elements in the thickness direction (fine "50 μm" mesh).

The polycrystalline plasticity model has ingredients to simulate non-proportional loadings, in particular the interaction matrices between slip systems (cross hardening). After parameter calibration, the load-displacement curve for the second loading is well simulated (SIM-PC, Fig. 9).

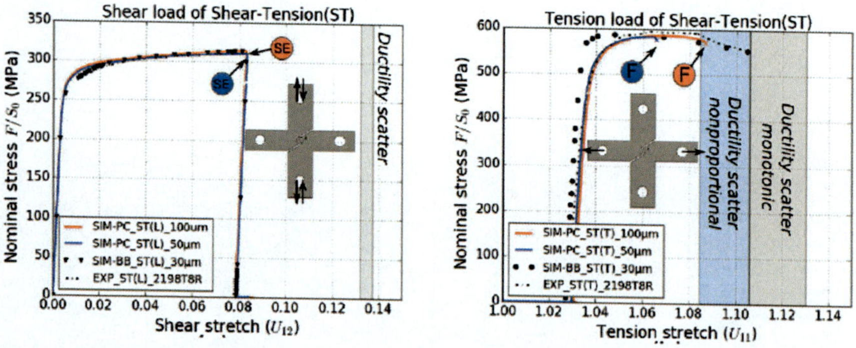

Fig. 9. 2198-T8R aluminum alloy. Tension curves after a first loading in shear. Blue and red curves correspond to different element size. F are the failure points (crack initiation), the **red** one (elements twice the size) falls in the experimental ductility scatter. Large dots correspond to macroscopic plasticity model simulations (Bron-Besson model, without damage). Small dots correspond to the experimental curves.

In particular, the hardening capability of the model is not saturated by the first loading, as it is the case with the macroscopic model (SIM-BB). In 2198-T8R aluminum alloy, the main damage mechanisms are microscopic voids and grain-related smooth, flat fracture in shear. With porous plasticity and the Coulomb **ductile** fracture model, neglecting the modeling of submicrometric voids, it was possible to calibrate the Coulomb damage parameters to obtain a good modeling of crack initiation and propagation for all loadings: T, S, TS and ST (**red** curve). The **blue** curve with a twice smaller element size shows the mesh size effect in ductile fracture. The simulated damage fields are in agreement with the damage vizualization by in-situ synchrotron laminography [9].

3 Conclusions

Numerical calibration of a **reduced texture** (usually only 3 orthotropic texture components and 12 crystallographic orientations), initiated by Rousselier and Barlat [1], enables the use of mean field polycrystalline plasticity with a drastic reduction in computational cost. Numerous simulations of laboratory specimens with more than 100,000 dof are affordable. To be more precise, a comprehensive laboratory study with experiments and simulations requires tens or hundreds of simulations and computational cost reduction is imperative. Besides, polycrystalline plasticity is more effective than macroscopic plasticity for anisotropic materials and complex loadings. Moreover, polycrystalline plasticity is the natural framework of damage and failure modeling because damage takes place at various scales: macroscopic scale, grain scale or slip system scale, obviously in aluminum alloys but also in steels, e.g. submicrometric voids initiated on carbides [7].

The extension to numerical modeling of processes and structural behavior is not so easy because numerical calibration of reduced texture parameters and damage parameters also is a long task to be performed before final finite element simulations. Damage models also have requirements for element size that may be too small for structural simulations. The problem of mesh size dependence (not addressed in this paper) also requires numerical developments that may make applications more difficult. But this goal can be achieved in the longer term.

References

1. Rousselier, G., Barlat, F.: Modeling of plastic anisotropy with reduced polycrystalline models. Application to aluminum alloys. In: 11th ESAFORM Conference on Material Forming, MS04 paper 17 (2008)
2. Rousselier, G., Barlat, F., Yoon, J.W.: A novel approach for anisotropic hardening modeling – Part I: Theory and its application to finite element analysis of deep drawing. Int. J. Plasticity **25**, 2383–2409 (2009)
3. Luo, M., Rousselier, G.: Modeling of large strain multiaxial deformation of anisotropic metal sheets with strength-differential effect using a Reduced Texture Methodology. Int. J. Plasticity **53**, 66–89 (2014)
4. Rousselier, G., Luo, M.: A fully coupled void damage and Mohr-Coulomb based ductile fracture model in the framework of a Reduced Texture Methodology. Int. J. Plasticity **55**, 1–24 (2014)

5. Rousselier, G., Quilici, S.: Combining porous plasticity with Coulomb and Portevin-Le Chatelier models for ductile fracture analyses. Int. J. Plasticity **69**, 118–133 (2015)
6. Rousselier, G., Morgeneyer, T., Ren, S., Mazière, M., Forest, S.: Interaction of the Portevin-Le Chatelier phenomenon with ductile fracture of a thin aluminum specimen: experiments and simulations. Int. J. Fracture **206**, 95–122 (2017)
7. Rousselier, G.: Porous plasticity revisited: Macroscopic and multiscale modeling. Int. J. Plasticity **136**, 102881 (2021)
8. Rousselier, G.: Lode-dependent second porosity in porous plasticity for shear-dominated loadings. Int. J. Plasticity **159**, 103446 (2022)
9. Kong, X., Morgeneyer, T., Missoum-Benziane, D., Rousselier, G.: A polycrystalline damage model applied to an anisotropic aluminum alloy 2198 under non-proportional load-path changes. Int. J. Plasticity **168**, 103674 (2023)
10. Z-set software (2023) Non-linear material & structure analysis suite. http://www.zset-software.com
11. Nahshon, K., Hutchinson, J.: Modification of the Gurson model for shear. Eur. J. Mech. A/Solids **27**, 1–17 (2008)
12. Madou, K., Leblond, J.-B.: A Gurson-type criterion for porous ductile solids containing arbitrary ellipsoidal voids-I: limit-analysis of some representative cell. J. Mech. Phys. Solids **60**, 1020–1036 (2012)

Inverse Identification of the YLD2000-2D Yield Locus Exponent for Stainless Steel 1.4301 Using a Time-Dependent Optimization Method

Konrad Barth[1](✉), Mohamadreza Afrasiabi[1], and Markus Bambach[2]

[1] Inspire AG, 8005 Zurich, Switzerland
konrad.barth@inspire.ch

[2] Advanced Manufacturing Laboratory, ETH Zurich, 8005 Zurich, Switzerland

Abstract. The non-quadratic yield locus YLD2000-2D is widely used in sheet metal forming to describe anisotropy. In this yield function, the exponent, often denoted by *m*, is a material parameter that determines the yield locus shape and is typically defined by the grain structure ($m = 6$ for body-centered cubic (bcc) and $m = 8$ for face-centered cubic (fcc)). However, recent studies have shown that this assumption is not always the best choice for reproducing the strain distribution of standard experiments, such as the Nakazima tests. In the present work, an inverse time-dependent optimization is proposed to determine the parameter *m* for the metastable austenitic stainless steel 1.4301. The material is subject to the Transformation Induced Plasticity (TRIP) effect, and the microstructure transforms from austenite (fcc) to martensite (bcc) during cold forming. The strain distributions along cross-sections of different Nakazima experiments are evaluated at different drawing depths for the inverse optimization. The FE simulation results are compared with the experimental cross-section data in order to determine the optimal values of *m*. The outcome is analyzed to quantify the influence of the Nakazima specimen and the drawing depth. Despite the increase in martensite content in the material, the overall optimal value of *m* for the stainless steel 1.4301 was determined to be $m = 8$.

Keywords: YLD2000-2D · M-value · Stainless steel · Inverse fitting · Time-dependent

1 Introduction

Determination of a yield locus is an essential part of materials constitutive modeling in sheet metal forming simulation. The more sophisticated the yield locus is, the more input parameters are usually required. These parameters are usually determined from experimental data. In most cases, tensile tests in 0°, 45°, and 90° with respect to the rolling direction are performed to determine the yield strength and Lankford parameters. Additionally, a bulge test can be conducted for the biaxial yield strength and Lankford parameter. A commonly used yield locus for sheet metal forming applications is the non-quadratic yield locus YLD2000-2D proposed by Barlat et al. [1]. It requires eight

K. Mocellin et al. (Eds.): ICTP 2023, LNME, pp. 351–362, 2024.
https://doi.org/10.1007/978-3-031-40920-2_37

basic input parameters to be identified from tensile tests and bulge test. With this input, the exponent m for the non-quadratic yield locus remains unknown. The exponent m influences the shape of the yield locus in the areas without experimental inputs. In particular, the value of m affects the position of the plane strain stress point, which is the most critical point in the forming limit curve (FLC). In order to obtain realistic finite element (FE) simulation results, especially for plane-strain conditions, m should be determined precisely.

Logan and Hosford [2] proposed to set the value of m according to the grain structure. They suggested choosing $m = 6$ for body-centered cubic (bcc) and $m = 8$ for face-centered cubic (fcc). However, recent studies such as Harsch [3], Pilthammar et al. [4], and Hippke et al. [5] show that this choice of the m-value is not always the best. These studies used different materials and methods to find the optimal value of m. Harsch [3] used the strain distribution before necking from three different Nakazima specimens to determine the value of m for the stainless steel 1.4301. m was optimized to fit the same failure point (major and minor strain) of the Nakazima experiment. Harsch showed that the optimal value of m can differ depending on the investigated Nakazima specimen. Pilthammar et al. [4] used Nakazima experiments in plane strain and biaxial direction. The major and minor strain distribution along two cross-sections (rolling direction and transversal direction) were determined at a defined drawing depth and compared to the simulated results. The best result was achieved with $m = 5$ for the investigated CR4 steel (bcc) instead of $m = 6$. Both Harsch and Pilthammar used only a single point in time of the Nakazima experiments to determine the exponent m. Possible deviations of the strains between experiment and simulation before or afterward cannot be detected this way. Hippke et al. [5] compared the usage of crystal plasticity (CP) and cruciform tension (CT) experiments on the example of an AA6014-T4 aluminum alloy (fcc). The yield strengths at different stress states were determined and used to fit the YLD2000 yield locus and, therefore, to find the best value of m. $m \approx 6$ was found to be optimal and was able to reduce the error in predicting the major strain by more than half. The disadvantages of this method are the time-consuming experiments and the special equipment required to perform the tests.

Due to the discussion about the value of m for non-quadratic yield loci, Hora and Barth [6] and Barth et al. [7] investigated the use of the Vegter yield locus. They proposed the time-dependent method for investigating the position of the plane strain stress point. The method, similar to Pilthammar et al. [4], used the strain distribution of Nakazima experiments at different drawing depths measured by digital image correlation (DIC). Based on the comparison with the FE simulation, the optimal Vegter parameters for the plane strain stress point position were determined.

The metastable austenitic stainless steel 1.4301 is an interesting material for investigating the value of m. Due to the TRIP effect, the basic grain structure of austenite (fcc) transforms to martensite (bcc) during cold forming. According to the findings of Logan and Hosford [2], this would lead to a transition from $m = 8$ to $m = 6$ based on the current martensite content of the material. When Harsch [3] investigated the value of m for the stainless steel, he found $m = 6.5$ for the 50 mm (direction of tensile test) and the 200 mm (biaxial direction) wide Nakazima specimen and $m = 8$ for the 90 mm (near plane strain) wide specimen. However, Harsch further used $m = 8$ in his study, as this value was validated by Krauer using a rectangular cup [8].

In this paper, 24 simulations and three Nakazima experiments are utilized to investigate the most appropriate value of m in the YLD2000 yield locus for the metastable austenitic stainless steel 1.4301. The inverse time-dependent optimization method is used for the analysis, as all necessary data from standard Nakazima experiments were available. The value of m is examined for the influence of the selection of the Nakazima specimen and the drawing depth. The latter is also used to draw conclusions about the influence of the martensite content on determining the optimal value of m.

2 Theory

The non-quadratic yield locus YLD2000-2D from Barlat [1] was used in this paper to represent the anisotropy of the stainless steel. Eight input arguments ($\sigma_0, \sigma_{45}, \sigma_{90}, \sigma_{biax}, r_0, r_{45}, r_{90}, r_{biax}$) and the exponent m are prerequisite to defining the yield locus. Equations (1), (2), and (3) state the basic equations of the yield condition of YLD2000.

$$\phi = \phi' + \phi'' = 2\overline{\sigma}^m \tag{1}$$

$$\phi' = \left|X_1' - X_2'\right|^m \tag{2}$$

$$\phi'' = \left|2X_2'' + X_1''\right|^m + \left|2X_1'' + X_2''\right|^m \tag{3}$$

Here, $\overline{\sigma}$ is the effective stress, and the material-dependent exponent m controls the shape of the yield locus. The higher the value of m, the closer the yield locus to the shape of a Tresca yield locus.

The hardening behavior of the stainless steel 1.4301, including the TRIP effect, is represented by a Hänsel-type model [9]. The model consists of two parts, which account for:

- describing the martensite formation rate as a function of the temperature T and the martensite content V_M itself through this expression:

$$\frac{dV_m}{d\varepsilon} = a \cdot e^{Q/T} \cdot \left(\frac{1 - V_M}{V_M}\right)^b \cdot V_M^p \cdot 0.5 \cdot (1 - \tanh(C + D \cdot T)) \tag{4}$$

  where a, b, C, D, p, Q are material-dependent parameters.
- representing the yield behavior, which includes a Hockett-Sherby yield curve, a temperature influence, and an additional hardening term due to the martensite formation.

$$\sigma = \left[B_{HS} - (B_{HS} - A_{HS}) \cdot \exp\left(-m_{HS} \cdot \varepsilon^{n_{HS}}\right)\right] \cdot \left[1 - K \cdot \left(T - T_{ref}\right)\right] + \Delta k_f^{\gamma \to \alpha'} \cdot V_M^z \tag{5}$$

The material model was implemented in a user material subroutine of LS-DYNA for the finite element simulations.

3 Materials and Methods

3.1 Materials

All experiments were performed on metastable austenitic stainless steel 1.4301 sheets with a thickness of 0.8 mm. As the TRIP effect causes the transformation of austenite in martensite, the Hänsel model was used to describe the flow curve of the material. The parameters were calibrated by the author and are listed in Table 1. Figure 1 shows the isothermal evolution of the hardening curve and the corresponding formation of martensite at different temperatures.

Table 1. Parameters for the TRIP material model

Parameter	Value	Unit
a	0.0823	-
b	0.7253	-
C	−25.94	-
D	0.0775	1/K
p	1.3974	-
Q	1267.0	K
ε_S	0.0176	-
$\Delta\varepsilon$	0.0304	-
n	1.1161	-
V_{M0}	0.37	%
T_{Ref}	353	K
A_{HS}	276.07	MPa
B_{HS}	1429.3	MPa
m_{HS}	1.9368	-
n_{HS}	0.9764	-
K	0.0023	1/K
$\Delta k_f^{\gamma\rightarrow\alpha\prime}$	377.38	MPa
z	1.4461	-

The input parameters for the YLD2000 yield locus were determined at room temperature using a series of tensile tests along 0°, 45°, and 90°of the rolling direction and a biaxial bulge experiment. The corresponding parameters are listed in Table 2. The resulting yield locus shapes for $m = 4.5$ and 8 are shown in Fig. 2. The influence of m on the position of the plane strain stress point can be seen. A high value of m causes little change of the yield locus around the plane strain stress point, while a lower value of m causes the yield locus normal to change more rapidly in this region. The higher the

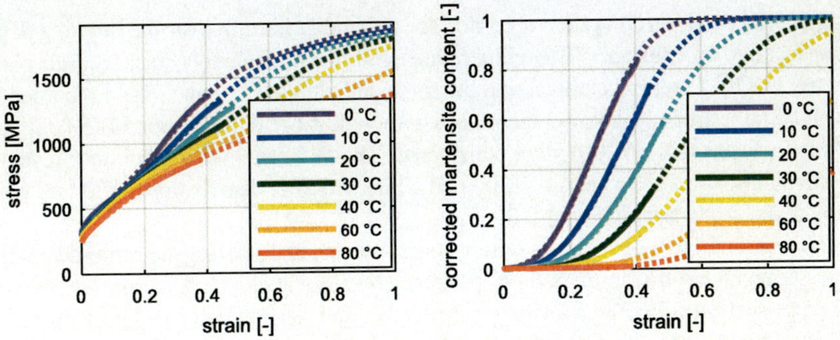

Fig. 1. Material behavior of 1.4301 for different isothermal temperatures (solid line: experimental data; dotted line: model) **a** Hardening behavior, **b** Martensite formation

value of *m*, the higher the resulting equivalent stress for the same principal stress, and this is accompanied by earlier yielding.

Table 2. Parameters for the YLD2000 yield locus without *m*

σ_0	σ_{45}	σ_{90}	σ_{biax}
287.5 MPa	270.5 MPa	265.1 MPa	287.9 MPa
r_0	r_{45}	r_{90}	r_{biax}
0.926	1.411	0.901	0.8784

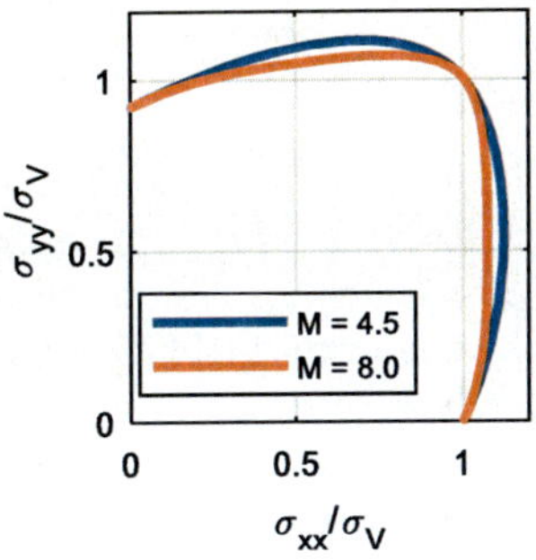

Fig. 2. YLD2000 yield locus with different exponent *m*

3.2 Methods

Nakazima experiments were conducted on the material with a coolable punch to ensure a blank temperature at the pole close to 20 °C. To ensure low-friction contact between blank and punch three layers of grease separated through PE foil was used. A digital image correlation (DIC) system recorded the strain distribution over the drawing depth at consecutive time instants. The temperature evolution on the upper surface of the blank was tracked by a thermal camera, and the force and path of the punch were recorded by the machine sensors. Nakazima specimens with widths of 50 mm (named W50), 80 mm (W80), and 200 mm (W200) were considered. The W50 and W80 specimens revealed negative minor strains, whereas W200 had a biaxial stress distribution. W80 was close to the plane strain state ($\Delta\varepsilon_2 = 0.0$).

The optimal value of m for the stainless steel was evaluated using the 'time-dependent cross-sectional optimization method' proposed by Hora et al. [6]. Accordingly, the central cross-section along the specimen was determined for the major and minor strain at different drawing depths. For this study, the drawing depths of 10, 20, 30, and 40 mm (named H10, H20, H30, and H40, respectively) were chosen. The cross-sections were also limited to the central 40 mm to ensure better comparability between the drawing depths.

The explicit finite element solver LS-DYNA was used to simulate the Nakazima experiments. In the first step, the friction coefficient and the heat transfer coefficient (HTC) between the punch and the blank were identified by comparing them to the experimental punch force and the temperature evolution, respectively. In the second step, simulations with variable m values ranging from 4.5 to 8 were set up for the three Nakazima experiments.

The difference between the strain along the cross-section of the experiment and the simulation was calculated according to

$$\Delta\varepsilon = \varepsilon_{Exp}(x) - \varepsilon_{Sim}(x) \tag{6}$$

wherein x denotes the distance along the cross-section. Subsequently, the root mean square error (RMSE) can be computed from:

$$RMSE = \sqrt{\frac{\int_{x_{min}}^{x_{max}} \Delta\varepsilon^2 dx}{x_{max} - x_{min}}} \tag{7}$$

The overall RMSE for a given value of m is the mean of all considered RMSEs. These can consist of all three Nakazima specimens, all four drawing depths, and all three principal strains. In total, 36 individual RMSEs were calculated for each value of m. The thinning strain was calculated by means of volume constancy.

The influence of the Nakazima specimen and the drawing depth were analyzed separately. Furthermore, the connection of the drawing depth with regard to the increasing strains to the increasing martensite content was established.

4 Results

The calibration of the friction coefficients resulted in $\mu = 0.02$ between punch and blank and $\mu = 0.04$ between die/binder and blank. The maximal force deviation between the experiment and simulations at 40 mm drawing depth was 10% for W200. The simulated temperature evolution on the pole with $HTC = 1.0\frac{kW}{m^2K}$ lies within 1 K compared to the experiment.

Using these input parameters, the resulting strain distributions of the Nakazima simulations for W50, W80, and W200 are shown in Fig. 3, Fig. 4, and Fig. 5, respectively. A good agreement between simulations and experiments was observed. However, the predicted 40 mm drawing depth strains deviate from the experimentally measured values. The deviation of the major strain at the central cross-section point (drawing depth of 40 mm) between the experiment and the simulation with $m = 8$ are 0.0296 for W50, 0.0261 for W80, and 0.0037 for W200. The differences in the strains due to the variation of m were largest for W200.

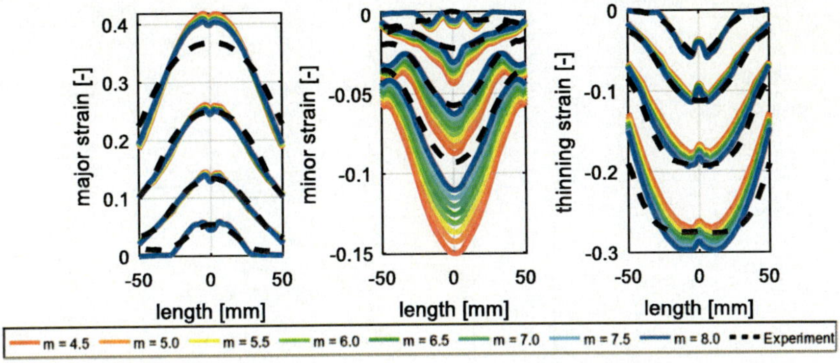

Fig. 3. Major, minor, and thinning strains along cross-sections at different drawing depths for 50 mm specimen

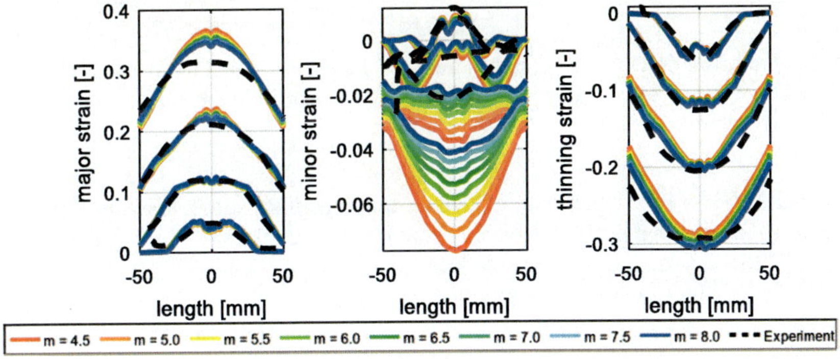

Fig. 4. Major, minor, and thinning strains along cross-sections at different drawing depths for 80 mm specimen

As shown in Fig. 6, the punch forces and the effective plastic strain hardly fluctuate when the value of m changes. The largest deviation was found for W200. The martensite formation in all three specimens is exemplified in Fig. 6 for $m = 8$. Only for W200 a clear difference between the values of m could be observed. No experimental martensite data were available as the measurement of martensite and DIC could not be done at the same time.

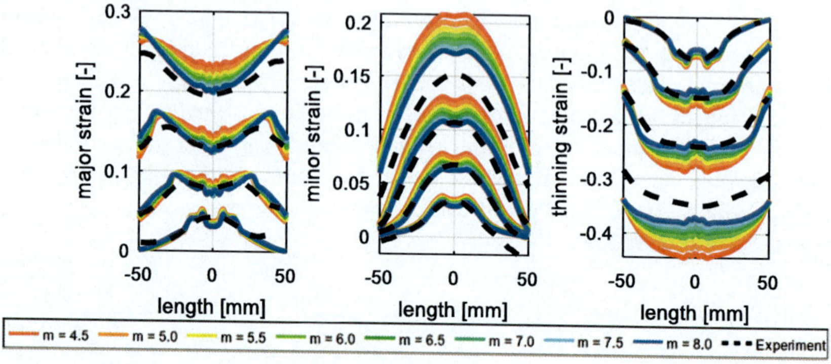

Fig. 5. Major, minor, and thinning strains along cross-sections at different drawing depths for 200 mm specimen

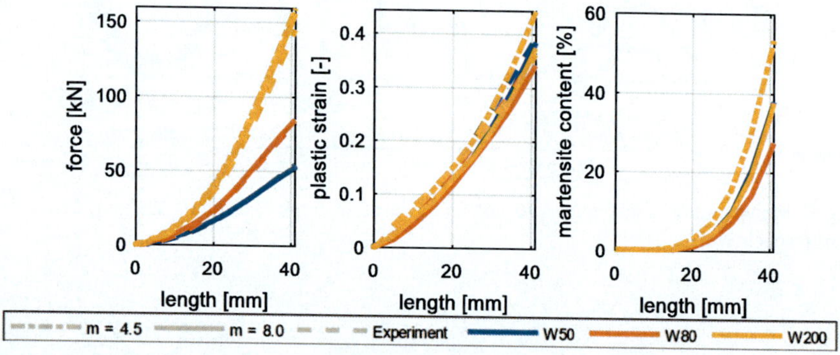

Fig. 6. Comparison of experiment (dashed line) and simulations (solid lines) for all Nakazima specimens with $m = 8.0$ and $m = 4.5$ only for W200 **a** punch force **b** effective plastic strain **c** martensite content

The RMSE for each combination of Nakazima specimen, strain, drawing depth, and the value of m was analyzed. The overall result is shown in black in Fig. 7. No minimum for the RMSE was detected within the investigated range from $4.5 \leq m \leq 8$. The best result was obtained with $m = 8$. Divided among the Nakazima specimens, a similar behavior of the RMSEs as for the overall result emerged (Fig. 7a). W200 showed the highest error for all three specimens, especially for lower values of m. The distribution of the RMSE according to the drawing depth showed that the error became larger with increasing depth. For H10, no influence of the value of m was recognizable ($m_{min} = 7$) and for H20 there was a lower plateau at $m \geq 6.5$ ($m_{min} = 8$). The level of the RMSEs for 10 to 30 mm drawing depth were similar, especially towards $m = 8$. H40 stoud out as the error was at least twice as high as the other three.

Figure 8 shows the RMSEs of the drawing depths divided among the individual Nakazima specimens. W50 shows an optimal value of $m = 8$ for all four depths. For W80, H10 had the optimal value of $m = 5.5$. Above this, the optimal value of m was

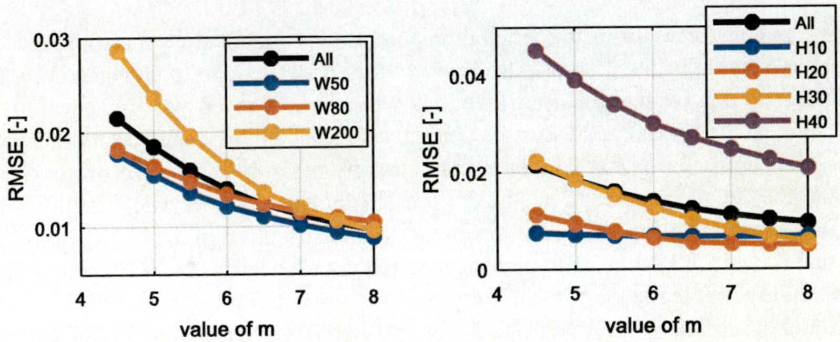

Fig. 7. RMSE for overall and **a** each investigated Nakazima specimen; **b** each evaluated drawing depth

always $m = 8$, and the gradient became steeper with increasing height. The minimal error for W200 developed from 7 for H10 over 6.5 for H20 to 8 for H30 and H40.

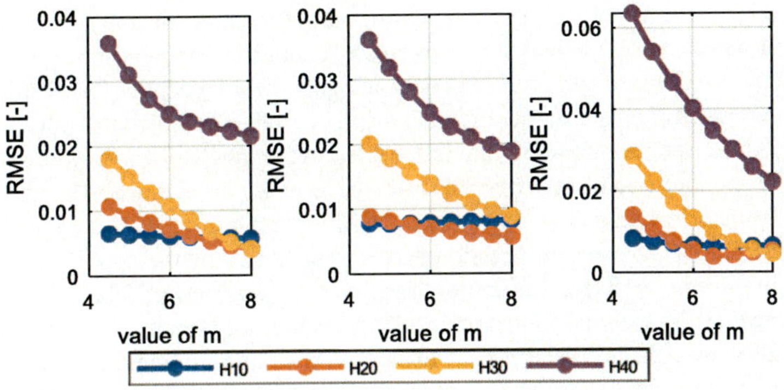

Fig. 8. RMSE for each drawing depth and **a** W50 specimen; **b** W80 specimen; and **c** W200 specimen

5 Discussion

The determined friction and heat transfer parameters allow the model to represent the experimental strain distribution, the temperature at the pole, and the punch force well. A strong influence of the friction on the strain distribution was observed in the same way as by Pilthammar [4]. For better determination of the friction coefficient, the thinning strain and, thus, the position of the crack was added as a further criterion in addition to the major and minor strain. The choice of the friction coefficient between punch and blank influences especially the major strain distribution. Therefore, determining the friction coefficient has to be done carefully to gain realistic parameters and not influence the latter analysis of the optimal value of m. The direct influence of friction on the optimal value of m should be investigated in a further study.

The punch forces from the simulations are almost identical even if the value of m varies. This indicates that despite the variation in strains due to changes in m, the equivalent strains seem to be insensitive to it, which is confirmed by Fig. 6.

Although the biaxial yield stress is perfectly represented by the YLD2000-2D yield locus, the strains for W200 fluctuate more strongly regarding the value of m. This is because the yield locus around the biaxial stress point is strongly curved, thus the direction of the normal changes rapidly. Since the strain distribution of the Nakazima W200 specimen covers this range up to plane strain, as shown by Barth et al. [7], the influence of m can be seen clearly.

The distribution of the major strain for H10, H20, and H30 agrees well with the experimental data. The variations of the strains become larger with increasing drawing depth. The minor strains vary more strongly as it is directly influenced by the value of m. Despite the larger error of major and minor strains for 40 mm, the thinning strains for W50 and W80 agree well with the experiment. The two strains balance each other in magnitude as the thinning strain is the negative sum of the two. For W200, the major strain is acceptable for $m = 8$, but the error for minor and thinning strains is larger. Higher major strains in the simulation at the same drawing depth have already been described by Harsch for W50 [3]. Comparing the Nakazima specimens by the time-dependent optimization method shows the faster strain development in the simulation for the other investigated specimens but only at higher drawing depths. Small strains can be misrepresented if only a higher drawing depth is used to optimize the value of m.

The good agreement up to 30 mm is reflected in the low RMSEs for these drawing depths. The optimum value of m deviates from the three curves mentioned in the Results, but the absolute variation is very small. For W200, a trend from $m \approx 6.75$ to $m = 8.0$ can be observed. However, the direction of this trend is exactly the opposite to the idea of changing from $m = 8$ (fcc: austenite) to $m = 6$ (bcc: martensite). The differentiation of the RMSE by the specimen and drawing depth reveals only small differences between them. W200 shows the highest error due to the largest strain variations. It is interesting to note that the depths of 20 and 30 mm for all three geometries are better reproduced with $m = 8$ than the ones at 10 mm. One reason for this may be that the initial local deformation state of the experiment cannot be reproduced well by the simulation. However, the differences, at least for $m = 8$, are minimized for the following drawing depths. The overall best value of m, which is $m = 8$, supports the earlier value for the YLD2000 exponent from the literature. The results of $m = 6.5$ for W50 and W200 from Harsch [3] cannot be confirmed using the time-dependent optimization method.

For H40, the major and minor strains are overestimated by the simulation. As these strains become smaller with increasing value of m, the minimal error is determined for the largest investigated m, which is $m = 8$. The large deviation of 40 mm drawing depth affects the overall result clearly. Evaluating the RMSEs without 40 mm, the optimal value of m results still in 8, but the gradient towards it is nearly halved. Since the deviations for 40 mm occur in all three samples and the other drawing depths are represented well, it is evident that the cause is to be found in the simulation model. Changing the friction model from uniform Coulomb friction toward a more complex friction model, e.g., depending on the contact pressure as studied by Filzek [10], could lead to a better representation of the initial and final strain distribution. Likewise, the material model has a strong influence on the simulation. The yield behavior and the martensite formation were only fitted based on tensile tests. The optimization could be extended to include the parameters of the Hänsel model. Since the experimental strains cannot be mapped satisfactorily over the entire drawing depth with a constant value of m, the use of a variable m depending on the equivalent plastic strain can also be considered, as was already done by Wang et al. for an aluminum [11]. Alternatively, a different yield locus, such as the Vegter yield locus [12], can also improve the strain prediction over all drawing depths.

An influence of the martensite volume on the optimum value of m could not be determined during the study. The tendency toward $m = 8$ can be seen in most of the metamodels and does not change with increasing martensite content. The simulation rather points to an opposite influence. The more the value of m can influence the equivalent strain, the more the martensite formation is altered, which can be seen at W200.

6 Conclusion and Outlook

The optimal value for the YLD2000-2D exponent m for the metastable austenitic stainless steel was investigated in this paper. The following conclusions can be drawn:

- The choice of the value of m influences the strain distribution in simulations. The higher the strains, the higher the differences; therefore, the choice of m must be considered.
- The overall optimum value of m for 20 °C is $m = 8$ and confirms the previously taken value in the literature for the YLD2000-2D exponent. All Nakazima specimens show similar behavior for the metamodel.
- No influence of the increasing martensite content on the value of m could be found for the material 1.4301.

The resulting $m = 8$ applies to the simulation with a blank temperature around 20 °C. As the stainless steel is strongly temperature dependent and the yield locus parameters were determined at room temperature, the optimum value of m should also be investigated for other temperatures.

References

1. Barlat, F., et al.: Plane stress yield function for aluminum alloy sheets - Part 1: Theory. Int. J. Plast. **19**(9), 1297–1319 (2003)
2. Logan, R.W., Hosford, W.F.: Upper-bound anisotropic yield locus calculations assuming ⟨111⟩-pencil glide. Int. J. Mech. Sci. **22**(7), 419–430 (1980)
3. Harsch, D.: Digitale Zwillingsmodelle zur optimalen Auslegung und Regelung von bedingt robusten Blechumformprozessen. ETH Zürich (2018)
4. Pilthammar, J., Banabic, D., Sigvant, M.: BBC05 with non-integer exponent and ambiguities in Nakajima yield surface calibration. Int.J. Mater. Form. **14**(4), 577–592 (2020). https://doi.org/10.1007/s12289-020-01545-0
5. Hippke, H., Hirsiger, S., Berisha, B., Hora, P.: Optimized and validated prediction of plastic yielding supported by cruciform experiments and crystal plasticity. Int.J. Mater. Form. **13**(5), 841–852 (2020). https://doi.org/10.1007/s12289-020-01569-6

6. Hora, P., Barth, K.: Impact of advanced measurement methods on the development and setup of new (complex) constitutive models. Form. Technol. Forum (2019)
7. Barth, K., Berisha, B., Hora, P.: Time-dependent method for the inverse evaluation of yield locus using nakazima experiments. In: Inal, K., Levesque, J., Worswick, M., Butcher, C. (eds.) NUMISHEET 2022. The Minerals, Metals & Materials Series. Springer, Cham (2022). https://doi.org/10.1007/978-3-031-06212-4_25
8. Krauer, J.: Erweiterte Werkstoffmodelle zur Beschreibung des thermischen Umformverhaltens metastabiler Stähle. ETH Zürich (2010)
9. Hänsel, H.C.A.: Nichtisothermes Werkstoffmodell für die FE-Simulation von Blechumformprozessen mit metastabilen austenitischen CrNi-Stählen. ETH Zürich (1998)
10. Filzek, J., Ludwig, M., Groche, P.: Improved FEM simulation of sheet metal forming with friction modelling using laboratory tests. In: Proceedings of. IDDRG, Bilbao, Spain, pp. 5–8 (2011)
11. Wang, H., Wan, M., Wu, X., Yan, Y.: The equivalent plastic strain-dependent Yld 2000–2d yield function and the experimental verification. Comput. Mater. Sci. **47**(1), 12–22 (2009)
12. Vegter, H., Van Den Boogaard, A.H.: A plane stress yield function for anisotropic sheet material by interpolation of biaxial stress states. Int. J. Plast. **22**(3), 557–580 (2006)

Capability Testing of Layered Specimens in Plane Strain Compression Method

Dániel Gy. Szőke, Martin L. Kölüs, Richárd Borbély, and Gábor J. Béres(✉)

GAMF Faculty of Engineering and Computer Science, John Von Neumann University, 6000 Kecskemét, Hungary

beres.gabor@gamf.uni-neumann.hu

Abstract. This study presents the experimental results of layer-compression tests in plane strain state as well as the role of the obtained parameters on the simulated thinning tendency of a cylindrical deep drawn part. The work hardening law has a special importance in numerical simulations, thus it is mostly expected to be known over a large range of deformation. Tensile test, which counts as a fundamental method for material characterization, is therefore often complemented by other experiments, such as hydraulic bulge test, or plane strain compression test in sheet metal forming. The main drawbacks of the plane strain compression test are the lateral tool misalignment, the stress concentration at the tool edges and the undesirable effect of the friction. The thinner the applied sheets, the first two things are more pronounced due to the geometrical constraints that ensure the plane strain state. To overcome these disadvantages, layered specimens were investigated with rounded-edge tools in the present research. Our results show that the shear contribution in the compression zone can be effectively reduced by two- or three-layered sheets, with a minimum of two times higher deformation than it occurs in tension. Besides, a better lubricant condition is also easier to achieve in the multi-layered arrangement. The effect of the obtained work hardening parameters is also briefly outlined using numerical simulations and the Taguchi method. It was initially observed that none of the hardening parameters have more significance on thinning than others.

Keywords: Plane Strain Compression · Layered Specimens · Large Strain Work Hardening

1 Introduction

The knowledge of large strain flow curve is essential for numerous types of material processing technologies realized by forming. Although the basic method of tension test is widely accepted, standardized and relatively simple to implement in any laboratory, it does not provide direct information to the large strain deformations. Nowadays, the advanced data processing techniques (e.g., digital imagine correlation) help the users to exploit most of the data during tension for this purpose, but other methods are also widespread. One of such alternatives in the sheet metal forming is the plane strain compression test.

K. Mocellin et al. (Eds.): ICTP 2023, LNME, pp. 363–372, 2024.
https://doi.org/10.1007/978-3-031-40920-2_38

This method, like other possibilities for work hardening characterization that do it other ways than tension, suffers by its special drawbacks. These drawbacks are mostly due to imperfections resulting from a single reason, the inhomogeneous deformation. Readers can find examples on this in the literature, whether we are talking about compression test [1, 2] or hydraulic bulge test [3, 4]. At the plane strain compression test, the non-homogenous deformation can be developed because of the shear contribution and friction [5]. Moreover, the deformation inhomogeneity can be also interpreted in the thickness direction, as proposed in [6]. Interestingly, this kind of inhomogeneity is ignored in most of the research dealing with the problematic of work hardening.

Turning to the primarily causes of inhomogeneous deformation, rounded-edged tools are suggested by Chermette et al. [7] and Banabic. et al. [8] to reduce the shear contribution. This solution is useful in this regard, beyond any doubt. However, the smaller the punch width, the higher the risk of shearing in general. This can be explained by the higher risk for tool misalignment and the difficulties in the precise tool machining. In order to increase the punch width, the sheet thickness has to be increased, too. This is necessary to meet the plane strain state condition. Since the sheet thickness is given, and in most cases, it is very low in sheet metal forming, the only option for punch width increasing is the application of multi-layered specimens. The applicability of this kind of specimen arrangement is already validated and recommended for uniaxial compression by Merklein [9] and Coppieters [10, 11], but is not yet tested in plane strain conditions, at least according to the authors best knowledge.

In the case of friction, it is self-evident that the undesired effect of the additional shear stresses can be reduced by good lubricant between the tools and the workpiece, however, the situation is bit more complicated when multi-layered specimens are applied. In this respect, we relied on literature [12] during the experiences.

This paper presents in the following sections how multi-layered specimens can be interpreted for the plane strain compression method, with the aim of work hardening characterization at large strains. The description of the experimental tests as well as the results' discussion supplemented by finite element modelling are shown in the next sub-chapters.

2 Plane Strain Compression Tests

The plane strain compression tests (PSCT) were carried out by in-house manufactured tool sets, taking into consideration the geometrical features matching the plane strain state in every cases. To this end, different tool sets were manufactured for the single specimens' tests and for the multi-layered specimens' tests. The reason of it was that the increased (multi-layered) sheet thickness allowed increased punch dimensions, too.

According to the original idea of the PSCT, a sheet with *w*' width, *l*' length and *h*' height is compressed by flat punches (Fig. 1a). The geometry features of the two punches, namely the width (*w*) and the length (*l*) that together with the sheet geometry responsible for the plane strain state are also given in the same figure, schematically. Note that *w* does not include the radii on the edges, as shown in Fig. 1a. At the same time, Fig. 1b shows the real tool arrangement without workpiece, indicating only the importance of the punch axis' coincidence.

If the shear contribution and the friction are negligible or compensated, a homogeneous deformation zone under the punches develops, in which the normal stress (or equivalently the deformation resistance K_k) and the normal (true) strain (ε^h) can be computed as follows:

$$\varepsilon^h = ln\left(\frac{h_1'}{h_0'}\right) \tag{1}$$

$$K_k = \frac{F}{w \cdot w\prime}. \tag{2}$$

In Eq. (1) and (2), subscripts *1* and *0* refer to the final and the initial state, as well as F means the compression force. Assuming the Tresca yield criterion (referred as TMG later), K_k is identical to the equivalent flow stress, σ_f, while ε^h is identical to the equivalent plastic strain, ε^{pl}. In turn, considering the plane strain state by means of the Huber-Mises-Hencky yield criterion (referred as von Mises criterion in the rest of the paper), the equivalent stress and strain will differ from the previous results:

$$\varepsilon^{pl} = \frac{2}{\sqrt{3}}\varepsilon^h \tag{3}$$

$$\sigma_f = \frac{\sqrt{3}}{2}K_k. \tag{4}$$

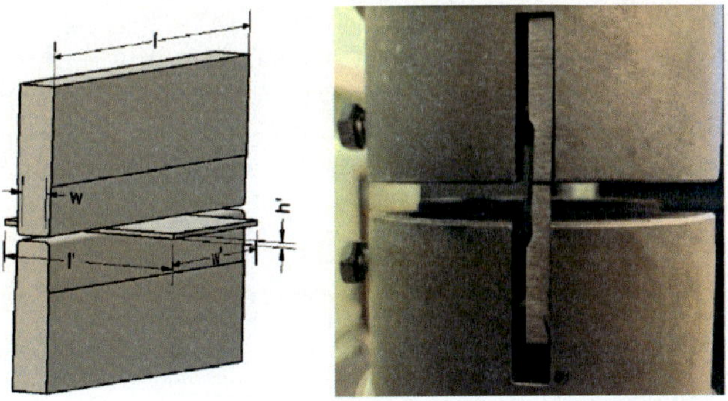

Fig. 1. Schematic view (a) and real arrangement of the plane strain compression tool set (b).

If we also take into account the fact of plastic anisotropy, anisotropic yield criterion should be applied. Given that the Hill'48 [13] criterion is well suited for analytical calculations, it can be deduced that the equivalent units will be changed what is described in Eq. (5) and (6), for the anisotropic case.

$$\varepsilon^{pl} = \frac{1+R}{\sqrt{1+2R}}\varepsilon^h \tag{5}$$

$$\sigma_f = \sqrt{1+\alpha^2 - \frac{2R}{1+R} \cdot \alpha \cdot K_k}. \tag{6}$$

Here R denotes the average Lankford coefficient and α is the ratio of the minor and the major, principal stresses. Using the associated flow rule, α can be determined after assuming that the ratio of the principal strain rates (β) is equal to zero.

$$\alpha = \frac{(1+R)\cdot\beta + R}{1+R+R\beta} \tag{7}$$

To make the stress and strain evaluation more accurately, Banabic et al. (Eq. (8)) and Chermette et al. (Eq. (9)) suggested the friction and the shear compensation in similar, but different forms:

$$\sigma_f = K_k \cdot \left[1 + \frac{\mu w}{2h'} + \frac{h'}{4w}\right]^{-1} \tag{8}$$

$$\sigma_f = K_k \cdot \left[\frac{h'}{\mu w} \cdot \left(e^{\mu\cdot\frac{w}{h'}} - 1\right) + \frac{1}{4}\frac{h'}{w}\right]^{-1}. \tag{9}$$

From this it can be observed that only the Coulomb friction coefficient needs to be defined or evaluated if accurate geometry measurements are available. In this study, all the mentioned yield criteria as well as the compensation suggested by Banabic et al. were used for further calculations.

2.1 Applied Material

The specimens were manufactured from commercial DC04 steel sheets. This type of material is developed for production by sheet forming operations, e.g., by deep drawing. The clearly ferritic microstructure provides good formability and thanks to the high value of normal anisotropy, deep components can be formed from this material in the press shop. The average, basic mechanical properties obtained by standard (ISO6892) tensile tests in 0°, 45° and 90° to the rolling direction are listed in Table 1.

Table 1. Tension-based mechanical properties of the investigated DC04 sheet.

YS (MPa)	UTS (MPa)	A_{80} (%)	A_g (%)	R (-)	K (MPa)	ε_0 (-)	n (-)
229	336	37.9	20.4	1.706	578	0.017	0.220

Here *UTS* and *YS* refer to the ultimate tensile and the yield strength, and A_{80}, A_g are the total and uniform elongations. The *R*-value was measured by an Instron advanced video extensometer and has been averaged between 0.1 plastic strain and the deformation associated with the A_g point. The Swift-type isotropic hardening law was embodied by the K, ε_0 and n parameters, and was fitted by the least squares method ($R^2 = 0.998$ for tension). The values are averages of three parallel measurements.

2.2 Single Specimens' Tests

Both the single and the multi-layered specimens' tests were carried out on the same Instron 5900R (100 kN) material tester equipment, as the tensile tests previously. 40 × 34 × 1 mm ($l' \times w' \times h'$) sheet samples were cut out, in which the l'direction coincided with the rolling direction. The compression was provided by dies with a total width of 2.8 mm, using combined (oil and polytetrafluoroethylene) lubricant on both sides of the sheet. To accurately measure the sheet thickness, the tests were interrupted sequentially at different force values: 60, 80 and 90 kNs. The thicknesses then were measured by microscopic tests, with a Zeiss Imager M2m equipment.

Our experiments showed that in lubricated state, the thickness distribution was relatively homogenous at the middle of the deformed region, however the traces of shear appearance were visible at the edges, presumably due to the tools' misalignment. One example for the shear contribution after 80 kN compression is depicted by Fig. 2. We observed that the more efficient the lubrication, the greater the twisting of the specimen under compression. Thus, the following tests of single specimens were examined in non-lubricated conditions, and a more significant friction compensation had to be applied during the calculations.

Fig. 2. Microscopic picture of the shear occurring in single specimen after 80 kN compression.

The isotropic hardening characteristic using the tensile and the single-layer compression data (without lubrication) from three parallel measurements is shown by Fig. 3. The stress state- and the friction compensation were calculated using Eq. (1)–(7) and Eq. (8). Coulomb friction coefficient with the value of 0.12 was assumed after Kraus et al. [14], due to the non-lubricated tool surfaces.

Authors know that the friction compensation should be validated by measurements and/or finite element simulations, but this will be investigated in the future. With this assumed coefficient of friction (and compensation strategy), it seems that the stress state compensation by the von Mises and the TMG yield theories slightly overestimates the tensile data points.

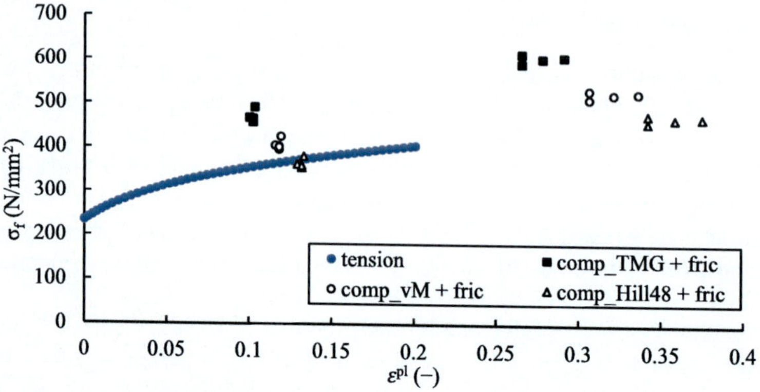

Fig. 3. Isotropic hardening characteristic obtained by tensile and PSC tests, for single-layered specimen. Both stress state and friction compensation were considered.

2.3 Multi-Layered Specimens' Tests

The multi-layered specimens proposed here to be an opportunity to avoid shear appearance and tools' misalignment. The assumption is based on the simple fact that wider tools are allowed for higher sheet thicknesses. Here we doubled the width of the dies, which led to 5.6 mm. The measurement method was the same that has been mentioned above, but samples were investigated in every 5 kNs above 60 kN. The wider tools allowed the usage of the combined lubrication between the tools and the workpieces. However, dry condition was maintained between the sheet layers. Representative samples for two- and three-layered specimens compressed by 90 kN are shown in Fig. 4. The corresponding work hardening characteristics are indicated in Fig. 5.

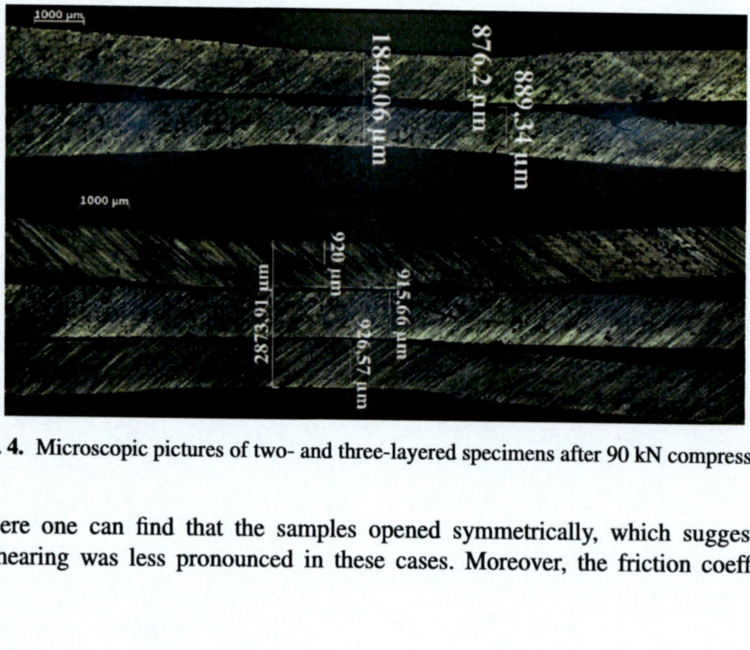

Fig. 4. Microscopic pictures of two- and three-layered specimens after 90 kN compression.

Here one can find that the samples opened symmetrically, which suggest that the shearing was less pronounced in these cases. Moreover, the friction coefficient was proposed to be 0.05 with the applied lubrication, which fits the reference [10]. Another advantage of the multi-layered condition is that the achievable strains were approximately twice higher than the deformation of the single specimens.

The Swift equation related to the tension-compression data yielded the relative error (R^2) of 0.938…0.987 (Fig. 5). A most precise hardening behavior can certainly be achieved with the combination of a power- and a saturation hardening law, since the former seems to be more suitable for describing the beginning, and the latter one is for the end of the curve. The observed fluctuation of the Swift parameters is: $K = 470_{(Hill48)}$ … $590_{(TMG)}$; $\varepsilon 0 = 0.001_{(Hill48)}$ … $0.025_{(TMG)}$; $n = 0.125_{(Hill48)}$ … $0.252_{(TMG)}$.

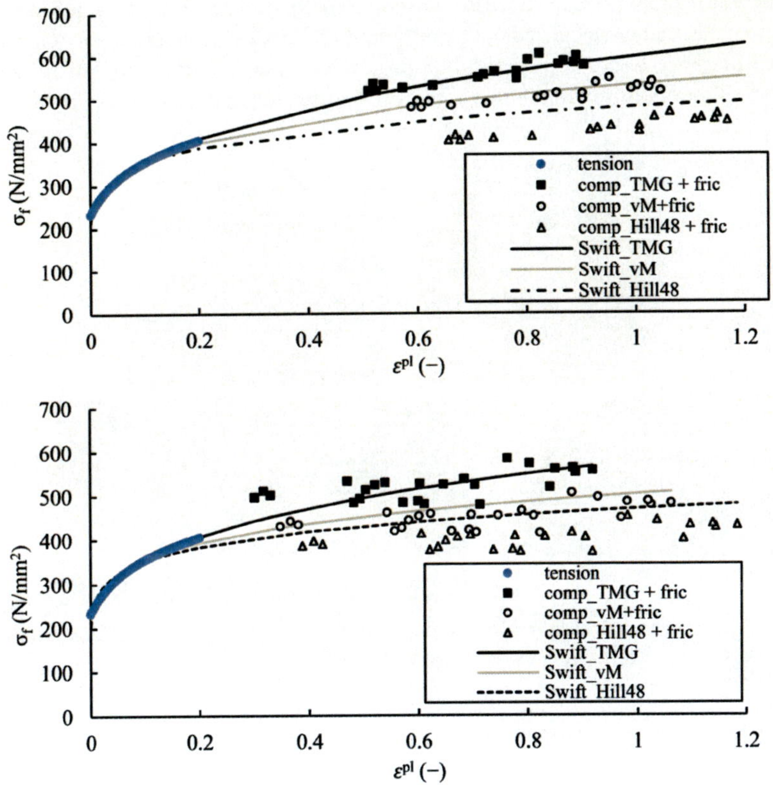

Fig. 5. Isotropic hardening characteristics obtained by tensile and PSC tests, for two- (above) and three-layered (below) specimens. Both stress state and friction compensation were considered

3 Discussion

The stress strain characteristics of the two- and three-layered specimens are similar to the single specimens' results in the sense that the TMG yield criterion evaluates the higher and the Hill'48 yield criterion shows the lowest stresses again. Naturally, the measurement of the friction coefficient and a more precise yielding behavior identification can clarify this behavior somewhat, but the role of the yield criterion continues to be significant. Consequently, we tried to reveal the effect of the hardening parameters with modelling the thinning of a deep-drawn part. Thinning, and the subsequent necking and tearing has of paramount importance in sheet metal forming, that is why it is in the focus now.

The experimental deep drawing tests were caried out by an Erichsen 142–40 universal sheet metal testing equipment, with the next technical parameters: punch diameter: Ø33 mm; punch radius: 5 mm; die clearance: 1.2 mm; initial blank dimensions: 1 mm × Ø66 mm; blank holder force: 8kN; lubrication type: polyethylene foil and oil on the die-side of the blank; punch speed 20 mm/min. The blanks had grid mesh for the strain measurements that were carried out by a Vialux® in-process camera system. One deep-drawn workpiece with the mesh grid on its surface as well as the steps of the surface generation after deformation in the AutoGrid® software are presented in Fig. 6.

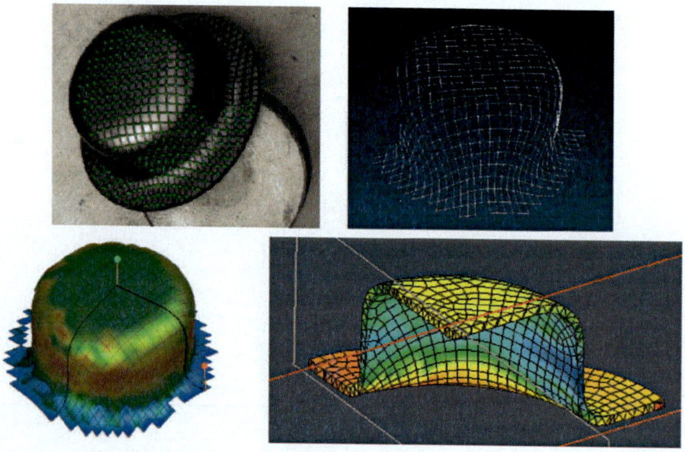

Fig. 6. Deep-drawn workpiece with the mesh grid on the surface, as well as the deformation field generation in the practice and in the modeled workpiece.

The drawing process simulation was built up in the Simufact Forming® software, and the thinning was investigated in the cross sections coincident with, and perpendicular to the rolling direction (Fig. 6). The initial element size used to set up the simulation was 1.4 mm, which came along with three layers in the thickness direction. This resulted in a total of 6696 initial elements. The coefficient of friction between the sheet and the die was counted as 0.05, while the value of 0.20 was used between the sheet and the punch. This was because of the punch-side of the blank was un-lubricated.

To specify the most important parameters, systematic modeling series based on the Taguchi L27 (3^8) design were executed with the changing of the input variables. Besides the hardening parameters (Swift-law), the initial sheet thickness, the elasticity modulus, the flow stress data, and the R-value (r_0 and r_{90}) were varied, too. The initially applied yield criterion was the Hill'48 theory, since the plane stress values calculated by this and by Abspoel et al. [15] have shown a good consistency. This is not possible to go into this in detail due to the limited content of the paper, but based on our first observations and at first glance, none of the hardening parameters have more of an impact on thinning than other.

4 Conclusion

The capability of multi-layered specimens in plane strain compression test was investigated in this study. Our results show that both the friction and the shear contribution can be reduced by using stacked specimens and rounded-edge tools at the same time. The deformation homogeneity can be developed even if efficient lubrication is applied, and an approximately two-times higher deformation is available with the increased sheet thickness. However, the obtained results are sensitive to the compensation methodology resulted from the plane strain conditions. To clarify the relevance of the compensation strategy through different yield criteria, a case study was performed on the thinning tendency of a DC04 steel sheet, using numerical simulations. The variation of the input parameters was designed according to the Taguchi theorem. Based on our first observations, the Swift-law hardening parameters have similar significance to each other as well as other variables, like the R-value or the thickness fluctuation do not show outstanding results, too.

References

1. Lange, K.: Handbook of Metal Forming. SME, Dearborn, Michigan, USA (1985)
2. Osakada, K., Kawasaki, T., Mori, K.: A method of determining flow stress under forming conditions. Ann. CRIP **30**(1), 135–138 (1981)
3. Slota, J., Spisak, E.: Determination of flow stress by hydraulic bulge test. Metalurgija **47**(1), 13–17 (2008)
4. Pearce, R.: Sheet metal forming. AdamHilger Series on new manufacturing processes and materials. IOP Publising Ltd. (1991)
5. Yabe, S., Terano, M., Yoshino, M.: Plane strain compression test and simple shear test of single crystal pure iron. Proc. Eng. **81**, 1342–1347 (2014). https://doi.org/10.1016/j.proeng.2014.10.154
6. Masaki, K., et al.: Experimental simulation of recrystallized microstructure in friction stir welded Al alloy using a plane-strain compression test. Scripta Mater. **58**, 355–360 (2008)
7. Chermette, C., Unruh, K., Peshekhodov, I., Chottin, J., Balan, T.: A new analytical method for determination of the flow curve for high-strength sheet steels using the plane strain compression test. Int.J. Mater. Form. **13**(2), 269–292 (2019). https://doi.org/10.1007/s12289-019-01485-4
8. Dorel Banabic, H.-J., Bunge, K., Pöhlandt, A.E., Banabic, D. (eds.) Formability of Metallic Materials. Springer, Berlin (2000). https://doi.org/10.1007/978-3-662-04013-3
9. Merklein, M., Kuppert, A.: A method for the layer compression test considering the anisotropic material behavior. Int. J. Mater. Form. **2**(1), 483–486 (2009)
10. Coppieters, S., et al.: Influence of hydrostatic pressure shift on the flow stress in sheet metal. Proc. Manuf. **47**, 1245–1249 (2020)
11. Coppieters, S., et al.: Determination of the flow stress and contact friction of sheet metal in a multi-layered upsetting test. J. Mater. Process. Technol. **210**, 1290–1296 (2010)

12. Coppieters, S., et al.: Large strain flow curve identification for sheet metal. J. Mater. Process. Tech. **308**, 117725 (2022)
13. Hill, R.: A theory of the yielding and plastic flow of anisotropic metals. Hydrodyn. Non-Newton. Fluids **I**, 281–297 (1947)
14. Kraus, M., Lenzen, M., Merklein, M.: Contact pressure-dependent friction characterization by using a single sheet metal compression test. Wear **476**, 1203679 (2021)
15. Abspoel, M., et al.: A new method for predicting advanced yield criteria input parameters from mechanical properties. J. Mater. Process. Tech **248**, 161–177 (2017). https://doi.org/10.1016/j.jmatprotec.2017.05.006

Measurement and Analysis of Elasto-Plastic Deformation Characteristics of Aluminum Alloy Sheet Subjected to Non-linear Stress Paths Without Intermediate Elastic Unloading

Shunsuke Asari(✉) and Toshihiko Kuwabara

Tokyo University of Agriculture and Technology, 2-24-16 Naka-Cho, Koganei-shi, 184-8588 Tokyo, Japan

asari@st.go.tuat.ac.jp

Abstract. The objective of this study is to investigate whether the material model identified using linear stress path (LSP) experiment can predict the deformation behavior of the material subjected to nonlinear stress paths (NLSPs). First, many LSPs are applied to a test sample (6000-series aluminum alloy sheet A6116-T4) with a nominal thickness of 1.1 mm to measure the contours of plastic work and the directions of the plastic strain rates, $\mathbf{D}^p$. Second, proper yield functions that reproduce the stress points forming counters of plastic work and the directions of $\mathbf{D}^p$ are determined. Third, a NLSP is applied to the test sample; the NLSP consists of several linear stress paths without intermediate elastic unloading. Forth, the elasto-plastic deformation behavior of the test sample subjected to the NLSP is compared with that predicted using the yield function determined from the LSP experiment. It is concluded that although the measured $\mathbf{D}^p$ slightly deviated from those predicted from the normality flow rule, the deformation behavior of the test sample predicted for the NLSP using the model determined from the LSP experiment is generally consistent with the measurement within the experimental conditions of the present study.

Keywords: Aluminum Alloy Sheet · Biaxial Stress Test · Non-linear Stress Path · Yield Function · Material Modeling · Normality Flow Rule

1 Introduction

The application of aluminum alloy sheets to transportation components is expanding to reduce the body weight. However, aluminum alloy sheets are less ductile and have smaller Young's modulus than mild steel sheets [1]. Therefore, forming defects, such as fracture and springback, frequently hinder efficient production. In order to predict defects in advance and quickly determine appropriate process design, it is necessary to perform accurate forming simulation. In press forming, materials are subjected to nonlinear stress paths (NLSPs). Therefore, applying a material model that is robust for NLSPs is necessary to improve the accuracy of forming simulations. Material models

K. Mocellin et al. (Eds.): ICTP 2023, LNME, pp. 373–380, 2024.
https://doi.org/10.1007/978-3-031-40920-2_39

are usually determined from linear stress path (LSP) test [2–4]. However, there are few experimental studies [5–9] that investigated how accurately a material model determined from LSP test can reproduce the deformation behavior of the material subjected to NLSPs.

The objective of this study is to clarify whether the deformation behavior of a test sample (A6116-T4; nominal thickness: 1.1 mm) under a NLSP can be reproduced using the material model determined from a LSP experiment. First, LSP tests are conducted using uniaxial tensile tests and multiaxial tube expansion tests [10] to create a material model. Next, a NLSP test is performed to compare the experimental data of the elasto-plastic deformation behavior of the test sample with that calculated using the material model identified from the LSP test. And the predictability of the deformation behavior of the specimen in NLSP using the material model determined from the LSP test is verified.

2 Experimental Methods

Hereafter, the rolling (RD), transverse (TD), and thickness directions (ND) of the material are referred to as the x-, y-, and z-axes, respectively. First, the material model was identified using LSP tests with σ_x:σ_y = 0:1, 1:4, 1:2, 3:4, 1:1, 4:3, 2:1, 4:1, and 1:0. The σ_x:σ_y = 0:1 and 1:0 tests were performed using uniaxial tensile test specimens

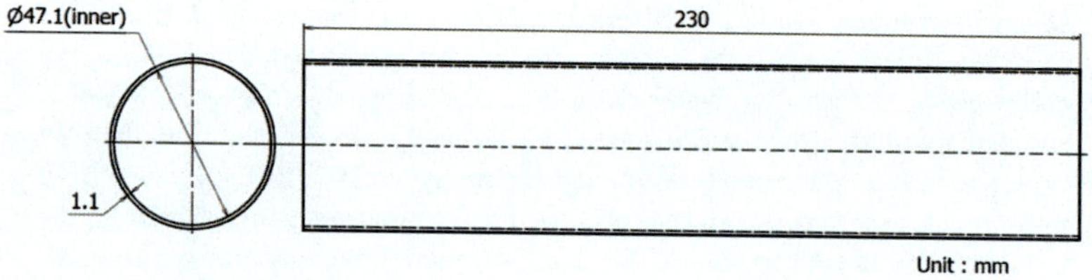

Fig. 1. Sample geometry of specimens used for the BTET.

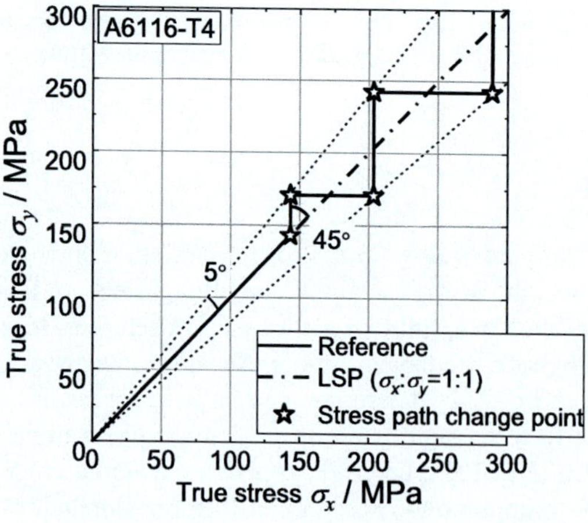

Fig. 2. Non-linear stress path applied to the test sample.

(JIS 13B). For other stress paths the biaxial tube expansion test (BTET) method was used. The geometry of the tubular specimen used in the BTET is shown in Fig. 1. The specimens were fabricated by uniformly bending a sheet sample and welding the sheet edges. The specimens were annealed so that premature rupture in the heat-affected zone could be suppressed. For details of the BTET method including test apparatus, see Ref. 10). The biaxial true stress-plastic strain curves were used to determine a proper material model for the test sample.

Next, a NLSP test was performed by applying the nonlinear stress path shown in Fig. 2 to a tubular specimen. The first stress path was the equibiaxial tension $((\sigma_x, \sigma_y) = 1{:}1)$ up to a stress point of $(\sigma_x, \sigma_y) = (143, 143)$ MPa, which corresponds to a von Mises equivalent plastic strain of 0.005. Then, the stress increment, $d\sigma_x = d\sigma_y$, was abruptly changed without intermediate elastic unloading to $(0, d\sigma_y)$, successively followed by $(d\sigma_x, 0)$, $(0, d\sigma_y)$, and $(d\sigma_x, 0)$. The amplitude of the direction of stress vector, $\varphi(= \tan^{-1}(d\sigma_x/\, d\sigma_y))$, during the NLSP test was $\pm 5°$ with respect to the stress path of $(\sigma_x{:}\sigma_y) = 1{:}1$. The star marks in Fig. 2 indicate the positions of abrupt stress path change.

To evaluate the elasto-plastic deformation behavior of the test sample under the linear and nonlinear stress paths, counters of plastic work and the direction of the plastic strain rate, $\mathbf{D}^{\mathrm{p}}$, were measured. $\mathbf{D}^{\mathrm{p}}$ was calculated as an average of the instantaneous gradient, $d\varepsilon_y^{\mathrm{p}}(t)/d\varepsilon_x^{\mathrm{p}}(t)$, for a time interval of 3 s.

3 Experimental Results

3.1 LSP Experiments and Material Modeling

The stress points forming counters of plastic work measured in the LSP test are shown in Fig. 3, and In Fig. 3 a group of stress points forming a work counter corresponding to $\varepsilon_0^{\mathrm{p}}$ were normalized by the uniaxial true stress σ_0 corresponding to $\varepsilon_0^{\mathrm{p}}$. Work contours up to $\varepsilon_0^{\mathrm{p}} = 0.092$ could be measured for all LSPs. As $\varepsilon_0^{\mathrm{p}}$ evolves, the normalized work contours changed in shape from a von Mises-type yield locus to a Tresca-type yield locus. Thus, the material exhibited differential hardening (DH).

Figure 4 shows the measured directions of $\mathbf{D}^{\mathrm{p}}$, β. β was almost constant during every LSP test with the increase of $\varepsilon_0^{\mathrm{p}}$; the maximum variation in β was 1.8° for $\sigma_x : \sigma_y = 4{:}3$.

The yield loci and β calculated using the von Mises and the Yld2000-2d [12] yield functions are also shown in Fig. 3 and Fig. 4, respectively. The Yld2000-2d yield function for isotropic hardening (IH) model was determined from the measured data for $\varepsilon_0^{\mathrm{p}} = 0.092$; the exponent M and parameters $\alpha_i (i = 1-6)$ of the Yld2000-2d yield function were determined to minimize the mean square error between the experimental and calculated values of the work contour and β using the genetic algorithm (GA) method [11].

The procedure for determining the DH model is as follows [11]. First, the values of M and $\alpha_i (i = 1-6)$ of the Yld2000-2d yield function were determined to minimize the mean square error between the experimental and calculated values of the work contour and β using GA for every $\varepsilon_0^{\mathrm{p}}$ for $0.005 \le \varepsilon_0^{\mathrm{p}} \le 0.09$ with a strain increment of $\Delta\varepsilon_0^{\mathrm{p}} = 0.005$. Next, the change in M with the increase of $\varepsilon_0^{\mathrm{p}}$ was approximated using Eq. (1), and $\alpha_i (i = 1-6)$ was determined by Levenberg-Marquardt Method for every $\varepsilon_0^{\mathrm{p}}$ using the approximated values of M. The resulting variation of α_i with the increase of ε_0^p was also approximated using Eq. (1).

$$M, \alpha_i = A(\varepsilon_0^p)^2 + B(\varepsilon_0^p) + C + D\exp(-E\varepsilon_0^p) \tag{1}$$

where, A, B, C, and D are the parameters to be determined. The evolution of the work contours with the increase of ε_0^p was accurately reproduced by the DH model based on the Yld2000-2d yield function. β was nearly constant during each LSP test with the increase of ε_0^p. As is clearly seen in Figs. 2 and 3, it was confirmed that the work contours and β reproduced by the IH and DH models based on the Yld2000-2d yield function were consistent with the measurement.

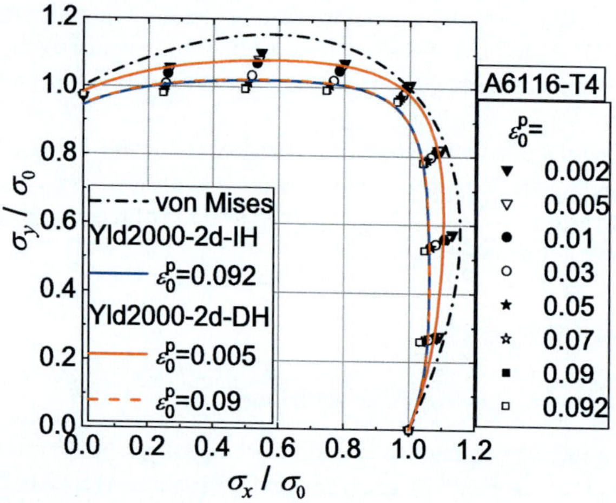

Fig. 3. Stress points forming counters of plastic work, normalized by the σ_0 belonging to respective work contours, compared with the yield loci calculated using selected yield functions.

3.2 NLSP Experiments

To evaluate the work contours for the NLSP, the experimental and calculated values of the stress vector φ with the increase of ε_0^p are shown in Fig. 5(a). The maximum deviation of the calculation from the experiment is 2.2° ($\varepsilon_0^p = 0.04$) for the IH model and 1.8° ($\varepsilon_0^p = 0.019$) for the DH model. Therefore, the Yld2000-2d yield function determined considering the DH of the test sample (hereafter, the Yld2000-2d (DH)) has more accurate reproducibility of the work hardening behavior of the test sample subjected to the NLSP than that determined assuming IH (hereafter, the Yld2000-2d (IH)).

Figure 5(b) the experimental and calculated values of β with the increase of ε_0^p for the NLSP. Again, the prediction using the Yld2000-2d (DH) yield function was closer to the experiment than the Yld2000-2d (IH); the maximum deviation of the prediction from the experiment was 20.1° at $\varepsilon_0^p = 0.023$ for the Yld2000-2d (IH) and 12.3° at

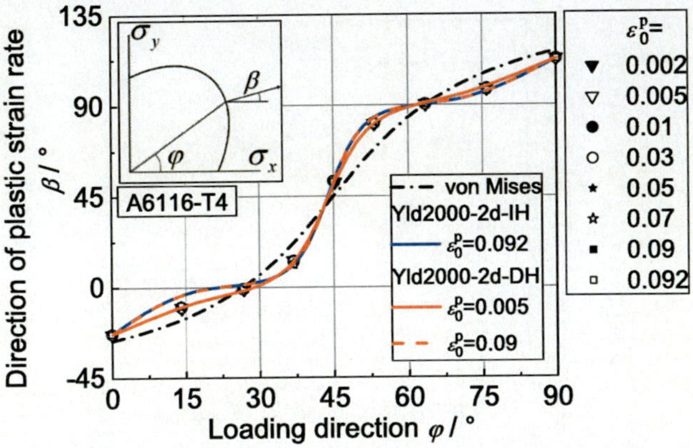

Fig. 4. Measured directions of $\mathbf{D}^p$ compared with those calculated using selected yield functions.

$\varepsilon_0^p = 0.08$) for the Yld2000-2d (DH). For the IH model the maximum deviation from the experiment occurred at an early stage of plastic deformation because the IH model was determined using the measured data at a larger plastic strain of $\varepsilon_0^p = 0.092$.

Figure 6 (a) shows the deviation, $d\beta$, of the measured β shown in Fig. 5 (b) from those calculated using the Yld2000-2d (DH) and (IH). The calculation of β was based on the normality flow rule with respect to the yield locus. $d\beta$ is defined as positive when the experimental $\mathbf{D}^p$ inclines toward the moving direction of the stress point. β in the NLSP test was calculated as an average of the instantaneous gradient, $d\varepsilon_y^p(t)/d\varepsilon_x^p(t)$, for a time interval of 3 s. Therefore, the data of the $d\beta$ immediately before each abrupt stress path change point in Fig. 6(a) are missing.

It is clearly seen in Fig. 6 (a) that $d\beta$ was negative immediately after the stress path change points. That means that the measured $\mathbf{D}^p$ was inclined toward the opposite direction of the stress point movement. Then, as ε_0^p increased by 0.002–0.005, $d\beta$ rapidly increased to become positive, and then gradually decreased; β got closer to the direction predicted by the normality flow rule. The Yld2000-2d (IH) yield function had a larger deviation from the normality flow rule than the Yld2000-2d (DH). The maximum values of $d\beta$ were 18.4° for the IH model and 10.5° for the DH model.

In summary, the work contours in the NLSP were generally predictable by the material models identified from the LSP test results. The Yld2000-2d (DH) yield function was superior in reproducibility to the Yld2000-2d (IH) yield function under the range of experimental conditions. On the other hand, the measured $\mathbf{D}^p$ in the NLSP slightly deviated from the normality flow rule; the deviation was larger immediately after the stress path change than in the later stage. The behavior of $\mathbf{D}^p$ is similar to those observed in the abrupt strain path change experiments by other authors [6, 7].

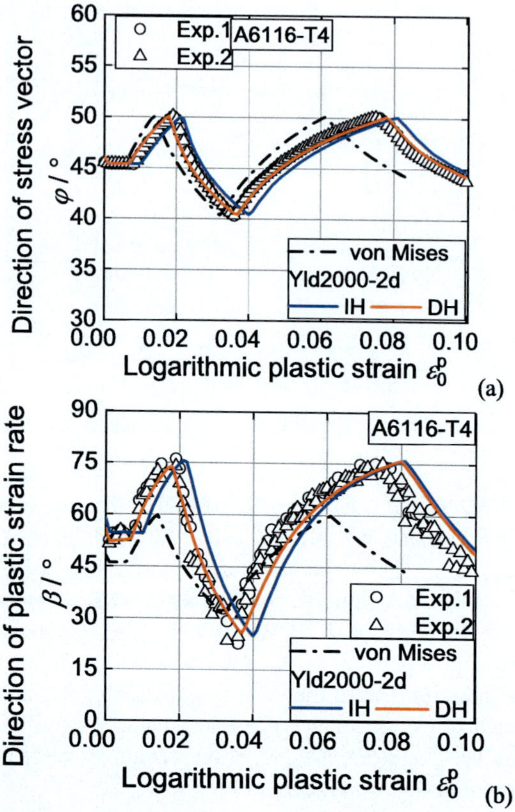

Fig. 5. Measured directions of (a) stress vector and (b) plastic strains compared with those predicted using the selected yield functions.

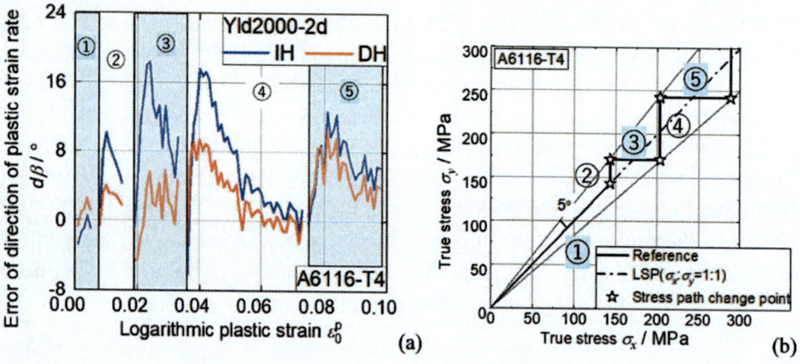

Fig. 6. Deviation of the $\mathbf{D}^{\mathbf{p}}$, $d\boldsymbol{\beta}$, from the normality flow rule in the NLSP experiment.

4 Conclusions

A LSP experiment was performed using uniaxial and BTET, with a 6000-series aluminum alloy sheet (A6116-T4, nominal thickness: 1.1 mm), to precisely measure the contours of plastic work and the directions of the plastic strain rates, β. Using the Yld2000-2d yield function, a material model that accurately reproduces the experimental data was identified. Furthermore, a NLSP experiment without intermediate elastic unloading was performed to compare the measured work hardening behavior and β with those calculated using the material models identified from the LSP experiment. The following findings were obtained.

(1) The contours of plastic work of the test sample subjected to LSPS changed in shape; the test sample exhibited DH. The deformation behavior was well reproduced by the Yld2000-2d yield function (DH).
(2) The work hardening behavior measured in the NLSP experiment was compared with that calculated using the material models determined from the LSP experiment. We found that the Yld2000-2d yield function (DH) accurately reproduced the experimental data.
(3) The Yld2000-2d (IH) yield function had a larger deviation from the normality flow rule than the Yld2000-2d (DH). The maximum values of $d\beta$ were 18.4° for the IH model and 10.5° for the DH model.
(4) Further research is needed to elucidate the application limits of the material model determined from LSP experiments to NLSPs.

References

1. Hu, J., Ishikawa, T.: Texture and plastic anisotropy in aluminum alloy sheet. J. Japan Inst. Light Metals **48**(11), 144–150 (1998)
2. Kuwabara, T.: Advances in experiments on metal sheets and tubes in support of constitutive modeling and forming simulations. Int. J. Plast. **23**(3), 385–419 (2007)
3. Barlat, F., Kuwabara, T.: Anisotropic yield conditions in mathematical theory of plasticity. J. JSTP **57**(662), 230–243 (2016)
4. Banabic, D., Barlat, F., Cazacu, O., Kuwabara, T.: Advances in anisotropy of plastic behaviour and formability of sheet metals. Int. J. Mater. Form. **13**(5), 749–787 (2020). https://doi.org/10.1007/s12289-020-01580-x
5. Hecker, S.S.: Experimental investigation of corners in the yield surface. Acta Mech. **13**, 69–86 (1972)
6. Kuwabara, T., Kuroda, M., Tvergaard, V., Nomura, K.: Use of abrupt strain path change for determining subsequent yield surface: experimental study with metal sheets. Acta Mater. **48**(9), 2071–2079 (2000)
7. Yoshida, K., Tsuchimoto, T.: Plastic flow of thin-walled tubes under nonlinear tension-torsion loading paths and an improved pseudo-corner model. Int. J. Plast. **104**, 214–229 (2018)
8. Hama, T., Yagi, S., Tatsukawa, K., Maeda, Y., Maeda, Y., Takuda, H.: Evolution of plastic deformation behavior upon strain-path changes in an A6022–T4 Al alloy sheet. Int. J. Plast. **137**(102913), 1–23 (2021)
9. Takada, Y., Kuwabara, T.: Nonlinear biaxial tensile stress path experiment without intermediate elastic unloading for validation of material model. Int. J. Solids Struct. **257**(5), 1–9 (2022)

10. Kuwabara, T., Sugawara, F.: Multiaxial tube expansion test method for measurement of sheet metal deformation behavior under biaxial tension for a large strain range. Int. J. Plast. **45**, 103–118 (2013)
11. Hakoyama, T., Kuwabara, T.: Effect of biaxial work hardening modeling for sheet metals on the accuracy of forming limit analyses using the Marciniak-Kuczynski approach. In: Altenbach, H., Matsuda, T., Okumura, D. (eds.) From Creep Damage Mechanics to Homogenization Methods, pp. 67–95. Springer, Switzerland (2015)
12. Barlat, F., et al.: Plane stress yield function for aluminum alloy sheets–part 1: theory. Int. J. Plast. **19**(9), 1297–1319 (2003)

Characterization of the Flow Behavior of a Ti6Al4V Alloy Considering the Temperature Change of the Specimen During Cylinder Compression

Su Min Ji[1], Mohd Kaswandee Razali[2], Jeong Muk Choi[3], and Man Soo Joun[2(✉)]

[1] School of Mechanical and Aerospace Engineering, Gyeongsang National University, Jinju 52828, South Korea

[2] Engineering Research Institute, School of Mechanical and Aerospace Engineering, Gyeongsang National University, Jinju 52828, South Korea

msjoun@gnu.ac.kr

[3] Jinhap Co. Ltd., Daejeon 34302, South Korea

Abstract. There are several methods of revealing temperature effect in hot cylinder compression test on actual flow curves. However, few methods were verified by the direct comparison between the experimental and predicted compression load-stroke curves. In this study, the systematic and scientific method of obtaining the flow curves formulated by the generalized *C–m* model considering the actual test environments is applied to the temperature-sensitive Ti6Al4V alloy in the temperature range of warm forming. Finite element analyses of the cylinder compression tests are conducted at various sample temperatures and strain rates using the fitted flow function, revealing that the predicted compression load-stroke curves are consistent with the experimental results.

Keywords: Titanium alloy · Cylinder compression · Flow curve · Temperature effect · Temperature-compensation

1 Introduction

The Ti6Al4V alloy has been widely utilized in the aerospace, defense, energy, and chemical industries because of its high specific strength, good mechanical properties at elevated temperatures, and high resistance to corrosion and creep [1, 2]. However, it is characterized by low heat capacity, low thermal conductivity, and high flow stress. The alloy thus undergoes remarkable temperature changes during its forming processes [3, 4] which may cause severe macroscopic instability of deformation [5, 6], especially at high strain rate [7]. Accurate acquisition of the flow behavior of the alloy considering the actual material test environments is thus essential. Particularly, for metal forming of the Ti6Al4V alloy, accurate flow characterization is of great significance since flow stress is also highly temperature-sensitive. Flow behavior greatly influences microstructural evolution and macroscopic flow phenomena, which govern failure in metal forming (e.g., the failure of a bolt near its head) [8, 9].

K. Mocellin et al. (Eds.): ICTP 2023, LNME, pp. 381–393, 2024.
https://doi.org/10.1007/978-3-031-40920-2_40

A few researchers have studied the flow characterization of the Ti6Al4V alloy at the temperature range (550–1050 °C) for its warm forming and at the various sample strain rates. In most previous studies of Ti6Al4V alloy characterization, the strain rates and temperatures investigated did not adequately represent the actual industrial conditions, although Majorell et al. [10] and Ji et al. [9] covered a wide temperature range for its warm forming considering the temperature effect on flow behavior with non-optimized flow basis functions. Even though Luan et al. [11] employed a systematic way of compensating for both friction and temperature during flow characterization, they did not apply it to the material with great flow stress and low thermal conductivity like the Ti6Al4V alloy.

Recently, Ji et al. [12] obtained the temperature-compensated flow curve (TCFC) of the Ti6Al4V alloy at the warm forming temperature range using some improved schemes. In this study, the flow characterization schemes are introduced with an emphasis on the flow basis functions. Improved TCFC curve is first verified by analyzing the cylinder compression test using the finite element method. A general-purpose implicit rigid- or elastic-thermoviscoplastic FE package AFDEX (Altair APA) [13] was used, based on the tetrahedral MINI-element scheme [14–16] with automatic adaptive remeshing [17]. The elastic deformation was assumed to be negligible.

2 Reference Flow Curves and Flow Model

2.1 Experiments and Thermal Information

The flow behavior of a Ti6Al4V alloy at warm forming temperatures (800 °C) was previously characterized using experimental flow data obtained via hot cylinder compression tests (Gleeble 3500 system) at the sample temperatures of 500, 600, 650, 700, 750, and 800 °C and strain rates of 0.1, 1.0, 5.0, 10.0, and 15.0 s^{-1}, as shown in Fig. 1 [9].

The initial specimens had a diameter of 7.7 mm and a height of 11.6 mm. The initial material, a coil rod with a diameter of 14.1 mm, was fully annealed at 750 °C. It was drawn in three consecutive drawing stages with conical dies. Each drawing stage reduced the rod area by 15% and its diameter was 11.05 mm just before final annealing and drawing. The drawn rod was annealed following Aerospace Material Specifications [18] and then drawn to 10.88 mm (a 3% diameter reduction) to improve its tolerance.

Figure 2 shows the experimental flow curves, called reference flow curves (RFCs) before temperature compensation (RFC-BTCs) exhibiting greater flow stress sensitivity to temperature than to strain rate. The strain rate effect, however, increased with temperature. At lower strain rates and temperatures and at higher strain rates and temperatures, flow behaviors look quite unstable, with excessive disordering of flow curves over a wide range of fixed strains (Fig. 2(a) and Fig. 2(d), respectively). The flow curves at strains of over 0.4 revealed extreme instability, especially at higher strain rates, because of the high strength and low thermal capacity of the alloy. As shown in Fig. 2, lower temperatures of samples were associated with earlier onsets of metallurgical instability at the large strain. Consequently, the flow curves revealed a generally narrow range of metallurgical stable forming conditions in the test range of the state variables. Note that an ambiguous and tool-dependent part of compression load at early stroke was properly eliminated (See Fig. 4 or 6 where the part was shown).

It was assumed that the friction coefficients at the interfaces between the specimens and tools were 0.3 [21]. It is noted that the friction effects in the cylinder compression tests may not be considerable because the friction-induced deformation heterogeneity causing the compression load can lead to its decrease owing to the decreased contact area between the material and tool [22]. Table 1 summarizes the thermal capacity and conductivity [20] of the titanium alloy and die or tool materials, which are used in all the finite element simulations in this study.

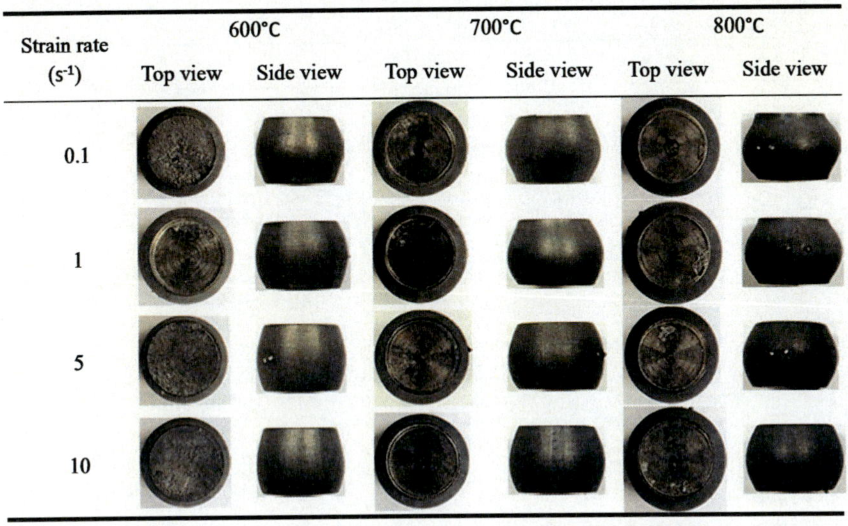

Fig. 1. Deformed specimens after cylinder compression test [9].

2.2 Flow Characterization Using the Generalized *C-m* Model

The flow patterns of the Ti6Al4V alloy are diverse and peculiar. Therefore, a traditional or even optimized flow model may not be sufficiently accurate even in the flow model case for narrow ranges of hot temperature and low strain rate. Recently, a general extended *C–m* model [23] was successfully applied to characterize several magnesium alloys with higher complexity and diversity than titanium alloys. It was thus employed to describe the experimental flow curves of the Ti6Al4V alloy as functions of state variables using a piecewise bilinear description scheme of flow parameters as follows:

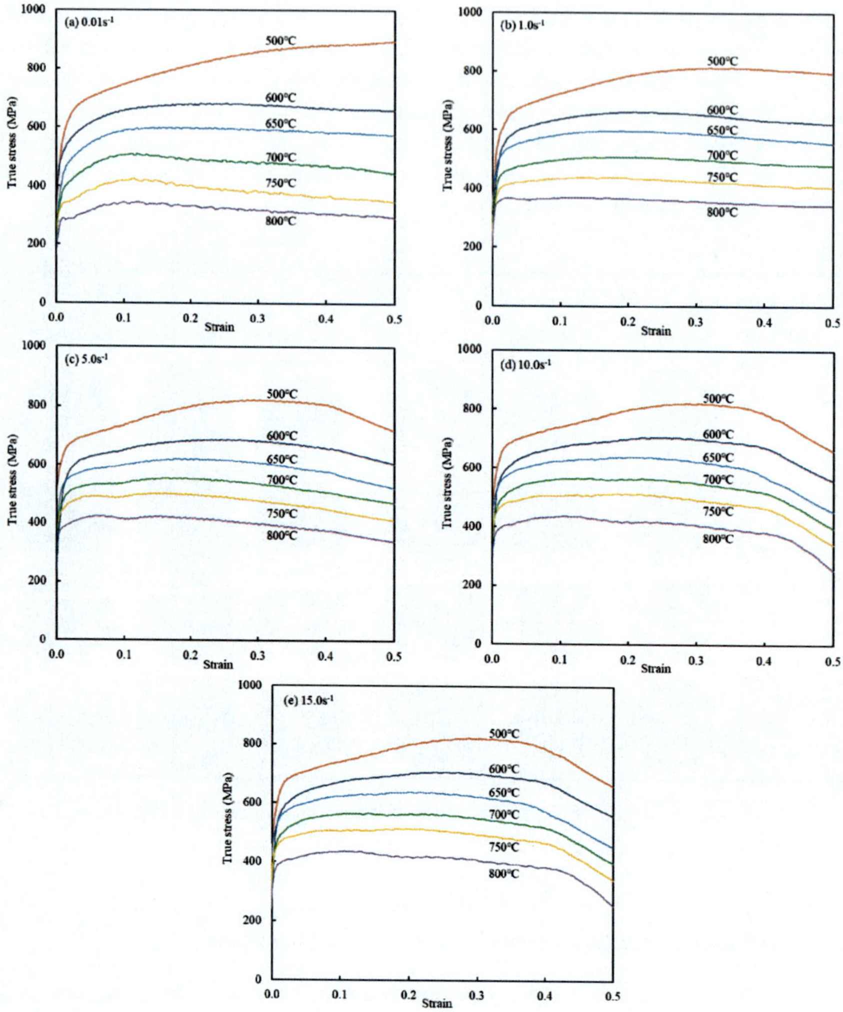

Fig. 2. Experimental flow curves

$$\sigma = C(\varepsilon, \dot{\varepsilon}, T)\dot{\varepsilon}^{m(\varepsilon, \dot{\varepsilon}, T)} \tag{1}$$

where

$$C(\varepsilon, \dot{\varepsilon}, T)C_0 + C_1 \ln(1 + \dot{\varepsilon}) \tag{2}$$

$$m(\varepsilon, \dot{\varepsilon}, T)m_0 + m_1 \ln(1 + \dot{\varepsilon}) \tag{3}$$

Table 1. Thermal properties and conditions

Initial material temperature (°C)	500–800
Heat transfer coefficient to ambient (W/mm^2 °C) [19]	2.95×10^{-6}
Heat radiation coefficient to ambient (W/mm^2 °C^4) [19]	3.97×10^{-14}
Heat transfer coefficient between material and die (W/mm^2 °C) [19]	0.01
Thermal capacity of material Ti6Al4V alloy (Ws/mm^3 °C) [20]	0.0029 (500 °C) 0.0033 (995 °C)
Thermal capacity of dies WC-24Co (Ws/mm^3 °C)	0.0050
Thermal conductivity of material Ti6Al4V alloy (W/mm °C) [20]	0.0127 (500 °C) 0.0227 (995 °C)
Thermal conductivity of dies WC-24Co (W/mm °C)	0.0950

where C_0, C_1, m_0, and m_1 are all flow constants defined at fixed sample strains and temperatures. Note that we preferred the basis flow function $\ln(1 + \dot{\varepsilon})$ to $\ln(\dot{\varepsilon})$ [9] for increased stability especially at the small strain rate.

Table 2. Flow constants of the fitted RFC-BTCs of the Ti6Al4V alloy.

T (°C)	ε	C_0	$C1$	m_0	m_1
500	0.03	668.9	−9.7	0.0007	0.0159
	0.05	695.7	−11.9	0.0006	0.0155
	0.10	736.3	−12.3	0.0000	0.0145
	0.20	800.9	−13.3	0.0005	0.0131
	0.30	837.8	−17.1	0.0000	0.0116
	0.40	855.1	−29.4	0.0000	0.0065
	0.50	790.5	0.0	−0.0531	0.0000
600	0.03	575.0	17.6	0.0000	0.0007
	0.05	609.0	1.9	0.0000	0.0090
	0.10	650.6	−12.6	0.0005	0.0159
	0.20	669.0	2.1	0.0000	0.0078
	0.30	669.6	−4.4	0.0011	0.0100
	0.40	658.8	0.0	−0.0032	0.0000
	0.50	596.0	0.0	−0.0430	0.0000
700	0.03	437.6	46.2	0.0097	−0.0103
	0.05	458.9	37.5	0.0013	−0.0022

(continued)

Table 2. (*continued*)

T (°C)	ε	C_0	$C1$	m_0	m_1
	0.10	506.4	−4.1	0.0010	0.0205
	0.20	491.8	28.1	0.0032	0.0006
	0.30	489.9	15.5	0.0130	0.0023
	0.40	475.9	13.7	0.0075	−0.0006
	0.50	471.0	14.5	0.0333	−0.0474
800	0.03	333.4	44.6	0.0632	−0.0406
	0.05	298.8	94.5	0.0000	−0.0406
	0.10	332.4	57.0	0.0000	−0.0173
	0.20	316.3	79.6	0.0000	−0.0365
	0.30	303.9	81.4	0.0000	−0.0402
	0.40	293.2	77.6	0.0000	−0.0440
	0.50	268.7	116.0	0.0000	−0.1298

All the flow constants were defined at each sample strain and temperature and calculated using an optimization technique [9], minimizing the squared errors between the fitted and experimental RFC-BTCs. They were listed in Table 2. The calculated flow stresses at the sample points are marked on the flow curves. These flow curves are compared with the experimental results in Fig. 3, where the dots represent the fitted flow stresses at their associated sample points. The calculated results are consistent with the experimental results. The mean absolute relative error (%) and root-mean-square error (MPa) are 1.3 and 4.3, respectively.

Figure 3 shows that the dashed flow curves (i.e., the fitted RFC-BTCs) are very close to the experimental flow curves. However, the FE predictions by the RFC-BTCs exhibit some substantial difference from the experimental compression load-stroke curves (CLSCs), as shown in Fig. 4. It should be noted that as the sample strain rate increases, the difference becomes more remarkable. This fact implies that the heat conduction has a certain relationship with the difference.

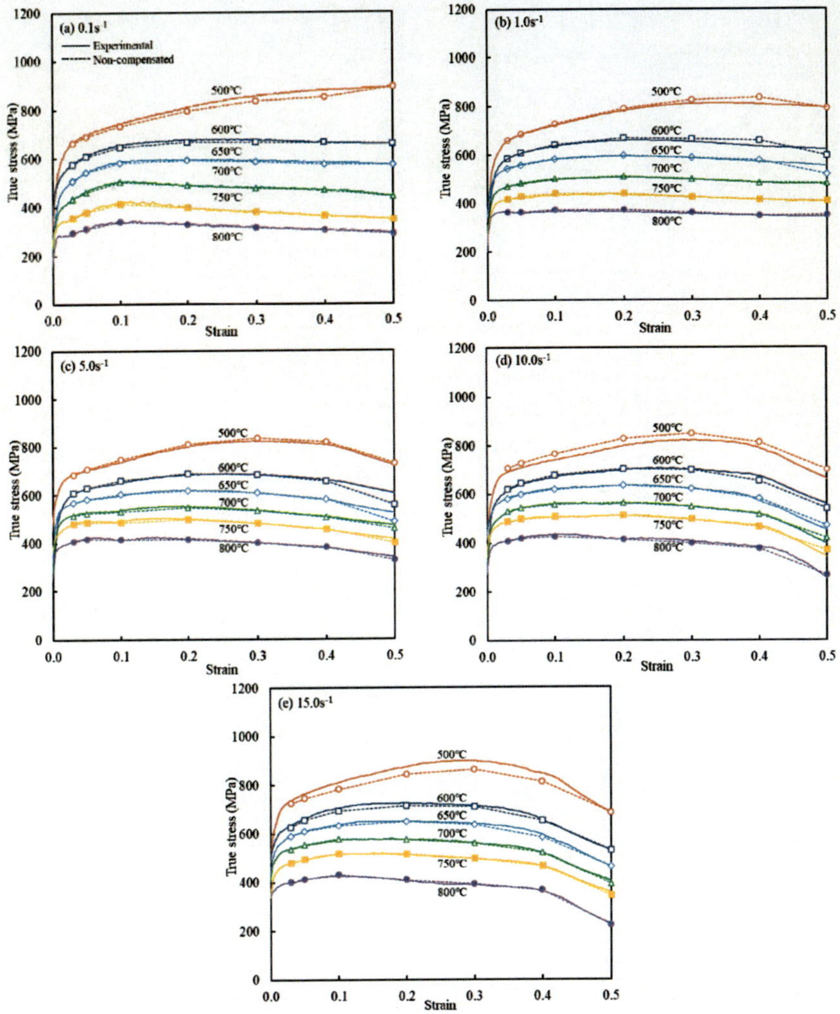

Fig. 3. Comparisons of non-compensated flow curves with the RFCs.

Figure 3 shows that the Ti6Al4V alloy does not exhibit strong flow stress dependence on strain. However, all flow curves (except those at small strain rates and 500 °C) exhibit distinct peak stress points over a relatively wide range of strains.

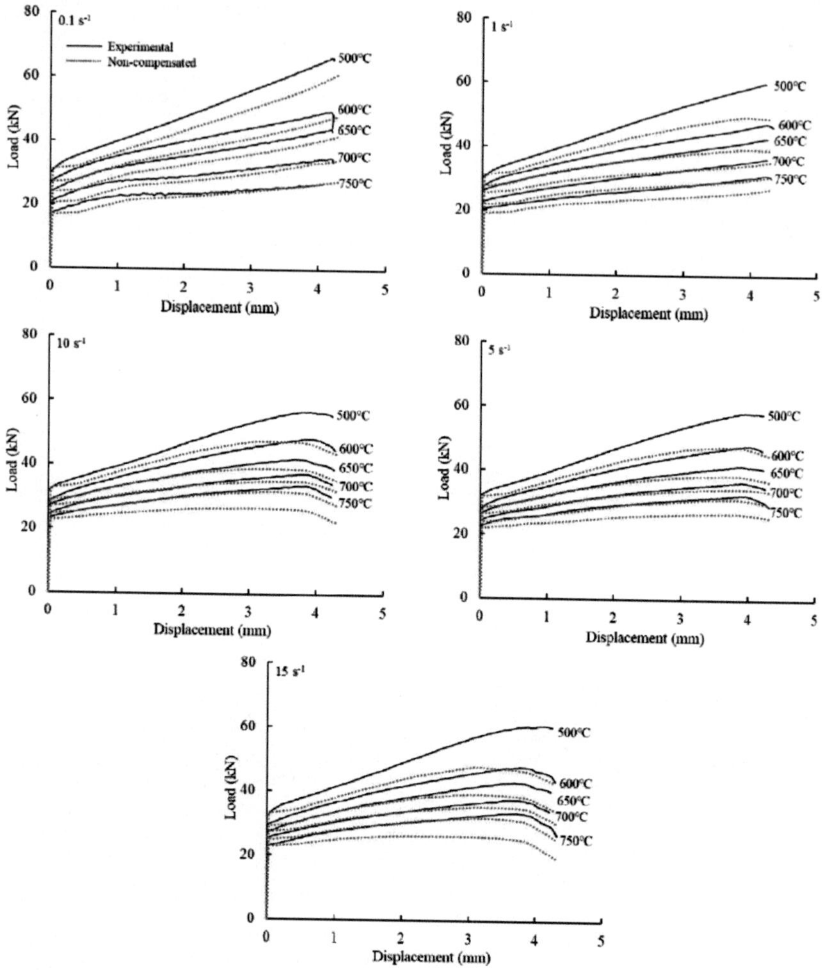

Fig. 4. Comparison of the predicted and experimental compression load–stroke curves.

2.3 Flow Characterization with an Emphasis on Temperature Compensation

The flow curves of titanium alloys in the temperature range of warm or hot forming with a wide range of strain rate depend sensitively on the temperature and strain rate because of the low thermal conductivity and small ratio of heat capacity to flow stress. The experimental flow information does not reflect any viscous heating and a kind of temperature compensation is thus needed.

The flow stress at a point on the RFC-BTCs does not reflect any increase in temperature caused by energy dissipation owing to the plastic work. This emphasizes the need for temperature compensation of RFCs. The experimental flow curves underestimate the actual flow stresses at large strains. The errors increase as the strain and strain rate increase. To cope with this matter, Joun et al. [22] presented the temperature-compensation method using a generalized *C-m* flow model (See Eqs. (1), (2) and (3)) where the C_{ij} and m_{ij}-values defined at the *i*-th sample strain and *j*-th sample temperature, that is, the flow constants of the generalized *C-m* model are systematically modified using the approximately estimated temperature increase.

When the sample temperatures and strains are systematically selected during the temperature compensation, the temperature increase at the *i*-th sample strain ε_i and *j*-th sample temperature T_j (ΔT_{ij}) is approximated, using the energy dissipation of the material, as follows:

$$\Delta T_{ij} = \int_0^{\varepsilon_i} \frac{\xi\sigma}{\rho c} d\varepsilon \approx \xi \frac{\overline{\sigma}\varepsilon_i}{\overline{\rho c}} \tag{4}$$

where ρc ($\overline{\rho c}$), σ $\overline{\sigma}$ and ξ are the volumetric thermal capacity, flow stress (their mean values) and instantaneous temperature correction parameter (TCP), respectively.

The TCP, ξ, is dependent on the unit system employed, the ratio of energy dissipation to the stress power, the strain rate, and the thermal conductivity. It means the fraction of the plastic work affecting the temperature rise at its generated spot. It thus depends much on the test speed and thermal conductivity of the material. Notably, the TCP is approximately 0.9 not only when $[\sigma]$ is provided in MPa and $[\overline{\rho c}]$ is provided in J·mm^{-3}·°C^{-1} but also when the strain rate is high and the thermal conductivity is low. To the contrary, it may approach to zero when the material should be heated during the compression test at low speed to maintain the selected sample temperature.

The fitted RFC-BTCs employed for FE predictions were calculated under an isothermal assumption. After several trials, the TCPs were set as follows:

(1) $\xi = 0$ for a strain rate of 0.1 s^{-1}.
(2) $\xi = 0.7, 0.8, 0.9$, and 0.9 for strain rates of 1, 5, 10, and 20 s^{-1}, respectively.

Note that the above TCP values indicate the existence of a critical strain rate at which temperature softening accelerates; this rate is less than 1 s^{-1}. The temperature compensation scheme was used when drawing the compensated flow curves of Fig. 5, shown with the experimental and fitted RFC-BTCs. Notably, the compensated flow stresses are much higher than the non-compensated stresses, particularly at large strains (except at small strain rates).

For example, the temperature increase of 120 °C was calculated when the strain at the specimen center reaches 0.6 during the cylinder compression test at an initial temperature of 500 °C and strain rate of 1 s^{-1} (TCP = 0.8), emphasizing that the temperature rise during the compression test is so great in the case of the titanium alloy.

Figure 6 indicates that the RFC-AFCs predicted quite accurate CLSCs compared to the experiments. The maximum and mean errors of the predicted CLSCs are 5.2% (at the 0.1 s^{-1}-strain rate and 500 °C-temperature) and 2.1% (Table 3), respectively.

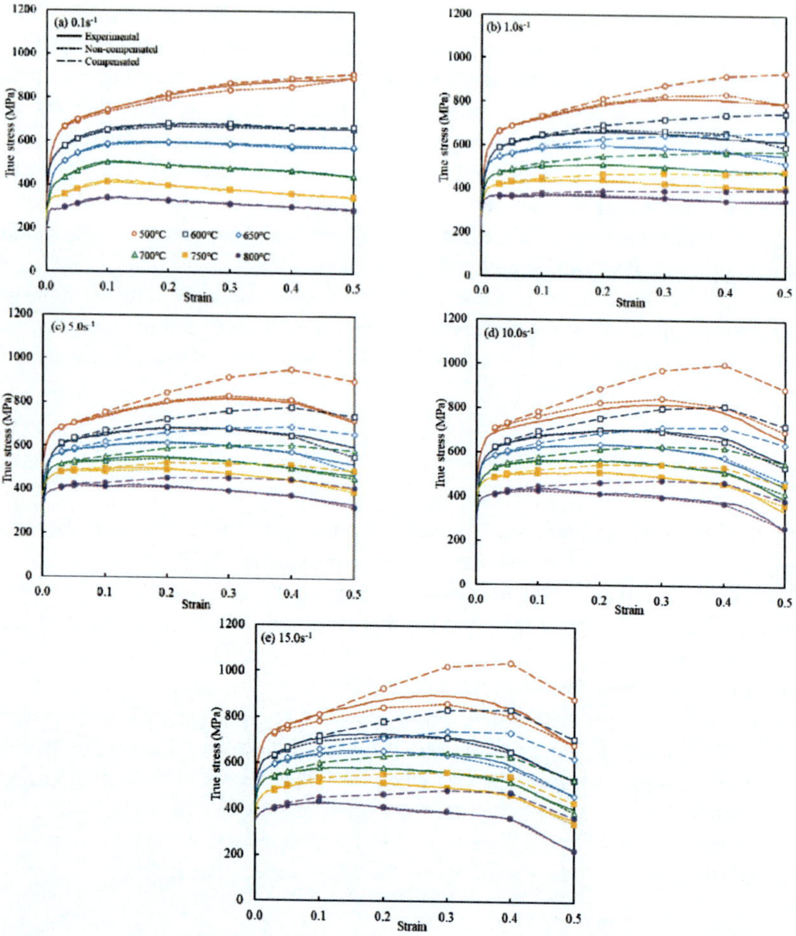

Fig. 5. Comparison non-compensated and compensated flow curves with the RFCs.

Table 3. Errors of predicted CLSCs relative to experiments

$\dot{\varepsilon}$ (s^{-1})	500 °C	600 °C	650 °C	700 °C	750 °C	Mean
0.1	4.5%	3.1%	2.8%	2.0%	1.9%	2.9%
1	3.9%	2.5%	2.4%	2.1%	1.5%	2.5%
5	2.4%	2.1%	2.0%	1.8%	1.3%	1.9%
10	3.0%	1.0%	1.8%	1.4%	1.2%	1.7%
15	2.9%	1.2%	1.7%	1.2%	1.0%	1.6%
Mean	3.3%	2.0%	2.1%	1.7%	1.4%	2.1%

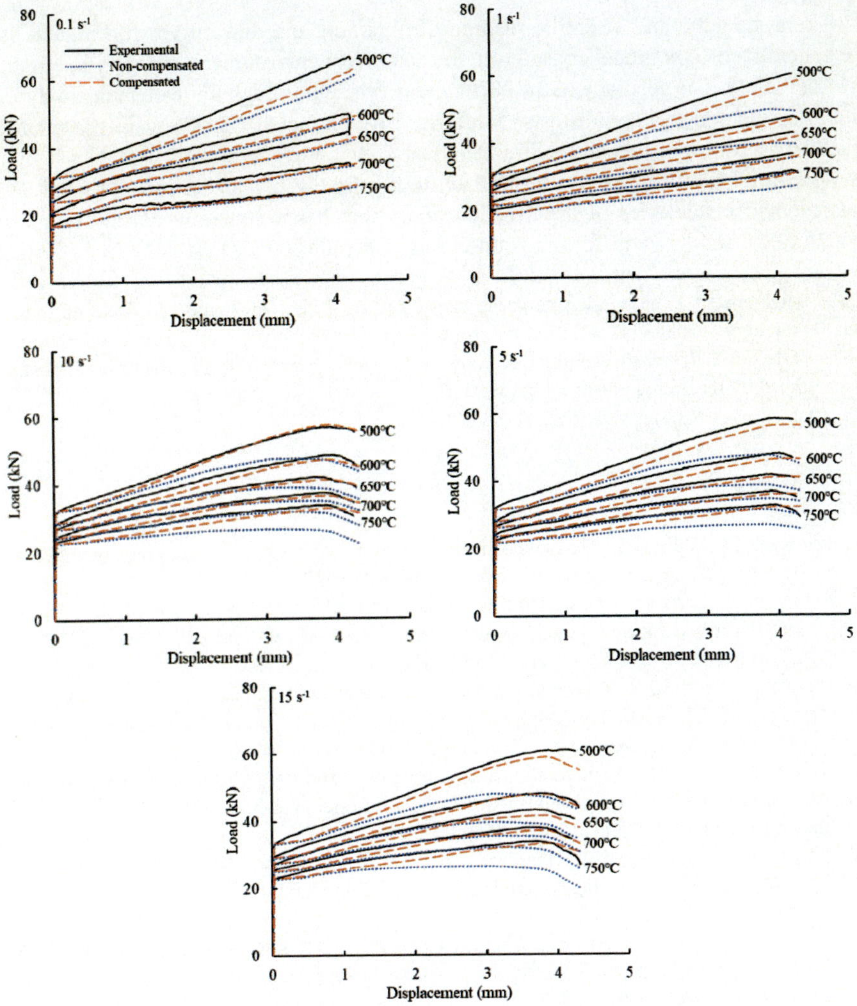

Fig. 6. Comparison of the predicted and experimental compression load–stroke curves.

3 Conclusion

The detailed procedure of characterizing the flow behavior of a Ti6Al4V alloy in the warm forming temperature range (500–800 °C) and moderate strain rates of 0.1 s^{-1} to 15.0 s^{-1} was introduced with an emphasis on its verification.

The experimental flow curves showed the effects of high ratios of flow stress to thermal capacity and conductivity, which inevitably require their temperature-compensation. All the flow curves were accurately fitted using the generalized *C–m* model. Compared to the experimental flow curves in the strain range of 0.03–0.5, the mean errors of the fitted flow curves were less than 1.3%.

The systematic and scientific method of obtaining the flow curves formulated by the generalized *C–m* model considering the actual test environments was applied to the Ti6Al4V alloy. The accuracy of the TCFCs was verified through finite element analyses of the cylinder compression tests. Moreover, the predicted CLSCs were reasonably consistent with experimental CLSCs, demonstrating a mean error of 2.1%.

The fitted flow curves can be utilized to accurately reveal the metallurgical and macroscopic instabilities of the Ti6Al4V alloy that has a generally narrow range of metallurgical stable warm-forming conditions of the alloy.

Acknowledgement. This work was partly supported by Korea Evaluation Institute of Industrial Technology (10081334) and Korea Institute of Energy Technology Evaluation and Planning (20214000000520, Human Resource Development Project in Circular Remanufacturing Industry) grant funded by the Korea government (MOTIE).

References

1. Banerjee, D., Williams, J.C.: Perspectives on titanium science and technology. Acta Mater. **61**(3), 844–879 (2013). https://doi.org/10.1016/j.actamat.2012.10.043
2. Semiatin, S.L.: An overview of the thermomechanical processing of α/β titanium alloys: current status and future research opportunities. Metall. Mater. Trans. A. **51**(6), 2593–2625 (2020). https://doi.org/10.1007/s11661-020-05625-3
3. Basak, D., Overfelt, R.A., Wang, D.: Measurement of specific heat capacity and electrical resistivity of industrial alloys using pulse heating techniques. Int. J. Thermophys. **24**, 1721–1733 (2003). https://doi.org/10.1023/B:IJOT.0000004101.88449.86
4. Kulakov, M., Rahimi, S., Semiatin, S.L.: Effect of deformation heating on microstructure evolution during hot forging of Ti-6Al-4V. Metall. Mater. Trans. A. **53**(2), 407–419 (2021). https://doi.org/10.1007/s11661-021-06493-1
5. Niu, L., Zhang, Q., Wang, B., Han, B., Li, H., Mei, T.: A modified Hanse-Spittel constitutive equation of Ti-6Al-4V during cogging process. J. Alloys Compd. **894**, 162387 (2022). https://doi.org/10.1016/j.jallcom.2021.162387
6. Zhang, Z.X., Qu, S.J., Feng, A.H., Shen, J., Chen, D.L.: Hot deformation behavior of Ti-6Al-4V alloy: effect of initial microstructure. J. Alloys Compd. **718**, 170–181 (2017). https://doi.org/10.1016/j.jallcom.2017.05.097
7. Skripnyak, V.V., Skripnyak, V.A.: Localization of plastic deformation in Ti-6Al-4V alloy. Metals **11**(11), 1745 (2021). https://doi.org/10.3390/met11111745
8. Jha, A.K., Singh, S.K., Kiranmayee, M.S., Sreekumar, K., Sinha, P.P.: Failure analysis of titanium alloy (Ti6Al4V) fastener used in aerospace application. Eng. Fail. Anal. **17**(6), 1457–1465 (2010). https://doi.org/10.1016/j.engfailanal.2010.05.007
9. Ji, S.M., Jang, S.M., Lee, Y.S., Kwak, H.M., Jm, C., Joun, M.S.: Characterization of ti-6Al-4V alloy in the temperature range of warm metal forming and fracture analysis of the warm capping process. J. Mater. Res. Technol. **18**, 1590–1606 (2022). https://doi.org/10.1016/j.jmrt.2022.03.066
10. Majorell, A., Srivatsa, S., Picu, R.C.: Mechanical behavior of Ti6Al4V alloy at high and moderate temperatures—Part I: experimental results. Mater. Sci. Eng. A **326**, 297–305 (2002). https://doi.org/10.1016/S0921-5093(01)01507-6
11. Luan, J., Sun, C., Li, X., Zhang, Q.: Constitutive model for AZ31 magnesium alloy based on isothermal compression test. Mater. Sci. Technol. **30**, 211–219 (2014). https://doi.org/10.1179/1743284713Y.0000000341

12. Ji, S.M., Choi, J.M., Joun, M.S.: Evaluation of the macroscopic formability of metallic materials using a cylinder compression test. J. Mater. Res. Technol. **23**, 2798–2809 (2023). https://doi.org/10.1016/j.jmrt.2023.01.173
13. Joun, M.S., Lee, M.C., Joun, E.J.G., Lee, M.C., Ecom, J.G.: Intelligent metal-forming simulation. Int. Manuf. Sci. Eng. Conf. **44304**, 161–168 (2011). https://doi.org/10.1115/MSEC2011-50128
14. Arnold, D.N., Brezzi, F., Fortin, M.: A stable finite element for the Stokes equations. Calcolo **21**, 337–344 (1984). https://doi.org/10.1007/BF02576171
15. Perchat, E.: Mini-élément et factorisations incomplètes pour la parallélisation d'un solveur de Stokes 2D. Application au forgeage. [D.Sc. dissertation]. Ecole Nationale supérieure des Mines, Paris (2000). https://pastel.archives-ouvertes.fr/tel-00005654
16. Lee, M.C., Chung, S.H., Jang, S.M., Joun, M.S.: Three-dimensional simulation of forging using tetrahedral and hexahedral elements. Finite Elem. Anal. Des. **45**(11), 745–754 (2009). https://doi.org/10.1016/j.finel.2009.06.002
17. Lee, M.C., Joun, M.S., Lee, J.K.: Adaptive tetrahedral element generation and refinement to improve the quality of bulk metal forming simulation. Finite Elem. Anal. Des. **43**(10), 788–802 (2007). https://doi.org/10.1016/j.finel.2007.05.006
18. Aerospace SAE: Titanium alloy bars, wire, forgings, rings, and drawn shapes 6Al-4V annealed. AMS 4928, Aerospace Material Specification (2007)
19. Lee, S.W., Lee, J.M., Joun, M.S.: On critical surface strain during hot forging of lubricated aluminum alloy. Tribol. Int. **141**, 105855 (2020). https://doi.org/10.1016/j.triboint.2019.105855
20. Lu, X., et al.: Finite element analysis and experimental validation of the thermomechanical behavior in laser solid forming of Ti-6Al-4V alloy. Addit. Manuf. **21**, 30–40 (2018). https://doi.org/10.1016/j.addma.2018.02.003
21. Joun, M.S., Moon, H.G., Choi, I.S., Lee, M.C., Jun, B.Y.: Effects of friction laws on metal forming processes. Tribol. Int. **42**(2), 311–319 (2009). https://doi.org/10.1016/j.triboint.2008.06.012
22. Joun, M.S., et al.: Characterization of AZ31B, AZ61A and AZ80A magnesium alloys with an emphasis on temperature compensation for their application to a hot forging. J. Manufact. Proc. **84**, 764–785 (2022). https://doi.org/10.1016/j.jmapro.2022.10.054
23. Joun, M.S., Razali, M.K., Yoo, J.D., Kim, M.C., Choi, J.M.: Novel extended C-m models of flow stress for accurate mechanical and metallurgical calculations and comparison with traditional flow models. J. Magnes. Alloy. **10**(9), 2516–2533 (2021). https://doi.org/10.1016/j.jma.2021.08.018

Flow Behavior of an A6082 Alloy at Elevated Temperature and Precision Finite Element Prediction of Its Hot Forging Process of an Automotive Part

Jeong Hwi Park[1], Nurhidayah Abd Hamid[2], Jong Bok Byun[3], Su Min Ji[2], Jeong Muk Choi[4], and Man Soo Joun[3](✉)

[1] Material Manufacturing Engineering Team, Hyundai WIA, Uiwang 16082, South Korea
[2] School of Mechanical and Aerospace Engineering, Gyeongsang National University, Jinju 52828, South Korea
[3] Engineering Research Institute, School of Mechanical and Aerospace Engineering, Gyeongsang National University, Jinju 52828, South Korea
msjoun@gnu.ac.kr
[4] Jinhap Co. Ltd., Daejeon 34302, South Korea

Abstract. During forging, temperature effect is relatively weak in the case of an A6082 alloy to magnesium and titanium alloys and both friction and temperature compensations are thus necessary to accurately characterize it. This point is quite different from other materials for which friction-compensation can be neglected. The characterized flow information of the A6082 alloy is verified by comparing the predicted and experimental stroke-compression load curves. The improved flow information is utilized to simulate a hot forging process of an A6082 alloy automotive part with high accuracy. The predictions are compared with the experiments, showing a good agreement with each other. The effect of friction on the flow pattern of the aluminum forging process is also investigated to practically find the ac-acceptable frictional condition.

Keywords: A6082 · Hot Cylinder Compression · Friction Effect · Temperature Effect · Hot Forging · Lateral Arm

1 Introduction

Aluminum alloys are one of the attractive metallic materials satisfying the demands of the aircraft and automobile industries, particularly for weight reduction of the vehicle. They have good mechanical and metallurgical features, including lightweight, weldability and machinability, and recycling capability [1–3]. Aluminum alloy parts for structural members with strength and integrity have traditionally been forged. Notably, the aluminum billets, fabricated mostly by extrusion of cast materials with limited reduction of area, have relatively poor metal flow lines to rolled steel billets with distinct metal flow lines. The forging of aluminum alloys contributes to improving metal flow lines that are useful to their strength [4].

K. Mocellin et al. (Eds.): ICTP 2023, LNME, pp. 394–405, 2024.
https://doi.org/10.1007/978-3-031-40920-2_41

Aluminum alloys exhibit unique flow behaviors, which affect much on not only the shapes of forgings but also metal flow lines, surface quality, etc. For example, they do not experience distinct strain hardening and significant change in their flow stresses owing to increased temperature [5–7] compared to the other commercial metals [8]. Understanding their flow behaviors and mathematical description with high accuracy is thus essential [9]. During the experimental acquisition of flow curves using a specific material testing method, the effects of friction and temperature are pretty influential. They should be corrected for the flow curves with acceptable accuracy [10]. In addition, mathematically describing the flow curves with accuracy, flexibility, and practicability is also important because the flow behavior exhibits high non-linearity with strain, strain rate, temperature, etc.

The flow behaviors of aluminum alloys are too complicated to be simply formulated in a mathematical form. Even though many kinds of research on their flow characterization have been metallurgically or phenomenologically conducted for several decades, based on specific mathematical model equations [11], few works could meet the need of the industries because of limited accuracy.

It should be noted that the flow information at an instant has traditionally been obtained under the assumption of isothermal and homogeneous cylinder compression. Acquisition of accurate flow information from the material testing should thus be preceded by the compensation of the friction and temperature effects on the flow test [10].

Recently a systematic approach to compensating for the effects of friction and temperature on flow characterization was presented by Joun et al. [12] for the magnesium alloys and it was applied to the A6082 aluminum alloy [13] with an emphasis on the friction compensation. In this study, the procedure of accurately characterizing the flow behavior of the A6082 alloy is introduced. The flow information with high accuracy is employed to simulate a practical aluminum forging process to reveal the tribological phenomenon.

For all the finite element analyses of the cylinder compression tests, an implicit thermoviscoplastic FE package AFDEX (Altair APA) [14, 15] was used, based on the tetrahedral MINI-element scheme [16, 17] with automatic adaptive remeshing [18]. The die or tool deformation and elastic deformation of material were assumed to be negligible. The material-tool interface was formulated by the law of Coulomb friction.

2 Material Characterization

The A6082 alloy material was manufactured by a continuous casting process accompanied by a peeling process to improve its surface. Solid cylinder compression tests were conducted using a Gleeble 3500 system. The initial diameter and height of the specimen were 10 and 15 mm, respectively. The sample test temperatures (°C) and strain rates (s^{-1}) were 350, 400, 450, 500, and 550 and 0.1, 1.0, 5.0, 10.0, and 20.0, respectively. A tantalum foil lubricated the interface between the specimen and the test tool. Some selected specimens at the stoke of around 6.3 mm are shown in Fig. 1. Note that the measured maximum diameter of the compressed specimen was 13.7 mm at the final stroke of 6.3 mm.

Solid lines in Fig. 2 describe the experimental compression load-stroke curves (experimental CLSCs). The dashed lines with square or circle marks in Fig. 2 represent the CLSCs obtained by isothermal and non-isothermal flow analyses of the cylinder compression test, respectively, using the primitive flow curves directly calculated from the experimental CLSCs (see the RFCs-BFTC in Fig. 3). Figure 2 indicates that the difference between isothermal and non-isothermal predictions increases either as the strain rate increases or as the temperature decreases, revealing that there is a distinct effect of temperature owing to viscous heating even though it is not remarkable compared to the magnesium or titanium alloys [8, 12]. Notably, the oscillation of the compression load of the specimen with the initial temperature (550 °C) near the melting temperature could be clearly observed in Fig. 2 as the strain rate and strain increased. The maximum measured temperature at the barreled surface reached 557 °C, implying that internal local melting might occur during the compression test at the strain rate of 20 s^{-1}.

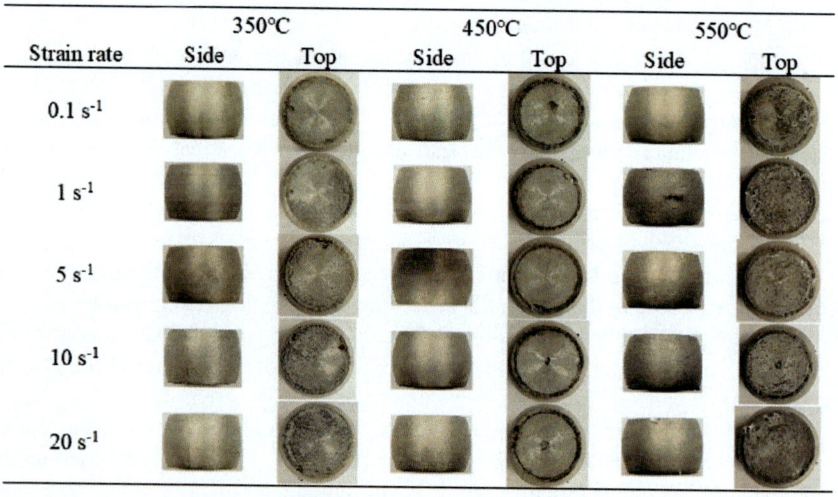

Fig. 1. Selected A6082 alloy specimens.

Solid lines in Fig. 3 exhibit the approximate, primitive flow curves for the sample temperatures and strain rates, termed the reference flow curves before friction- and temperature-compensation (RFCs-BFTC), which were calculated under the assumption of isothermal and homogeneous cylinder compression, viz., without any compensation for friction and temperature effects on flow behavior.

The friction and temperature compensation algorithm [12] was applied to obtain the friction- and temperature-compensated reference flow curves (RFCs-AFTC) from the RFCs-BFTC. The RFCs-AFTC were given in Fig. 3 together with the RFCs-BFTC, notably showing that they are almost the same. It is interesting to note that the reason why there is a negligible difference lies in the fact that the friction effect compensates for the temperature effect on occasion [13, 19].

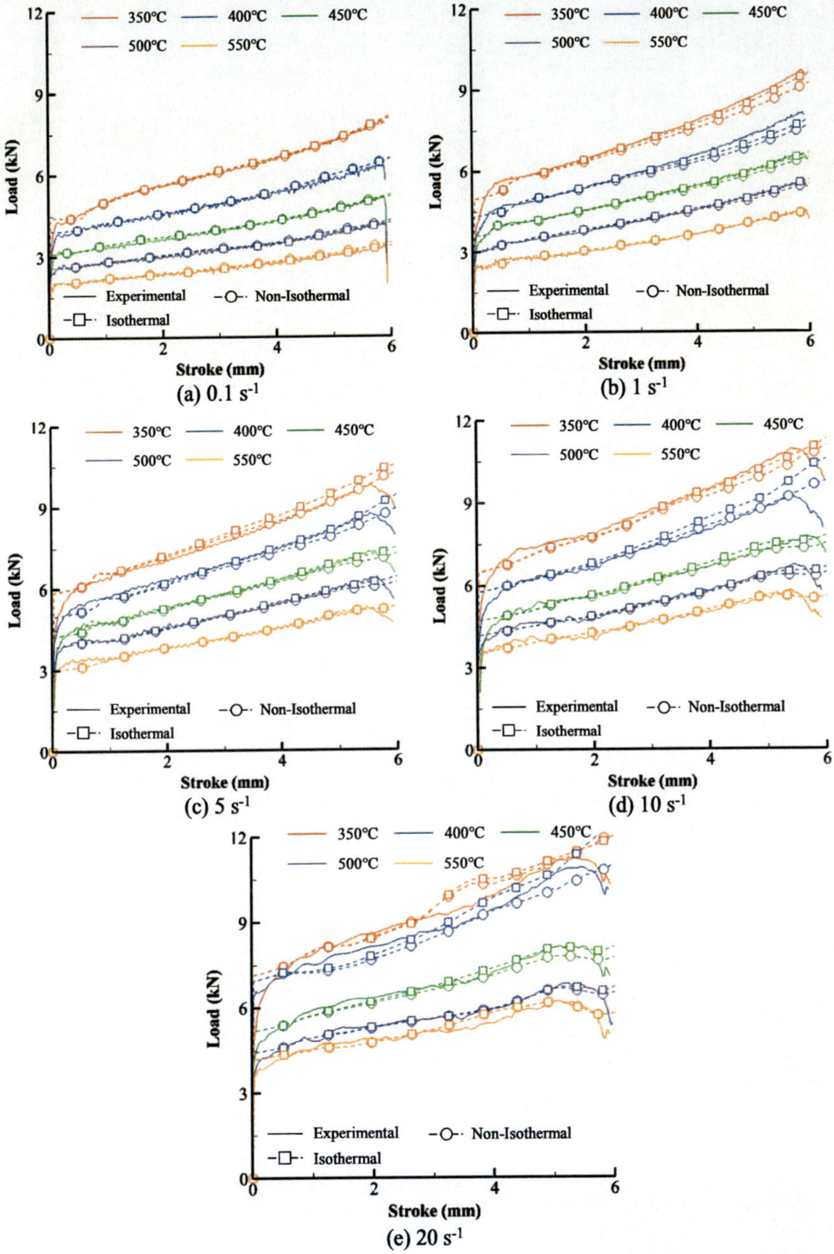

Fig. 2. Experimental CLSCs and their predictions by the RFCs-BFTC in Fig. 3.

350℃ 400℃ 450℃
500℃ 550℃

RFCs-BFTC By RFCs-AFTC

(a) 0.1 s^{-1}

350℃ 400℃ 450℃
500℃ 550℃

RFCs-BFTC By RFCs-AFTC

(b) 1 s^{-1}

350℃ 400℃ 450℃
500℃ 550℃

RFCs-BFTC By RFCs-AFTC

(c) 5 s^{-1}

350℃ 400℃ 450℃
500℃ 550℃
RFCs-BFTC By RFCs-AFTC

(d) 10 s^{-1}

350℃ 400℃ 450℃
500℃ 550℃
RFCs-BFTC By RFCs-AFTC

(e) 20 s^{-1}

Fig. 3. Experimental CLSCs and their predictions by RFCs-AFTC and RFCs-BFTC.

350℃ 400℃ 450℃
500℃ 550℃

Experimental By RFCs-AFTC

(a) 0.1 s^{-1}

350℃ 400℃ 450℃
500℃ 550℃

Experimental By RFCs-AFTC

(b) 1 s^{-1}

350℃ 400℃ 450℃
500℃ 550℃

Experimental By RFCs-AFTC

(c) 5 s^{-1}

350℃ 400℃ 450℃
500℃ 550℃

Experimental By RFCs-AFTC

(d) 10 s^{-1}

350℃ 400℃ 450℃
500℃ 550℃

Experimental By RFCs-AFTC

(e) 20 s^{-1}

Fig. 4. Experimental CLSCs and their predictions by RFCs-AFTC.

Figure 4 compares the experimental CLSCs with their predictions obtained using the RFCs-AFTC. Table 1., summarizing the errors of the predicted CLSCs by RFCs-AFTC with respect to the experimental CLSCs, shows that the friction- and temperature-compensation scheme remarkably reduced the errors of the improved and fitted flow behaviors.

The previous study [13] showed that only either friction-compensation or temperature-compensation may deteriorate the accuracy of the modified flow curves. However, the comparison of the experimental and predicted CLSCs summarized in Table 1. implies that the friction and temperature effects on its flow curve compensate for each other. Notably, this fact is one of the essential flow characteristics of a few aluminum alloys, including the A6082 alloy. This phenomenon causes by the fact that the A6082 alloy has a particular characteristic of a relatively weak effect of temperature and strain on the variation of flow stress, compared to other commercial metals [19].

Table 1. Error analysis of the CLSCs predicted by RFCs-AFTC

	Average error for all sample strain rates				
Temp.	Mean	Max.	Temp.	Mean	Max.
350	2.03	4.28	400	2.16	4.43
450	1.65	4.05	500	1.36	4.12
550	1.55	4.59			

3 Precision Hot Aluminum Forging Simulation with an Emphasis on Friction

The hot aluminum forging process analyzed in this study was aimed at fabricating an automobile lateral arm with reduced weight. The lateral arm is characterized by its great length compared to the cross-sectional area. The process is composed of two stages but only the first stage which has three under-filling defects was selected for the comparison between the experiments and predictions with an emphasis on the defects. The purpose is to identify the tribological condition since it is the next important factor in the flow behavior affecting the macroscopic phenomena. The initial material and die geometries were given in Fig. 5.

The initial temperatures of the material and dies were 540 and 250 °C, respectively. The thermal properties and conditions are summarized in Table 2. [20, 21]. A mechanical press with an operating speed of 50 rpm, a crank diameter of 400 mm, a rod length of 1350 mm, and a load capacity of 4000 tons were utilized for the experiments.

Table 2. Thermal properties and condition of the die and materials

Heat transfer coefficient to ambient ($W/mm^2 \cdot °C$)	2.95×10^{-6}
Heat transfer coefficient between material and die ($W/mm^2 \cdot °C$)	0.01
Heat radiation coefficient ($W/mm^2 \cdot °C^4$)	3.97×10^{-14}
Thermal conductivity ($W/mm \cdot °C$)	$0.16\ W/mm \cdot °C$
Heat capacity ($Ws/mm^3 \cdot °C$)	$0.00244\ Ws/mm^3 \cdot °C$

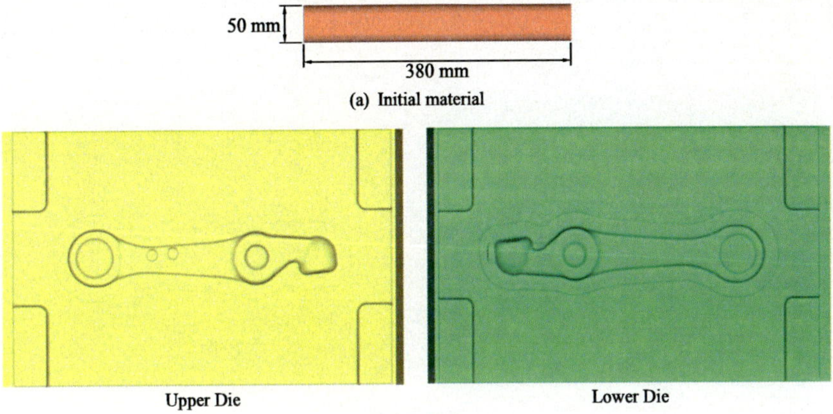

(a) Initial material

(b) Die geometries

Fig. 5. Initial material and die geometries.

The material was not coated by any lubricant but the water with graphite was sprayed onto the surfaces. The quantification of the friction is thus very difficult. Therefore, several friction conditions using the law of Coulomb friction with various friction coefficients ($\mu = 0.05, 0.1, 0.15, 0.2, 0.25, 0.3, 0.35$) were tried to determine the acceptable friction condition. Figures 6 and 7 globally and locally show the predictions of the aluminum hot forging process in the case of $\mu = 0.1$ and experiments, respectively, revealing that they are quite close to each other on the macro-scale. Notably, the three under-filling defects were clearly predicted, which are the major control points in this study.

To compare the predictions with experiments in a quantitative way, the lengths of the defect cavities defined as A and B in Fig. 7 were measured and summarized in Table 3. where the number in the bracket means percentage error. It is seen that the variation of them with the friction coefficient is not remarkable if the friction coefficient is between 0.05 and 0.2, indicating that the friction in this hot forging process of the lateral arm is not much sensitive in terms of the under-filling defect formation. It is owing to the negligible material flow along the longitudinal direction in the hot forging of the long billet. This phenomenon is quite different from that occurring in the hot forging of aluminum pistons [21] in which the material was coated by the lubricant and a critical surface strain governing the frictional phenomena was observed.

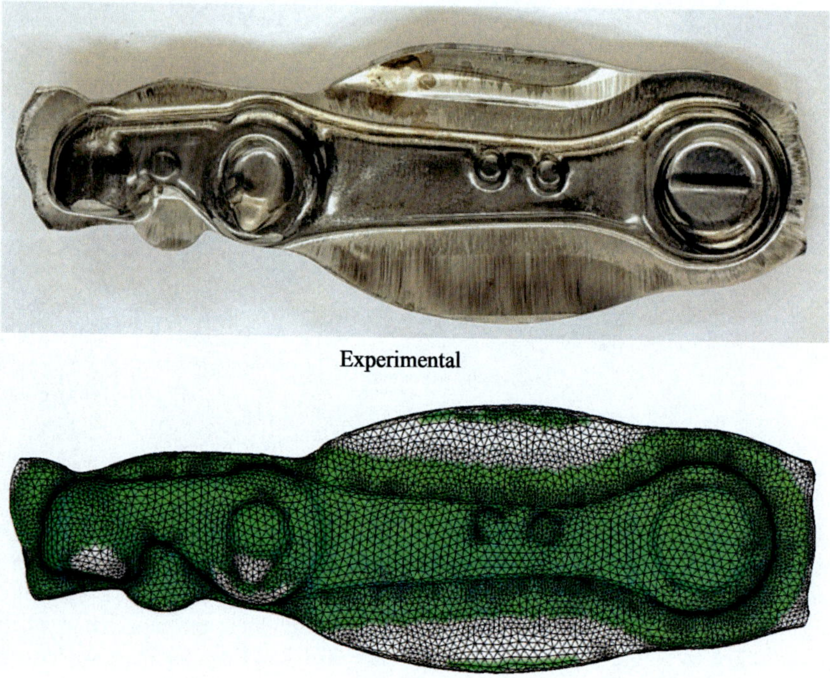

Experimental

Predicted

Fig. 6. Global comparison of the experiments and predictions ($\mu = 0.1$)

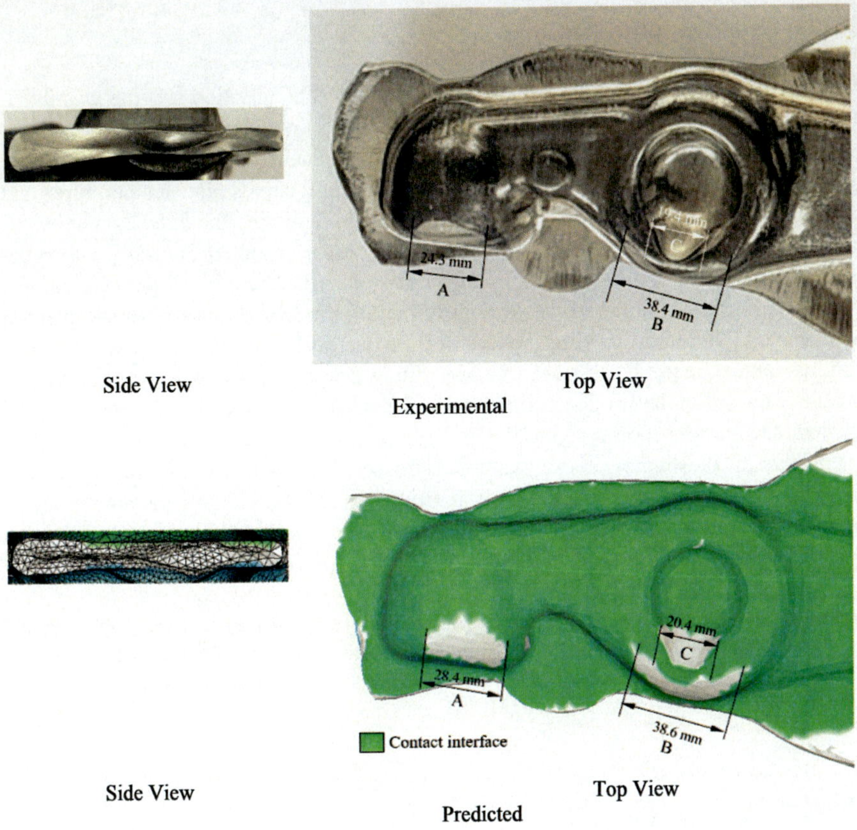

Fig. 7. Detailed comparison of the experiments and predictions ($\mu = 0.1$)

Table 3. Comparison of the under-filling defect lengths

Point	Experiment	Coulomb Friction						
		$\mu = 0.05$	$\mu = 0.10$	$\mu = 0.15$	$\mu = 0.20$	$\mu = 0.25$	$\mu = 0.3$	$\mu = 0.35$
A (mm)	24.3	27.91 (14.86%)	28.42 (16.95%)	30.47 (25.39%)	32.55 (33.95%)	30.92 (27.24%)	32.55 (33.95%)	30.89 (27.12%)
B (mm)	38.4	41.12 (7.08%)	38.56 (0.42%)	44.97 (17.11%)	45.02 (17.24%)	46.06 (19.95%)	45.8 (19.27%)	42.56 (10.83%)
C (mm)	19.4	19.59 (10.77%)	20.44 (5.36%)	19.85 (2.32%)	21.22 (9.38%)	20.48 (5.57%)	22.3 (14.95%)	20.00 (3.09%)

4 Conclusion

A systematic and practical friction- and temperature-compensation scheme was introduced to characterize the flow behavior of the A6082 alloy. The importance of friction-compensation during the flow characterization was emphasized, compared with titanium and magnesium alloys. It was emphasized that the flow curves after temperature- and friction-compensations exhibit improved accuracy in predicting the finite element solutions of the hot cylinder compression test. It is, however, noted that only either the temperature- or friction-compensation may deteriorate the accuracy of the flow curve.

Based on the accuracy of the flow information, the friction was observed using a practical hot aluminum forging process of an automobile part called lateral arm. The comparison of the predicted sizes of under-filling defects for the various friction coefficients shows that the friction coefficient in the range from 0.05 to 0.20 predicted the acceptable solutions, compared to the experimental under-filling defect size. This fact means that the plastic flow in this example is not sensitive to the friction since the material flow in the longitudinal direction is severely constrained by the geometrical nature of the long shaft or bar forging like the lateral arm in which a plane strain assumption can be adopted for simple calculation of plastic deformation at the cross-section.

Acknowledgment. This work was partly supported by Korea Evaluation Institute of Industrial Technology (10081334) and Korea Institute of Energy Technology Evaluation and Planning (20214000000520, Human Resource Development Project in Circular Remanufacturing Industry) grant funded by the Korea government (MOTIE).

References

1. Miller, W.S., et al.: Recent development in aluminium alloys for the automotive industry. Mater. Sci. Eng. A **280**(1), 37–49 (2000). https://doi.org/10.1016/S0921-5093(99)00653-X
2. Fridlyander, I.N., Sister, V.G., Grushko, O.E., Berstenev, V.V., Sheveleva, L.M., Ivanova, L.A.: Aluminum alloys: promising materials in the auto-motive industry. Metal Sci. Heat Treat. **44**(9/10), 365–370 (2002). https://doi.org/10.1023/A:1021901715578
3. Gronostajski, J., Marciniak, H., Matuszak, A.: New methods of aluminium and aluminium-alloy chips recycling. J. Mater. Process. Technol. **106**, 34–39 (2000)
4. Kim, M.C., Chung, S.H., Joun, M.S.: Optimal process design in hot forging in terms of grain flow quality. Int. J. Automot. Technol. **20**, 45–46 (2019)
5. Rudnytskyj, A., Simon, P., Jech, M., Gachot, C.: Constitutive modelling of the 6061 aluminium alloy under hot rolling conditions and large strain ranges. Mater. Des. **190**, 108568 (2020)
6. Lin, Y.C., Ding, Y., Chen, M.S., Deng, J.: A new phenomenological constitutive model for hot tensile deformation behaviors of a typical Al–Cu–Mg alloy. Mater. Des. **52**, 118–127 (2013)
7. Trimble, D., O'Donnell, G.E.: Constitutive modelling for elevated temperature flow behaviour of AA7075. Mater. Des. **76**, 150–168 (2015)
8. Ji, S.M., Jang, S.M., Lee, Y.S., Kwak, H.M., Choi, J.M., Joun, M.S.: Characterization of Ti-6Al-4V alloy in the temperature range of warm metal forming and fracture analysis of the warm capping process. J. Mater. Res. Technol. **18**, 1590–1606 (2022). https://doi.org/10.1016/j.jmrt.2022.03.066

9. Park, J.H., Razali, M.K., Yoo, J.D., Kim, M.C., Joun, M.S.: Modelling of hot flow behaviors of aluminum alloy Al6061 at elevated temperature: A review. In: International Joint Conference on Materials & Manufacturing Technology and Simulation IJCAMTS 2021, Malaysia (2021)
10. Luan, J., Sun, C., Li, X., Zhang, Q.: Constitutive model for AZ31 magnesium alloy based on isothermal compression test. Mater. Sci. Technol. **30**, 211–219 (2014)
11. Joun, M.S., Razali, M.K., Yoo, J.D., Kim, M.C., Choi, J.M.: Novel extended Cm models of flow stress for accurate mechanical and metallurgical calculations and comparison with traditional flow models. J. Magnes. Alloy. **10**, 2516–2533 (2021)
12. Joun, M.S., et al.: Characterization of AZ31B, AZ61A and AZ80A magnesium alloys with an emphasis on temperature compensation for their application to a hot forging. J. Manufact. Proc. **84**, 764–785 (2022)
13. Park, J.-H., Ji, S.-M., Choi, J.-M., Joun, M.-S.: Accurate flow characterization of A6082 for precision simulation of a hot metal forming process. Materials **15**(23), 8656 (2022). https://doi.org/10.3390/ma15238656
14. Joun, M.S., Lee, M.C., Eom, J.G.: Intelligent metal-forming simulation. Int. Manuf. Sci. Eng. Conf. **44304**, 161–168 (2011)
15. Joun, M.: Recent advances in metal forming simulation technology for automobile parts by AFDEX. IOP Conf. Ser. Mater. Sci. Eng. **834**(1), 012016 (2020). https://doi.org/10.1088/1757-899X/834/1/012016
16. Arnold, D.N., Brezzi, F., Fortin, M.: A stable finite element for the Stokes equations. Calcolo **21**, 337–344 (1984)
17. Razali, N.A., Chung, S.H., Chung, W.J., Joun, M.S.: Implicit elastoplastic finite element analysis of tube-bending with an emphasis on springback prediction. Int. J. Adv. Manuf. Technol. **120**(9–10), 6377–6391 (2022). https://doi.org/10.1007/s00170-022-09073-8
18. Lee, M.C., Joun, M.S., Lee, J.K.: Adaptive tetrahedral element generation and refinement to improve the quality of bulk metal forming simulation. Finite Elem. Anal. Des. **43**, 788–802 (2007)
19. Ji, S.M., Choi, J.M., Joun, M.S.: Evaluation of the macroscopic formability of metallic materials using a cylinder compression test. J. Mater. Res. Technol. **23**, 2798–2809 (2023)
20. Yoo, J.D., Hwang, T.H., Joun, M.S.: Finite element prediction of temperature variation of material and dies in aluminum forging. Key Eng. Mater. **830**, 93–100 (2020)
21. Lee, S.W., Lee, J.M., Joun, M.S.: On critical surface strain during hot forging of lubricated aluminum alloy. Tri. Int. **141**, 105855 (2020)

Measurement and Analysis of the Strength Differential Effect of 5000 Series Aluminum Alloy Sheet

Kaisei Akiyama[1](✉), Ren Tachibana[1], Toshihiko Kuwabara[2], and Takeo Sakurai[3]

[1] Department of Mechanical Systems Engineering, Graduate School of Engineering, Tokyo University of Agriculture and Technology, 2-24-16 Naka-cho, Koganei-shi, 184-8588 Tokyo, Japan
ks-akiyama@st.go.tuat.ac.jp

[2] Division of Advanced Mechanical Systems Engineering, Institute of Engineering, Tokyo University of Agriculture and Technology, 2-24-16 Naka-cho, Koganei-shi, 184-8588 Tokyo, Japan

[3] The Japan Institute of Light Metals, 4-2-15, Ginza, Chuo-ku, 104-0061 Tokyo, Japan

Abstract. Tension-Compression Asymmetry (TCA) of a 1.5-mm-thick 5000-series aluminum alloy sheet, A5083-O, is measured using a uniaxial tensile test and in-plane compression test. In particular, the TCA observed for annealed materials is referred to as the Strength Differential Effect (SDE). The in-plane compressive flow stress is 2–6% higher than the uniaxial tensile flow stress; therefore, the material exhibits the SDE. Moreover, a stacked compression test in the thickness direction of the test sample is also performed to measure the hydrostatic stress dependency of the yield stress. The uniaxial compressive flow stress in the thickness direction is almost identical to the equibiaxial tensile flow stress measured using a cruciform equibiaxial tension test (ISO 16842) and a hydraulic bulge test; therefore, the hydrostatic stress dependency of the yield stress has not been confirmed. Hence, it is concluded that the SDE observed in the A5083-O is not caused by the hydrostatic stress dependency of the yield stress.

Keywords: Tension-Compression Asymmetry · Equibiaxial stress test · In-plane compression test · Stacked compression test · Hydrostatic stress

1 Introduction

Aluminum alloy sheets are used for manufacturing automotive parts to reduce the weight of the car body. Automotive parts are mainly fabricated by press forming, which has superior productivity. During press forming, materials are subjected to compressive stress as well as tensile stress. For some metals, the magnitude of a tensile true stress is different from that of compressive stress at the same magnitude of plastic strain. The phenomenon is referred to as Tension-Compression Asymmetry (TCA). In particular, the TCA observed for annealed materials is referred to as the Strength Differential Effect (SDE).

K. Mocellin et al. (Eds.): ICTP 2023, LNME, pp. 406–414, 2024.
https://doi.org/10.1007/978-3-031-40920-2_42

Richmond and Spitzig [1, 2] performed uniaxial tension and compression tests under hydrostatic pressure on both iron-based materials, including Fe single crystals, and 1100 aluminum to find that these materials exhibit the SDE, and demonstrated that the cause of the SDE is attributed to the fact that the flow stress of these materials is dependent on the hydrostatic pressure. Kuwabara et al. [3] performed an in-plane compression test (IPCT) on a 5182-O aluminum alloy sheet and two low-carbon steel sheets using comb-shaped dies. They found that the flow stresses were higher in compression than in tension for the steel sheets, whereas the aluminum alloy sheet did not exhibit the SDE. Kuwabara et al. [4] concluded that the SDE in mild steel sheet is also due to the hydrostatic stress dependence of the flow stress. Holmen et al. [5] investigated the SDE of four 6000-series aluminum alloys (one was rolled and other three were extrude) in several tempers with yield strengths varying from 27 MPa to 373 MPa: a total of thirteen different material configurations. They observed that the axial stresses measured in compression tests were significantly higher than corresponding tensile stresses for nearly all material configurations. Ku et al. [6] experimentally verified the TCA in rolled aluminum alloy 7056 in the overaged condition (T7) with two tempers and reported that the TCA was attributed to the microstructural evolution. However, no experimental verification of the hydrostatic stress dependency of aluminum alloy sheets has been reported, except for the work of Spitzig and Richmond [1, 2].

This study investigates the SDE of a 5000-series rolled aluminum alloy sheet (A5083-O) with a nominal thickness of 1.5 mm. In-plane tension and compression tests were performed to evaluate the SDE. Moreover, an equibiaxial tension test using a cruciform specimen [7], hydraulic bulge test, and a uniaxial compression test in the thickness direction of the test sample [4] were performed to clarify whether the flow stresses were affected by hydrostatic stress. The work contributes experimental evidence that the test sample exhibits the SDE. However, the observed SDE is not attributed to pressure sensitivity to the flow stress.

2 Effect of Hydrostatic Stress on Flow Stress

Hereafter, the rolling (RD), transverse (TD), and thickness directions (ND) of the material are referred to as the x-, y-, and z-axes, respectively. When the hydrostatic stress, $-\sigma_b$, is superimposed on the equibiaxial tensile stress state, σ_b, in the xy-plane, the equibiaxial tensile stress σ_b is cancelled by the hydrostatic stress $-\sigma_b$ and is equivalent to a uniaxial compressive stress state, $-\sigma_b$, in the z-axis (Fig. 1). Therefore, assuming that the flow stress of the material is independent of hydrostatic stress, the $\sigma_b - \left|\varepsilon_z^p\right|$ (ε_z^p: plastic strain in the z-axis direction) curve obtained from the equibiaxial tensile test should be identical to the $|\sigma_{zC}| - \left|\varepsilon_z^p\right|$ curve obtained from the uniaxial compression test in the thickness direction. On the other hand, if the flow stress is affected by hydrostatic stress, the two curves should be different from each other.

3 Experimental Method

3.1 Test Material

The test sample is a 5000-series aluminum alloy sheet (A5083-O) with a nominal thickness of 1.5 mm. The mechanical properties of the sample are shown in Table 1.

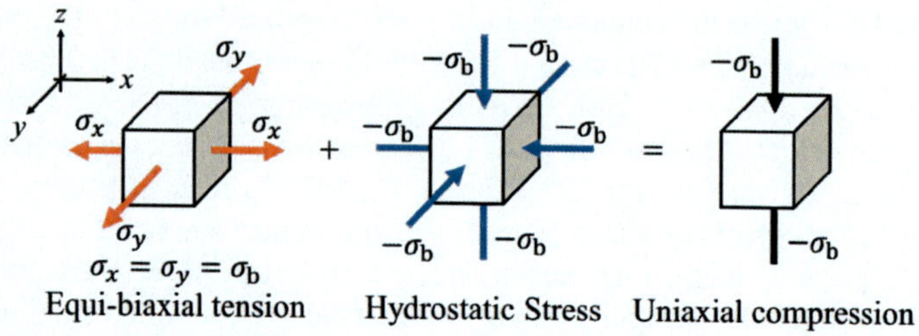

Fig. 1. Equivalence of the equibiaxial tension with the uniaxial compression with an assumption that the flow stress is independent of hydrostatic stress

Table 1. Mechanical properties of A5083-O

Angle from RD /°	Stress direction*	E /GPa	$\sigma_{0.2}$/MPa	r	σ_B/MPa	ε_{TS}^{p}**	a*** /MPa	b*** /MPa	c***
0	T	72.2	157	0.682	312	0.171	401	254	12.5
	C	71.4	161	0.640			380	227	15.8
90	T	72.0	159	0.613	305	0.155	389	243	12.6
	C	72.0	165	0.717			382	226	14.8

* "T" and "C" mean tension and compression, respectively.

** ε_{TS}^{p} is a logarithmic plastic strain giving the maximum tensile load

*** Approximated using $\sigma = a - b exp(-c\varepsilon^{p})$ at $\varepsilon^{p} = 0.002 - \varepsilon_{TS}^{p}$ (tension) $\varepsilon^{p} = 0.002 - 0.1$ (compression)

3.2 In-Plane Compression Test

Figures 2a and b show the IPCT apparatus used in this study. This apparatus was first developed in [3], modified in [8], and used in [9–11]. Figure 2a shows an overview of the dies used for applying in-plane compression to a sheet specimen. Figure 2b shows an overview of the test apparatus. Lower die 1 is bolted to the lower plate of the die set. Lower die 2 is bolted to the slide rail, so that the die can move in the horizontal direction without friction. A specimen is placed on lower dies 1 and 2 with both ends clamped by clamping jigs (not shown in the figure). Upper dies 1 and 2 are placed on the specimen.

The positioning pins (not shown in the figure) fixed to upper dies 1 and 2 align with holes in lower dies 1 and 2. Accordingly, upper die 1 is stationary, and the motion of upper die 2 is synchronized with that of lower die 2. Lower die 2 is connected to a load cell to measure the tension-compression force during a test. It is actuated by hydraulic cylinder A to apply an in-plane tension-compression force to the specimen. Hydraulic cylinder B exerts a blank holding force on the specimen to prevent the buckling of the specimen during a test. The blank holding force is kept constant during a test via a hydraulic control valve. Steel cylindrical rollers are inserted between the blank holding platens and upper dies 1 and 2 so that the blank holding force can be transmitted to the specimen during an IPCT without friction. In this study, the blank holding force applied to the specimen was 1.2% of 0.2% yield stress (=1.92 MPa). The strain was measured using a strain gauge (YFLA-2-1LJC-F, Tokyo Measuring Instruments Laboratory Co. Ltd.). The strain rate was approximately 5×10^{-4} s^{-1} and the testing temperature was room temperature for both tension and compression.

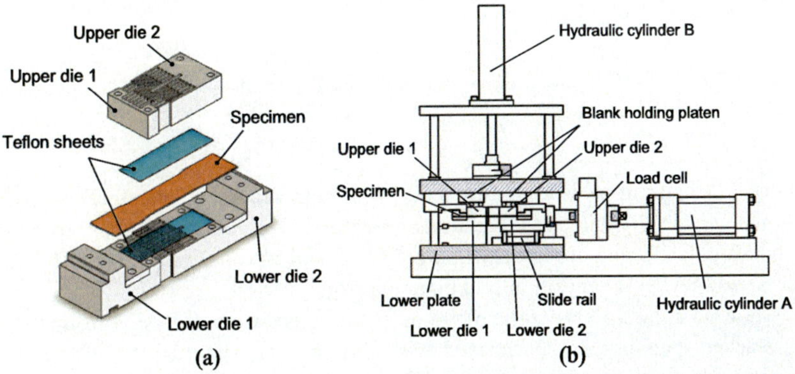

Fig. 2. Experimental apparatus for IPCT [8]: (a) upper and lower dies and (b) overview of test apparatus

3.3 Uniaxial Compression Test of Stacked Specimen [4]

Figure 3a shows a schematic diagram of the uniaxial compression testing device for a stacked specimen. This device was incorporated into a regular tensile testing machine (AUTOGRAPH AG-100kNX, Shimadzu Corporation). Figure 3b shows the details of the compression die. The compression dies have a thickness of 4 mm (a 3-mm-thick super steel base plate coated with 1-mm-thick sintered polycrystal diamond, PCD) and a diameter of 19 mm. The PCD is effective for reducing the friction between the specimen and the compression dies and thus suppresses the non-uniform deformation of the specimen [12]. The compression dies were positioned on the upper and lower supporting plates using a clamping plate. The supporting plates were bolted to the upper and lower die plates. Since the initial thickness of the specimen is 1.5 mm, the amount of compressive displacement due to compression in the thickness direction is small,

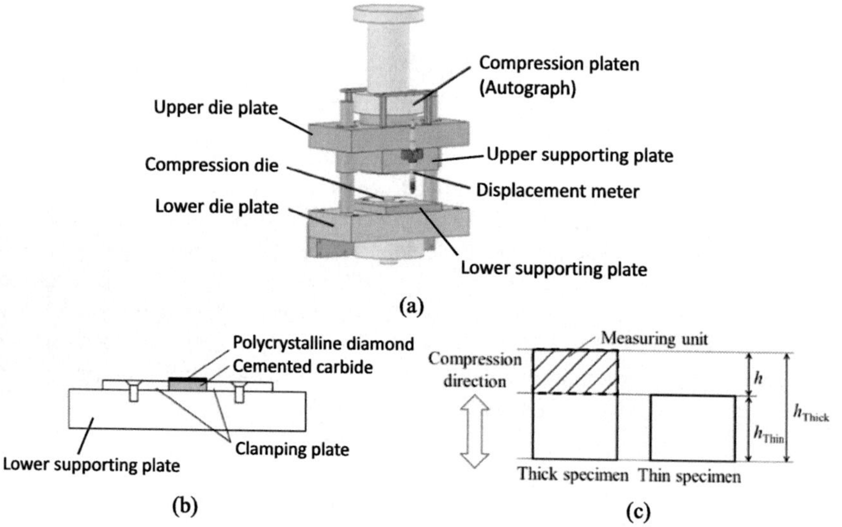

Fig. 3. SCT device [4]: (a) overview of test apparatus, (b) details of lower part of compression die, and (c) specimen geometry

making it difficult to measure plastic strain with high accuracy. Therefore, we stacked the specimens in the thickness direction and increased the initial thickness to improve the measurement accuracy. The sheets were fixed with an adhesive (CEMEDINE Metal Lock Series Y610) to prevent interlaminar slippage.

Figure 3c shows a schematic diagram of the uniaxial compression testing device for a stacked specimen. We performed compression tests using two stacked specimens with the same cross-sectional area but different heights, h_{Thick} and h_{Thin}, and measured the nominal stress-displacement for each specimen using a displacement meter (GT2-S5, KEYENCE Co. Ltd.). By denoting the difference in displacement between the two stacked specimens for a given nominal stress as Δu and calculating the compressive true strain, ε_z, as $\ln\left(1-\frac{\Delta u}{h_{\mathrm{Thick}}-h_{\mathrm{Thin}}}\right)$, we can determine the true stress-true strain curve, $|\sigma_z c| - |\varepsilon_z|$, for a specimen with an initial stack heights of $h(= h_{\mathrm{Thick}} - h_{\mathrm{Thin}})$. This approach enables us to exclude the influences of the elastic deformation of the compression device and potential friction between the specimen and compression platens on the measurement of $|\varepsilon_z|$. Thus, Young's modulus is calculated from the slope at unloading, and we can determine the compressive true stress vs. compressive plastic strain curve, $|\sigma_z c| - \left|\varepsilon_z^{\mathrm{p}}\right|$.

4 Experimental Results

4.1 Measured Result of the SDE

Uniaxial tensile test and in-plane compression test were performed in the RD and TD to quantitatively evaluate the SDE of the specimen. The true stress-logarithmic plastic strain ($|\sigma| - |\varepsilon^{p}|$) curves of the tension and compression in the RD and TD are shown in Fig. 4. (average of two measurements in each direction).

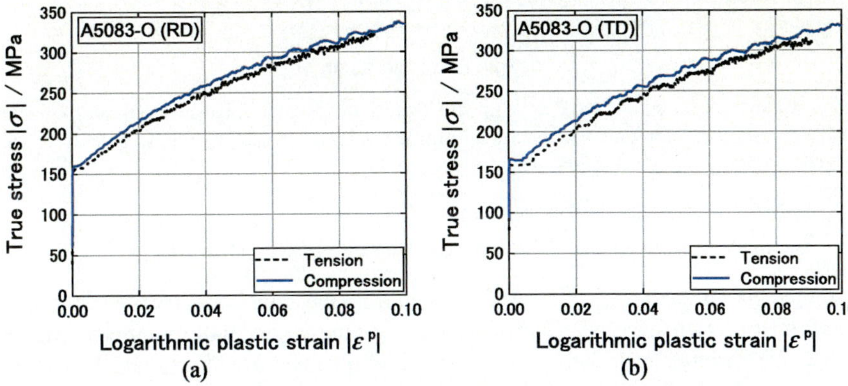

Fig. 4. Comparison of true stress - logarithmic plastic strain curves measured using uniaxial tensile test and in-plane compression test for (a) RD and (b) TD

The compressive stress $|\sigma_{C}|$ exceeds the tensile stress σ_{T} in both RD and TD. Therefore, the SDE was clearly observed. In order to quantitatively evaluate the magnitude of the SDE in the uniaxial stress state, the increase rate in the flow stress, $\beta_{SDE-uni}$ is defined by Eq. (1) using the values of σ_{T} and $|\sigma_{C}|$ at an arbitrary plastic strain ε^{p}.

$$\beta_{SDE-uni} = 2\frac{|\sigma_{C}| - \sigma_{T}}{|\sigma_{C}| + \sigma_{T}} \tag{1}$$

The results of both uniaxial tensile test and in-plane compression test were averaged over two tests and approximated by Voce's law to eliminate the influence of the Portevin–LeChatelier effect. The variation of $\beta_{SDE-uni}$ with ε^{p} for the RD and TD is shown in Fig. 5. $\beta_{SDE-uni} > 0$ was always observed from the early stage of plastic deformation for both the RD and TD; therefore, the test sample clearly exhibits the SDE.

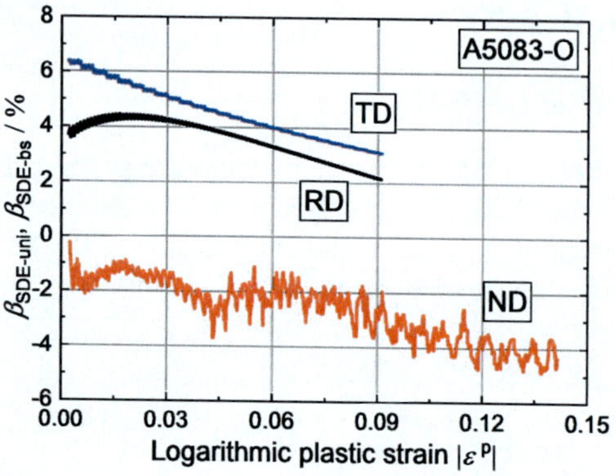

Fig. 5. Change in $\boldsymbol{\beta}_{\mathbf{SDE-uni}}$ and $\boldsymbol{\beta}_{\mathbf{SDE-bs}}$ with logarithmic plastic strain measured using uniaxial tensile test, in-plane compression test, cruciform biaxial tension test, hydraulic bulge test, and stacked compression test for A5083-O

4.2 Verification of the Hydrostatic Stress Dependency of Flow Stress

A comparison of the $\sigma_b - |\varepsilon_z^p|$ curves obtained from the equibiaxial tensile test using a cruciform specimen (average of two measurements) and the hydraulic bulge test (average of three measurements) with the $|\sigma_{zC}| - |\varepsilon_z^p|$ curve obtained from the uniaxial compression test in the thickness direction using stacked specimens is shown in Fig. 6. The $\sigma_b - |\varepsilon_z^p|$ curve measured using the equibiaxial tensile test with a cruciform specimen was combined with that of the hydraulic bulge test to make an accurate equibiaxial tensile test result. It is clearly seen that the $\sigma_b - |\varepsilon_z^p|$ and $|\sigma_{zC}| - |\varepsilon_z^p|$ curves were almost identical to each other. In addition, the compressive stress $|\sigma_{zC}|$ does not exceed the tensile stress σ_b. The result indicates that the flow stress is not affected by hydrostatic stress at least for a strain range of $0 \leq |\varepsilon_z^p| \leq 0.075$.

In order to quantitatively evaluate the effect of hydrostatic stress on the equibiaxial flow stress, the plastic flow stress increase rate, β_{SDE-bs} is defined by Eq. (2) using σ_b and $|\sigma_{zC}|$ at any thickness plastic strain ε_z^p.

$$\beta_{SDE-bs} = 2\frac{|\sigma_{zC}| - \sigma_b}{|\sigma_{zC}| + \sigma_b} \tag{2}$$

The variation of β_{SDE-bs} with ε_z^p for ND is shown in Fig. 5 (the result of $\beta_{SDE-uni}$ is also shown for reference). β_{SDE-bs} is in the range from 0 to -4% and it is clear that $\beta_{SDE-bs} \leq 0$. Therefore, it is concluded that the SDE observed in the uniaxial stress state was not caused by the hydrostatic stress dependency of the flow stress. The cause of the SDE of the test sample is a subject of future research.

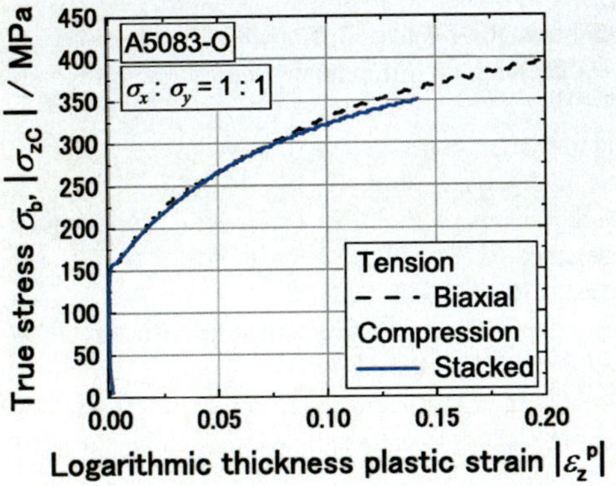

Fig. 6. Comparison of true stress – logarithmic thickness plastic strain curve measured using cruciform biaxial tension test, hydraulic bulge test, and stacked compression test for A5083-O

5 Conclusion

Uniaxial tensile test and in-plane compression test were performed on a 5000-series aluminum alloy sheet (A5083-O) to measure the SDE. In addition, an equibiaxial tensile test using a cruciform specimen, a hydraulic bulge test, and uniaxial compression tests in the thickness direction using stacked specimens were performed to verify the hydrostatic stress dependency of the flow stress. The experimental results are summarized follows.

(1) The SDE was observed for the test sample; $\beta_{\mathrm{SDE-uni}} = 2.5\%$ and 3.4% at $|\varepsilon^{\mathrm{p}}| = 0.08$ in the RD and TD, respectively.
(2) The $\sigma_{\mathrm{b}} - |\varepsilon_z^{\mathrm{p}}|$ curve observed in the equibiaxial tensile test and $|\sigma_{z\mathrm{C}}| - |\varepsilon_z^{\mathrm{p}}|$ curve observed in the uniaxial compression test in the thickness direction almost identical to each other. Therefore, the flow stress of the test sample was not affected by hydrostatic stress.
(3) Based on the above two results, we conclude that the SDE of the test sample was not caused by the hydrostatic stress dependency of the flow stress.

References

1. Richmond, O., Spitzig, W.A.: Pressure dependence and dilatancy of plastic flow. In: IUTAM Conference, Theoretical and Applied Mechanics, Proceedings of 15th International Congress of Theoretical and Applied Mechanics, pp. 377–386. North-Holland Publishers, Amsterdam (1980)
2. Spitzig, W.A., Richmond, O.: The effect of pressure on the flow stress of metals. Acta Metall. **32**, 457–463 (1984)
3. Kuwabara, T., Morita, Y., Miyashita, Y., Takahashi, S.: Elastic-plastic behavior of sheet metal subjected to in-plane reverse loading. In: Tanimura, S., Khan, A.S. (eds.) Proceedings Plasticity 1995, The Fifth International Symposium on Plasticity and its Current Applications, pp. 841–844. Gordon and Breach Publishers (1995)

4. Kuwabara, T., Tachibana, R., Takada, Y., Koizumi, T., Coppieters, S., Barlat, F.: Effect of hydrostatic stress on the strength differential effect in low-carbon steel sheet. Int. J. Mater. Forming **15**, 13 (2022)
5. Holmen, J.K., Frodal, B.H., Hopperstad, O.S., Børvik, T.: Strength differential effect in age hardened aluminum alloys. Int. J. Plast. **99**, 144–161 (2017)
6. Ku, A.Y., Khan, A.S., Gnaupel-Herold, T.: Quasi-static and dynamic response, and texture evolution of two overaged Al 7056 alloy plates in T761 and T721 tempers: experiments and modeling. Int. J. Plast. **130**, 102679 (2020)
7. ISO 16842:2021 Metallic materials –Sheet and strip – Biaxial tensile testing method using a cruciform test piece
8. Kuwabara, T., Kumano, Y., Ziegelheim, J., Kurosaki, I.: Tension-compression asymmetry of phosphor bronze for electronic parts and its effect on bending behavior. Int. J. Plast. **25**, 1759–1776 (2009)
9. Noma, N., Kuwabara, T.: Specimen geometry optimization for in-plane reverse loading test of sheet metal and experimental validation. J. Jpn. Soc. Technol. Plast. **53**(617), 574–579 (2012)
10. Noma, N., Hashimoto, K., Maeda, T., Kuwabara, T.: Enhancement of accuracy of spring-back simulation using the material model considering the strength differential effect. J. JSTP **61**(708), 20–25 (2020). (in Japanese)
11. Maeda, T., Noma, N., Kuwabara, T., Barlat, F., Korkolis, Y.P.: Measurement of the strength differential effect of DP980 steel sheet and experimental validation using pure bending test. J. Mater. Process. Technol. **256**, 247–253 (2018)
12. Koizumi, T., Kuroda, M.: Evaluation of tension-compression asymmetry of a low-carbon steel sheet using a modified classical compression test method. IOP Conf. Ser. J. Phys. **1063**, 012167 (2018)

Application of Barlat's Yld 2000-2d Yield Stress Function for Modeling the Anisotropic Plastic Behaviour and the Forming Limit Strain Curve

José Divo Bressan[1(✉)] and Mauricio Vicente Donadon[2]

[1] Department of Mechanical Engineering, University of Santa Catarina State, UDESC, Campus Universitario, CEP 89.219-710, SC Joinville, Brazil
josedivo.bressan@udesc.br

[2] Aeronautical Engineering Division, Technological Institute of Aeronautics, ITA, DCTA CEP, 12.228-901, São José Dos Campos, SP, Brazil

Abstract. The objective of present paper is to examine the plastic anisotropy behaviour of steel sheet, employing the Barlat´s Yld 2000-2d yield stress criterion and the corresponding non-associated plastic flow rule. New Barlat´s coefficients of anisotropy were defined and calibrated from material experimental data of simple uniaxial tension and equal biaxial stress tests. The new set of coefficients calculated from the experimental Lankford anisotropy coefficients (r-values), normalized yield stress (s-values), equal biaxial stress parameters (r_b and σ_b) were numerically obtained using the Newton-Raphson method. The investigated metal was the highly anisotropic AISI 439 steel sheets found in the literature. In the results analysis and discussion, the new coefficients of anisotropy of the Barlat´s non-associated plastic flow rule were calculated and validated by plotting on the same graph the predicted r-value and s-value curves and the experimental data for the anisotropic steel sheets. The correlations have revealed that the Barlat´s yield criterion and the plastic flow stress potential were not coincident. Furthermore, the predicted limit strain curve of 439 steel correlated better with the experimental FLC_{TD} transverse curve when using the shear stress fracture criterion and the non-associated plastic potential than the associated flow rule. Therefore, the Barlat´s Yld 2000-2d non-associated plastic flow rule provides a better fit with the experimental Lankford and equal biaxial coefficients of anisotropy and the FLC_{TD} curve results of AISI 439 steel sheets.

Keywords: Plastic anisotropy · Barlat´s Yld 2000-2d non-associated plastic flow rule · FLC · AISI 439 steel

1 Introduction

In sheet metal forming operations practice, it is well-known that the Formability of the blank material is greatly influenced by the strain and strain-rate hardening and the plastic anisotropy coefficients, which can be measured by the Lankford's coefficients of anisotropy and the equal biaxial anisotropy coefficient. Metal blanks as received exhibits

K. Mocellin et al. (Eds.): ICTP 2023, LNME, pp. 415–426, 2024.
https://doi.org/10.1007/978-3-031-40920-2_43

plastic anisotropy with orthotropic symmetry originated from the cold rolling. The three orthotropic axes of plastic anisotropy are mutually perpendicular and are defined in the rolling direction RD, transverse TD and normal or thickness ND directions. These are assumed as the principal axes of orthotropic plastic anisotropy.

One first work to investigate an anisotropic yield stress criterion to describe the plasticity behaviour of sheet metals with orthotropic symmetry was the paper presented by Rodney Hill in 1948 [1]. The phenomenological Hill's 48 yield stress criterion was an extension of von Mises's isotropic yield stress criterion. Since then, several other phenomenological anisotropic yield stress criteria have been proposed. A brief review of yield stress criteria can be found in the articles by Barlat et al. [2], Yoon et al. [3] and Bressan et al. [4].

Barlat's Yld 2000-2d anisotropic yield stress function [5] has been widely applied in numerical simulations and analytical modelling of cup ears and forming limit strains of aluminium and steel alloys in sheet metal shaping operations in plane stress conditions. This yield function was developed from findings of previous works by Hershey [6], Hosford [7] yield stress criteria for isotropic material and the Yld96 anisotropic yield stress criterion proposed by Barlat et al. [8]. In 2003, the Yld 2000-2d anisotropic yield stress criterion applied to sheet metal forming for plane stress condition, pressure independent metals, orthotropic anisotropy was proposed by Barlat et al. [5] and was defined as the sum of two scalar non-quadratic functions, which are generated by two linear transformations of the deviatoric stress tensor.

1.1 Non-associated Plastic Flow Stress Potential

In the modelling of plastic flow behavior of metal, it is commonly postulated and accepted the associated plastic flow rule or normality rule, by which the plastic flow stress potential function is assumed to be coincident with the yield stress criterion. Thus, it is usually assumed that both the yield stress surface and the metal plastic strain increments or strain rate are calculated from a unique yield stress criterion function, which must be a convex function.

In the recent literature, the non-associated plastic flow rule approach has been frequently employed in numerical simulations of sheet metal forming processes. The material yield stress criterion and plastic flow stress potential were defined by two distinct functions [9, 10]. One first work to investigate a non-associated plastic stress potential was presented by Stoughton [9]. The author argued that the associated plastic flow rule postulation is not rigorously true, revealed by comparing the Hill's 48 yield stress criterion function for both the associated and the non-associated flow rule approaches for the experimental data for mild steel, Al 2008-T4 and Al 2024-T3 aluminium alloys. Employing the tensile test experimental data for uniaxial yield stress and the Lankford coefficient of anisotropy, r-value, at 0°, 45° and 90° to the RD to calibrate the non-associated flow rule modelling, the author found good fittings for other points for different experimental tensile test angles with the predicted curves for yield stress and r-values. The curves fitting was poor or failed to correctly predict the uniaxial yield stress and r-value variations with the angle to RD, using the Hill's 48 yield criterion as the plastic stress potential or the associated flow rule.

Numerical simulations of cylindrical cup drawing and ears formation and its height profile of highly anisotropic Al 2090-T3 and Al 5042 aluminium sheets were studied by Park and Chung [10]. The authors used the associated and non-associated Barlat's Yld2000-2D plastic potential. Numerical simulations, using m = 8 and the non-associated plastic potential showed good correlations with the experimental results of ear numbers and height profiles for both materials; better cylindrical cup earing profiles prediction than the associated flow rule and the Hill 48 yield criterion.

The FLC of AISI 439 steel sheet was investigated by Lian et al. [11], employing the non-associated Hill's 48 plastic flow potential and the yield stress criterion, and the modified maximum force criterion to predict the FLC. The non-associated flow rule model presented better correlation with the experimental FLC curve than the Hill's 48 associated flow rule model. As well the predicted punch force versus displacement curve of the Nakajima equal biaxial stress test showed good correlation with the experimental data. The authors also presented a short review of recent references that deal with the application of non-associated plastic potentials.

Safei et al. [12] analyzed the non-associated Barlat's Yld2000-2D flow stress potential model, applying to deformation processes of the highly anisotropic Al 2090-T3 alloy and DC06 steel sheets. The predicted s-value and r-value curves as function of angle to the rolling direction RD have shown good correlation with the experimental data. Although, the numerical simulation of cylindrical cup deep drawing of Al 2090-T3 alloy showed the same number of ears, the simulations revealed poor fitting with the experimental results of the earing height profile.

The objective of present work is to examine the Barlat´s Yld 2000-2d stress function as the anisotropic non-associated plastic flow potential or the flow rule to better describe the plastic anisotropy coefficients of highly anisotropic AISI 409 steel sheets. New generalised and exact equations to calculate the eight coefficients of plastic anisotropy a_i of Barlat´s Yld 2000-2d yield stress function have been proposed by Bressan [4]. Thus, to calibrate the Barlat´s coefficients of plastic anisotropy a_i, eight experimental tests and eight material plastic anisotropy parameters are needed. The plastic anisotropic parameters required are defined as Lankford coefficients of anisotropy r_0, r_{45}, r_{90}; yield stress $\sigma_0, \sigma_{45}, \sigma_{90}$; balanced biaxial coefficient of anisotropy r_b and balanced biaxial yield stress, σ_b. Newton-Raphson numerical method with relaxation factor were employed to numerically calculate the a_i-values with high accuracy for assumed minimum error. The calculated eight a_i coefficients of anisotropy were validated and calibrated by comparing the predicted r-values and s-values with the experimental data for aluminium alloys [4] and present steel sheets.

2 The Non-Associated Barlat's Yld 2000-2d Anisotropic Plastic Flow Stress Potential

Defining $f\left(\sigma_{ij}, \overline{\varepsilon}, \dot{\overline{\varepsilon}}\right)$ as the yield stress function or yield stress surface and $\Phi(\sigma_{ji})$ the yield stress criterion, the general loading condition for the onset of plastic straining is,

$$f\left(\sigma_{ij}, \overline{\varepsilon}, \dot{\overline{\varepsilon}}\right) = \Phi\left(\sigma_{ij}\right) - \overline{\sigma}\left(\varepsilon_o, \overline{\varepsilon}, \dot{\overline{\varepsilon}}\right) = 0 \tag{1}$$

and commonly is a first degree homogeneous function of the Cauchy stress tensor, $\sigma_{ji} = (1/3)\,\sigma_{kk}\,\delta_{ij} + S_{ij}$, which defines the yield surface shape (S_{ij} is the stress deviator), $\overline{\sigma}$ is the current equivalent flow stress and is identified as scalar function of the equivalent plastic strain $\overline{\varepsilon}$, pre-strain ε_o and strain rate $\dot{\overline{\varepsilon}}$. The equivalent flow stress represents the yield stress surface size and material hardening law. The modified Swift equation is: $\overline{\sigma}\left(\varepsilon_o, \overline{\varepsilon}, \dot{\overline{\varepsilon}}\right) = \sigma_o(1 + \overline{\varepsilon}/\varepsilon_o)^n\left(\dot{\overline{\varepsilon}}/\dot{\varepsilon}_o\right)^M$, where σ_o is the material yield stress limit. Assuming the existence of a plastic flow potential, the plastic strain increments $d\varepsilon_{ij}$ can be calculated from the plastic flow stress potential $\Psi(\sigma_{ij})$ as,

$$d\varepsilon_{ij} = d\lambda\,\frac{\partial\,\Psi\left(\sigma_{ij}\right)}{\partial\,\sigma_{ij}} \tag{2}$$

where $d\lambda$ is the plastic multiplier increment and σ_{ij} are the stress tensor components. Usually, it is assumed that the plastic potential coincides with the yield stress criterion: $\Psi(\sigma_{ij}) \equiv \Phi(\sigma_{ij})$.

As mentioned above, the Barlat's Yld2000-2d anisotropic yield stress criterion [5] applied to sheet metal forming for plane stress condition and orthotropic anisotropy was defined as the sum of two scalar non-quadratic functions, which are generated by two linear transformations of the deviatoric stress tensor $\mathbf{S_{ij}}$ or the Cauchy true stress tensor σ_{ij}: $S'_{ij} = C'\,S_{ij} = L'_{ij}\,\sigma_{ij}$ and $S''_{ij} = C''\,S_{ij} = L''_{ij}\,\sigma_{ij}$,

$$\Phi = \Phi' + \Phi'' = \left|S'_1 - S'_2\right|^m + \left|2S''_1 + S''_2\right|^m + \left|2S''_2 + S''_1\right|^m = 2\,\overline{\sigma}^m \tag{3}$$

where $\overline{\sigma}$ is the equivalent stress, S'_j and S''_k are the principal values of the transformed stress tensors S'_{ij} and S''_{ij}. Also, the proposed function Φ is convex for exponent $m > 1$ [5]. These principal values, in Eq. (3), can be expressed in terms of the Cauchy stress tensor components by the definition of proposed new coefficients of plastic anisotropy, a_i, assuming (x,y)-axis as the material principal cartesian axis of orthotropic *anisotropy* in the sheet plane and the two matrices $\mathbf{C'}$ and $\mathbf{C''}$ expressed as a function of eight coefficients of anisotropy, thus [4],

$$\sigma'_1 = 2S''_1 + S''_2 = \left(a_1\,\sigma_{xx} + a_2\,\sigma_{yy}\right)/2 + \sqrt{\Delta'} \tag{4a}$$

$$\sigma'_2 = 2S''_2 + S''_1 = \left(a_1\,\sigma_{xx} + a_2\,\sigma_{yy}\right)/2 - \sqrt{\Delta'} \tag{4b}$$

$$\sigma''_1 - \sigma''_2 = S'_1 - S'_2 = 2\sqrt{\Delta''} \tag{4c}$$

where,

$$\begin{aligned}\Delta' &= \left(\frac{a_3\,\sigma_{xx} - a_4\,\sigma_{yy}}{2}\right)^2 + \left(a_8\tau_{xy}\right)^2; \\ \Delta'' &= \left(\frac{a_5\,\sigma_{xx} - a_6\,\sigma_{yy}}{2}\right)^2 + \left(a_7\tau_{xy}\right)^2\end{aligned} \tag{4d,e}$$

where a_i are the proposed eight new coefficients of material plastic anisotropy [4], which are related to the original linear transformation coefficients α_i. Material plastic isotropy is when $a_i = \alpha_i = 1$. The two anisotropy matrices L'_{ij} and L''_{ij} can be expressed as a function of eight new coefficients of material plastic anisotropy [4],

$$a_{ij} = \begin{bmatrix} a_1 & a_2 & 0 & 0 \\ a_3 & a_4 & 0 & 0 \\ a_5 & a_6 & 0 & 0 \\ 0 & 0 & a_7 & 0 \\ 0 & 0 & 0 & a_8 \end{bmatrix} = \begin{bmatrix} 3(L''_{11}+L''_{21}) & 3(L''_{22}+L''_{12}) & 0 & 0 \\ (L''_{11}-L''_{21}) & (L''_{22}-L''_{12}) & 0 & 0 \\ (L'_{11}-L'_{21}) & (L''_{22}-L''_{12}) & 0 & 0 \\ 0 & 0 & L'_{66} & 0 \\ 0 & 0 & 0 & L''_{66} \end{bmatrix} \tag{5}$$

Substituting Eqs. (4a-e) into Eq. (3), Barlat's Yld 2000-2d anisotropic yield stress function can be written in terms of Cauchy stress tensor components and the coefficients of plastic anisotropy. The non-associated Barlat's Yld 2000-2d flow potential is assumed to be similar to the yield stress criterion function. However, the anisotropy parameters a_i-values will be substituted by $\tilde{a}_i$-values [4], assuming material with $M = 0$,

$$\begin{aligned} \Psi(\sigma_{xx}, \sigma_{yy}, \tau_{xy}, \tilde{a}_i) = \frac{1}{2^m}\left| \tilde{a}_1\sigma_{xx} + \tilde{a}_2\sigma_{yy} + \left[\left(\tilde{a}_3\sigma_{xx} - \tilde{a}_4\sigma_{yy}\right)^2 + \left(2\,\tilde{a}_8\tau_{xy}\right)^2\right]^{1/2}\right|^m \\ + \frac{1}{2^m}\left| \tilde{a}_1\sigma_{xx} + \tilde{a}_2\sigma_{yy} - \left[\left(\tilde{a}_3\sigma_{xx} - \tilde{a}_4\sigma_{yy}\right)^2 + \left(2\,\tilde{a}_8\tau_{xy}\right)^2\right]^{1/2}\right|^m \\ + \left|\left(\tilde{a}_5\sigma_{xx} - \tilde{a}_6\sigma_{yy}\right)^2 + \left(2\,\tilde{a}_7\tau_{xy}\right)^2\right|^{m/2} = 2\bar{\sigma}^m = 2\sigma_o^m(1+\bar{\varepsilon}/\varepsilon_o)^{nm} \end{aligned} \tag{6}$$

2.1 Coefficients of Anisotropy from the Non-associated Plastic Potential

For a standard tensile test specimen cut at angle θ to RD, loading in uniaxial tensile stress σ_θ in the x'-axis of tensile direction and y'-axis the width direction, for plane stress condition, the stress components and strain increments in the anisotropy (x,y)-axis or (RD,TD)-axis can be obtained. For a standard tensile test specimen under a uniaxial tensile stress σ_θ, the equivalent stress can be expressed as $\overline{\sigma} = A\,\sigma_\theta$, where $A = \{2^m F + |B + C|^m + |B - C|^m\}^{1/m}/2^{1+1/m}$. Then, assuming Barlat´s non-associated flow rule and applying the Euler's identity theorem for a homogeneous stress function of degree 'm' to the Barlat's yield stress criterion function $\Psi\left(\sigma_{xx}, \sigma_{yy}, \tau_{xy}, \tilde{a}_i\right)$, after rearranging, the Lankford´s coefficient of anisotropy and the equal biaxial stress coefficient of anisotropy can be calculated by [4],

$$r_\theta = \frac{d\varepsilon_{width}}{d\varepsilon_{thickness}} = \frac{\partial\Psi/\partial\sigma_{y'y'}}{\partial\Psi/\partial\sigma_{zz}} = \frac{2mA^m\sigma_\theta^{m-1}}{\frac{\partial\Psi}{\partial\sigma_{xx}} + \frac{\partial\Psi}{\partial\sigma_{yy}}} - 1 \tag{7}$$

$$r_b = \frac{d\varepsilon_{TD}}{d\varepsilon_{RD}} = \frac{d\varepsilon_{yy}}{d\varepsilon_{xx}} = \frac{\partial\Psi/\partial\sigma_{yy}}{\partial\Psi/\partial\sigma_{xx}} \tag{8}$$

The derivatives of the stress function $\Psi\left(\sigma_{xx}, \sigma_{yy}, \tau_{xy}, \tilde{a}_i\right)$ seen in Eqs. (7–8) are obtained from Eq. (6), considering the derivative chain rule and the derivative sign for the modulus function, thus, they are calculated by [4],

$$\frac{\partial \Psi}{\partial \sigma_{xx}} = \left\{ 2mE\ \tilde{a}_5 + \frac{m}{2^m}(B+C)^{m-1}(\tilde{a}_1 + D\tilde{a}_3) + \frac{m}{2^m}(B-C)^{m-1}(\tilde{a}_1 - D\tilde{a}_3) \right\} \sigma_\theta^{m-1} \tag{9}$$

$$\frac{\partial \Psi}{\partial \sigma_{yy}} = \left\{ -2mE\ \tilde{a}_6 + \frac{m}{2^m}(B+C)^{m-1}(\tilde{a}_2 - D\tilde{a}_4) + \frac{m}{2^m}(B-C)^{m-1}(\tilde{a}_2 + D\tilde{a}_4) \right\} \sigma_\theta^{m-1} \tag{10}$$

Finally, substituting into Eqs. (7–8), the generalized Lankford's coefficient of anisotropy r_θ and the equal biaxial coefficient of anisotropy r_b are calculated by,

$$r_\theta = \frac{2^m F + |B+C|^m + |B-C|^m}{2^{m+1}E(\tilde{a}_5 - \tilde{a}_6) + (B+C)^{m-1}(A + D_{34}) + (B-C)^{m-1}(A - D_{34})} - 1 \tag{11}$$

$$r_b = \frac{2^m|\tilde{a}_5 - \tilde{a}_6|^m + |(\tilde{a}_1+\tilde{a}_2) + |\tilde{a}_3 - \tilde{a}_4||^m + |(\tilde{a}_1+\tilde{a}_2) - |\tilde{a}_3 - \tilde{a}_4||^m}{2^{m+1}LB\tilde{a}_5 + G_1^{m-1}(\tilde{a}_1 + H_{34}\tilde{a}_3) + G_2^{m-1}(\tilde{a}_1 - H_{34}\tilde{a}_3)} - 1 \tag{12}$$

Where,

$$A = \tilde{a}_1 + \tilde{a}_2;\ B = \tilde{a}_1 \cos^2\theta + \tilde{a}_2 \sin^2\theta \tag{13a,b}$$

$$C = \left[\left(\tilde{a}_3\cos^2\theta - \tilde{a}_4\sin^2\theta\right)^2 + 4\,\tilde{a}_8^2\sin^2\theta\cos^2\theta\right]^{1/2} \tag{13c}$$

$$D = \left(\tilde{a}_3\cos^2\theta - \tilde{a}_4\sin^2\theta\right)/C;\ D_{34} = D(\tilde{a}_3 - \tilde{a}_4) \tag{13d,e}$$

$$E = \left(\left(\tilde{a}_5\cos^2\theta - \tilde{a}_6\sin^2\theta\right)^2 + 4\,\tilde{a}_7^2\sin^2\theta\cos^2\theta\right)^{(m/2)-1}\left(\tilde{a}_5\cos^2\theta - \tilde{a}_6\sin^2\theta\right) \tag{13f}$$

$$F = \left|\left(\tilde{a}_5\cos^2\theta - \tilde{a}_6\sin^2\theta\right)^2 + 4\tilde{a}_7^2\sin^2\theta\cos^2\theta\right|^{m/2} \tag{13g}$$

$$G_1 = (\tilde{a}_1 + \tilde{a}_2) + |\tilde{a}_3 - \tilde{a}_4|;\ G_2 = (\tilde{a}_1 + \tilde{a}_2) - |\tilde{a}_3 - \tilde{a}_4| \tag{13h,i}$$

$$LB = (\tilde{a}_5 - \tilde{a}_6)^{m-1};\ H_{34} = (\tilde{a}_3 - \tilde{a}_4)/|\tilde{a}_3 - \tilde{a}_4| \tag{13k,l}$$

For a tensile test specimen under uniaxial tensile stress σ_θ, the equivalent stress of Eq. (6) is $\overline{\sigma} = A\,\sigma_\theta$. Consequently, the normalized yield stress at any angle θ to RD is,

$$\frac{\sigma_\theta}{\sigma_o} = \frac{2^{1+1/m}(1+\overline{\varepsilon}/\varepsilon_o)^n}{\left\{2^m F + |B+C|^m + |B-C|^m\right\}^{1/m}} \tag{14}$$

where σ_o is the material yield stress limit. Assuming the current equivalent flow stress, $\overline{\sigma} = \sigma_o(1+\overline{\varepsilon}/\varepsilon_o)^n$, at any current yielding condition, hence, the normalized equal biaxial stress, $\sigma_{xx} = \sigma_{yy} = \sigma_b$ *and* $\tau_{xy} = 0$, is calculated by,

$$\frac{\sigma_b}{\sigma_o} = \frac{2^{1+1/m}(1+\overline{\varepsilon}/\varepsilon_o)^n}{\left\{2^m(a_5 - a_6)^m + |(a_1+a_2) + |a_3 - a_4||^m + |(a_1+a_2) - |a_3 - a_4||^m\right\}^{1/m}} \tag{15}$$

Substituting into Eq. (6) the boundary conditions for a uniaxial tensile test specimen under loading at 0° RD, the stress components are $\sigma_{xx} = \sigma_0$ and $\sigma_{yy} = \tau_{xy} = 0$, thus,

$$|2a_5|^m + |a_1 + a_3|^m + |a_1 - a_3|^m - 2^{1+m}(1 + \bar{\varepsilon}/\varepsilon_o)^n = 0 \tag{16}$$

Similarly, for a specimen under pure shear stress test, the stress components are $\sigma_{xx} = \sigma_{yy} = 0$ and $\tau_{xy} = \tau_o = k$, where $\tau_0 = k$ is the material shear yield stress limit, the Eq. (6) is reduced to: $a_7 = \pm 2^{(1/m)-1}(\sigma_o/\tau_o) = \pm 2^{(1/m)-1}\sigma_o/k$.

2.2 The FLC Strain Path in the MPLS Map

The linear strain path β to calculate the FLC_{RD} curve in the Map of Principal Surface Limit Strains (MPLS) is obtained from Eqs. (9–10). Assuming the principal coordinate axis coincident with the principal direction of anisotropy axis, (1,2)-axis = (x,y)-axis = (RD,TD), and defining the stress ratio $X = \sigma_2/\sigma_1$, the linear strain path is [4],

$$\beta = \frac{d\varepsilon_2}{d\varepsilon_1} = \frac{\partial\Psi/\partial\sigma_{yy}}{\partial\Psi/\partial\sigma_{xx}} = \frac{-2^{m+1}L\,\tilde{a}_6 + (K_1)^{m-1}(\tilde{a}_2 - M_{34}\,\tilde{a}_4) + (K_2)^{m-1}(\tilde{a}_2 + M_{34}\,\tilde{a}_4)}{2^{m+1}L\,\tilde{a}_5 + (K_1)^{m-1}(\tilde{a}_1 + M_{34}\,\tilde{a}_3) + (K_2)^{m-1}(\tilde{a}_1 - M_{34}\,\tilde{a}_3)} \tag{17}$$

where,

$$L = |\tilde{a}_5 - \tilde{a}_6X|^{m-2}(\tilde{a}_5 - \tilde{a}_6X);\ K_1 = (\tilde{a}_1 + \tilde{a}_2X) + |\tilde{a}_3 - \tilde{a}_4X|;$$
$$K_2 = (\tilde{a}_1 + \tilde{a}_2X) - |\tilde{a}_3 - \tilde{a}_4X|; M_{34} = (\tilde{a}_3 - \tilde{a}_4X)/|\tilde{a}_3 - \tilde{a}_4X| \quad (18a,b,c,d)$$

2.3 The Shear Stress Fracture Criterion to Predict FLC_{TD} in the MPLS Map

The Bressan-Williams shear stress fracture criterion [13] was proposed firstly in 1983 to predict ductile local necking in sheet metal forming. It is an instability stress point of maximum in a critical shear stress just when local necking initiate by shear bands formation. This maximum shear stress is a critical value that acts in a plane inclined through the sheet thickness and in the direction of *pure shear strain*, namely τ_{cr}, that is a material property which depends on temperature and strain rate. Considering the experimental observations of fast shear rupture without visible local necking in sheet metal shaping of strain-rate-independent materials, M = 0, this approach was modified by Bressan and Barlat [14], using the anisotropic Barlat´s Yld-2000-2d yield criterion.

However, assuming the maximum loading axis of specimen coincident with the *transverse direction* of anisotropy axis, (2,1)-axis = (TD,RD)-axis, and defining the stress ratio $X_T = \sigma_1/\sigma_2$, linear strain path is $\beta_T = de_1/de_2$, the governing equation of fast shear stress fracture criterion to calculate FLC_{TD} curve in the MPLS map is [14],

$$\sigma_2 = \left|(2 + \beta_T)/\sqrt{1 + \beta_T}\right| \tau_{cr} \tag{19}$$

where $\sigma_2 > 0$ is the maximum principal in-plane true stress coincident with TD, τ_{cr} the critical shear stress. Assuming materials exhibiting planar anisotropy and strain hardening behavior only, strain rate sensitivity exponent is neglected, M = 0, the equivalent flow stress for linear deformations is described by the Swift's law, $\overline{\sigma} = K(\varepsilon_o + \overline{\varepsilon})^n$, where K is the material strength coefficient, $\overline{\varepsilon}$ the equivalent strain, ε_o the pre-strain, n the material strain hardening exponent. Assuming the equivalent plastic strain increment calculated from the definition of equivalent plastic work increment, using the previous definition of strain path β_T and stress ratio X_T, is obtained,

$$d\overline{\varepsilon} = (\sigma_2/\overline{\sigma})(1 + X_T \beta_T)d\varepsilon_2 \tag{20}$$

Assuming linear plastic strain path up to the transverse FLC_{TD} curve, thus, integrating Eq. (20), the predicted major true plastic strain limit ε_{2f} is calculated by [14],

$$\varepsilon_{2f} = \left\{\frac{1}{2}\left(\left|\frac{\sigma_1'' - \sigma_2''}{\sigma_2}\right|^m + \left|\frac{\sigma_1'}{\sigma_2}\right|^m + \left|\frac{\sigma_2'}{\sigma_2}\right|^m\right)\right\}^{\frac{1+n}{mn}} \left|\frac{2+\beta_T}{\sqrt{1+\beta_T}}\right|^{\frac{1}{n}} \frac{(\tau_{cr}/K)^{\frac{1}{n}}}{(1+X_T\beta_T)} - \left\{\frac{1}{2}\left(\left|\frac{\sigma_1'' - \sigma_2''}{\sigma_2}\right|^m + \left|\frac{\sigma_1'}{\sigma_2}\right|^m + \left|\frac{\sigma_2'}{\sigma_2}\right|^m\right)\right\}^{\frac{1}{m}} \frac{\varepsilon_o}{(1+X_T\beta_T)} \tag{21}$$

The non-dimensional critical shear stress parameter τ_{cr}/K is a material property, which can be obtained and calibrated from experimental bulge test or the plane strain point, FLC_o, in the forming limit strain curve point at the minor true strain $\varepsilon_1 = 0$.

3 Materials and Experimental Anisotropy Parameters

The experimental data of AISI 439 steel available in the literature [11] for simple uniaxial tension and equal biaxial stress tests are shown in Tables 1.. The plastic anisotropy parameters used to calibrate the model were: Lankford coefficient of anisotropy r-values, normalized tensile yield stress s-values ($=\sigma_\theta/\sigma_{0.19}$), and equal biaxial stress parameters r_b and σ_b at strain $\varepsilon = 0.19$, according to Barlat et al. [6, 8].

Table 1. Mechanical plastic anisotropy parameters of studied AISI 439 steel [11].

Metal alloy	Anisotropy coefficients, r-values							
	ro	r15	r30	r45	r60	r75	r90	rb
AISI 439 [11]	1.60	1.45	1.16	1.03	1.45	2.25	2.53	1.32
Normalized yield stress calibrated at $\varepsilon = 0.19$, s-values ($= \sigma_\theta/\sigma_0$ and σ_b/σ_0)								
	σ_0	σ_{15}	σ_{30}	σ_{45}	σ_{60}	σ_{75}	σ_{90}	σ_b
AISI 439 [11]	1.00	1.008	1.031	1.059	1.059	1.027	1.008	1.238

The plastic strain hardening law of the AISI 439 steel sheet of thickness 1.0 mm were obtained from experimental tests, which was applied to calculate the theoretical transverse FLC_{TD} curves: the average tensile tests: $\overline{\sigma} = 732(0.009 + \overline{\varepsilon})^{0.235}$ MPa, according to the uniaxial experimental results by Lian et al. [11].

4 Results and Discussions

The system of exact non-linear equations, Eq. (11), Eq. (12), Eq. (14), Eq. (15) and Eq. (16), constitute a set of eight equations with eight experimental anisotropy properties and the unknown variables: the anisotropy coefficients $\tilde{a}_i$, i = 1,..8. The details of Newton-Raphson numerical method were presented in the paper by Bressan and Donadon [4]. Therefore, it is necessary to choose among these, eight independents equations and eight experimental parameters to form a system of non-linear equations to find numerically the anisotropy coefficients $\tilde{a}_i$. The Newton-Raphson method with a relaxation factor is a robust, accurate and reasonably fast numerical method which was employed to solve this system of non-linear equations within a given tolerance.

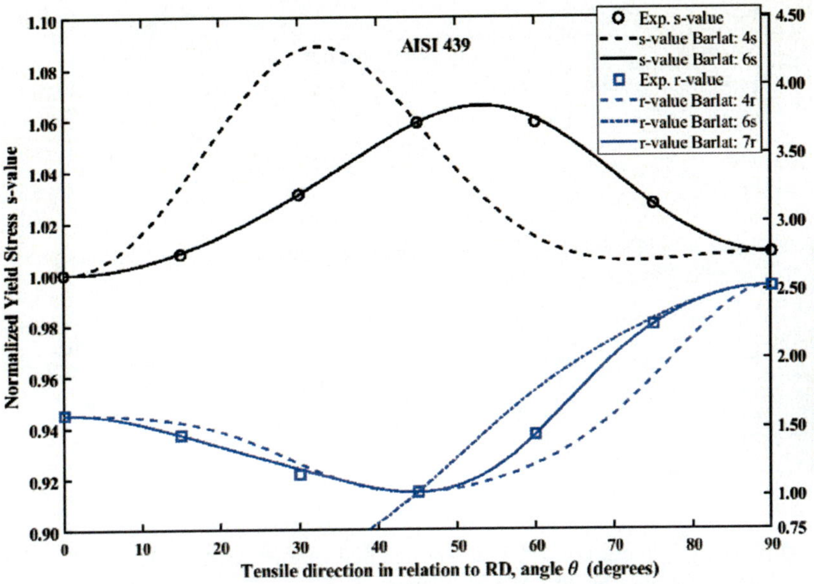

Fig. 1. Comparisons of predicted r-ralue and s-value curves for the associated and the non-associated Barlat's Yld 2000-2d yield stress with experimental results for tensile tests of AISI 439 [11], exponent m = 6, using Newton-Raphson method with experimental calibration parameters: a) associated rule, 4s- and 4r-values: σ_0, σ_{45}, σ_{90}, σ_b, r_o, r_{45}, r_{90}, r_b and b) non-associated plastic potential, 6s- and 2r-values: σ_o, σ_{30}, σ_{45}, σ_{75}, σ_{90}, σ_b, r_{45}, r_{90}; c) non-associated plastic potential, 7r-values and 1s-value: r_0, r_{15}, r_{45}, r_{60}, r_{75}, r_{90}, r_b, σ_{45}.

In Fig. 1, comparison of predicted curves for r-values and s-values versus angle θ to RD with experimental points of AISI 439, using the Barlat's Yld 2000-2d yield criterion and corresponding associate flow rule and employing the recommended material selection of calibration parameters: 4 r-values and 4 s-values (r_0, r_{45}, r_{90}, r_b, σ_0, σ_{45}, σ_{90} and σ_b), plotted in the same graph. Although the curve fitting passes exactly through the previously selected 4 r-values and 4 s-values points, for others experimental points the correlation is poor, particularly for s-values points, indicating the necessity of better curve fit. The curve fit is noticeably increased when selecting 6 s-values points for calibration of the normalized yield stress and 7 r-values points for calibration of the Lankford coefficients of anisotropy. Therefore, the yield stress criterion and plastic potential functions are not coincident, but are two distinct functions. The calculated a_i and $\tilde{a}_i$ plastic anisotropy coefficients of Barlat Yld 2000-2d stress functions for curve fitting of Fig. 1 are seen in Table 2.. Comparison between the calculated and experimental balanced biaxial plasticity parameters are shown in Table 3..

Table 2. The calculated plastic anisotropy coefficients for the Barlat Yld 2000-2d yield stress criterion $\Phi(\sigma_{ij}, a_i)$ and the independent non-associated Barlat Yld 2000-2d plastic stress potential $\Psi(\sigma_{ij}, \tilde{a}_i)$ of AISI 439 steel, using m = 6 and experimental points.

Exper	a_1	a_2	a_3	a_4	a_5	a_6	a_7	a8
Φ:4s + 4r	0.067310	1.443596	0.926687	0.657302	1.120455	0.907724	0.906021	1.180186
Φ:6s + 2r	0.046876	1.215132	1.545069	1.004261	1.077962	0.583045	0.917654	1.214749
Exper	$\tilde{a}_1$	$\tilde{a}_2$	$\tilde{a}_3$	$\tilde{a}_4$	$\tilde{a}_5$	$\tilde{a}_6$	$\tilde{a}_7$	$\tilde{a}_8$
Ψ: 7r + 1s	-1.457668	0.408759	0.096955	2.187854	−1.356585	−1.091135	.000663	0.506318

Table 3. The comparison between the experimental and calculated balanced biaxial plasticity parameters of AISI 439 steel for the numerical cases.

Experimental points	m	$r_{b\,exp}$	$\sigma_{b\,exp}$	r_b calculated	σ_b calculated
$\Psi \equiv \Phi$: 4s + 4r	6			1.320	1.238
Φ: 6s + 2r	6	1.32	1.238	0.038	1.239
Ψ: 7r + 1s	6			1.32	0.871

Comparisons of the predicted and experimental FLC_{TD} of AISI 439 steel [11] are seen in Fig. 2. The Bressan-Barlat critical shear stress criterion was calibrated with the experimental data at the plane strain condition point: $FLC_o = 0.271$ and m = 6. The calibrated normalized critical shear stress obtained was: for associated $\tau_{cr}/K = 0.532$ and $\tau_{cr}/K = 0.335$ for the and non-associate Barlat's plastic potential case. The non-associated Barlat´s Yld 2000-2d plastic flow potential calibrated with 7 r-values provided a better fit with the experimental FLC_{TD} transverse curve of AISI 439 steel.

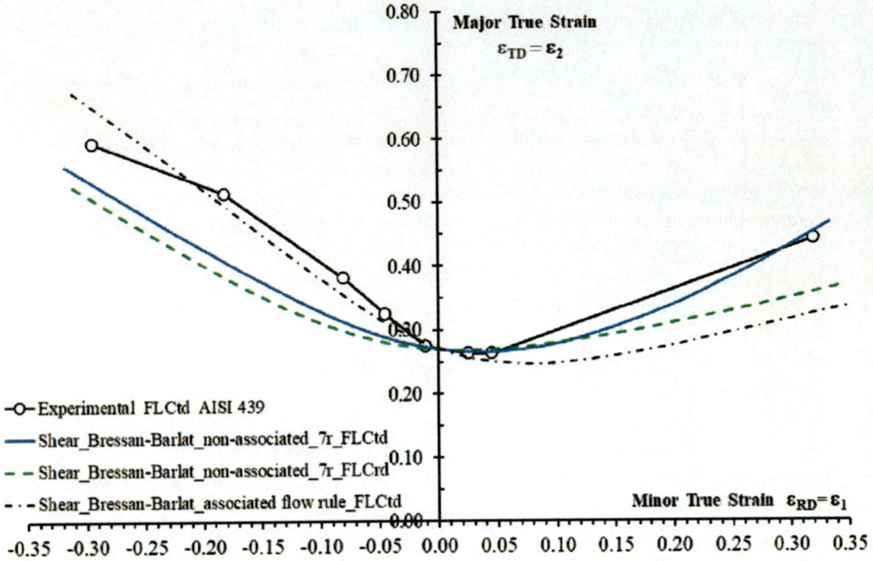

Fig. 2. Comparisons of the predicted and experimental FLC_{TD} of AISI 439 steel sheet [11], using the Bressan´s shear stress fracture criterion. The Bartlat´s Yld 2000-2d yield stress function was employed for the associated (4r) and the non-associated plastic stress potential (7r).

5 Conclusions

Prediction of the forming limit strain curve of AISI 439 steel was improved substantially when calibrating the non-associated Barlat´s Yld 2000-2d plastic potential with 7 r-values and using the Bressan-Barlat shear stress fracture criterion. Furthermore, Barlat´s Yld 2000-2d non-associated flow rule provides a better fit with the experimental Lankford and equal biaxial coefficients of anisotropy than the associated rule.

References

1. Hill, R.: A theory of the yielding and plastic flow of anisotropic metals. Proc. R. Soc. London A **193**, 281–297 (1948)
2. Barlat, F., Whan, J., Yoon, J.W., Cazacu, O.: On linear transformations of stress tensors for the description of plastic anisotropy. Int. J. Plast. **23**, 876–896 (2007)
3. Yoon, J.W., Lou, Y., Yoon, J., Glazoff, M.V.: Asymmetric yield function based on the stress invariants for pressure sensitive metals. Int. J. Plast. **56**, 184–202 (2014)
4. Bressan, J.D., Donadon, M.V.: An improved anisotropic non-associated plastic potential based on the Barlat's Yld 2000–2d yield stress criterion. J. Mater. Eng. Perform., ASTM, online March/2023 (2023). https://doi.org/10.1007/s11665-023-07799-4
5. Barlat, F., et al.: Plane stress yield function for aluminum alloy sheets - Part 1: theory. Int. J. Plast. **19**, 1297–1319 (2003)
6. Hershey, A.V.: The plasticity of an isotropic aggregate of anisotropic face centred cubic crystals. J. Appl. Mech. Trans. ASME **21**, 241–249 (1954)

7. Hosford, W.F.: A generalized isotropic yield criterion. J. Appl. Trans. ASME **39**, 607–609 (1972)
8. Barlat, F., et al.: Yield function development for aluminium alloy sheets. J. Mech. Phys. Solids **45**, 1727–1763 (1997)
9. Stoughton, T.B.: A non-associated flow rule for sheet metal forming. Int. J. Plast. **18**, 687–714 (2002)
10. Park, T., Chung, K.: Non-associated flow rule with symmetric stiffness modulus for isotropic-kinematic hardening and its application for earing in circular cup drawing. Int. J. Solids Struct. **49**, 3582–3593 (2012)
11. Lian, J., et al.: An evolving non-associated Hill48 plasticity model accounting for anisotropic hardening and r-value evolution and its application to forming limit prediction. Int. J. Solids Struct. **151**, 20–44 (2017)
12. Safaei, M., Yoon, J.W., De Waele, W.: Study on the definition of equivalent plastic strain under non-associated flow rule for finite element formulation. Int. J. Plast. **58**, 219–238 (2014)
13. Bressan, J.D., Williams, J.A.: The use of a shear instability criterion to predict local necking in sheet metal deformation. Int. J. Mech. Sci. **25**(3), 155–168 (1983)
14. Bressan, J.D., Barlat, F.: A shear fracture criterion to predict limit strains in sheet metal forming. Int. J. Mater. Form. **3**(1), 235–238 (2010)

The Role of Local Crystallographic Texture on Strain Localization at Weld Seams in Al-Mg-Si Porthole Die Extrusions

Andrew Zang[1], Jean-François Béland[2], Yu Wang[3], Nick Parson[4], and Warren J. Poole[1](✉)

[1] The University of British Columbia, Vancouver, BC V6T 1Z4, Canada
warren.poole@ubc.ca
[2] National Research Council Canada (NRC), Saguenay, Québec, Canada
[3] University of Waterloo, Waterloo, ON N2L 2G1, Canada
[4] Rio Tinto Aluminium, Saguenay, Québec, Canada

Abstract. The use of hollow aluminum extrusions fabricated using porthole dies in automotive applications is increasing. In a porthole die, the aluminum feed splits into multiple paths to pass around the bridges which support the internal mandrel and then recombines in the weld chamber of the die. A challenge for these products is the spatial variation of microstructure near the weld seam, in particular crystallographic texture. In the current study, an Al-Mg-Si alloy with a high density of Mn/Cr containing dispersoids was extruded through a simple die with two portholes and a single bridge into a rectangular strip. The high density of dispersoids minimized recrystallization during and after the extrusion process. The microstructure in the profile was analyzed using electron backscatter diffraction (EBSD) maps. It was observed that the microstructure could be segmented into a number of regions with different crystallographic textures. Using the information on local texture, the visco-plastic self-consistent (VPSC) polycrystal plasticity code was used to predict the local anisotropic plastic response in each region. The VPSC results were used to fit a Barlat YLD2004-18p flow rule which was employed in an LS-Dyna finite element method (FEM) model to predict the spatial variation of plastic strain near the weld seam. The FEM simulations were validated with tensile testing where the local strain was characterized using digital image correlation (DIC). Good agreement was found between the DIC and FEM results.

Keywords: Aluminum extrusion · porthole dies · weld lines · crystallographic texture · digital image correlation

1 Introduction

The use of hollow profile aluminum extrusions is increasing in automotive applications due to the need for vehicle weight reduction in either traditional internal combustion engine or battery electric vehicles [1, 2]. Hollow profiles can be produced using dies where the internal mandrel is supported outside the die (thereby limiting the number

K. Mocellin et al. (Eds.): ICTP 2023, LNME, pp. 427–434, 2024.
https://doi.org/10.1007/978-3-031-40920-2_44

of internal cavities) or by porthole dies where the aluminum splits into different ports which then weld back together after the die bridges before exiting the die. It is of interest to understand the role of the complex strain path in porthole dies and how this affects the mechanical response of the extrudate during forming or crash scenarios. Recently, Wang and co-workers have proposed that the modification of crystallographic texture near the weld line due to flow through the porthole die affects the tensile response of the extrudate [3, 4]. In cases where the microstructure is fully recrystallized after extrusion, the weld line region is slightly stronger than the surrounding material and localization of strain when tested in tension perpendicular to the weld line is pushed outside the weld line region. In contrast, unrecrystallized extrudates exhibit a softer region near the weld line and localization occurs in this region. The objective of this paper is to examine the relationship between the experimentally measured local crystallographic texture, its predicted anisotropic plastic response and the resulting spatial distribution of plastic strain for loading perpendicular to the weld line.

2 Methodology

Aluminum billets with a diameter of 101 mm and a composition of 0.7 Mg, 1.0 Si, 0.2 Fe, 0.5 Mn and 0.15 Cr were direct chill cast by Rio Tinto Aluminium. The billets were homogenized by heating at a rate of 250 °C/h to 500 °C and then a lower heating rate of 50 °C/h to a final soak temperature of 550 °C, held for 2h and then water quenched. The billets were then re-heated to 400 °C and extruded on a laboratory extrusion press. In the die, the aluminum billet was split into two portholes which were separated by a streamlined bridge. The two streams rejoined in the weld chamber, which had a length of 12.5 mm, before finally exiting through a 2.5 mm × 50 mm die opening (see Fig. 1 [3]). The extruded samples were artificially aged in an oil bath for 4h at 180 °C, i.e. to the peak, T6, strength.

Tensile tests were conducted on the extruded strips where the tensile axis was perpendicular to the weld line as shown in Fig. 2. Prior to testing, a black and white speckle was spray painted on the surface of the tensile sample so that the distribution of deformation along the gauge length could be quantified using digital image correlation (DIC). Optical images of the sample surface were acquired with a Titanar 2 8/50 digital camera at 6 Hz. The images were analyzed using the Aramis v6 DIC software in order to calculate the strain component in the tensile axis.

Electron backscatter diffraction (EBSD) was used to characterize the microstructure using a Zeiss-Σigma SEM equipped with a Nikon high speed camera. The EBSD maps were analyzed using the TSL OIM and MTEX software packages. Samples for EBSD were metallographically prepared in the usual manner using a final polishing step with 0.05 μm colloidal silica solution. Selected areas of interest from the EBSD were characterized in terms of their crystallographic texture by calculating orientation distribution functions (ODF). The ODF was used to create an input texture file for the visco-plastic self-consistent (VPSC, version 7d) polycrystal plasticity code [5]. The material constitutive law for VPSC was determined by a fit to a tensile test (tensile axis parallel to the transverse direction) from an extruded strip which did not have a weld line. A modified

Voce law was used, i.e.

$$\tau^s = \tau_0^s + (\tau_1^s + \theta_1^s \Gamma\left[1 - \exp\left(-\Gamma\left|\frac{\theta_0^s}{\tau_1^s}\right|\right)\right] \quad (1)$$

where τ_0^s is the critically resolved shear stress for the onset of plasticity, τ_1^s is the scaling stress, θ_0^s is the initial work hardening rate and θ_1^s is the large strain work (Stage IV) work hardening rate. The fit parameters are summarized in Table 1. A rate sensitivity of 0.02 and a grain interaction parameter (n_{eff}) of 10 were employed for the calculations. Calculations of the initial flow stress were conducted for tensile loading at angles of 0, 15, 30, 45, 60, 75 and 90° with respect to the transverse direction of the extrudate (and rotating around the extrusion direction). Using the VPSC data, a yield surface was constructed using the Barlat YLD2004-18p function.

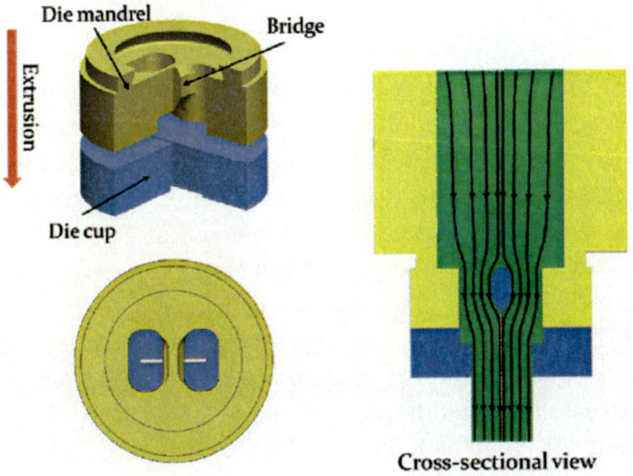

Fig. 1. Schematic diagram of porthole die and a plane view showing the flow streamlines through the die.

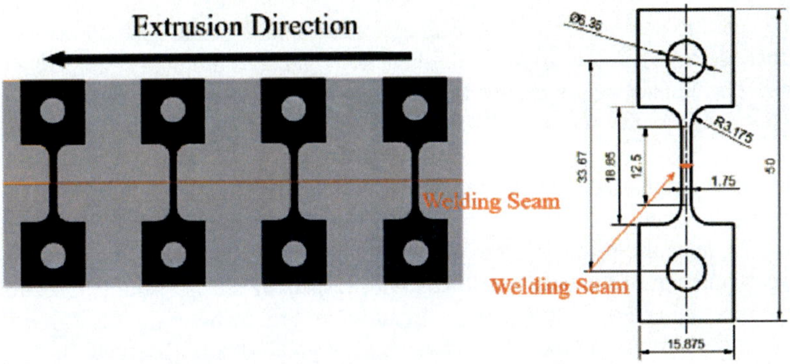

Fig. 2. Illustration of tensile samples and their orientation for tests perpendicular the weld seam.

The LS-Dyna 3D (explicit solver) was then used to conduct finite element method calculations on a tensile sample that was meshed to include regions with different material properties (as described by their different yield surface). The overall sample dimensions for the FEM simulation were 20 × 8 × 2.5 mm with a weld line region of 0.6 mm at the centre of the sample.

Table 1. Summary of fit parameters used for the constitutive law in the VPSC simulations.

Condition	τ_0^s (MPa)	τ_1^s (MPa)	θ_0^s (MPa)	θ_1^s (MPa)
T6	123	16	150	0

3 Results and Discussion

Figure 3 shows the EBSD inverse pole figure (IPF) maps for the end-on view of the extrusion where the microstructure has been delineated for the region around the weld line and a region near the ¼ width of the sample. It can be seen qualitatively that there is no evidence for recrystallization in either area and the grains near the weld line have a different crystallographic orientation to the ¼ width region. The ideal texture components have been reported previously [3] and are summarized in Table 2. It can be seen that the ¼ width region texture exhibits a typical high temperature plane strain deformation texture (consistent with the deformation path through the die for this region) texture with Copper, Brass and S components [4]. The region near the weld line is more complicated having a spatial distribution of different texture components. The data averaged over this area reported in Table 2 shows significant components of Cu, S and $Cube_{ED}$ components. This reflects the complex strain path the material follows as it is split into the portholes, rejoins in the weld chamber and then undergoes a very large deformation as it exits the die [4].

Figure 4 shows the DIC results for the distribution of strain in the tensile direction as a function of position along the gauge length (the centre of the sample is at the "0" position, i.e. the position of the weld line) for a far field strain of 0.05. The results show that, as expected, the strain is relatively homogeneous along the gauge length for the sample without a weld line. In contrast, the sample with a weld line exhibits strain localization adjacent to the weld line with a strain amplitude ~ 2.8 times the far-field value.

In order to evaluate the effect of the different local crystallographic textures, VPSC simulations were conducted using input files calculated from ODFs for each region. Figure 5 presents the results of the VPSC simulations for the yield stress and the R-value as a function of the tensile test axis with respect to the TD sample direction for both the ¼ width and weld line regions. In addition, Fig. 5 shows the fits of the Barlat YLD2004-18p surfaces to the VPSC data. A very good fit was observed for the yield stress and R-value for the ¼ width region and a reasonable fit was obtained for the more complicated weld line region. It can be seen from these data that both regions show considerable plastic anisotropy.

Figure 6 plots the YLD2004-18p yield surfaces in the TD-ND plane of the extrudate. Here, it can be observed that the flow stress for a tensile axis of 0° (parallel to the TD

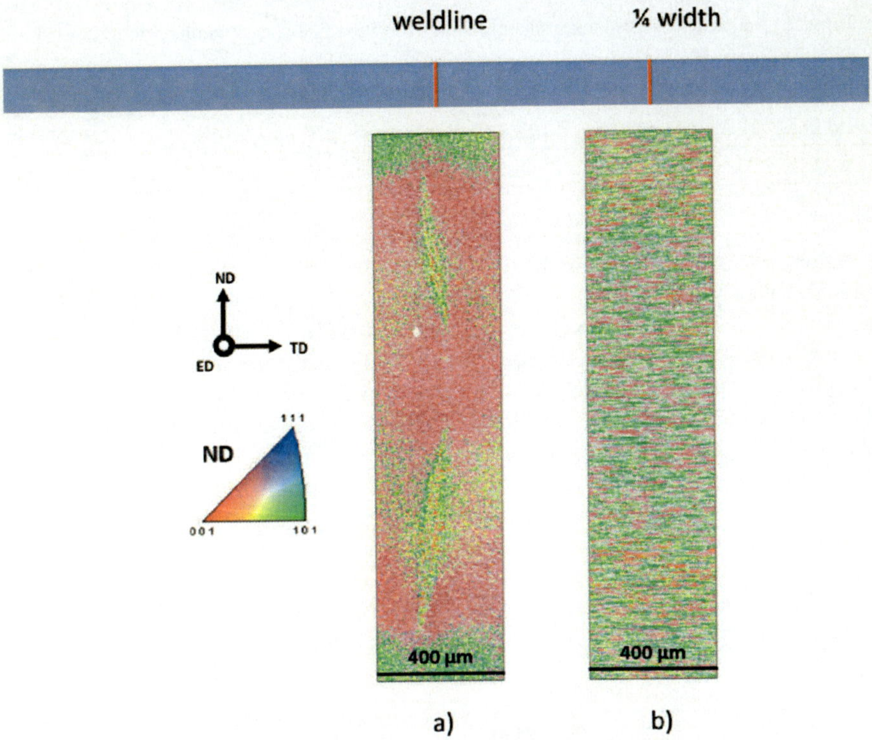

Fig. 3. EBSD IPF maps of the grain structure a) weld line and b) quarter width.

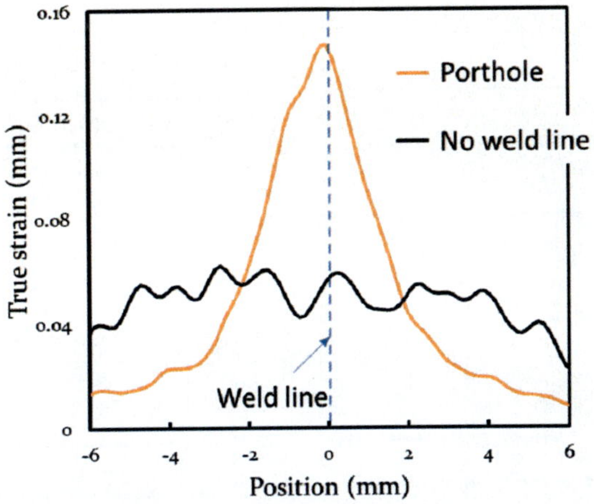

Fig. 4. Digital image correlation results for true strain distribution along a tensile sample with a far field strain of ~ 0.05. Sample was loaded in uniaxial tension perpendicular to the weld line.

Table 2. Summary of ideal texture components for weld line and ¼ width regions (%) [3].

	Cu	Br	S	Cube	Goss	CubeED	other
Weld line	36.7	4.5	22.5	3.7	8.1	14.5	10.0
¼ width	8.0	35.2	48.5	1.4	0.1	3.0	3.8

direction) is ~ 20 MPa lower in the weld line region. However, the shapes of the yield surfaces have significant differences and, in some regions, the weld line is softer than the ¼ region and in other regions it is stronger (however, it should be noted that the regions where the weld line is stronger are regions of near plane strain and equi-biaxial stress states for which the yield surface was not explicitly fit). In order, to evaluate these differences in plastic response, a FEM model was built with the mesh shown in Fig. 7a. The model consisted of two regions with different material behavior, i.e. the blue region for the weld line and the red region for the rest of the sample. Figure 7b presents the results for effective plastic strain at a far field strain of ~ 0.05, showing that a complex localized strain state develops in the weld line region. The largest value of strain is ~ 0.1, i.e. ~ twice the far field strain. This is consistent with the DIC results shown in Fig. 4, supporting the hypothesis that the variations in local crystallographic texture can account for the strain localization behavior seen experimentally.

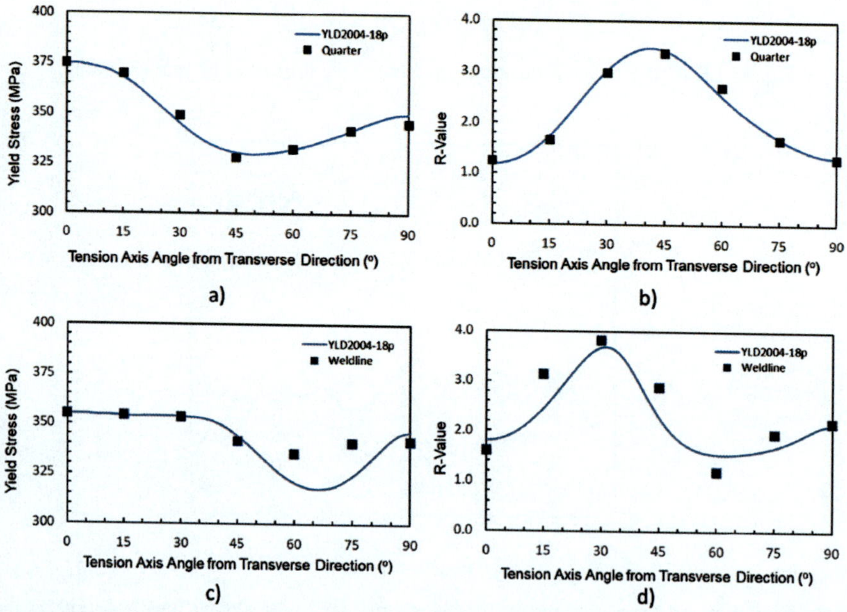

Fig. 5. Results for the orientation dependence of the yield stress (left figures) and R-value (right figures) for uniaxial tensile loading at different angles to the transverse direction. a) and b) for quarter width, c) and d) for the weld line. Solid symbols are from VPSC simulations and blue lines are the fit of the Barlat YLD2004-18p yield surface.

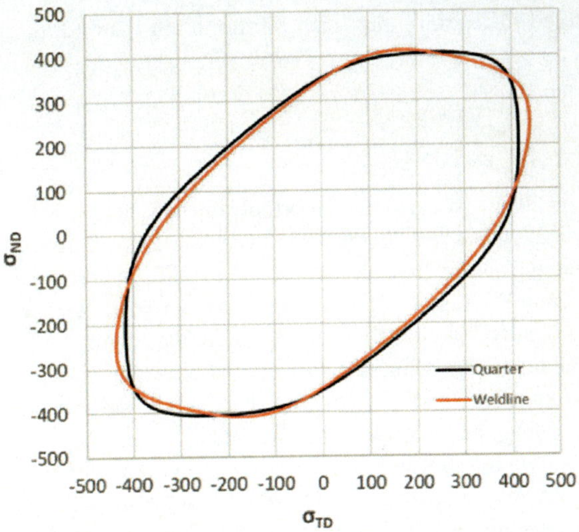

Fig. 6. Barlat YLD2004-18p yield surfaces fit using results from VPSC simulations for the weld line and quarter width regions.

The FEM calculations show a lower level of strain localization compared to the DIC results, i.e. 2 times vs 2.8 times the far field strain, respectively. However, it is noted that the use of one single region to represent the material behavior in the weld line region is a simplification. It can be observed from Fig. 3 that there is a more complex spatial distribution of crystallographic texture (the green, red and purple regions to first approximation). Currently, more detailed FEM calculations are being conducted to account for these local heterogeneities. In addition, higher resolution DIC experiments

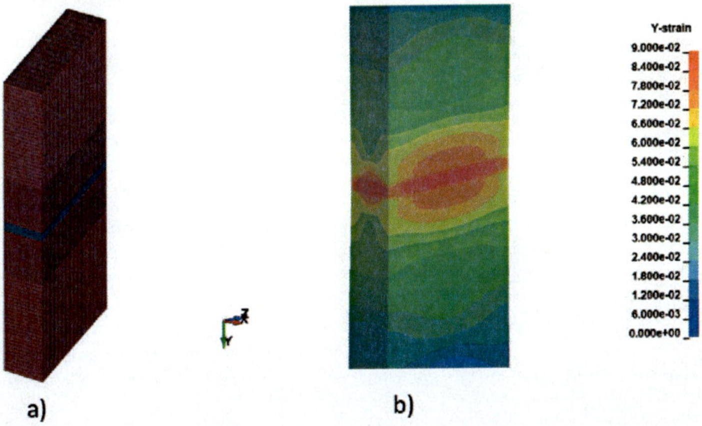

Fig. 7. Results from FEM simulations where weld line region and rest of sample were given different yield surfaces (far field strain of ~ 0.05). a) shows the FEM mesh and b) the strain parallel to the tensile axis.

are being performed to examine the local plastic strain distribution in the weld line region.

4 Summary

Aluminum extrusion through an idealized porthole die have been conducted under conditions where recrystallization was suppressed. It was found that the local crystallographic texture varied spatially in the cross-section of the profile. In particular, the region near the weld line was found to have a different texture compared to the quarter width. Digital image correlation results from tensile tests perpendicular to the weld seam showed that plastic strain localized in the weld line. Polycrystal plasticity simulations using VPSC for the weld line and quarter width regions found that they both exhibited an anisotropic plastic response. Using the VPSC results, Barlat YLD2004-18p yield surfaces were constructed for the two regions and then incorporated into a LS-Dyna 3D model of the tensile sample where the geometry was simplified into one region for the weld line and one for the rest of the sample. The FEM simulations indicated that plastic strain localized in the weld line region suggesting that the details of the local crystallographic texture and its associated anisotropic plastic response could explain the localization of strain in this region.

Acknowledgements. This work was undertaken, in part, thanks to funding from the Canada Research Chair program (Poole). Support of Rio Tinto Aluminium and the Natural Sciences and Engineering Research Council of Canada (NSERC) is gratefully acknowledged.

References

1. Scamans, G.: Electric vehicle spike demand for high strength aluminium extrusions. Light Metal Age **76**, 6–12 (2018)
2. Joost, W.J.: Reducing vehicle weight and improving U.S. Energy efficiency using integrated computational materials engineering. JOM **64**, 1032–1038 (2012)
3. Wang, Y., Zang, A., Wells, M., Poole, W.J., Parson, N.: Strain localization at longitudinal weld seams during plastic deformation of Al–Mg–Si–Mn–Cr extrusions: the role of microstructure. Mat. Sci. Eng. A **849**, 143454 (2022)
4. Wang, Y., Zang, A., Mahmoodkhani, Y., Wells, M., Poole, W., Parson, N.: The effect of bridge geometry on microstructure and texture evolution during porthole die extrusion of an Al–Mg–Si–Mn–Cr Alloy. Metall. Mater. Trans. A. **52**(8), 3503–3516 (2021)
5. Lebensohn, R.A., Tomé, C.N.: A self-consistent anistropic approach for the simulation of plastic deformation and texture development of polycrystals: Application to zirconium alloys. Acta Metall. Mater. **41**, 2611–2624 (1993)

Integration of Local Formability Limits of AHSS into Local/Global Formability Map

Hyun-ho Bok(✉) and Jongwon Choi

Steel Solution Research Lab, POSCO, Incheon, South Korea
bokhh@posco.com

Abstract. In accordance with the emergence of new automotive steel types, such as 3rd Generation advanced high strength steel, the reasonable estimation of their local ductility (tight-radius bendability, edge stretchability, etc.) has become of great importance to the automobile industry. Regarding this, the local/global formability map, expressed by true fracture (or thickness) strain and uniform strain, has recently been introduced to present intrinsic local and global formability of steels simultaneously. In this study, an improved form of the local/global formability map is proposed in combination with limit strain models. Isolines of tight-radius bending limit and hole expansion ratio at arbitrary intervals are superimposed on the existing map, thereby enabling designers of stamping parts to easily confirm whether their material selections are appropriate for concerned part geometries, or vice versa.

Keywords: AHSS · Local/Global Formability Map · Tight-Radius Bendability · Edge Stretchability

1 Introduction

The local ductility (an ability to withstand severe deformation) of advanced high strength steels (AHSS) has drawn great attention from both the automobile and raw material sectors as the strength level of AHSS increases. This could limit part shapes or reduce opportunities to apply higher strength AHSS, specifically, of which the minimum guaranteeing strength exceeds 780 MPa. As a local ductility measure, true fracture strain ($TFS = \ln(A_0/A_f)$) or true thickness strain ($TTS = \varepsilon_{3f} = \ln(t_0/t_f)$) at fracture can be displayed together with uniform strain (ε_u) through the local/global formability map (L/G map) [1, 2] to represent intrinsic local/global formability characteristics of concerned materials. Using an L/G map, materials can be roughly classified into four categories, i.e. global, balanced global, balanced local and local groups, of which the boundaries are divided by TFS/ε_u ratios of 5, 10 and 20, respectively.

Meanwhile, commercial software specialized in sheet metal forming such as Autoform R10 (AF R10), Pam-Stamp2G, etc. have also introduced relevant functions into their material models, through which fracture behaviors can be simulated properly for AHSS. For example, AF R10 requires additional material parameters named Max Surf Strain (MSS) and Max Edge Strain (MES) to predict the initiations of bending and edge

K. Mocellin et al. (Eds.): ICTP 2023, LNME, pp. 435–441, 2024.
https://doi.org/10.1007/978-3-031-40920-2_45

stretching failures, respectively. In this, MSS can be regarded as limit bending strain (LBS) on the outer surface of a specimen being bent with tight radius, while MES can be uniaxial strain corresponding to the hole expansion ratio (HER).

It is, however, difficult for users to obtain limit strains by the material formability classification. In other words, conversions of the intrinsic data into the limit strains, using appropriate models, are required. Nevertheless, it is possible for an L/G map to be combined with any model as long as it is expressed with the same variables, i.e. fracture strains and ε_u. For instance, Kim and Hance [3] have presented a model combined with an L/G map to predict the maximum bending angle from the three-point bending based on VDA238-100 [4].

In order to further improve the functionality of L/G maps, this study proposes an additional method, in which a combined L/G map with local formability limits as LBS and HER of various AHSS is presented.

2 Local Ductility Models

Firstly, a data set of ε_u and TTS from 50 specimens has been prepared through conducting tensile tests. The materials are DP780 (Galvanized/1.2t), CP780 (Uncoated/0.9t, Galvanized/1.2t), TRIP780 (Galvannealed/1.2t, 1.4t), DP980 (Uncoated/1.6t), TRIP1180 (Galvanized/1.6t), etc. The specimen geometry for ε_u is referred to 13 B specimen (gauge length = 50 mm, width of parallel region = 12.5 mm) of KS B 0801 standard [5], while that for TTS has notches in the center as depicted in Fig. 1 to induce deformation in a manner similar to the plane strain mode prior to fracture [6]. The tests are performed in rolling, diagonal and transverse directions which are denoted as RD, DD and TD, respectively. Five specimens are used for each test. In this, only the data from the specimens aligned in TD is involved in the modeling explained in following paragraphs.

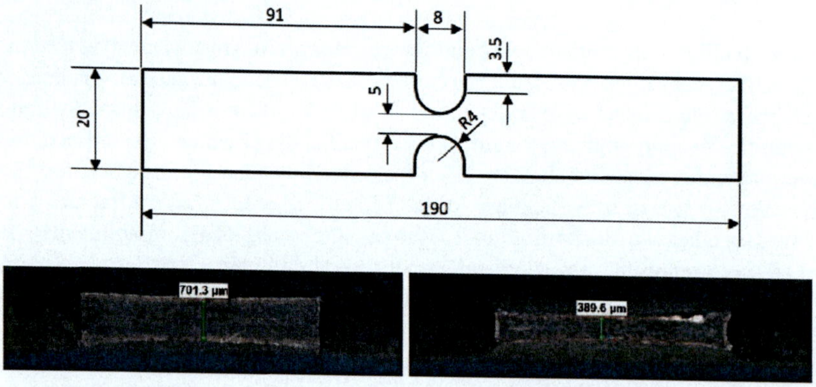

Fig. 1. Design of notched specimen to measure TTS and exmaples of fractured faces after tensile tests in transverse direction (Lower left: Galvanized DP780/1.2t, right: Galvanized CP780/1.2t).

Corresponding HER and LBS have also been gathered by relevant tests complying with ISO 16630 [7] and VDA238-100 standards, respectively.

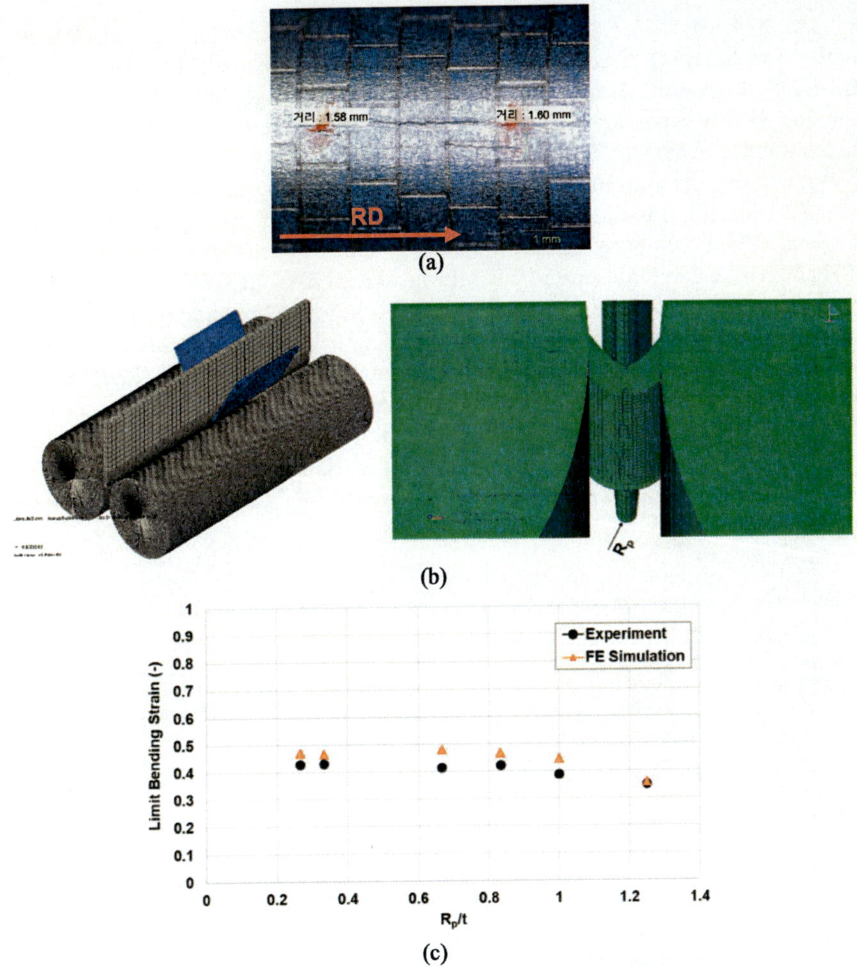

Fig. 2. Three-point bending based on VDA238-100 obtaining limit bending strain (LBS): (a) Surface of a bent DP780 specimen, on which square grids are marked for measuring, is shown by a stereoscopic microscope (two apparent cracks are visible on the surface), (b) Corresponding fracture simulation using ABAQUS/Explicit with ductile damage model (Element size: 0.1 mm × 0.1 mm × 0.1 mm), (c) Comparison of LBS between measured and simulated results (DP780, thickness = 1.2, 1.5 mm)

In this, LBS values are obtained from measuring the maximum strain on the surface after unloading caused by cracks. The bending axis is parallel with the rolling direction so it can be consistently correlated with ε_u and TTS by maintaining the same loading direction, namely, TD. To measure LBS, square grids of 1 mm × 1 mm are applied on the surfaces of the specimens prior to bending (Fig. 2a). Although the basic tip radius of the bending punch, R_p, is 0.4 mm as VDA238-100 stipulates, two more tip radii (1.0 and 1.5 mm) are additionally used to find the minimum values in the given range of radius-to-thickness ratio, R_p/t (Fig. 2c). Fig. 2c, in which the bending until crack is also simulated with ABAQUS/Explicit applying a ductile damage model [8], shows that the LBS of DP780 sheets (1.2 and 1.5 mm) representing the average values of 5 tests for each condition slightly decreases as R_p/t increases in the applied R_p/t range in the experiment and the simulation as well. The other materials show a similar trend (not presented here).

With the data set from 50 specimens acquired, a popular machine learning (ML) tool, the scikit-learn [9], is used to obtain regression models for HER and LBS as a function of ε_u and TTS. 80% of the data set (40 data) is allocated as training data and the rest (10 data) is set aside for testing in this ML, in which PolynomialFeatures class creates new features for a polynomial regression model followed by the training of the model. The degree option of the class is set to be 2 leading to a quadratic regression model in order to avoid overfitting problems. Obtained features are X0, X1, $X0^2$, X0X1 and $X1^2$ for both HER and LBS, where X0 and X1 are none other than ε_u and TTS, respectively. Then the coefficients to the features are fitted employing scikit-learn's LinearRegression class. Training scores in this ML are 0.90 and 0.72 for HER and LBS, respectively. These models of which the common form is $Y = Intercept + \sum Coefficient_i \times Feature_i$ are summarized in Table 1.

Table 1. Quadratic polynomial models for HER and LBS created by machine learning manner

Model	Feature	Coefficient	Intercept
HER (%)	X0	−182.5907	−11.6482
	X1	184.83621	
	$X0^2$	2452.30522	
	X0X1	−807.06705	
	$X1^2$	−52.121	
LBS (−)	X0	4.18807	−0.24182446
	X1	1.53498	
	$X0^2$	0.24052	
	X0X1	−7.62456	
	$X1^2$	−0.59825	

Figure 3a and 3b graphically show the resultant performance of the models. The coefficients of determination (R^2 values) for HER and LBS models are 0.9015 and 0.668, which shows a consistent trend with the training scores. The insufficient accuracy in the LBS prediction can be attributed to the lack of data quantity and features to express the bending. For instance, if thickness variable as an additional feature (X2) is involved, R^2 of the LBS model can be improved as shown in Fig. 3c. This study, however, adopts material properties solely to exclude the geometry effect from the thickness difference in constructing a new L/G diagram, combined with the HER and LBS models.

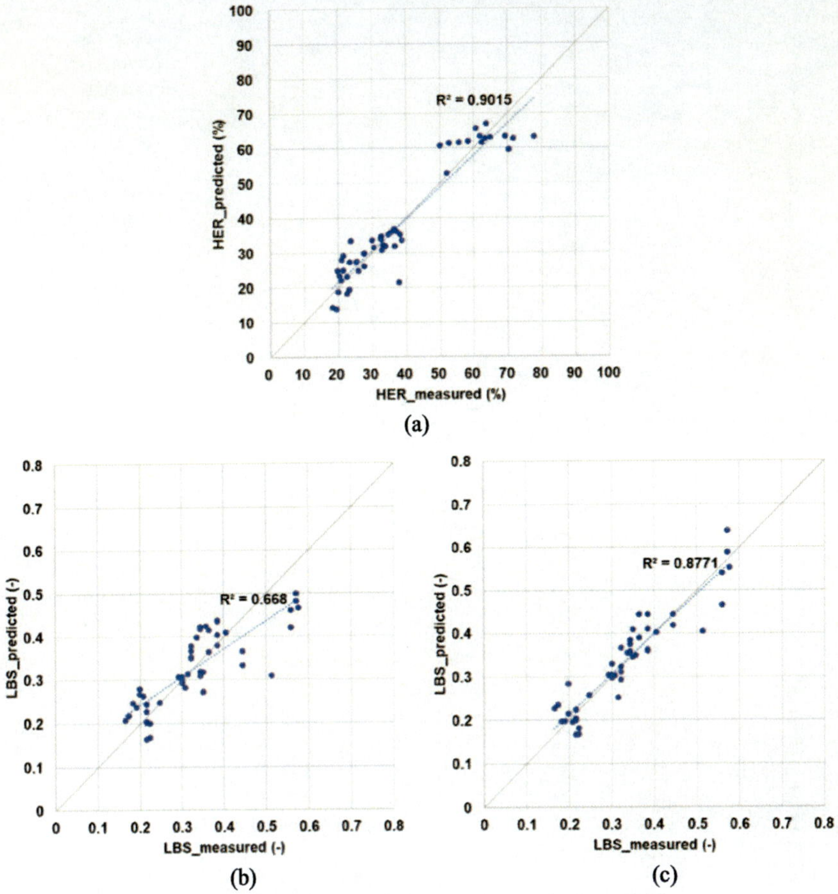

Fig. 3. Comparison of measured and predicted local ductility: (a) Hole expansion ratio (HER) model, (b) Limit bending strain (LBS) model and (c) LBS modeled with additional feature, X2 (Thickness)

3 Combined Local/Global Formability Map

The quadratic equations of the HER and the LBS models are solved against X1 (= TTS) to be integrated into an L/G map, sharing the same axes with the models. Resultant solutions are shown as the isolines of HER and LBS in Fig. 4a and 4b (black solid lines), respectively. In the case of the isolines with negative curvatures, for example, the lines for HER, which are larger than 30%, they cannot be plotted in higher ε_u-TTS space attributing to mathematical reason (the discriminants of the quadratic equations gaining negative values). This may depend on the form and coefficients of a model.

The scatter plot of ε_u and TTS on the background is constructed using the entire tensile data tested along RD, TD and DD. The scatters in the map, therefore, indicate intrinsic formability in all directions leading to rather dispersed distributions for each

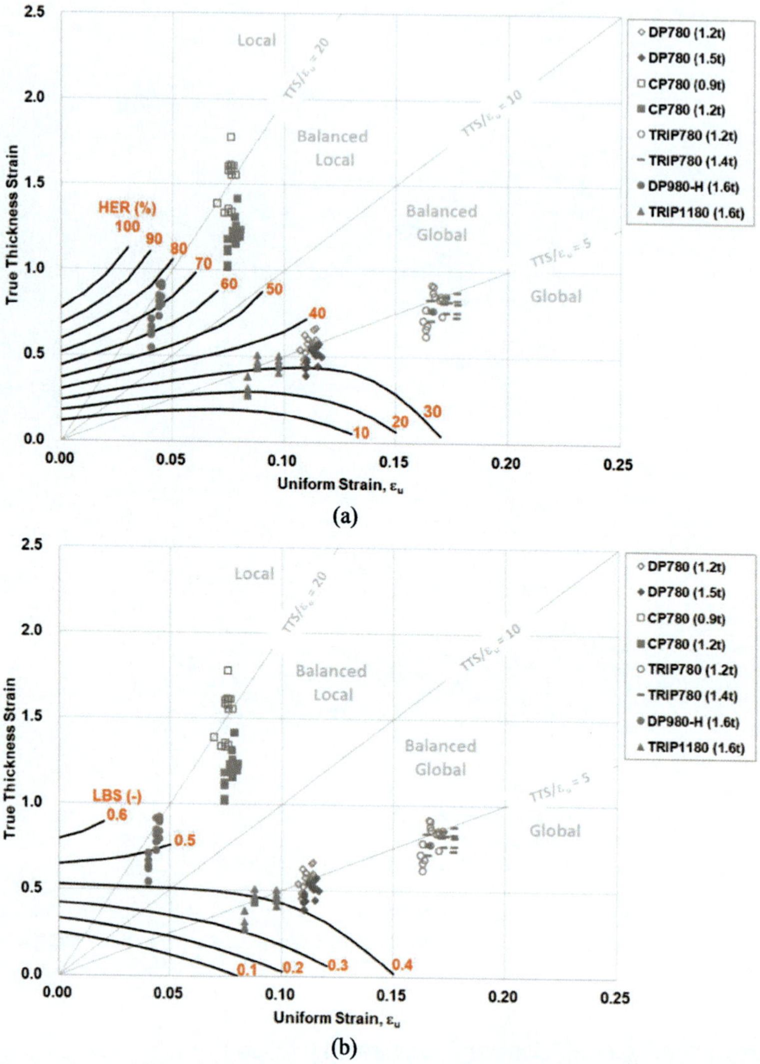

Fig. 4. Predicted isolines in different local ductility levels integrated into local/global formability map (L/G map), in which uniform strain (ε_u) and true thickness strain (TTS) tested in three directions (rolling, diagonal and transverse directions) are also scattered for various AHSS: (a) Combined L/G map with HER and (b) that with LBS

material, while the predicted isolines representing the local ductility are correlated with the material properties in TD as previously explained in Sect. 2. Through this procedure, it is shown that specific local ductility, e.g. edge stretchability and tight-radius bendability as HER and LBS, can be effectively displayed in the ε_u-TTS space, shared by the L/G map.

4 Conclusion

A method which incorporates forming limit strains into an L/G map is proposed in order to construct a combined form of the L/G map. The edge stretchability and tight-radius bendability (HER and LBS) of several AHSS, of which the minimum tensile strength is larger than 780 MPa, are measured as the forming limit strains to be incorporated. Then a quadratic regression model, in which uniform strain and true thickness strain at fracture are set as common features with the L/G map, is obtained by a machine learning tool aiming to predict HER or LBS of concerned AHSS.

As a result, it is shown that the predicted isolines of HER and LBS can be successfully integrated into the L/G map as they are expressed in common space. Further improved forms of a combined map can be expected depending on the volume of involved material data in combination with more advanced models.

References

1. Hance, B., Davenport, M.D.: Advanced high strength steel: deciphering local and global formability. In: Proceedings of International Automotive Body Congress (2016)
2. Larour, P., Freudenthaler, J., Weissböck, T.: Reduction of cross section area at fracture in tensile test: measurement and applications for flat sheet steels. J. Phys.: Conf. Ser. **896**, 012073 (2017)
3. Kim, J.Y., Hance, B.: The combined influence of global and local formability on the VDA 238-100 bending angle in advanced high-strength steels. In: Proceedings of International Automotive Body Congress (2022)
4. VDA238-100 Plate bending test for metallic materials (2010)
5. KS B 0801 Test pieces for tensile test for metallic materials (2007)
6. Gruenbaum, M., Aydin, G., Dettinger, T., Heibel, S.: Local formability of AHSS: measurement technique, specimen types and robustness. Mater. Sci. Eng. **651**, 012056 (2019)
7. ISO 16630 Metallic materials – Sheet and strip – Hole expanding test (2017)
8. Dassault Systems Homepage. https://www.3ds.com/products-services/simulia/products/abaqus/
9. Scikit learn Homepage. https://scikit-learn.org/stable/

Characterization of the Dynamic Recrystallization Kinetics Using Directly Flow Stress Model and Its Application to 42CrMo Steel

Mohd Kaswandee Razali[1], Suk Hwan Chung[2], Missam Irani[3], Jeong Muk Choi[4], and Man Soo Joun[1](✉)

[1] Engineering Research Institute, School of Mechanical and Aerospace Engineering, Gyeongsang National University, Jinju 52828, South Korea
msjoun@gnu.ac.kr
[2] Metal Forming Research Corporation (MFRC), Jinju 52818, South Korea
[3] Institute of Metal Forming, Technische Universität Bergakademie Freiberg, 09599 Freiberg, Germany
[4] Jinhap Co. Ltd, Daejeon 34302, South Korea

Abstract. In order to improve and enhance the solution accuracy and practicability in prediction of dynamic recrystallization kinetics, we present a direct method to characterize the flow-related DRX parameters including two essential parameters when predicting DRX microstructural evolution which are the peak strain and the strain at 50% recrystallization. In this method, flow curves are described accurately using the general and improved C-m models in which C and m are defined as functions of strain rate at varying strains and temperatures. It eliminates the need for mathematical modeling of these parameters in order to predict the volume fraction of DRX grains and increases the practical utility of microstructural prediction by a great deal. The DRX behavior of extruded 42CrMo high-strength steel is investigated with a finite element approach using the direct method and the Avrami kinetic model. Our present method predicts the DRX kinetics volume fraction and grain size during microstructural evolution with remarkable accuracy when compared to other approaches and experiments. DRX Microstructural predictions are greatly enhanced by this method since it eschews mathematical modeling of the peak strain and the strain at 50% recrystallization in Avrami kinetic model.

Keywords: Solution Accuracy · Dynamic Recrystallization Kinetics · Flow Stress

1 Introduction

Metallic materials have crucial microstructures that determine their properties, including their strength, ductility, elongation, corrosion resistance, fatigue, and wear resistance [1, 2]. Through the use of well-known mathematical models for flow curves and kinetics of

K. Mocellin et al. (Eds.): ICTP 2023, LNME, pp. 442–454, 2024.
https://doi.org/10.1007/978-3-031-40920-2_46

microstructural evolution, FE simulations have been used widely to predict microstructures of metallic materials during hot forming processes [3]. This implies that simulation accuracy greatly depends on the mathematical models of flow curves and microstructural evolution [4].

A number of years have been spent researching dynamic recrystallization (DRX) kinetics models [5–20] and their characterization techniques. The influence of deformation behavior on DRX process has been studied by McQueen and Ryan [10] by introducing a creep equation combining dynamic activation energy, strain rate, and temperature, and Wang et al. built another physically-based constitutive model which is a hyperbolic-sine equation to express the relationship among peak stress, temperature, and strain rate. However, the model coefficients can only be determined with enough experimental data, whereby, linear or nonlinear fitting is performed to the experimental data at different deformation conditions [11]. Due to early metallurgists' contributions [2, 12, 13], a global methodology was established based on the Avrami kinetic model or another. In general, Avrami kinetic model [14–16] or other models [17–19] have been used to describe DRX kinetics of metallic materials. The hot deformation behavior of metallic materials has been studied extensively with this method since then. Although the Avrami kinetics model for commercial metallic materials has been simplified much, the practical acquisition of ideal parameters is still complicated, costly, sometimes mistaken, and time-consuming. Due to these factors, it has been quite difficult to apply the methodology to mathematical model equations and finite element (FE) simulations.

This work presents a direct method (DM) that directly calculates the flow-related DRX kinetics parameters peak strain (ε_p) and strain at 50% DRX ($\varepsilon_{0.5}$) for Avrami's model from flow curves without any mathematical modeling and kinetics constants α_d, β_d and k_d treated as the functions of state variables. Moreover, this method is supported by the general and improved *C-m* model of Joun et al. [20], which provides an accurate description of flow stress. We demonstrate how to acquire the constants in practice and prove the DM's validity. Its greatest strength is its mathematical flexibility when modifying its key parameters in order to incorporate the effect of state variables on flow stress and eventually on DRX evolution [3].

2 Experimental Overview

Experimental data from hot compression tests of 42CrMo steel alloy (chemical composition in wt%: C, 0.450; Si, 0.280; Cr, 0.960; Mn, 0.630; Mo, 0.190; P, 0.016; S, 0.012; Cu, 0.014; Fe to 100 wt%) were used in this study [1], which were considered comprehensive and relatively reliable in the development of a mathematical model, as illustrated in Fig. 1. In their work, a Gleeble test has been used to acquire flow stress behaviors and microstructure characteristics with various sample temperatures (1123, 1198, 1273, and 1348 K) and strain rates (0.01, 0.1, 1, and 10 s^{-1}) during hot-cylinder compression test, as sketched in Fig. 2. The specimens measured 10 mm in diameter and 12 mm in height, then compressed with a 60% reduction in height. Following the test, all specimens were quenched immediately in order to preserve a deformed or evolved microstructure.

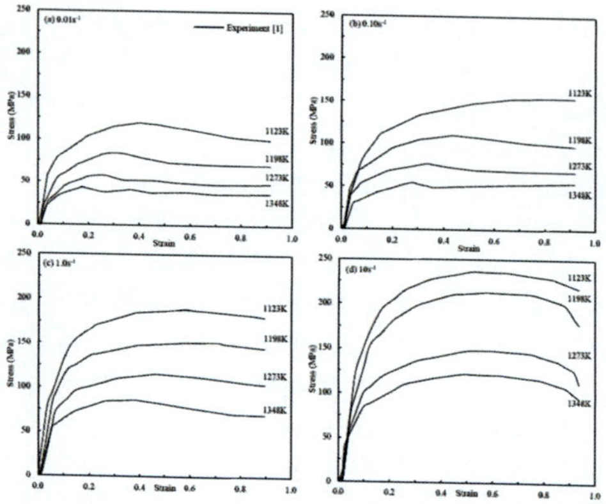

Fig. 1. Flow stress curves of 42CrMo steel at different deformation temperatures with strain rates [1]. (a) 0.01 s^{-1}, (b) 0.1 s^{-1}, (c) 1 s^{-1} and (d) 10 s^{-1}.

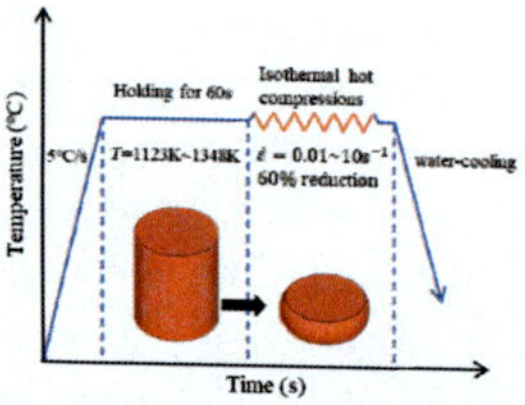

Fig. 2. Heating process diagram [1].

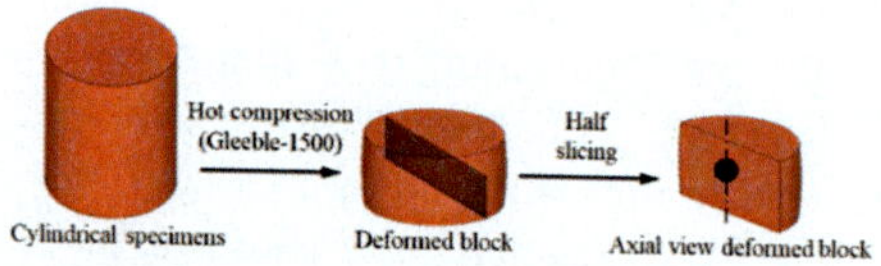

Fig. 3. Schematic diagram of sectioning the specimen for optical microstructure analysis.

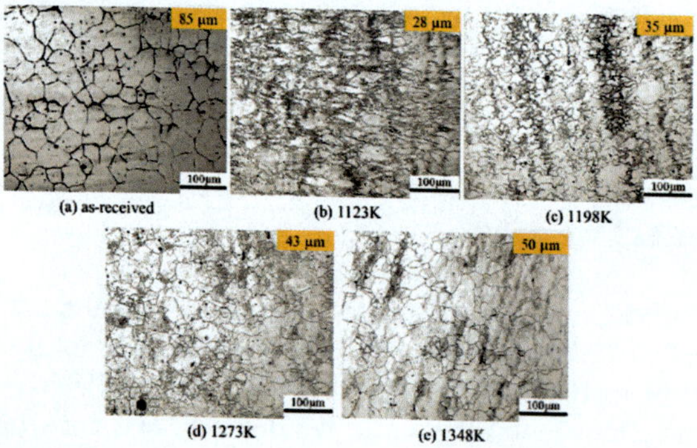

Fig. 4. Microstructures at the strain rate of 0.1 s^{-1} and different deformation temperatures: (a) as-received, (b) 1123 K, (c) 1198 K, (d) 1273K and (e) 1348 K.

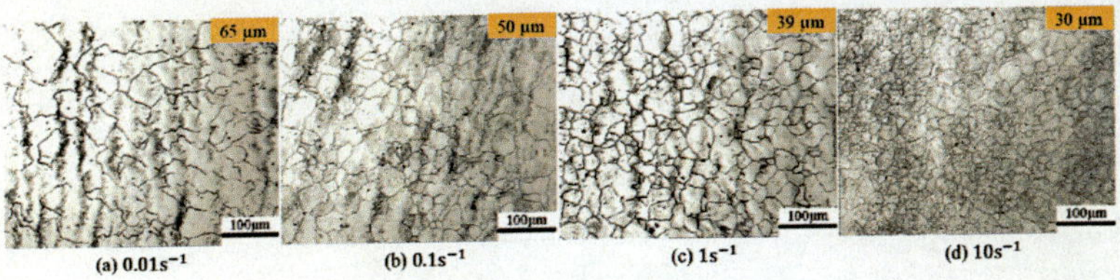

Fig. 5. Microstructures at the temperature of 1348 K and different strain rates: (a) 0.01 s^{-1}, (b) 0.1 s^{-1}, (c) 1 s^{-1}, (d) 10 s^{-1}.

For metallographic examination, all specimens were sectioned perpendicular to the longitudinal compression axis, as schematically shown in Fig. 3. The sections were polished and etched in saturated picric acid. In the center region of the section plane, optical microstructures were analyzed. The measured grain sizes and their corresponding volume fraction were summarized in Fig. 4 and Fig. 5.

3 Modeling and Its Prediction

3.1 Flow Stress Model

The piecewise *C-m* model of Joun et al. [20] was applied to commercial steel 42CrMo and had a systematic formulation and practical identification scheme, as follows:

$$\sigma = C(\varepsilon, \dot{\varepsilon}, T)\dot{\varepsilon}^{m(\varepsilon,\dot{\varepsilon},T)} \tag{1}$$

where

$$C(\varepsilon, \dot{\varepsilon}, T) = C_0 + C_1 ln(1+\dot{\varepsilon}) \tag{2}$$

$$m(\varepsilon, \dot{\varepsilon}, T) = m_0 + m_1 ln(1+\dot{\varepsilon}) \tag{3}$$

where C_i [MPa], m_i, and γ_i represent all material properties defined at the fixed sample strains and temperatures. Determination of these properties involves minimizing the difference between experimental and fitted flow stress functions. A detailed list of the material constants can be found in Table 1. Then, the predicted flow stress is compared with the experimental results as shown in Fig. 6, and there is good agreement, whereby its maximum error is 3.4%. A specific temperature and strain rate can be calculated using this flexible flow stress model. For given temperatures and strain rates, it is therefore possible to precisely determine the strain values at which DRX occurs.

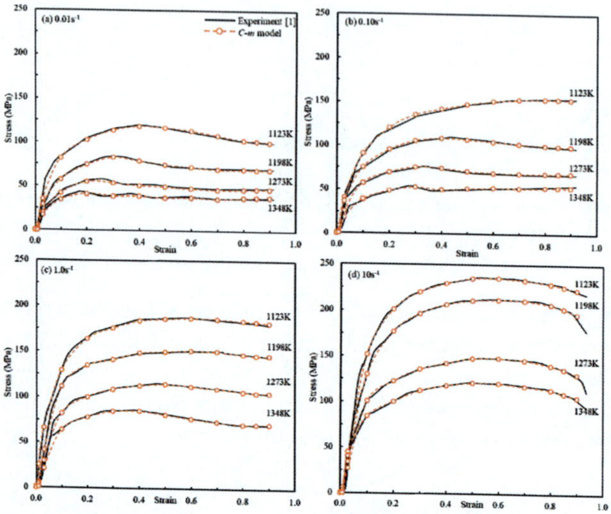

Fig. 6. Comparison of flow stress on 42CrMo at different deformation temperatures with strain rates [1]. (a) $0.01\ s^{-1}$, (b) $0.1\ s^{-1}$, (c) $1\ s^{-1}$ and (d) $10\ s^{-1}$.

3.2 X_{DRX} Kinetics Model

DRX occurs at a critical strain and is affected by temperature and strain rate. It can be seen on the flow stress curve as a decrease in the flow stress from peak stress to steady-state stress [3, 21]. To describe the volume fraction of DRX (X_{DRX}) in this study, Avrami's kinetics model was used. This model relies on four major parameters of β_d, k_d, ε_p and $\varepsilon_{0.5}$, as follows:

$$X_{DRX,prediction} = 1 - \exp\left[-\beta_d\left(\frac{\varepsilon - \alpha_d\varepsilon_p}{\varepsilon_{0.5}}\right)^{k_d}\right] \tag{4}$$

where β_d, k_d, ε_p, $\varepsilon_{0.5}$ and α_d are the model parameters but α_d is usually given 1.0.

Based on the previous work by Razali et al. [3, 21], ε_p and $\varepsilon_{0.5}$ can be calculated from the flow stress curves by fitting the improved *C-m* model. We can directly determine peak stress (σ_p) and its strain (ε_p), steady-state stress (σ_s) and its strain (ε_s) and the strain of 50% recrystallization ($\varepsilon_{0.5}$) from the flow curve. Conceptually, the procedure was summarized in Fig. 7. In general, Razali et al. [3] procedure for calculating flow-related DRX kinetics parameters can be easily programmed for general purposes. It is important to realize that this simple but powerful scheme could be made possible only using a general and accurate description of the flow curves [20, 21].

Then, using an optimization technique, kinetic rate constant (β_d) and Avrami exponent (k_d) were calculated using a high degree of accuracy for each pair of sampled temperature and strain rate by minimizing the error between $X_{DRX,prediction}$ and $X_{DRX,experiment}$ with high accuracy [3]. There is a noteworthy aspect of this research: most researchers assumed the β_d and k_d were constant over the entire range of strain rate and temperature which distorted Avrami's kinetic model solution accuracy and flexibility. Interpolation

Table 1. Flow stress information matrix of 42CrMo steel.

T (K)	ε	C_0	$C1$	m_0	m_1
1123	0.001	2.47	−0.52	0.30	−0.20
	0.01	4.93	29.74	0.00	−0.44
	0.03	34.57	45.17	0.00	−0.23
	0.10	85.88	63.26	0.01	−0.09
	0.20	129.76	49.05	0.05	−0.06
	0.30	154.38	30.82	0.07	−0.03
	0.40	163.70	29.17	0.07	−0.03
	0.50	186.25	0.00	0.10	0.00
1198	0.001	196.25	−10.90	0.12	0.01
	0.01	208.18	−29.58	0.14	0.04
	0.03	205.97	−27.67	0.14	0.03
	0.10	205.48	−28.14	0.14	0.03
	0.20	94.12	0.04	0.00	0.14
	0.30	0.26	0.00	0.07	0.00
	0.40	7.08	−1.07	0.25	-0.16
	0.50	47.96	−9.02	0.11	0.00
1273	0.001	89.49	31.98	0.10	-0.09
	0.01	115.68	27.66	0.10	-0.05
	0.03	133.22	11.77	0.10	-0.01
	0.10	149.93	−1.80	0.14	0.01
	0.20	155.47	−7.38	0.16	0.01
	0.30	153.27	−2.37	0.16	0.00
	0.40	149.34	2.35	0.17	−0.01
	0.50	143.66	5.98	0.16	−0.02
1123	0.001	140.80	8.10	0.15	−0.02
	0.01	62.21	2.03	0.00	0.18
	0.03	0.03	0.00	0.09	0.00
	0.10	3.02	2.90	0.38	-0.61
	0.20	34.56	−6.99	0.11	0.03
	0.30	74.97	10.48	0.13	−0.05
	0.40	81.08	27.78	0.09	−0.07
	0.50	103.89	7.04	0.14	-0.04

of these parameters was then carried out using piecewise bilinear functions in relation to strain rate and temperature [3, 22].

Additionally, the experimental X_{DRX}, denoted as $X_{\mathrm{DRX,experiment}}$, can be calculated through flow curve analysis as follows:

$$X_{DRX,experiment} = \left(\frac{\sigma_p - \sigma_{DRX}}{\sigma_p - \sigma_s} \right) \tag{5}$$

where σ_{DRX} is the stress at the sample strain in the DRX region. Subsequently, the four major parameters of the Avrami's kinetic model at various temperatures and strain rates were optimally acquired and disclosed in Table 2. An analysis of the experimental and fitted S-curves for 42CrMo steel is described in Fig. 8. It can be seen that the predicted X_{DRX} curves are well congruent with the experimental X_{DRX} curves. Moreover, a quantitative comparison between the experiments and DM predictions together with Quan et al. [1] are shown in Fig. 9. In terms of absolute relative error (*AARE*) and coefficient of determination (R^2) of X_{DRX} using the DM (*AARE* = 2.18%, R^2 value = 0.98), the DM was substantially better than Quan et al. (*AARE* = 9.27%, R^2 value = 0.89). In addition to its superiority, the DM also benefits from direct acquisition of the ε_{p} and $\varepsilon_{0.5}$ values from the fitted flow curves. This dominance is also explained by the description of β_{d} and k_{d} as functions of temperature and strain rate.

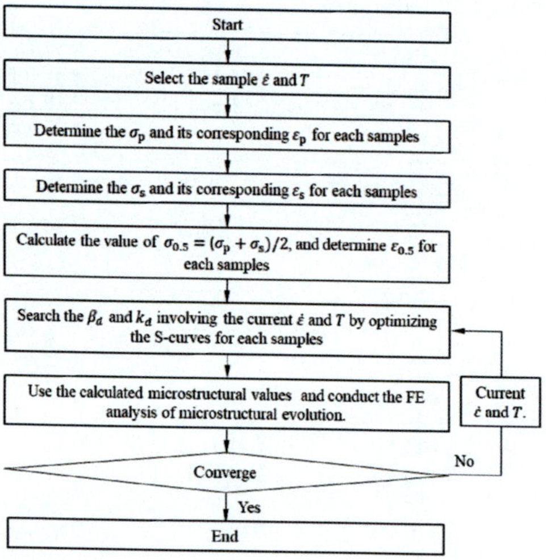

Fig. 7. Flow chart on how flow-related DRX kinetic parameters are calculated at different temperatures and strain rates.

3.3 DRX Grain Size ($d_{\mathrm{DRX,average}}$) Kinetics Model

In order to calculate the $d_{\mathrm{DRX,average}}$, we need to perform the following calculations using the calculated X_{DRX} and initial grain size (d_0):

$$d_{DRX,average} = x_{DRX} d_{DRX} + d_0(1 - x_{DRX}) \tag{6}$$

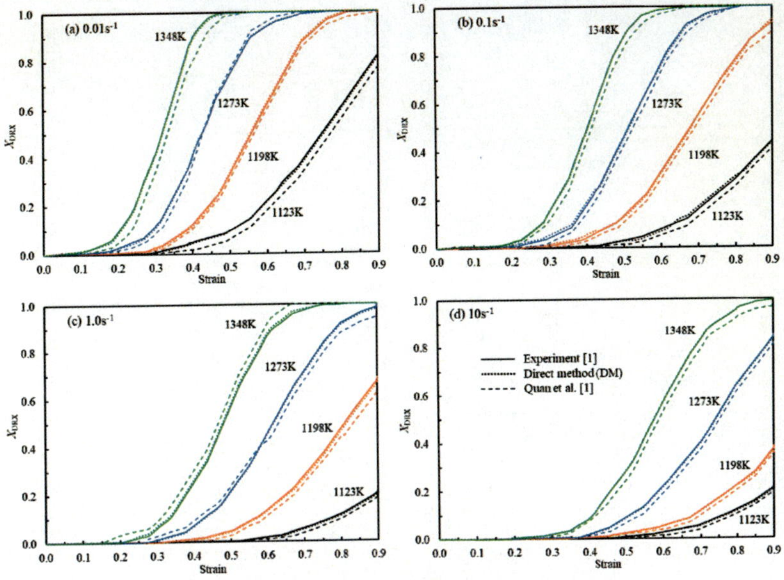

Fig. 8. The S-curve of the 42CrMo steel.

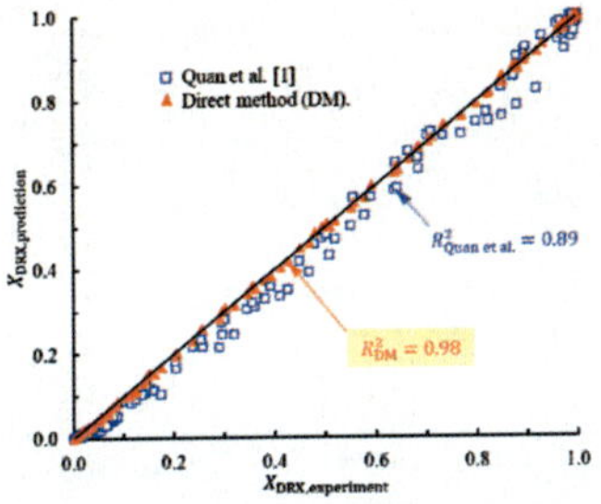

Fig. 9. Comparison of R^2 value between the direct and Quan et al.'s methods.

where the local DRX grain size (d_{DRX}) is calculated from the following function:

$$d_{DRX} = p d_0^h \varepsilon^n \dot{\varepsilon}^m \exp(Q/RT) \tag{7}$$

where the constants *p*, *h*, *n*, *m* and *Q* were obtained via a practical method of a combined FE analysis and optimization technique [3, 11, 23] and *R* is the given value of the universal gas constant, as conveyed in Table 3.

This DM was then incorporated into commercial FE method software [23] and applied to the prediction of DRX behaviors of 42CrMo steel during hot upsetting. An axisymmetric FE analysis model was constructed by 2000 quadrilaterals as in Fig. 10 and applied a coefficient of Coulomb friction of 0.15 at the contact surface between the specimen and dies. A series of FE analyses were conducted at various initial temperatures of 1123 to 1348 K and strain rates of 0.01 to 10 s^{-1}. Noted that, die and specimen

Table 2. Microstructural parameters of the AZ91D alloy for Avrami's kinetics DRX model.

T(K)	$\dot{\varepsilon}\left(s^{-1}\right)$	ε_p	$\varepsilon_{0.5}$	β_d	k_d
1123	0.01	0.298	0.149	1.32	1.67
	0.1	0.351	0.176	1.24	1.31
	1.0	0.432	0.216	1.97	1.58
	10.0	0.454	0.227	1.72	1.63
1198	0.01	0.236	0.118	1.23	1.06
	0.1	0.282	0.141	1.12	1.23
	1.0	0.338	0.169	1.86	1.52
	10.0	0.369	0.185	1.46	1.77
1273	0.01	0.181	0.091	0.85	0.9
	0.1	0.208	0.104	1.16	1.16
	1.0	0.261	0.131	1.81	1.61
	10.0	0.278	0.139	1.39	1.4
1348	0.01	0.056	0.028	0.85	0.9
	0.1	0.088	0.044	1.16	1.16
	1.0	0.149	0.075	1.81	1.61
	10.0	0.193	0.097	1.39	1.4

Table 3. d_{DRX} constants of the 42CrMo steel.

Q	R	d_0	p	h	n	m
−82.175	8.314	85.0	0.417	0.263	0.721	0.287

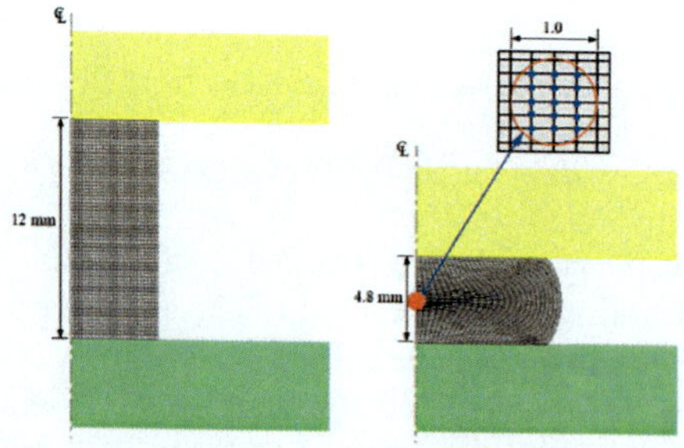

Fig. 10. The FE upsetting process and $d_{DRX,average}$ measurement points.

were assumed to be rigid and rigid-thermoviscoplastic, respectively. After upsetting the cylindrical specimen at 1348 K, the microstructure of 42CrMo steel using a measured grain sizes depicted in Fig. 5 were investigated. The measurement method [3, 22] for $d_{DRX,average}$ can be found in Fig. 10, and the FE prediction were obtained at various temperatures and strain rates for the verification process of DM.

Figure 11 shows the $d_{DRX,average}$ prediction of 42CrMo steel at different reductions and temperatures at a strain rate of 0.1 s^{-1}. The $d_{DRX,average}$ clearly increases with increasing temperature and reduction. It also can be seen that the $d_{DRX,average}$ at the measured point increased from about 28 to 50 μm as the temperature increased from 1123 K to 1348 K, respectively. The characteristic changes of $d_{DRX,average}$ at different temperatures (Fig. 11) are consistent with the flow stress curves shown in Fig. 1 and the metallographic observations in Fig. 4(b–e). On the other hand, the higher the sample strain rates, the smaller the $d_{DRX,average}$, as predicted (Fig. 12) and observed experimentally (Fig. 5). Figure 12 shows the $d_{DRX,average}$ at different reductions and strain rates for temperature 1348 K. It shows that the AVGDRX decreased from about 65um to 30μm as the strain rate increased from 0.01 to 10 s^{-1}. Subsequently, the FE predictions are in agreement with the flow stress prediction shown in Fig. 1 and with the microstructural observation shown in Fig. 5(a–d).

In general, it can be postulated that the $d_{DRX,average}$ of the most-deformed region started and was completed earlier than that of less-deformed regions (Figs. 11 and 12). On the specimen, the region with the smallest $d_{DRX,average}$ was located in the middle, while the DRX on the upper and lower edges, as well as on the right side, were generally larger than the center. The $d_{DRX,average}$ distribution in hot cylinder compression tests is unequal since frictional effects (dead metal zones) and barreling are inevitable. As the microstructure at the center was less affected by friction, the metal flow resistance decreased and the radial expansion increased, resulting in large plastic deformations. In this way, the microstructure mostly experienced DRX and a large number of fine and equiaxed grains were formed.

A comparison is shown in Fig. 13 between the predicted $d_{DRX,average}$ by Quan et al. [1] and DM. Figure 13 (a) exhibits a change of the experimental and predicted $d_{DRX,average}$ with increase in the temperature at strain rates of 0.1 s^{-1} and Fig. 13 (b) shows that with the increase in the strain rates at the temperature of 1348 K. Overall, the numerical predictions of $d_{DRX,average}$ are in good agreement with the experimental results, which demonstrates the coherence of the established DM in predicting the DRX evolution with *C-m* flow stress model approach.

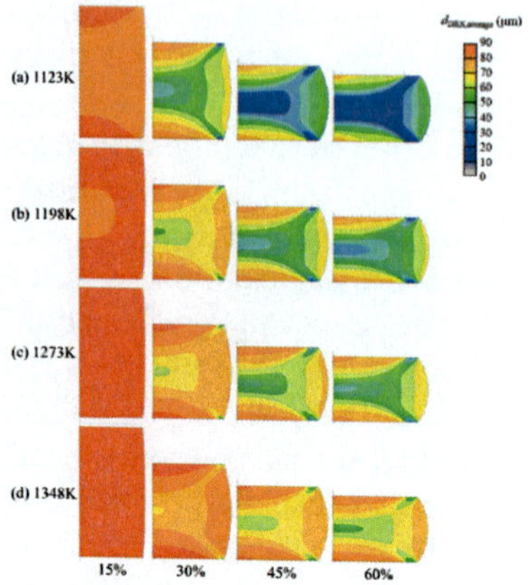

Fig. 11. Predicted $d_{\mathrm{DRX,average}}$ evolution at a strain rate of 0.1 s^{-1}.

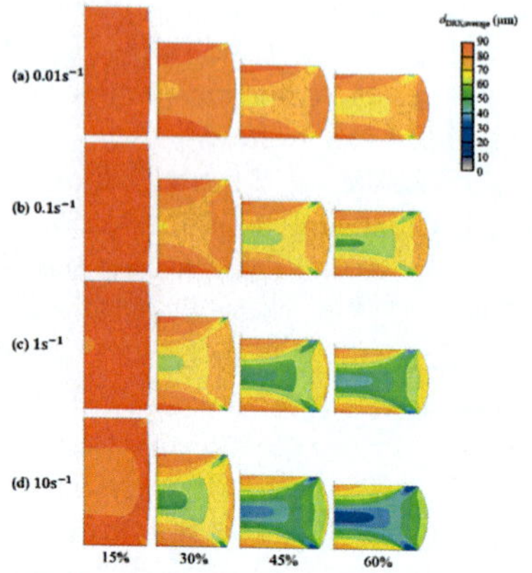

Fig. 12. Predicted $d_{\mathrm{DRX,average}}$ evolution at a temperature of 1348 K.

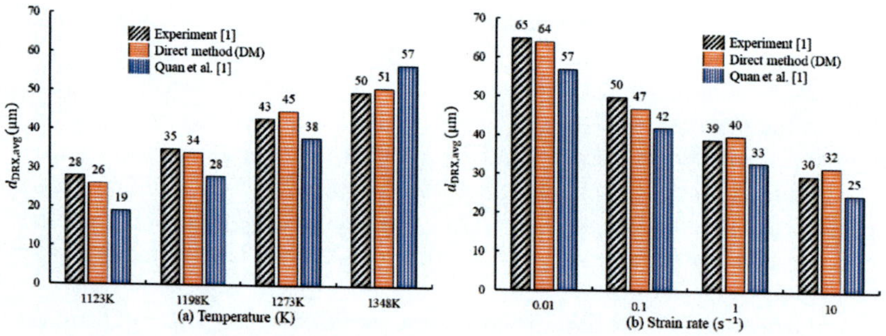

Fig. 13. Comparison of $d_{\mathrm{DRX,average}}$ between the experiments and predictions.

4 Conclusion

In this paper, DRX kinetic parameters related to flow curve are directly calculated using a DM and treated as function of temperature and strain rate. Basically, this method uses the general and improved *C-m* model to describe flow curves accurately. The parameters were determined by using FE characterization coupled with optimization. A significant advantage of this method is the fact that there is no need to generate any additional mathematical models for Avrami kinectic model parameters such as the stress and strain at peak, steady-state, and strain values at 50% DRX, i.e., $\varepsilon_{0.5}$. In spite of this, it is quite accurate and predicted X_{DRX} values significantly improved. Based on the *AARE* and R^2, the DM is also very reliable, 2.18% and 0.98, respectively.

It was also revealed that the microstructure analysis at the evaluated conditions showed that $d_{\mathrm{DRX,average}}$ get smaller as the temperature decreased and strain rate increased, indicating that fine grains gradually grew with each subsequent change. It is evident from the results that there is excellent agreement between the measured and predicted $d_{DRX,average}$, underscoring the validity of the suggested approaches for modeling flow stress curves and DRX kinetics. As demonstrated in the applied example case, the average and maximum $d_{DRX,average}$ errors were 4.2% and 5.6%, respectively, emphasizing that the DM enhances accuracy, reliability and practicability of FE prediction of microstructural evolution with the impact of good correlation in X_{DRX} prediction.

Acknowledgement. This work was partly supported by Korea Evaluation Institute of Industrial Technology (KEIT) (10081334, Development of Commercialization Technology for High Strength Ti Alloy Aircraft Fasteners using Warm Forming) and Korea Institute of Energy Technology Evaluation and Planning (KETEP) (20214000000520, Human Resource Development Project in Circular Remanufacturing Industry) grant funded by the Korea government (MOTIE).

References

1. Quan, G.Z., Li, G.S., Chen, T., Wang, Y.X., Zhang, Y.W., Zhou, J.: Dynamic recrystallization kinetics of 42CrMo steel during compression at different temperatures and strain rates. Mater. Sci. Eng., A **528**(13–14), 4643–4651 (2011). https://doi.org/10.1016/j.msea.2011.02.090
2. Sakai, T., Jonas, J.J.: Overview no. 35 Dynamic recrystallization: mechanical and microstructural considerations. Acta Metall. **32**(2), 189–209 (1984). https://doi.org/10.1016/0001-6160(84)90049-X
3. Razali, M.K., Joun, M.S.: A new approach of predicting dynamic recrystallization using directly a flow stress model and its application to medium Mn steel. J Mater. Res. Technol. **11**, 1881–1894 (2021). https://doi.org/10.1016/j.jmrt.2021.02.026
4. Gourdet, S., Montheillet, F.: An experimental study of the recrystallization mechanism during hot deformation of aluminium. Mater. Sci. Eng. A **283**, 274–288 (2000). https://doi.org/10.1016/S0921-5093(00)00733-4
5. Sun, X., Zhang, M., Wang, Y., Sun, Y., Wang, Y.: Kinetics and numerical simulation of dynamic recrystallization behavior of medium Mn steel in hot working. Steel. Res. Int. **91**, 1900675 (2020). https://doi.org/10.1002/srin.201900675
6. Sellars, C.M.: Computer modelling of hot-working processes. Mater. Sci. Technol. **1**, 325–332 (1985). https://doi.org/10.1179/mst.1985.1.4.325
7. Liu, J., Cui, Z., Ruan, L.: A new kinetics model of dynamic recrystallization for magnesium alloy AZ31B. Mater. Sci. Eng. A **529**, 300–310 (2011). https://doi.org/10.1016/j.msea.2011.09.032
8. Chen, M.S., Lin, Y.C., Ma, X.S.: The kinetics of dynamic recrystallization of 42CrMo steel. Mater. Sci. Eng. A **556**, 260–266 (2012). https://doi.org/10.1016/j.msea.2012.06.084
9. Mirzadeh, H., Parsa, M.H., Ohadi, D.: Hot deformation behavior of austenitic stainless steel for a wide range of initial grain size. Mater. Sci. Eng. A **569**, 54–60 (2013). https://doi.org/10.1016/j.msea.2013.01.050
10. McQueen, H.J., Ryan, N.D.: Constitutive analysis in hot working. Mater. Sci. Eng. A **322**, 43–63 (2002). https://doi.org/10.1016/S0921-5093(01)01117-0
11. Irani, M., Joun, M.S.: Determination of JMAK dynamic recrystallization parameters through FEM optimization techniques. Comput. Mater. Sci. **142**, 178–184 (2018). https://doi.org/10.1016/j.commatsci.2017.10.007
12. Mecking, H., Kocks, U.F.: Kinetics of flow and strain-hardening. Acta Metall. **29**, 1865–1875 (1981). https://doi.org/10.1016/0001-6160(81)90112-7
13. Fanfoni, M., Tomellini, M.: The Johnson-Mehl-Avrami-Kolmogorov model: A brief review. Nuovo Cim Della Soc Ital Di Fis D - Condens Matter, At Mol Chem Physics, Biophys **20**, 1171–82 (1998). https://doi.org/10.1007/BF03185527

14. Cho, J.R., Jeong, H.S., Cha, D.J., Bae, W.B., Lee, J.W.: Prediction of microstructural evolution and recrystallization behaviors of a hot working die steel by FEM. J. Mater. Process. Technol. **160**, 1–8 (2005). https://doi.org/10.1016/j.jmatprotec.2004.01.001
15. Chen, F., Cui, Z.: Mesoscale simulation of microstructure evolution during multi-stage hot forging processes. Model. Simul. Mater. Sci. Eng. **20**, 045008 (2012). https://doi.org/10.1088/0965-0393/20/4/045008
16. Li, X., Duan, L., Li, J., Wu, X.: Experimental study and numerical simulation of dynamic recrystallization behavior of a micro-alloyed plastic mold steel. Mater. Des. **66**, 309–320 (2015). https://doi.org/10.1016/j.matdes.2014.10.076
17. Zhang, C., Zhang, L., Xu, Q., Xia, Y., Shen, W.: The kinetics and cellular automaton modeling of dynamic recrystallization behavior of a medium carbon Cr-Ni-Mo alloyed steel in hot working process. Mater. Sci. Eng. A **678**, 33–43 (2016). https://doi.org/10.1016/j.msea.2016.09.056
18. Wan, Z., Sun, Y., Hu, L., Yu, H.: Experimental study and numerical simulation of dynamic recrystallization behavior of TiAl-based alloy. Mater. Des. **122**, 11–20 (2017). https://doi.org/10.1016/j.matdes.2017.02.088
19. Xu, Y., Chen, C., Zhang, X., Dai, H., Jia, J., Bai, Z.: Dynamic recrystallization kinetics and microstructure evolution of an AZ91D magnesium alloy during hot compression. Mater. Charact. **145**, 39–52 (2018). https://doi.org/10.1016/j.matchar.2018.08.028
20. Joun, M.S., Razali, M.K., Yoo, J.D., Kim, M.C., Choi, J.M.: Novel extended C-m models of flow stress for accurate mechanical and metallurgical calculations and comparison with traditional flow models. J. Magnes. Alloys **10**(9), 2516–2533 (2021). https://doi.org/10.1016/j.jma.2021.08.018
21. Razali, M.K., Irani, M., Joun, M.: General modeling of flow stress curves of alloys at elevated temperatures using bi-linearly interpolated or closed-form functions for material parameters. J. Mater. Res. Technol. **8**(3), 2710–2720 (2019). https://doi.org/10.1016/j.jmrt.2019.04.007
22. Razali, M.K., Abd Ghawi, A.A., Irani, M., Chung, S.H., Choi, J.M., Joun, M.S.: Practical approach for determining material parameters when predicting grain size after static recrystallization. J. Market. Res. (2023). https://doi.org/10.1016/j.jmrt.2023.02.054
23. Joun, M.S., Lee, M.C., Eom, J.G.: Intelligent Metal-Forming Simulation. In: ASME 2011 International Manufacturing Science and Engineering Conference, ASMEDC, pp. 161–8, vol. 1 (2011). https://doi.org/10.1115/MSEC2011-50128

Forming Limit Prediction of Multi-layered Metal-Polymer Sheet Using Enhanced Marciniak-Kuczynski Model with Non-associated Hill48 Plasticity

Yong Hou[1], Chanmi Moon[1], Qi Hu[2], Jung-Hyeon Park[1], Aishwary Gupta[1], Junhe Lian[3], and Myoung-Gyu Lee[1](✉)

[1] Department of Materials Science and Engineering and RIAM, Seoul National University, 1 Gwanak-Ro, Seoul 08826, South Korea
myounglee@snu.ac.kr

[2] Chair of Metal Forming and Casting, Technical University of Munich, Walther-Meißner-Straße 4, 85748 Garching, Germany

[3] Advanced Manufacturing and Materials, Department of Mechanical Engineering, Aalto University, Puumiehenkuja 3, 02150 Espoo, Finland

Abstract. Formability evaluation of the laminated polymer-metal films is important for manufacturing the pouch type Li-ion battery cell. In this work, the Marciniak-Kuczynski (M-K) model was enhanced to predict the forming limit of the multi-layered films used for the battery pouch. A rate-dependent Swift hardening law and anisotropic Hill48 yield criterion under the non-associated flow rule were adopted to accurately describe the plastic deformation of each layer. By assuming the iso-strain condition and perfect bonding between layers for the simplicity of modeling, the M-K model was re-formulated with the unknown variables such as strain increment and strain ratios in the groove zone based upon the compatibility condition and force equilibrium. In order to experimentally verify the predictive accuracy of the developed model, the calculated forming limit curve (FLC) was compared with the experimental FLC of a three-layer AA5182-O/Polypropylene/AA5182-O (AA/PP/AA) sandwich sheet. The results demonstrate the high accuracy of the proposed enhanced M-K method. In addition, based on the established modeling framework, the effect of each layer and its thickness on the formability of the laminated film were investigated for a nylon/AA/PP sandwich film for the battery pouch.

Keywords: Marciniak-Kuczynski model · Laminated film · Formability · Hill48 plasticity

1 Introduction

Li-ion batteries are important components as power sources in a variety of electrical appliances. Recently, Li-ion batteries are widely used in electric vehicles. The development of high-performance Li-ion batteries promotes the research on the packaging

K. Mocellin et al. (Eds.): ICTP 2023, LNME, pp. 455–462, 2024.
https://doi.org/10.1007/978-3-031-40920-2_47

material of pouch batteries. Multi-layered aluminum-polymer ultra-thin films, as shown in Fig. 1, are important for pouch battery packaging due to its light weight and excellent corrosion resistance. The stable formability is essential for the aluminum-polymer pouches. However, the formability of aluminum-polymer film still needs to be improved for the increasing area and depth of pouch to accommodate the battery cell with high energy density. Forming of the aluminum-polymer film is complicated due to the difference in the mechanical properties of aluminum alloy and polymer layers. Yu et al. [1] experimentally studied the effect of film properties, friction coefficient, and manufacturing processes on the formability of aluminum-polymer pouches. The results showed that lower friction coefficient and a stable elongation property of the polymer film (PP in Fig. 1) improved the formability. On the other hand, the development of predictive modeling to calculate the FLC of aluminum-polymer sheets significantly boosts the application of multi-layered sheet materials. Moreover, it is helpful for designing the multi-layered structures and the forming processes. Kim et al. [2] predicted the FLC of a three-layer AA5182/polypropylene/AA5182 sandwich sheet based on the Marciniak-Kuczynski (M-K) model and plastic strain-rate potentials. The results showed that formability of the sandwich sheet has been improved compared to the single AA5182 sheet, attributed to more ductile of polypropylene core layer than the AA5182 sheet.

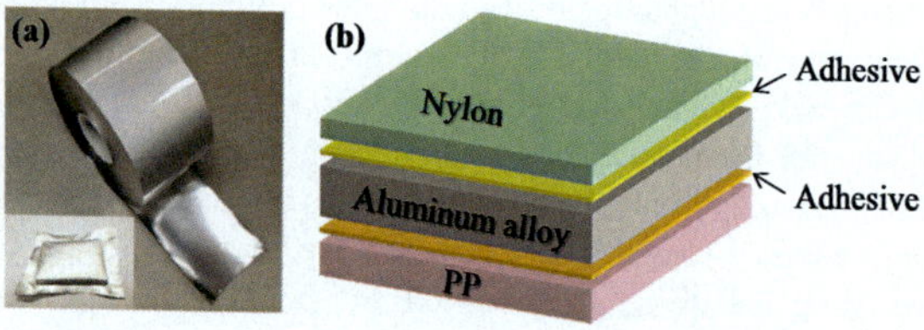

Fig. 1. (a) The aluminum-polymer pouch film and deformed pouch film, and (b) a typical structure of aluminum-polymer pouch film with multiple layers.

In this work, to accurately predict the FLC of multi-layered film, the Marciniak-Kuczynski method was enhanced: (1) the unknowns were set as the strain increment and two strain ratios based on the assumption of the iso-strain condition; (2) the anisotropic Hill48 yield criterion under the non-associated flow rule was implemented into the M-K model. The proposed predictive method was applied to calculate the formability of a three-layer AA5182/polypropylene/AA5182 sandwich sheet, and the predicted FLC was compared with the experimental results obtained from Kim et al. [2]. The results showed the high accuracy of the enhanced M-K method. After the validation, the established modeling framework was used to study the effect of each layer and its thickness on the formability of laminated film (see Fig. 1b). It is shown that the nylon layer significantly improves the formability compared with the single aluminum alloy sheet, and the thicker the nylon layer, the higher the increased FLC.

2 Enhanced Marciniak-Kuczynski (M-K) Model

As the localization-based formability model, the M-K model has been widely used to predict the FLC of sheet metals. In the M-K model, the strain increment ($\Delta\varepsilon_{11}^{b}$) and two stress ratios (α^{b} and γ^{b}) in the narrow groove (region '*b*' in Fig. 2) as shown in Eq. (1) are generally set as the unknowns, which can be solved through the Newton-Raphson (N-R) method in accordance with the equilibrium and compatibility equations.

$$\Delta\varepsilon_{11}^{b}, \alpha^{b} = \frac{\sigma_{22}^{b}}{\sigma_{11}^{b}}, \gamma^{b} = \frac{\sigma_{12}^{b}}{\sigma_{11}^{b}} \tag{1}$$

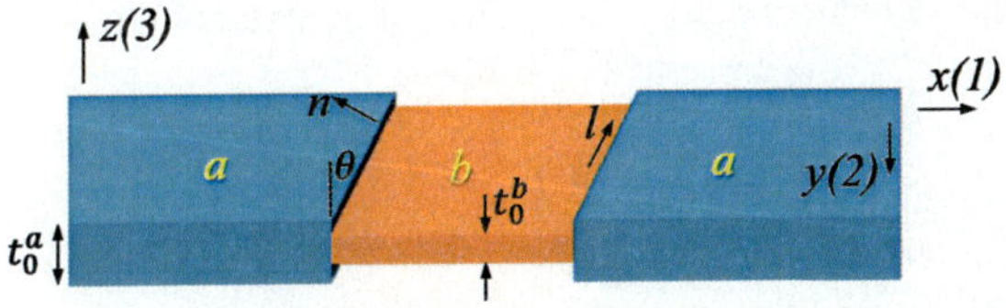

Fig. 2. Schematic representation of the initial imperfection band (region '*b*') in the M-K model.

An increment method for the M-K model was developed by Hu et al. [3] through analyzing the relationship between the internal stress variables in the groove. To enhance the computational efficiency, a modified increment method was further established by combining modified N-R method and increment method. The corresponding equations and procedure in details can be found in Hu et al. [3].

In this work, the robust algorithm for M-K model developed by Hu et al. [3] was extended to predict the FLC of a general multi-layered (with *N* layers) sheet. For the simplicity of modeling, the assumptions of iso-strain and perfect bonding between layers were employed. Therefore, the stress of each layer is different due to the distinctive mechanical properties of each layer. The stress ratios cannot be set as the unknows in the N-R iterations. To solve the problem, two strain ratios (β^{b} and δ^{b} as shown in Eq. (2)) and the strain increment were set as the unknows.

$$\Delta\varepsilon_{11}^{b}, \beta^{b} = \frac{\Delta\varepsilon_{22}^{b}}{\Delta\varepsilon_{11}^{b}}, \delta^{b} = \frac{\Delta\varepsilon_{12}^{b}}{\Delta\varepsilon_{11}^{b}} \tag{2}$$

The proportional deformation path is assumed as β^{a} = constant (−0.5–0). In each strain increment of $\Delta\varepsilon_{11}^{a} = 0.001$, the three unknowns can be obtained by solving the three equations, Eqs. (3–5). The geometric compatibility condition is Eq. (3), and the force equilibrium conditions are Eqs. (4) and (5).

$$\Delta\varepsilon_{11}^{b}\left(\tan^{2}\theta + \beta^{b} + 2\delta^{b}\tan\theta\right) = \Delta\varepsilon_{11}^{a}\left(\tan^{2}\theta + \beta^{a}\right) \tag{3}$$

$$\sum_{i=1}^{N}\left[f_i\frac{h\left(\bar{\varepsilon}^{bi}+\Delta\bar{\varepsilon}^{bi}\right)\left(\Delta\bar{\varepsilon}^{bi}\right)^{m_i}}{\varphi^{bi}}\left(1+\alpha^{bi}\tan^2\theta-2\gamma^{bi}\tan\theta\right)t_0^{ai}\right]=\Gamma_1^a \tag{4}$$

$$\sum_{i=1}^{N}\left[f_i\frac{h\left(\bar{\varepsilon}^{bi}+\Delta\bar{\varepsilon}^{bi}\right)\left(\Delta\bar{\varepsilon}^{bi}\right)^{m_i}}{\varphi^{bi}}\left[\left(\alpha^{bi}-1\right)\tan\theta+\gamma^{bi}\left(\tan^2\theta-1\right)\right]t_0^{ai}\right]=\Gamma_2^a \tag{5}$$

where $\Delta\varepsilon_{11}^a$ (=0.001) and $\Delta\varepsilon_{11}^b$ are the strain increments in region '*a*' and '*b*'. The angle between the negative *y*-axis and the narrow groove direction is θ. The subscript *i* is the number to identify each layer, 1 ~ *N*. f_i is the imperfection factor of each layer, *h* is the rate-independent flow stress, m_i is the strain rate sensitivity exponent, φ^{bi} is the ratio of the effective stress to the stress along *x*-axis, and t_0^{ai} is the initial thickness of each layer. $\bar{\varepsilon}^{bi}$ and $\Delta\bar{\varepsilon}^{bi}$ are the equivalent plastic strain and its increment in region '*b*', respectively. The two term Γ_1^a and Γ_2^a are expressed as follows.

$$\Gamma_1^a=\sum_{i=1}^{N}\left[\frac{h\left(\bar{\varepsilon}^{ai}+\Delta\bar{\varepsilon}^{ai}\right)\left(\Delta\bar{\varepsilon}^{ai}\right)^{m_i}}{\varphi^{ai}}\left(1+\alpha^{ai}\tan^2\theta\right)t_0^{ai}\right] \tag{6}$$

$$\Gamma_2^a=\sum_{i=1}^{N}\left[\frac{h\left(\bar{\varepsilon}^{ai}+\Delta\bar{\varepsilon}^{ai}\right)\left(\Delta\bar{\varepsilon}^{ai}\right)^{m_i}}{\varphi^{ai}}\left[\left(\alpha^{ai}-1\right)\tan\theta\right]t_0^{ai}\right] \tag{7}$$

where $\bar{\varepsilon}^{ai}$ and $\Delta\bar{\varepsilon}^{ai}$ are the equivalent plastic strain and its increment in region '*a*', respectively, and φ^{ai} is the ratio of the effective stress to the stress along *x*-axis in region '*a*'. Once solutions are obtained for each time increment, the onset of strain localization is determined by comparing the average equivalent plastic strains in region '*a*' and 'b', as follows.

$$\frac{\langle\bar{\varepsilon}^b\rangle}{\langle\bar{\varepsilon}^a\rangle}=\frac{\sum t_0^{ai}\bar{\varepsilon}^{bi}}{\sum t_0^{ai}\bar{\varepsilon}^{ai}}>10 \tag{8}$$

The forming limit strains are calculated for a given strain ratio β^a as the ultimate strain $(\varepsilon_{22}^a, \varepsilon_{11}^a)$ in region '*a*' at the moment of Eq. (8).

3 Plasticity Modeling

There have been numerous anisotropic models for describing the plastic anisotropy of sheet metals [4–9]. In this study, the Hill48 yield criterion based on the non-associated flow rule was adopted as follows.

$$Y(\sigma)=\left[\left(G_y+H_y\right)\sigma_{11}^2+\left(F_y+H_y\right)\sigma_{22}^2-2H_y\sigma_{11}\sigma_{22}+2N_y\sigma_{12}^2\right]^{1/2} \tag{9}$$

$$P(\sigma)=\left[\left(G_p+H_p\right)\sigma_{11}^2+\left(F_p+H_p\right)\sigma_{22}^2-2H_p\sigma_{11}\sigma_{22}+2N_p\sigma_{12}^2\right]^{1/2} \tag{10}$$

where Y and P are the yield stress and plastic potential functions, σ_{11} and σ_{22} are the stress components along the rolling (RD) and transverse directions (TD), respectively, and σ_{12} is the in-plane shear stress. And $F_{y/p}$, $G_{y/p}$, $H_{y/p}$, $N_{y/p}$ are the corresponding anisotropy parameter, which can be analytically determined as follows.

$$F_y = -\frac{1}{2} + \frac{1}{2}\left(\frac{1}{\hat{\sigma}_{90}}\right)^2 + \frac{1}{2}\left(\frac{1}{\hat{\sigma}_b}\right)^2, G_y = \frac{1}{2} - \frac{1}{2}\left(\frac{1}{\hat{\sigma}_{90}}\right)^2 + \frac{1}{2}\left(\frac{1}{\hat{\sigma}_b}\right)^2 \tag{11a}$$

$$H_y = \frac{1}{2} + \frac{1}{2}\left(\frac{1}{\hat{\sigma}_{90}}\right)^2 - \frac{1}{2}\left(\frac{1}{\hat{\sigma}_b}\right)^2, N_y = 2\left(\frac{1}{\hat{\sigma}_{45}}\right)^2 - \frac{1}{2}\left(\frac{1}{\hat{\sigma}_b}\right)^2 \tag{11b}$$

$$F_p = \frac{r_0}{r_{90}(1+r_0)}, G_p = \frac{1}{1+r_0}, H_p = \frac{r_0}{1+r_0}, N_y = \frac{(r_0+r_{90})(2r_{45}+1)}{2r_{90}(1+r_0)} \tag{11c}$$

where $\hat{\sigma}_{45}$ and $\hat{\sigma}_{90}$ are the normalized yield stress along 45° and 90° to the RD under uniaxial tension by the yield stress along the RD under uniaxial tension, and $\hat{\sigma}_b$ is the normalized yield stress under equi-biaxial tension. r_0, r_{45}, and r_{90} are the Lankford coefficient along 0°, 45°, and 90° to the RD under uniaxial tension.

The relationship between the strain ratios, stress ratios and anisotropic parameters can be expressed based on the non-associated Hill48 plasticity as follows.

$$\beta = \frac{\Delta\varepsilon_{22}}{\Delta\varepsilon_{11}} = \frac{\partial P(\sigma)/\partial\sigma_{22}}{\partial P(\sigma)/\partial\sigma_{11}} = \frac{(F_p + H_p)\alpha + H_p}{G_p + H_p - H_p\alpha} \tag{12}$$

$$\delta = \frac{\Delta\varepsilon_{12}}{\Delta\varepsilon_{11}} = \frac{\partial P(\sigma)/\partial\sigma_{12}}{\partial P(\sigma)/\partial\sigma_{11}} = \frac{2N_p\gamma}{G_p + H_p - H_p\alpha} \tag{13}$$

$$\varphi = \frac{Y(\sigma)}{\sigma_{11}} = \left[G_y + H_y - 2H_y\alpha + (F_y + H_y)\alpha^2 + 2N_y\gamma^2\right]^{1/2} \tag{14}$$

$$\phi = \frac{\Delta\varepsilon_{11}}{\Delta\overline{\varepsilon}} = \frac{\partial P(\sigma)}{\partial\sigma_{11}} = \frac{G_p + H_p - H_p\alpha}{\left[G_p + H_p - 2H_p\alpha + (F_p + H_p)\alpha^2 + 2N_p\gamma^2\right]^{1/2}} \tag{15}$$

In this work, the Swift hardening law was used to capture the hardening curves $h(\overline{\varepsilon})$ of each layer as follows.

$$h(\overline{\varepsilon}) = K(\overline{\varepsilon} + \varepsilon_0)^n \tag{16}$$

where K, ε_0 and n are material parameters, and $\overline{\varepsilon}$ is the equivalent plastic strain.

4 Model Verification

For the verification of the proposed predictive model based on M-K method, the experimental results of a three-layer AA5182/PP/AA5182 sandwich sheet (see Fig. 3a) from Kim et al. [2] were utilized. The mechanical properties of the AA5182 and PP sheets are summarized in Table 1. Each layer was assumed to share the same initial imperfection ratio f_0 within the whole groove, which was determined as 0.991 from the measured FLC0 value of 0.33. Note that the mechanical properties of the EVA adhesive between the AA and PP sheet were neglected. The predicted FLC was compared with the measured FLC in Fig. 3b. The predicted value matches with the measurement with reasonable accuracy.

Table 1. The mechanical properties of the aluminum and polypropylene sheets.

Layer	r_0	r_{45}	r_{90}	$\hat{\sigma}_0$	$\hat{\sigma}_{45}$	$\hat{\sigma}_{90}$	$\hat{\sigma}_b$	K [MPa]	ε_0	n	m	t_0 [mm]
AA	1.11	1.38	1.03	1.00	0.97	0.97	0.96	647.7	0	0.32	0.02	0.4
PP	2.98	2.98	2.98	1.00	1.00	1.00	1.00	31.3	0	0.13	0.05	0.8

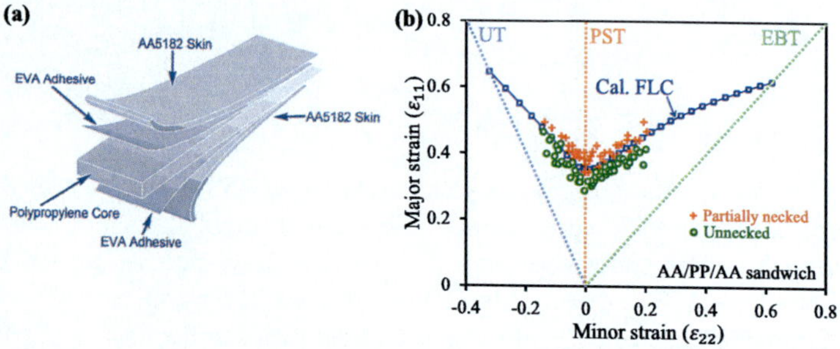

Fig. 3. (a) Structure of the AA/PP/AA sandwich sheet from Kim et al. [2]; (b) comparison of the predicted and experimental FLC of the AA/PP/AA sandwich sheet.

5 Results and Discussion

Based on the verified M-K framework for a general multi-layered sheet, the formability of the pouch material, e.g. the nylon/AA/PP structure, was investigated. The two adhesive layers were neglected in Fig. 1b. The mechanical properties of each layer were given artificially in accordance with the features of aluminium alloy and polymers. The corresponding stress vs. strain curves are shown in Fig. 4a.

Table 2. The given (artificial) mechanical properties of nylon, AA and PP layers.

Layer	$r_{0/45/90}$	$\hat{\sigma}_{0/45/90/b}$	K [MPa]	ε_0	n	m	t_0 [mm]	f_0
Nylon	1.00	1.00	200	0.002	0.50	0.05	0.04	1.000
AA	1.00	1.00	200	0.002	0.10	0.02	0.06	0.998
PP	2.00	1.00	20	0.002	0.01	0.05	0.08	1.000

5.1 Effect of Polymer Layers on the Formability

Figure 4b shows the predicted FLC of films with different structures. It can be seen that the FLC of nylon/AA is much higher than that of single AA film. The FLC of AA/PP is not much different from the single AA film. Note that the initial imperfection ratio f_0 of nylon and PP are both 1.000, and these two polymer materials have relatively high strain rate sensitivity exponents. Hence, it can be concluded that nylon film plays a vital role in improving the formability of pouch battery packaging materials, which is related to the high slope ('hardening') of stress vs. strain curve of nylon materials (Table 2).

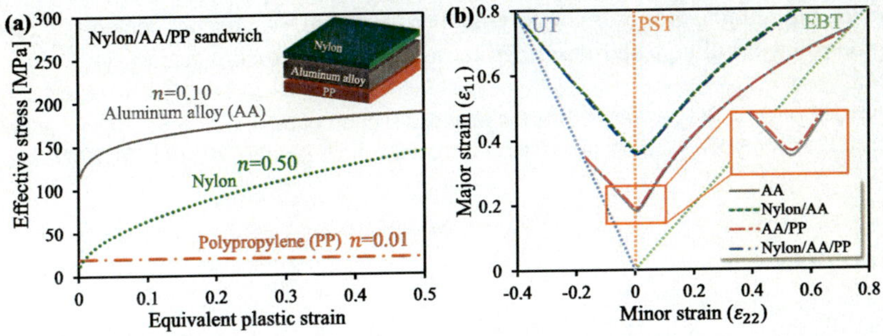

Fig. 4. (a) The hardening curves of nylon, AA and PP; (b) the predicted FLCs of AA, nylon/AA, AA/PP, and nylon/AA/PP films.

5.2 Effect of Thickness of Nylon Layer on the Formability

Figure 5a compares the FLCs of nylon/AA/PP sandwich films with different thicknesses of nylon film (from 0 to 0.08 mm) predicted by the proposed M-K model. As the thickness of the nylon film increases, the FLC of the sandwich structure increases. As shown in Fig. 5b, FLC0, that is the forming limit strain under plane strain tension, increases linearly with the thicknesses of nylon layer.

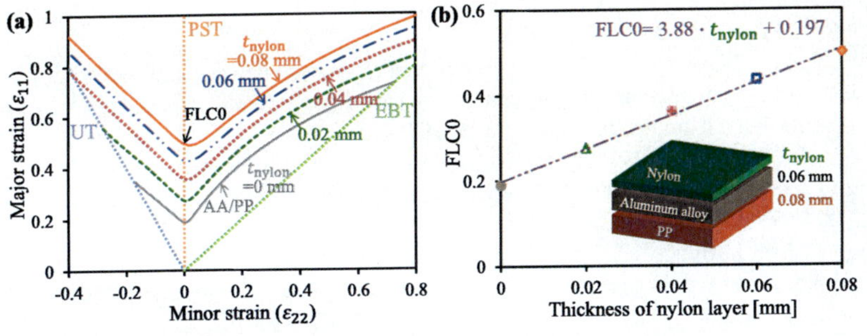

Fig. 5. (a) FLC of the nylon/AA/PP film with various thickness of nylon layer; (b) the relationship between FLC0 and the thickness of nylon layer in the nylon/AA/PP sandwich film.

6 Conclusions

A predictive model based on the M-K method and the non-associated Hill48 plasticity was developed to predict the formability of a general multi-layered sheet. Experimental results in the literature validated the accuracy of the predictions. Based on the verified predictive model, the formability of pouch material with nylon/AA/PP sandwich structure was theoretically investigated. The nylon film with strong hardening behavior (stress-strain curve with high slope) plays a crucial role in improving the formability of pouch sheet, which increases linearly with the thickness of the nylon layer. As a future work, the mechanical properties of polymer layers, which have been modeled as viscous hyper-elasticity, will be better described to improve the prediction accuracy.

Acknowledgments. YH acknowledges the financial support of the BK21 Four program at Seoul National University. The authors are grateful for grant from NRF (No. 2022R1A2C2009315).

References

1. Yu, M., Song, M., Kim, M., Kim, D.: Experimental study on the formability of aluminum pouch for lithium polymer battery by manufacturing processes. J. Mech. Sci. Technol. **33**(9), 4353–4359 (2019). https://doi.org/10.1007/s12206-019-0831-y
2. Kim, D., et al.: Modeling of forming limit for multilayer sheets based on strain-rate potentials. Int. J. Plast. **75**, 63–99 (2015)
3. Hu, Q., Li, X., Chen, J.: New robust algorithms for Marciniak-Kuczynski model to calculate the forming limit diagrams. Int. J. Mech. Sci. **148**, 293–306 (2018)
4. Hou, Y., Min, J., Stoughton, T.B., Lin, J., Carsley, J.E., Carlson, B.E.: A non-quadratic pressure-sensitive constitutive model under non-associated flow rule with anisotropic hardening: Modeling and validation. Int. J. Plast. **135**, 102808 (2020)
5. Hou, Y., Min, J., Lin, J., Lee, M.-G.: Modeling stress anisotropy, strength differential, and anisotropic hardening by coupling quadratic and stress-invariant-based yield functions under non-associated flow rule. Mech. Mater. **174**, 104458 (2022)
6. Hou, Y., Du, K., Abd El-Aty, A., Lee, M.-G., Min, J.: Plastic anisotropy of sheet metals under plane strain loading: A novel non-associated constitutive model based on fourth-order polynomial functions. Mater. Des. **223**, 111187 (2022)
7. Hou, Y., Lee, M.-G., Lin, J., Min, J.: Experimental characterization and modeling of complex anisotropic hardening in quenching and partitioning (Q&P) steel subject to biaxial non-proportional loadings. Int. J. Plast. **156**, 103347 (2022)
8. Lee, H.-R., Lee, M.-G., Park, N.: Incremental tube forming process with a novel free rotating bearing tool tip: Experiment and FE modeling with anisotropic plasticity model. Met. Mater. Int. **28**, 2356–2370 (2022)
9. Choi, Y., Lee, J., Bong, H.J., Lee, M.-G.: Hole expansion characteristics of W-tempered 7075 aluminum alloy sheet in comparison with peak aged T6 tempered alloy sheet. Met. Mater. Int. **29**, 157–167 (2023)

Determination of Optimized Biaxial Cruciform Specimens of Mild Steels, SS 316L, and Aluminum Alloys

Dilip K. Banerjee(✉), Mark A. Iadicola, and Evan Rust

NIST, Material Measurement Laboratory, Gaithersburg, MD 20899, USA
Dilip.Banerjee@nist.gov

Abstract. A methodology is described for obtaining optimum biaxial cruciform specimen designs by combining equibiaxial tension test data with finite element analysis (FEA) in conjunction with the application of an appropriate optimization strategy. This approach is applied for developing optimum specimen designs for an ASTM A1008 steel, a SS 316L, and an AA 6xxx-T4 alloy. This methodology requires in sequence the (a) selection of an appropriate initial biaxial specimen design and development of an equivalent, verified FEA model, b) optimization of the verified FEA models, (c) fabrication of cruciform specimens according to the optimized dimensions and testing under appropriate loading conditions, and (d) simulating the validation test with an equivalent FEA model with boundary conditions obtained from the tests. This study shows that such a combined "test-FEA-optimization" approach can be successfully applied to develop optimum cruciform specimen designs for advanced lightweighting materials.

Keywords: Cruciform · finite element analysis · optimization · biaxial tension test

1 Introduction

Biaxial testing of alloys is being widely used in the sheet metal forming industries to determine the mechanical properties of materials, because most forming operations include some biaxial deformation. It is preferred over the traditional uniaxial testing since sheet metal is usually highly anisotropic and biaxial testing data are expected to provide more complete information about the multidirectional deformation behavior of materials. Since advanced lightweighting alloys with relatively unknown forming behavior are increasingly being used in the automotive industries to enhance fuel efficiency of vehicles, biaxial testing of these alloys is becoming more widespread. However, there are significant challenges in developing optimum specimen designs for these materials as specimens can often fail prematurely in regions away from the gauge area.

In-plane biaxial testing of cross-shaped or cruciform specimens has become popular for determining materials' multiaxial forming behavior [1]. These specimens offer the possibility of obtaining various biaxial stress states by changing proportion of

K. Mocellin et al. (Eds.): ICTP 2023, LNME, pp. 463–474, 2024.
https://doi.org/10.1007/978-3-031-40920-2_48

loads/displacements of two axes [2]. An effective biaxial testing system relies on a good design of a biaxial test specimen. Such a specimen must experience both elastic and plastic deformation and eventual failure in the gauge area. Many specimen designs have been used by researchers. They generally fall in four main categories: (a) cut, (b) reduced section, (c) strip and slit, and (d) strip and slit with reduced central section [3]. However, a cruciform specimen design has not been universally accepted [4, 5].

In this work an approach is presented for developing optimum designs of biaxial cruciform specimens for three materials (ASTM A1008 steel, stainless steel SS 316L, and aluminum alloy AA 6xxx-T4) after starting with suitable initial designs. The initial finite element analysis (FEA) models were verified by comparing predicted displacement and strain fields with those from digital image correlation (DIC) technique. Constitutive material property (isotropic) data for the steel specimens were developed from uniaxial tensile tests conducted at various strain rates, and those for the aluminum alloy specimens were developed from static uniaxial tensile tests of specimens fabricated with three different orientations relative to the sheet rolling direction. The influence of uncertainties in hardening model parameters on material deformation behavior has been described elsewhere [6]. An anisotropic yield condition (Hill'48 [7]) was used for the models constructed for the aluminum alloy specimens, which incorporated the plastic strain ratios, r-values, (based on Lankford [8]).

2 Initial FEA Model Development and Verification

For each material, the initial cruciform specimen design was chosen based on a literature review. FEA models of these specimens were developed using above mentioned constitutive models along with boundary conditions obtained from cruciform tensile tests conducted. 1/2 symmetry models (symmetry in z or thickness direction) were created for FEA verification and validation, since test displacement/force data on all 4 arms are not exactly the same. 1/8 symmetry models were used for optimization, since 1/2 symmetry model is computationally prohibitive.

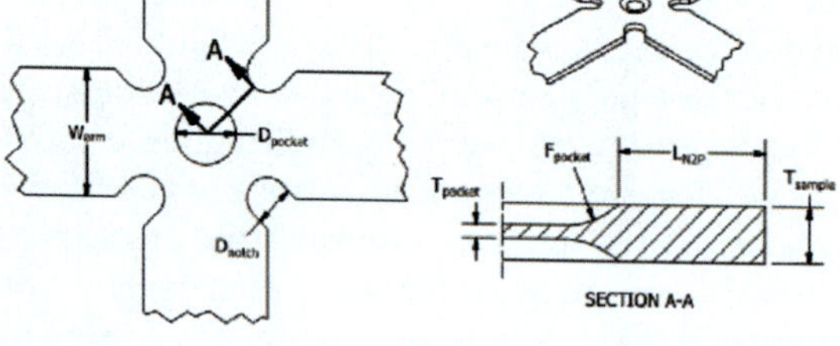

Fig. 1. A schematic of the cruciform geometry used for the ASTM A1008 steel with identification of key design parameters.

2.1 ASTM A1008 Steel Specimen

The initial cruciform specimen geometry is the N225D635 design [9] (Fig. 1). This design was chosen because this specimen was reported to fail in the central pocket region, although failure occurred when the specimen center was heated to an elevated temperature. This specimen has a central thinned-down gauge (pocket) area. Key geometric features for the design (see Fig. 1 and Fig. 9 of [10]) are the radius of the pocket flat area (R_{gauge}), pocket fillet radius (F_{pocket}), reduced thickness of the pocket flat region (T_{pocket}), corner notch radius (R_{notch}), arm width (W_{arm}), and specimen thickness, (T_{sample}). R_{pocket} is the radius of the pocket. Specimens were tested with a multiaxial test frame that has four independently controllable actuators with a ±500 kN capacity and ±50 mm stroke. Additional details are in [11]. Specimens were cut from a 3 mm thick sheet using an abrasive waterjet. The central pocket thickness reduction and pocket fillet were fabricated with electro-discharge machining (EDM). The dimensions of key specimen geometric features are listed in the "Sample 2" column in Table 1 of [12] and were used to develop the initial FEA model of the specimen.

A FEA model of the specimen was built in Abaqus [13][1]. Fully integrated, hexahedral elements (C3D8) were used for discretization. Potential issues related to the disparity seen in strain outputs from FEA software employing different types of elements are discussed elsewhere [14]. An isotropic, strain rate-dependent hardening evolution behavior was constructed from uniaxial tensile tests conducted at three different strain rates [15]. von Mises yielding behavior was assumed. Measured history-dependent, displacement and displacement gradient boundary conditions (obtained with DIC) were specified at about 59 mm from the specimen center on each arm (Fig. 2). The FEA verification methodology is similar to the one described in [16, 17] and was conducted using DIC image closest to 90% of the von Mises effective strain at peak force (hereinafter referred as t_{90}). As discussed in [12] verification was conducted by comparing FEA calculated and DIC measured values of: (a) average effective strain (ε_{VM}) in the region from the pocket center to half of R_{gauge}, (b) ε_{VM} along an arc midway between the notch and pocket (arc D in Fig. 1 of ref [12]), (c) maximum principal strains (ε_1) in the pocket along two diagonals through the center, and (d) ε_1 along diamond lines (Fig. 3). Figure 4 shows a comparison of measured and FEA ε_1 along two diamond lines. The overall agreement is good even though FEA strain profiles had slightly lower maxima. Figure 5 shows a comparison of the ε_1 (corresponding to t_{90}) along a diagonal in the central pocket. Again, the agreement is quite good between the model and measured plots. Other comparisons of different variables have been made and show good agreement between measured and computed values but are not shown here. The FEA model was refined until no noticeable change in FEA predictions was observed. This refined model was used for further optimization.

---

[1] Certain equipment, instruments, software, or materials, commercial or non-commercial, are identified in this paper in order to specify the experimental procedure adequately. Such identification is not intended to imply recommendation or endorsement of any product or service by NIST, nor is it intended to imply that the materials or equipment identified are necessarily the best available for the purpose.

2.2 SS 316L Specimen

The verified cruciform geometry from ASTM A1008 alloy was also used as the SS 316L verified geometry except that a few key dimensions were scaled [18] based on sheet thickness of 1.2 mm. The width of the specimen arms was such that it fit the test frame at the University of New Hampshire (UNH). Additional details are in ref. [10].

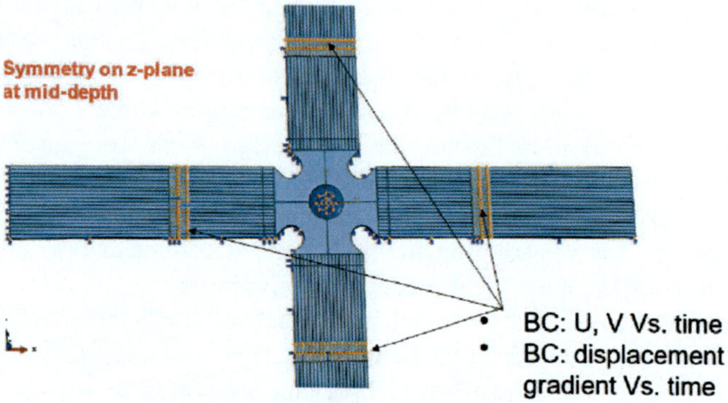

Fig. 2. A 1/2 symmetry FEA part model used for FEA verification (steel specimens) in optimization showing location of displacement and displacement gradient boundary conditions.

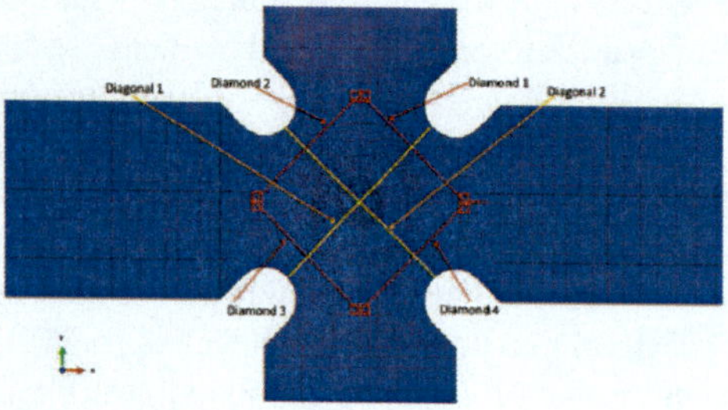

Fig. 3. FEA model for steel specimens showing diagonal and diamond lines used for model verification and validation.

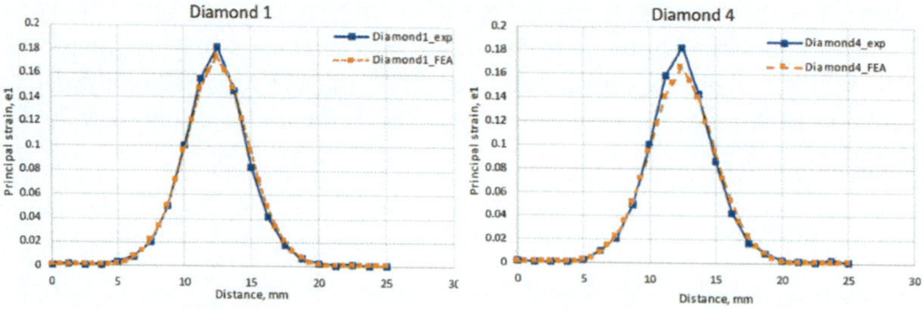

Fig. 4. Measured and FEA maximum principal strain (ε_1) along two of the four diamond lines at a step corresponding to 90% of strain at maximum force (t_{90}) for the ASTM A1008 steel.

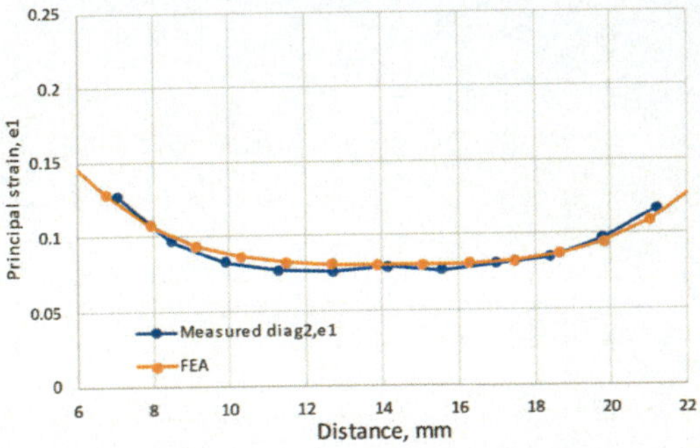

Fig. 5. Measured and FEA ε_1 in the pocket area along a diagonal through the center (from notch to notch) at t_{90} used for verification prior to the optimization study for the ASTM A1008 steel.

2.3 AA 6xxx-T4 Specimen

The planar shape of the aluminum cruciform specimen was cut from an AA 6xxx-T4 alloy sheet by an abrasive waterjet according to the design in Fig. 6. Relevant dimensions in Table 1 for the "Initial Value" column were used. The center pocket was machined (by end mill) from both faces of the specimen. Initial specimen dimensions are very close to the "optimized specimen 5" in [19] corresponding to the sheet thickness of 2.48 mm, but use a simple fillet to transition from the full thickness down to the pocket thickness. Isotropic hardening was assumed, and these data were obtained from a uniaxial RD tensile test. Since the biaxial tests result in strains beyond the uniform strain in uniaxial tests, stress-strain data were extrapolated to strains beyond uniform deformation. An approximation method was chosen that uses the DIC data sampled across the width of the specimen at the axial location of the eventual localized neck and failure. True axial strain was calculated as the average axial strain across the width for each point in time. The true stress was calculated as the applied end force divided by the current cross-sectional area at the same location and time. No correction for stress triaxiality was used. The model was built using both C3D8 and C3D8R elements. Force vs. time values for each of the four arms (from the testing machine actuators) were used as boundary conditions. A comparison of measured and FEA mean Hill'48 equivalent strain (ε_{H48}) as a function of image number in the region from specimen center to $R_{gauge}/2$ is shown in Fig. 7. Also shown is the image number corresponding to the 75% strain at maximum force point. Very good agreement is seen up to the 75% strain point. Plots of ε_{VM} along diamond and diagonal lines are not shown for brevity. After some mesh refinement and when results did not change appreciably further, the FEA model was then used further for optimization.

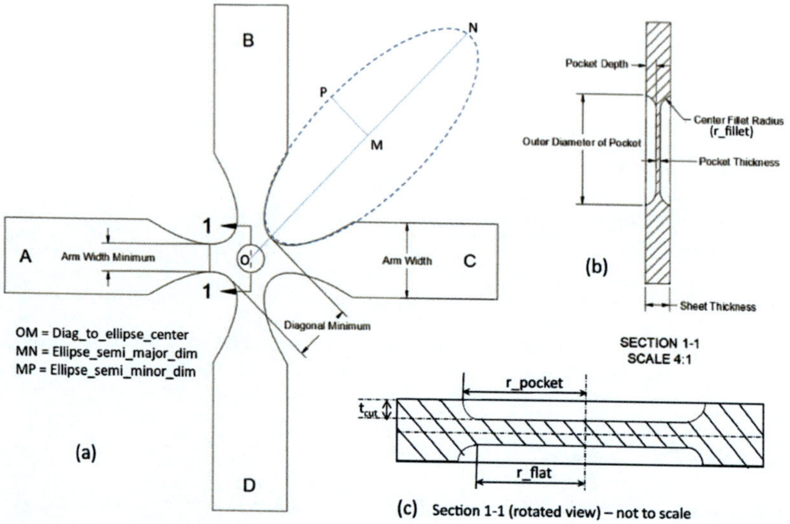

Fig. 6. A diagram of the AA 6xxx-T4 specimen with design dimensions labeled as variables.

Table 1. Values of design parameters during optimization and fabrication of the specimen (AA 6xxx-T4).

Parameters	Initial Values	DOE (#56)	LSGRG (#55)	Fabricated
Ellipse center to specimen center, L_{MO} (mm)	73.19	70.48	70.479	70.48
Ellipse semi major radius, L_{MN} (mm)	61.62	57.11	57.11	57.11
Ellipse semi minor radius, L_{MP} (mm)	22.4	14.13	14.13	14.13
Central pocket outer radius, r_{pocket} (mm)	5.46	6.857	6.86	6.88
Central pocket fillet radius, r_{fillet} (mm)	1.63	3.044	3.048	2.87
Pocket cut depth, t_{cut} (mm)	1.035	0.8802	0.898	0.886
Max. value of objective function, peeq_max	-	0.2511	0.2611	-
(std_dev - threshold)	-	0.0038	0.0039	-

3 Specimen Optimization

3.1 ASTM A1008 Steel Specimen Optimization

Optimization was conducted using the verified FEA model described in Sect. 2.1 in Isight software [20] using three different techniques: Large Scale Generalized Reduced Gradient (LSGRG), Sequential Quadratic Programming (NLPQLP), and design of experiments (DOE) [20]. All three techniques produced similar results. Average displacements of each arm as a function of time obtained from the verification mechanical test were used as boundary conditions and specified on each arm end. Design parameters were the following: fillet radius (F_{pocket}), radius of the pocket (R_{pocket}), and thickness of the specimen that was cut in the central pocket, T_{cut}, which equals $0.5*(T_{\text{sample}} - T_{\text{pocket}})$. At the beginning of the optimization, initial values of these parameters and minimum

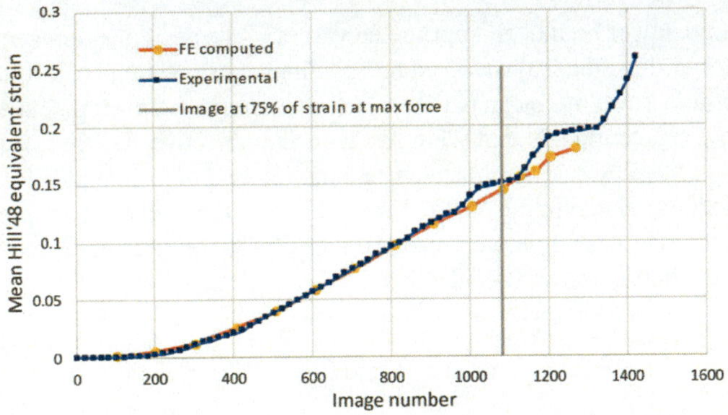

Fig. 7. A plot of mean measured and computed Hill'48 equivalent strain as a function of image number in the region from specimen center to $R_{\text{gauge}}/2$ (AA 6xxx-T4).

and maximum values of the bounds were specified. Multiple FEA runs in a loop were conducted by varying the design parameter values each time and computing objective function values and evaluating the constraint condition (see below). This process continued until the desired objective and constraint conditions were achieved. The following were used as the objective function and constraint condition:

$$\overline{\varepsilon_{VM}} = \frac{1}{n}\sum_{i=1}^{n}\left[\varepsilon_{\text{VM}}\left(\sqrt{x_i^2 + y_i^2} \leq \frac{R_{\text{pocket}}}{2}\right)\right] \tag{1}$$

$$\text{Maximize } [\overline{\varepsilon_{VM}}]$$

$$\text{While}\left[0.99\overline{\varepsilon_{\text{VM}}} \leq \left(\varepsilon_{\text{VM}}\left(\sqrt{x_i^2 + y_i^2} \leq \frac{R_{\text{pocket}}}{2}\right)\right) \leq 1.01\overline{\varepsilon_{\text{VM}}}\right] \tag{2}$$

where, ε_{VM} "overbar" signifies the mean value in the central region of the pocket, x_i and y_i are nodal coordinates, respectively, of the nodes in the undeformed configuration, and n is the number of nodes. The objective function and the constraint condition were determined based on findings described in [12], which indicated variability of strains in the central pocket of the optimized specimen. Equations (1) and (2) were evaluated at a step corresponding to t_{90}.

3.2 SS 316L Steel Specimen Optimization

The design goals were (1) maximized ε_{VM} in the central flat gauge area (3 mm radius), (2) moderate strain uniformity (±2.5% of the mean ε_{VM}) in the gauge area, (3) sufficient flat gauge area for material characterization, (4) a machinable geometry, and (5) a design suitable for the UNH apparatus. Key specimen geometry features are same as those given in Sect. 2.1. T_{sample} (sheet thickness) was 1.2 mm and T_{pocket} was set to 50% of the sheet thickness or 0.6 mm. The lower bound of F_{pocket} was set to 0.25 mm based on practical machinability limit of end mills. Specimen width was set to 180 mm. See Table 5 of [10] for lower bound, upper bound, and optimized values of the design parameters. A 2-level DOE approach with Isight [20] was used for specimen optimization [10]. A total of 32 × 32 (1024) combinations of design parameters (based on the Latin Hypercube sampling) were used to produce models that were run with the objective function and constraint conditions (±2.5% ε_{VM} variation) given in Eqs. (3) and (4) respectively [10]. These conditions were evaluated in the center 3 mm radius of the pocket. Isotropic hardening and von Mises yield condition were used in DOE FEA simulations. From these 1024 simulations, the optimum run was identified.

$$(\overline{\varepsilon_{VM}})_{R\leq 3} = \frac{1}{n}\sum_{i=1}^{n}\left[\varepsilon_{VM}\left(\sqrt{x_i^2 + y_i^2} \leq 3\right)\right] \tag{3}$$

$$\text{Maximize}\left[(\overline{\varepsilon_{VM}})_{R\leq 3}\right]$$

$$\text{While}\left[0.975(\overline{\varepsilon_{VM}})_{R\leq 3} \leq \left(\varepsilon_{VM}\left(\sqrt{x_i^2 + y_i^2} \leq 3\right)\right) \leq 1.025(\overline{\varepsilon_{VM}})_{R\leq 3}\right] \tag{4}$$

3.3 AA 6xxx-T4 Specimen Optimization

Six design parameters along with initial values and allowable ranges of values were identified (following [19]) corresponding to sheet thickness of 2.48 mm. They are: fillet radius (r_{fillet}), outer radius of the thinned down central pocket (r_{pocket}), thickness of the specimen that was cut from each face to form the central pocket (T_{cut}), distance from the specimen center to the ellipse center (L_{MO}), ellipse semi-major axis radius (L_{MN}), and ellipse semi-minor axis radius (L_{MP}) (Fig. 6). Maximization of the mean values of ε_{H48} from center to $r_{\text{flat}}/2$ (r_{flat} is distance from center to the tangent point of the fillet) was chosen as the first objective function. The second objective is the minimization of the difference between the standard deviation ε_{H48} and a threshold value to ensure uniformity of strain in the gauge area:

$$\text{Maximize}\left\{\text{Mean}\left[\varepsilon_{\text{H48}}\left(\sqrt{x^2 + y^2} \leq \frac{r_{\text{flat}}}{2}\right)\right]\right\} \tag{5}$$

$$\text{Minimize}\left[\text{Stdev}\left\{\varepsilon_{\text{H48}}\left(\sqrt{x^2 + y^2} \leq \frac{r_{\text{flat}}}{2}\right)\right\} - \zeta\right] \tag{6}$$

where, x, and y are nodal coordinates in the undeformed FEA model and ζ is the threshold value (which was taken as 1% of the mean ε_{H48} obtained at the time step corresponding to the maximum force point). Equations (5) and (6) were evaluated at a step (N_75) that is 75% of the step corresponding to the maximum force point step. A 2-step approach (DOE followed by the LSGRG method) was used for an efficient exploration of the design space. Figure 8 (a) shows a comparison of the evolution of FEA mean ε_{H48} (in the $r_{\text{flat}}/2$ region or the area of interest, AOI) as a function of frame number for the initial verified FE model, the best DOE model, and the final DOE & LSGRG optimized model. Figure 8(b) shows a similar plot for the standard deviation of ε_{H48} in AOI. The frame corresponding to N_75 is shown by a vertical line. The optimized model (LSGRG4_55) had the highest maximum strain, while the strain variability was very close to the best DOE run (DOE_56). At frame N_75, the optimized model shows over double the mean ε_{H48} in the initial verified model and less than half the standard deviation of ε_{H48} from the initial verified model. This fulfills both objectives of this study. It can also be seen that beyond frame 1225, the optimized model shows more variability due to large associated strains (post-necking) at that time. The optimized model had lower values for the L_{MO}, L_{MP}, L_{MN}, and t_{cut} parameters and larger values of the r_{pocket} and r_{fillet} parameters in relation to the initial verified model (Table 1).

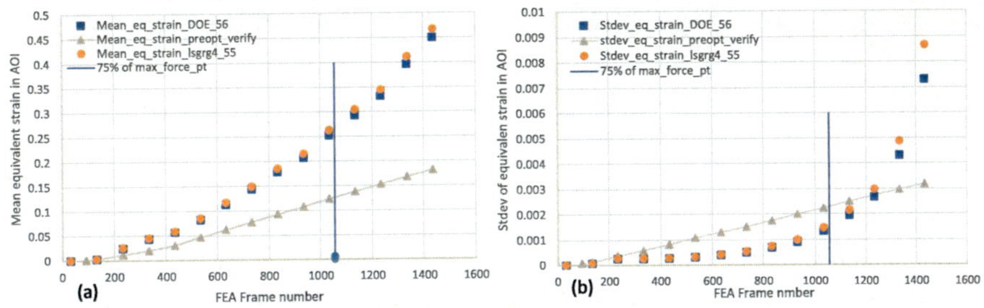

Fig. 8. (a) A plot of the evolution of FE computed mean equivalent strains (in AOI) Vs. Frame number for the initial verified FE model, the best DOE model, and the final LSGRG optimized model; (b) a plot of the standard deviation of the equivalent strains in AOI (AA 6xxx-T4).

4 Validation of the Optimized Model

4.1 ASTM A1008 Steel Specimen

Finally, a verification study was conducted by fabricating a specimen according to the optimized parameter values, testing this specimen under equibiaxial loading, and comparing measured results with those from an equivalent FEA model. The FEA model was developed in half symmetry. Displacement boundary conditions at each arm end were taken from the mechanical test conducted. The comparison is made for an image, which is just before the maximum force is reached in the specimen. Figure 9(b) shows the contour plots of FEA computed ε_{VM}. These trends match quite well with measured contour plots in Fig. 9(a). Comparisons of ε_{VM} plots along diagonal and diamond lines are not shown here for brevity.

4.2 SS 316L Steel Specimen

The optimized cruciform specimen was tested in four loading conditions (ratio of x- and y-displacements = 1:1, 2:1, 8:1, and 1: free), which resulted in deformation close to equibiaxial through plane strain to near uniaxial strain evolution conditions. Equivalent FE analyses were conducted with an 1/8 symmetry model using both von Mises isotropic

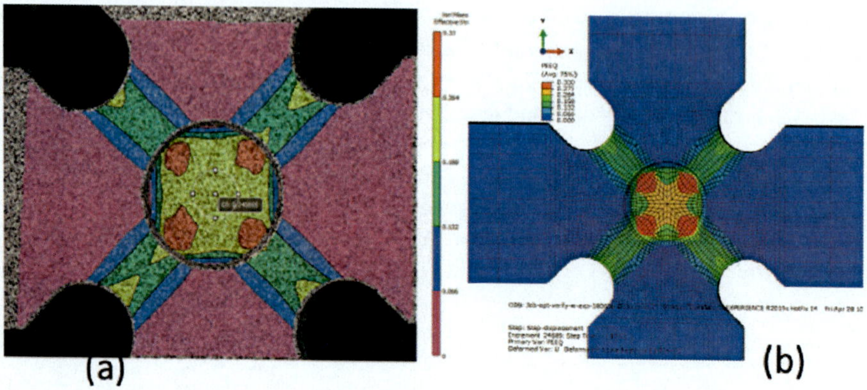

Fig. 9. Contour plots of (a) measured and (b) FEA computed equivalent strain at a frame just before the onset of localization in the validation experiment for ASTM A1008 steel.

yield function and an anisotropic yield function (Yld2004-18p($\bar{\varepsilon}$)) with a Hockett-Sherby strain hardening model [10]. The anisotropic yield function model predicted force-displacement curves matched the measured plots quite well. Principal strain predictions for both isotropic and anisotropic models compared reasonably well with those from tests under all four loading path conditions [10].

4.3 AA 6xxx-T4 Specimen

A cruciform specimen with dimensions of design parameters in the "Fabricated" column of Table 1 (which are close to the values in the LSGRG column) was fabricated and tested in equibiaxial tensile stretch conditions. An equivalent FE model using the as-fabricated specimen dimensions was also built. The model was built in half symmetry. Displacement vs. time data from each actuator of the mechanical testing machine was used in the FE model as boundary conditions. The FE model development procedure and computation methodology are same as the ones described in Sect. 2.3 and are not repeated here. Two FE simulations were conducted: one with C3D8 and the other with C3D8R elements. The validation efforts are ongoing where strain profiles along diamond and diagonal lines before the onset of localization are being compared with those from test data. Contour plots of the measured, FEA (C3D8 element), and FEA (C3D8R element) ε_{H48} at a frame just before the onset of localization are shown in Fig. 10. A good agreement was achieved. It is anticipated that failure would initiate from the region near the base of the fillet at four corners and progress inward.

5 Summary

This study reviews a systematic approach for developing optimum biaxial cruciform specimen designs for three materials (ASTM A1008 steel, SS 316L stainless steel, and AA 6xxx-T4). For each material, the starting geometry was chosen after conducting a literature review. In each case, an initial FEA model was first verified by comparing model

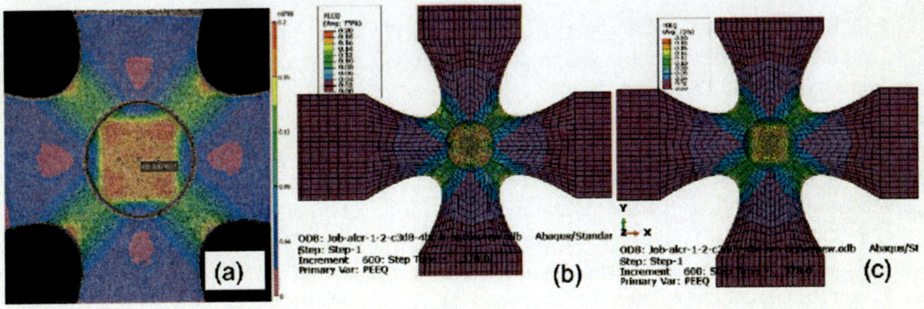

Fig. 10. Plots of (a) measured, (b) FEA computed (C3D8), and (c) FEA computed (C3D8R) equivalent strain at a frame before the onset of localization for the AA 6xxx-T4 validation test.

results with measured data in the gauge region of specimens tested under equibiaxial loading conditions. For ASTM A1008 steel, three separate optimization methodologies were used yielding similar results. For SS 316L, a 2-level DOE was used. DOE followed by LSGRG was used for AA 6xxx-T4. Validation of optimized specimens was conducted by fabricating specimens with optimized specimen dimensions, testing under equibiaxial tension, and comparing test results in the gauge area with those from equivalent FEA simulations. The approach allows to obtain the following objectives: a) strain uniformity in the gauge area [10], b) much higher equivalent strain (e.g., 0.246 for optimized Vs. 0.156 for unoptimized ASTM A1008 steel at t_{90} [12]), c) elimination of premature failure in regions away from gauge area and d) absence of any undesirable load sharing among horizonal and vertical arms. Hence, stress-strain data to much larger strains can be obtained from tests with optimum specimen designs than those that can be obtained from unoptimized specimens.

Acknowledgment. The authors appreciate the collaboration they had with E. Mamros, J. Ha, and B. Kinsey, University of New Hampshire for the SS 316L study. The Al cruciform work was partially funded by a Fulbright U.S. Research Award (Banerjee, 2019/APE-R/104).

References

1. Demmerle, S., Boehler, J.P.: Optimal design of biaxial tensile cruciform specimens. J. Mech. Phys. Solids **41**(1), 143–181 (1993)
2. Wu, X.D., Wan, M., Zhou, X.B.: Biaxial tensile testing of cruciform specimen under complex loading. J. Matl. Proc. Tech. **168**(1), 181–183 (2005)
3. Hannon, A., Tiernan, P.: A review of planar biaxial tensile test systems for sheet metal. J. Mater. Process. Technol. **198**(1), 1–13 (2008)
4. Lin, S.B., Ding, J.L., Zbib, H.M., Aifantis, E.C.: Characterization of yield surfaces using balanced biaxial tests of cruciform plate specimens. Scr. Metall. Mater. **28**(5), 617–622 (1993)
5. Lin, S.B., Ding, J.L.: Experimental study of the plastic yielding of rolled sheet metals with the cruciform plate specimen. Int. J. of Plast. **11**(5), 583–604 (1995)
6. Prasad, K., Kumar, D., Krishnaswamy, H., Banerjee, D.K.: Uncertainties in the swift hardening law parameters and their influence on the flow stress and the hole expansion behavior of dual-phase (DP600) steel specimens. J. Matl. Eng. Perf. (2023). https://doi.org/10.1007/s11665-022-07793-2

7. Hill, R.: The Mathematical Theory of Plasticity. Oxford University Press, New York (1950)
8. Lankford, W.T.: New criteria for predicting the press performance of deep drawing sheets. Trans. ASM **42**, 1197–1232 (1950)
9. Abu-Farha, F., Hector, L.G., Khraisheh, M.: Cruciform-shaped specimens for elevated temperature biaxial testing of lightweight materials. JOM **61**(8), 48–56 (2009)
10. Mamros, E.M., Mayer, S.M., Banerjee, D.K., Iadicola, M.A., Kinsey, B.L., Ha, J.: Plastic anisotropy evolution of SS316L and modeling for novel cruciform specimen. Int. J. Mech. Sci. **234**, 107663 (2022). https://doi.org/10.1016/j.ijmecsci.2022.107663
11. Iadicola, M.A., Creuziger, A.A., Foecke, T.: Advanced biaxial cruciform testing at the NIST center for automotive lightweighting. In: Stress, R. (ed.) Thermomechanics & Infrared Imaging, Hybrid Techniques and Inverse Problems, vol. 8, pp. 277–285. Springer, Cham (2014). https://doi.org/10.1007/978-3-319-00876-9_34
12. Creuziger, A., Iadicola, M.A., Foecke, T., Rust, E., Banerjee, D.: Insights into cruciform sample design. JOM **69**(5), 902–906 (2017). https://doi.org/10.1007/s11837-017-2261-6
13. Abaqus Unified FEA - SIMULIA Homepage.https://www.3ds.com/products-services/simulia/products/abaqus/. Accessed 09 Aug 2022
14. Iadicola, M., Banerjee, D.: A comparison of strain calculation using digital image correlation and finite element software. J. Phys.: Conf. Ser. **734**, 032013 (2016). https://doi.org/10.1088/1742-6596/734/3/032013
15. Banerjee, D.K., Iadicola, M.A., Creuziger, A.A.: Understanding deformation behavior in uniaxial tensile tests of steel specimens at varying strain rates. J. Res. Nat. Inst. Stand. Technol. **126**, 126050 (2021). https://doi.org/10.6028/jres.126.050
16. Banerjee, D., Iadicola, M., Creuziger, A., Foecke, T.: An experimental and numerical study of deformation behavior of steels in biaxial tensile tests. In: TMS2015 Supplemental Proceedings, pp. 279–288. Springer, Cham (2016). https://doi.org/10.1007/978-3-319-48127-2_35
17. Banerjee, D., Iadicola, M., Creuziger, A., Foecke, T.: Finite element modeling of deformation behavior of steel specimens under various loading scenarios. Key Eng. Mater. **651**, 969–974 (2015)
18. Mamros, E.M., Eaton, M.C., Ha, J., Kinsey, B.L.: Numerical analysis of SS316L biaxial cruciform specimens under proportional loading paths. In: ASME International Manufacturing Science and Engineering Conference (2021). https://doi.org/10.1115/MSEC2021-59877
19. Baptista, R., Claudio, R.A., Reis, L., Madeira, J.F.A., Guelho, I., Freitas, M.: Optimization of cruciform specimens for biaxial fatigue loading with direct multi search. Theoret. Appl. Fract. Mech. **80**, 65–72 (2015)
20. Isight & SIMULIA Execution Engine by Dassault Systèmes Homepage.https://www.3ds.com/products-services/simulia/products/isight-simulia-execution-engine/. Accessed 09 Aug 2022

Examination of Bending Stress Superposition Effect on Martensite Transformation in Austenitic Stainless Steel 304

Elizabeth M. Mamros[1], Lenard A. Polec[2], Fabian Maaß[2], Till Clausmeyer[2], A. Erman Tekkaya[2], Jinjin Ha[1](✉), and Brad L. Kinsey[1]

[1] Mechanical Engineering, University of New Hampshire, Durham, NH 03824, USA
jinjin.ha@unh.edu

[2] Institute of Forming Technology and Lightweight Components, TU Dortmund University, Dortmund, Germany

Abstract. Uniaxial tension is a universal material characterization experiment. However, studies have shown that increased formability can be achieved with simultaneous bending and unbending of the material. This so-called continuous bending under tension process is an example of bending stress superposition to a uniaxial tension process. In this research, experiments are conducted on stainless steel 304 to investigate the effects of bending stress superposition on the austenite to martensite phase transformation. Two vortex tubes are mounted to the carriage of the machine and used to decrease the temperature in a localized region of the specimen to evaluate two temperature conditions. The *in-situ* strain and temperature fields are captured using 3D digital image correlation and infrared cameras. The deformation induced α′-martensite volume fraction is measured at regular intervals along the deformed gauge length using a Feritscope. The number of cycles that the rollers traverse the gauge length, corresponding to the strain level, is also varied to create five conditions. The deformed specimens revealed heterogeneous martensite transformation along the gauge length due to the non-uniform temperature fields observed for each test condition. Decreasing the temperature and increasing the number of cycles led to the highest amount of phase transformation for this bending-tension superposed process. These results provide insight on how stress superposition can be applied to vary the phase transformation in more complex manufacturing processes, such as incremental forming, which combines bending, tension, and shear deformation.

Keywords: Martensite Transformation · Stainless Steel · Stress Superposition

1 Introduction

Incremental forming is a rapid prototyping process that can be used to manufacture customized sheet metal parts without requiring specialized tooling. In incremental forming, a tool, typically hemispherical, traverses a piece of clamped sheet metal following a layer-by-layer toolpath to form the desired geometry. The conventional process is known

K. Mocellin et al. (Eds.): ICTP 2023, LNME, pp. 475–485, 2024.
https://doi.org/10.1007/978-3-031-40920-2_49

as single point incremental forming (SPIF) and results in increased material formability compared to stamping and deep drawing [1]. Three dominant forming mechanisms are observed during incremental forming: bending, tension, and shear [2], i.e., stress superposition. Stress superposition is defined as the incorporation of additional stresses into a primary forming operation during a single process step [3]. This innately stress-superposed process can be manipulated to affect the properties of the formed part [4], but complex modeling efforts are required to inform manufacturing decisions, e.g., process parameter selection.

To assist with modeling incremental forming, a similar, simpler manufacturing process, known as continuous bending under tension (CBT), can be investigated. The CBT process is more complex than uniaxial tension because bending, via rollers, is superposed to a uniaxial process but less complex than incremental forming, due to the absence of shear deformation. A numerical model of the CBT process was introduced by Hadoush et al. [5] to better understand the effects of cycling in sheet metal forming, specifically in relation to incremental forming and deformation beyond the forming limit curve. The deformation mechanisms in the CBT process have been likened to those of SPIF, including the bending depth to the wall angle and the roller speed of the carriage to the feed rate and step size. A CBT machine designed with an additional roller, which is subjected to a constant compression force, to superpose compression has been used to investigate the deformation mechanisms in double-sided incremental forming (DSIF) [6].

Several variations of CBT machines exist, including one at the University of New Hampshire (UNH) [7]. Previous studies have shown that the CBT process increases elongation to failure, i.e., the formability of the material, compared to uniaxial tension and decreases the required forming load. This is consistent with two of the three main objectives of using the stress superposition strategy, which are increased formability [8], reduced forming loads, and enabling customization [9], which will be demonstrated in this work through the manipulation of phase transformation. CBT experiments have been conducted for aluminum [10], magnesium [11], and dual phase steels [12] using the UNH machine to validate various material models used in finite element analyses.

Austenitic stainless steels undergo an irreversible strain-induced phase transformation from γ-austenite (face-centered cubic, FCC) to α'-martensite (body-centered cubic, BCC). Alternatively, the phase transformation can occur from γ-austenite to ε-martensite (hexagonal close packed, HCP) to α'-martensite. This phase transformation is a function of several parameters including temperature [13], strain level [14], strain rate [15], strain path [16], stress state [17], crystallographic texture [18], stacking fault energy [19], and chemical composition [4]. Related to these parameters, certain grades of stainless steel (SS) transform at different rates when subjected to the same temperature and loading conditions, e.g., SS316L transforms significantly less than SS304 and SS301LN [20]. This variance dependent upon the material and a combination of parameters during experiments makes it difficult for martensitic transformation kinetics models to be transferable between materials and processes without modification.

As one example, a model proposed by Olson and Cohen in 1975 [14] has been adapted several times to include additional parameters in order to attempt to fit other materials. Their model, based on data from Angel [21], predicts the martensitic transformation as a function of equivalent plastic strain and temperature. In 1992, Stringfellow et al.

incorporated stress triaxiality into Olson and Cohen's model [22] to compare with the experimental results presented by Young [23]. Several years later, Tomita and Iwamoto amended the previous version of the model to include the effect of strain rate [24]. In 2011, Beese and Mohr expanded the model to incorporate the Lode angle parameter as another means to define the stress state in addition to the stress triaxiality [17]. The following year, Beese and Mohr proposed an anisotropic version of their model to account for the differences in phase transformation observed previously for rolling direction and transverse direction SS301LN uniaxial tension tests [25]. For additively manufactured SS304L, Wang and Beese proposed a stress state dependent model that also accounts for the crystallographic texture and chemical composition of the material [18]. Future numerical analyses of the CBT process will include martensitic transformation kinetics modeling for the chosen material.

In this work, continuous bending under tension experiments are conducted on stainless steel 304, to investigate the effects of bending stress superposition on the austenite to martensite phase transformation. Two temperature conditions and five strain levels, i.e., varied numbers of cycles, are investigated to determine the effects of localized cooling and strain level on the phase transformation of this material. A Feritscope is used to measure the α'-martensite volume fraction along the gauge length, and the results are validated by electron backscatter diffraction (EBSD) analyses of select samples. In future work, x-ray diffraction analyses will be conducted to determine the effect of temperature and strain level variation on the residual stress development due to phase transformation. Finite element analyses will also be performed and validated by experimental results.

2 Material Characterization

2.1 Material

Uniaxial tension experiments were conducted using the geometry described in the ASTM E8 standard and the procedure detailed in [26] to characterize 1.2 mm thick sheets of SS304, which are fully austenite in the undeformed condition (i.e., zero α'-martensite volume fraction). The chemical composition provided by the manufacturer is shown in Table 1. The resulting true stress-strain curves from specimens oriented along the rolling direction (RD) is shown in Fig. 1. Note that starting at ~0.20 true strain, the strain hardening behavior of SS304 increases, which is an indication of the α'-martensite phase transformation occurring.

Table 1. Chemical composition provided by manufacturer for SS304.

	C	Si	Mn	P	S	Cr	Mo	Ni	N	Cu	Fe
SS304	0.07	1.00	2.00	0.045	0.030	16.71	–	10.12	–	–	Bal

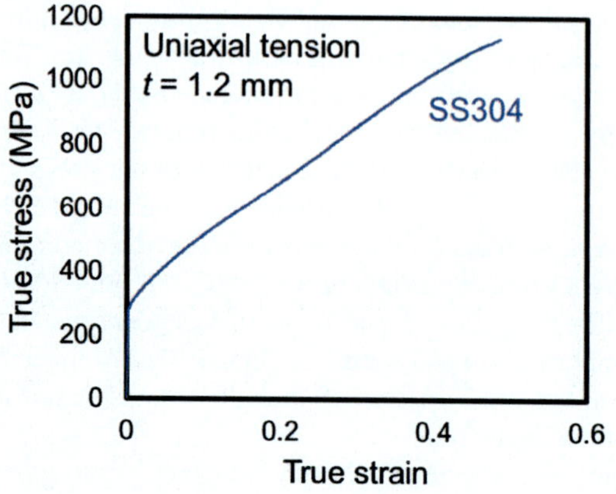

Fig. 1. True stress-strain curve for SS304 from uniaxial tension experiment oriented along the RD.

2.2 Specimen Preparation

The ASTM E8 geometry with an extended gauge length of 200 mm was used for the CBT experiments as shown in Fig. 2. The specimens were cut from the sheets along the RD by laser. The top surfaces of the specimens were prepared with layers of white paint, a black speckle pattern, and matte clear paint for detection in the digital image correlation post-processing software.

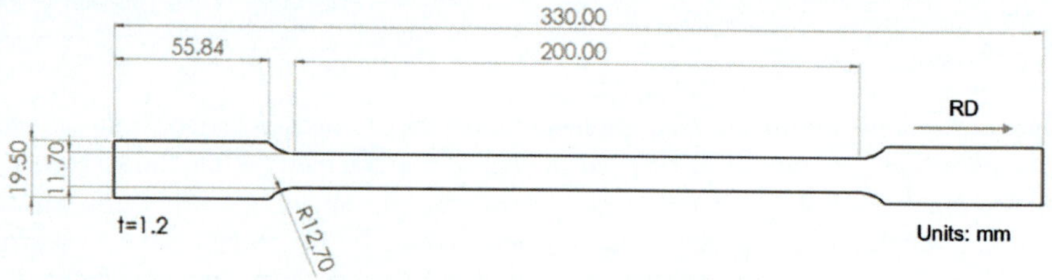

Fig. 2. Specimen geometry for CBT experiments.

3 Experiments

3.1 Experimental Setup

A custom CBT machine at UNH [7, 27] was used for experiments and is shown in Fig. 3. The CBT specimen was clamped in the grips, and the center roller of the three stationary roller system was displaced downward to create the bending condition. During the experiments, the cylinder was displaced to apply axial tension to one end of the specimen while the opposite side of the specimen remained fixed. The entire carriage assembly was driven by a ball screw and motor, which caused the specimen to traverse through the stationary rollers over the gauge length of the specimen while superposing bending to the tensile process. To capture the strain and temperature *in-situ*, a stereo (3D) digital image correlation (DIC) system (Correlated Solutions Inc.) with two 2.3 megapixel cameras (FLIR Grasshopper2) and 17 mm lenses (Schneider) was coupled with an SC-645 infrared camera (FLIR). The three cameras were fixtured to the carriage, using aluminum extrusion and 3D-printed camera mounts as shown in Fig. 3, to allow the cameras to remain focused on the same area of interest throughout the experiment. Subset, step, and filter sizes of 4 pixels, 9 pixels, and 5 were used for all analyses (VIC-3D, Correlated Solutions Inc.).

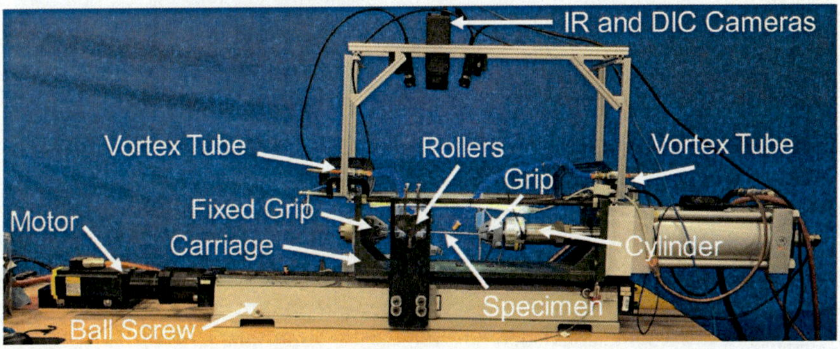

Fig. 3. CBT machine at the University of New Hampshire.

Based on previous experiments using this machine, the speed of the carriage, which is the so-called roller speed, was set to 66 mm/s, and the pulling speed, which is the speed of the cylinder, was set to 1.3 mm/s. The bending depth, which is the ratio of the bend depth from the rollers to the material thickness, was 2 for all experiments.

3.2 Temperature Control

To locally control the temperature, two vortex tubes (Vortec) were mounted to the carriage with 3D-printed fixtures, and the attached nozzles were aimed at two locations along the gauge length in the DIC area of interest (Fig. 4). Note that the cooling effect was extremely localized, and two controlled zones were created in the gauge area. The amount of hot air exhausted from the vortex tubes was adjusted to create two temperature conditions, low and high temperatures.

3.3 Phase Transformation Measurements

The γ-austenite to α'-martensite phase transformation in stainless steel was measured using an FMP30C Feritscope (Fischer Technology) on the unpainted, bottom surface of the specimens. The fixed end of the specimen was designated as the reference location, and measurements were taken every 12.7 mm along the gauge length after 10 cycles. The strain and temperature data were extracted at the same locations.

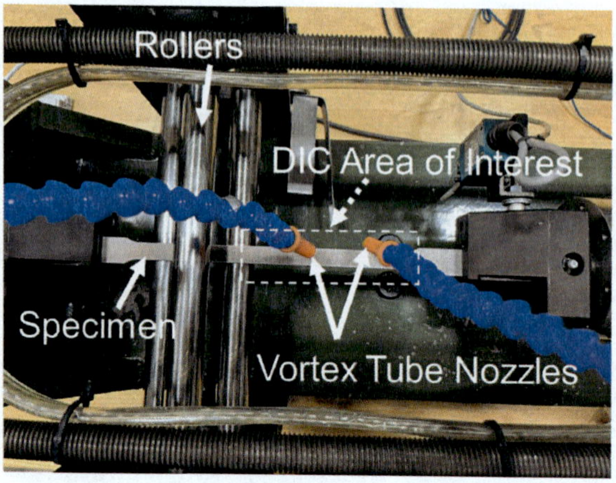

Fig. 4. Close-up of experimental setup from cameras' perspective.

To validate the α'-martensite measurements from the Feritscope, electron backscatter diffraction (EBSD) analyses were performed on select samples extracted near P3 after 10 cycles (Table 2). A Tescan Lyra3 GMU focused ion beam scanning electron microscope equipped with an Edax EBSD detector was used for these analyses. The following parameters were selected: magnification: 750x; beam intensity: 19.1; acceleration voltage: 10 kV; sample tilt: 70°; binning: 8 × 8; and step size: 0.7 μm.

Table 2. Comparison of Feritscope and EBSD α'-martensite measurements near P3.

Material	Temperature Condition	α'-Martensite Volume Fraction	
		Feritscope	EBSD
SS304	Low	0.89	0.82
	High	0.75	0.72

4 Results

4.1 Strain

The true strain evolution along the pulling direction in the gauge length of a SS304 specimen deformed at the low temperature condition with different number of cycles is shown in Fig. 5. The strain field is relatively uniform over the gauge length throughout the forming process and reaches a true strain value of approximately 0.35 after 10 cycles.

4.2 Temperature

The evolution of the temperature in the gauge length with increasing deformation, i.e., number of cycles, is shown for a SS304 low temperature experiment in Fig. 6. The two

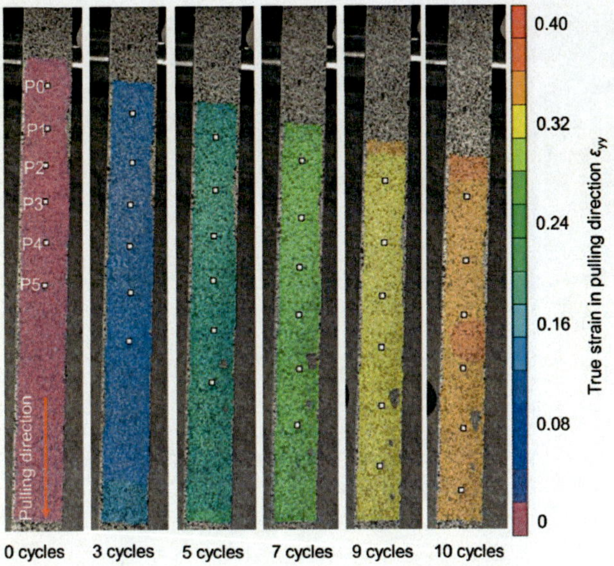

Fig. 5. Evolution of true strain in pulling direction (ε_{yy}) recorded in gauge length of SS304 low temperature CBT specimen at 0 (undeformed), 3, 5, 7, 9, and 10 cycles.

controlled zones created by the vortex tubes are distinguishable by their reduced temperature compared to the rest of the area of interest. Note that the controlled zones appear to remain stationary, but, due to elongation, different regions of the gauge area passed through the controlled zones as the experiment progressed. Throughout the experiment shown in Fig. 6, the temperature in the center of the controlled zones increases ~4°. In the surrounding regions, the temperature increases ~11 °C since the cooling effect of the vortex tubes is lessened significantly at distances farther away from the nozzle. At the extreme captured by the thermal camera, i.e., near the bottom of Fig. 6, the temperature increases to nearly 30 °C due to the lack of temperature control and deformation-induced heating.

Comparing Figs. 5 and 6, it appears that the localized cooling did not cause significant inhomogeneity in the true strain in the pulling direction in the region of interest. However, the DIC images of each cycle were captured after the rollers had traversed the area of interest. Some localized effects on the strain may occur when the rollers pass through the controlled zones and superpose bending, but, unfortunately, these regions were obscured from the cameras' view.

4.3 Martensitic Transformation

From Olson and Cohen [14], it is known that increased martensitic transformation occurs at decreased temperatures and increased strain levels. Both trends were observed for the α'-martensite volume fraction of SS304 when comparing the low temperature and high temperature conditions at varying numbers of cycles as shown in Fig. 7. The average volume fraction over the area of interest for three specimens at each condition is plotted

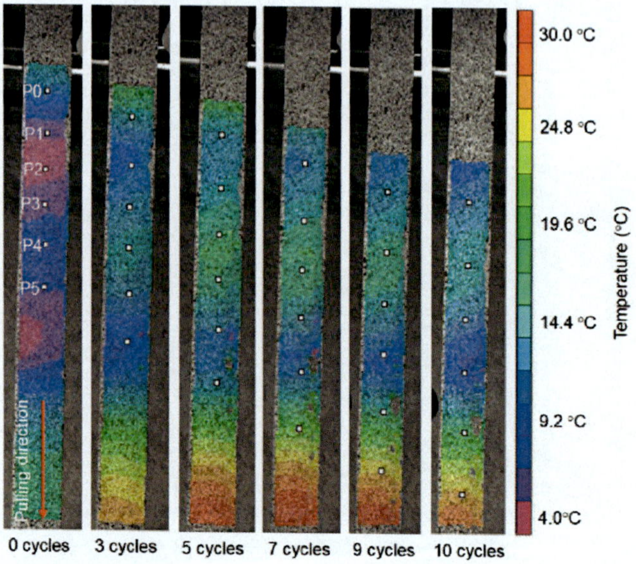

Fig. 6. Evolution of temperature recorded in gauge length of SS304 low temperature CBT specimen at 0 (undeformed), 3, 5, 7, 9, and 10 cycles.

with vertical error bars representing the maximum and minimum values from the three specimens' averages and horizontal error bars representing the true strain values recorded by the DIC after the designated number of cycles.

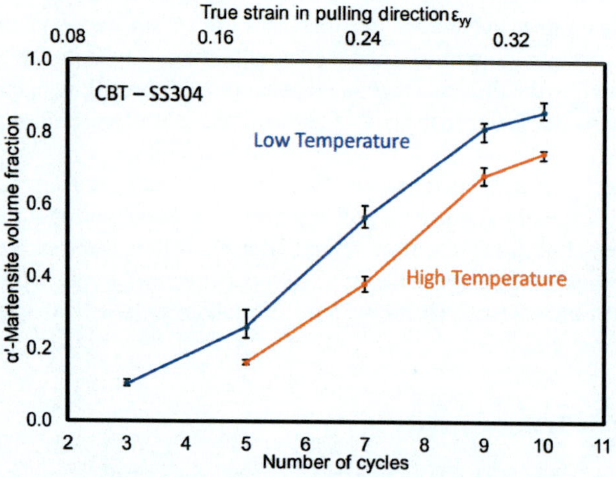

Fig. 7. Average α′-martensite volume fraction in the area of interest of SS304 low temperature and high temperature CBT experiments with respect to true strain in pulling direction and number of cycles.

The α′-martensite volume fraction and average temperature at the analysis points in the area of interest vary slightly along the gauge length of the specimen. An example from a SS304 low temperature, 10 cycles experiment is shown in Fig. 8. The α′-martensite volume fraction varies ~0.05, and the temperature varies ~3°. Again, the analysis points in the area of interest moved through the cooling regions created by the vortex tubes as the specimen was elongated, so different locations with respect to the reference were exposed to varying amounts of cooling throughout the experiment. Note that the accuracy of the Feritscope is not guaranteed above 97.1% volume fraction [28].

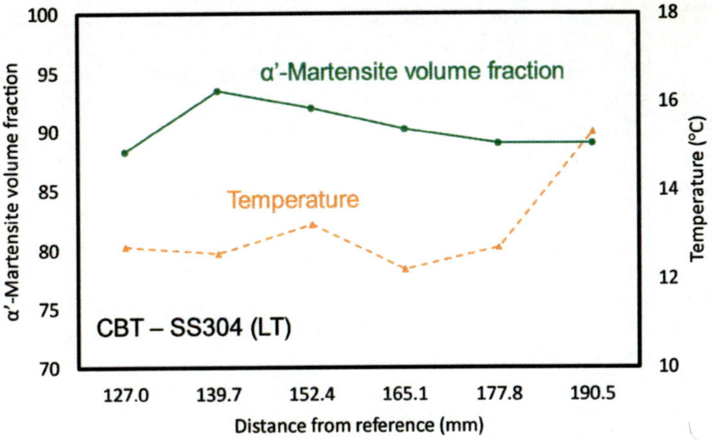

Fig. 8. SS304 low temperature, 10 cycle CBT experiment: α′-martensite volume fraction and average temperature measurements in the area of interest.

5 Conclusions

SS304 specimens were subjected to continuous bending under tension experiments conducted at high and low temperature conditions, which were created using vortex tubes. The α′-martensitic phase transformation resulting from the stress superposition of bending onto a tension process was evaluated using a Feritscope. Five strain levels, corresponding to 3, 5, 7, 9, and 10 cycles, were investigated.

Increased strain levels and decreased temperature conditions led to increased phase transformation. The strain in the pulling direction remained relatively homogeneous throughout the experiments, while the temperature varied due to the presence of two controlled zones created by the vortex tubes and deformation-induced heating. The variation in material properties along the gauge length, e.g., volume fraction, supports that stress superposition can be used in manufacturing processes to achieve functionally graded materials.

Acknowledgements. Support for the New Hampshire Center for Multiscale Modeling and Manufacturing of Biomaterials (NH BioMade) project is provided by the US National Science Foundation (NSF) EPSCoR award (#1757371). This research is also supported by the German Academic

Exchange Service (DAAD) Research Internships in Science and Engineering (RISE) Worldwide. Special thanks to Dirk Hoffmann from IUL for his assistance with specimen preparation. The SEM/FIB used is managed by the University Instrumentation Center (UIC) at UNH and was purchased with funds awarded from the US National Science Foundation (NSF) (MRI Grant 1337897) with additional funds from UNH.

References

1. Martins, P.A.F., Bay, N., Skjoedt, M., Silva, M.B.: Theory of single point incremental forming. CIRP Ann. **57**, 247–252 (2008). https://doi.org/10.1016/j.cirp.2008.03.047
2. Allwood, J.M., Shouler, D.R., Tekkaya, A.E.: The increased forming limits of incremental sheet forming processes. Key Eng. Mater. **344**, 621–628 (2007). https://doi.org/10.4028/www.scientific.net/KEM.344.621
3. Tekkaya, A.E., Groche, P., Kinsey, B.L., Wang, Z.G.: Plasticity and future opportunities of stress superposition in metal forming. CIRP Ann. (under revision) (2023)
4. Feng, Z., Mamros, E.M., Ha, J., et al.: Modeling of plasticity-induced martensitic transformation to achieve hierarchical, heterogeneous, and tailored microstructures in stainless steels. CIRP J. Manuf. Sci. Technol. **33**, 389–397 (2021). https://doi.org/10.1016/j.cirpj.2021.04.006
5. Hadoush, A., Van den Boogaard, T., Huetink, J.: Stable incremental deformation of a strip to high strain. Int. J. Mach. Tool Manuf. **344**(615), 620 (2007). https://doi.org/10.4028/www.scientific.net/KEM.344.615
6. Peng, W., Ou, H.: Deformation mechanisms and fracture in tension under cyclic bending plus compression, single point and double-sided incremental sheet forming processes. Int. J. Mach. Tools **184**, 103980 (2023). https://doi.org/10.1016/j.ijmachtools.2022.103980
7. Roemer, T.J., Kinsey, B.L., Korkolis, Y.P.: Design of a continuous-bending-under-tension machine and initial experiments on Al-6022-T4. Am. Soc. Mech. Eng. Digit. Collect. (2015). https://doi.org/10.1115/MSEC2015-9440
8. Barrett, T., Kinsey, B.L., Knezevic, M., Korkolis, Y.P.: Numerical and experimental investigation of formability enhancement during continuous-bending-under-tension (CBT) of AA6022-T4. Procedia Eng. **207**, 1940–1945 (2017). https://doi.org/10.1016/j.proeng.2017.10.965
9. Maaß, F., Hahn, M., Tekkaya, A.E.: Adjusting residual stresses by flexible stress superposition in incremental sheet metal forming. Arch. Appl. Mech. **91**(8), 3489–3499 (2021). https://doi.org/10.1007/s00419-021-01929-x
10. Zecevic, M., Knezevic, M.: Origins of improved elongation to fracture in cyclic bending under tension of AA6022-T4 sheets as revealed using crystal plasticity modeling. Mech. Mater. **177**, 104546 (2023). https://doi.org/10.1016/j.mechmat.2022.104546
11. Matukhno, N., Kljestan, N., Vogel, S.C., Knezevic, M.: Cyclic bending under tension of alloy AZ31 sheets: influence on elongation-to-fracture and strength. Mater. Sci. Eng. A **857**, 144127 (2022). https://doi.org/10.1016/j.msea.2022.144127
12. Poulin, C.M., Korkolis, Y.P., Kinsey, B.L., Knezevic, M.: Over five-times improved elongation-to-fracture of dual-phase 1180 steel by continuous-bending-under-tension. Mater. Des. **161**, 95–105 (2019). https://doi.org/10.1016/j.matdes.2018.11.022
13. Mamros, E.M., Bram Kuijer, M., Davarpanah, M.A., Baker, I., Kinsey, B.L.: The effect of temperature on the strain-induced austenite to martensite transformation in SS 316L during uniaxial tension. In: Daehn, G., Cao, J., Kinsey, B., Tekkaya, E., Vivek, A., Yoshida, Y. (eds.) Forming the Future. TMMMS, pp. 1853–1862. Springer, Cham (2021). https://doi.org/10.1007/978-3-030-75381-8_155

14. Olson, G., Cohen, M.: Kinetics of strain-induced martensitic nucleation. Metall. Trans. A **6**, 791–795 (1975). https://doi.org/10.1007/BF02672301
15. Talonen, J., Nenonen, P., Pape, G., Hänninen, H.: Effect of strain rate on the strain-induced $\gamma \rightarrow \alpha'$-martensite transformation and mechanical properties of austenitic stainless steels. Metall. Mater. Trans. A **36**, 421–432 (2005). https://doi.org/10.1007/s11661-005-0313-y
16. Darzi, S., Adams, M.D., Roth, J.T., et al.: Manipulating martensite transformation of SS304L during double-sided incremental forming by varying temperature and deformation path. CIRP Ann. Manuf. Technol. (2023)
17. Beese, A.M., Mohr, D.: Effect of stress triaxiality and Lode angle on the kinetics of strain-induced austenite-to-martensite transformation. Acta Mater. **59**, 2589–2600 (2011). https://doi.org/10.1016/j.actamat.2010.12.040
18. Wang, Z., Beese, A.M.: Stress state-dependent mechanics of additively manufactured 304L stainless steel: Part 1 – characterization and modeling of the effect of stress state and texture on microstructural evolution. Mater. Sci. Eng. A **743**, 811–823 (2019). https://doi.org/10.1016/j.msea.2018.11.094
19. Neding, B., Tian, Y., Ko, J.Y.P., Hedström, P.: Correlating temperature-dependent stacking fault energy and in-situ bulk deformation behavior for a metastable austenitic stainless steel. Mater. Sci. Eng. A **832**, 142403 (2022). https://doi.org/10.1016/j.msea.2021.142403
20. Katajarinne, T., Kivivuori, S.: Strain induced martensite in incremental forming - formation, effect and control. MSF **773–774**, 119–129 (2013). https://doi.org/10.4028/www.scientific.net/MSF.773-774.119
21. Angel, T.: Formation of martensite in austenitic stainless steels, effect of deformation, temperature and composition. J. Iron Steel Inst. **177**, 165–174 (1954)
22. Stringfellow, R.G., Parks, D.M., Olson, G.B.: A constitutive model for transformation plasticity accompanying strain-induced martensitic transformations in metastable austenitic steels. Acta Metall. Mater. **40**, 1703–1716 (1992). https://doi.org/10.1016/0956-7151(92)90114-T
23. Young, C.-C.: Transformation toughening in phosphocarbide-strengthened austenitic steels. Massachusetts Institute of Technology (1988)
24. Tomita, Y., Iwamoto, T.: Constitutive modeling of trip steel and its application to the improvement of mechanical properties. Int. J. Mech. Sci. **37**, 1295–1305 (1995). https://doi.org/10.1016/0020-7403(95)00039-Z
25. Beese, A.M., Mohr, D.: Anisotropic plasticity model coupled with Lode angle dependent strain-induced transformation kinetics law. J. Mech. Phys. Solids **60**, 1922–1940 (2012). https://doi.org/10.1016/j.jmps.2012.06.009
26. Mamros, E.M., Mayer, S.M., Banerjee, D.K., et al.: Plastic anisotropy evolution of SS316L and modeling for novel cruciform specimen. Int. J. Mech. Sci. **234** (2022). https://doi.org/10.1016/j.ijmecsci.2022.107663
27. Roemer, T.: Experimental apparatus for continuous-bending-under-tension and experiments on A6022-T4. Master's theses and capstones (2016)
28. Talonen, J., Aspegren, P., Hänninen, H.: Comparison of different methods for measuring strain induced martensite content in austenitic steels. Mater. Sci. Technol. **20**, 1506–1512 (2004). https://doi.org/10.1179/026708304X4367

Temperature-Dependent Plasticity and Fracture Properties of Modern BCC Steels

Fuhui Shen[1,2](✉), Hao Xu[1], Sebastian Münstermann[1], and Junhe Lian[2]

[1] Steel Institute, RWTH Aachen University, Intzestraße 1, 52072 Aachen, Germany
fuhui.shen@iehk.rwth-aachen.de
[2] Department of Mechanical Engineering, Aalto University, Otakaari 4, 02150 Espoo, Finland

Abstract. For some modern steels with a body-centered cubic (bcc) crystal structure, it is observed that both tensile strength and ductility are significantly improved with decreasing temperature, which motivates the exploration of the cryogenic formability and fracture properties of these materials. The temperature-dependent plasticity and fracture phenomena of a modern bainitic steels with the bcc structure have been investigated by performing a comprehensive experimental program and finite element simulations, covering a broad range of loading conditions. Uniaxial tensile tests have been performed at different temperatures along three loading directions. Tensile tests using flat specimens with various geometries, including shear, central hole and notched dog bone, have been performed along the rolling direction at room temperature and –196 °C. An advanced non-associated constitutive plasticity model is used to describe the temperature-dependent strength and hardening properties of the material. The local critical stress and strain variables extracted from finite element simulations of different fracture tests have been used to calibrate a unified fracture criterion, which considers the stress state dependence. The effects of temperature on the plasticity and stress state dependent fracture behavior of the modern bcc steels have been quantitatively determined.

Keywords: Anisotropy · Stress State · Cryogenic Fracture · FE Simulation

1 Introduction

Recent studies show that the formability of several aluminum alloys (with a face-centered cubic structure) at cryogenic temperatures is superior to room temperature, which motivates the exploration of the cryogenic formability of other metallic materials [1–3]. For metallic materials with a body-centered cubic (bcc) crystal structure, brittle fracture is typically observed in toughness tests when the testing temperature is below the ductile to brittle transition temperature. However, only a narrow range of stress states with high triaxiality is considered in the toughness tests. The fracture properties of modern bcc steels under low to moderate triaxiality conditions have not been fully explored. Recent studies show that significant amount of plasticity can occur in bainitic pipeline steels with a bcc structure at extremely low temperatures when the local stress triaxiality is below a threshold value [4, 5].

K. Mocellin et al. (Eds.): ICTP 2023, LNME, pp. 486–493, 2024.
https://doi.org/10.1007/978-3-031-40920-2_50

The effects of stress states on ductile fracture have been extensively investigated for various metallic materials [6–9]. Several uncoupled ductile fracture criteria have been proposed and applied to describe the stress state dependence of failure strain, which is typically formulated as the weight function of stress triaxiality and Lode angle parameter [10–12]. There are several successful applications of these criteria for the prediction of ductile fracture at elevated temperatures as well. However, the impacts of temperature on the plasticity and fracture properties of modern bcc steel over a wide range of stress states have rarely been investigated for temperatures lower than room temperature. For materials with anisotropic plasticity properties, different advanced constitutive models are available for the accurate description of plastic deformation at various temperatures [13–15]. Therefore, the temperature influence on the plasticity and fracture behavior of a modern bcc steel with anisotropic effects has been investigated under low to moderate triaxiality conditions at cryogenic temperatures.

2 Materials and Experiments

2.1 Materials and Testing Program

The plasticity and fracture properties of a modern bainitic steel have been investigated by performing tensile tests using different specimen geometries at room temperature and −196 °C under quasi-static conditions. The microstructure of the investigated bcc steel is shown in Fig. 1a. The plasticity of the material is characterized by performing uniaxial tensile tests using smooth dog bone (SDB) specimens along three different loading directions (00°, 45°, and 90° with respect to the rolling direction). The biaxial flow behavior is characterized by compression tests through the thickness direction. During uniaxial tensile tests at room temperature, the strains along both longitudinal and width directions have been measured to calculate the r-value. Tensile tests using fracture specimens, including notched dog bone (NDB), central hole (CH-R3) and shear (SH) geometries, have been performed along the rolling direction at room temperature and −196 °C. The gauge length of 40 mm is used for all fracture specimens. The detailed descriptions of the experimental setup are provided in previous studies [4, 16].

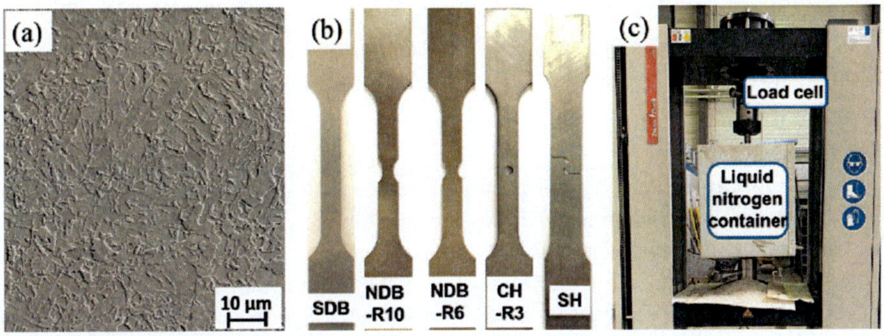

Fig. 1. Microstructure of the investigated material (a); Overview of all tensile specimens (b) and setup of the machine for tensile tests at −196 °C.

2.2 Experimental Results

The anisotropic engineering stress and strain curves obtained from uniaxial tensile tests at room temperature and −196 °C are shown in Fig. 2. Both yield strength and tensile strength are significantly increased at the cryogenic testing temperature than room temperature due to thermal softening effects. In addition, the tensile ductility at −196 °C is also significantly improved than at room temperature. The r-value along three loading directions has been measured continuously during tensile deformation at room temperature, as shown in Fig. 2d. In addition to the orientation-dependent flow stress, the anisotropic effects on the r-value are more pronounced. For instance, the r-value along 00° and 90° is lower than one. The r-value measured along different loading directions is also changing during plastic deformation.

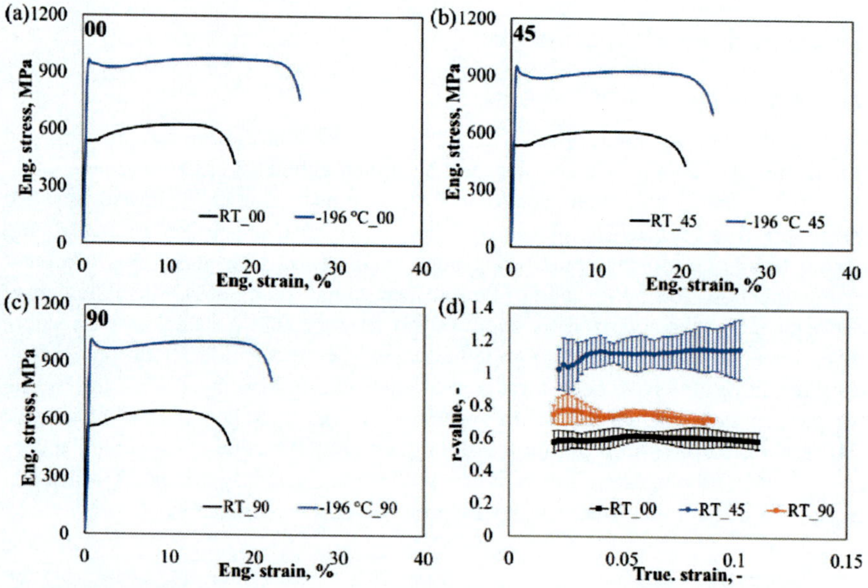

Fig. 2. Anisotropic uniaxial tensile properties of the investigated bcc steel at room temperature and −196 °C. Engineering stress and strain curves along (a) 00°, (b) 45°, and (c) 90°; (d) Evolution of r-value at room temperature.

Significant plastic deformation occurs in tensile tests using fracture specimens at both temperatures, as shown in the following section. However, the underlying failure mechanisms are different, depending on the loading stress states and temperature. For instance, shear failure occurs in the SH specimens at both room temperature and −196 °C. For the NDB specimens, cleavage fracture occurs after significant plastic deformation at −196 °C while ductile fracture takes place at room temperature.

3 Material Models

3.1 Plasticity Model

The non-associated flow rule has been adopted to overcome the drawback of the conventional Hill48 model, which provides limited accuracy in the simultaneous description of anisotropic strength and r-values [15, 17]. The features of anisotropic hardening and r-value evolution have been further considered in the evolving non-associated Hill48 (enHill48) model to improve the accuracy in describing the anisotropic plasticity [18]. The yield function f and flow potential g are described using separate equations.

$$f = \bar{\sigma}_\sigma(\boldsymbol{\sigma}|F_\sigma, G_\sigma, H_\sigma, L_\sigma, M_\sigma, N_\sigma) - \sigma_Y\left(\bar{\varepsilon}^p\right) \leq 0 \tag{1}$$

$$g = \bar{\sigma}_r(\boldsymbol{\sigma}|F_r, G_r, H_r, L_r, M_r, N_r) - \sigma_Y\left(\bar{\varepsilon}^p\right) \leq 0 \tag{2}$$

The same quadratic Hill48 equivalent stress $\bar{\sigma}(\boldsymbol{\sigma})$ is adopted in the yield function f and flow potential g, where the anisotropic parameters are calibrated independently. Anisotropic flow stress is used to calibrate parameters $(F_\sigma, G_\sigma, H_\sigma, N_\sigma)$ in yield function f while anisotropic r-values are used to determine parameters (F_r, G_r, H_r, N_r) in flow potential g.

$$\begin{aligned}\overline{\sigma}(\boldsymbol{\sigma}) = \{&\frac{1}{2}[F(\sigma_{22} - \sigma_{33})^2 + G(\sigma_{33} - \sigma_{11})^2 + H(\sigma_{11} - \sigma_{22})^2] \\ &+ L\sigma_{23}^2 + M\sigma_{13}^2 + N\sigma_{12}^2\}^{\frac{1}{2}}\end{aligned} \tag{3}$$

According to the non-associated flow rule, the equivalent plastic strain rate $\dot{\bar{\varepsilon}}^p$ is calculated in Eq. (4), where a correction factor is added to the typical non-negative plastic multiplier $\dot{\lambda}$.

$$\dot{\bar{\varepsilon}}^p = \dot{\lambda} \cdot \frac{\bar{\sigma}_r}{\bar{\sigma}_\sigma} \tag{4}$$

3.2 Fracture Criterion

The unified fracture criterion proposed by Shen et al., [4, 5] is used in this study, where the Bai-Wierzbicki [10] phenomenological expression of fracture initiation strain is adopted. The anisotropic damage mechanics model has been used to describe the fracture properties of high-strength bainitic steels at different temperatures [4, 16]. The general expression of stress state using stress triaxiality η and Lode angle parameter $\overline{\theta}$ is adopted in the fracture criterion, which are derived from stress invariants (I_1, J_2, J_3).

$$\eta = I_1/\sqrt{27 \cdot J_2} \tag{5}$$

$$\overline{\theta} = 1 - \frac{6}{\pi}\theta = 1 - \frac{2}{\pi}\cos^{-1}(\sqrt{27/4} \cdot J_3 \cdot {J_2}^{-3/2}) \tag{6}$$

The stress state effects on failure strain are quantified by a phenomenological function with four parameters ($C_{1\sim4}$) for proportional loading conditions according to Eq. (8). To consider the non-proportional loading effects, the average values of stress triaxiality and Lode angle parameter ($\eta_{\mathrm{avg}}, \overline{\theta}_{\mathrm{avg}}$) are used to describe the overall stress state. Fracture occurs when the fracture indicator I_{f} reaches the unity.

$$\overline{\varepsilon}_{\mathrm{f}}(\eta, \overline{\theta}) = \left(C_1 exp^{-C_2\eta} - C_3 exp^{-C_4\eta}\right)\overline{\theta}^2 + C_3 exp^{-C_4\eta} \tag{7}$$

$$\eta_{\mathrm{avg}} = \frac{1}{\overline{\varepsilon}_{\mathrm{f}}}\int_0^{\overline{\varepsilon}_{\mathrm{f}}} \eta\, d\overline{\varepsilon}^{\mathrm{p}}, \overline{\theta}_{\mathrm{avg}} = \frac{1}{\overline{\varepsilon}_{\mathrm{f}}}\int_0^{\overline{\varepsilon}_{\mathrm{f}}} \overline{\theta}\, d\overline{\varepsilon}^{\mathrm{p}} \tag{8}$$

$$I_{\mathrm{f}} = \int_0^{\overline{\varepsilon}^{\mathrm{p}}} \frac{1}{\overline{\varepsilon}_{\mathrm{f}}(\eta, \overline{\theta})} d\overline{\varepsilon}^{\mathrm{p}} \tag{9}$$

4 Simulation Results

4.1 Calibration of Plasticity Model

The orientation-dependent distribution of flow stress and r-value at the equivalent plastic strain of 0.04 can be well captured by the enHill48 plasticity model for both temperatures, as shown in Fig. 3a and Fig. 3b. Assuming the anisotropy of the investigated material is not affected by temperature, the r-value is considered to be independent of temperature [4, 19]. The anisotropic hardening behavior of the material is described by fitting the individual flow curves along 00°, 45°, 90° and the biaxial flow curve using the combined Swift-Voce function. The evolution of the r-value along three loading directions during plastic deformation is described using the Voce function. The fitting of the flow curve and r-value evolution function along the rolling direction is shown in Fig. 3c and Fig. 3d.

4.2 Critical Stress and Strain Values

Finite element simulations have been performed to extract the local critical stress and strain variables in different specimens. The material model is implemented as a user-defined VUMAT subroutine. Half-thickness models with solid elements have been used for all geometries during simulations using the ABAQUS/EXPLICIT software. The evolution of equivalent plastic strain with stress state variables in the critical element of CH-R3 and NDB-R6 specimens is demonstrated in Fig. 4 for both room temperature and −196 °C. Solid symbols represent the fracture point. It is observed that significant plastic deformation occurs at −196 °C during tensile tests along the rolling direction using these specimens. It can be seen that the fracture strain at room temperature is slightly higher than at −196 °C, indicating the temperature effects on fracture strain are not very pronounced under the investigated conditions.

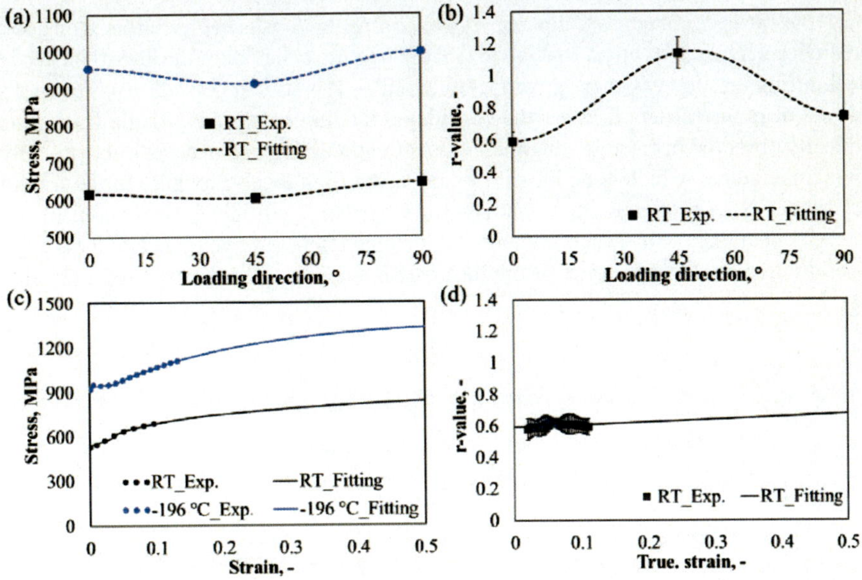

Fig. 3. Calibration of anisotropic plasticity model at room temperature and −196 °C: Anisotropic description of flow stress (a) and r-value (b) at the equivalent plastic strain of 0.04; Fitting of flow curve (c) and r-value evolution function (d) along the rolling direction.

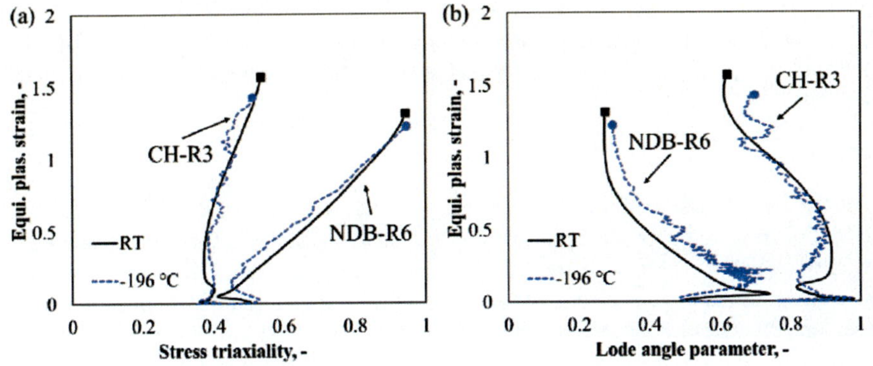

Fig. 4. The evolution of equivalent plastic strain with stress triaxiality (a) and Lode angle parameter (b) at critical position in CH-R3 and NDB-R6 specimens during tensile tests at room temperature and −196 °C along the rolling direction. Solid symbols represent fracture point.

4.3 Fracture Prediction

Considering the marginal effects of temperature on fracture strain, one set of fracture parameters ($C_{1\sim4}$) has been calibrated to describe the fracture behavior of the investigated bcc steel during tensile tests at room temperature and −196 °C along the rolling direction. The predicted force and displacement curves during tensile tests using CH-R3 and NDB-R6 specimens are compared with experimental results in Fig. 5. Based on good prediction accuracy, it can be concluded that the fracture behavior of the investigated bcc steel during tensile tests at room temperature and −196 °C can be well captured using one set of failure criterion under the considered loading conditions. Brittle fracture is typically observed in fracture toughness tests of engineering bcc materials at extremely low temperatures, which is mainly attributed to the high local triaxiality in the testing conditions. Recent studies show that significant plastic deformation can occur in bcc steels by delaying the cleavage fracture at cryogenic temperatures, far below the ductile to brittle transition temperature, if the local stress triaxiality is below a threshold value [4, 5].

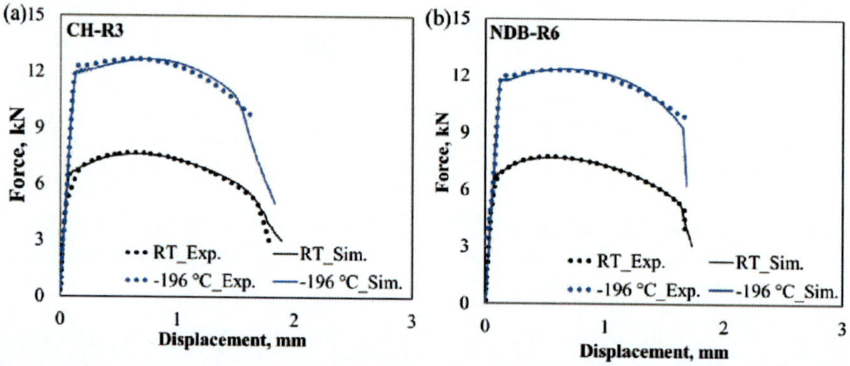

Fig. 5. Experimental and numerical results of force and displacement curves during tensile tests using CH-R3 (a) and NDB-R6 (b) specimens at room temperature and −196 °C along the rolling direction.

By comparing the fracture behavior at room temperature and −196 °C, it confirms that the local fracture strain of the investigated bcc steel at −196 °C is comparable to that of room temperature under the considered stress states below the threshold triaxiality, which is approximately 1 for the same grade of steel [4]. More systematic investigations should be performed to quantify the temperature effects on failure strain at other temperatures, particularly within the ductile to brittle transition region. However, these observations indicate that the local formability of modern bcc steels under low to moderate triaxiality conditions needs more exploration at cryogenic temperatures.

5 Conclusions

The effects of temperature on the plasticity and fracture properties of a bcc steel have been investigated by performing tensile tests using various specimen geometries at room temperature and −196 °C. The fracture properties during tensile tests at different temperatures have been simulated using an evolving plasticity model combined with a strain-based fracture criterion. The key findings are summarized as follows:

- Significant plasticity occurs at −196 °C in the investigated bcc steel during tensile tests using various specimen geometries, where the local triaxiality is below a threshold value.
- The temperature has weak impacts on the failure strain of the investigated bcc steel under the considered loading conditions.

References

1. Wang, X., Fan, X., Chen, X., Yuan, S.: Cryogenic deformation behavior of 6061 aluminum alloy tube under biaxial tension condition. J. Mater. Process. Technol. **303** (2022)
2. Yuan, S., Cheng, W., Liu, W.: Cryogenic formability of a solution-treated aluminum alloy sheet at low temperatures. J. Mater. Process. Technol. **298** (2021)
3. Liu, W., Hao, Y.: Damage and fracture prediction of 7075 high-strength aluminum alloy during cryogenic stamping process. Mech. Mater. **163** (2021)
4. Shen, F., Münstermann, S., Lian, J.: A unified fracture criterion considering stress state dependent transition of failure mechanisms in bcc steels at –196 °C. Int. J. Plast. **156** (2022)
5. Shen, F., Münstermann, S., Lian, J.: Cryogenic ductile and cleavage fracture of bcc metallic structures – influence of anisotropy and stress states. J. Mech. Phys. Solids, 105299 (2023)
6. Mu, L., Jia, Z., Ma, Z., Shen, F., Sun, Y., Zang, Y.: A theoretical prediction framework for the construction of a fracture forming limit curve accounting for fracture pattern transition. Int. J. Plast. **129** (2020)
7. Bao, Y., Wierzbicki, T.: On fracture locus in the equivalent strain and stress triaxiality space. Int. J. Mech. Sci. **46**(1), 81–98 (2004)
8. Shen, F., et al.: Local formability of medium-Mn steel. J. Mater. Process. Technol. **299** (2022)
9. Zhang, Y., et al.: Ductility prediction of HPDC aluminum alloy using a probabilistic ductile fracture model. Theor. Appl. Fract. Mech. **119** (2022)
10. Bai, Y., Wierzbicki, T.: A new model of metal plasticity and fracture with pressure and Lode dependence. Int. J. Plast. **24**(6), 1071–1096 (2008)
11. Lou, Y., Huh, H., Lim, S., Pack, K.: New ductile fracture criterion for prediction of fracture forming limit diagrams of sheet metals. Int. J. Solids Struct. **49**(25), 3605–3615 (2012)
12. Mohr, D., Marcadet, S.J.: Micromechanically-motivated phenomenological Hosford-Coulomb model for predicting ductile fracture initiation at low stress triaxialities. Int. J. Solids Struct. **67–68**, 40–55 (2015)
13. Barlat, F., Aretz, H., Yoon, J.W., Karabin, M.E., Brem, J.C., Dick, R.E.: Linear transfomation-based anisotropic yield functions. Int. J. Plast. **21**(5), 1009–1039 (2005)
14. Barlat, F., et al.: Plane stress yield function for aluminum alloy sheets—Part 1: theory. Int. J. Plast. **19**(9), 1297–1319 (2003)
15. Stoughton, T.B., Yoon, J.W.: Anisotropic hardening and non-associated flow in proportional loading of sheet metals. Int. J. Plast. **25**(9), 1777–1817 (2009)
16. Shen, F., Münstermann, S., Lian, J.: Investigation on the ductile fracture of high-strength pipeline steels using a partial anisotropic damage mechanics model. Eng. Fract. Mech. **227**, 106900 (2020)
17. Stoughton, T.B.: A non-associated flow rule for sheet metal forming. Int. J. Plast. **18**(5–6), 687–714 (2002)
18. Lian, J., et al.: An evolving non-associated Hill48 plasticity model accounting for anisotropic hardening and r-value evolution and its application to forming limit prediction. Int. J. Solids Struct. **151**, 20–44 (2018)
19. Shen, F., Münstermann, S., Lian, J.: An evolving plasticity model considering anisotropy, thermal softening and dynamic strain aging. Int. J. Plast **132**, 102747 (2020)

Improving Formability of Titanium Bipolar Plate via Hot Stamping

Xianglu Zhang[1], Nan Guo[1], Wenyao Wang[1], Zhuoqi Yan[1], Daijun Yang[2,3], Junying Min[1](✉), Pingwen Ming[2,3], and Cunman Zhang[2,3]

[1] School of Mechanical Engineering, Tongji University, Shanghai 201804, China
junying.min@tongji.edu.cn
[2] Clean Energy Automotive Engineering Center, Tongji University, Shanghai 201804, China
[3] School of Automotive Studies, Tongji University, Shanghai 201804, China

Abstract. Due to the limited formability of ultra-thin titanium sheet at room temperature, it is difficult to form fine flow channels with increasing geometrical complexity of titanium bipolar plate. Premature occurrence of cracks is commonly observed in conventional room-temperature stamping process, restricting further improvement of the performance of PEMFC. In this work, hot stamping using an on-site resistance heating system is applied to ultra-thin titanium sheet to improve formability of titanium bipolar plate. It is demonstrated that fracture limits of ultra-thin titanium sheet at 700 °C are obviously higher than those at room temperature. Finally, a crack-free titanium bipolar plate with fine flow channels is exemplarily formed via hot stamping process.

Keywords: Formability · Hot stamping · Titanium bipolar plate

1 Introduction

To meet the rising demands on the proton exchange membrane fuel cells (PEMFC) with high power density, low weight and long service life, titanium bipolar plates are receiving more and more attention due to their high specific strength, high electrical conductivity and excellent corrosion resistance [1]. However, ultra-thin titanium sheets exhibit relatively poor formability at room temperature, which is insufficient to form fine flow channels with high aspect ratio through the conventional room-temperature stamping method [2], restricting the improvement of power density of PEMFC. Hence, several processes including multistage stamping [3], rubber forming [4–6], electromagnetic forming [7] as well as additive manufacturing [8] were developed to enhance the formability of titanium bipolar plates. To date, the aspect ratio of channels can be ascended to 0.67 through a three-stage stamping process proposed by Xu et al. [3], which is still difficult to fulfill the requirements of high-powered PEMFC. Thus, it is an urgent need to develop a novel forming method, which can further improve the formability of ultra-thin titanium sheets.

K. Mocellin et al. (Eds.): ICTP 2023, LNME, pp. 494–502, 2024.
https://doi.org/10.1007/978-3-031-40920-2_51

Base on the fact that metallic materials generally exhibit increased plasticity at elevated temperature, hot stamping process is widely used to enhance the formability of sheet metals. Müller et al. [9] applied the hot stamping process to form stainless steel bipolar plates and successfully enlarged the ultimate channel depth. Guo et al. [10] revealed that the forming accuracy of stainless steel bipolar plates can also be obviously improved, indicating that hot stamping can be a potential solution for high precision forming of metallic bipolar plates.

In this study, the fracture limits of ultra-thin titanium sheets at high temperatures were investigated using the equipment reported previously in Guo et al. [10]. Finally, titanium bipolar plates with an aspect ratio of 0.73 was exemplarily formed via hot stamping process.

2 Material and Methods

2.1 Material

A commercially pure titanium sheet with a thickness of 0.1 mm was employed in this study. Its chemical compositions that were tested through X-Ray Fluorescence (XRF) analysis are listed in Table 1. Besides, its mechanical properties along three orientations to the rolling direction were measured via uniaxial tensile testing and are summarized in Table 2.

Table 1. Chemical compositions of the ultra-thin commercially pure titanium sheet (wt.%).

Element	Ti	Fe	Mg	Na	Ni	Si	Others
Content/%	99.5	0.04	0.08	0.23	0.02	0.02	Bal

Table 2. Mechanical Properties of the ultra-thin titanium sheet along various orientations.

Orientation	Young's modulus (GPa)	Yield Strength (MPa)	Tensile Strength (MPa)	r-value	Elongation (%)
RD	86.8	205.0	297.5	1.89	39.2
DD	98.0	197.5	283.2	3.67	40.2
TD	81.2	189.9	297.5	6.18	36.2

2.2 Experimental Setup for Hot Stamping

In order to achieve a uniform temperature field through a rapid heating process and simultaneously avoid the transfer process from heating device to forming tool, which will result in a significant temperature drop, resistance heating was considered as a promising heating method for hot stamping of ultra-thin metallic sheets [11]. Figure 1 shows a lab-scale hot stamping platform with an on-site resistance heating system, which enable the titanium sheets to be heated inside the forming tool. The supplied electric power and the heating time can be adjusted through a control panel, thus achieving accurate temperature control of the titanium sheets. The average temperature and the uniformity of temperature distribution were monitored via an Optris PI400 infrared thermal imager in a preliminary resistance heating experiment. More details about the hot stamping platform can be found in the previous study [10].

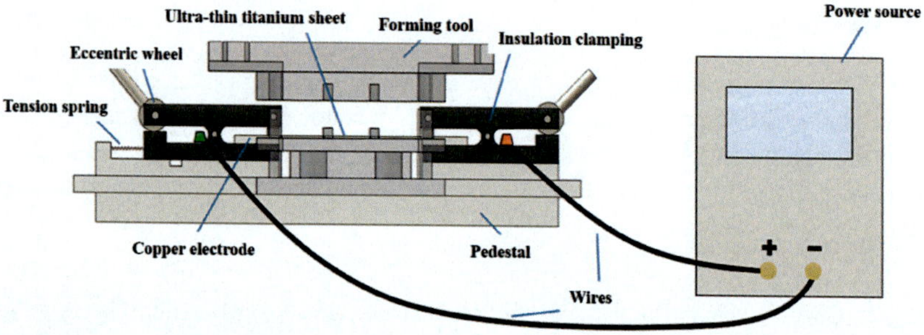

Fig. 1. Schematic diagram of the hot stamping platform

In this study, rectangular specimens with the dimensions of 230×60 mm^2 were prepared for the hot stamping experiments to investigate the fracture limits at elevated temperature. The heating time was set to 3 s, and the upper and lower tools were closed immediately (within 0.1 s) after power-off. The heating parameters to achieve the target temperatures of 700, 800 and 900 °C were summarized in Table 3.

Table 3. Heating parameters of ultra-thin commercially pure titanium sheet

Target temperature (°C)	Electric power (kW)	Heating time (s)
700	12.5	3
800	15.0	3
900	17.5	3

2.3 Method to Obtain Fracture Forming Limit Diagram (FFLD)

Fracture forming limit diagram (FFLD) is a quantitative evaluation method of formability of sheet metals, which is obtained by connecting their fracture limits under typical stress states. However, there are several issues when applying conventional testing approaches to micro/meso-scale forming. Since the feature size is reduced into the submillimeter range, the diameter of circular grids for strain calculation should be decreased to less than 0.2 mm, which is challenging to produce. Besides, the traditional Marciniak or Nakazima tests require a specimen size of 180 mm. The reliability of their results may be incorrect in micro forming due to the possible size effect.

To accurately characterize the fracture limits in micro/meso-scale forming, a novel testing method based on a simple stamping process was proposed [12]. As shown in Fig. 2, three types of features including straight channels, truncated circular cones and semispheres with various dimensions can be formed simultaneously through stamping, enabling the tested metallic sheet to deform under different strain states. In-plane strains can be computed subsequently through the measured thickness reduction. JD-3028 boron nitride spray was used as lubricant to reduce the effect of friction.

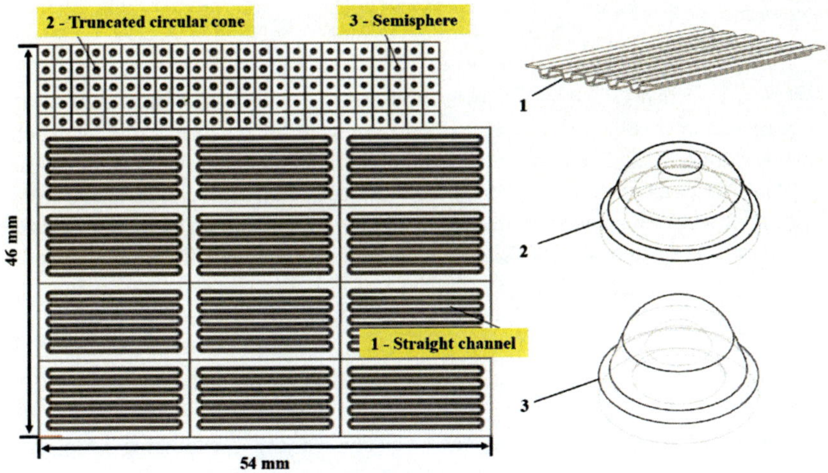

Fig. 2. Typical features with various dimensions of the stamping die

Since strain localization tend to occur in corner areas, strain states of corners were investigated for the research of fracture limits. For each type of features, ratios of minor and major strain $\boldsymbol{\beta}$ in the corner areas were calculated using the finite element method (FEM). Finite element analyses (FEA) were conducted in the commercial finite element software LS-DYNA. Shell element was used for both blank and tools, where element size was $0.015 \times 0.015\ mm^2$ for blank and $0.1 \times 0.1\ mm^2$ with refined grid at corners for tools. By considering that the specimens were deformed under quasi-proportional loadings, the FEA employed the classic Hill48 anisotropic yield model and Hockett-Sherby isotropic hardening law, which allowed for considerable simulation accuracy with only a few experiments. The parameters for the models were obtained from the uniaxial tensile tests along different directions at room temperature, assuming that the strain ratio does not vary with temperature. As shown in Fig. 3, strain paths are quasi-linear for all feature types, indicating a nearly constant strain ratio **β** during the entire forming process. Thus, the **β**-values can be derived from the fitting straight-lines, which were −0.02, 0.40 and 0.89 for straight channels, truncated circular cones and semispheres, respectively.

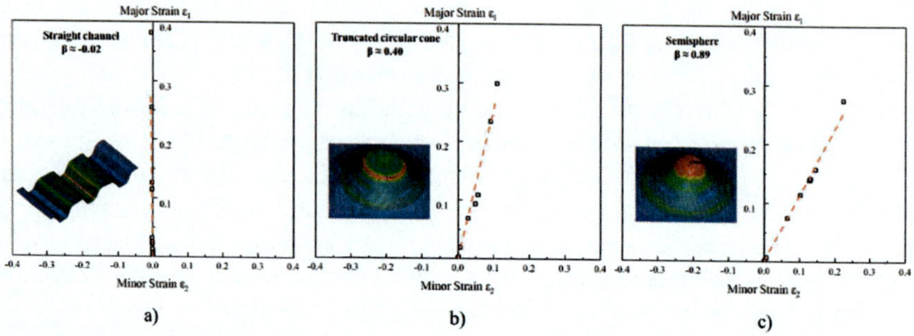

Fig. 3. Strain paths of points in the corner areas of a) straight channels, b) truncated circular cones and c) semispheres

Besides, it is revealed that **β** shows little change, as long as the corner radii of features remain unchanged. Hence, the corner radii were set to a fixed value, while other dimension parameters varied and were designed via FEM, ensuring that the in-plane strains can vary in a wide range covering both unfractured and fractured states, without variation of strain ratio. Based on the volume invariant hypothesis, the major strain $\boldsymbol{\varepsilon_1}$ and minor strain $\boldsymbol{\varepsilon_2}$ can be derived from the strain in the thickness direction $\boldsymbol{\varepsilon_t}$ as follow:

$$\varepsilon_1 = -\varepsilon_t/(1+\beta) \tag{1}$$

$$\varepsilon_2 = -(\beta \cdot \varepsilon_t)/(1+\beta) \tag{2}$$

Strain in the thickness direction $\boldsymbol{\varepsilon_t}$ can be calculated from the thinning rate $\boldsymbol{\eta}$, which is determined by measuring the minimum sheet thickness $\mathbf{t_1}$ in the corner area.

$$\varepsilon_t = \ln(1+\eta) \tag{3}$$

$$\eta = (t_1 - t_0)/t_0 \cdot 100\% \tag{4}$$

where t_0 is the initial thickness of the tested metallic sheet (i.e. 0.1 mm). Note that the thickness in the fracture location should be measured instead in case of breakage, as shown in Fig. 4.

Afterwards, labeled with either "Crack-free" or "Fracture", the measured points can be filed in a coordinate system, where x-axis represents minor strain $\boldsymbol{\varepsilon_2}$ and y-axis represents major strain $\boldsymbol{\varepsilon_1}$. Finally, the right hand side of the fracture limit diagram can be obtained by a linear regression from the data.

Fig. 4. Measuring the minimum sheet thickness $\mathbf{t_1}$ in case of fracture

3 Result and Discussion

3.1 Increased Fracture Limits at Elevated Temperature

Rectangular specimens with a size of 54 × 46 mm^2 stamped at different temperatures are compared in Fig. 5. As can be seen, less cracks occurred by elevating the forming temperature from room temperature to 700 °C, indicating an enhanced formability of ultra-thin titanium sheet (Fig. 5a, 5b). Nevertheless, more cracks as compared to cold forming were observed by further raising the temperature to 800 °C and 900 °C, which denotes that overhigh temperature can in turn cause a reduction of fracture limit of ultra-thin titanium sheet (Fig. 5c, 5d). Thus, forming at a temperature above 800 °C should be avoid in the hot stamping process, whose effect consequently will not be discussed in detail.

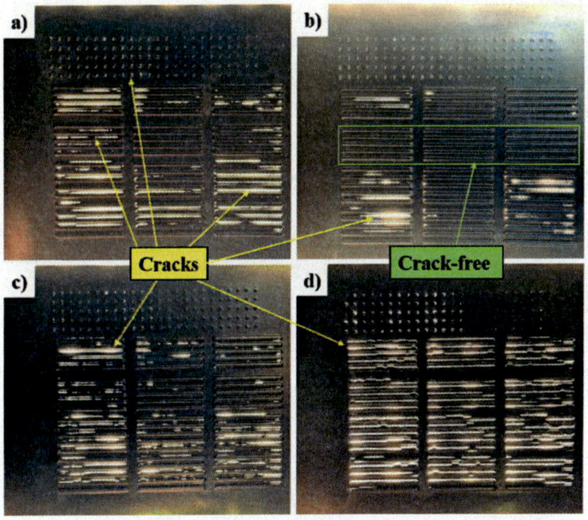

Fig. 5. Specimens formed at a) room temperature, b) 700 °C, c) 800 °C and d) 900 °C

To quantify the improvement of formability, strain-based FFLDs of ultra-thin titanium sheets formed at room temperature and at 700 °C are compared in Fig. 6. Significantly, a wide "safety-fracture" interlaced zone can be observed in the FFLD for cold-formed specimen as shown in Fig. 6a, which denotes that the titanium sheet exhibits a larger fluctuation of formability at room temperature. It may be attributed to the size effect induced localized deformation [13], since the ultra-thin titanium sheet exhibits a t/d ratio (thickness/average grain size) of only approximately 8. In contrast, a more homogeneous deformation may be achieved by means of hot forming, as more slip systems are activated at elevated temperatures.

The fracture forming limit curves (FFLCs) in both forming conditions are compared and summarized in Fig. 6b. Since no crack on the semispheres was found for both specimens, the FFLCs were derived from the fracture points with minimum equivalent strain on the straight channels and truncated circular cones. Hence, an obvious increasement in limit strain at 700 °C can be perceived, indicating an enhancement of about 20% in plane strain state.

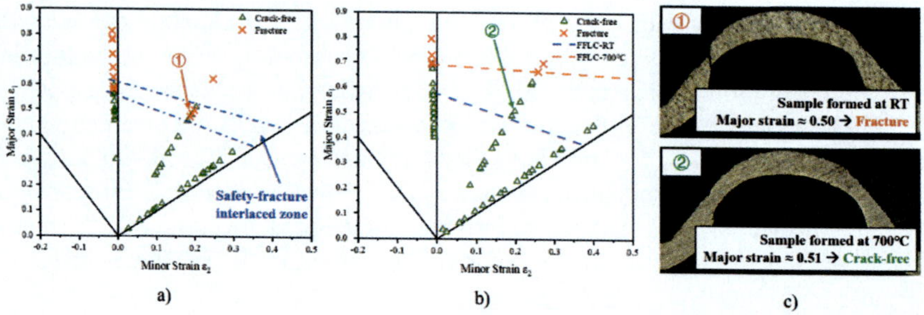

Fig. 6. Strain-based fracture limit diagrams of ultra-thin titanium sheets formed a) at room temperature and b) at 700 °C as well as c) cross sections of selected truncated circular cones formed under both conditions

3.2 Improved Formability of Titanium Bipolar Plate Through Hot Stamping

To validate the feasibility and effect of the introduced process, a lab-scale titanium bipolar plate was experimentally formed via hot stamping. Tool geometry of the flow channel and design of the bipolar plate are illustrated in Fig. 7a and 7b, respectively. Note that the aspect ratio **AR** of the flow channel is defined as:

$$AR = c/w \tag{5}$$

where c is the channel depth and w is the length of trapezoidal median line of the channel, indicating a high aspect ratio of approximately 0.73.

Bipolar plates formed at room temperature and at 700 °C are compared in Fig. 7c. For the cold-stamped specimen, cracks occur both in curved sections of the channels as well as in the area designed for structural strength, whereas titanium bipolar plate without cracks can be formed via hot stamping at 700 °C. It can be therefore concluded that hot stamping process could be an applicable and economical solution to form high precise titanium bipolar plates with finer flow channels, since only one forming step is required.

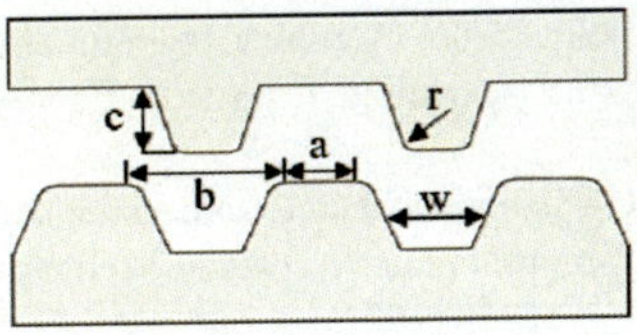

Parameter		Values [μm]
Rib width	a	500
Channel width	b	877
Channel depth	c	350
Corner radius	R	100
Length of trapezoidal median line	w	480

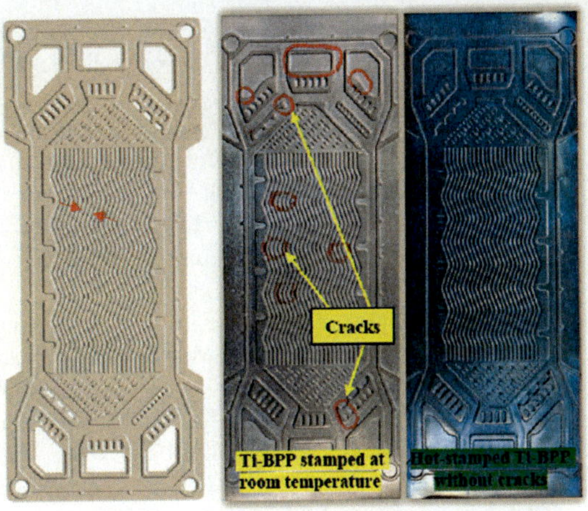

a) b) c)

Fig. 7. a) Tool geometries of the flow channel, b) design of bipolar plate and c) comparison of titanium bipolar plates formed at room temperature and at 700 °C

4 Conclusions

In the present work, the feasibility of applying hot stamping process to titanium bipolar plates is studied and validated. The main conclusions can be summarized as follow:

1) Fracture limit of the ultra-thin titanium sheet is obviously improved by increasing the forming temperature to 700 °C, indicating an increase of approximately 20% in plane strain state.
2) Based on the enhanced fracture limit at elevated temperatures, titanium bipolar plate with a high aspect ratio of 0.73 can be successfully formed by using only one forming step, thus the feasibility and advantages of hot stamping process are validated.

Acknowledgement. This study is supported by the National Key Research and Development Program of China (No. 2022YFE0207500).

References

1. Karacan, K., Celik, S., Toros, S., et al.: Investigation of formability of metallic bipolar plates via stamping for light-weight PEM fuel cells. Int. J. Hydrog. Energy **45**(60), 35149–35161 (2020)
2. Nakagaki, N.: The newly developed components for the fuel cell vehicle, Mirai. SAE Technical Paper (2015)
3. Xu, Z., Li, Z., Zhang, R., et al.: Fabrication of micro channels for titanium PEMFC bipolar plates by multistage forming process. Int. J. Hydrog. Energy **46**(19), 11092–11103 (2021)
4. Lee, K., Jin, C., Kang, C., et al.: Fabrication of titanium bipolar plates by rubber forming process and evaluation characteristics of TiN coated titanium bipolar plates. Fuel Cells **15**(1), 170–177 (2015)

5. Jin, C., Jeong, M., Kang, C.: Fabrication of titanium bipolar plates by rubber forming and performance of single cell using TiN-coated titanium bipolar plates. Int. J. Hydrog. Energy **39**(36), 21480–21488 (2014)
6. Liu, Q., Wang, X., Yin, D., et al.: Mechanical properties, corrosion resistance, and rubber pad forming of cold differential speed-rolled pure titanium for bipolar plates of proton-exchange membrane fuel cells. Int. J. Hydrog. Energy **47**(40), 17737–17748 (2022)
7. Dong, P., Li, Z., Feng, S., et al.: Fabrication of titanium bipolar plates for proton exchange membrane fuel cells by uniform pressure electromagnetic forming. Int. J. Hydrog. Energy **46**(78), 38768–38781 (2021)
8. Celik, S., Timurkutluk, B., Aydin, U., et al.: Development of titanium bipolar plates fabricated by additive manufacturing for PEM fuel cells in electric vehicles. Int. J. Hydrog. Energy **47**(89), 37956–37966 (2022)
9. Müller, C., Janssen, H.: Hot forming of metallic bipolar plates using conductive heating. In: FC^3 Fuel Cell Conference Chemnitz 2021 (2022)
10. Guo, N., Zhang, X., Hou, Z., et al.: Hot stamping of ultra-thin stainless steel sheets for bipolar plates. J. Mater. Process. Technol. **317**, 117987 (2023)
11. He, Z., Ming, P., Guo, N., et al.: Heating methods for hot stamping of ultra-thin stainless steel metal. In: AWMFT&APSTP, 2021 (2021)
12. Min, J., Ming, P., Zhang, C., et al.: A test method of forming limit. China patent CN113128001B (2022)
13. Fu, M., Chan, W.: A review on the state-of-the-art microforming technologies. Int. J. Adv. Manuf. Technol. **67**, 2411–2437 (2013)

Effect of the Choice of Data Used for Analytical Identification of Orthotropic Criteria for Aluminum Extrusions

Oana Cazacu(✉) and Benoit Revil-Baudard

Department of Mechanical and Aerospace Engineering, University of Florida, REEF, Shalimar, FL 32579, USA
cazacu@reef.ufl.edu

Abstract. Recently, Cazacu [1] derived the expression in terms of stress invariants of the Yld91 yield function [2]. This enables the determination of the anisotropy coefficients analytically and makes possible the investigation of the influence of the data subset used for identification on the predictions. In this paper, we conduct such a study for AA 7079 extrusion. It is shown that if Yld91 and respectively Hill's parameters are obtained analytically using the same data subset, similar predictions of the anisotropy in uniaxial tension are obtained. However, there are distinct differences in terms of the predicted yield surface evolution with the level of shear.

Keywords: Orthotropy · Analytical Parametrization · Yield Surface Shape · Influence of Dataset · Material Variability · Aluminum Extrusion

1 Introduction

For anisotropic materials, the quality of the finite-element (FE) predictions is strongly dependent on the quality of the parametrization used for the anisotropic yield functions. The main reason is that for non-quadratic anisotropic yield functions, the expression of the anisotropy coefficients in terms of mechanical properties is not known. Therefore, the choice of the parameterization to be used for a specific application is strongly dependent on the experience of the analyst. Moreover, comprehensive mechanical data sets are available only for materials in sheet form while for extrusions, generally only uniaxial data along directions belonging to a single symmetry plane are reported (e.g., see [3]). Recently, Cazacu [1] derived the expression of the Hershey-Hosford yield function in terms of stress invariants. This in turn enabled the derivation of the expression of Yld91 for both FCC (e.g. aluminum and its alloys) and BCC materials (e.g. steel alloys) in terms of all the six stress components. Moreover, it was shown that as in the case of Hill's criterion, the anisotropy coefficients of Yld91 can be expressed solely in terms of yield stresses or solely in terms of r-values.

In this paper, we investigate the influence of the data subset used for identification of Hill and Yld 91 criterion on the predictions of the anisotropy of an AA 7079 extrusion

K. Mocellin et al. (Eds.): ICTP 2023, LNME, pp. 503–512, 2024.
https://doi.org/10.1007/978-3-031-40920-2_52

material. Specifically, we used subsets of yield stresses and r-values calculated from the uniaxial tensile stress-strain curves reported in [4] for a total of 12 orientations in all three coordinate planes θz, rz, and rθ of the cylindrical specimen (see Fig. 1).

2 Modeling the Anisotropy of AA 7079

2.1 Analytical Identification of the Yld91 and Hill Yield Function Using Only Six Uniaxial Yield Stresses

Yld91 is an orthotropic yield function that was obtained by replacing in the expression of the Hershey-Hosford yield function the principal values of the stress deviator **s** with the principal values $\tilde{S}_1$, $\tilde{S}_2$ $\tilde{S}_3$ of a transformed tensor $\tilde{\mathbf{S}}$, defined as

$$\tilde{\mathbf{S}} = \mathbf{C}\mathbf{s} \tag{1}$$

In Eq. (1), **C** is a fourth-order tensor which is represented in the axes of orthotropy (**x, y, z**) by a 6 × 6 matrix with six non-zero components, *a, b, c, f, g, h* (see [2]), i.e.

$$\mathbf{C} = \frac{1}{3}\begin{bmatrix} b+c & -c & -b & 0 & 0 & 0 \\ -c & c+a & -a & 0 & 0 & 0 \\ -b & -a & a+b & 0 & 0 & 0 \\ 0 & 0 & 0 & 3f & 0 & 0 \\ 0 & 0 & 0 & 0 & 3g & 0 \\ 0 & 0 & 0 & 0 & 0 & 3h \end{bmatrix} \tag{2}$$

In [3], it was deduced the expression of Yld91 in terms of all stress components. In particular, for FCC metals (m = 8), the expression of Yld91 is:

$$\text{Yld91} = \left(\tilde{S}_1 - \tilde{S}_2\right)^8 + \left(\tilde{S}_2 - \tilde{S}_3\right)^8 + \left(\tilde{S}_1 - \tilde{S}_3\right)^8 = 258\tilde{J}_2^4 - 648\tilde{J}_2\tilde{J}_3^2 \tag{3}$$

with

$$\begin{aligned}\tilde{J}_2 =& f^2\sigma_{yz}^2 + g^2\sigma_{xz}^2 + h^2\sigma_{xy}^2 + \left[(b+c)\sigma_{xx} - c\sigma_{yy} - b\sigma_{zz}\right]^2/18 \\ &+ \left[-c\sigma_{xx} + (c+a)\sigma_{yy} - a\sigma_{zz}\right]^2/18 + \left[-b\sigma_{xx} - a\sigma_{yy} + (a+b)\sigma_{zz}\right]^2/18\end{aligned} \tag{4}$$

$$\begin{aligned}\tilde{J}_3 =& 2(fgh)\,\sigma_{xy}\sigma_{xz}\sigma_{yz} + \frac{1}{27}\left[\begin{matrix}\left[(b+c)\sigma_{xx} - c\sigma_{yy} - b\sigma_{zz}\right]\cdot\left[-c\sigma_{xx} + (c+a)\sigma_{yy} - a\sigma_{zz}\right]\cdot \\ \left[-b\sigma_{xx} - a\sigma_{yy} + (a+b)\sigma_{zz}\right]\end{matrix}\right] \\ &- f^2\sigma_{yz}^2\left[(b+c)\sigma_{xx} - c\sigma_{yy} - b\sigma_{zz}\right]/3 - g^2\sigma_{xz}^2\left[-c\sigma_{xx} + (c+a)\sigma_{yy} - a\sigma_{zz}\right]/3 \\ &- h^2\sigma_{xy}^2\left[-b\sigma_{xx} - a\sigma_{yy} + (a+b)\sigma_{zz}\right]/3\end{aligned} \tag{5}$$

To simplify writing, let us denote the uniaxial yield stress along the z-axis as σ_0. Using the expression of Yld91 given by Eq. (3)–(5), it can be shown that for the AA 7079 extrusion, the anisotropy coefficients *a, b, c* can be determined analytically by solving an algebraic system of three independent relations involving only the uniaxial tensile yield stresses along the *r*, θ, *z*-axis of the cylindrical specimen, i.e. denoted, σ_r, σ_θ, σ_0.

The anisotropy coefficient h is expressed in terms of the coefficients a and b and the uniaxial yield stress in tension at $\alpha = 45°$ in the θ z plane, denoted $\sigma_{45}^{\theta z}$; the anisotropy coefficient g is expressed in terms of the coefficients a and c and the uniaxial yield stress in tension at $\beta = 45°$ in the rz plane, the respective yield stress being denoted as σ_{45}^{rz}; the coefficient f is expressible in terms of b and c and the uniaxial yield stress at $\gamma = 45°$ in the $r\theta$ plane, denoted $\sigma_{45}^{r\theta}$ (see Fig. 1).

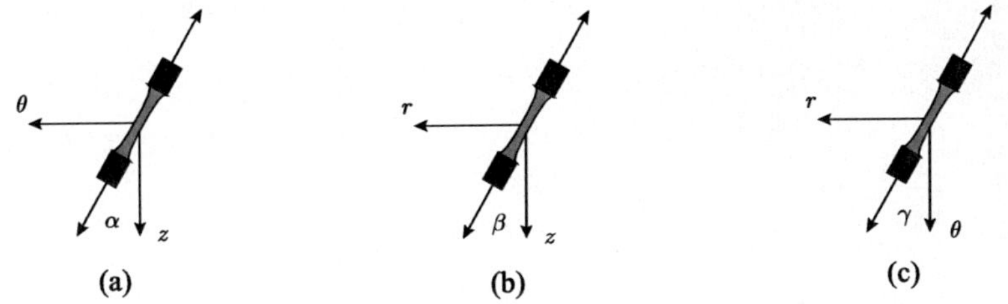

Fig. 1. Orientation of the tensile specimens in the: (a) θz, (b) rz, or (c) rθ plane of the extrusion material; the extrusion direction is along the z-axis.

The six anisotropy coefficients of Hill's criterion can be obtained analytically using the same six experimental yield stresses. Specifically, the anisotropy coefficients F, G, H can be obtained analytically using three independent relations involving only the uniaxial yield stresses along the orthotropy axes, i.e. σ_0, σ_θ, σ_r; the coefficient N is expressible solely in terms of the anisotropy coefficients F and G and the experimental yield stress $\sigma_{45}^{\theta z}$; the coefficient M is expressible solely in terms of the anisotropy coefficients F and H and the experimental yield stress σ_{45}^{rz}; the coefficient L is expressible solely in terms of the anisotropy coefficients G and H and the experimental yield stress $\sigma_{45}^{r\theta}$.

Figure 2 shows a comparison between the predictions of the anisotropy in uniaxial tensile yield stresses according to the two yield criteria identified using analytical formulas and the same data subset consisting only of yield stresses. Although only 6 datapoints were used for the determination of the parameters involved in each model, namely the tensile yield stresses along the z-axis ($\alpha = 0°$ which is the same test as $\beta = 0°$), θ-axis (for $\alpha = 90°$ which is the same test as $\gamma = 0°$), and the r-axis (for $\beta = 90°$ which is the same test as $\gamma = 90°$) and the yield stresses for $\alpha = 45°$, $\beta = 45°$ and $\gamma = 45°$, the predicted yield stresses for all the other orientations are within 5% of the experimental values. It is worth noting that irrespective of the symmetry plane, Yld91and Hill criterion predict the same variation of the uniaxial tensile yield stresses with the loading orientation (see Fig. 2).

For any other plane stress loadings with principal directions aligned with the r, θ, z-directions, Hill's criterion predicts slightly higher yield stresses than Yld91 (see Fig. 3 (a)–(c) for the comparison between the projections of the respective yield surfaces in the symmetry planes).

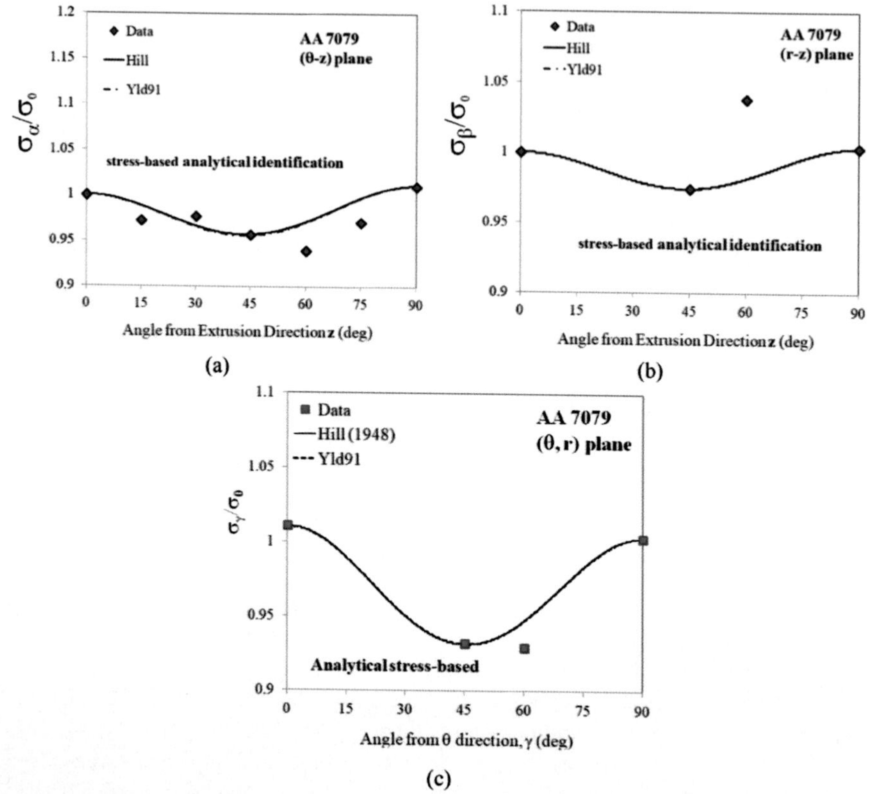

Fig. 2. Comparison between the predictions for aluminum alloy AA 7079 of the anisotropy in uniaxial tensile yield stresses according to the Hill and Yld91 identified using analytical formulas involving only 6 tensile yield stresses and data (symbols): (a) θz plane (b) rz plane (c) rθ plane.

The Lankford coefficient, r_σ corresponding to the orientation α with respect to the z-direction in the θz plane is given by:

$$r_\alpha = -\frac{\left(sin^2\alpha\right)\dot{\varepsilon}^p_{zz} - sin(2\alpha)\dot{\varepsilon}^p_{\theta z} + \left(cos^2\alpha\right)\dot{\varepsilon}^p_{\theta\theta}}{\dot{\varepsilon}^p_{zz} + \dot{\varepsilon}^p_{\theta\theta}} \tag{6}$$

The Lankford coefficient, r_β corresponding to the orientation β with respect to the z-direction in the rz plane is given by

$$r_\beta = -\frac{\left(sin^2\beta\right)\dot{\varepsilon}^p_{zz} - sin(2\beta)\dot{\varepsilon}^p_{rz} + \left(cos^2\beta\right)\dot{\varepsilon}^p_{rr}}{\dot{\varepsilon}^p_{zz} + \dot{\varepsilon}^p_{rr}} \tag{7}$$

The Lankford coefficient, r_γ corresponding to the orientation γ with respect to the θ-direction in the rθ plane is given by

$$r_\gamma = -\frac{\left(sin^2\gamma\right)\dot{\varepsilon}^p_{\theta\theta} - sin(2\gamma)\dot{\varepsilon}^p_{r\theta} + \left(cos^2\gamma\right)\dot{\varepsilon}^p_{rr}}{\dot{\varepsilon}^p_{\theta\theta} + \dot{\varepsilon}^p_{rr}} \tag{8}$$

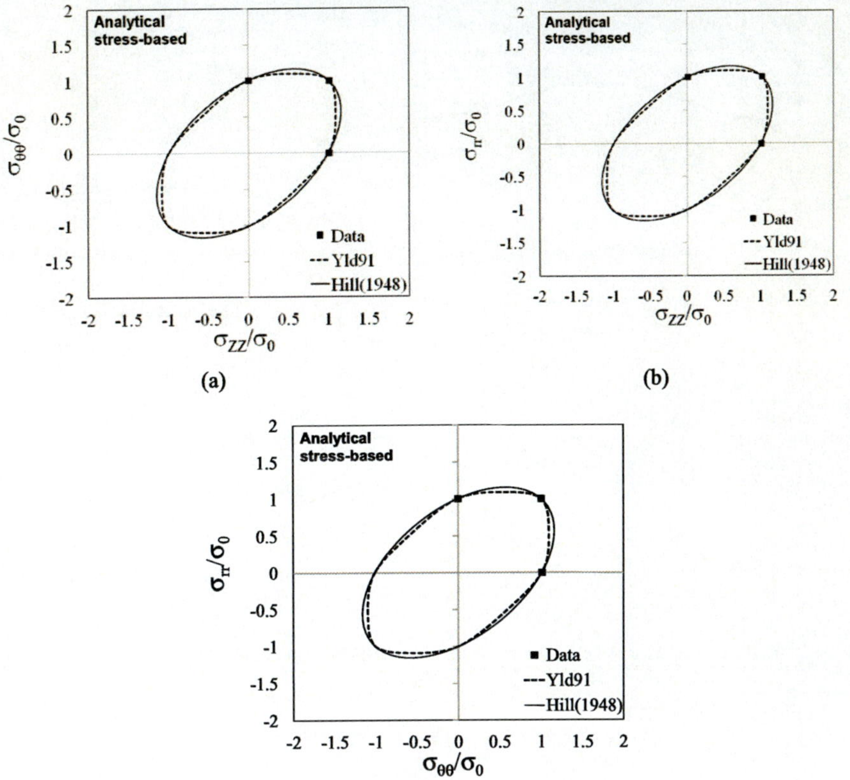

(c)

Fig. 3. Projection of yield surfaces according to the Hill and Yld91 yield criterion identified using analytical formulas involving only $\sigma_{a=0}$, $\sigma_{a=90}$, $\sigma_{45}^{\theta z}$, σ_{45}^{rz}, $\sigma_{45}^{r\theta}$: (a) ($\sigma_{\theta\theta}$, σ_{zz}) plane (b) (σ_{zz}, σ_{rr}) plane, (c) ($\sigma_{\theta\theta}$, σ_{rr}) plane. Stresses are normalized by σ_0 , the yield stress in uniaxial tension along the *z*-axis; data represented by symbols.

It is to be noted that since $\alpha = 0°$ and $\beta = 0°$ correspond to uniaxial tension along the z-direction in the respective planes, per above equations, we have: $r_{\beta=0} = 1/r_{\alpha=0}$; likewise $\alpha = 90°$ and $\gamma = 0°$ correspond to uniaxial loading along θ-direction, $r_{\gamma=0} = 1/r_{\alpha=90}$; $\beta = 90°$ and $\gamma = 90°$ correspond to uniaxial loading along the r-direction, so $r_{\gamma=90} = 1/r_{\beta=90}$. The r-values represented by symbols in the subsequent figures were calculated using Eq. (2)–(4) from the stress-strain curves corresponding to the uniaxial tensile tests conducted in the θz plane, rz plane, and rθ plane that were reported in [4].

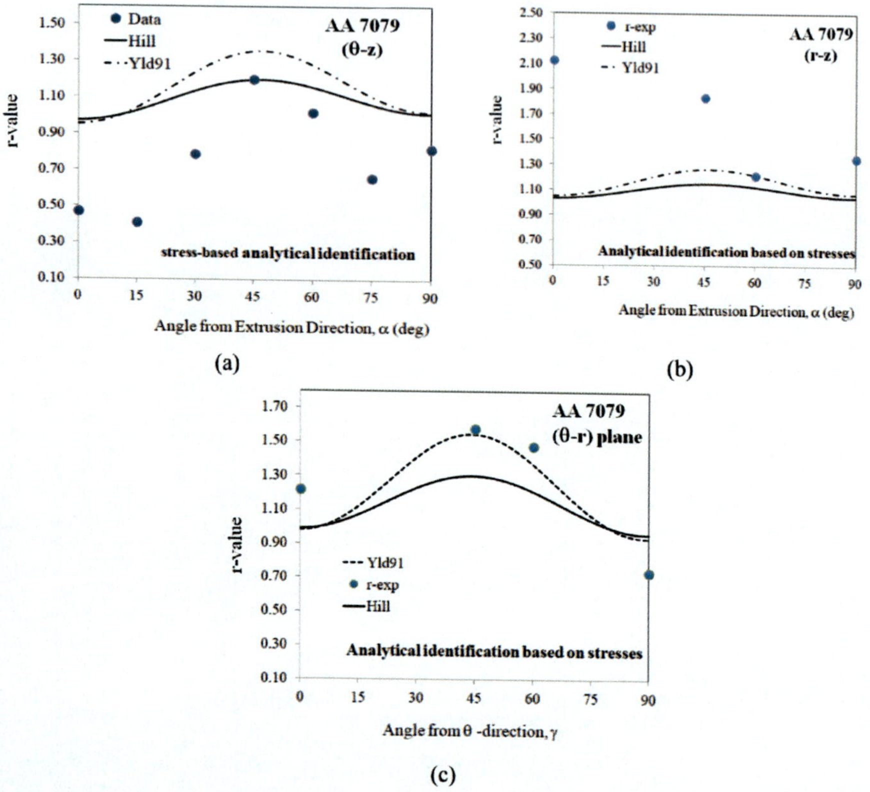

Fig. 4. Comparison between the predictions of the anisotropy in r- values for AA 7079 obtained with Hill and Yld91 criteria identified using analytical formulas involving only 6 tensile yield stresses: (a) θz plane variation, (b) rz plane variation, (c) rθ plane variation.

Figure 4 shows a comparison between the predictions of the anisotropy in r-values according to the two yield criteria identified using analytical formulas and the same data subset consisting of 6 tensile yield stresses. Note that the two models predict almost identical Lankford coefficients in the r, θ, z-directions and the same trends for the anisotropy evolution in the respective planes. While both criteria overpredict the experimental values, the Yld91 predicts larger r-values than Hill's criterion. For example, in the θz plane both models predict a maximum at $\alpha = 45°$ with the Lankford coefficients increasing for $0 < \alpha < 45°$ and then decreasing for $45° < \alpha < 90°$. However, the experimental r-values are largely overestimated for α close to 0° or 90°, hence the predicted curvature of the respective yield loci for stress states close to uniaxial tension along the z-axis and θ-axis is not accurate (see also Fig. 3).

2.2 Analytical Identification of the Yld91 and Hill Yield Function Using Six Experimental Lankford Coefficients

Let us denote r_0 the Lankford coefficient corresponding to $\alpha = 0°$ (z-direction), r_{90} the Lankford coefficient corresponding to $\alpha = 90°$ (θ-direction) and by $r_{45}^{\theta z}$, $r_{45}^{r\,z}$, $r_{45}^{r\,\theta}$, the Lankford coefficients corresponding to the tensile tests at a direction at 45°in the rz, rθ, and θz planes, respectively (see Fig. 1). Using the expression of Yld91 given by Eq. (3)–(5) and associated flow rule, it can be shown that the anisotropy coefficients *a, b, c* are expressible only in terms of the Lankford coefficients along the *z*-axis and the θ–axis of the cylindrical specimen, i.e. r_0 and r_{90}. To be noted that the parameters *F*, *G*, *H*, which are the anisotropy coefficients associated with the normal stress components in the expression of Hill's criterion are expressible only in terms of same Lankford coefficients, r_0 and r_{90}. It can also be shown that the anisotropy coefficient *h* of Yld91 is expressed in terms of the anisotropy coefficients *a* and *b* and $r_{45}^{\theta z}$ and likewise, the anisotropy coefficient *N* of Hill's criterion is expressible solely in terms of the anisotropy coefficients *F* and *G* and $r_{45}^{\theta z}$. The Yld91 anisotropy coefficient *g* is expressible solely in terms of the anisotropy coefficients *a* and *c* and $r_{45}^{r\,z}$ while the anisotropy coefficient *f* is expressible solely in terms of the anisotropy coefficients *b* and *c* and $r_{45}^{r\,\theta}$. Likewise, the Hill's parameter *M* is expressible solely in terms of the parameters *H* and *F* and $r_{45}^{r\,z}$. The parameter *L* is expressible solely in terms of the parameters *H* and *G* and $r_{45}^{r\,\theta}$. As an example, in Fig. 5 are presented comparisons between the predictions of the material's anisotropy in the θz plane which were obtained with Yld91 and Hill criterion identified analytically using the same five r-values, i.e. (r_0, r_{90}, $r_{45}^{\theta z}$ $r_{45}^{r\,z}$, $r_{45}^{r\,\theta}$). Note that irrespective of the loading orientation a, the predictions of the Lankford coefficients obtained with Hill's and Yld91 yield criterion are nearly the same (see Fig. 5(b)).

Note that the anisotropy in r-values in the θz plane is well described, the experimental r-values being only slightly overpredicted for the orientations other than those used for the identification of the anisotropy coefficients of either criterion (see Fig. 5(b)). Concerning the predictions of the anisotropy in stresses presented in Fig. 5 (a) and Fig. 5(c), it is important to note that none of the datapoints shown in these plots were used for the identification of the anisotropy coefficients. Note that both models predict the same trends for the yield stress anisotropy in the θz plane, namely the experimental values for $\alpha > 45°$ are overpredicted with the difference between the highest and lowest yield stresses is 18.07% according to Hill and 10.05% according to the Yld91 criterion, against 1.07% experimentally. Also, note that the yield locus according to Yld91 is interior to the one according to Hill's criterion (see Fig. 5(c)). Similar results are obtained concerning the predicted anisotropy in the rz plane and the rθ plane, respectively. It is also worth comparing the predicted evolution of the shape of the yield loci with the level of shear for Yld91 and Hill's, respectively. In Fig. 6 are shown the cross-sections in the (rθ) plane

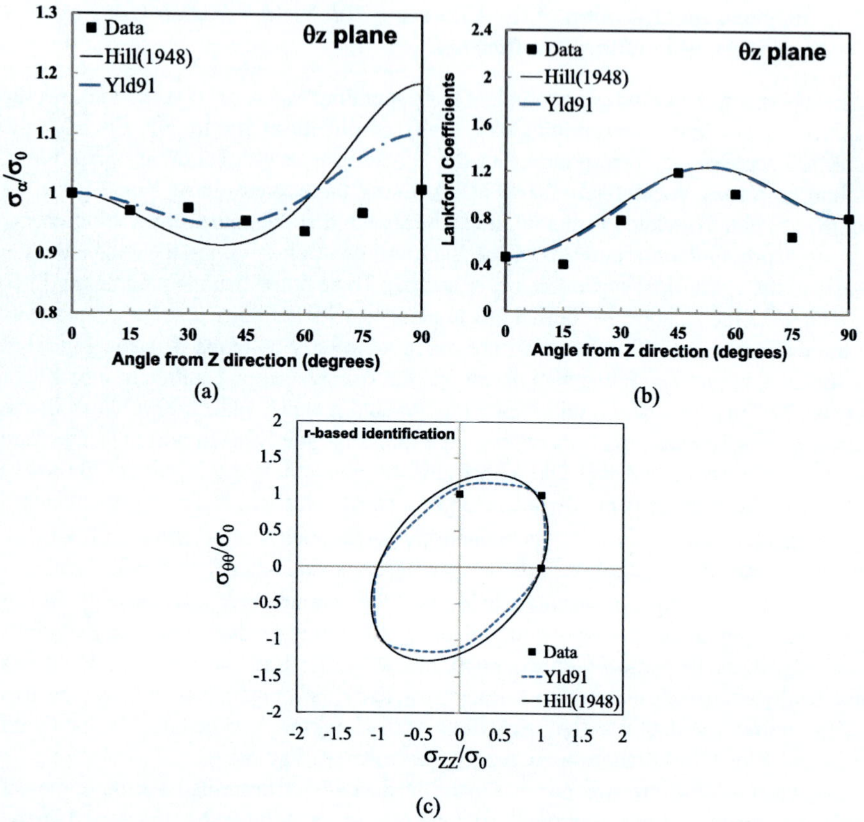

Fig. 5. Comparison between data and the predictions of the anisotropy for AA 7079 in the θz plane obtained with Hill's and Yld91 criteria identified using analytical formulas involving only 5 Lankford coefficients (a) tensile yield stress variation, (b) r-values variation, (c) yield surfaces in the biaxial plane; stresses are normalized by σ_0, the yield stress along z- axis. Data reported in [4] are represented by symbols.

using the anisotropy coefficients identified with the same five r-values. It is interesting to note that the Yld91 isosurfaces are not self-similar. Note that even a small level of shear induces a significant change in curvature with respect to the shape corresponding to zero shear. On the other hand, the Hill's surfaces are self-similar. This trend is true for cross-sections in the (rz) and (θz) planes, respectively.

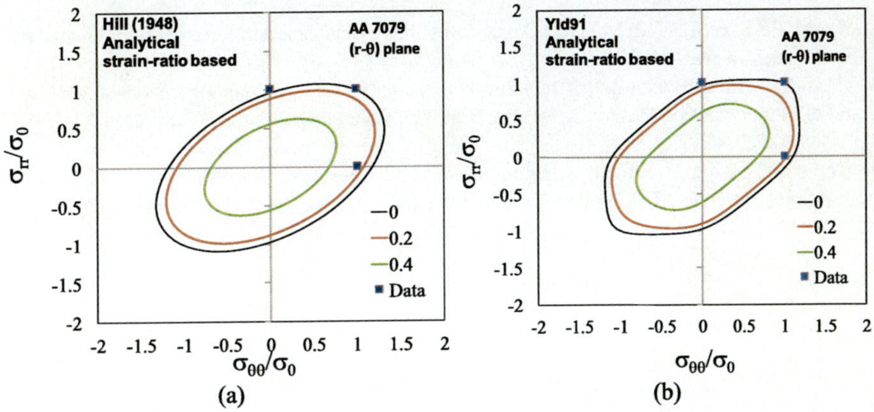

Fig. 6. Cross-sections of the yield surfaces for AA 7079 extrusion corresponding to fixed levels of shear stresses obtained identified using analytical formulas involving only 5 r-values in the ($\sigma_{\theta\theta}$, σ_{zz}) plane for (a) Hill criterion; (b) Yld91. Stresses are normalized by σ_0, the yield stress in uniaxial tension along the z-axis. Data represented by symbols.

3 Conclusions

Based on the new mathematical results obtained in [1], namely the derivation of equivalent expressions in terms of all stress components of 6-parameters anisotropic yield surfaces such as Yld91, it is possible to obtain the anisotropy coefficients analytically. In this paper, the benefits are shown for the case of the aluminum 7079 alloy extrusion. Most importantly, a clear understanding of the influence of the type of data used for parameterization on the predictions of the anisotropy of the material was provided. As in the case of Hill's criterion, three of the parameters involved in Yld91, namely a, b, c are expressible solely in terms of uniaxial tensile properties in the axes of symmetries of the extruded material while each of the other three shear-related parameters are expressible in terms of a single tensile uniaxial stress in an off-axis direction in the respective symmetry plane. It is important to note is that like in the case of Hill's criterion all coefficients of Yld91 can be expressed analytically only in terms of 6 yield stresses or alternatively only in terms of five r-values. The relative advantages of Yld91 and Hill criteria were analyzed. It was concluded that for 7079 aluminum extrusion, if Yld91 and Hill's model parameters are obtained analytically using the same datapoints, their predictions of the material's anisotropy for uniaxial yield stresses and r-values in each of the symmetry are close. For multi-axial loadings involving shear, there are distinct differences in terms of the evolution of the shape of the yield loci. Specifically, according to Yld91 the yield surface shape is strongly influenced by the level of shear whereas according to the Hill's criterion the yield surfaces are self-similar.

References

1. Cazacu, O.: New mathematical results and explicit expressions in terms of the stress components of Barlat et al.(1991) orthotropic yield criterion. Int. J. Solids Struct. (176–177), 86–95 (2019)

2. Barlat, F., Lege, D.J., Brem, J.C.: A six-component yield function for anisotropic materials. Int. J. Plast. **7**, 693–712 (1991)
3. Cazacu, O., Revil-Baudard, B., Chandola, N.: Plasticity-Damage Couplings: From Single Crystal to Polycrystalline Materials, 1st edn. Springer, Cham (2019). https://doi.org/10.1007/978-3-319-92922-4
4. Corona, E., Kramer, S.L.B., Scherzinger, W.M., Jones, A.R.: Anisotropic plasticity model forms for extruded Al 7079: Part I, calibration. Int. J. Solids Struct. **213**, 135–147 (2021)

Influence of the Initial Microstructure on the Mechanical Behavior During Forming for Inline Manufacturing Process Routes

Avantika Jhanji[1](✉), Benjamin Sydow[1], Tom-Eric Adams[2], Stefan Habisch[2], and Sebastian Härtel[1]

[1] Brandenburg University of Technology, Cottbus-Senftenberg, Cottbus, Germany
jhanji@b-tu.de

[2] Technical University of Munich, Munich, Germany

Abstract. With a view of sustainability and the rising energy costs currently, manufacturing processes of metals are becoming increasingly focused on optimizing process parameters such as energy and time consumption. A conventional hot-forming process route currently involves casting an ingot, letting it cool down, and heating it up again for the hot-forming process (see Fig. 1a). In order to implement the combination of casting and forging, avoiding the reheating cycle and using less energy, by utilizing the casting heat (see Fig. 1b), a methodology was developed within the present work to quantify the influence of the resulting microstructure as a function of the cooling rate on the forming and recrystallization behavior (see Fig. 1c).

For this purpose, AISI 301 austenitic stainless-steel cast samples with different cast cooling rates were generated. An in-situ high-temperature microscope is used to determine the holding time and the heating rate. Dilatometer tests are performed to characterize the interaction between initial microstructure and the flow curves to verify the determination method (see Fig. 1d). The aim was to demonstrate whether the microstructure evolution and mechanical behavior is affected by the initial microstructure. The flow curves and the post-forming microstructure show a higher degree of recrystallization in fast-cooled microstructure than slow-cooled microstructure. Hence, it was found that the initial microstructure and the associated temperature history does have an impact on the mechanical properties.

Keywords: Inline process · mechanical behavior · temperature history

1 Introduction

There is a significant demand to have more efficient processes to reduce energy costs and reduce energy consumption. A way to reduce energy consumption is by reducing the number of steps in a manufacturing process. Currently, stainless steel manufacture is done by first, casting a material, then cooling the material to room temperature. Next, the sample is then reheated to a forming temperature and then formed in a die, eventually letting the part cool to room temperature. To optimize this process, the heat from the warm ingot can already be used for the forming procedure, as an inline process.

K. Mocellin et al. (Eds.): ICTP 2023, LNME, pp. 513–524, 2024.
https://doi.org/10.1007/978-3-031-40920-2_53

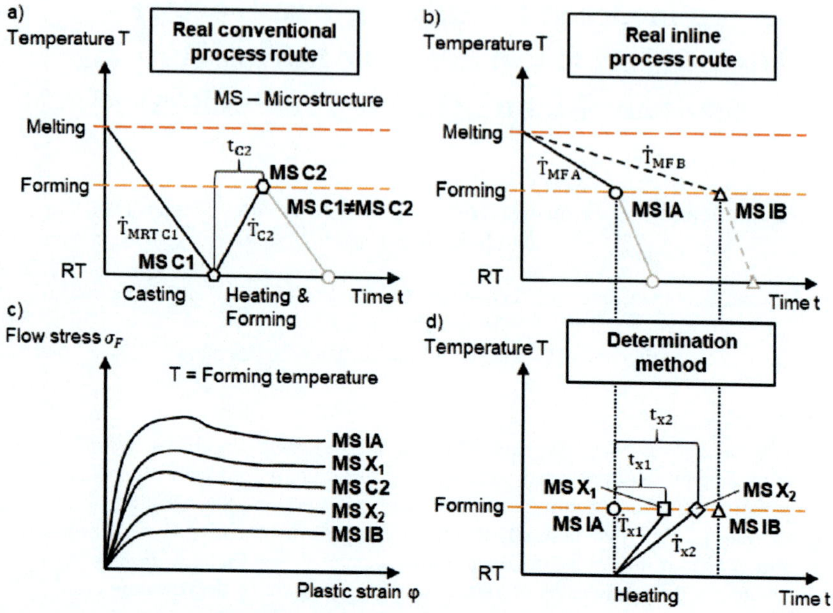

Fig. 1. Temperature-microstructure relationship curve during processing for hot forming operations for a) conventional process route and b) inline process route c) flow curves based on different microstructures, d) determination method to model the microstructure evolution during an inline process route

There are several manufacturing techniques that make use of this phenomenon such as Hot Form Quench (HFQ) technology, used in the manufacture of aluminum sheet metal forming, which involves heating a blank to a solution heat treatment temperature leading to a supersaturated solid solution and then subsequently transferring it to a press to be stamped and simultaneously quenched by cooled tools. Here, the heat from the warm blank is utilized in the hot forming process [1, 2]. The advantages of this process include higher formability, less energy required due to the reduction in reheating steps, and fewer opportunities for contamination of metal within each process [3].

Another example of a process combination is the new inline process called Weld forming, which is used to improve the mechanical properties of weld seams. As the name suggests, this process combines welding and a forming process (e.g. rolling) to homogenize the microstructure in the heat-affected zone, which greatly degrades the toughness of the material in this area [4].

Work by Dedov et al., combines the casting-forging process for the production of automotive parts. This work exhibits a direct transition from casting to forging operation using the same tool, so that the heat loss decreases and no additional heat treatment between these operations is necessary, leading to a higher process efficiency [5]. Other similar examples include liquid forging, rheo forging [5, 6], and strip casting and roller forming [7]. However, the research data is limited to aluminium alloys, and extensive literature on the influence of the initial stainless-steel cast microstructure on the mechanical properties is not readily available yet.

Hence, the motivation for this study is to explore the feasibility of casting and forming combined to determine mechanical properties, before developing an in-line process and making an investment into designing tools and purchasing materials and machinery. A way to conceptualize this idea is by using a determination method as shown in Fig. 1d. This involves casting ingots in-house with different tool geometries to provide different microstructures. Samples from this ingot will then be sent for analysis under an in-situ microscope to determine a heating rate, temperature, and holding time under which the microstructure does not change significantly. The objective here is to characterize the material in the as-cast state (depending on the cooling rates) in subsequent dilatometer tests. For solution heat treatment, the microstructure dissolves and faster cooling is required to prohibit grain growth [8, 9]. Taking these exact determined conditions, and simulating them in a dilatometer, with subsequent forming will provide an indication of the effect of the initial microstructure. If there is no difference in mechanical properties based on the initial microstructure, this suggests that the inline process can be easily implemented without consideration of cast conditions. However, if there is a difference in the mechanical properties and the recrystallization behavior, this suggests that further work may be required to optimize the determination method. Whilst, the potential of the inline process development is huge, the limitation is that the results are only applicable to a set material with set conditions, in this case, AISI 301 with set heating rate and temperature. However, a big advantage is that the determination method is relatively simple to conduct and requires much less investment and effort than setting up an inline process machine.

The analysis in this study is done by examining the recrystallization behavior and flow curves using hot compression testing of AISI 301, austenitic stainless steel with excellent toughness, formability, and high mechanical properties [10, 11]. During this process when a sample is hot worked at a temperature range of 0.5–0.9 Tm, dynamic recrystallization (DRX) can occur, a mechanism that greatly impacts the microstructure and therefore mechanical properties such as flow behavior of the sample. Conventional DRX consists of nucleation and then grain growth. At low strains, strain hardening and softening by dynamic recovery (DRV) are the deformation-controlling mechanisms. The presence of the DRV promotes equilibrium [12].

2 Material and Methods

2.1 As-Cast Specimen Production

Plates of AISI 301 material are subsequently cut into smaller sections to facilitate an efficient melting procedure. The chemical composition is listed below in Table 1.

The AM500 electric-arc melting furnace from Edmund Buehler is used to produce the as-cast samples, using inert gas pressure of 1.0 bar. The material is heated up by the arc beyond the melting point and then sucked into the suction-casting copper molds with a diameter of 40 mm (Fig. 2a) and 6 mm (Fig. 2b) to produce slow-cooled and fast-cooled samples in Fig. 2c and d, respectively. The basic idea is that casting smaller

Table 1. Chemical composition (%)

Element	C	Si	Mn	P	S	Cr	Mo	N	Ni
min	0.05	–	–	–	–	16.00	–	–	6.00
max	0.15	2.00	2.00	0.045	0.015	19.00	0.8	0.11	9.50

diameter samples lead to higher cooling rates due to the favorable surface-to-volume ratio.

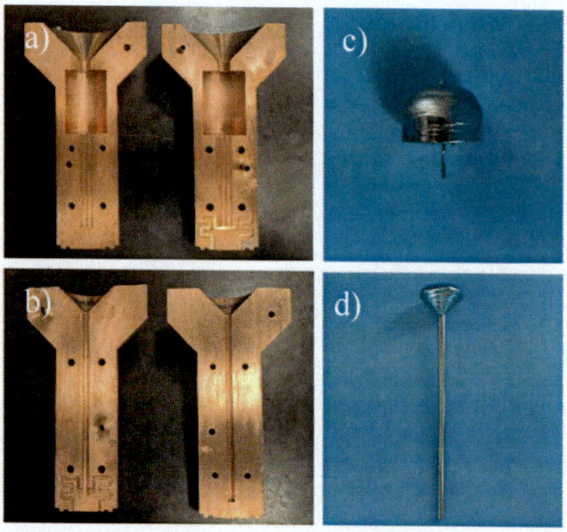

Fig. 2. Copper cast mold a) 40 mm diameter (slow-cooling) and b) 6 mm diameter (fast-cooling) c) 40 mm diameter cast-sample d) 6 mm diameter cast sample

2.2 In-Situ Microstructure Analysis

In-situ microstructure analysis is performed at Technical University of Munich, Germany with a Yonekura High Temperature Confocal Scanning Laser Microscope (HT-CSLM) and the ultra-high temperature module (SVF17SP). Cylindrical samples of dimensions Ø5 mm × 2 mm are obtained from the half-height of the 6 mm as-cast sample. The microscope chamber is filled with 400 ml/min Argon. Heating is performed with a ramp function. The first stage slowly heats the sample up to 150 °C. The second stage uses a heating rate of 100 K/s up to approx. 900 °C with a subsequent slower heating rate up to 1200 °C. The holding time at 1200 °C is 5 min. The microstructure evolution is video recorded during the whole analysis.

2.3 Hot Compression Testing

Cylindrical dilatometer samples (Ø5 mm × 10 mm) are machined from the two as-cast geometries, as shown in Fig. 2. The heating rate of 100 K/s was adapted as in the in-situ microscope, however, with the dilatometer testing constant heating rate without ramp function is used until the testing temperature is reached. The samples are then isothermally deformed, using a quenching and deformation dilatometer DIL805A/D/T (TA Instruments), at temperatures 1000 °C, 1100 °C and 1200 °C, with the strain rate of 0.1 s^{-1} and a total strain of 0.7. The samples are heated under an Argon atmosphere with a heating rate of 100 K/s, with a holding time of 30 s to obtain a homogeneous temperature distribution before compression testing. The Si3N4-punches are used in the compression process. The samples are deformed between molybdenum plates of ~1 mm thickness.

2.4 Microstructure Analysis

The microstructure of the as-cast samples and the deformed samples is observed by means of optical microscopy. The samples are mounted with cut cross-sections on top and ground flat with successively finer grades of silicon carbide (SiC) papers. The surface of the samples is polished with a 0.05 μm OP-S silica solution. Aqueous sodium hydroxide solution agent etching was used to reveal the microstructure.

2.5 Simulations

FEM simulations are performed with the software Simufact Forming. The model was simplified such that the pouring of the molten material into the mold was not simulated and only the cooling phase of the casting procedure within each casting mold was considered. The cast molds are modeled as shown in Fig. 2. The model is set with a 2D-axisymmetric simulation, where the initial temperature is set to 1400 °C based on the approx. Melt temperature of AISI 301. The heat transfer principles applied here can be defined via Newton's law of convection and the Stefan-Boltzmann law of radiation.

These are especially applicable for the heat transfer during any shrinkage that occurs during the cooling phase of casting. The contact heat transfer coefficient used was 10000 $W/m^2/K$ [13] and the ambient temperature of 20 °C was used. The simulation run time was 200 s. Material data was taken from the existing material library in Simufact.

3 Results and Discussion

3.1 Simulation of Cooling Rate in Casting Geometry Analysis

Temperature measurement was not possible with the setup described in Sect. 2.1. Hence, cooling rates were evaluated by simulations, and microstructural analysis and compared with the literature for validation. The simulation setup was based on the investigations in a welding+forming study conducted by Härtel et al. [14]. In the paper mentioned, the simulations for comparable cooling simulations were validated with experimental investigations.

Figure 3a shows the temperature vs time curve for the marked locations in the 6 mm (fast-cooling rate), 40 mm tool (slow-cooling rate). The positions 1, 2 and 3 indicate the location of the samples extracted from the cast material and the samples analyzed

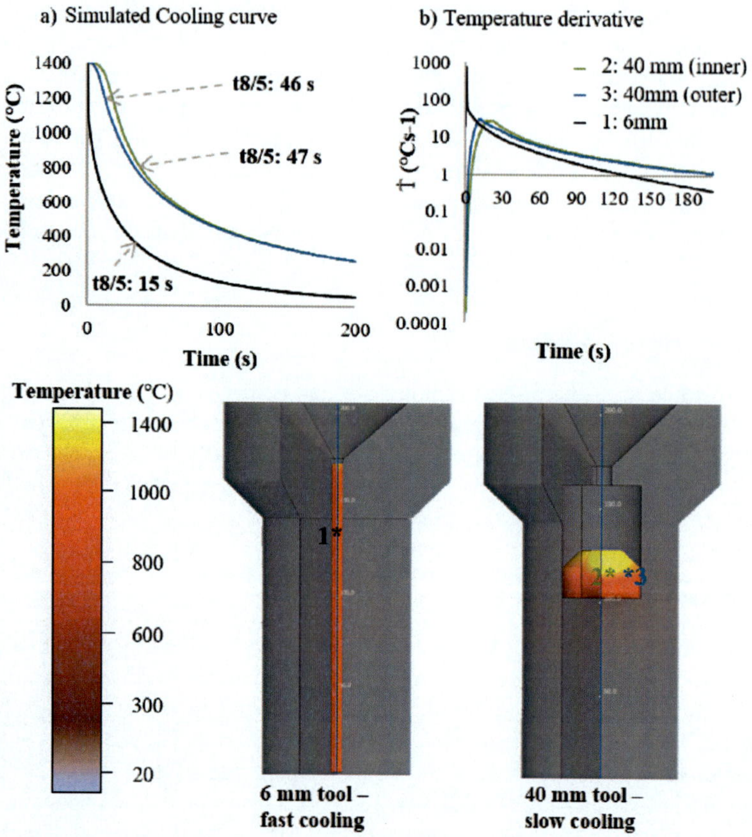

Fig. 3. Simulation data: a) Temperature vs. time curve for each tool at the marked location for 6 mm (1*), 40 mm inner (2*) and 40 mm outer (3*) b) Temperature derivative for each marked point

for microstructure analysis. The $t_{8/5}$ for the 6 mm sample is 15.4 s and for the $t_{8/5}$ for 40 mm is 45.8 s for position 2 (inner) and 47 s for position 3 (outer). This is expected due to a thinner geometry of the cast ingot, where the heat flux by conduction [15] cools the edges rapidly, with a small shrinkage in the outer layer of the cast sample. Thus, the decrease in the cooling rate is attributed to the gap between the tool and the sample.

From the study conducted by Härtel et al. [14], whilst the study is on mild steel, a typical dendritic microstructure is noted with fast cooling rates, with a $t_{8/5}$ of 17.85 s, similar to the fast-cooled values obtained here. This is further confirmed by the 6 mm microstructure in Fig. 4a with δ-ferrite (dendritic) phase and fine grains, present during fast cooling rates or in welding [16]. The darker, vermicular, thin lines represent the δ-ferrite phase with the light areas being the austenite matrix. The formation of these thick arrowhead dendrites occurs when grains bond together early during the solidification process to resist solidification cracking [16]. From the temperature derivative curves in Fig. 4b, it can be seen that the temperature decreases rapidly in the first 5 s for the 6 mm tool and does not reach a constant level to room temperature in the first 200 s. For the 40 mm tool the curves for temperature derivative show a slow cooling, as also reflected in the $t_{8/5}$ time. The 40 mm sample also did not reach a constant level of cooling at room temperature.

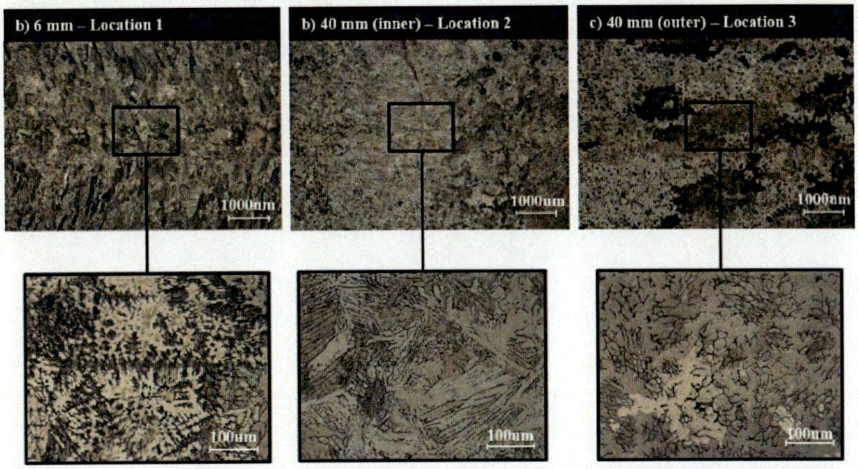

Fig. 4. Optical microscope images of the initial cast microstructure before forming for a) 6 mm, b) 40 mm (inner) and c) 40 mm (outer) samples. The top 3 images are at 100x magnification and zoomed in images are at 500x magnification

There is a higher percentage of dendrites (δ-ferrite), visually noted for fast cooling than slow cooling microstructure (Fig. 4b and c) as a result of suppression of the transformation of ferrite to austenite [15]. The outer section of the 40 mm tool shows a more homogenous microstructure with coarse grains in Fig. 4a than the middle of the ingot (40 m inner) in Fig. 4b, where the left side of the sample has more columnar grains and the middle of the sample showing larger grains. A study done by Bai et al., on the phase transformation during slow cooling of AISI 301 show a ferrite core and ferrite lamellae with austenite Matrix structure [17], which are noted in Fig. 5b and c. The δ-ferrite (dendritic) phase is normally enveloped by the precipitation of austenite at a slow cooling rate [17].

3.2 In-Situ Microscope Analysis

Figure 5 shows images obtained from the in-situ microscope video of the heating of the cast material. With a heating rate of 100 K/s at a temperature of 1200 °C and a subsequent holding time of ca. 30 s, no substantial change in microstructure occurs (see Fig. 5a and b). It is only after a longer holding time of 300 s at 1200 °C (see Fig. 5c), dissolution of dendrites (δ-ferrite) and grain boundaries become visible. Hence, the conditions of 1200 °C, 100 K/s with a holding time of 30 s were selected for the subsequent hot compression testing. The circled area at 0 s in Fig. 5a could indicate dendrites (δ-ferrite phase) in the 6 mm sample (fast-cooled), within the lighter austenitic matrix, as mentioned in Sect. 3.1

Fig. 5. In-situ microscope video showing microstructure at a) 0 s at room temperature, b) 30 s holding time at 1200 °C, and c) 300 s holding time at 1200 °C (where the microstructure change occurs)

3.3 Flow Stress Analysis

The obtained flow curves have been analysed for the slow and fast cooled cast samples and are graphed in Fig. 6. Generally, it can be noted that the flow stress decreases as the temperature increases. Initially, the flow stress increases rapidly due to work hardening.

All flow curves show that recrystallisation occurred in all samples, where a decline at the end in flow stress is an indication of DRX (Dynamic Recrystallisation), except at 1000 °C 40 mm outer and inner curves, shown in Fig. 6. Furthermore, the difference in flow behaviour for the 40 mm inner and outer becomes negligible at higher temperatures.

Austenitic stainless steels can have transformation induced plasticity (TRIP) effect. Transformation from austenite to martensite can also occur by two other mechanisms such as spontaneous transformation at low temperatures and stress-assisted nucleation [18]. However, TRIP effect tends to occur more so at room temperature i.e. during cold forming. Furthermore, the microstructure post forming at 1100 °C also does not show any signs of martensite α' [19] in Fig. 7 a, b and c. Hence, it can be assumed that the martensite concentration is low enough to not have an impact on the mechanical properties.

For 1100 °C, the evidence of recrystallisation can be seen in Fig. 7a. This suggests that the initial microstructure has an impact on mechanical properties. The fine grains and the higher δ-ferrite content in the initial microstructure for the fast cooling sample could be the reason for higher flow stress in the 6 mm sample as compared to the 40 mm ones. A study on the microstructure of welded AISI 301 austenitic stainless-steel joints confirms that formation of very fine grains and delta ferrite structure in weld region, due to high cooling rates leads to superior tensile properties [20]. A study done on the characterisation of grain size and yield strength of AISI 301 shows a relationship of the larger the grain size, the lower the yield strength [21].

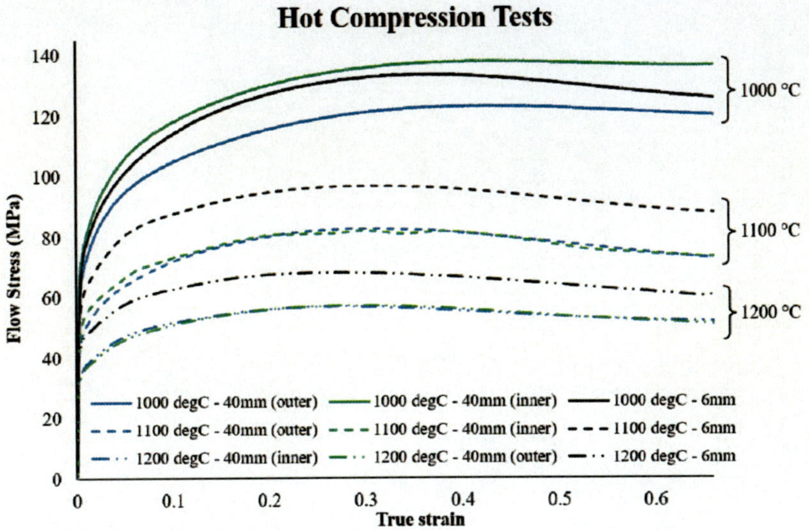

Fig. 6. Flow curves for temperatures 1000 °C, 1100 °C and 1200 °C for each tool at 0.1 s-1 strain rate and 0.7 strain.

The 40 mm inner and outer samples show very little recrystallisation and mostly flow localisation [22]. Extensive literature on flow curves for AISI 301 was not found; however, a study on 316L austenitic steel gives an idea on the expected trend for the three different temperature flow curves.

Overall, the hot compression tests show that the initial microstructure (dependent on the casting cooling rate) does have an impact on the flow behaviour. The impact is more noticeable at higher temperatures. The fast cooling rate for the initial cast microstructure, leads to an increased formation of delta-ferrite dendritic structure and fine grains, ultimately leading to a higher flow stress and a higher degree of recrystallisation as compared with the slow-cooled samples. The slow-cooled samples show variation in the initial microstructure within the samples. However, this difference does not impact the mechanical behavior at higher temperatures of 1100 °C and 1200 °C. The slow cooled samples show more homogenous, coarser grains and generally lead to lower flow stresses and a reduction in the degree of recrystallisation.

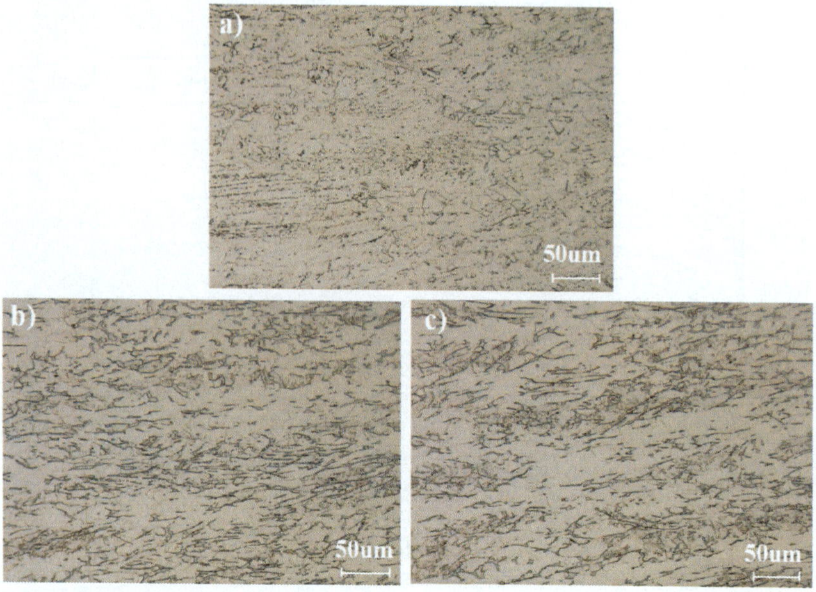

Fig. 7. Microstructure of samples post forming at 1100 °C a) 6 mm b) 40 mm (inner) c) 40 mm (outer)

4 Conclusions

In order to implement the combination of casting and forging by utilizing the casting heat, a methodology was developed within the present work to quantify the influence of the resulting microstructure as a function of the cooling rate on the forming and recrystallization behavior. For this purpose, casting samples with different cooling rates were generated.

The aim of this study was to demonstrate whether the microstructure evolution and mechanical behavior is affected by the initial microstructure and heating time. This was successfully simulated using a determination method (based on an in-situ high-temperature microscope) involving heating the sample at a high heating rate and a set temperature and holding time, to avoid any microstructural changes. The subsequent testing involved emulating these conditions in the dilatometer and forming at different temperatures. This study has shown:

- The slow-cooled material (40 mm) has a coarse initial microstructure with large grains. This is reflected by a t8/5 of 45 −47 s and a lower flow stress than the fast-cooled material (6 mm). Recrystallisation occurred at all temperatures, except at 1000 °C forming temperature.
- On the other hand, a fast-cooled material (6 mm) has a fine grain and dendritic microstructure, with a t8/5 of 15 s. The hot compression tests show higher flow stress values and thus, more recrystallized grains than the slow-cooled microstructure.
- Thus, the initial microstructure has an impact on the mechanical properties of the sample. This is especially noticeable at higher deformation temperatures.

Hence, future work will be required to optimize the determination method. An ideal solution might be to have an in-situ casting and dilatometer process. Furthermore, there is data still required for an in-depth recrystallisation behavior analysis and grain size calculations to quantify the degree of recrystallisation.

References

1. Zheng, K., et al.: The effect of hot form quench (HFQ®) conditions on precipitation and mechanical properties of aluminium alloys. Mater. Sci. Eng. A, 1–13 (2019)
2. Jiang, Y., Ding, H., Cai, M., Chen, Y., Liu, Y., Zhang, Y.: Investigation into the hot forming-quenching integrated process with cold dies for high strength alumnium alloy. Mater. Charact. **16**(10), 1–9 (2019)
3. Mohamed, M.S.K.: An investigation of hot forming quench process for AA6082 aluminium alloys. Mechanical Engineering PhD theses, p. 30 (2010)
4. Adams, T.-E., Härtel, S., Hälsig, A., Awiszus, B., Mayr, P.: WeldForming: a new inline process combination to improve weld seam properties. Weld. World **64**(4), 601–610 (2020). https://doi.org/10.1007/s40194-020-00856-9
5. Dedov, S., Lehmann, G., Kawalla, R.: Application of combined casting-forging process for production of durable lightweight alumnium parts. Key Eng. Mater., 264–273 (2013)
6. Hong-Min, G., Xiang-Jie, Y.: Rheoforging of wrought aluminum alloys. Solid State Phenom. **07**(07), 271–276 (2008)
7. Raabe, D.: Texture and microstructure evolution during cold rolling of a strip cast and of a hot rolled austenitic stainless steel. Institut fiir Metallkunde und Metallphysik **5**(06), 1137–1151 (1996)
8. Scott, J.A., Spruiell, J.E.: Grain refinement of wrought austenitic stainless steels by rapid heating. Metall. Mater. Trans. B. **5**(1), 255–259 (1974). https://doi.org/10.1007/BF02642949
9. Elmer, J.W.: The influence of cooling rate on the microstructure of stainless steel alloys, Doctorate thesis, pp. 28–29 (1998)
10. A. International: Austenitic Stainless Steels, in Stainless Steels for Design Engineers (2008)
11. Al-Fadhalah, K.J., Al-Attal, Y., Rafeeq, M.A.:Microstructure refinement of 301 stainless steel via thermomechanical processing. Metals **10**(12), 2–23 (2022). https://doi.org/10.3390/met12101690
12. Varela-Castro, G., Cabrera, J.M., Prado, J.M.: Critical strain for dynamic recrystallisation. The particular case of steels. Metals **10**, 2–58 (2020). https://doi.org/10.3390/met10010135
13. Nayak, R.K., Sundarraj, S: Selection of initial mold–metal interface heat transfer coefficient values in casting simulations—a sensitivity analysis. Metall. Mater. Trans. B (41), 151–160 (2010)
14. Härtel, S., Adams, T.-E., Hoefer, K., Awiszus, B., Mayr, P.: A novel method for improving weld seam properties through inline coupling of welding and forming. Materials **13**, 271 (2020). https://doi.org/10.3390/ma13020271
15. Cardoso, W., et al.: Influence of heat in the microstructure of AISI 301 stainless steel cast zone AISI 301. Ifes ciencia **7**(1), 1–15 (2021). https://doi.org/10.36524/ric.v7i1.1202
16. Yu, P., Thompson, K.J., McCarthy, J., Kou, S.: Microstructure evolution and solidification cracking in austenitic stainless steel welds. Weld. J., 301–314 (2018). https://doi.org/10.29391/2018.97.026
17. Bai, L., Ma, Y., Xing, S., Zhang, J., Wang, B.: Research on non-directional solidification process of AISI 301 Austenite stainless steel. In: TMS2015 Annual Meeting Supplemental Proceedings, Shanghai (2015)

18. Montepagano, D., Citi, I., Guerra, R., Nunzio, P.D., Ruffini, F.: Enhancement of ductility of work hardened strips in AISI 301 austenitic stainless steel. La Metallurgia Italiana, 7–10 (2022)
19. Mukarati1, TW, Mostert RJ, Siyasiya, C.W.: The direct observation of surface martensite formation upon cooling to temperatures close to ambient in a heat treated AISI 301 stainless steel. In: Conference of the South African Advanced Materials Initiative (CoSAAMI-2018), Pretoria (2018)
20. Gnanasekaran, S., Kumar, S.S., Venugopal, N., Upadhyaya, M.: Effect of laser power on microstructure and tensile properties of pulsed Nd:YAG laser beam welded AISI 301 austenitic stainless steel joints. Mater. Today **37**, 934–939 (2021). https://doi.org/10.1016/j.matpr.2020.06.145
21. Aghaie-Khafri, M., Honarvar, F., Zanganeh, S.: Characterization of grain size and yield strength in AISI 301 stainless steel using ultrasonic attenuation measurements. J. Non-destr. Eval. (2012). https://doi.org/10.1007/s10921-012-0134-z
22. Chen, L., et al.: Investigation on the hot deformation behavior of 316L stainless steel using 3D processing map. Trans. Indian Inst. Met. **72**(12), 2997–3006 (2019). https://doi.org/10.1007/s12666-019-01674-4

A Robust Method to Determine True Stress–true Strain Curves for Sheet Metals at High Temperatures

Ruiqiang Zhang(✉), Siyi Chen, and Jianguo Lin

Department of Mechanical Engineering, Imperial College London, London SW7 2AZ, UK
r.zhang17@imperial.ac.uk

Abstract. Thermo-mechanical uniaxial tensile tests on sheet metals have been widely carried out to characterize the mechanical behavior of the materials at high temperatures. Because of nonuniform temperature along gauge length, however, nonuniform deformation occurs from the beginning of deformation, leading to challenges to determine stress–strain curves. In the present study, a robust method for determining true stress–true strain curves has been proposed; in this method, true strains are obtained within a local area in which fracture occurs, and true stresses are computed based on the engineering strains within this local area. Uniaxial tensile tests on aluminum alloy AA6082 sheets under hot stamping conditions have been carried out, with the strain measurement using the digital image correlation (DIC) technique. The onset of necking in those tests has been determined using the spatio-temporal method. Then the data before the necking have been analyzed and the robust method has been applied to determine the associated true stress–true strain curves. The true stresses computed using the robust method have also been benchmarked against those computed using the standard method. It concludes that this robust method is able to accurately determine the true stress–true strain curves at high temperatures.

Keywords: Hot stamping · Uniaxial tensile test · True stress–true strain curve · Necking · Temperature nonuniformity

1 Introduction

To overcome the low formability of sheet metals in cold forming, advanced warm or hot forming technologies such as hot stamping of steels [1] and aluminum alloys [2], have been developed. Uniaxial tensile tests at high temperatures have been commonly conducted to determine materials thermo-mechanical properties [3]. In those tests, however, the temperature distribution in the gauge area maybe nonuniform [4]. For example, in uniaxial tensile tests using the Gleeble thermal-mechanical simulation system, temperature usually decreases along specimen length direction with increasing distance from the specimen center due to the heat loss to the specimen grips [5]. This results in nonuniform deformation along the specimen gauge length from the beginning of the deformation. Therefore, the engineering stress–engineering strain curves for materials

K. Mocellin et al. (Eds.): ICTP 2023, LNME, pp. 525–532, 2024.
https://doi.org/10.1007/978-3-031-40920-2_54

obtained from the tests are considerably dependent on the predefined gauge lengths [6]. Due to the nonuniformity of the deformation along the gauge length, it is also challenging to determine true stress–true strain curves uniquely.

In the present study, a robust method for determining the true stress–true strain curves for sheet metals at high temperatures has been proposed. Uniaxial tensile tests on AA6082 sheets under hot stamping conditions have been carried out, and the strain fields in the gauge area have been obtained and analyzed. The robust method has been applied to determine the true stress–true strain curves for the material.

2 Methodology

Aluminum alloy AA6082 sheets with a thickness of 1.5 mm are used in the present study. Figure 1a shows the temperature cycle of the material that mimics industrial hot stamping. As can be seen, the material is heated to 535 °C for solution heat treatment, and after soaking for 60 s at this temperature, the material is quenched at a rate of 60 °C/s to the forming temperature T_F for deformation under certain conditions, as listed in Table 1. To reproduce this temperature cycle, a Gleeble 3800 thermo-mechanical simulator system, as shown in Fig. 1b, is used, which provides electric current to heat specimens using the direct resistance heating method. Figure 1c shows the specimen geometry and dimensions; all specimens are cut along the rolling direction.

During the testing, full-field strains are measured using the digital image correlation (DIC) technique. For this purpose, random speckle patterns are generated on the specimen surfaces by painting a black background and white dots. Also, a high-speed camera (Photron FASTCAM Mini delivering 1,280 by 1,024 pixel resolution) together with the cross-polarization technique (Fig. 1b) proposed in [7], is used to record the deformation of specimens. The cross-polarization technique which uses orthogonal linear polarizers on light source and camera [8], could eliminate the shinning effects caused by the exposure of AA6082 due to tiny cracking of the patterns during deformation at high temperatures. The frame rate is 250 frames per second, and the distance of the camera from the specimens is ~2 m. The whole specimen gauge area (~80 mm) is recorded,

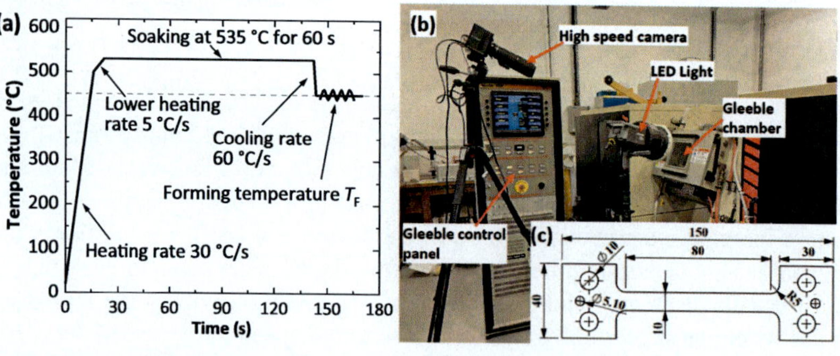

Fig. 1. Experimental setup for uniaxial tensile tests on AA6082 sheets under hot stamping conditions, (a) temperature profile for the material, (b) setup of Gleeble, high-speed camera and LED light source, and (c) geometry and dimensions (unit: mm) of dog-bone specimen.

so the pixel/millimeter ratio is ~16. All deformation images are then processed in the commercial software GOM Correlate 2018.

Table 1. Testing conditions for uniaxial tensile tests on AA6082 in the present study.

Cases	Temperature (°C)	Strain rate (s^{-1})	Effective gauge length (mm)
1	400	0.5	20
2	500	0.5	20

3 Results

3.1 Nonuniformity of Strain Fields Along Tensile Direction

Figure 2a and Fig. 2b show the major strain fields in the gauge area of specimens deformed at a forming temperature of 400 °C and 500 °C, respectively. The values 25%, 65%, 75%, 85% and 95% are the deformation times of the strain fields from the beginning of deformation, which are normalized by the time at fracture. It can be seen that the deformation localized significantly near the specimen center with increasing the deformation time, in particular at the normalized time of 95%. This is because the

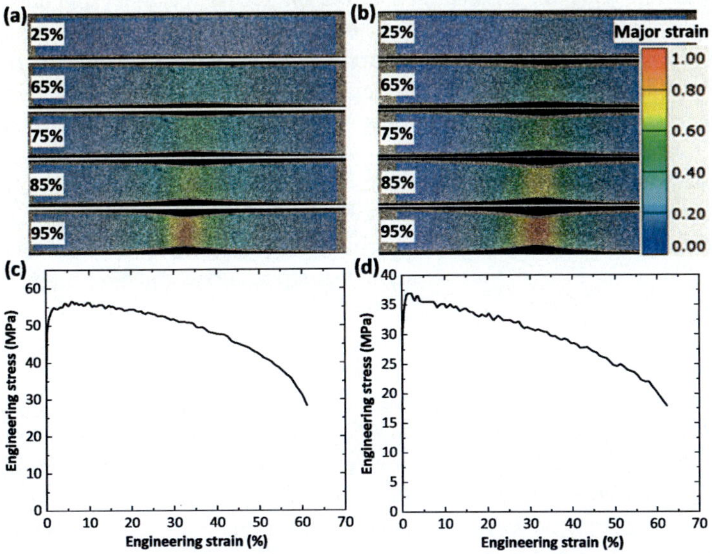

Fig. 2. Strain fields measured using the DIC technique, and the associated engineering stress engineering strain curves of the material deformed at two forming temperatures, (a) strain fields at 400 °C, (b) strain fields at 500 °C, (c) stress–strain curve at 400 °C and (d) stress–strain curve at 500 °C.

temperature distribution along the long direction is nonuniform, and the specimen center has highest temperature, leading to the localized deformation even from the beginning of deformation [6].

Figure 2c and Fig. 2d show the engineering stress engineering strain curves of the material at the forming deformation of 400 °C and 500 °C, respectively. The curves were determined using the average strains within a predefined effective gauge length of 20 mm. As can be seen, the engineering stress reaches the ultimate tensile strength (UTS) quickly from the beginning of deformation, and then decreases with increasing the deformation time. For example, in the test at 500 °C, the engineering strain at UTS is only 1.84%, while the strain at fracture is 62.37%. This indicates that the conventional method of plotting true stress–true strain curves using the data before UTS (i.e. before necking) does not work in these tests anymore. In other words, in these tests, the onset of necking cannot be simply determined as the time of UTS.

3.2 Determination of Necking Using the Spatio-Temporal Method

In the present study, the onset of necking in the tests at hot stamping temperatures was determined using the spatio-temporal method first developed by Zhang et al. [9, 10]. Figure 3a and Fig. 3b show the determination of necking for the tests deformed at 400 °C and 500 °C, respectively. A base zone (BZ) and a reference zone (RZ) were selected around the fracture initiation location, as indicated in the inset in Fig. 3b. At the beginning of deformation, along the length direction, BZ and RZ have a dimension of 3 mm ($2 \times h$) and 6 mm ($4 \times h$), respectively, where h is the initial thickness of the specimens, while along the width direction, BZ and RZ have a dimension of 6 mm. As shown in Fig. 3a and Fig. 3b, by plotting the thickness reduction within RZ and RZ, two lines (i.e. red and green dashed lines) were fitted using the thickness reduction data in the initial and later stages of each deformation. According to the intersection of two fitted lines, the normalized times at the onset of necking in the tests deformed at 400 °C and 500 °C are 83.3% and 79.6%, respectively. Figure 3c and Fig. 3d show the limit strain ($\varepsilon_2, \varepsilon_1$) at the onset of necking, and the associated strain path until fracture, where ε_2 is the minor strain and ε_1 is the major strain.

According to the onset of necking determined above, the experimental data before the necking were used to plot the true strain–true strain curves and three effective gauge lengths of 10, 20 and 30 mm (inset in Fig. 4b) were chosen for plotting those curves; true stresses are computed using Eq. (1). The results are presented in Fig. 4a and Fig. 4b for the tests conducted at 400 °C and 500 °C, respectively. As can be seen in both figures, the true stress–true strain curves are dependent on the effective gauge length, particularly the curves in the later stages of deformation. Specifically, a smaller gauge length results in higher true strains and higher true stresses. This dependency is because the strain distribution in specimen gauge length along the tensile direction is nonuniform from the beginning of each deformation, as presented in Fig. 2a and Fig. 2b. The curves in Fig. 4a and Fig. 4b present different material properties such as hardening behavior, and also affect the calibration of material constitutive models significantly. A robust method to determine true stress–true strain is therefore needed.

$$\sigma = F/A_0 \times (1 + e) \tag{1}$$

where σ is the true stress, F is the force, A_0 is the initial cross-sectional area of the specimen, and e is the engineering strain. It is worth noting that Eq. (1) is derived based on the following assumption [11]:

$$A_0 \times L_0 = A \times L \tag{2}$$

where L_0 is the initial gauge length, A is the instantaneous cross-sectional area of the specimen, and L is the instantaneous gauge length.

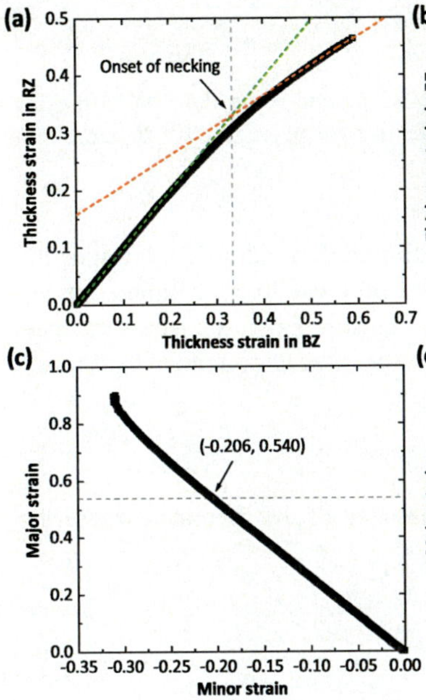

Fig. 3. Determination of the onset of necking and the limit strains at necking using the spatio-temporal method, (a) onset of necking at 400 °C, (b) onset of necking at 500 °C, (c) limit strains at 400 °C, and (d) limit strains at 500 °C.

3.3 A Robust Method Determining True Stress–true Strain Curves

In the present study, a robust method is proposed to determine true stress–true strain curves. In this method, the true strains are determined using the average strains within a local area where the fracture occurs, e.g. the BZ (Fig. 3b), and the true stresses are computed using Eq. (1); all those data are selected before the onset of necking which is determined using the spatio-temporal method [9, 10]. The reason to use the average strain within the local area is that according to previous studies [6, 12], in the tests at hot stamping temperatures using the Gleeble, both the strain rate and the temperature are nonuniformly distributed in the gauge area along the tensile direction, and only the strain rate and the temperature near the specimen center can be directly controlled to match the desired values.

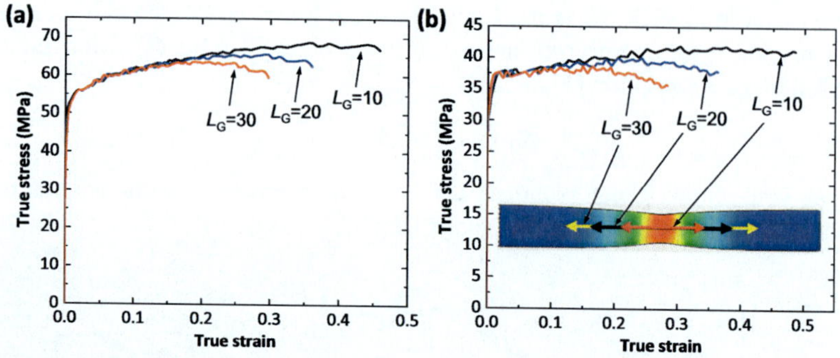

Fig. 4. True stress–true strain curves for AA6082 plotted using the data before the onset of necking and the average strains within three effective gauge lengths 10, 20 and 30 mm, for the tests conducted at (a) 400 °C and (b) 500 °C.

One may suspect that due to the nonuniform deformation near the specimen center, Eq. (2) may not be entirely accurate, which could lead to the computation of incorrect true stress data using Eq. (1). To eliminate those doubts, the true stresses computed using Eq. (1) were benchmarked against the true stresses computed using the following relationship:

$$\sigma = F/A \tag{3}$$

where A is the instantaneous cross-sectional area of the specimen where the fracture occurs (namely the specimen center).

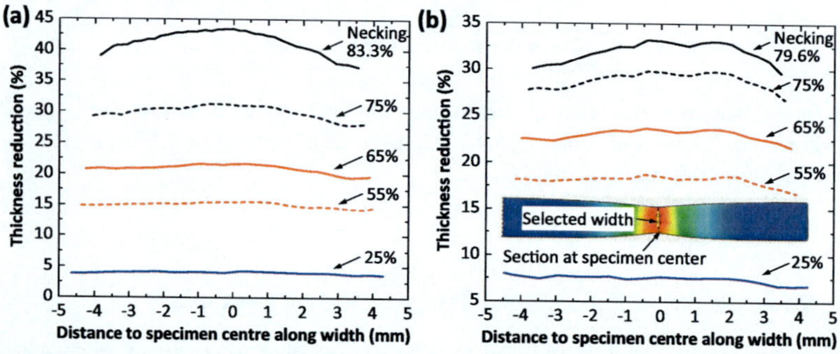

Fig. 5. Thickness reduction at the specimen center where the fracture occurs, along the specimen width direction at different normalized times before the onset of necking, for the tests conducted at (a) 400 °C and (b) 500 °C.

Figure 5a and Fig. 5b show the thickness reduction along the section (as illustrated in the inset in Fig. 5b) at the specimen center, which is computed based on volume constancy assumption, at different normalized times right before the onset of necking (e.g. 25%, 55%, and 65%) for the tests conducted at 400 °C and 500 °C, respectively. It is worth noting that in practice strain fields near the edges of the specimen gauge area cannot be measured using DIC, as demonstrated in Fig. 2a and Fig. 2b. As can be seen from Fig. 5a and Fig. 5b, the thickness reduction along the specimen width direction is almost uniform, particularly in the initial stages of each deformation. In the present study, the instantaneous cross-sectional area A was therefore computed using the following equation:

$$A = W \times H \tag{4}$$

where W is the instantaneous value of a selected width at the specimen center which is shorter than the initial width of the specimen, and H is the average instantaneous thickness of the specimen center along the selected width. In the present study, the value of W at the beginning of deformation is 6 mm (inset in Fig. 5b).

Based on the true stresses computed using Eqs. (1) and (3), and the average strains within the BZ, the true stress–true strain curves for AA6082 for the tests conducted at 400 °C and 500 °C were obtained; the results are plotted in Fig. 6a and Fig. 6b, respectively. As can be seen, the true stress values computed using the two methods are almost equal. This indicates that although deformation near specimen center before the onset of necking is nonuniform, Eq. (2) is still valid, enabling accurate true stress values to be computed using Eq. (1). Therefore, the robust method proposed in the present study can be used to determine true stress–true strain curves accurately.

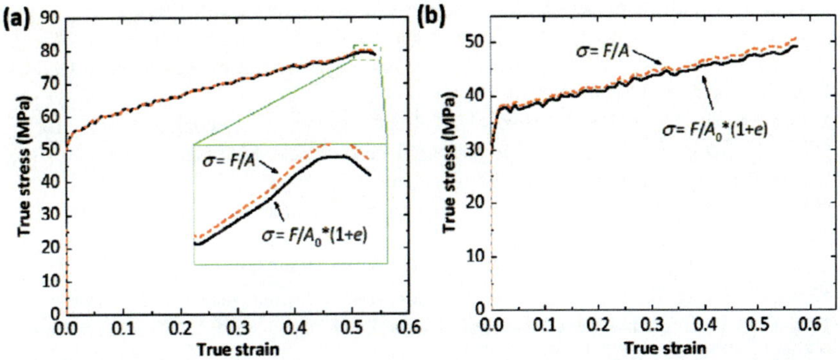

Fig. 6. True stress–true strain curves for AA6082, where the true stresses are computed using Eqs. (1) and (3), and the true strains are the average strains within the BZ, for the tests conducted at (a) 400 °C and (b) 500 °C.

4 Conclusions

In the present study, uniaxial tensile tests on AA6082 sheets under hot stamping conditions have been carried out, and the associated true stress–true strain curves have been determined by analyzing the data before the onset of necking detected using the spatio-temporal method. According to the data analyses, the following conclusions can be drawn:

1) Since the deformation in the specimen gauge area along the tensile direction is nonuniform from the beginning of the deformation due to temperature nonuniformity, true stress–true strain curves determined from the tests are significantly dependent on predefined effective gauge length.
2) A robust method for determining true stress–true strain curves for tests at high temperatures has been proposed. This method uses average strains within a local area where the fracture occurs, instead of a predefined gauge area, and computes true stress using the formula $\sigma = F / A_0 \times (1 + e)$.
3) The true stress values computed using $\sigma = F / A_0 \times (1 + e)$ have been benchmarked against those computed using $\sigma = F / A$. It has been found that the data computed using the former method are accurate, although the deformation around the local area where the fracture occurs is nonuniform.

References

1. GB1490535A, Manufacturing a hardened steel article, United Kingdom 'Patent' GB 1490535 A, (1977), 2 November 1977
2. Foster, A., Dean, T.A., Lin, J.: Process for forming aluminium alloy sheet components, United States 'Patent' US 2012/0152416 A1 (2012), 21 June 2012
3. ASTM International, Standard test methods for elevated temperature tension tests of metallic materials, ASTM International, E21–09 (2009). https://doi.org/10.1520/E0021-20
4. He, Z., Wang, Z., Lin, Y., Fan, X.: Hot deformation behavior of a 2024 aluminum alloy sheet and its modeling by fields-backofen model considering strain rate evolution, Metals **9** (2019). https://doi.org/10.3390/met9020243
5. Ganapathy, M., Li, N., Lin, J., Abspoel, M., Bhattacharjee, D.: A novel grip design for high-accuracy thermo-mechanical tensile testing of boron steel under hot stamping conditions. Exp. Mech. **58**(2), 243–258 (2017). https://doi.org/10.1007/s11340-017-0333-8
6. Zhang, R., Shao, Z., Lin, J., Dean, T.A.: Measurement and analysis of heterogeneous strain fields in uniaxial tensile tests for boron steel under hot stamping conditions. Exp. Mech. **60**(9), 1289–1300 (2020). https://doi.org/10.1007/s11340-020-00658-6
7. Chen, S., Zhang, R., Shi, Z., Lin, J.: An improved strain field measurement method using DIC and XPOL for sheet metals under hot stamping conditions (2023) (Under review)
8. LePage, W.S., Daly, S.H., Shaw, J.A.: Cross polarization for improved digital image correlation. Exp. Mech. **56**(6), 969–985 (2016). https://doi.org/10.1007/s11340-016-0129-2
9. Zhang, R., Shi, Z., Shao, Z., Dean, T.A., Lin, J.: A novel spatio-temporal method for determining necking and fracture strains of sheet metals. Int. J. Mech. Sci. **189**, 105977 (2021). https://doi.org/10.1016/j.ijmecsci.2020.105977
10. Zhang, R., Shi, Z., Shao, Z., Yardley, V.A., Lin, J.: An effective method for determining necking and fracture strains of sheet metals. MethodsX **8**, 101234 (2021). https://doi.org/10.1016/j.mex.2021.101234
11. Tu, S., Ren, X., He, J., Zhang, Z.: Stress–strain curves of metallic materials and post-necking strain hardening characterization: a review. Fatigue Fract. Eng. Mater. Struct. **43**, 3–19 (2019). https://doi.org/10.1111/ffe.13134
12. Zhang, R., Jiang, J., Lin, J., Yardley, V.A.: Investigation of variability in apparent values of materials properties in thermo-mechanical uniaxial tensile tests on sheet metals. J. Manuf. Processes **101**, 737–754 (2023). https://doi.org/10.1016/j.jmapro.2023.06.005

A Framework for Analytical Cup Height Computation in Multi-stage Deep Drawing

Holger Aretz(✉)

Speira GmbH, Research and Development, Georg-von-Boeselager-Str. 21, 53117 Bonn, Germany
holger.aretz@speira.com

Abstract. In the present work, an analytical framework for the computation of the final height of cylindrical cups produced by multi-stage deep drawing operations is proposed. Its capabilities of making accurate predictions are demonstrated by means of an application example.

Keywords: Sheet Forming · Cup Drawing · Anisotropy

1 Introduction

In order to overcome the problem of the maximum allowable drawing ratio in deep drawing operations, it is common practice to distribute the total forming over several sub-critical forming operations. In other words, single-stage forming is replaced by multiple consecutive forming operations. In this way, large deformations and complex final product geometries can be realized. However, due to economical reasons it is desirable to minimize the total number of involved forming operations. As a prominent example, the manufacturing of classical 330 ml aluminium beverage cans, possessing a form factor of (height h/radius R) ≈ 3.4, involves two consecutive cup forming operations, viz. drawing and redrawing (see e.g. Hosford & Duncan [5]). Other examples are cylindrical battery housings: the recently introduced cell format "4680" possesses the same form factor like a classical 330 ml beverage can, whereas the form factor of the cell format "18650" is even $h/R \approx 7$.

In the present work, an analytical framework for the computation of the final cup height in multi-stage deep drawing is proposed. Although the developed framework is sufficiently general to be readily used in a broader field of applications, considerations in this work are confined to the drawing of *cylindrical* cups.

Some notational conventions used in the present work are as follows. The symbols R_b, R_p, R_{pp}, R_d, R_{dp}, t_0 denote the blank radius, the punch radius, the punch profile radius, the die opening radius, the die profile radius and the initial sheet thickness, respectively. The drawing ratio is $\beta = R_b/R_p$. The angle relative to the rolling direction is φ. Unless stated otherwise, h and R denote the height and radius of the respective geometric entities. Symbols not mentioned here will be explained in the text.

K. Mocellin et al. (Eds.): ICTP 2023, LNME, pp. 533–540, 2024.
https://doi.org/10.1007/978-3-031-40920-2_55

2 Geometric Approach for Cup Height Computation

The most important assumptions made in the present work are: a) isochoric and rigid-plastic material behavior; b) forming occurs frictionless. The point of departure is a multi-stage deep drawing process to produce cylindrical cups, as depicted in Fig. 1.

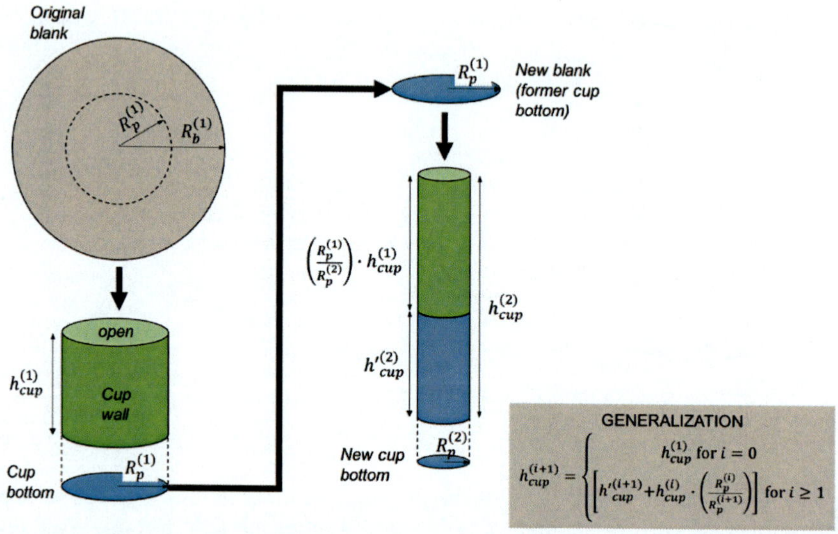

Fig. 1. Geometric derivation of the framework for cup height calculation in multi-step drawing operations.

In the *first* cup drawing operation (cf. Fig. 1) the originally round blank is formed to a cup, which resembles a tube with one closed end. In order to understand what happens in the *second* cup drawing operation (i.e. the redraw) it is helpful to conceptually cut the cup obtained from the first drawing into two parts, more specifically into the cup bottom and the cup wall. Both parts are considered *separately* in the second drawing operation. For the second drawing, the former cup bottom becomes the new blank which is drawn to a new cup. Simultaneously, the former cup wall, which is now a double-sided open tube, is essentially reduced in its diameter and stretched in axial direction, according to the drawing ratio of the second operation (assuming no change of wall thickness). Let $V_{tube}^{(i)} \approx (2\pi \cdot R_{tube}^{(i)} \cdot t_0) \cdot h_{tube}^{(i)}$ be the (approximate) volume of the thin-walled tube pertaining to stage no. i. Volume constancy states that $V_{tube}^{(i)} = V_{tube}^{(i+1)}$, which gives

$$h_{tube}^{(i+1)} = h_{tube}^{(i)} \cdot \left(R_{tube}^{(i)} / R_{tube}^{(i+1)} \right) \tag{1}$$

After the second drawing, the final cup height simply amounts to the height or length of the deformed open tube plus the cup height that results from the solo-drawing of the former flat cup bottom to a cup. For illustration, the verbally described process is cast into the sketch shown in Fig. 1.

The process outlined in Fig. 1 may be generalized to an arbitrary number of drawing operations. The associated generalized scheme to calculate the cup height, h_{cup}, reads as follows (cf. Fig. 1):

- For the *first* drawing stage:

$$h_{cup}^{(1)} = h_{cup}^{\prime(1)}, \quad \beta^{(1)} = R_b^{(1)} / R_p^{(1)} \tag{2}$$

- For the *subsequent* drawing stages ($i \geq 1$):

$$h_{cup}^{(i+1)} = h_{cup}^{\prime(i+1)} + h_{cup}^{(i)} \cdot \beta^{(i+1)}, \quad \beta^{(i+1)} = R_p^{(i)} / R_p^{(i+1)} \tag{3}$$

In the present work, the cup height h'_{cup} is predicted using the analytical model summarized in the next section.

The formulation given above is only first-order accurate. A more detailed look on the outlined situation reveals that the tube analogy should be confined to the *cup wall*, whose height is $h'_{wall} = h'_{cup} - R_{pp} - t_0$, which has been done in the present work. The previous approach may be extended to cup drawing processes involving *ironing*, which is very common in industrial manufacturing.

3 Analytical Cup Height Model

In the present work, the analytical earing/cup height model of van Houtte *et al.* [7] was used. Its predictive capabilities using both circular and non-circular blanks were demonstrated in [2,3]. Unfortunately, due to space limitations a detailed derivation of the model must be omitted here, which can be found in [3]. Compared to the original formulation in [2–4], two improvements were made to the earing model: a) a new definition of the stress ratio s has been derived; b) a new equation to calculate the punch profile radius contribution to the cup height has been derived.

According to the earing model of van Houtte *et al.* [6,7] the cup height is

$$h_{cup} = t_0 + \Delta h_{corner} + \Delta h_{flange} \tag{4}$$

with the flange contribution

$$\Delta h_{flange} = \frac{R_b^{q(s)+1} - (R_p + t_0)^{q(s)+1}}{(q(s)+1) \cdot \left(R_p + \frac{t_0}{2}\right)^{q(s)}} \tag{5}$$

and a contribution due to the presence of a rounded punch corner

$$\Delta h_{corner} = \left(2 - \frac{\pi}{2}\right) \cdot \left(R_{pp} + \frac{t_0}{2}\right) \tag{6}$$

Assuming *proportional straining* in the flange, the strain ratio $q(s)$ can be derived as (cf. [2,3])

$$q(s) = -\left(\frac{\partial \bar{\sigma}}{\partial \sigma_{rr}} \Big/ \frac{\partial \bar{\sigma}}{\partial \sigma_{tt}}\right)\Bigg|_{\sigma_{rr}=s,\sigma_{tt}=1} \tag{7}$$

Herein, $\bar{\sigma}$, σ_{rr} and σ_{tt} are the equivalent stress, the radial and the tangential stress components, respectively. If the material is anisotropic, then $q(s)$ as well as the cup height vary with the angular coordinate φ. The stress ratio $s = \sigma_{rr}/\sigma_{tt}$ is not known *a priori* and depends on the drawing ratio [6,7]. Based on a purely theoretical analysis assuming isotropic material, van Houtte *et al.* [7] derived the following relationship for the *local* stress ratio $\hat{s}(r)$:

$$\hat{s}(r) = -(R_b/r - 1), \quad R_p \leq r \leq R_b \tag{8}$$

The *average* stress ratio in the flange of the cup during the drawing process is obtained by averaging Eq. (8) using the mean-value theorem:

$$s(\beta) = \frac{1}{R_b - R_p} \cdot \int_{R_p}^{R_b} \hat{s}(r)\, dr = 1 - \left(\frac{\beta}{\beta - 1}\right) \cdot \ln(\beta) \tag{9}$$

This stress ratio gives similar results to the one used in [2–4], but possesses the beauty of being derived based on a closed-form theory.

In the *original* model formulation Δh_{corner} was calculated based on excessive *length* in the corner region, as explained in detail by Schouwenaars *et al.* [6]. A more consistent approach is to consider excessive *volume* instead of length. This analysis involves Guldin's rule to calculate the volume of the rounded corner region. Its final result is:

$$\Delta h_{corner} = \left(2 - \frac{\pi}{2}\right) \cdot \left(R_{pp} + \frac{t_0}{2}\right) + \frac{\left(R_{pp} + \frac{t_0}{2}\right)^2}{R_p} \cdot \left(\frac{\pi}{2} - \frac{3}{2}\right) \tag{10}$$

As one may see, the first term is identical to the one in the original model, but a second additive term appears, which results in a slightly higher cup than the original model predicts. However, in the present work the second term provides only a negligible contribution, which may be readily verified by the reader using the considered process data summarized in Table 1.

4 Validation Example

4.1 Input Data and Constitutive Model

In order to validate the proposed modeling approach, it is applied to the cup drawing and the subsequent redrawing step of the Numisheet 2016 benchmark [8]. All the material and process data used in the present work were adopted from the benchmark description of Watson *et al.* [8] and are compiled in Table 1.

The considered material (cf. Table 1) is a strongly anisotropic aluminium sheet made of the alloy AA5352 with unknown temper condition. Since the equi-biaxial yield stress Y_b was not explicitly mentioned in [8], reverse engineering was employed using the parameters for the Hill48 and Yld89 yield functions provided in [8]. This revealed a ratio of the equi-biaxial yield stress to the yield stress in uniaxial tension in rolling direction of $Y_b/Y_{0°} \approx 1.141$.

Table 1. Material and process data of the Numisheet 2016 benchmark problem. All yield stresses (labeled as $Y_{\square}$) are in MPa, while the r-values (labeled as $r_{\square}$) are dimensionless. All geometric quantities pertaining to the process data (die radii and sheet thickness t_0) are in mm. Except of Y_b, all data were adopted from Watson *et al.* [8].

Material data of AA5352							
$Y_{0°}$	$Y_{15°}$	$Y_{30°}$	$Y_{45°}$	$Y_{60°}$	$Y_{75°}$	$Y_{90°}$	Y_b
197.59	193.39	196.62	193.94	197.03	196.89	198.74	(225.51)
$r_{0°}$	$r_{15°}$	$r_{30°}$	$r_{45°}$	$r_{60°}$	$r_{75°}$	$r_{90°}$	r_b
0.535	0.465	0.655	1.105	1.415	1.595	2.270	0.620

Process data						
	R_b	t_0	R_p	R_{pp}	R_d	R_{dp}
Cup drawing	81.4832	0.2790	58.4200	4.0640	58.7225	2.5400
Redraw	58.4200	0.2790	43.9295	4.0640	44.2445	2.5400

The yield surface of the considered material was described using the yield function "Yld2011-27p" proposed by Aretz & Barlat [1]. Its parameters were fitted to the material data in Table 1 by means of non-linear optimization. The yield function exponent and reference yield stress were selected as $m = 12$ and $Y_{\text{ref}} \equiv Y_{0°}$, respectively. The calculated Yld2011-27p parameters are documented in Table 2 and Fig. 2 shows the corresponding yield surface.

Table 2. Computed Yld2011-27p parameters for AA5352, using an exponent of $m = 12$.

Parameters C'_{ij}		
$C'_{12} = -0.207580\text{E}+01$,	$C'_{13} = -0.644787\text{E}+00$,	$C'_{21} = 0.405947\text{E}+00$,
$C'_{23} = -0.152111\text{E}+01$,	$C'_{31} = 0.155244\text{E}+01$,	$C'_{32} = 0.153879\text{E}+01$,
$C'_{44} = 0.100000\text{E}+01$,	$C'_{55} = 0.100000\text{E}+01$,	$C'_{66} = -0.130234\text{E}+01$
Parameters C''_{ij}		
$C''_{12} = 0.205928\text{E}+01$,	$C''_{13} = 0.222635\text{E}+01$,	$C''_{21} = -0.147313\text{E}+01$,
$C''_{23} = -0.341526\text{E}+00$,	$C''_{31} = -0.795468\text{E}-01$,	$C''_{32} = -0.177653\text{E}+00$,
$C''_{44} = 0.100000\text{E}+01$,	$C''_{55} = 0.100000\text{E}+01$,	$C''_{66} = 0.952689\text{E}+00$
Parameters C'''_{ij}		
$C'''_{12} = -0.673947\text{E}+00$,	$C'''_{13} = 0.113179\text{E}+01$,	$C'''_{21} = -0.167562\text{E}+01$,
$C'''_{23} = 0.637574\text{E}+00$,	$C'''_{31} = -0.109894\text{E}+01$,	$C'''_{32} = 0.110714\text{E}+01$,
$C'''_{44} = 0.100000\text{E}+01$,	$C'''_{55} = 0.100000\text{E}+01$,	$C'''_{66} = -0.143155\text{E}+01$

The associated predictions in terms of characteristic uniaxial and equi-biaxial yield stresses as well as related r-values are documented in Table 3. The quality of the predictions made by Yld2011-27p is visualized in Fig. 3, which can be judged to be very good. As an alternative quality measure one may use the root mean square error, RMS,

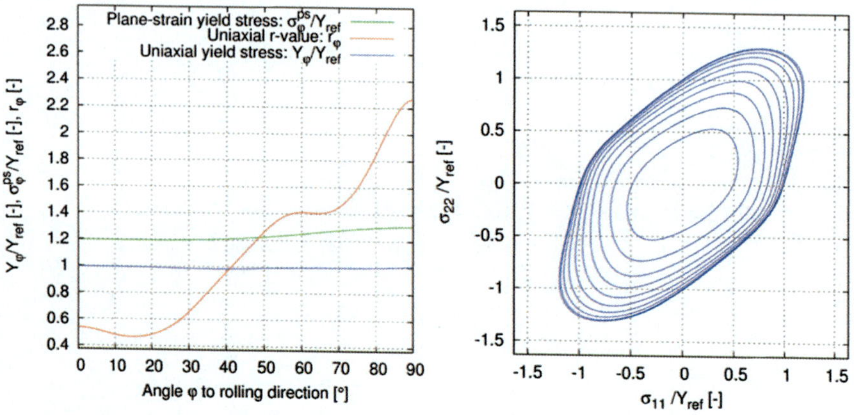

Fig. 2. Yld2011-27p predictions for AA5352: directional yield stresses (plane-strain and uniaxial) as well as r-values (left); yield surface in iso-level representation (i.e. levels of $\sigma_{12} = \text{const.}$) (right). Stresses are normalized by the reference yield stress $Y_{\text{ref}} \equiv Y_{0^\circ}$.

Table 3. Characteristic data predicted by the utilized yield function Yld2011-27p. Stresses were normalized according to the definition $Y'_\square \equiv Y_\square / Y_{0^\circ}$.

Normalized yield stresses							
Y'_{0°	Y'_{15°	Y'_{30°	Y'_{45°	Y'_{60°	Y'_{75°	Y'_{90°	Y'_b
0.993	0.992	0.985	0.986	0.996	0.998	1.005	1.141
r-values							
r_{0°	r_{15°	r_{30°	r_{45°	r_{60°	r_{75°	r_{90°	r_b
0.534	0.466	0.653	1.110	1.412	1.596	2.266	0.620

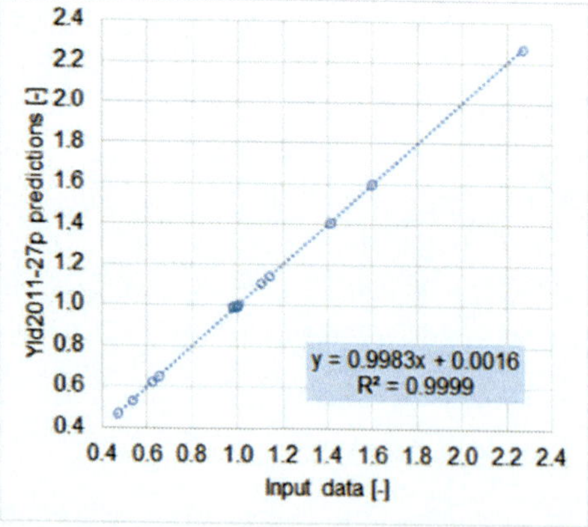

Fig. 3. Comparison of the Yld2011-27p predictions with the provided input data (including all yield stresses and r-values) from Table 1. All stress quantities were normalized with respect to the uniaxial yield stress in rolling direction.

defined here as RMS $\equiv \sqrt{(\Delta \mathbf{X}^T \cdot \Delta \mathbf{X})/N}$, with $\Delta \mathbf{X} \equiv \mathbf{X}_{pred} - \mathbf{X}_{meas}$. Herein, $\mathbf{X}_{pred}, \mathbf{X}_{meas} \in \mathbb{R}^N$ denote the column matrices storing the N predicted and measured data, respectively. Using stress values normalized by $Y_{0°}$, Yld2011-27p displays a dimensionless error of RMS $= 5.051\text{E-}03$ for the considered material, which again suggests a very good quality of its predictions.

4.2 Results and Discussion

The results after drawing and redrawing are shown in Fig. 4. The predicted *average cup height* was calculated using the mean-value theorem according to

$$\bar{h}_{pred} = \frac{1}{2\pi} \cdot \int_0^{2\pi} h(\varphi)\, d\varphi \tag{11}$$

Its value after redrawing is $\bar{h}_{pred} = 52.79\,\text{mm}$, which can be compared to the measured value of $\bar{h}_{meas} = 52.05\,\text{mm}$ as documented in the benchmark results report [8, p. 42]. The associated error is $|\bar{h}_{pred} - \bar{h}_{meas}|/\bar{h}_{meas} \cdot 100\,\% \approx 1.42\,\%$, which permits the judgment that the model prediction agrees fairly well with the experimental result. Furthermore, Fig. 4 shows that eight ears after redrawing are predicted, which is in agreement with the benchmark results [8, p. 42]. Figure 4 does also reveal that the earing profile characteristics between drawing and redrawing remain unchanged. The predicted *amplitude* after redrawing is $A_h \equiv h_{max} - h_{min} = (54.05 - 51.40)\,\text{mm} = 2.65\,\text{mm}$, whereas the benchmark report mentions a measured value of $A_h = 4.13\,\text{mm}$ [8, p. 42]. This discrepancy may be explained by the fact that the benchmark report documents *pinching* of the largest ears [8, p. 42], which is associated with an additional stretching of the affected areas, thereby biasing the amplitude value to an unknown extent. In practice, pinching can be avoided by reducing the blankholder force at the end of the drawing.

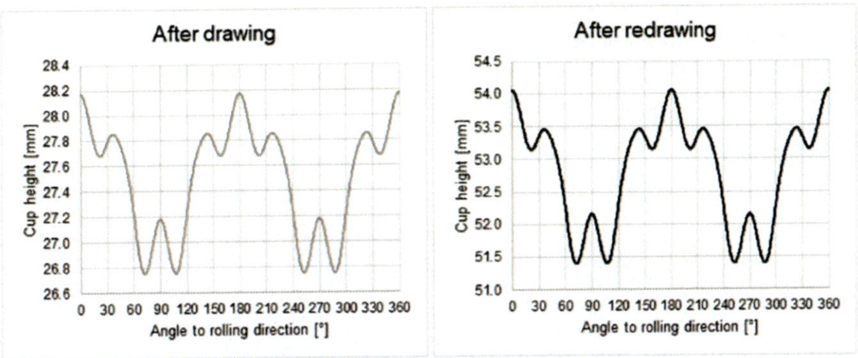

Fig. 4. Cup height profile after drawing (left) and after redrawing (right). The average cup heights are 27.50 mm and 52.79 mm, respectively.

5 Conclusions

Based on geometric considerations a framework for cup height calculation in multi-stage deep-drawing was developed. Application of the proposed framework to the Numisheet 2016 benchmark revealed very good predictive capabilities at very low and hardly measurable computation times. In practice, the developed model is particularly useful to obtain accurate estimates e.g. for blank design or stage layout, which may be subsequently refined using more sophisticated, but computationally expensive simulation tools like the finite element method.

Acknowledgements. This work is dedicated to Professor Frédéric Barlat, who was always a great inspiration to me, on the occasion of his retirement. I would also like to express my gratitude and respect to the organizers and the courageous participants of the Numisheet 2016 benchmark, respectively.

References

1. Aretz, H., Barlat, F.: New convex yield functions for orthotropic metal plasticity. Int. J. Non-Linear Mech. **51**, 97–111 (2013)
2. Aretz, H., Engler, O.: Accuracy assessment of analytical earing models. Eur. J. Mech. A Solids **78**, 103839 (2019)
3. Aretz, H., Engler, O.: Accuracy analysis of earing compensation procedures. Int. J. Solids Struct. **191–192**, 418–433 (2020)
4. Engler, O., Kalz, S.: Simulation of earing profiles from texture data by means of a visco-plastic self-consistent polycrystal plasticity approach. Mater. Sci. Eng. A **373**, 350–362 (2004)
5. Hosford, W.F., Duncan, J.L.: The Aluminum Beverage Can. Scientific American, pp. 48–53 (1994)
6. Schouwenaars, R., van Houtte, P., van Bael, A., Winters, J., Mols, K.: Analaysis and prediction of the earing behaviour of low carbon steel sheet. Textures Microstruct. **26–27**, 553–570 (1996)
7. van Houtte, P., Clarke, P., Saimoto, S.: A quantitative analysis of earing during deep drawing. In: Morris, J.G., Merchant, H.D., Westerman, E.J., Morris, P.L. (eds.) Aluminum Alloys for Packaging, pp. 261–273. The Minerals, Metals & Materials Society (1993)
8. Watson, M., Dick, R., Huang, Y.H., Lockley, A., Cardoso, R., Santos, A.: Benchmark 1 - failure prediction after cup drawing, reverse redrawing and expansion – Part A: benchmark description. J. Phys. Conf. Ser. **734**, 022001 (2016). https://doi.org/10.1088/1742-6596/734/2/022001

Computing Sheet Rolling Instabilities with a Shell Finite Element Model

A. Cometa(✉), H. J. M. Geijselaers, J. Havinga, and A. H. van den Boogaard

Faculty of Engineering Technology, Chair of Nonlinear Solid Mechanics, University of Twente, 7500 AE Enschede, The Netherlands
a.cometa@utwente.nl

Abstract. In this paper a novel model for 3D finite element sheet metal rolling calculations is presented. A global model which represents the behavior and stress state of the strip outside the roll bite is coupled to a local model which represents in detail the mechanics of deformation within the roll bite. Shell finite elements are used for the global model of the rolled sheet, while 2D plane strain elements are used for the local model of the strip and the roll. The coupling is made via an equivalent roll bite model, incorporated into the shell model to represent the physics of the roll bite. The rolling velocity, the zero out-of-plane position and the thickness strain are enforced at the roll line via a set of constraint equations. The amount of prescribed thickness reduction is determined based on the local tensions and friction coefficient. A metamodel which provides the relation between these conditions and the local thinning is obtained from 2D off-line rolling calculations. The proposed model can be used instead of a full 3D rolling model, as it is computationally less expensive, especially for thin strip rolling simulations. It is shown how the developed model can be applied to analyze instability phenomena in cold rolling processes, specifically strip buckling due to disruptions in the process conditions.

Keywords: Thin strip rolling · Steel sheets · Flatness defects

1 Introduction

3D finite element models of strip rolling are used to simulate the deformation and stress distribution in a rolling process. With these models sensitivity analyses can be performed to evaluate the effect of changes in the process settings on the rolling behavior. Sensitivity analyses can help to identify critical parameters that affect the stability of the rolling process and the quality of the final products. For example, 3D FEM models are a very useful tool to investigate the occurrence of flatness defects in rolling processes [1,7,8,13]. Flatness defects in sheet rolling are caused by inhomogeneous deformations in the roll bite across the width of the sheet. These can be caused by a number of factors, including variations in the roll gap (thermo-elastic roll deformation) and variations in the material properties of the sheet. Inhomogeneous deformations can result in

K. Mocellin et al. (Eds.): ICTP 2023, LNME, pp. 541–552, 2024.
https://doi.org/10.1007/978-3-031-40920-2_56

non-uniform stress distributions within the sheet, with local compressive stresses which can induce buckling [5]. These phenomena can lead to a variety of defects in the rolled sheet, such as waviness (e.g. center buckles and wavy edges [7,10]), wrinkles [1], or pinching [3,4]. Pinching in sheet metal rolling refers to the occurrence of defects, that appear as repetitive ripples, waviness, surface marks and local ruptures. Pinching can also cause damage to the rolling mill itself, leading to downtime and increased maintenance costs. It was shown experimentally that pinching defects can be induced by non-uniform frictional conditions between the sheet and the rolls, due to disruptions in the lubrication during rolling [4]. In an industrial rolling mill, there are several potential causes of disruptions in the lubrication, that result in a change of the frictional conditions between the strip and the roll. These can include inadequate lubricant application or distribution, changes in the viscosity or composition of the lubricant [3], or contamination of the lubricant with foreign particles [14]. It was observed that due to disturbances in the lubrication conditions buckles may appear in front of the roll bite. These buckles get trapped between the rolls, resulting in folds in the metal, which will finally appear as repetitive ripples. To capture the mechanism of formation of flatness defects, both a detailed local model of the roll bite and a global model of the strip are needed. Moreover, the conditions outside the roll bite (stress distribution and onset of instabilities) have a significant interaction with the conditions within the roll bite. Therefore, it is crucial to ensure the coupling between the conditions inside and outside the roll bite [1,2]. However, it is computationally very expensive to perform 3D rolling simulations based on conventional FEM models. The high computational cost of 3D FEM rolling models is due to the size of the simulation domain, which must cover the full thickness and width and enough length of the rolled sheet, and to the resolution of the mesh, which needs to be high in the roll bite, where high gradients of deformation occur. This leads to very large models, especially for thin strip rolling [6,12].

To reduce the computational cost of 3D rolling simulations a novel strategy is presented in this paper. The sheet is modeled using shell finite elements. In this shell model the mechanical interaction between the strip and the work rolls is reproduced via an equivalent roll bite model. A set of constraints is applied to the shell finite elements crossing the roll line, to enforce the rolling velocity, the zero out-of-plane position and the thickness reduction. A metamodel is used to couple the prescribed thickness reduction to the local tensions and friction coefficient. By using the equivalent roll bite model, sheet rolling calculations can be performed more efficiently compared to full 3D FEM simulations. The proposed method is applied to model instability phenomena which arise from disruptions in the lubrication conditions during cold rolling. It is shown that, by using the equivalent roll bite model for shell rolling calculations, it is possible to simulate flatness defects in sheet metal rolling. Various scenarios are analyzed to study the behavior of the rolled sheet when different types of disruptions in the lubrication conditions are applied.

2 Equivalent Roll Bite Model

A graphical overview of the equivalent roll bite model is given in Fig. 1. A global shell finite element model of the strip is considered in the x–y plane, being x the rolling direction, as shown in Fig. 1a. Three-dimensional 3-node Discrete Kirchhoff Triangles are used for the finite element discretization. In the global model, the strip is subjected to the entry and exit tension forces, T_{in} and T_{out}. A roll line is defined parallel to the y-axis. A numerical algorithm is used to enforce a set of constraint conditions to the shell elements crossing the roll line. These constraints allow to reproduce in the global model the mechanics of the roll bite, as described in Sect. 2.1. The strip is seen as a collection of adjacent slices. Each slice is represented as a plane-strain problem, using a local 2D ALE model, as shown in Fig. 1b. Input parameters for the 2D model are: the geometrical parameters (roll radius R, roll center distance D and strip's entry thickness H), the roll and strip material parameters and the local conditions occurring in each slice, that are the specific entry and exit tensions, t_{in} and t_{out}, and the friction coefficient μ. The global and the local models are coupled to ensure a continuous update of the conditions inside and outside the roll bite, as explained in Sect. 2.2.

2.1 Constraint Conditions

The finite element rolling simulation is performed using the incremental analysis, based on the Updated Lagrangian method. The conditions of rolling velocity, zero out-of-plane position and thickness strain are enforced to the elements crossing the roll line in the current time increment n. The configurations of the shell element at the beginning and at the end of the calculation step n are shown in Fig. 2.

Rolling Velocity. At the beginning of the current step, for any element passing over the roll line, the intersection points, P_1 and P_2, are first identified. The incremental x-displacements of those points can be expressed as a function of the incremental nodal x-displacements $\{\mathbf{d}_x\}$ via the shape functions matrix $[\mathbf{N}]$. At the intersection points the rolling velocity d_{R} is prescribed. The constraint equation to be satisfied is:

$$\mathbf{c}_{\mathrm{R}}(\{\mathbf{d}_x\}) = [\mathbf{N}]\{\mathbf{d}_x\} - d_{\mathrm{R}}\{\mathbf{1}\} = \mathbf{0} \tag{1}$$

The rolling velocity constraint equation corresponds to two linearly independent scalar equations, one for each intersection point.

Zero Out-of-Plane Position. The zero out-of-plane position is enforced at the roll line, which allows to restrain the strip's z-displacement due to the presence of the rolls. The current z-position of the two intersection points, represented by the vector $\{\mathbf{z}_{\mathrm{P}}\}$, can be obtained by interpolating the current nodal z-coordinates via the shape functions. The constraint equation to be satisfied is:

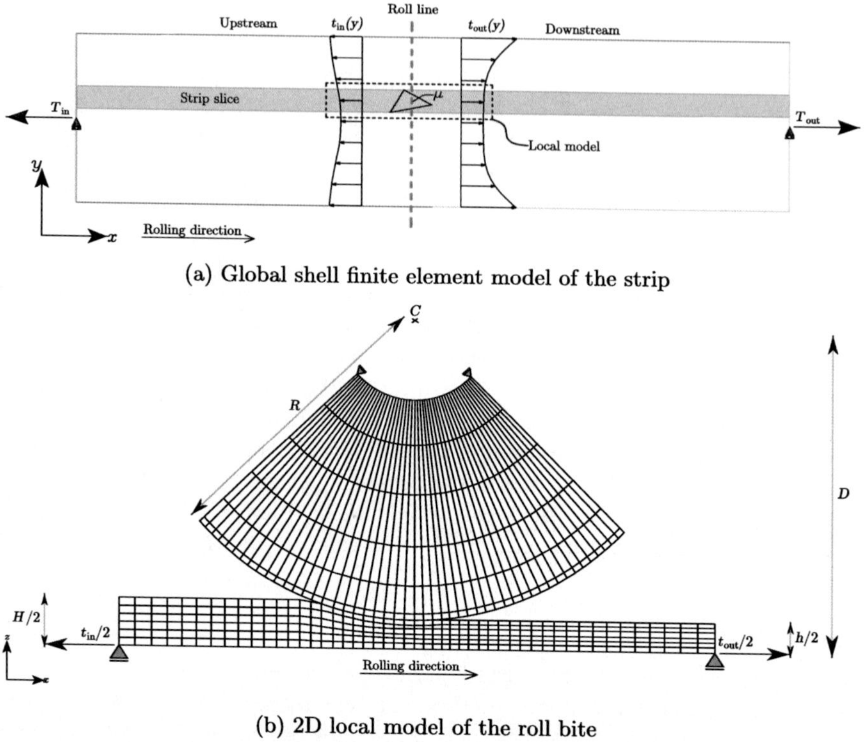

Fig. 1. Overview of the equivalent roll bite model.

$$\mathbf{c}_z(\{\mathbf{d}\}) = \{\mathbf{z}_P\}(\{\mathbf{d}\}) = \mathbf{0} \tag{2}$$

where $\{\mathbf{d}\}$ is the set of nodal incremental displacements in x, y, and z directions.

Thickness Strain. The thickness strain constraint allows to enforce a certain thickness strain ε_{zz} by changing the element area. The area of the shell element crossing the roll line is increased by an amount which is proportional to the fraction of the element area that is rolled in the current step, A^n_{rol}, and to the area strain, ε^n_{A}. The latter one can be defined as function of the thickness strain. Here, for simplicity, the area strain is approximated as:

$$\varepsilon^n_{\text{A}} \approx -\varepsilon^n_{zz} \tag{3}$$

The constraint equation to be satisfied is:

$$c_{\text{A}}(\{\mathbf{d_{xy}}\}) = A^{n+1}(\{\mathbf{d_{xy}}\}) - [A^n + A^n_{\text{rol}} \cdot \varepsilon^n_{\text{A}}] = 0 \tag{4}$$

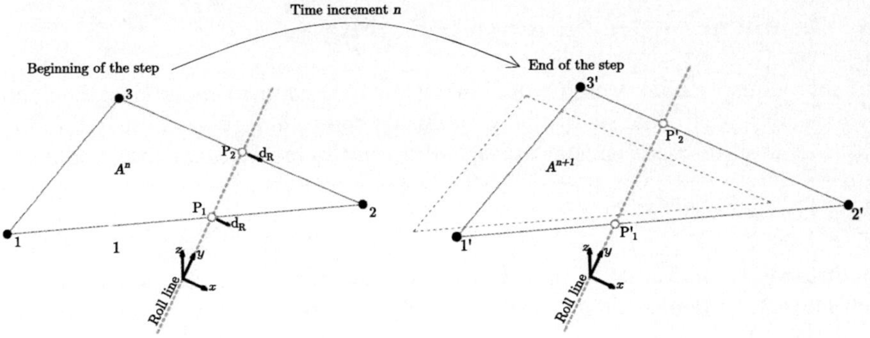

Fig. 2. Shell element crossing the roll line during the current time increment n.

where $\{\mathbf{d_{xy}}\}$ is the set of incremental nodal displacements in x and y directions. The prescribed change of the element area results in both a longitudinal strain (strain in the rolling direction) and a transversal strain. This allows to account for material later flow during rolling.

To enforce the constraints given by Eqs. (1), (2) and (4), the penalty method is used. For any element currently crossing the roll line, this results in a modified stiffness matrix and force vector. In this way, a modified global system of equations is solved at each time increment.

2.2 Metamodel

The plastic deformations occurring in the roll bite affect the in-plane stress state outside the roll bite and vice-versa. Therefore, to achieve accurate results, the reciprocal influence between the global and the local mechanical behavior of the rolled sheet must be accounted for. This is of particular importance when modeling the effect of disruptions on the rolling behavior. When a local disturbance occurs during the process (e.g. variations in the lubrication conditions, in the profile of the strip or the roll, or in the material properties), the local deformation of the strip being rolled changes. This results in a stress re-distribution outside the roll bite, which will in turn affect the in-bite deformations. Therefore, in the proposed model, the thickness strain prescribed at the roll line is obtained as a function of the local tensions and friction coefficient. The coupling between in-bite and out-of-bite conditions is enabled via a metamodel, which is built from off-line rolling calculations, using a refined 2D ALE model of the roll bite, as shown in Fig. 1b. A set of rolling simulations is performed for a specific rolling process (stand geometry, strip entry thickness and material properties), considering a range of different entry and exit tensions and different friction coefficients. The results of these simulations are used to define the thickness strain as function of the local tensions and friction coefficient.

3 Application to Thin Strip Rolling

The developed model is used to perform thin strip rolling simulations. The aim is to analyze the behavior of the strip during the rolling process, including the occurrence of instabilities due to disruptions in the lubrication conditions.

Model Input. A sheet of low carbon packaging steel is rolled, undergoing a thickness reduction of 22.35% (at the strip center line). The input parameters for the model are summarized in Table 1. An isotropic elasto–plastic material model with Swift hardening is used for the strip. The yield stress is expressed as $\sigma_y(\bar{\varepsilon}^{\mathrm{p}}) = \sigma_{y0} + C \cdot (\bar{\varepsilon}^{\mathrm{p}} + \varepsilon_0)^n$, where $\bar{\varepsilon}^{\mathrm{p}}$ is the equivalent plastic strain, σ_{y0} is the offset yield stress, ε_0 is the offset yield strain, C is the hardening modulus and n is the Swift power law variable. For the considered rolling case two metamodels are constructed which give the thickness reduction as a function of the entry and exit local tensions, for two levels of friction: one for $\mu_{\mathrm{low}} = 0.025$ and one for $\mu_{\mathrm{high}} = 0.1$. The lower friction coefficient, which is within a typical range for a well lubricated cold rolling process [11], leads to a value of the total roll force approximately equal to 550 kN. On the other hand, the higher friction coefficient is assumed for a non-lubricated rolling process.

Table 1. Input parameters of the shell rolling model.

Process parameters		
Strip's width	W	180 mm
Strip's length	L	1440 mm
Strip's entry thickness	H	0.255 mm
Roll radius	R	198.5 mm
Roll center distance	D	198.527 mm
Entry tension force	T_{in}	4.5 kN
Exit tension force	T_{out}	5.5 kN
Material parameters		
Elasticity modulus	E	200 GPa
Poisson's ratio	ν	0.3
Offset yield stress	σ_y	397 MPa
Offset yield strain	ε_0	0.0588
Hardening modulus	C	630 MPa
Swift power law variable	n	0.1632

First the rolling simulation is performed considering uniform good lubrication conditions ($\mu_{\mathrm{low}} = 0.025$) over the strip's width. The simulation is continued until steady-state conditions are reached. Thereafter, a disturbance is introduced, by applying $\mu_{\mathrm{high}} = 0.1$, on a portion of the strip width. This is shown in Fig. 3, where the lubricated and non-lubricated patches are highlighted on top of the shell finite element mesh. Two scenarios are considered, to study the sensitivity of the results to the width of the lubricated/non-lubricated patches. In the second scenario being analyzed, there is a disruption in the lubrication conditions that is the opposite of the first scenario. In the first scenario, the lubricated patch has a width of 60 mm. In the second scenario, the lubricated patch has a width of 120 mm.

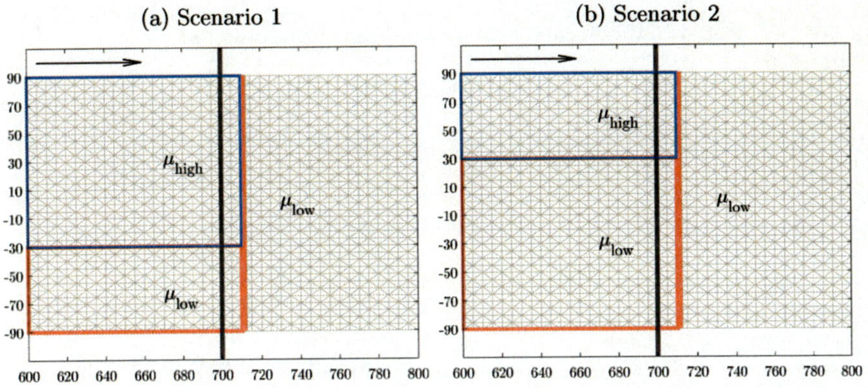

Fig. 3. Application of lubrication disturbance during the rolling simulation.

The first scenario is chosen as a representative example of one of the experimental observations reported in the previous work [4]. Figure 4 shows two consecutive photo frames taken during the pinching experiments from the upstream side of the roll bite. When there is uneven lubrication, with the roll bite becoming mostly dry and local traces of lubricant being present on the strip, the sheet experiences waviness upstream of the roll bite. By simulating the first scenario, it is possible to verify whether the developed model can be used to identify the pinching mechanism observed during the experiments. The second scenario is chosen as a more technologically relevant case, in which insufficient lubrication can locally occur, due to several factors, like nozzle clogging, presence of contaminants or impurities, or changes in the lubricant composition [3] or viscosity.

Simulation Results. Figure 5 shows how the distribution of the longitudinal stress (stress acting in the rolling direction) changes from the steady-state to the state after disruption. In both scenarios, it can be seen that due to the applied disturbance, local compressive stresses develop mainly upstream of the roll line, while local higher tensile stresses occur mainly downstream of the roll line. The most compressive stresses are localized in the lubricated patch, while the higher tensile stresses appear near the line of transition between the lubricated and the non-lubricated patch.

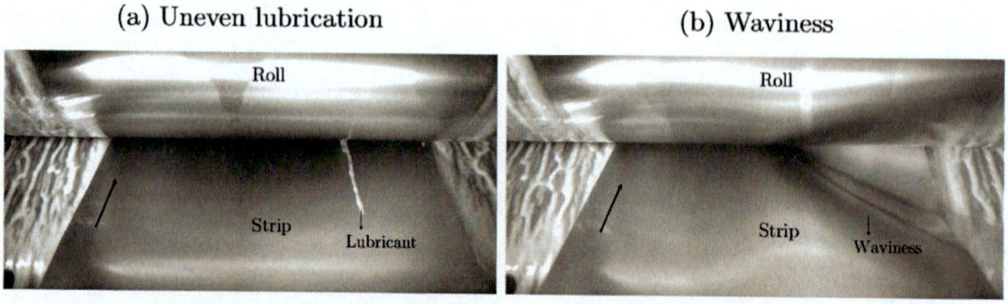

Fig. 4. Experimental observations from a pinching test [4].

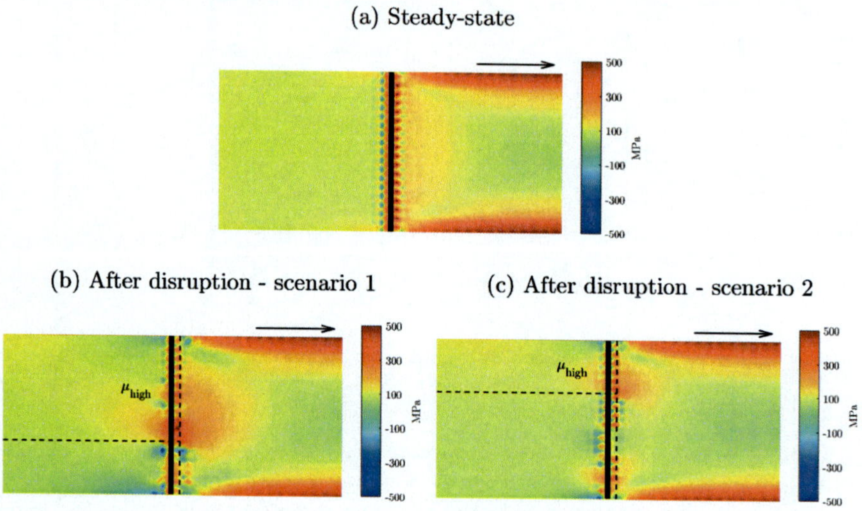

Fig. 5. Longitudinal stress σ_{xx}

The likelihood of buckling in the sheet depends on the most compressive principal stresses. Figure 6 shows the distribution of these stresses upstream of the roll line for both scenarios at the moment before stress relaxation occurs due to the first buckles appearing on the upstream side.

The magnitudes of the compressive principal stresses are higher in the first scenario. If the compressive principal stress in the sheet exceeds a critical value, buckling occurs. The critical buckling stress depends on the geometry (thickness and width) of the sheet, on its material properties, and on the boundary conditions. The out-of-plane displacements induced by the applied disruptions are shown in Fig. 7. It must be noted that there is no steady-state after the disruption is applied, and the results are taken at a specific moment in time shortly after application of the disruption. In both scenarios the strip experiences waviness, not only downstream but also upstream of the roll line. The buckling behavior is shown in Fig. 8. The buckles upstream of the roll line are confined to the part of the strip where the lower friction coefficient is applied.

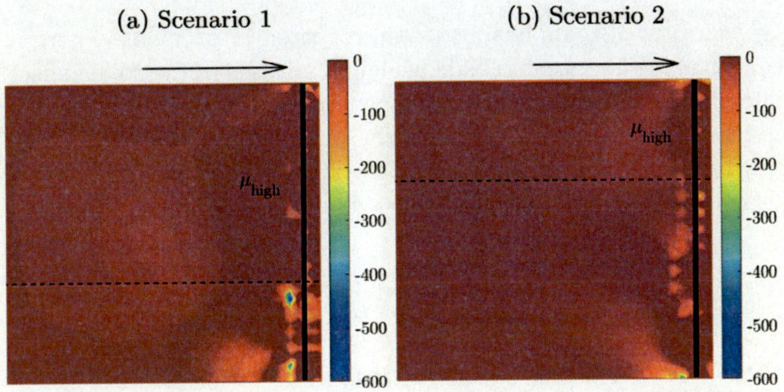

Fig. 6. Compressive principal stress after disruption (before stress relaxation), upstream of the roll line.

However, in the occurrence of the waviness upstream of the roll line, a difference exists between the two scenarios, that has to do with the different level of compressive principal stresses, which is lower in the second scenario than in the first one. In the second scenario, as the rolling process continues, these stresses accumulate and eventually result in buckling, which occurs later than in the first scenario. This difference can be better expressed in terms of length of the sheet being rolled from the moment of disruption to the appearance of buckles on the upstream side. This length is equal to 2.2 mm in the first scenario and to 3.2 mm in the second scenario.

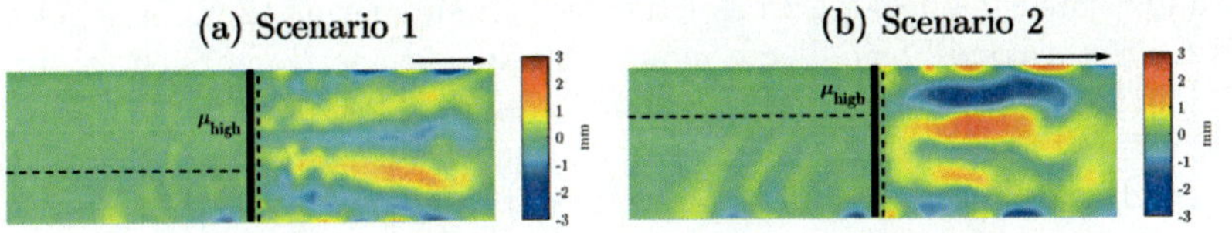

Fig. 7. Out-of-plane displacements induced by the disruption.

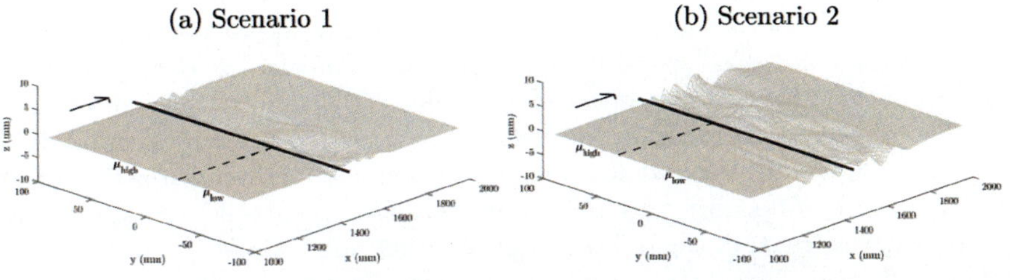

Fig. 8. Buckling behavior of the strip after disruption.

The appearance of tears has also been reported in previous experiments [4]. These are caused by excessive tensile principal stresses in the rolled sheet. These stresses are shown in Fig. 9 for the two scenarios. Consistently in the two cases, the highest magnitudes appear close to the line of transition between the lubricated and the non-lubricated patch. In the first scenario, these stresses are higher and approximately aligned with the most compressive stresses upstream of the roll line.

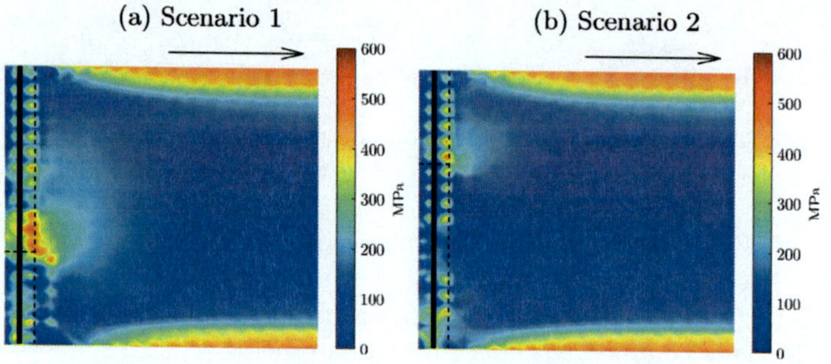

Fig. 9. Tensile principal stress after disruption, downstream of the roll line.

4 Discussions

3D finite element models are an effective tool for simulating the rolling behavior under different conditions. These models can be specifically applied to analyze instabilities in thin strip rolling, which can arise due to various factors. The proposed method for 3D sheet rolling calculations was used to study the effect of uneven lubrication conditions on the stress state and stability of the rolled sheet. The simulation results showed how a local disruption in the frictional conditions affects the rolling process, providing some insights into the complex interplay between lubrication, rolling conditions, and sheet behavior. It was shown that a local variation of the friction coefficient across the strip width causes a stress re-distribution in the rolled sheet. The most significant effects of this disruption are the local lower tensile and even compressive stresses upstream of the roll bite and the local higher tensile stresses downstream of the roll bite. Consistently for the two analyzed cases, the most compressive principal stresses occur locally in the part of the sheet where a lower friction coefficient is applied. If the most compressive principal stress reaches the critical buckling stress, waviness appears in the sheet. Differently from what is commonly described in literature concerning the occurrence of flatness defects [1,9], it was shown that waviness can appear not only downstream of the roll bite, but also upstream. Moreover, unlike other types of flatness defects (e.g. center buckles and wavy edges), the simulated buckles are not always fully aligned with the rolling direction. The compressive principal stresses can take more or less time during the process to build up and lead to buckles, depending on the type of disruption, as shown in the comparison between the two scenarios. This can be one of the factors influencing the distance between repetitive ripples in pinched strips. The deformed configuration of the buckled strip resembles qualitatively the behavior observed during the pinching experiments, as reported in previous work [4]. Another result of interest that was shown is the local high concentration of tensile principal stress, downstream of the roll line, around the boundary between the lubricated and the non-lubricated patches. The most tensile principal stress is higher than the yield stress of the material in both the two analyzed scenarios. If the tensile principal stress is higher than the ultimate tensile stress of the material, the sheet will experience tears as observed in some pinching experiments. The similarity between simulation results and experimental observations suggests that the proposed model is able to capture the underlying mechanism of pinching or similar types of defects. In this way, the model can contribute to a better understanding of instability phenomena in thin strip rolling. However, it is important to note that the proposed method has a limitation, due to not accounting for the elastic bending of the rolls. This deformation can affect the final shape and thickness of the sheet metal, with a significant effect on its flatness.

5 Conclusions

3D FEM models are a useful tool to simulate and analyze the sheet metal rolling process. These models allow to predict stress and strain distributions in the sheet metal. These predictions can be used to optimize the process parameters and improve the quality of the final products. In this paper a strategy to reduce the computational cost of 3D sheet rolling calculations has been presented. The deformations within the roll bite and the stress distribution outside the roll bite influence each other. This interplay has a critical role in determining the stress and strain fields in the rolled sheet. To account for the mutual influence between in-bite and out-of-bite conditions, an equivalent roll bite model is developed to be connected to a shell model of the sheet. The local deformations within the roll bite are coupled to the local tensions and frictional conditions. The shell model describes the deformation and stress state of the entire sheet. It is shown that a qualitative study of instability phenomena in thin strip rolling can be performed using the proposed model. The rolling behavior which arises from disruptions in the lubrication conditions is analyzed, with a focus on the occurrence of pinching phenomena. Hence, the proposed model has the potential to provide valuable insights into certain complex phenomena which can occur during the rolling process, like those leading to flatness defects.

Acknowledgments. This research was carried out under project number S16039 in the framework of the Partnership Program of the Materials innovation institute M2i and the Technology Foundation TTW, which is part of the Netherlands Organization for Scientific Research.

References

1. Abdelkhalek, S., Montmitonnet, P., Legrand, N., Buessler, P.: Coupled approach for flatness prediction in cold rolling of thin strip. Int. J. Mech. Sci. **53**(9), 661–675 (2011). https://doi.org/10.1016/j.ijmecsci.2011.04.001
2. Abdelkhalek, S., Zahrouni, H., Potier-Ferry, M., Legrand, N., Montmitonnet, P., Buessler, P.: Coupled and uncoupled approaches for thin cold rolled strip buckling prediction. Int.J. Mater. Form. **2**(Suppl. 1), 833–836 (2009). https://doi.org/10.1007/s12289-009-0547-0
3. Cometa, A., et al.: Experimental investigation of pinching phenomena in cold rolling of thin steel sheets. In: AIP Conference Proceedings, vol. 2113 (2019). https://doi.org/10.1063/1.5112560
4. Cometa, A., Geijselaers, H.J.M., van den Boogaard, A.H., Wentink, D.J., Hol, C.W.J., Jacobs, L.J.M.: Experimental study on the mechanism of pinching in cold-rolling processes. Steel Res. Int. **93**(8) (2022). https://doi.org/10.1002/srin.202100812
5. Fischer, F., Rammerstorfer, F., Friedl, N.: Residual stress-induced center wave buckling of rolled strip metal. J. Appl. Mech. Trans. ASME **70**(1), 84–90 (2003). https://doi.org/10.1115/1.1532322
6. Linghu, K., et al.: 3D FEM analysis of strip shape during multi-pass rolling in a 6-high CVC cold rolling mill. Int. J. Adv. Manuf. Technol. **74**, 1733–1745 (2014). https://doi.org/10.1007/s00170-014-6069-z
7. Moazeni, B., Salimi, M.: Investigations on relations between shape defects and thickness profile variations in thin flat rolling. Int. J. Adv. Manuf. Technol. **77**(5–8), 1315–1331 (2015). https://doi.org/10.1007/s00170-014-6544-6
8. Nakhoul, R., Montmitonnet, P., Legrand, N.: Manifested flatness defect prediction in cold rolling of thin strips. Int. J. Mater. Form. **8**(2), 283–292 (2015). https://doi.org/10.1007/s12289-014-1166-y
9. Nakhoul, R., Montmitonnet, P., Potier-Ferry, M.: Multi-scale method for modeling thin sheet buckling under residual stresses in the context of strip rolling. Int. J. Solids Struct. **66**, 62–76 (2015). https://doi.org/10.1016/j.ijsolstr.2015.03.028
10. Roberts, W.L.: Cold Rolling of Steel. Marcel Dekker Inc., New York and Basel (1978)
11. Tieu, A., Zhu, H., Lu, C., You, C., Jiang, Z., D'Alessio, G.: Modelling of friction coefficient in cold strip rolling. Mater. Sci. Forum **505-507**(Part 2), 1285–1290 (2006). https://doi.org/10.4028/0-87849-990-3.1285
12. Wang, Q.L., Sun, J., Liu, Y.M., Wang, P.F., Zhang, D.H.: Analysis of symmetrical flatness actuator efficiencies for UCM cold rolling mill by 3D elastic-plastic FEM. Int. J. Adv. Manuf. Technol. **92**, 1371–1389 (2017). https://doi.org/10.1007/s00170-017-0204-6
13. Wang, X.C., Yang, Q., Jiang, Z.Y., Xu, J.W.: Research on the improvement effect of high tension on flatness deviation in cold strip rolling. Steel Res. Int. **85**, 1560–1570 (2014). https://doi.org/10.1002/srin.201400048
14. Yan, X., Sun, J., Xiong, S.: Effects of lubricants on the rolling performances of cold rolled copper strips. Procedia Eng. **207**, 2227–2232 (2017). https://doi.org/10.1016/j.proeng.2017.10.986

In-Plane Torsion Test - Analysis of the Tool Design

Fabian Stiebert(✉), Heinrich Traphöner, and A. Erman Tekkaya

Institute of Forming Technology and Lightweight Components (IUL), TU Dortmund University, Baroper Street 303, 44227 Dortmund, Germany
Fabian.Stiebert@iul.tu-dortmund.de

Abstract. The in-plane torsion test is more frequently used in various research facilities to obtain flow curves and the Bauschinger-coefficients of sheet materials. The execution and the experimental setup of the test currently depend on the experience, available machines as well as tools at the respective testing laboratories. For this reason, the author's aim is to standardize the in-plane torsion test in order to create a common basis for carrying out the tests and to enable the characterization of comparable material parameters. The shape and structure of the inner clamping are often selected based on experience without knowing the exact effects on the process. Although the clamps appear less relevant, they are the only component in contact with the specimen and have a major influence on the process limits of the in-plane torsion test, such as wrinkling and slippage. To provide such a base, the shape and the surface structure of the clamping on the process limit of slipping are investigated. Results show that a ring-shaped clamping surface can transmit up to 50% more torque at the same clamping force compared to a full-circle clamping surface. It is derived analytically that torque transmission through plain clamping surfaces is not possible for a part of thin sheets ($t = 0.5$ and 3.0 mm). For such cases, structured clamping surfaces were used to enable successful torque transmission.

Keywords: Standardization · Material characterization · In-plane torsion test

1 Introduction

The in-plane torsion test (Fig. 1) is successfully used for the characterization of sheet metal with large deformations for more than 60 years [1]. Knowledge of how to carry out in-plane torsion tests is based on the experience of individual research institutions [2, 3]. Most works focus on the application, evaluation and results [4] but not the test setup and description of the testing procedure. Hence, standardization of the in-plane torsion test is the next step towards commercialization of the test. For this reason, the aim of the authors is to prepare a test standard for testing thin sheets ($t = 0.5$ and 3.0 mm) using the in-plane torsion test.

The inner and outer clamps as the main transmitters of torque, are considered as one of the most critical components regarding the process window of the in-plane torsion test.

K. Mocellin et al. (Eds.): ICTP 2023, LNME, pp. 553–561, 2024.
https://doi.org/10.1007/978-3-031-40920-2_57

Process boundaries, such as the specimen slipping, limit the use of the test, especially testing high-strength materials [5]. Stress superposition due to the compression force of the clamping and the associated transverse contraction of the material also influence the ideal stress state in the test area [6]. The test can only be carried out successfully, if the geometric design of the clamps is appropriate, the clamping strategy and clamping force are selected accordingly. This work focuses on the tool design and clamping forces for the inner clamping, since the outer clamping is not critical regarding the torque that has to be transmitted. The aim is to define a tool shape and clamping conditions, which transmit the required torque to characterize high-strength materials.

2 Transmittable Torque – Design of the Inner Clamping

High-strength materials with simultaneously high sheet thickness provide a challenge to material characterization. In the in-plane torsion test, this process boundary is shown by the slippage of the specimen under the internal clamping, since the maximum transmittable torque is limited. Due to the used machine, the clamping force of the punch is limited, such that the transmittable torque can only be influenced by the contact area, i.e. the size and shape and of the punch.

2.1 Shape of the Contact Surface

In order to establish an analytical relationship between transmittable torque M and applied clamping force F_C, the tangential force F_θ that can be transmitted into the specimen by friction, must be derived. The shape of the inner clamping surface is defined by the radius of the inner clamping r_I and a minimum radius r_{min}, which describes the left-out pocket in the center of the clamping surface, as shown in Fig. 1.

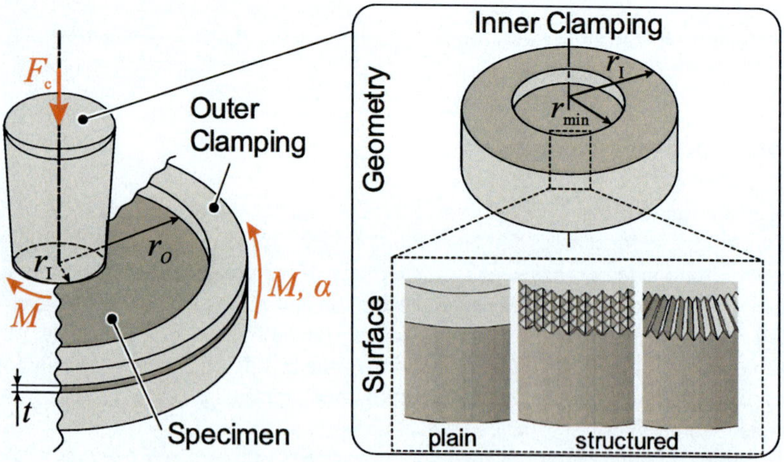

Fig. 1. Schematic layout and geometric dimensions in the in-plane torsion test

The normal stress which acts on the clamping surface can be calculated from the clamping force F_C and the contact area A by

$$\sigma_z = \frac{F_C}{A} = \frac{F_C}{\pi \cdot \left(r_I^2 - r_{min}^2\right)}. \tag{1}$$

Assuming Coulomb friction between the punch with a flat contact surface and the specimen, this normal stress σ_z leads to a tangential force F_θ which is given by

$$\begin{aligned} dF_\theta &= \sigma_\theta \cdot dA \\ &= \sigma_\theta \cdot r \cdot d\theta \cdot dr \text{ with } \sigma_\theta = \mu \cdot \sigma_z = \text{const.} \end{aligned} \tag{2}$$

by integration over the area between the inner radius r_{min} and the outer radius of the inner clamp r_I, the transmittable torque of two surfaces with different friction coefficients μ_1 and μ_2 can be calculated by.

$$M = \frac{2}{3} \cdot \frac{\left(r_I^3 - r_{min}^3\right)}{\left(r_I^2 - r_{min}^2\right)} \cdot (\mu_1 + \mu_2) \cdot F_C. \tag{4}$$

To compare the transmittable torque of an ring-shaped ($r_{min} \neq 0$) and a full ($r_{min} = 0$) clamping surface, the torque-ratio can derived by

$$\frac{M_{ring}}{M_{full}} = \frac{\left(r_I^3 - r_{min}^3\right)}{r_I \cdot \left(r_I^2 - r_{min}^2\right)}. \tag{5}$$

Through a limit value consideration according to

$$\lim_{r_{min} \to r_I} \left(\frac{\left(r_I^3 - r_{min}^3\right)}{r_I \cdot \left(r_I^2 - r_{min}^2\right)} \right) = 1.5 \tag{6}$$

it can be derived that the ring-shaped clamping surface can transmit up to 50% more torque with the same clamping force and radius r_I, compared to a full clamping surface. In reality, this maximum value is never achieved, as the tool or the workpiece fail due to excessive compressive forces caused by the decreasing clamping surface. These high compressive stresses can, similar to upsetting tests, lead to material flow in radial direction in the specimen and negatively influence the stress state in the test area close to the inner clamping. Another limitation is the maximum clamping force that can be applied, which is limited by the machine used.

2.2 Structure of the Contact Surface – Plain

Besides the dimensions, the structure of the clamping surfaces also has a significant influence on the transmittable torque. The simplest surface structure is a smooth clamping surface. In the following, the suitability of such plain clamping surfaces for torque transmission in fine sheet metals is investigated. Therefore, a circular, ring-shaped clamping surface with the radius r_I is considered, which is shown in Fig. 1. The transmittable torque M for two acting clamping surfaces (top and bottom) can be described by Eq. (4), where the friction coefficients for the top and bottom surfaces are given by μ_t and μ_b respectively. To minimize material flow in the radial direction, the clamping force F_C is limited as a function of the initial yield stress $\sigma_{f,0}$ of the tested material. By taking the initial yield stress $\sigma_{f,0}$ of the material into account for the maximum clamping force F_C, the maximum transmissible torque M_{trans} is given by

$$M_{trans} = \frac{2\pi}{3} \cdot \left(r_I^3 - r_{min}^3\right) \cdot (\mu_t + \mu_b) \cdot \sigma_{f,0}. \tag{7}$$

In order to determine the real coefficient of friction when performing the in-plane torsion test, investigations were carried out with a smooth ring-shaped clamping surface. The maximum torque at the transition from static to dynamic friction M_{max} was investigated at different clamping forces and lubrication conditions (Fig. 2). In addition to the friction without lubricant, the friction coefficient was also determined with wet and dried paint, as used in the test for optical strain measurement.

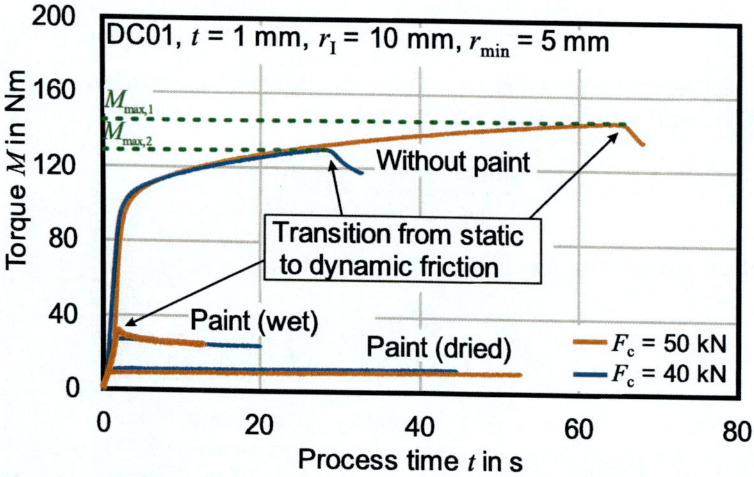

Fig. 2. Torque characteristics for determining the transition point between static and dynamic friction and deriving the maximum transmissible torque.

For each condition, three repetitions were carried out for statistical coverage. The current friction coefficient μ was then calculated with the maximum achieved torque M_{max}, by converting Eq. (2) to μ. The values for r_I and r_{min} were measured with a tactile measurement device. The friction coefficients of the individual configurations are summarized in Table 1.

The friction coefficient for the sheet without lubrication is the highest with $\mu = 0.101$. The lowest friction value $\mu = 0.009$ is found for dried paint. The higher friction coefficient with non-dried paint ($\mu = 0.021$) can be explained by the fact that a large part of the paint under the clamping is pressed away during the loading step and only a small layer of paint remains. In order to consider the real test conditions, it must be noted that the torque transmission in the in-plane torsion test is applied to the specimen via

Table 1. Friction values for different surface conditions in the plane torsion test.

Surface condition	Friction coefficient μ	Standard deviation
Blank	0.101	5,7%
Paint (wet)	0.021	7,5%
Paint (dried)	0.009	5,9%

one top-surface with almost wet paint and one bottom-surface without paint, resulting in a total friction coefficient $\mu_{total} = \mu_t + \mu_b = 0.11$. In order to evaluate whether a certain material can be characterized without structured clamping surfaces up to a certain point without slipping, the ratio of the torque transmittable by friction M_{trans} and a required torque M_{req} was calculated. The required torque M_{req} depends on the yield stress $\sigma_{f,max,tensile}$ needed for a certain strain in the flow curve, the internal clamping diameter $r_{I,}$ and the sheet thickness t. The ratio is given by

$$\frac{M_{req}}{M_{trans}} = \frac{\sqrt{3} \cdot \sigma_{f,max} \cdot t}{(\mu_t + \mu_b) \cdot r_I \cdot \sigma_{f,0} \cdot \left(1 - \frac{r_{min}}{r_I}^3\right)}. \quad (8)$$

As long as the maximum transmittable torque is greater than the required torque, i.e. the ratio M_{req}/M_{trans} is less than 1.0, the transmission with plain clamping surfaces is possible. Figure 3 shows the torque ratio over the sheet thickness. In order to represent different material behaviors, materials with different work hardening behaviors were analyzed. The ratio between the maximum flow stress from the tensile test before necking $\sigma_{f,UTS}$ and the initial yield stress $\sigma_{f,0}$ was formed for consideration. One material with strong work-hardening ($\sigma_{f,UTS}$ / $\sigma_{f,0} = 5$) and one material with low work-hardening ($\sigma_{f,UTS}$ / $\sigma_{f,0} = 1.25$) are shown in Fig. 3.

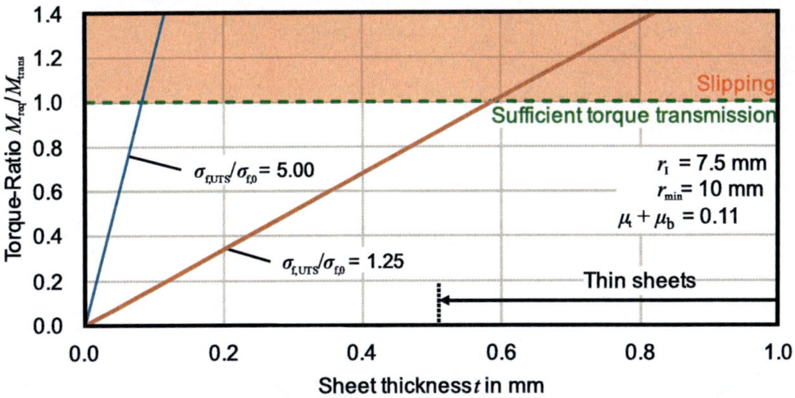

Fig. 3. Torque ratio over the sheet thickness for one material with high work-hardening ($\sigma_{f,UTS}$/ $\sigma_{f,0} = 5.0$) and one material with low work-hardening ($\sigma_{f,UTS}$/ $\sigma_{f,0} = 1.25$)

As shown, the required torque for the material with high work hardening can only be transmitted for sheets with a thickness lower than 0.08 mm for the given clamping geometries and clamping conditions. In contrast, materials with weak work hardening can be tested up to sheet thicknesses of almost 0.6 mm, which can be explained by the comparatively higher possible clamping forces, resulting from the high initial yield strength. By forming the ratio of the two yield stresses, a dimensionless quantity is given, which allows an evaluation independent of the tested material. The ratio of the initial flow stress and the flow stress required for a certain degree of deformation should be taken into account. This considered degree of deformation should be higher than the maximum degree of deformation achieved in the tensile test in order to fully benefit from the potential of the in-plane torsion test. In order to realize the torque transmission also for materials with higher sheet thicknesses, the clamping surface needs to be structured.

2.3 Structure of the Contact Surface - Radial Linear Serrations

In the literature, different surface structures are used to clamp the specimens in the in-plane torsion test. Marciniak [1] used radial serrations, while Yin [4] used cross-shaped serrations. In addition, Yin [6] investigated the use of additional pins through the sheet metal, which turned out to be disadvantageous. The aim is always to transmit the greatest possible tangential force with a low clamping force and at the same time homogeneous strains around the circumference of the specimen. Radial linear serrations, as shown in Fig. 4a, meet these requirements. A defined number of serrations n are arranged at regular angular intervals $\Delta\alpha = 2\pi/n$ over the circumference of the clamps. Figure 4b shows the geometry of individual serrations and the forces acting on the serrations for an infinitesimal element. The axial force F_n on the serrations results in pressure on the flanks, which leads to plastic deformation of the material (Fig. 4a).

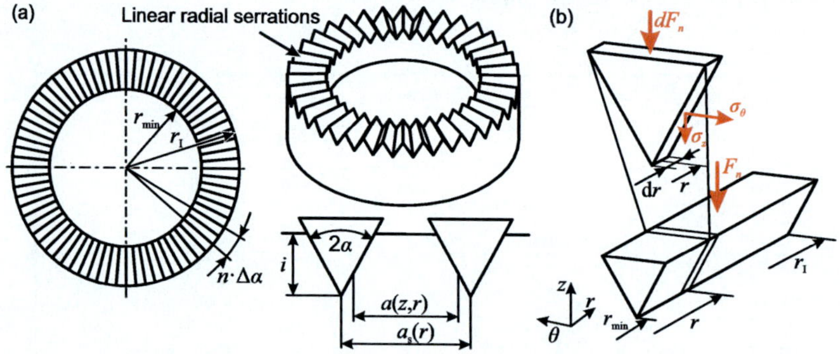

Fig. 4. (a) Inner clamping with radial serrations (b) simplified dimensions of the serrations

In numerical simulations of the indentation and subsequent torsion of a single serration (Fig. 5a), the material bulges on the sides as the serration penetrates the material. The surface is no longer flat. The indentation depth i is measured from the upper edge of the bulged material to the tip of the serration and is therefore larger than the axial feed of the serration. With a subsequent tangential load F_θ at constant axial force F_z, the serration penetrates deeper into the material. The indentation depth is approximately doubled because only one flank of the serration is in contact with the material. The simulation ends with the material shearing off, which is interpreted as the clamping slipping through.

Transmittable Torque and Required Clamping Force. In the following calculations, plane stress and strain conditions are assumed. The slip line theory is used to calculate the penetration depth i as a function of the clamping force F_C. The slip line field used for the indentation and subsequent torsion is shown in Fig. 5b.

A blunt tip of the serration with width b_s is assumed. The surface pressure at the tip and flank is calculated from the illustrated slip line field:

$$p_{1,II} = -2k(1+\alpha) \text{ and } p_{1,IV} = -2k\left(1+\frac{\pi}{2}\right). \tag{9}$$

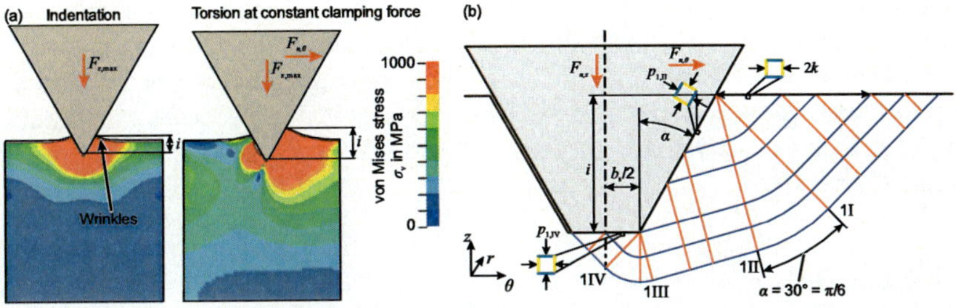

Fig. 5. (a) Numerical simulation of indentation and subsequent torsion for one serration and (b) slip line field for torsion with constant axial force.

The force on a single serration as a function of the indentation depth i is

$$F_n = \frac{F_C}{n} = p_{1,\mathrm{IV}} \cdot b_s \cdot (r_\mathrm{I} - r_{\min}) + 2p_{1,\mathrm{II}} \cdot i \cdot (r_\mathrm{I} - r_{\min}) \cdot \tan(\alpha) \tag{10}$$

and the penetration depth i results as

$$i(\mathrm{F}_C) = \frac{1}{4(1+\alpha)\cdot n \cdot (r_\mathrm{I} - r_{\min}) \cdot \tan(\alpha)} \cdot \frac{F_C}{k} - \frac{\left(1+\frac{\pi}{2}\right)}{2(1+\alpha)\cdot\tan(\alpha)} \cdot b_s. \tag{11}$$

The increase in indentation depth i is proportional to an increase in clamping force F_C. The equilibrium of the forces in tangential direction θ for a radial increment dr is

$$dF_\theta = 2k \cdot (1+\alpha) \cdot i \cdot n \cdot dr. \tag{12}$$

Assuming a constant radial distribution of the force $dF = \text{const.}$, it follows for the resulting maximum torque and two contact areas:

$$\begin{aligned} M_{\max} &= 2 \int_{r=r_{\min}}^{r_I} r \cdot dF_\theta(r) = 2k \cdot (1+\alpha) \cdot i \cdot n \cdot \left(r_I^2 - r_{\min}^2\right) \\ &= \frac{\left(r_I^2 - r_{\min}^2\right)}{\left(r_I - r_{\min}\right) \cdot \tan(\alpha)} \cdot F_C - \frac{2 \cdot \left(1 + \frac{\pi}{2}\right) \cdot \left(r_I^2 - r_{\min}^2\right)}{\tan(\alpha)} \cdot k \cdot n \cdot b_s \end{aligned} \tag{13}$$

For many materials, however, the technologically achievable torque is not necessary, so that lower clamping forces are sufficient. The minimum required clamping force $F_{C,\min}$ results to

$$F_{C,\min} = \frac{2 \cdot \pi \cdot \left(r_I - r_{\min}\right) \cdot \tan(\alpha) \cdot r_{\tau,\max}^2}{\left(r_I^2 - r_{\min}^2\right)} \cdot t \cdot \tau_{\max} + 2 \cdot \left(1 + \frac{\pi}{2}\right) \cdot k \cdot n \cdot b_s \cdot \left(r_I - r_{\min}\right). \tag{14}$$

$r_{\tau,\max}$ is the radius of maximum shear stress, which differs for the planar specimen and grooved specimen. $\tau_{\max}$ is the maximum shear yield stress occurring in the material. The right part of Eqs. (13) and (14) is a kind of power dissipation for blunt tips. For ideal sharp serrations ($b_s = 0$), the torque that can be transmitted is independent of the shear flow stress k of the material and only dependent on the geometry of the punch and the clamping force F_C.

3 Conclusion

Analytical and experimental analysis of the shape of the inner clamping in the in-plane torsion test showed that ring-shaped clamping surfaces can transmit up to 50% more torque than full clamping surfaces for the same clamping force. This is particularly advantageous in the case of technologically limited clamping forces. If the clamping force is not technologically limited, on the other hand, the use of full clamping surfaces leads to the transmission of higher maximum torques. Further investigations showed that a frictional torque transmission at the inner clamping surface is not sufficient for most thin sheets with a thickness of 0.5 to 3.0 mm. Structuring of the clamping surface is therefore necessary in most cases. The structure on the surface of the clamps (serrations) should be adapted to the direction of the load. For this reason, radial serrations are well suited, which create a homogenous deformation zone around the circumference of the specimen. For radial serrations, the necessary clamping force could be estimated analytically as a function of the punch geometry and the material properties. For ideal serrations, without blunt tips, the transmittable torque is material-independent.

Future work will focus on the influence of the clamping surface geometry on the process limit of wrinkling. The aim is to enable the future user to design the clamping tools in such a way that a test can be carried out without slipping and wrinkling.

Acknowledgments. The authors acknowledge the support of the research project IGF 21137 / VP1320 by the Research Association for Steel Application (FOSTA), which is supported by the Federal Ministry of Economic Affairs and Climate Action through the German Federation of Industrial Research Associations (AiF) as part of the program for promoting industrial cooperative research (IGF) on the basis of a decision by the German Bundestag.

References

1. Marciniak, Z.: Influence of the sign change of the load on the strain hardening curve of a copper test piece subject to torsion. Archiwum Mechaniki Stosowanej **13**, 743–752 (1961)
2. Grolleau, V., Roth, C.C., Mohr, D.: Design of in-plane torsion experiment to characterize anisotropic plasticity and fracture under simple shear. Int. J. Solids Struct. **236–237**, 111341 (2022). https://doi.org/10.1016/j.ijsolstr.2021.111341
3. Wagner, L., Gross, T., Gruber, P.G., Grillenberger, M., Schagerl, M.: Application of the in-plane torsion test in an industrial environment - recent advances and remaining challenges. In: Proceedings of the Forming Technology Forum 2019, Herrsching am Ammersee, Germany (2019)
4. Traphöner, H., Clausmeyer, T., Tekkaya, A.E.: Material characterization for plane and curved sheets using the in-plane torsion test – an overview. J. Mater. Process. Technol. **257**, 278–287 (2018). https://doi.org/10.1016/j.jmatprotec.2018.02.030
5. Tekkaya, A.E., Pöhlandt, K., Lange, K.: Determining stress-strain curves of sheet metal in the plane torsion test. CIRP Ann. Manuf. Technol. **31**, 171–174 (1982)
6. Yin, Q.: Verfestigungs- und Schädigungsverhalten von Blechwerkstoffen im ebenen Torsionsversuch. Dr.-Ing.-Dissertation, Technische Universität Dortmund, Dortmund, Germany (2014)

Non-iterative Stress Projection Method for Rate-Independent Plasticity

Seongyong Yoon[1,2], Shin-Yeong Lee[1], and Frédéric Barlat[1](✉)

[1] Graduate Institute of Ferrous and Energy Materials Technology, Pohang University of Science and Technology, 77 Cheongam-Ro, Nam-Gu, Pohang, Gyeongbuk 37673, Republic of Korea
f.barlat@postech.ac.kr

[2] Now at Max-Planck-Institut Für Eisenforschung GmbH, Max-Planck-Street 1, 40237 Düsseldorf, Germany

Abstract. A fully-explicit stress integration method is comprehensively discussed via finite element analyses using three different classes of continuum plasticity models, i.e., anisotropic yield function, anisotropic hardening, and continuum damage model. A unified but straightforward formulation covers a wide range of rate-independent plasticity models. High accuracy, numerical robustness, and great efficiency are achieved through the use of generalized effective plastic strain and stress projection equations. In order to estimate the stress integration quality, an associated variable mapping technique, namely, the precision map, is introduced. Based on the precision map study, the current fully-explicit integration exhibits excellent accuracy. Furthermore, this non-iterative stress calculation reduces the computation cost by about 50% compared to conventional iterative stress update algorithms.

Keywords: Computational plasticity · Anisotropic yield function · Anisotropic hardening · Continuum damage model

1 Introduction

Numerical simulations involving metal plasticity significantly contribute to the analysis and optimization of manufacturing processes in various industries. The modeling of advanced materials demands complex formulations to reproduce unique plastic deformation mechanisms that are not well described by classical approaches. However, inefficient computation, poor numerical robustness (or stability), and low-stress integration quality are obstacles to more widespread use of these advanced formulations.

Conventional stress integration methods (algorithms) show superb performances in solving non-linear problems involving plasticity. Iterative (implicit) solution-finding approaches, nevertheless, lead to conditional success in stress integration depending on the model formalisms, deformation modes, and simulation analysis parameters. In addition, the classical integration strategies are limited by the shortcomings of the root-finding process, namely, the Newton-Raphson method. Over the last decade, numerous efforts have been made such as the line-search method, the residual value control, etc.,

K. Mocellin et al. (Eds.): ICTP 2023, LNME, pp. 562–570, 2024.
https://doi.org/10.1007/978-3-031-40920-2_58

in order to enhance the numerical performance of these algorithms. However, such treatments are, in general, accompanied by increased computation costs.

Fully-explicit stress integration approaches are featured with two primary advantages such as high computation speed and numerical robustness. The critical drawbacks, e.g., poor integration quality and conditional stability mainly stem from the uncertain fulfillment of the yield condition. Rossi et al. [1] initially utilized a numerical technique, so-called stress projection, to get the unconditional yield condition satisfaction. Independently, Yoon and Barlat [2] suggested the non-iterative stress projection method and applied it to implicit as well as explicit finite element analyses resulting in impressive performances.

In this article, the non-iterative stress projection method is briefly introduced. The stress integration quality is comparatively validated by means of the mapping of a fidelity parameter on a yield locus. Finally, its performance is estimated via finite element simulations using a variety of advanced plasticity models.

2 Non-iterative Stress Projection Method

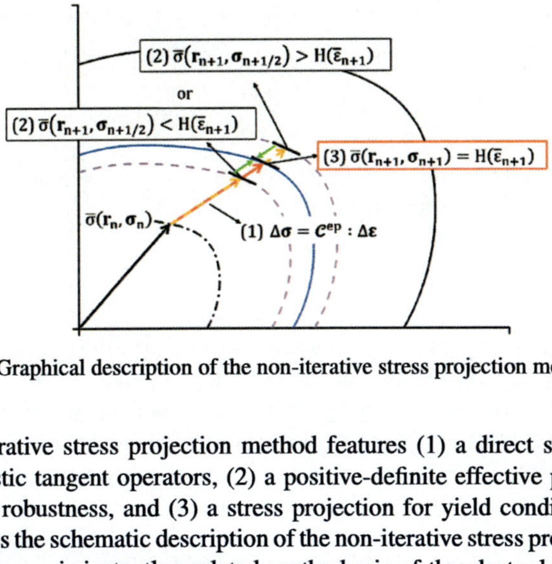

Fig. 1. Graphical description of the non-iterative stress projection method [2]

The non-iterative stress projection method features (1) a direct stress integration using elastoplastic tangent operators, (2) a positive-definite effective plastic strain for high numerical robustness, and (3) a stress projection for yield condition fulfillment. Figure 1 displays the schematic description of the non-iterative stress projection method.

The stress tensor is instantly updated on the basis of the elastoplastic constitutive law for the plastic deformation state.

$$\boldsymbol{\sigma}_{n+\frac{1}{2}} = \boldsymbol{\sigma}_n + \Delta\boldsymbol{\sigma}_{n+\frac{1}{2}} \tag{1a}$$

$$\Delta\boldsymbol{\sigma}_{n+\frac{1}{2}} = \mathcal{C}_n^{ep} : \Delta\boldsymbol{\varepsilon}_{n+1} \tag{1b}$$

where $\mathcal{C}_n^{ep}$ denotes an elastoplastic tangent modulus. The index (n + 1/2) indicates that the intermediate stress is not a final solution but needed to be corrected later. The plastic strain is approximated using the additive decomposition of the total strain, and the above-calculated stress increment.

$$\Delta\boldsymbol{\varepsilon}^{\mathrm{p}}_{\mathrm{n}+\frac{1}{2}} \approx \Delta\boldsymbol{\varepsilon}_{\mathrm{n}+1} - \boldsymbol{\mathcal{S}}^{\mathrm{e}} : \Delta\boldsymbol{\sigma}_{\mathrm{n}+\frac{1}{2}} \tag{2}$$

$\boldsymbol{\mathcal{S}}^{\mathrm{e}}$ represents the elastic compliance tensor. The positive-definite effective strain increment is acquired from the scalar product of the flow rule.

$$\Delta\overline{\varepsilon}_{\mathrm{n}+\frac{1}{2}} = \sqrt{\frac{\Delta\boldsymbol{\varepsilon}^{\mathrm{p}}_{\mathrm{n}+1/2} : \Delta\boldsymbol{\varepsilon}^{\mathrm{p}}_{\mathrm{n}+1/2}}{\frac{\partial\overline{\sigma}_{\mathrm{n}+1/2}}{\partial\boldsymbol{\sigma}_{\mathrm{n}+1/2}} : \frac{\partial\overline{\sigma}_{\mathrm{n}+1/2}}{\partial\boldsymbol{\sigma}_{\mathrm{n}+1/2}}}} \tag{3}$$

The effective plastic strain and internal state variables are explicitly updated making use of the effective strain increment (Eq. (3)).

$$\overline{\varepsilon}_{\mathrm{n}+1} = \overline{\varepsilon}_{\mathrm{n}} + \Delta\overline{\varepsilon}_{\mathrm{n}+\frac{1}{2}} \tag{4}$$

$$\mathbf{r}_{\mathrm{n}+1} = \mathbf{r}_{\mathrm{n}} + \frac{\mathrm{d}\mathbf{r}}{\mathrm{d}\overline{\varepsilon}}\Delta\overline{\varepsilon}_{\mathrm{n}+\frac{1}{2}} \tag{5}$$

Finally, the intermediate updated stress (Eq. (1a)) is projected onto the updated flow stress.

$$\boldsymbol{\sigma}_{\mathrm{n}+1} = \frac{\mathrm{H}(\overline{\varepsilon}_{\mathrm{n}+1})}{\overline{\sigma}\left(\mathbf{r}_{\mathrm{n}+1}, \boldsymbol{\sigma}_{\mathrm{n}+1/2}\right)}\boldsymbol{\sigma}_{\mathrm{n}+\frac{1}{2}} \tag{6}$$

Note that the current stress projection expression is available only when the yield condition is a homogeneous function.

3 Precision Map

In this section, the precision map is introduced to intuitively validate the stress integration quality of the non-iterative stress projection method (NSPM) for an imposed time (strain) increment through a comparative study with conventional iterative stress update algorithms, namely, the closest point projection method (CPPM) and the cutting plane method (CPM). For the accuracy validation, the R-value, a representative material anisotropy parameter, is valid only for uniaxial and balanced biaxial modes. Alternatively, the square root of the scalar product of the yield surface gradient is adopted as the fidelity parameter to cover entire deformation modes.

$$\Omega = \sqrt{\frac{\partial\overline{\sigma}}{\partial\boldsymbol{\sigma}} : \frac{\partial\overline{\sigma}}{\partial\boldsymbol{\sigma}}} \tag{7}$$

The precision map records the fidelity parameters on the normalized yield locus for numerous deformation modes. The corresponding plane stress tensors are defined by means of trigonometrical functions. To obtain the entire stress states on the outline of a yield surface ($\sigma_{12} = 0$), a primary angle (θ_1) which is the argument of the trigonometrical functions varies from 0° to 360°. The inside of a yield surface is filled out by rotating a plane stress tensor on the outline from 0° to 45° using a secondary angle (θ_2). Finally, 16,606 data points representing the fidelity parameter values are generated for a single map. Individual data points are determined after the stand-alone simulations where the major strain component is discretized by a prescribed boundary condition, and time increment, and the minor components are specified after the manipulation of the generalized Hooke's law and a reference stress tensor.

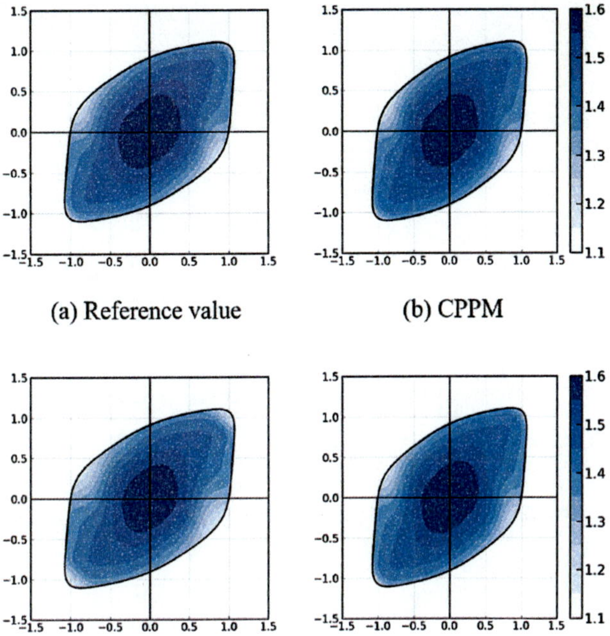

(a) Reference value (b) CPPM

(c) CPM (d) NSPM

Fig. 2. Fidelity parameter values (Eq. (7)) of Yld2000-2d calibrated for AA2090-T3 mapped on the yield locus where the values are obtained from (a) reference stress state, (b) CPPM (fully implicit), (c) CPM (semi-explicit), (d) NSPM (fully explicit)

An aluminum-lithium alloy (AA2090-T3) characterized by a non-quadratic anisotropic yield function (Yld2000-2d) was selected for this study [2]. For the precision mapping, a material point is subjected to 0.1 true strain ($\varepsilon_{true} = 0.1$) for each deformation mode and analyzed with a fixed time increment ($\Delta t = 0.01$). Thereby, the major strain increment ($\Delta\varepsilon_{major} = 0.001$) is evenly imposed at each analysis step. Note that the strain increment, in general, ranges from 10^{-6} to 10^{-3} for implicit finite element analysis.

Figure 2 shows the fidelity parameters at every plane stress state after corresponding simulations using Yld2000-2d. The magnitude of the parameter is marked blue varying its intensity. The reference value is obtained from the stress states prescribed by the trigonometrical functions before running simulations in virtue of the consistency of yield surface gradients. Due to the numerical error accumulated during a simulation, the desired stress tensors are frequently lost after passing stress integration processes. The discrepancy in the precision maps depends on the integration schemes. According to the mapping study, NSPM results in the exactly same fidelity parameter pattern as CPPM. The mapping pattern by CPM is apart from the others.

The distinction between the maps in Fig. 2 is visualized by mapping the relative errors on the yield condition, so-called the error map (See Fig. 3). The relative error is calculated below.

$$\mathrm{ERROR} = \frac{|\Omega_{\mathrm{ref}} - \Omega|}{\Omega_{\mathrm{ref}}} \times 100\ \% \tag{8}$$

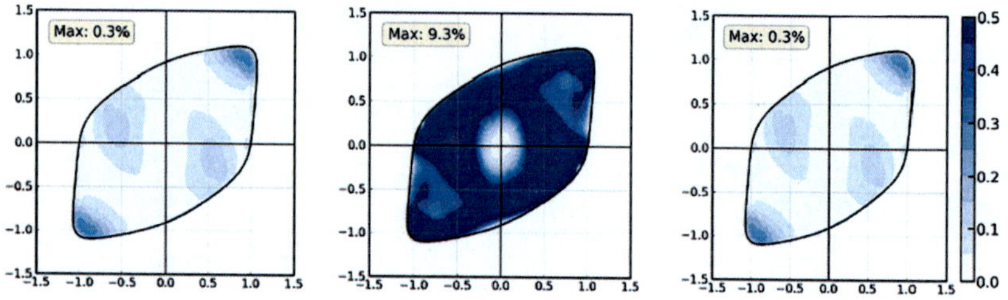

(a) CPPM (b) CPM (c) NSPM

Fig. 3. Relative error patterns which represent the gaps between the reference value and the fidelity values from (a) CPPM (fully implicit), (b) CPM (semi-explicit), and (c) NSPM (fully explicit)

NSPM shows the same maximum error value (0.3%) as that of CPPM. In contrast, CPM results in the highest error (9.3%). This implies that the stress integration quality of NSPM is very close to CPPM even with such a coarse strain increment ($\Delta\varepsilon_{\mathrm{major}} = 10^{-3}$). The error deviation between the considered integration methods may differ depending on the magnitude of the strain increment and the material anisotropy. In other words, for the more refined strain increment, CPM will give the same accuracy as the others. The precision map indirectly exhibits the accuracy of stress integration methods for a prescribed strain increment, which is very close to the displacement-controlled finite element analysis. The fidelity parameter can be replaced with other variables for various purposes. One may find the original source code at [7].

4 Finite Element Analysis

In this section, the numerical performance of NSPM is discussed via finite element simulations using three classes of plasticity models, i.e., anisotropic yield function (Yld200-2d), anisotropic hardening model (Homogeneous anisotropic hardening – HAH_{20}), and continuum damage model (Gurson-Tvergaard-Needleman – GTN). Detailed descriptions of model coefficients and finite element modeling are found in the works [2–4].

The plasticity models are implemented into the finite element program, namely, Abaqus 6.20/explicit through the vectorized user-defined material (VUMAT) subroutine. For objectivity, simulation results of NSPM are compared to those of CPPM and CPM in each sub-section. The used VUMAT subroutine code is available at [8].

4.1 Deep Drawing: Yld2000-2d

A non-quadratic anisotropic yield function (Yld2000-2d) is exploited to reproduce the material anisotropy of AA2090-T3 in the deep drawing simulation.

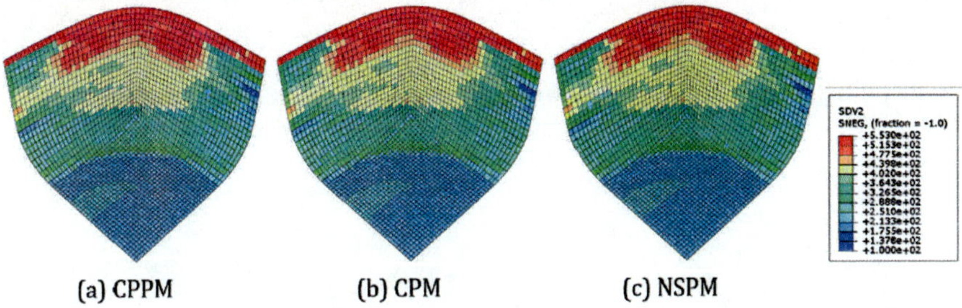

Fig. 4. Effective stress (SDV2) distribution of Yld2000-2d in deep drawing explicit finite element simulations integrated using (a) CPPM (fully-implicit), (b) CPM (semi-explicit), and (c) NSPM (fully-explicit)

Figure 4 shows the distribution of effective stress after explicit deep drawing simulation. The stress prediction by NSPM is identical to the other methods. The obtained earing profiles are completely overlapped with one another (See Fig. 5(a)). The computation time with NSPM is reduced by as much as 64% compared to CPPM although the latter requires only 1 or 2 iterations for algorithmic convergence (See Fig. 5(b)). The high efficiency of NSPM likely results from the simple formulation in addition to the zero iteration.

4.2 S-rail Forming: HAH_{20}

NSPM is also available for any type of anisotropic hardening when a morphological change of yield surface is incorporated in an elastoplastic tangent operator. The morphologic component contributes to the instant but an accurate calculation of the effective strain increment in the present method as well as other algorithms. A distortional plasticity hardening model (HAH_{20}) [5] is used for plasticity modeling of dual-phase steel (DP780) under non-proportional loading conditions and applied to an industrial scale simulation, namely, the S-rail forming.

NSPM results in an effective stress distribution in Fig. 6 in excellent agreement with those obtained from CPPM and CPM after the forming simulations. Figure 7(a) indicates that all the springback profiles are in excellent agreement as well. Figure 7(b) shows that the analysis time with NSPM is reduced by 46% without loss of accuracy after forming and springback simulations (See Fig. 7(b)). In this application, CPPM and CPM require 2 or 3 iterations to complete the stress integration.

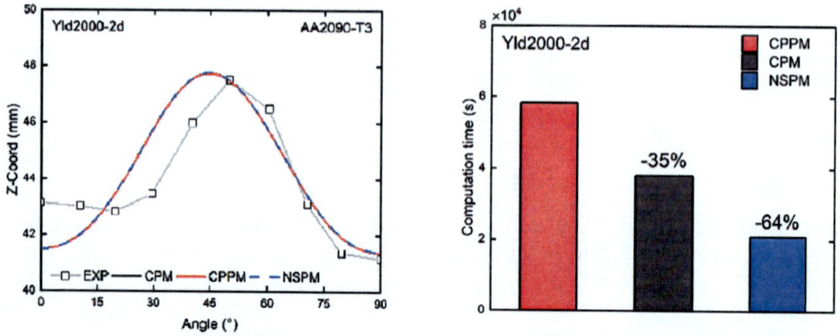

(a) Earing profiles (b) Computation costs

Fig. 5. (a) Earing profiles after deep drawing simulations and (b) computation costs obtained from CPPM (fully implicit), CPM (semi-explicit), and NSPM (fully explicit)

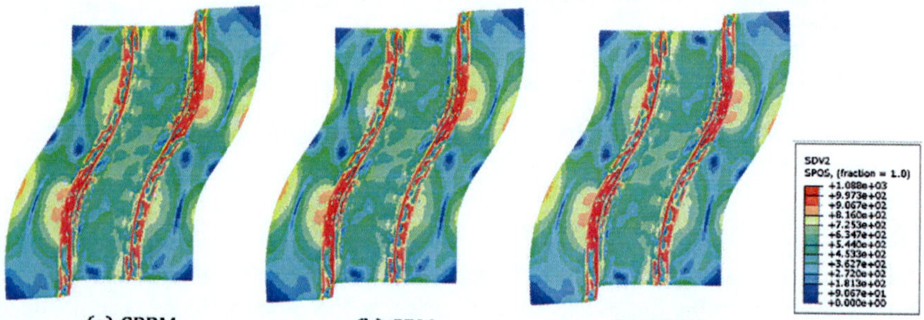

(a) CPPM (b) CPM (c) NSPM

Fig. 6. Effective stress (SDV2) distribution of HAH_{20} in S-rail forming explicit finite element simulations integrated using (a) CPPM (fully-implicit), (b) CPM (semi-explicit), and (c) NSPM (fully-explicit)

4.3 Hole Expansion: Modified GTN

The modified GTN model [6] is employed for failure prediction in a conical hole expansion simulation. The model coefficients are calibrated based on the fracture behavior of dual-phase steel (DP980).

Figure 8 exhibits the equivalent stress of the modified GTN when the strain localization initiates. The stress pattern and the localization positions show reasonable agreement. The deviations among the load-displacement curves in Fig. 9(a) might result from the extraordinarily large strain increment when the localization is initiated. CPPM demands 3 to 7 iterations for the algorithm convergence, and CPM 2 to 5. Consequently, NSPM boosts the simulation time by 60%.

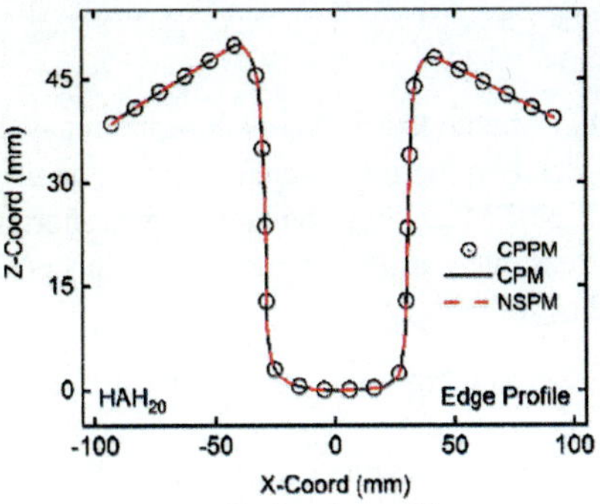

(a) Springback profiles

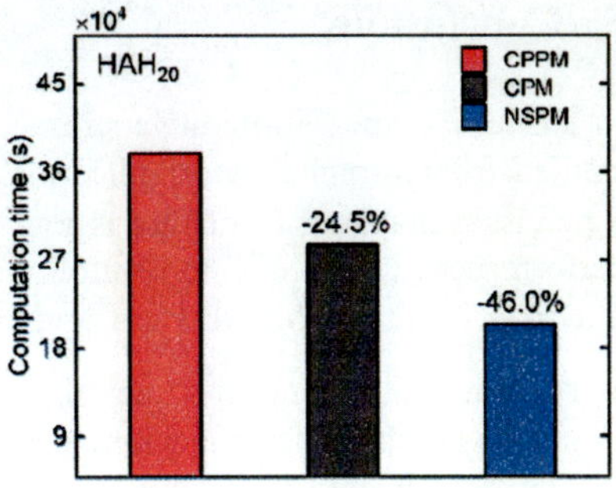

(b) Computation costs

Fig. 7. (a) Springback profiles after S-rail forming simulations and (b) computation costs obtained from CPPM (fully-implicit), CPM (semi-explicit), and NSPM (fully-explicit)

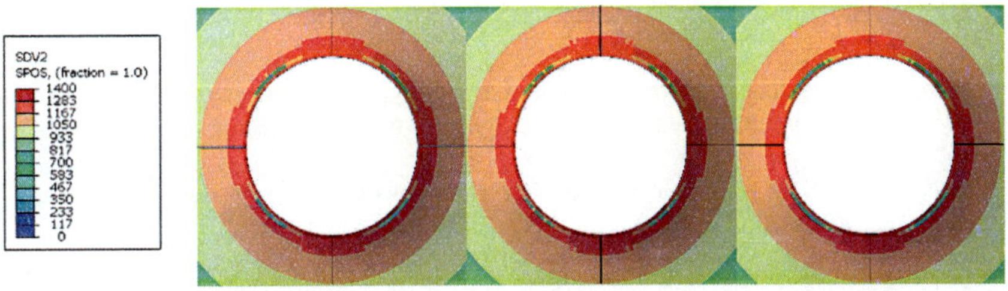

(a) CPPM (b) CPM (c) NSPM

Fig. 8. Effective stress (SDV2) distribution of the modified GTN in the hole expansion explicit finite element simulations integrated using (a) CPPM (fully-implicit), (b) CPM (semi-explicit), and (c) NSPM (fully-explicit)

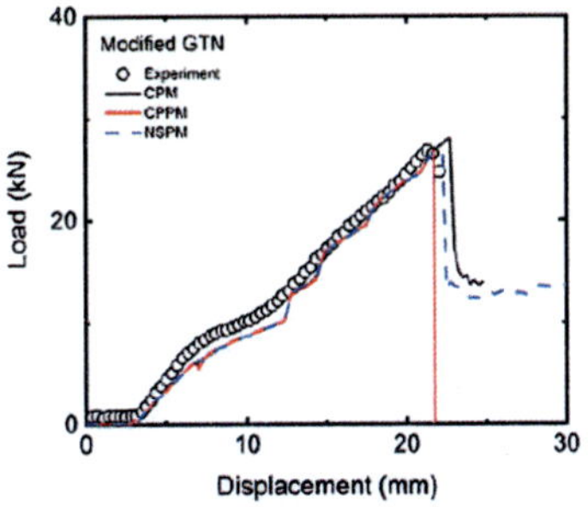

(a) Load-displacement curve

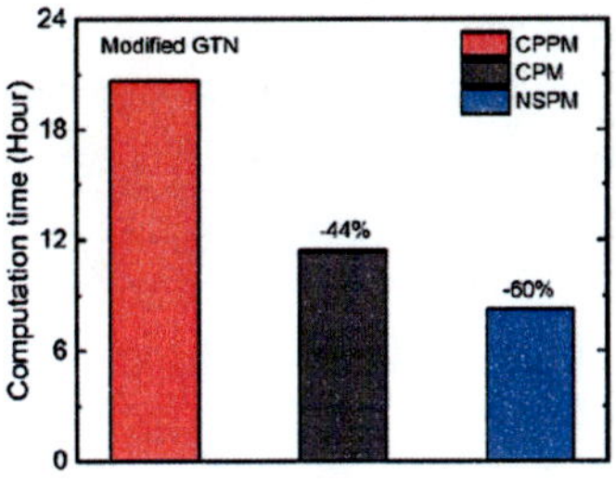

(b) Computation costs

Fig. 9. (a) Load-displacement curves in conical hole expansion simulations and (b) computation costs obtained with CPPM (fully implicit), CPM (semi-explicit), and NSPM (fully explicit)

5 Conclusions

In this article, the non-iterative stress projection method (NSPM) is comprehensively studied, from formulation to application. For a strict evaluation of the integration quality, a precision mapping technique is introduced. NSPM exhibits outstanding numerical performance in terms of computation speed, stress integration accuracy, and robustness. The main results of this work are summarized as follows:

- A chronic shortcoming of the fully-explicit stress integration method is resolved by employing the stress projection technique.
- The stress integration quality of NSPM is comparable to that of the return mapping-based stress update algorithm according to the precision map results.
- The excellent numerical performance of NSPM is qualitatively demonstrated based on a variety of finite element simulations and plasticity models.

References

1. Rossi, M., Lattanzi, A., Cortese, L., Amodio, D.: An approximated computational method for fast stress reconstruction in large strain plasticity. Int. J. Numer. Meth. Eng. **121**(14), 3048–3065 (2020)
2. Yoon, S., Barlat, F.: Non-iterative stress integration method for anisotropic materials. Int. J. Mech. Sci. **242**, 108003 (2023)
3. Yoon, S., Barlat, F.: Non-iterative stress projection method for anisotropic hardening. Mech. Mater. **183**, 104683 (2023b)
4. Lee, S., Barlat, F., Yoon, S.: Non-iterative stress projection method for continuum damage model. In: Preparation (2023)
5. Barlat, F., Yoon, S.Y., Lee, S.Y., Wi, M.S., Kim, J.H.: Distortional plasticity framework with application to advanced high strength steel. Int. J. Solids Struct. **202**, 947–962 (2020)
6. Nahshon, K., Hutchinson, J.: Modification of the Gurson model for shear failure. Eur. J. Mech-A/Solids **27**(1), 1–17 (2008)
7. https://github.com/theysy/Mapping_tools_public
8. https://github.com/theysy/mml_subroutine_public

Formability and Spring-Back of Light Metals at High Strain Rates

Shi-Hong Zhang(✉), Hao Li, Yong Xu, Shuai-Feng Chen, and Hong-Wu Song

Shi-changxu Innovation Center for Advanced Materials, Institute of Metal Research, Chinese Academy of Sciences, 72 Wenhua Road, Shenyang 110016, China
shzhang@imr.ac.cn

Abstract. Formability or plasticity of light metals, such as aluminum alloys and titanium alloys, are quite strain rate sensitive. For most normal plastic forming processes like forging, stamping, rolling and extrusion, their strain rates are usually at medium strain rates between 10^{-2} s^{-1} and 10^{2} s^{-1}, the formability becomes lower with increasing of strain rates. While at very low strain rates, the formability is often much higher, especially between 10^{-5} ~ 10^{-3} s^{-1}, at which super-plasticity is found for many metals and alloys. Spring-back is quite normal for sheet and tube forming, it is very small when the strain rate is lower than 10^{-3} s^{-1}, while it is usually very high when the strain rate is higher for medium strain rates (10^{-2} s^{-1} and 10^{2} s^{-1}), especially for titanium alloys. It has been found that there exists a range of strain rates around 1×10^{3}~$5 \times 10^{3} s^{-1}$ for many light metal alloys that the formability is increased remarkably, and the spring-back is also quite low. In this paper, variations of the formability and the spring-back with varied strain rates have been investigated, The research aims to find a suitable strain rate range with remarkable increased formability and lower spring-back, so that many sheet and tube parts of light metals can be formed with sufficient formability and very small spring-back at room temperature.

Keywords: Strain rate · Formability · Spring-Back · Light metal

1 Introduction

Light metals (aluminum alloys, magnesium alloys, titanium alloys, et al.) have excellent properties such as high specific strength, low density and corrosion resistance, they have various applications in aerospace, medical apparatus, automobile and other industries fields [1]. In view of the poor ductility of these alloys at room temperature, hot forming is commonly used in engineering production to improve the formability of components by improving the uniformity of microstructure [2]. However, the use of hot forming often leads to higher manufacturing costs and affects the forming efficiency of components. Therefore, it is necessary to develop a novel technology for enhancing forming accuracy and efficiency at room temperature.

Recently, it has been reported that the plasticity of light metals at room temperature can be significantly improved under high strain rate (HSR) tensile/compressive loadings

K. Mocellin et al. (Eds.): ICTP 2023, LNME, pp. 571–578, 2024.
https://doi.org/10.1007/978-3-031-40920-2_59

at room temperature [3]. Smerd et al. found that the elongation of 5754 aluminum alloy under high strain rate tensile loading at room temperature was 42% higher than that under low strain rate (LSR) tensile loading [4]. Li et al. carried out the electromagnetic expansion experiments with Ti-6Al-4V rings, and found that the uniform strain and the fracture strain of titanium alloy were both significantly increased when the strain rates exceeded 6935.6 s^{-1} [5]. In addition to uniaxial tensile loading under HSR, more and more experimental results show that the formability of light metal sheets can be significantly improved by using high strain rate forming technologies (electromagnetic forming, electro-hydraulic forming, impact hydraulic forming (IHF), explosive forming, et al.) [6]. Maris et al. found that the ultimate principal strain of 5182 aluminum alloy under electro-hydraulic forming mode was significantly higher than that under LSR loading [7]. Similarly, Yu et al. found that the forming limit height of DP600 high-strength steel sheet can be increased by 27% compared with LSR forming by using electro-hydraulic forming [8]. Further, they studied the formability improvement mechanism of high strength steel with electro-hydraulic forming, and found that the early necking of high strength steel samples could be avoided under HSR loading due to inertia effect, and the dislocation distribution changed from aggregating and tangling at the grain boundaries under LSR loadings to the uniform distribution in the grains, which promotes the coordinative deformation between grains [8]. Although the formability improvement of light metals at room temperature under HSR loadings has been proved, systematic studies on the plasticity evolution in wide strain rate range and the forming process window of light metal sheets under HSR loadings has been seldom reported.

In practical production, shape accuracy is also an important standard to evaluate the forming quality of components. For some light metals (such as titanium alloys), it is easy to produce severe spring-back after unloading at room temperature. At present, hot stamping [9], increment forming [10], multi-pass forming [11] et al. are commonly used to eliminate residual stress of components. However, the above methods also have the problems of low forming efficiency and high costs. According to recent reports, HSR loadings can also effectively reduce the spring-back of sheet metal at room temperature and improve the shape accuracy of components [12]. Golovashchenko et al. applied electro-hydraulic loading for correcting 6111-T4 aluminum U-shaped parts after unloading, and achieved high shape accuracy [13]. Similarly, Cui et al. performed electromagnetic assisted forming to correct the AA3033 arc-bent specimen, and also reduced the spring-back significantly [14]. Therefore, HSR loading has positive effect on reducing spring-back of light metal component. Whereas, there are still few reports on realizing spring-back reduction with one loading procedure under HSR.

In this paper, several typical aluminum alloys, Al-Li alloys and Ti-6Al-4V alloy sheets are selected as the tested materials. Their plasticity variation laws under wide strain rate loadings are studied by experiments. Furthermore, the plasticity enhancement mechanism under HSR loading is revealed based on 5A02 aluminum alloy sheet. Through conducting IHF experiments with 2B06 aluminum alloy sheets under different drawing ratios, the forming process window was established. In addition, the spring-back behaviors of Ti-6Al-4V sheets under IHF was tested and a Ti-6Al-4V complex component was formed with this technology by experiment.

2 Evolution of Plasticity of Light Metal Sheet Under Wide Strain Rates

In this work, wide strain rate tensile tests (1×10^{-3}~5×10^{3} s^{-1}) of light metals were performed at room temperature. Where, LSR tensile tests (1×10^{-3} ~ 1×10^{-1} s^{-1}) were conducted with Z150 tensile tester, and HSR tensile tests (1×10^{3} ~ 5×10^{3} s^{-1}) were based on the Split Hopkinson Tensile Bar (SHTB) equipment. The relationship between the fracture strain and the strain rate of several typical alloys obtained by experiments is shown in Fig. 1. It can be seen that the fracture strain of the tested materials under HSR is much higher than that under LSR (1×10^{3} s^{-1}). The percentages of the maximum amplification (MA) of fracture strain are shown in Table 1. According to the test results in Table 1, the maximum strain amplification of aluminum alloy is much larger than that of titanium alloy under HSR.

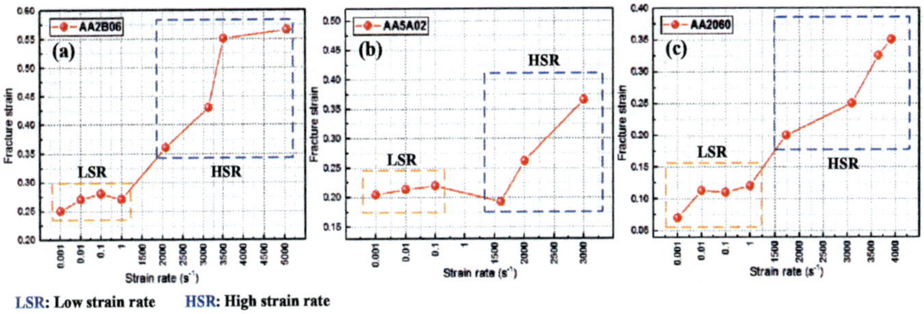

Fig. 1. Relationship between fracture strain and strain rate (a) AA2B06, (b) AA5A02, (c) AA2060

Table 1. Fracture strain amplification of the tested light metals under HSR

Material	2B06	5A02	5A06	2060	2195	Ti-6Al-4V
MA of strain %	126.0	78.92	86.67	400.0	111.45	56.95

AA5A02 sheet is selected to reveal the plasticity enhancement mechanism of light metals under HSR loadings. The fracture specimen with tensile strain rate of 2000 s^{-1} was selected to carry out EBSD measurement. According to the EBSD results, the average grain size near the fracture of the specimen does not change significantly under HSR loading (from 11.51 μm before tension to 11.44 μm after fracture). In addition, the average orientation difference angle of grains under HSR is significantly smaller than that of the undeformed specimen (from 34.65 ° before tension to 18.94 ° after fracture). It should be noted that the high-angle grain boundaries (HAGBs) ($\theta > 15$ °) have a significant reduction in proportion, and low-angle grain boundaries (LAGBs) (2 ° $< \theta$ < 5 °) have a significant increase in proportion, as shown in Fig. 2. Considering that continuous movable dislocation multiplication will produce a large number of LAGBs [15], a large number of dislocation multiplications occurred in the grain under HSR can be inferred. In addition, it was found that the distribution of LAGBs near the fracture is more uniform under HSR, which is quite different from the nonuniform distribution of LAGBs near the fracture of 5052 aluminum alloy specimen under LSR [15]. Above all, the multiplication of movable dislocations and the uniform distribution of dislocations are the main reasons for the plasticity improvement under HSR loading.

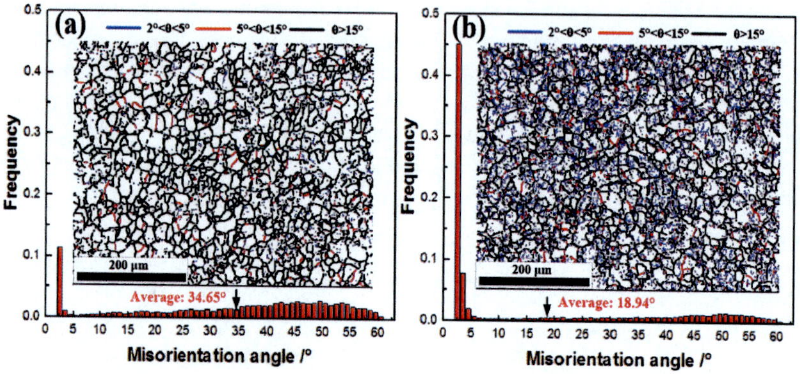

Fig. 2. Distributions of HAGBs and LAGBs of 5A02 aluminum alloy (a) microstructure before tension (b) microstructure near fracture under strain rate of 2000 s^{-1}.

3 Formability of Light Metal Sheet under HSR Loading

In this work, the formability of the 2B06 aluminum alloy sheet was evaluated with ordinary hydroforming and IHF experiments, respectively. The forming principles of the two kinds of loading methods are shown in Fig. 3. According to the experimental results shown in Fig. 4, the formability of AA2B06 cylindrical parts with impact forming of velocities of 10~30 m/s was remarkably improved compared with that under ordinary hydroforming. In order to reveal the mechanism of formability improvement under HSR loading, numerical simulation was carried out based on above the two loading methods. It was found that, the deformation begins from the central region to the edge of the workpiece under ordinary hydroforming, while for Impact Hydroforming the deformation begins from the sidewall edge to the central region. It will make more material flow from the edge to the deformation region, and further prevent the early occurrence of fracture. In addition, the strain rate at the sheet center under the loading velocity of 30 m/s is quantitatively calculated with simulation results. It is found that the maximum strain rate at the central region can reach to 7000 s^{-1}. Based on the relationship between the fracture strain and the strain rate shown in Fig. 1, it can be judged that the room temperature formability of 2B06 aluminum alloy sheet at this strain rate is in a formability improvement zone. Above all, the formability improvement of 2B06 aluminum alloy sheet under IHF at room temperature can be attributed to the change of deformation history of the sheet during forming process and the plasticity improvement effect under HSR loading.

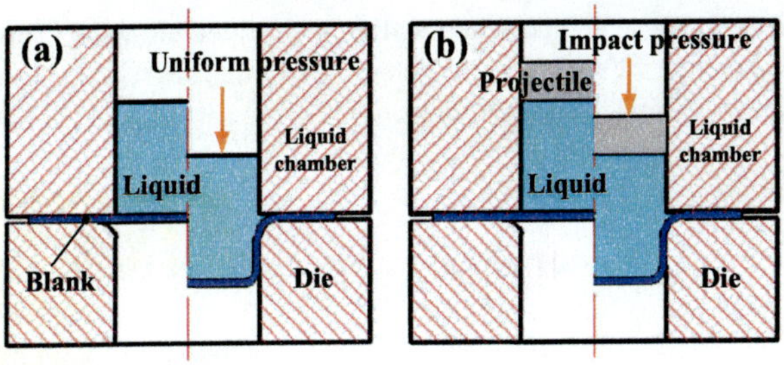

Fig. 3. Forming principle (a) ordinary hydroforming (b) impact hydroforming.

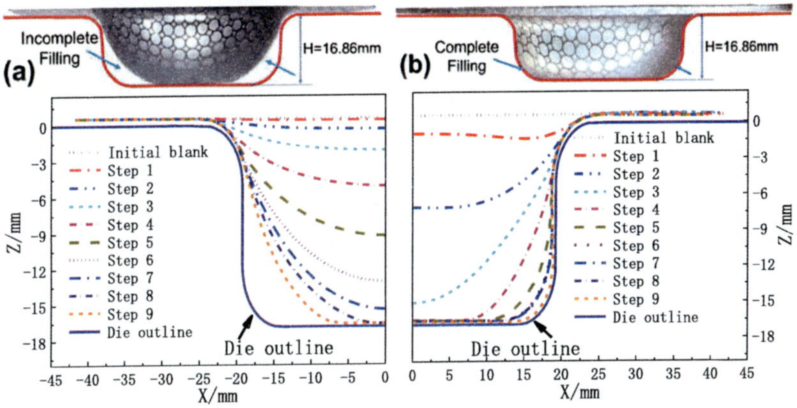

Fig. 4. Forming results: (a) ordinary hydroforming (0.1 m/s); (b) IHF (30 m/s).

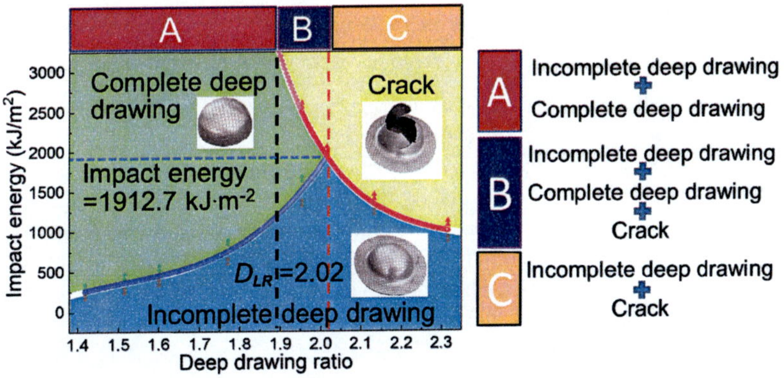

Fig. 5. Forming process window of 2B06 aluminum alloy sheet under IHF

The formability characterization of 2B06 aluminum alloy sheet was evaluated by experiments with different drawing ratios (1.42, 1.51, 1.60, 1.77, 1.95, 2.13 and 2.31) under IHF [16]. The obtained forming process window is shown in Fig. 5. It can be learned that the limit drawing ratio is 2.02 when realizing complete drawing. The corresponding impact energy is close to 2000 kJ/m^2. When the specimen size exceeds the limit drawing ratio and enhancing impact energy continuously, the tested specimen has the risk of fracture.

4 Spring-Back Behavior of Light Metal Sheet Under HSR Loading

Considering that the severe spring-back will occur when forming titanium alloy components with traditional methods at room temperature, the spring-back behaviors of Ti-6Al-4V sheet under IHF was tested. The length, the width and the thickness of sheet blank are 146 mm, 28 mm and 0.8 mm, respectively. The curvature radius and the depth of the die cavity are R80 mm and 17 mm. In order to find the spring-back characteristic of Ti-6Al-4V sheet under HSR, the conventional punch bending (CPB) and impact hydraulic bending (IHB) experiments were both carried out on the self-developed IHF equipment. The loading velocities used in the experiments were 0.1 m/s (CPB) and 43 m/s (IHB). According to the experimental results shown in Fig. 6, the spring-back amount of the bent specimen obtained under CPB is close to 100%. When the impact velocity is increased to 43 m/s, the spring-back amount is significantly reduced to 4.9%. Hence, it can be proved that the severe spring-back of titanium alloy sheet can be significantly reduced under HSR at room temperature.

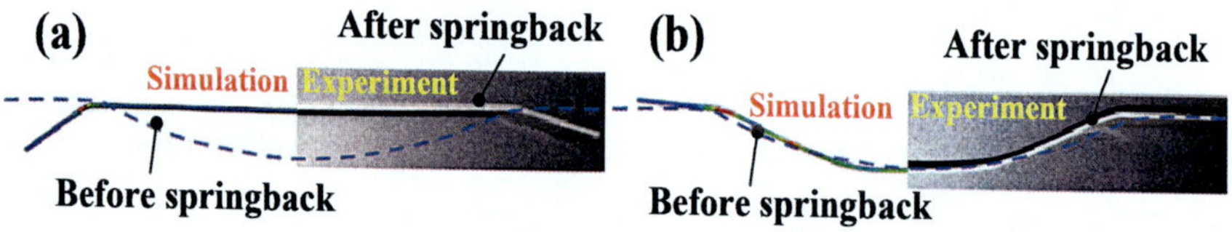

Fig. 6. Spring-back results of Ti-6Al-4V sheet: (a) CPB (0.1 m/s); (b) IHF (43 m/s)

In order to further quantitatively characterize the created strain rates at different zones during bending process under different loading velocities, numerical simulation was carried out. The simulation accuracy was verified by comparing with the experimental results, as shown in Fig. 6. For calculating the strain rate at the different regions of the sheet, the logarithmic strain is extracted from the simulation results. According to the calculation results, the maximum strain rate under CPB is 3.3 s^{-1}, and the maximum strain rate will reach 7040.46 s^{-1} when the impact velocity is up to 43 m/s. Additionally, it can be also learned from Fig. 7 that the maximum strain rate appears in the fillet region.

Complex thin-walled component of Ti-6Al-4V was formed by IHF at room temperature. The component obtained by experiment is shown in Fig. 8. By comparing the geometric profile of the obtained component with the geometric profile of the die cavity, it was found that the spring-back amount of the component under IHF is greatly reduced. It indicates that HSR forming can be used to improve the forming accuracy of the titanium alloy complex thin-walled components at room temperature.

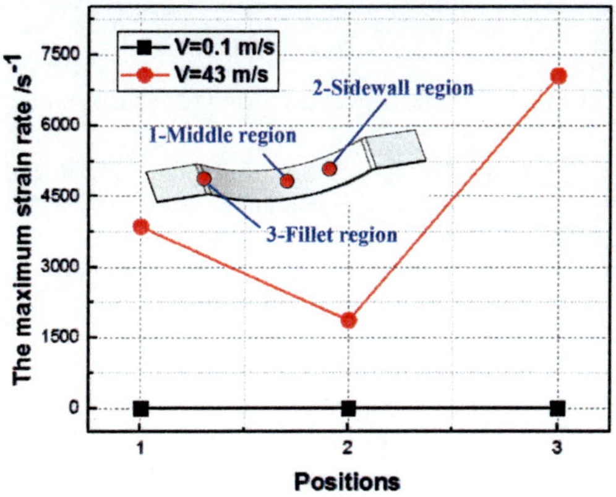

Fig. 7. The distributions of the maximum strain rates at different sheet regions at different loading velocities

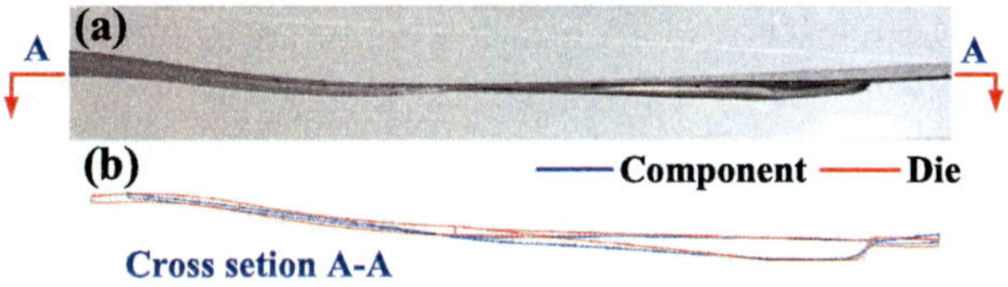

Fig. 8. Experimental results under IHF: (a) the obtained Ti-6Al-4V thin-walled component; (b) the die fitting result of the component obtained after unloading

5 Conclusions

(1) High Strain Rate loading can significantly improve the plasticity of light metal sheet at room temperature. The plasticity increment of aluminum alloy is much larger than that of titanium alloy under High Strain Rate loading.
(2) High Strain Rate can effectively improve the room temperature formability of light metal sheet. The proposed forming process window can be used to guide the design of precise forming of components and prevent the occurrence of cracks.
(3) High Strain Rate bending can significantly reduce the severe spring-back of titanium alloy sheet at room temperature. In addition, this technology can be further applied to realize the precise forming of titanium alloy of complex thin components at room temperature.

Acknowledgments. This work was supported by the National Natural Science Foundation of China (Grant No. 52111530293); and the Sino-Belarus Inter-Governmental Science and Technology Cooperation project (Grant No. CB02–01).

References

1. Banerjee, D., Williams, J.C.: Perspectives on titanium science and technology. Acta Mater. **61**, 844–879 (2013)
2. Odenberger, E.L., Oldenburg, M., Thilderkvist, P., Stoehr, T., Lechler, J., Merklein, M.: Tool development based on modelling and simulation of hot sheet metal forming of Ti-6Al-4V titanium alloy. J. Mater. Process. Technol. **211**, 1324–1335 (2011)
3. Zhang, H.R., Zhao, H., Wang, R., Liang, X.X., Zhu, Y.: Coupling influences of elevated temperature and strain rate on the behaviour of 6061–T6 aluminium alloy used as construction material. Structures **40**, 596–606 (2022)
4. Smerd, R., Winkler, S., Salisbury, C., Worswick, M., Lloyd, D., Finn, M.: High strain rate tensile testing of automotive aluminum alloy sheet. Int. J. Impact. Eng. **32**(1), 541–560 (2005)
5. Li, F.Q., Mo, J.H., Li, J.J., Huang, L., Fan, W., Fang, J.X.: Effects of deformation rate on ductility of Ti-6Al-4V. Material **81**, 754–759 (2014)
6. Zheng, Q., Yu, H., Cai, X.: Formability and deformation behavior of DP600 steel sheets during a hybrid quasi-static/dynamic forming process. Int. J. Adv. Manuf. Technol. **110**(7–8), 2169–2180 (2020). https://doi.org/10.1007/s00170-020-06014-1
7. Maris, C., et al.: Comparison of quasi-static and electrohydraulic free forming limitsfor DP600 and AA5182 sheets. J. Mater Process Technol. **235**, 206–219 (2016)
8. Zheng, Q.L., Yu, H.P.: Hyperplasticity mechanism in DP600 sheets during electrohydraulic free forming. J. Mater. Process Technol. **279**, 116582 (2020)
9. Maeno, T., Tomobe, M., Mori, K.I., Ikeda, Y.: Hot stamping of titanium alloy sheets using partial contact heating. In: 17th International Conference on Metal Forming. Metal Forming 2018, Procedia Manufacturing, vol. 15, pp. 1149–1155 (2018)
10. Jiang, B., Yang, W.B., Zhang, Z.Y., Li, X.F., Ren, X.P., Wang, Y.Q.: Numerical simulation and experiment of electrically-assisted incremental forming of thin TC4 titanium alloy sheet. Materials **13**, 1335 (2020)
11. Badr, O.M., Rolfe, B., Zhang, P., Weiss, M.: Applying a new constitutive model to analyse the spring-back behaviour of titanium in bending and roll forming. Int. J. Mech. Sci. **128–129**, 389–400 (2017)
12. Li, H., Xu, Y., Chen, S.F., Song, H.W., Zhang, S.H.: Spring-back inhibition of Ti-6Al-4V sheet with impact hydroforming at room temperature. In: Proceedings of the International Deep-Drawing Research Group Conference (IDDRG 2022), vol. 1238, p. 012076 (2022)
13. Golovashchenko, S.F., Gillard, A.J., Mamutov, A.V., Ibrahim, R.: Pulsed electrohydraulic spring-back calibration of parts stamped from advanced high strength steel. J. Mater. Process Technol. **214**, 2796–2810 (2014)
14. Du, Z.H., et al.: Spring-back control and large skin manufacturing by high-speed vibration using electromagnetic forming. J. Mater. Process Technol. **299**, 117340 (2022)
15. Zhao, G.H., Sun, M.X., Li, J., Li, H.Y., Ma, L.F., Li, Y.G.: Study on quasi-in-situ tensile microstructure evolution law of 5052-O aluminum alloy based on EBSD. Mater. Today Commun. **33**, 104572 (2022)
16. Chen, D.Y., et al.: A novel method to evaluate the high strain rate formability of sheet metals under impact hydroforming. J. Mater Process Technol. **287**, 116553 (2021)

A Material Model Optimization Approach for the Sheet Metal Forming Process Using the Hole Expansion Test

Trunal Bhujangrao(✉) and Toni Chezan

Tata Steel Research & Development, IJTC, PO Box 10000, NL-1970 CA IJmuiden, The Netherlands

trunal.bhujangrao@tatasteeleurope.com

Abstract. Sheet metal forming is an important manufacturing process widely used to produce complex stamped parts from flat sheet stock in industries such as automotive and packaging. Due to the global economic climate, these industries need to be highly competitive by reducing production costs and increasing process efficiency. Numerical simulation combined with sheet metal forming expertise is one of the technological innovations adopted to meet these requirements by reducing the traditional time-consuming and costly testing steps. With the progress of finite element simulation, questions about the accuracy or limitations of the type of material description adopted have become particularly important. The influence of the plasticity model is examined in this work by a numerical study using the hole expansion test. This work first presents the yield locus criterion adopted and developed by Tata Steel, which is hereafter referred to as the Tata Steel material model. Hole expansion tests are performed at different hole diameters and the results are compared with FE simulation. The simulations are performed with the finite element software Autoform R11 in which the yield criterion proposed by Abspoel & Scholting [1] has been implemented. The discussion therefore focuses on the influence of the material model on the numerical predictions and its accuracy based on the optimization of the different material parameters measured.

Keywords: hole expansion test · yield locus criteria · FE simulation · DIC technique

1 Introduction

Material flow control during the forming process plays an important role in the manufacture of complex automotive part geometries such as bodysides, door inners and tailgates. Such geometries pose a heavy demand on material stretchability and several strategies are necessary for optimizing the strain distribution. Holes are generally punched in the blank at door or window aperture areas in order to prevent excessive strains and part splitting. The location and shape of these relief holes is optimized in order to promote controlled material flow in the inner flange. The aim is to design a hole geometry that promotes linear strain paths in the flange and prevents strain localization as a result of

K. Mocellin et al. (Eds.): ICTP 2023, LNME, pp. 579–590, 2024.
https://doi.org/10.1007/978-3-031-40920-2_60

strain path changes. The material flow behavior at and in the neighborhood of the hole is dependent on the anisotropic properties of the sheet metal. It is therefore important to understand the influence of anisotropy on hole expansion in order to ensure linear strain paths in the flange. Therefore there is a high demand to develop an accurate material model to understand and to improve the predictive accuracy of finite element analysis of the deformation behavior in hole expansion tests. The selection of a material model that is able to reproduce the plastic deformation behavior of the material as accurately as possible is generally important. Several investigations of the strain distribution in biaxial stretching of holes has been reported in Parmar and Mellor [2] and other publications [3–8]. Beyond these works, a variety of anisotropic yield criteria, e.g., Hill48 [9, 10] is calibrated either on r-values (the plastic strain ratios) in three directions and the yield strength in rolling direction or on the yield stresses in those three directions and the biaxial yield stress. The function is too rigid to describe the other stress states correctly. Barlat 89 [11] is used mainly to describe aluminum materials and has as input the three strain ratios (r-values 0°, 45°, 90°) and an exponent m, and Yld2000-2D [12], or crystal plasticity, have been used in numerical simulations of hole expansion (HE) [13]. In some of these works, anisotropy was calibrated based on only uniaxial tensile tests at angles to the rolling direction (RD).

In this study, the strain behavior of zinc coated cold rolled sheet steel (DX57D + Z) subjected to hole expansion using a flat cylindrical punch has been investigated both experimentally and numerically. Experimental hole expansion tests have been undertaken with a circular punched hole in a circular blank. The hole is stretched using a cylindrical punch and the strain distribution at and near the hole is recorded using the GOM Aramis full-field online strain measurement system. In order to determine an appropriate anisotropic yield function for the test material, FEM simulations of the hole expansion were performed. The anisotropic yield functions used in the simulations are developed by Tata Steel compared with different yield criteria. The predictions of local responses have been precisely measured along the hole edge and compared with the simulation results. Consequently, sensitivity analysis of the mesh dependency on accuracy is also discussed.

2 Materials and Methods

2.1 Experiment

The hole expansion test with a circular punch hole was selected for these investigations. The experiment was performed using an Erichsen hydraulic press at a blank holder force of 250 kN). The same tooling setup as used for Nakajima tests (serrated blank holder and the levelling rings between the die and the lid) was used for this test with a 100 mm cylindrical punch diameter with punch head radius 10 mm as shown in Fig. 1.

The Aramis system (i.e. DIC system) was mounted on the Erichsen press for recording the strain evolution with the square grid size 0.45 mm. A typical forming steel in this study DX57D + Z (normally used for bodysides, door inners and tailgates) with thickness of 0.75 mm was selected for testing. The samples were manually centered in the Erichsen press in order to ensure concentricity of the hole and punch and the punch displacement and force were simultaneously recorded. The samples were then

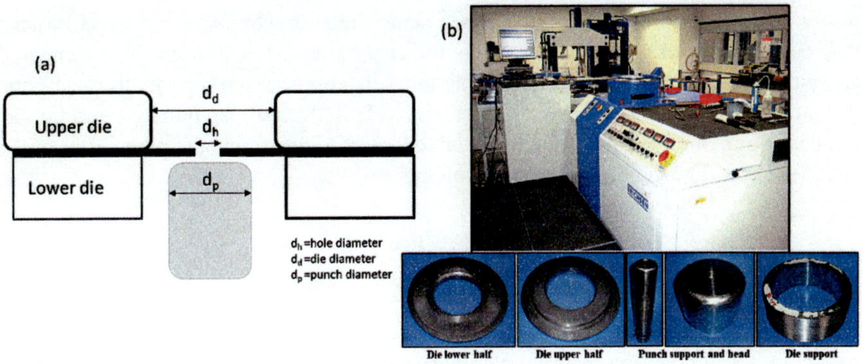

Fig. 1. (a) schematic representation of test equipment and (b) tool setup used for experimental testing in the Erichsen press

clamped with the maximum blank holder force which ensured proper clamping. The punch was then displaced at a constant velocity of 0.5 mm/s until visible sample failure wasobtained.

The influence of normal and planar anisotropy on material flow during hole expansion is studied. High normal anisotropy would promote material flow in the sheet plane. The magnitude of the deformation during hole expansion was recorded as a 3D measurement of the strain distribution at a constant rate of frames per second. The frame just before visible neck is considered as material limit. The experiment tests are performed for eight different hole diameters of 4, 6, 8, 10, 16, 20, 23 and 27 mm, to study the influence of the hole dimeter to the major strain. The hole expansion test stopped manually as soon as the failure occur.

3 Material Modelling

The material model consists of a hardening curve and a yield locus which are assumed to be independent, and isotropic hardening is considered. Where the plastic flow under multiaxial stress states is described by an yield criterion, which is derived directly from test data at the four key stress states: uniaxial stress, plane strain, biaxial stress and pure shear. To find the best yield surface description, several are mutually compared. With each description the number of parameters that need to be determined are stated.

3.1 Hill Yield Criteria

Hill48 yield criterion predicts yield locus based on a combination of the three principal stresses and their orientations. The Hill criterion is based on the assumption that the material fails when the maximum shear stress exceeds a critical value that depends on the principal stress orientations. It is one of the most commonly used yielding criteria, with a simple mathematical formulation and a few parameters. The criterion is expressed mathematically as follows:

$$f = (G + H)\sigma_{xx}^2 - 2H\sigma_{xx}\sigma_{yy} + (H + F)\sigma_{yy}^2 + 2N\sigma_{xy}^2 = \sigma^{-2} \tag{1}$$

where σ_{xx} and σ_{yy} are the normal stresses along the x direction (rolling direction) and y direction (transverse direction); σ_{xy} is the shear stress; F, G, H and N are material parameters. Where Hill90 [10] criterion assumes that the material is perfectly plastic and that the yield surface is an ellipsoid in stress space. Hill90 criterion requires one additional test consisting of either bi-axial tension or shear test to determine the parameters of the yield function. The yield criterion is given by:

$$\begin{aligned} f(\sigma) &= \left|\sigma_{xx} + \sigma_{yy}\right|^m + \left(\tfrac{\sigma_b}{\tau}\right)^m \left|\left(\sigma_{xx} - \sigma_{yy}\right)^m + 4\sigma_{xy}^2\right|^{\frac{m}{2}} \\ &+ \left|\sigma_{xx}^2 + \sigma_{yy}^2 + 2\sigma_{xy}^2\right|^{\frac{m}{2}} \left[-2a\left(\sigma_{xx}^2 - \sigma_{yy}^2\right) + b\left(\sigma_{xx} - \sigma_{yy}\right)^2\right] \\ &= (2\sigma_b)^m \end{aligned} \tag{2}$$

where, m, a and b are the material parameters with $m > 1$, σ_b and τ are yield stress in bi-axial tension and shear respectively.

3.2 Tata Steel Model (Vegter 2017 Criterion)

The Tata Steel model based on the Vegter 2017 yield criterion is a more recent development that is based on a generalized form of the von Mises yield criterion. The Vegter 2017 criterion assumes that the material exhibits strain hardening or softening. In order to obtain the yield locus several experimental tests are performed to derive the material model constants (tensile testing (ISO 6892–1 and ISO10113), plane strain testing (not standardised) biaxial testing (Bulge: ISO 16808); Cruciform specimen: ISO 16842; Compression: not standardised) and shear tests (not standardised)). Tata Steel has long been a proponent of physics-based hardening laws, as promoted by Bergström and adapted by Vegter & van Liempt [14, 15]. These laws are generally used in finite element simulations as tabulated data. The advantage of this law is that it considers the effect of strain rate and temperature on hardening behavior for high levels of strain. And for an accurate construction of the yield locus, stress and strain data in four key stress states are studied experimentally. In this paper, the yield locus is defined based on Vegter 2017. The development of this model and its explanation is given in [1]. The parameters required such as stress factor in uniaxial direction $F_{un\text{-}0} = \sigma_{un\text{-}0}/\ \sigma_{un\text{-}0} = 1$, $F_{un\text{-}45} = \sigma_{un\text{-}45}/\ \sigma_{un\text{-}0}$ $F_{un\text{-}90} = \sigma_{un\text{-}90}/\ \sigma_{un\text{-}0}$, and r_0, r_{45} and r_{90} are all easily derived from the tensile test and will be used to predict the stress factors for the biaxial, plane strain and shear points on the yield locus. Figure 2 indicates graphically the Vegter 2017 controls for the biaxial point, plane strain and shear strain.

To predict the biaxial stress factor biaxial (F_{bi}) with the uniaxial factors and the r-values, the biaxial point is first normalized with the average of the uniaxial factors and plotted versus the average of the r-values. Where $F_{un\_av} = (F_{un\text{-}0} + 2{\cdot}F_{un\text{-}45} + F_{un\text{-}90})/4$ and $r_{av} = (r_0 + 2{\cdot}r_{45} + r_{90})/4$, and the normalized biaxial stress factor is F_{bi}/F_{un_av}. As indicated in Fig. 2, there are materials with evolving yield loci and there are materials without or almost no yield locus evolution. So first the evolution of the biaxial factor for the entire data is given by following equation.

$$F_{bi} = F_{un\_av}\left[\frac{F_{bi\_min}}{1 + f_{dirac}} + \frac{F_{bi\_max}}{1 + \frac{1}{f_{dirac}}}\right] \tag{3}$$

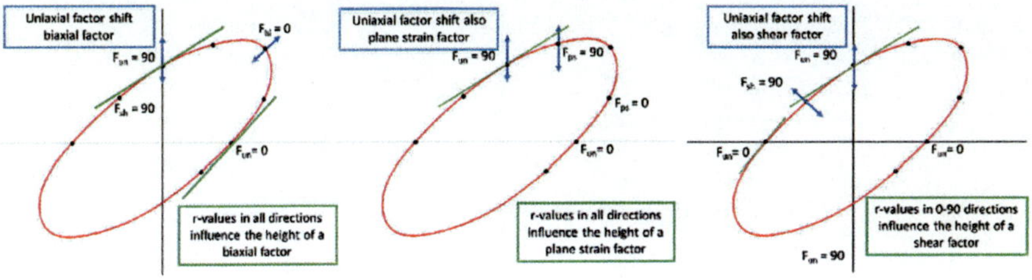

Fig. 2. Principle of how Vegter 2017 controls plane strain, shear and biaxial (each point individually)

Similarly the plane stress (ps) and shear stress (sh) factor are determined, f_{dirac} is Dirac function and can be defined for each stress factors [1].

$$F_{ps} = F_{un\_\varphi}[\frac{F_{ps\_min}}{1 + f_{dirac}} + \frac{F_{ps\_max}}{1 + \frac{1}{f_{dirac}}}] \tag{4}$$

$$F_{sh} = F_{un_{av0}-90}[\frac{F_{sh\_min}}{1 + f_{dirac}} + \frac{F_{sh\_max}}{1 + \frac{1}{f_{dirac}}}] \tag{5}$$

With these equations, a complex yield locus can now be predicted. Here, all the parameters are defined based on the average values. However, during the simulation, these parameters are optimized by considering the minimum and maximum parameters. The mechanical properties for this material grade is listed in Table 1.

Table 1. Mechanical properties of DX57D + Z with thickness $t_0 = 0.75$ mm

Grade: DX57D+Z														
Angle	Rp [MPa]	Rp-avg	Rm [MPa]	Rm-avg	Ag [%]	Ag-avg	A80 [%]	A80-avg	r [-]	r-avg	Δr	n [-]	n-avg
0°	154	161	287	291	24.2	23.1	47.1	45.4	2.268	2.04	0.444	0.231	0.224
45°	165		296		22.4		44.1		1.717			0.218	
90°	159		283		23.5		46.4		2.514			0.228	

In summary, the Hill48, Hill90 and the Tata Steel yield criteria are anisotropic yield criteria that are used to model the plastic behavior of the materials. However, the Tata Steel model criterion includes a term to account for strain hardening or softening and has large number of parameters. Moreover from Fig. 3, it is observed that the Hill90 and Tata Steel material models show quite similar yield locus. But it can provide a better representation of all measured data than with the Hill90 model.

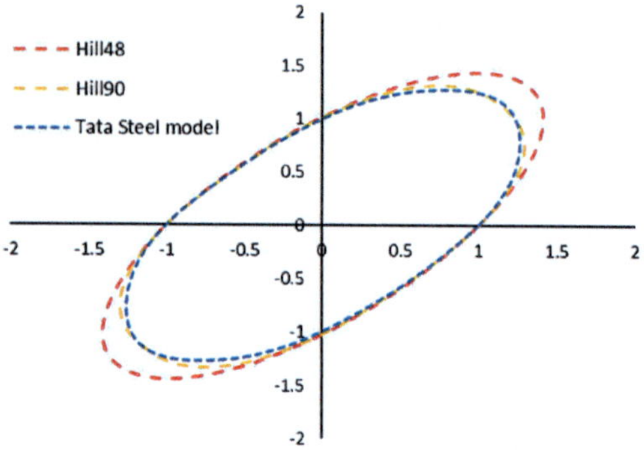

Fig. 3. Comparison of different yield locus for different yield criteria

4 Finite Element Analysis

FE-simulation was performed using Autoform R11 with the material models discussed above. Same tooling with the experiment, was modeled using non-deformable, analytical rigid body elements such as die and punch. The blank was constructed using deformable triangular element with constant size (0.45 mm) and taken with similar strain resolution as the Aramis measurements. The parameter 'heat generation' to include heat due to plastic work and friction is switched on. The tools and the meshed blank used for the FE simulation are shown in Fig. 4.

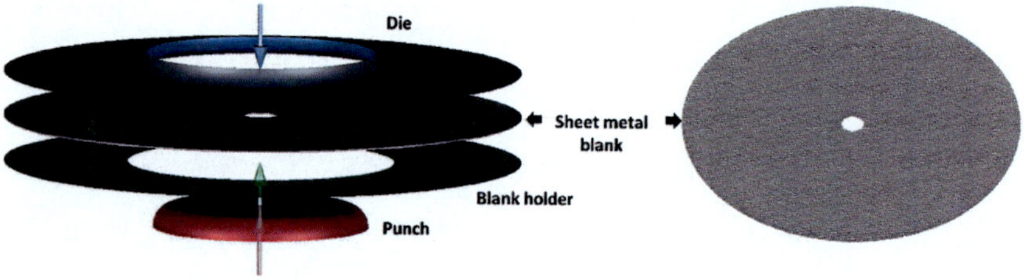

Fig. 4. Finite element model and the sheet metal blank mesh for the hole expansion simulation

Surface-to-surface contact was assumed for all contacts with the blank, and the Coulomb friction law was applied with a constant coefficient $\mu = 0.09$. The simulation results were compared with the experiment regarding the punch major strain in radial direction.

5 Results and Discussion

Force-punch displacement data recorded for all tests presented in Fig. 5. The trend of the force shows an increase with decreasing of the hole diameters. The tests are also repeated to check whether the results are in line with the trend. Each test is repeated 3 times for each hole diameter.

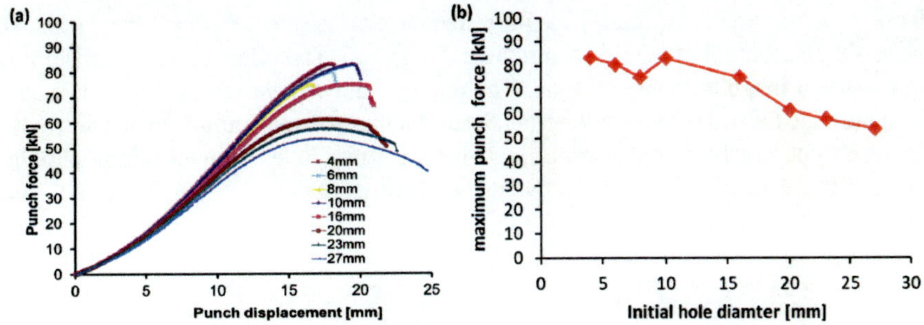

Fig. 5. (a) force-displacement curves for corresponding hole diameter. (b) max punch force as a function of the initial hole diameter for four materials.

In general the maximum punch force decreases with increasing initial hole diameter for the given material. The curves show a shift to a lower maximum force magnitude for initial diameter of 8 mm and lower. This effect appears to be related to the early failure.

5.1 Strain Analysis for Different Hole Diameters

The rectangular grid patterns were used for strain measurement, however there is a limit of grid measurement at the physical edge of the hole as shown in Fig. 6, there is missing information that cannot be analysed experimentally, therefore all the results are carried out at locations near to the hole edge.

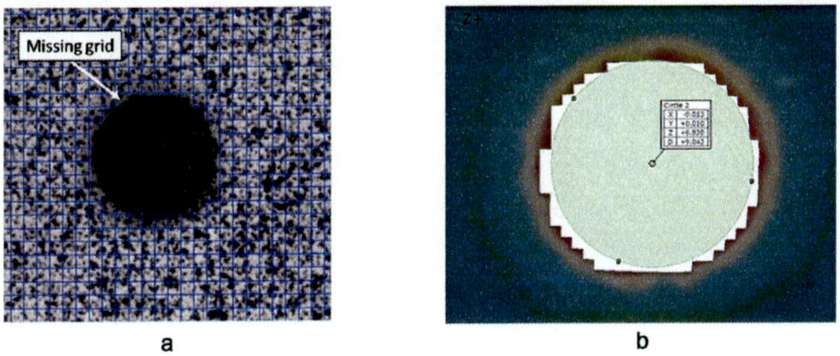

Fig. 6. Strain grid superimposed on test specimen showing limitation in strain computation at the physical hole edge (a) and the definition of the physical hole in the strain plot (b)

The strain state at a stage just before visible fracture is observed either in the bulk material or at the edge of the hole is studied for different initial hole diameters. The initial hole diameter determines the strain evolution in the circular region neighboring the hole whereas points farther away from the hole show lesser influence of the hole with far away points loaded in biaxial stretching. Circular sections were placed at radial distances of 1 mm and 5 mm from the deformed hole at a defined punch displacement. The maximum punch displacement was recorded to be 16 mm. The effect of initial hole diameter on the strain distribution for the tested material is shown in Fig. 7 (here it shown the strain distribution of 4,8,16,27 mm hole diameters). The effect of planar anisotropy on the strain magnitude can be seen. The major strain reduces as the location moves from the edge towards the bulk material. Strains at the location 5 mm from the hole edge are seen to be nearly in plane strain mode for the 4 mm hole diameter sample. Strains closer to the edge are observed to move in the direction of uniaxial stress loading for the 16 mm hole diameter.

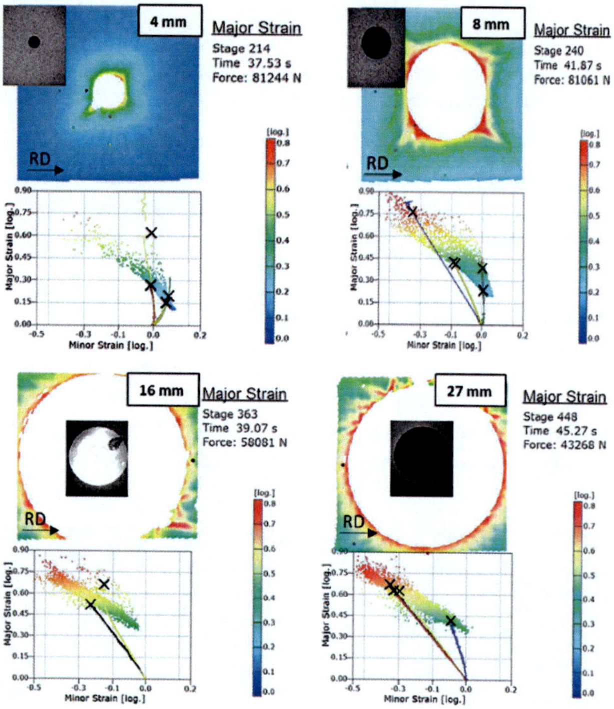

Fig. 7. Major strain obtained for the four different diameters of the hole at a punch displacement just before fracture

The sectional thinning profiles for the sections are also described below and is shown in Fig. 8. The influence of anisotropy on thinning can be observed with higher thinning in the diagonal directions. The anisotropy pattern is seen prominently for the section closer to the hole edge. Differences in thinning magnitude at the same angular location and at radial locations 1 mm and 5 mm are observed to be larger for the 4 mm samples compared to the 16 mm samples. Lower thinning is observed at location r + 5 mm (r is hole edge) in the case of 4 mm hole samples whereas it is seen to increase for the 16 mm samples.

The hole expansion FE simulation is assessed based on the radial plane strain distributions. Figure 9 shows the strain condition predicted at the boundary for all tested models. The strain distribution for the Tata Steel model and Hill90 is seen to be much

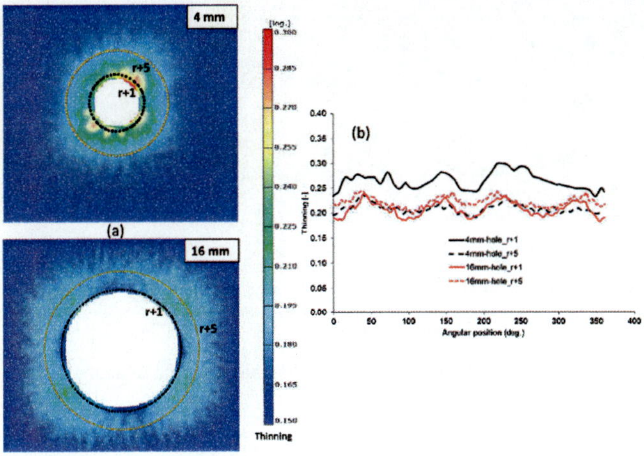

Fig. 8. Thickness distribution calculated from strain measurements for two hole expansion tests with 4 mm and 16 mm initial hole diameters.

more closer to the experimental results. In other words, an accurate description of the material behavior in especially near plane-strain is important for the accurate prediction of the strain variation around the hole which can lead to fracture.

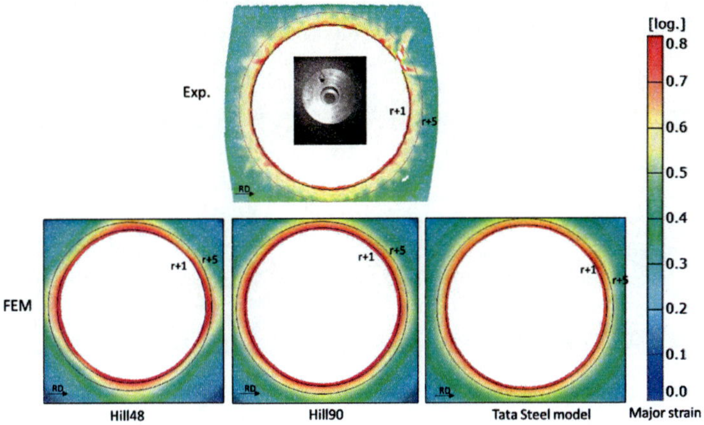

Fig. 9. Major strain distribution at the onset of fracture for a 16 mm initial hole diameter along the radial direction (experimental and FEM simulation) The color bars are identical for both cases.

The major strain distribution along the radial distance $r + 1$ and $r + 5$ shows that all material models are very close to the experimental results as shown in Fig. 10. However the strain distribution for Tata Steel and Hill 90 are more closer to the experimental results than Hill48.

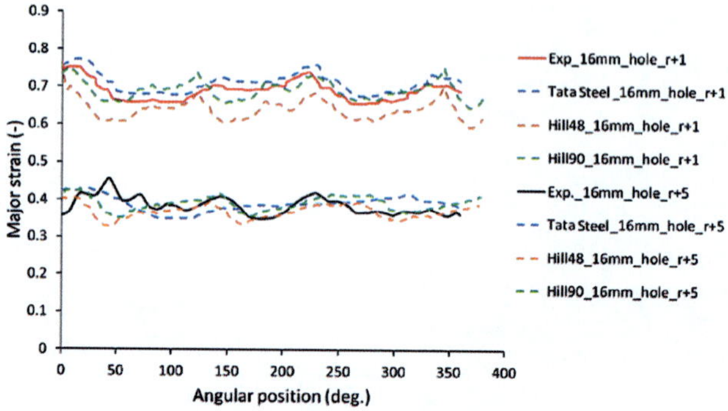

Fig. 10. Major strain distribution along radial distance

5.2 Mesh Sensitivity Analysis

A mesh sensitivity analysis was carried out to investigate the effect of element size on the predicated local response of the simulation. The influence of element size was considered using shell elements with three different sizes of 1mm, 0.50 mm (grid size used in experiment) and 0.25 mm with fully integrated formulation. The three different element sizes are chosen to verify the similarity in the experiment results. The mesh size of 1mm, which is higher than grid size used for experiment, slightly deviated from those of finer meshes. The element size 0.50 and 0.25 mm appeared to better fit to local responses of the material as shown in Fig. 11. It means that there is good agreement between the mesh size and experimental results when the mesh size is less than 0.50 mm.

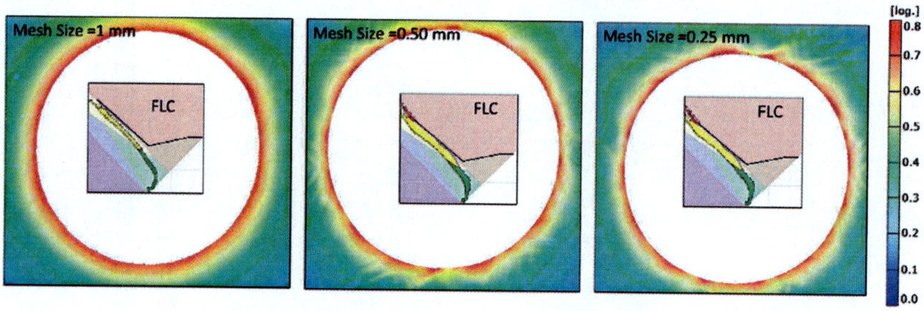

Fig. 11. Effect of mesh size on the predicted thinning simulation results

6 Conclusion

The experimental results of the hole expansion test are discussed in this paper and compared with the FE simulation. The following conclusions can be derived:

- Samples with larger hole diameters show lower forces and larger strains in the neighborhood of the deformed hole. Multiple shear bands are observed in such cases at the hole edge which incrementally reduce the loading capacity.

- Strain localization can occur either in the bulk or near the edge, depending on the stress state in the bulk. Strain localization does not necessarily lie along the diagonal (to rolling) direction and is dependent on the planar anisotropy.
- Thinning magnitudes near the edge are observed to decrease with larger hole and the difference in thinning magnitude near the edge and in the bulk is smaller for larger hole diameters.
- A comparison of experimental results and strain predictions with FE modelling was carried out using the flat hole expansion test. Hill90 criterion shows promising results whereas the Tata Steel model can be considered as a step forward for plane stress and predicts high accurate results, being most suited to fit all data under plane stress conditions. Moreover, for the sheet steel stretch forming part the onset of necking is predicted in a much better way with this model.
- The mesh sensitivity analysis in this paper confirms that the selected mesh element size of 0.50 and 0.25 mm with fully integrated formulation is appropriate for the proposed hole expansion test FE analysis for the localized zone.

The results of this work can help industry to select and calibrate an appropriate anisotropic yield function for different material. In particular, the comparison of predictions with HE experience reveals the ability of the yield function to capture the resulting deformation fields, including their local characteristics. This in turn is necessary for accurate predictions of failure in the form of necking and the resulting tearing, which is a prerequisite for the design of a reliable virtual process.

Acknowledgements. The authors would like to acknowledge experimental work carried out at Tata Steel R&D by Frank Schouten and Tushar Khandeparkar.

References

1. Abspoel, M., Scholting, M.E., Lansbergen, M., An, A., Vegter, H.: A new method for predicting advanced yield criteria input parameters from mechanical properties. J. Mater. Proc. **248**, 161–177 (2017)
2. Parmar, A., Mellor, P.B.: Plastic expansion of a circular hole in sheet metal subjected to biaxial tensile stress. Int. J. Mech. Sci. **20**, 707–720 (1978)
3. Hill, R.: Theoretical plasticity of textured aggregates. In: Mathematical Proceedings of the Cambridge Philosophical Society; Cambridge University Press, Cambridge, UK, vol. 85, pp. 179–191 (1979)
4. Durban, D.: Radial stressing of thin sheets with plastic anisotropy. Int. J. Mech. Sci. **28**, 801–813 (1986)
5. Durban, D.: On two stress concentration problems in plane-stress anisotropic plasticity. Int. J. Solids Struct. **23**, 469–484 (1987)
6. Durban, D., Birman, V.: On the elasto-plastic stress concentration at a circular hole in an anisotropic sheet. Int. J. Mech. Sci. **23**, 469–484 (1982)
7. Hosford, W.F.: A generalized isotropic yield criterion. J. Appl. Mech. **39**, 607–609 (1972)
8. Karafillis, A.P., Boyce, M.C.: A general anisotropic yield criterion using bounds and a transformation weighting tensor. J. Mech. Phys. Solids **41**, 1859–1886 (1993)
9. Hill, R.: A theory of the yielding and plastic flow of anisotropic metals. Proc. R. Soc. Lond. A Math. Phys. Eng. Sci. **193**, 281–297 (1948)

10. Hill, R.: Constitutive modelling of orthotropic plasticity in sheet metals. J. Mech. Phys. Solids **38**, 405–417 (1990)
11. Barlat, F., Lian, K. Plastic behavior and stretchability of sheet metals. Part I: a yield function for orthotropic sheets under plane stress conditions. Int. J. Plast. **5**, 51–66 (1989)
12. Vegter, H., Mulder, J., van Liempt, P., Heijne, J.: Work hardening descriptions in simulation of sheet metal forming tailored to material type and processing. Int. J. Plasticity **80**, 204–221 (2016)
13. Carleer, B.D.: Finite Element Analysis of Deep Drawing, Dissertation, University of Twente, The Netherlands (1997)
14. Bergström, Y.: A dislocation model for the stress strain behaviour of polycrystalline α-Fe with special emphasis on the variation of the densities of mobile and immobile dislocations. Mat. Sci. Eng. **5**, 179–192 (1969/70)
15. van Liempt, P.: Workhardening and substructral geometry of metals, In: Hartley, P. et al. (eds.) Proceedings Metal Forming 94, (Birmingham, UK, 1994), J. Mat. Proc. Tech. **45**, 459–464 (1994)

Influence of the Identification Procedures of the Material Model in Accurate Prediction of Incremental Sheet Forming Forces

Ehssen Betaieb[1], Jaekwang Shin[2], Youngrok Lee[2], Laurent Duchêne[1], Alan I. Taub[2,3], Mihaela Banu[2](✉), and Anne Marie Habraken[1]

[1] Urban and Environmental Engineering, University of Liège, Liège, Belgium
[2] Department of Mechanical Engineering, University of Michigan, Ann Arbor, USA
mbanu@umich.edu
[3] Department of Materials Science and Engineering, University of Michigan, Ann Arbor, USA

Abstract. Finite element simulations of sheet metal forming processes achieve high precision of the process model. Previous works [1] on the impact of the selection of the material law on the prediction accuracy of the formed results have shown that more sophisticated yield functions allow better prediction compared to von Mises or Hill 48 models. However, this choice increases the cost of calibration, as it requires extensive material testing, and in some cases, crystal plasticity modeling predictions. In this work, the selection of the material model is investigated with respect to the accuracy of the force prediction in single point incremental sheet forming (SPIF) simulations. It is shown that the calibration of the material model using shear tests instead of tensile tests can satisfy the required accuracy of SPIF force prediction. This study investigates the accuracy and computation cost of using only von Mises yield surface and an isotropic hardening model versus Hill 48 and Barlat Yld2004-18p yield criteria. AA7075-O cone parts are used to compare the force prediction results.

Keywords: Incremental Sheet Forming · Shear Testing · Finite Element Method · Material Parameters Calibration

1 Introduction

Single Point Incremental Forming (SPIF) is an alternative to conventional forming technologies such as stamping and hydroforming. It is suitable for low volume production because it does not require a dedicated die and press, it is easily customizable for various shapes, and requires a lower forming force. Cost analyses show that the incremental forming process is the most cost-efficient when applied to less than 1000 parts per year production [2, 3]. SPIF process is an attractive alternative for agile sheet metal forming processes targeting low volume, complex geometry, and low cost per part. However, there still exists challenges to increase the geometric accuracy of the forming process.

The issue of geometry accuracy has been addressed by many researchers [4] and today, the coupling with machine learning brings novelties [5]. In recent years, advancement in numerical modeling has allowed engineers to predict tool forces, final shape,

K. Mocellin et al. (Eds.): ICTP 2023, LNME, pp. 591–602, 2024.
https://doi.org/10.1007/978-3-031-40920-2_61

optimal tool path to reach the target geometry, part properties or material state with satisfactory accuracy. It lowered the number of iterations required to achieve precision in manufacturing. The finite element (FE) models played a key role in understanding the final material properties and the geometry of the formed part by capturing the deformation that occurs during the process [6, 7].

Many researchers have investigated different combinations of yield functions and hardening laws and their influence on the prediction results, such as the forming force [8]. The simplest case is the von Mises yield criterion coupled with an isotropic hardening law identified by a tensile test. For instance, Li et al. [9] showed that FE LS-Dyna simulations based on Swift type hardening law and von Mises yield criterion could explain the deformation mechanisms and predict the tool force in a truncated AA 7075-O cone. Robert et al. [10] investigated the same combination with Abaqus FE software and found the errors acceptable while computation cost was lower than more complex models. Flores et al. [11] emphasized the importance of material parameter identification. A biaxial tensile testing machine was used to obtain the coefficients of the Hill (1948) and Hosford (1979) yield criteria. Both tensile and shear test results were exploited for the constitutive law calibration.

Henrard et al. [12] investigated the effect of the type of FE element, of constitutive laws and their identification process on the accuracy of the SPIF predicted force of two cones. A combination of von Mises and Hill 48 yield functions with different hardening laws: isotropic Swift, isotropic Voce, kinematic Armstrong-Fredericks, and kinematic Ziegler hardening laws coupled or not, were used. The highest accuracy was reached with brick elements with a fine mesh with a material model combining von Mises yield locus with mixed isotropic-kinematic hardening model of Voce-Ziegler. Eyckens et al. [13] compared measured and predicted cone forces and strain fields. Their FE simulations were based on isotropic and anisotropic yield criteria combined with isotropic hardening and kinematic hardening laws and their validation results relied on Digital Image Correlation (DIC) and load cell measurements. The effect of the chosen 3D yield phenomenological function in SPIF simulations was also examined by Esmaeilpour et al. [14] while the results based on Yld 2004-18p yield function were compared by Barlat et al. [15] with the ones due to von Mises and Hill 48 yield functions. In Esmaeilpour et al. [16], the out-of-plane strain and stress components were provided using FE crystal plasticity simulations. The results showed that the thickness and strain distribution closely match data, while triaxiality and Lode parameters only deviated slightly.

For an Al SPIF cone simulation, the present paper analyses the effect of the material model on the force prediction: Von Mieses or Hill 48 or Barlat Yld2004-18p yield locus coupled with an isotropic Voce hardening law identified by a tensile and/or shear tests. Bauschinger effect was neglected being not significant for this material. The article also discusses the best way to identify Hill 48 according the target results. Three FE simulations are conducted, first a combination of von Mises yield locus and Voce hardening law based solely on tensile test results, second with the Hill 48 yield locus and Voce hardening law identified both by tensile and shear test results, and third with Barlat Yld2004-18p yield locus and Voce hardening law. Note that the possibility to replace the shear test in the identification methodology by a simple SPIF test to calibrate the kinematic hardening law is demonstrated in [17].

2 Material Models

The von Mises [18], Hill 48 [19], and Yld2004-18p [20] yield functions are three widely used material models in FE simulation for sheet metal forming. A brief remind of the three yield functions is provided below. Equation (1) provides Von Mises yield function:

$$\overline{\sigma}_{VM} = \frac{1}{\sqrt{2}}\sqrt{(\sigma_{11}-\sigma_{22})^2+(\sigma_{22}-\sigma_{33})^2+(\sigma_{33}-\sigma_{11})^2+6\sigma_{23}^2+6\sigma_{31}^2+6\sigma_{12}^2} \tag{1}$$

where $\overline{\sigma}_{VM}$ denotes von Mises equivalent stress. For Hill 48, the equation is:

$$\overline{\sigma}_{Hill} = \sqrt{F(\sigma_{22}-\sigma_{33})^2+G(\sigma_{33}-\sigma_{11})^2+H(\sigma_{11}-\sigma_{22})^2+2L\sigma_{23}^2+2M\sigma_{31}^2+2N\sigma_{12}^2} \tag{2}$$

Here the material parameters F, G, H, L, M, and N are constants characterizing the material anisotropy. They usually rely on Lankford coefficients (r-values) measured in tensile test conducted at 0°, 45°, and 90° from the rolling direction (RD), however they can also be determined from anisotropic yield stress ratio functions [21]. It should be noted that L, M, and N values are assumed to be the same since, it is extremely hard to perform experimental tests on sheet samples to measure through-thickness shear.

Yld2004-18p (Eq. (3)) yield function uses 18 parameters to describe anisotropic behaviour of metals. Their calibration requires uniaxial tension data for 0°, 15°, 30°, 45°, 60°, 75°, and 90° from RD, balanced biaxial and out-of plane stresses test data. Esmaeilpour et al. [16] employed FE crystal plasticity simulations to conduct virtual experiments to account for the out-of plane coefficients

$$\begin{aligned}\varphi = \varphi(\tilde{\mathbf{S}}', \tilde{\mathbf{S}}'') &= \left|\tilde{S}'_1-\tilde{S}''_1\right|^a+\left|\tilde{S}'_1-\tilde{S}''_2\right|^a+\left|\tilde{S}'_1-\tilde{S}''_3\right|^a+\left|\tilde{S}'_2-\tilde{S}''_1\right|^a \\ &+\left|\tilde{S}'_2-\tilde{S}''_2\right|^a+\left|\tilde{S}'_2-\tilde{S}''_3\right|^a+\left|\tilde{S}'_3-\tilde{S}''_1\right|^a+\left|\tilde{S}'_3-\tilde{S}''_2\right|^a+\left|\tilde{S}'_3-\tilde{S}''_3\right|^a = 4\overline{\sigma}^a\end{aligned} \tag{3}$$

$$\begin{aligned}\tilde{\mathbf{s}}' &= \mathbf{C}'\mathbf{s} = \mathbf{L}'\boldsymbol{\sigma} \\ \tilde{\mathbf{s}}'' &= \mathbf{C}''\mathbf{s} = \mathbf{L}''\boldsymbol{\sigma}\end{aligned} \tag{4}$$

In above yield function, a = 8 for FCC materials, $\overline{\sigma}$ is the flow stress, $\tilde{S}'_i$ and $\tilde{S}''_j$ are principal values of the tensors $\tilde{\mathbf{s}}'$ and $\tilde{\mathbf{s}}''$ which are defined by the two linear transformations C′ and C″ on the deviator $\boldsymbol{s}$ of the stress tensor $\boldsymbol{\sigma}$.

3 Material Tests and Model Calibration

Tensile tests (Fig. 1) of AA7075-O sheet metal samples at every 15° relative to the RD were performed with 3 repetitions based on the ASTM B557M standard with a MTS Insight 10 Mechanical Tester at university of Michigan. The 0.2% offset method was used to determine the yield point of the material.

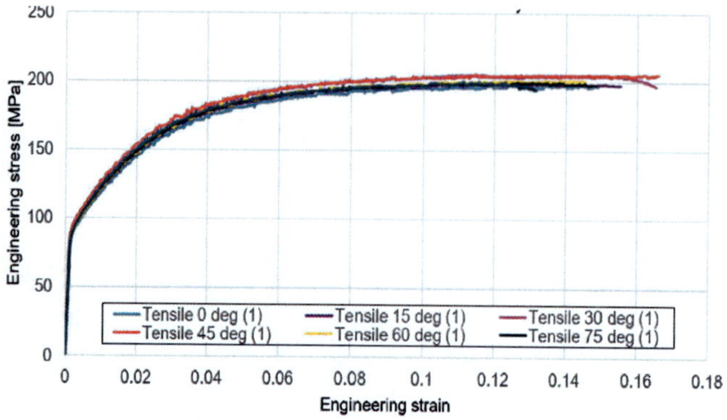

Fig. 1. Stress – strain curve under simple tension every 15° for AA7075-O

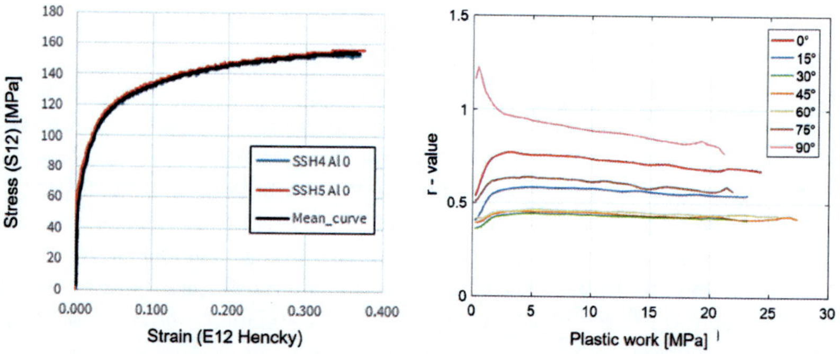

a) b)

Fig. 2. a) Stress-strain curve under simple shear for AA7075-O;b plasticity entrance, b) R-value evolution as a function of plastic work per unit volume

The shear tests were carried out using a home-made bi-axial machine with a capacity of 100 kN developed at the university of Liège. The test (Fig. 2a) was repeated two times and the 0.2% offset method was used to establish the plasticity entrance.

Figure 2b shows Lankford coefficients (the width-to-thickness strain ratio) function of plastic work used for the calibration of the material constitutive models. The thickness strain was calculated using the strain in axial and transverse directions by assuming material incompressibility.

The Voce isotropic hardening model, was selected for both tensile and shearing tests. The shear stress-strain diagram was converted to equivalent stress and strain using Eq. (5) and fitted as described in Eq. (6) to describe isotropic hardening properties of the Al alloy. The hardening material parameters are summarized in Table 1.

$$\overline{\sigma} = \sqrt{3}\sigma_{12}; \overline{\varepsilon} = \frac{2}{\sqrt{3}}\varepsilon_{12} \tag{5}$$

$$\sigma_{voce} = \sigma_0 + K\left(1 - e^{n\overline{\varepsilon}}\right) \tag{6}$$

Table 1. Material constants for Voce isotropic hardening model

	σ_0 [MPa]	K [MPa]	n
Tensile	91.30	149.34	26.71
Shear	115.72	145.02	15.91

In this study, the Hill 48 material parameters were calibrated based on both yield stresses and r-values by Eq. (7) with a weight factor η, choosing the higher impact of strain (r-value) or stress measurements. Equation (7) defines the cost function minimized:

$$\chi = \underbrace{\sum_{i=1,l} (1-\eta)\left[\frac{\sigma_{Theo\_i} - \sigma_{exp\_i}}{\sigma_y}\right]^2}_{\chi\sigma} + \underbrace{\sum_{j=1,m} \eta\left[\frac{r_{Theo\_j} - r_{exp\_j}}{\bar{r}}\right]}_{\chi r} \tag{7}$$

where the sub-index l defines the number of experimental tests used to measure stress points ($l = 8$: tensile tests in 7 different directions + shear test) and the sub-index m is the number of tests used to calculate r-values ($m = 7$ tensile tests in 7 different directions). The sub-indexes *Theo* and *exp* refer respectively to the theoretical and experimental values. Three cases were studied in this work involving different measurements ($\eta = 0$ strain ones; $\eta = 0.5$ stress and strain ones; $\eta = 1$ stress ones) (Tables 2 and 3).

Table 2. Anisotropy coefficients for the Hill 48 yield function

	F	G	H	N, L, M
$\eta = 0$	0.9635	1.1892	0.8108	1.8507
$\eta = 0.5$	1.0184	1.1831	0.8169	1.9538
$\eta = 1$	1.1686	1.1915	0.8085	2.6355

Table 3. Coefficients for Yld2004-18p included in C′ and C″

c'_{12}	c'_{13}	c'_{21}	c'_{23}	c'_{31}	c'_{32}	c'_{44}	c'_{55}	c'_{66}
0.8440	1.2864	1.2398	1.3860	1.3214	1.4549	0.9374	0.8194	1.2077
c''_{12}	c''_{13}	c''_{21}	c''_{23}	c''_{31}	c''_{32}	c''_{44}	c''_{55}	c''_{66}
0.8341	1.0716	0.8122	0.6433	0.4379	0.3502	0.9940	1.0418	0.7782

4 SPIF Setup and FE Simulation Features

Cones of 45° and 67° are manufactured using the SPIF setup (Fig. 3a) established on the Cincinnati HMC 400 EP CNC milling machine. An AA7075-O sheet of 1.63 mm thickness is clamped and a layer of MoS_2 multipurpose grease is applied on the surface to reduce friction in the tool-sheet interface. A hemispheric tool with a radius of 4 mm is used. It follows a spiral trajectory created with a Matlab code. The tool moves at 2540 mm/min with the spindle speed set at zero and in an unlocked condition. The step-down size between each revolution is set to 0.5 mm. Forming forces measured by a KISTLER force sensor are continuously recorded.

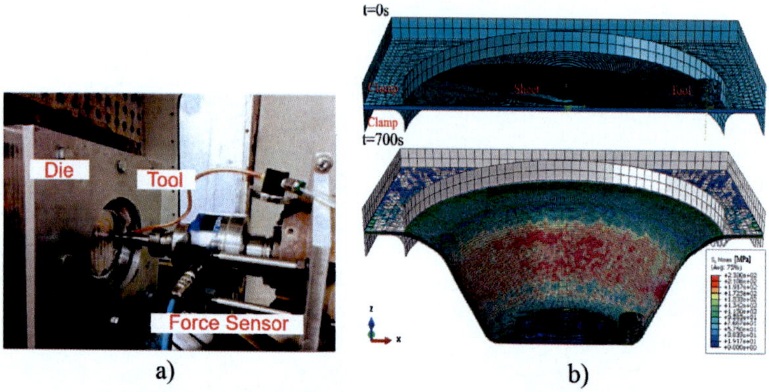

Fig. 3. a) Experimental setup of SPIF, b) FE simulation, initial and final geometry, isovalues of equivalent stress for the of 67° cone computed with von Mises yield function

Each component of the model has the same dimension and location as that of the experimental component and setup. The clamps have a circular opening with a radius of 85 mm. Figure 3b shows the established model of SPIF for 67° cone. The top/bottom clamp, die, and tool that are more rigid than the sheet are meshed with R3D4 rigid elements of Abaqus. The tool and the die are meshed sufficiently small to avoid any element distortion during the contact with the sheet. The sheet is meshed with a C3D8R general-purpose linear brick element with reduced integration. The circular meshing technique is implemented with the element size in the radial direction being 0.5 mm and one element every 1° of the circumference. This allows smooth deformation for axisymmetric parts such as cones. Five elements are placed in the thickness direction to sufficiently capture the stress distribution in the thickness direction without making the simulation computationally costly. A general contact algorithm provided by Abaqus defines the contact for three-dimensional surfaces and mechanical finite-sliding contact. Penalty method and friction coefficients of 0.1 and 0.09 are selected. For this quasi-static problem involving slow motion of the parts, a scaling factor of mass of 10^7 is used and provides a ratio of kinetic energy to internal energy KE/IE less than 1%.

5 Results and Discussions

Five cases for FE simulations defined in Table 4, all with Voce hardening law were conducted for the target geometry 45° and 67° cone to determine the accuracy of the force prediction and the computational costs for each material model (see Table 5). The predicted forces in the in-plane radial direction and the vertical z-direction were compared with the experimental forces (Fig. 4). The corresponding errors were measured based on the experimental results in the steady-state region where most of the deformation is taking place (Table 4). A synthetic bar plot containing the average error percentage depending on the calibration method, hardening law, and yield function is given in Fig. in Conclusion section. Hereafter, shear stress distribution was plotted for each wall angle in the bending, thinning and steady state regions respectively to investigate dominant deformation mode in SPIF and analyse the results (Fig. 5).

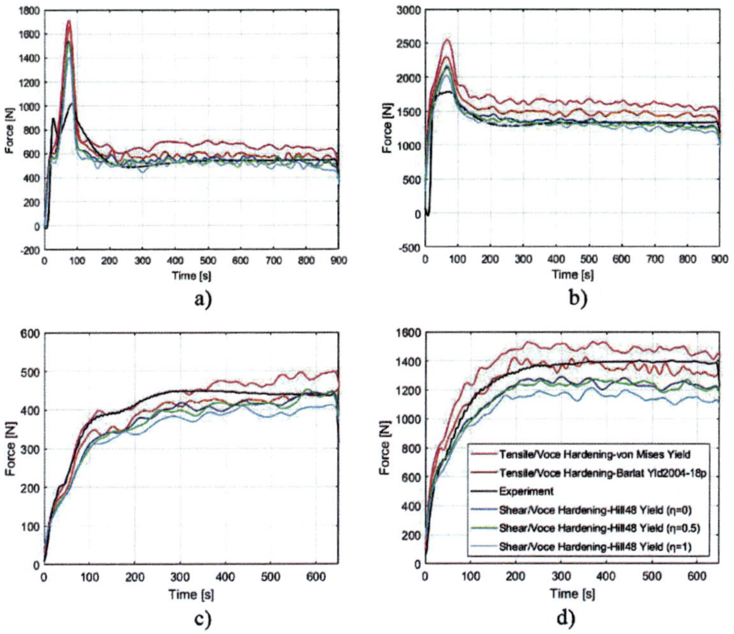

Fig. 4. Predicted and measured SPIF forming forces, a) SPIF 67° in-plane radial force, b) Z-direction vertical force, c) SPIF 45° in-plane radial force, d) Z-direction vertical force

The simulation results show that the conventional von-Mises yield criterion with Voce isotropic hardening identified solely on tensile test result gives fair force prediction with around 5% of error for the SPIF 45° wall angle. However, for the SPIF 67° wall angle, the in-plane and vertical forces are severely overpredicted which leads to around 15% error on average. The difference in force prediction for 45° and 67° wall angle can be explained by the distribution of shear stresses occurring during the forming. The presence of significant amount of in-plane and through-thickness shears in high

Table 4. Case definition and percentage of errors on the force prediction

	Calibration method/Yield function η = 0 only strain; η = 0.5 combining stress and strain; η = 1 only stress	SPIF 45° (%)	SPIF 67° (%)	Average (%)
In-plane Radial Force	Tensile/von Mises	5.05	28.38	16.72
	Tensile/Yld2004-18p	−4.33	7.43	1.55
	Tensile + Shear/ Hill 48 (η = 0)	−5.50	2.56	−1.47
	Tensile + Shear/ Hill 48 (η = 0.5)	−7.08	1.64	−2.72
	Tensile + Shear/ Hill 48 (η = 1)	−12.60	−5.45	−9.03
Z-Direction Vertical Force	Tensile/von Mises	6.18	22.58	14.38
	Tensile/Yld2004-18p	−3.52	7.43	1.96
	Tensile + Shear/ Hill 48 (η = 0)	−12.48	1.00	−5.74
	Tensile + Shear/ Hill 48 (η = 0.5)	−13.47	0.39	−6.54
	Tensile + Shear/ Hill 48 (η = 1)	−18.77	−3.83	−11.3

wall angle (highlighted by Fig. 5b, d, f) induces the need for a more complex model to accurately predict the forming forces when tensile test is used for calibration.

The Barlat's Yld2004-18p Voce isotropic hardening calibrated by tensile tests and crystal plasticity predicts the experimental forces with less than 10% error for all the cases. It proves to be an accurate material model compared to the conventional von Mises isotropic yield criterion. The out-of-plane components required for the calibration of the Yld2004-18p yield function parameter were found employing crystal plasticity in addition to the yield function nature able of taking the full stress state into account. This research confirms that the Yld2004-18p better predicts the material stress state compared to the isotropic yield locus. It enables to capture the deformation in both wall angles and to accurately predict the forming forces in SPIF simulation.

Lastly, Hill 48 yield function and Voce isotropic hardening identified with both tensile and shear tests were investigated and compared with the previous results. The simulations were conducted with three different approaches to identify Hill 48 anisotropy coefficients. It can be seen that the Hill 48 yield function with Voce hardening was able to capture both in-plane and vertical forces in SPIF 67° with great accuracy (less than 6%) for the three identification methodologies. This observation indicates that the shear-tensile-based calibrated Hill 48 yield function with isotropic hardening enables

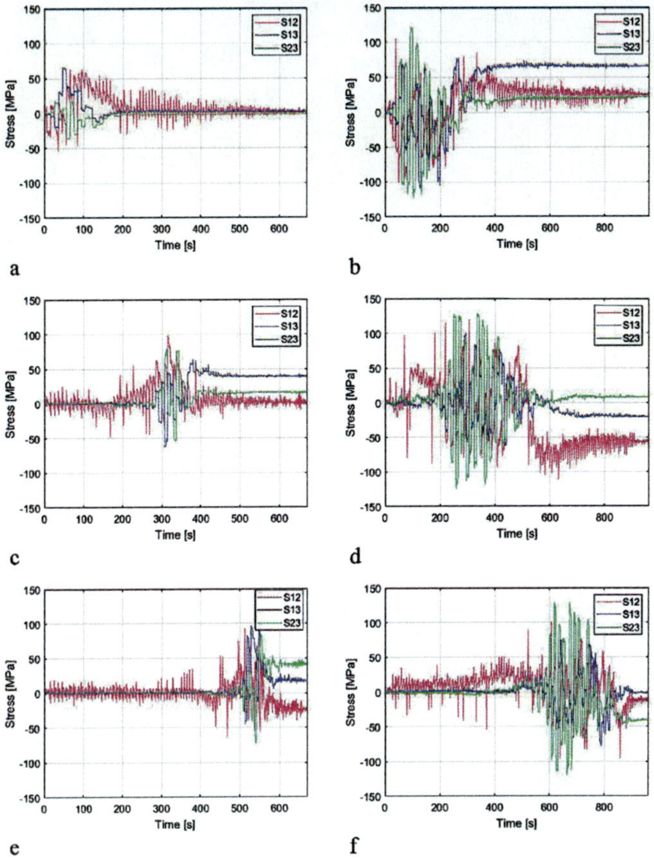

Fig. 5. Shear stress distribution in the bending, thinning, and steady-state region (isotropic von-Mises criterion with Voce isotropic hardening) in SPIF 45°: a), c), e) and SPIF 67°: b), d), f)

the calibration of the shear region of the yield locus. It allows considering the dominant in-plane and out-of-plane shears in high wall angle. Although this material model shows excellent prediction in the high wall angle case, it was not able to capture both in-plane and vertical forces in low wall angle: an increase of average errors on the force prediction is observed. It can be explained by the difference in deformation mode for each cone. The low wall angle cone does not include much shear deformation, e.g. in-plane and out-of-plane shears, compared to the high wall angle cone, it may still include deviation from the shear-based Hill 48 yield locus and more deviation in the prediction of forming forces.

Table 5. Comparison of the computational time for the material model in SPIF 45 and 67 cones

		Case 1	Case 2	Case 3	Case 4	Case 5
Targeted part shape		Truncated cone shape with 45° wall angle Abaqus/Explicit				
Material Model	Hardening Law	Voce type, Isotropic	Voce type, Isotropic	Voce type, Isotropic	Voce type, Isotropic	Voce type, Isotropic
	Yield locus	von Mises	Yld2004-18p	Hill 48 ($\eta = 0$)	Hill 48 ($\eta = 0.5$)	Hill 48 ($\eta = 1$)
Mass scaling		1.0E7	1.0E7	1.0E7	1.0E7	1.0E7
Friction coefficient		Penalty, 0.1	Penalty, 1.0	Penalty, 1.0	Penalty, 1.0	Penalty, 1.0
Total # of elements		291,700	291,700	291,700	291,700	291,700
Number of CPUs		108	108	108	108	108
CPU time		15:41:27	~48:00:00	15:45:48	15:48:49	15:51:56
Targeted part shape		Truncated cone shape with 67° wall angle Abaqus/Explicit				
Material Model	Hardening Law	Voce type, Isotropic	Voce type, Isotropic	Voce type, Isotropic	Voce type, Isotropic	Voce type, Isotropic
	Yield locus	von Mises	Yld2004-18p	Hill 48 ($\eta = 0$)	Hill 48 ($\eta = 0.5$)	Hill 48 ($\eta = 1$)
Mass scaling		1.0E7	1.0E7	1.0E7	1.0E7	1.0E7
Friction coefficient		Penalty,0.09	Penalty,0.09	Penalty,0.09	Penalty,0.09	Penalty,0.09
Total # of elements		191,340	191,340	191,340	191,340	191,340
Number of CPUs		108	108	108	108	108
CPU time h:m:s		35:41:40	~56:00:00	37:54:56	38:32:50	39:04:37

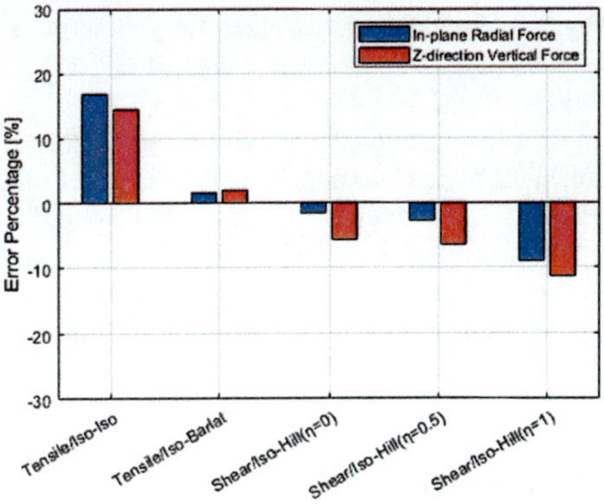

Fig. 6. Average error comparison depending on the hardening law and the yield function

6 Conclusions

In this work, the selection and identification procedures of the material model were investigated with respect to the accuracy of force prediction in SPIF simulation. An identification method was proposed to capture not only the deformation mode in SPIF process but also material anisotropy in a cost-effective way by complementing tensile tests with shear tests. While the error analysis shows that the combination of isotropic hardening and Yld2004-18p yield (Fig. 6) results in a minimum amount of average error, performing the mechanical or virtual tests required to calibrate this yield function is challenging and the computational cost is high (see Table 5). Thus, replacing it with a simpler yield function and less experiments (Hill 48, tensile and shear tests) can be a great alternative to predict the force in the SPIF process. It has been found that the Hill 48 yield function calibrated by tensile and shear tests has superior accuracy close to that of Yld2004-18p and can provide even better force predictions in high wall angle. In conclusion, it can be seen that the characteristics of the shear test are aligned with that of the deformation taking place, especially in the high wall angle in SPIF.

References

1. Shin, J., et al.: A comparative study of incremental sheet forming process to achieve optimal accuracy. In: Inal, K., Levesque, J., Worswick, M., Butcher, C. (eds.) NUMISHEET 2022: Proceedings of the 12th International Conference and Workshop on Numerical Simulation of 3D Sheet Metal Forming Processes, pp. 141–154. Springer, Cham (2022). https://doi.org/10.1007/978-3-031-06212-4_13
2. Giardini, C., Ceretti, E., Attanasio, A.: Innovative sheet forming processes: evaluation of the economic convenience of an actual case. In: Kuljanic, E. (ed.) AMST 2005 Advanced Manufacturing Systems and Technology, pp. 473–482. Springer, Vienna (2005). https://doi.org/10.1007/3-211-38053-1_46
3. Ambrogio, G., Di Lorenzo, R., Micari, F.: Analysis of the economical effectiveness of incremental forming processes: an industrial case study. In: Proceedings VI AITeM Conference, Gaeta, Italy (2003)

4. Lu, H., Liu, H., Wang, C.: Review on strategies for geometric accuracy improvement in incremental sheet forming. Int. J. Adv. Manuf. Technol. **102**(9–12), 3381–3417 (2019). https://doi.org/10.1007/s00170-019-03348-3
5. Carette, Y., Duflou, J.R.: Mastering the complexity of incremental forming: geometry-based accuracy prediction using machine learning. Eng. Proc. **26**, 12 (2022)
6. Jeswiet, J., Micari, F., Hirt, G., Bramley, A., Duflou, J., Allwood, J.: Asymmetric single point incremental forming of sheet metal. CIRP Ann. **54**(2), 88–114 (2005)
7. Eyckens, P., He, S., Van Bael, A., Van Houtte, P., Duflou, J.: Forming limit predictions for the serrated strain paths in single point incremental sheet forming. AIP Conf. Proc. **908**(1), 141–146 (2007)
8. Esmaeilpour, R., Kim, H., Park, T., Pourboghrat, F., Agha, A., Abu-Farha, F.: Effect of hardening law and process parameters on finite element simulation of single point incremental forming (SPIF) of 7075 aluminum alloy sheet. Mech. Ind. **21**, 302 (2020)
9. Li, Y., Daniel, W.J.T., Meehan, P.A.: Deformation analysis in single-point incremental forming through finite element simulation. Int. J. Adv. Manuf. Technol. **88**(1–4), 255–267 (2016). https://doi.org/10.1007/s00170-016-8727-9
10. Robert, C., Dal Santo, P., Delamézière, A., Potiron, A., Batoz, J.L.: On some computational aspects for incremental sheet metal forming simulations. Int. J. Mater. Form. **1**(SUPPL. 1), 1195–1198 (2008)
11. Flores, P., et al.: Model identification and FE simulations: effect of different yield loci and hardening laws in sheet forming. Int. J. Plast. **23**(3), 420–449 (2007)
12. Henrard, C., et al.: Forming forces in single point incremental forming: prediction by finite element simulations, validation and sensitivity. Comput. Mech. **47**(5), 573–590 (2011)
13. Eyckens, P., et al.: Strain evolution in the single point incremental forming process: digital image correlation measurement and finite element prediction. Int. J. Mater. Form. **4**(1), 55–71 (2011)
14. Esmaeilpour, R., Kim, H., Park, T., Pourboghrat, F., Mohammed, B.: Comparison of 3D yield functions for finite element simulation of single point incremental forming (SPIF) of aluminum 7075. Int. J. Mech. Sci. **133**, 544–554 (2017)
15. Barlat, F., Aretz, H., Yoon, J.W., Karabin, M.E., Brem, J.C., Dick, R.E.: Linear transfomation-based anisotropic yield functions. Int. J. Plast. **21**(5), 1009–1039 (2005)
16. Esmaeilpour, R., et al.: Calibration of barlat Yld 2004–18P yield function using CPFEM and 3D RVE for the simulation of single point incremental forming (SPIF) of 7075-O aluminum sheet. Int. J. Mech. Sci. **145**, 24–41 (2018)
17. Betaieb, E., Duchêne, L., Habraken, A.M.: Calibration of kinematic hardening parameters on sheet metal with a Computer Numerical Control machine. Int. J. Mater. Form. **15**, 69 (2022)
18. Von Mises, R.: Mechanics der plastischen Formänderung von Kristallen. Zeitschrift Angew Math. Mech. **8**, 161–85 (1928)
19. Hill, R.: A theory of the yielding and plastic flow of anisotropic metals. Proc. R Soc. **193**, 281–297 (1948)
20. Abaqus 6.14 manual/8.2.6 Anisotropic yield/creep
21. Barlat, F., Aretz, H., Yoon, J.W., Karabin, M.E., Brem, J.C., Dick, R.E.: Linear transformation - based anisotropic yield functions. Int. J. Plast. **21**, 1009–1039 (2005)

Vibrating Tool Path Design for New Multi-axial Vibration Assisted Incremental Sheet Forming

Zhidong Chang, Jamie Booth, and Hui Long(✉)

Department of Mechanical Engineering, The University of Sheffield, Sheffield, UK
h.long@sheffield.ac.uk

Abstract. In this study, a new multi-axial vibration assisted incremental sheet forming (MV-ISF) process is introduced by developing a vibrated toolpath and an ellipsoidal headed tool design. The process has been shown to improve the formability of tested materials due to the introduction of a low-frequency and high-amplitude vibration field. To understand the forming characteristics of the MV-ISF process, the geometrical accuracy, surface quality and formability in forming a conic geometry are investigated through experiments and finite element analysis using two aluminium alloys and compared with that formed using the conventional ISF tool path design. An analytical model of the material deformation under the vibration condition is also proposed to assess the enhanced formability of the MV-ISF process. The new vibrated tool path design combined with the use of the ellipsoidal headed tool provides a simple and potentially effective solution for improving the forming accuracy, surface quality and formability of ISF manufactured parts.

Keywords: Incremental Sheet Forming · Multi-axial Vibration · Tool Path Design

1 Introduction

Incremental sheet forming (ISF) is considered as a flexible sheet metal forming process, potentially suitable for low volume production of customised sheet metal parts. Considerable studies have been conducted to investigate the deformation mechanisms of ISF process and its potential industrial applications. Investigation on the forming quality is of importance to address current issues of ISF parts, including geometrical accuracy, surface quality, and formability.

Vibration-assisted ISF (VAISF) process was proposed with the aim of improving the formability and surface quality in ISF. Vahdati et al. [1] firstly introduced the ultrasonic vibration into ISF process and demonstrated that the VAISF process had advantages in reducing the forming force, geometrical deviation and surface roughness, which was also validated by other studies in [2, 3]. The forming force was reduced significantly for the materials with higher yield stress, by adopting higher vibration energy and lower feed speed [4, 5]. Amini et al. [6] and Yang et al. [7] demonstrated that the use of ultrasonic vibration had a positive effect on improving the formability of sheet metals.

K. Mocellin et al. (Eds.): ICTP 2023, LNME, pp. 603–612, 2024.
https://doi.org/10.1007/978-3-031-40920-2_62

Amini et al. [8] reported a distinct improvement of surface hardness under the effect of ultrasonic vibration. Li et al. [9] developed a hybrid constitutive model to describe the material behaviour under VAISF condition. Different from the ultrasonic vibration with high frequency, Lu et al. [10] developed low-frequency and high amplitude vibrations in ISF process through the design of single- and double-offset on ellipsoidal headed tools under rotation. A hard-to-form material, magnesium alloy AZ31B, was processed and laminated ultrafine-grained structures were generated by these vibrating tools [10].

In addition to the formability studies of sheet materials, the improvement of forming quality has also been investigated. The ISF tool path design has shown significant effect on the forming quality. Filice et al. [11] proposed a spiral tool path to eliminate scratch marks on the part surface to improve the surface quality. Wang et al. [12] added an equal-diameter spiral path to the conventional helix tool path along circumferential direction to improve the forming limit of the sheet material. To avoid the biaxial tension condition using the conventional toolpath, Wen et al. [13] applied the multi-direction tool path, in which the moving direction of each contour kept changing. Lu et al. [14] developed a feature-based tool path through calculating the equi-potential contour lines to improve the geometrical accuracy and surface quality. Chang et al. [15] proposed a point-contact tool path to reduce the circumferential friction of the tool and showed the benefits in reducing geometrical deviation and improving formability of the material.

The vibration-assisted forming methods described above applied a unidirectional vibration field along the tool axis. However the ISF process involves three-dimensional stress states, the multi-axial vibration field using the tool path design can provide further improvement in forming quality but also maintain process flexibility.

In this work, a new multi-axial vibration assisted incremental sheet forming (MV-ISF) with a low-frequency and high-amplitude vibration field was proposed by creating a new vibrated tool path design and using an ellipsoidal headed tool. It is applied to form two materials of aluminium alloy sheets, AA5251-H22 and AA6082-T6. The effect of the new tool path design on the geometrical accuracy, surface quality and formability are investigated and compared with the conventional ISF (CISF) process. The influence of vibration amplitude on the formability and forming quality is investigated through experiments, an analytical model and finite element method.

2 Multi-axial Vibration Assisted ISF (MV-ISF)

To further improve the formability of sheet metal under vibration fields, the MV-ISF process is proposed with the flexible and controllable 3D vibration field through combining a new vibrated tool path (VTP) with the use of the ellipsoidal headed tool. Different from the conventional ball head tool (T0), the ellipsoidal headed tool has an ellipsoidal head, named as T2, it exerts a radial vibration field with high amplitude and low frequency onto sheet metal, as shown in Fig. 1. In addition to the radial vibration caused by the ellipsoidal headed tool, an axial vibration is also applied to the sheet through employing a vibrated tool path. Employing this vibrated tool path, the forming tool moves up and down at a certain axial distance h while travelling along the circumferential direction of the conic part as shown in Fig. 1(c). This results in discontinuous contact between tool and sheet. The tool employing this new vibrating tool path thus introduces the axial vibration field into the sheet in MV-ISF process.

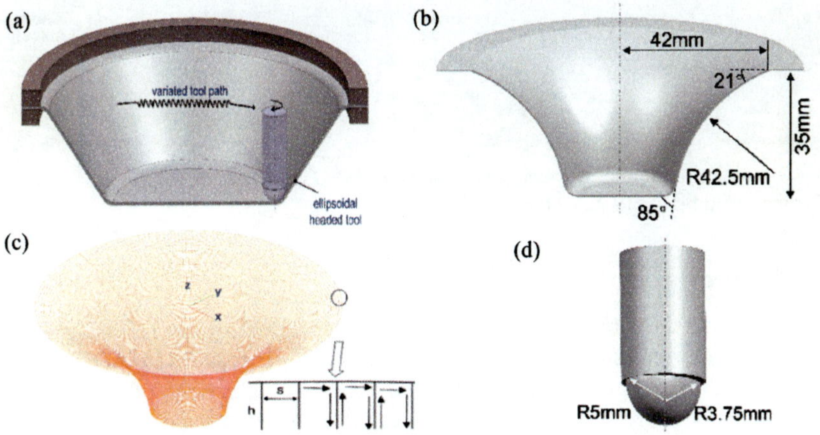

Fig. 1. Illustration of the MV-ISF (a) schematic of the MV-ISF process; (b) conic part geometry; (c) vibrating tool path; (d) ellipsoidal headed tool.

3 Analysis of MV-ISF Experimental Results on Forming Quality

3.1 Geometrical Accuracy

Figure 1(b) shows the conical shape with varied wall angle formed in the experiment. To compare the difference of geometrical accuracy of the formed parts, Fig. 2 shows the measured part profiles along the cross section of the conic part using different tools and tool paths with step depth of 0.3 mm using AA5251 with sheet thickness 1 mm. The vibration tool path (VTP) with different axial distances ($h = 0.1$ mm and 0.3 mm) is adopted and compared with the conventional tool path (CTP).

As shown in Fig. 2, the part profiles using VTP show a considerably improved geometrical accuracy, especially for VTP with $h = 0.3$ mm. It is possible that the MV-ISF may reduce the residual stress by introducing both axial and radial vibration fields, which will be discussed in the following finite element analysis.

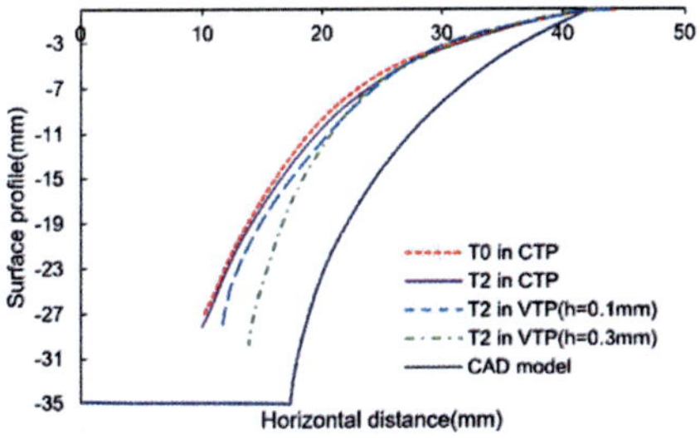

Fig. 2. Part profiles by using MV-ISF and CISF.

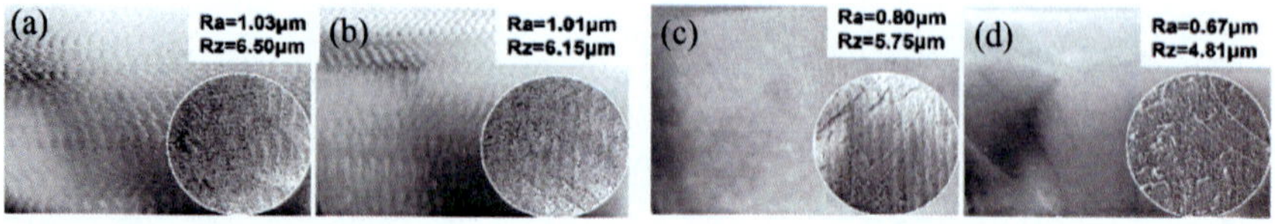

Fig. 3. Surface morphology by using MV-ISF and CISF: (a) T0 & CTP; (b) T2 & CTP; (c) T2 & VTP h = 0.1 mm; (d) T2 & VTP h = 0.3mm.

3.2 Surface Quality

Figure 3 shows the surface morphology on the inner surface of the inclined wall of the conic part by different tools and tool paths. The surface is scanned by Alicona and the surface roughness is measured along the circumferential direction of the conic part. In the MV-ISF experiment, sufficient oil lubricant is used between tool and sheet. For the conic part produced by the CTP with tool rotation, a fish-scale texture can be seen on the surface as shown in Figs. 3(a) and (b), resulting in a reduced surface quality. By employing the VTP in MV-ISF, the surface quality is improved with reduced surface roughness and no obvious fish-scale texture can be observed on the sheet surface, as shown in Figs. 3(c) and (d). It is worth noting that by using the VTP with h = 0.3 mm, the inner surface of the part as shown in Fig. 3(d) has achieved a smoother surface finishing which is difficult to achieve using conventional ISF processes. This is because by using the VTP, the fish-scale surface texture is repeatedly flattened by the forming tool under the action of the axial vibration field, thus achieving the improved surface quality.

3.3 Forming Limit

To investigate the formability, conical parts using AA5251 and AA6082 sheets are produced by employing different tools and tool paths. The thickness strain is obtained by measuring the sheet thickness at the fracture, and the fracture strain is calculated based on the assumptions of volume constancy and plane strain condition [16]:

$$\overline{\varepsilon_f} = \frac{2}{\sqrt{3}} \ln \frac{t_0}{t} \quad (1)$$

where t_0 is the original thickness, and t is the measured thickness.

Figure 4 shows an increased maximum thinning and fracture strain by the MV-ISF, when compared with that by the CTP. The fracture strain by VTP $h = 0.1$ mm is 8.6% higher than that by the CISF for AA5251 and 23% higher for AA6082. Compared with using T0 and the CTP, the fracture strain by using T2 and VTP $h = 0.3$ mm is increased by 28.1% for AA5251 and 16.1% for AA6082. Therefore, the sheet formability can be effectively improved by the developed MV-ISF process. It is widely recognised that the application of a vibration field can reduce the dislocation density of the sheet metal thus improving the formability. The dislocation density of AA6082, with a higher strength, is significantly higher than that of AA5251, therefore the volume effect caused by the vibration field is more obvious in AA6082, leading to the greater decreasing of dislocation density and significant improvement of formability compared with AA5251.

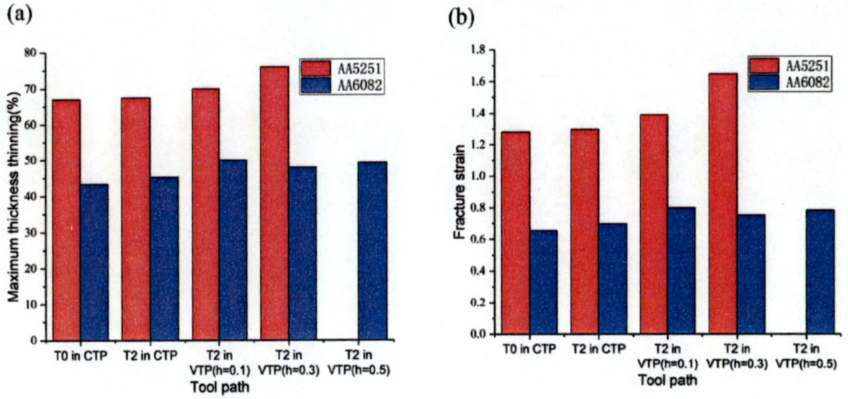

Fig. 4. Formability by using CTP and VTP: (a) sheet thinning; (b) fracture strain.

4 Analytical Model of Estimating Stress Reduction in MV-ISF

To investigate the effect of vibration field on the material plastic deformation, vibration amplitude and frequency generated by different tools and tool paths are measured by using an Eddy current sensor, with the highest measurement accuracy of 0.01 μm, to obtain the displacement of the sheet. When the tool contacts the sheet measurement point where the sensor is located, elastic-plastic deformation of the sheet is measured by displacement, as shown in Fig. 5. The fluctuations of displacement are considered being caused by the vibration field. The axial fluctuation due to the VTP will cause a reduction of feed rate and an increase of forming time. As shown in Fig. 5(a) and Fig. 5(b), in the same time period, the tool by the CTP contacts the sheet measurement point three times, while using the VTP the tool only contacts the sheet once, indicating an increase of forming time by using the VTP.

After applying Fourier processing of the measured displacement data in Fig. 5(c), the vibration amplitude and frequency of the sheet in the MV-ISF are obtained, as shown in Fig. 5(d). The MV-ISF process creates both axial and radial vibration fields, leading to a higher frequency. No obvious difference of radial vibration amplitude is observed by using different tool paths, but the axial amplitude by the VTP increases significantly, as shown in Fig. 5(d). The maximum amplitudes of CTP and VTP ($h = 0.1$ mm) are 25 μm and 32.6 μm respectively, while the amplitude generated by the VTP ($h = 0.3$ mm) is measured as 87.2 μm, which is much greater than other process conditions tested.

It has been reported that the vibration field shows significant effect on the material deformation, which can be explained by vibration softening, stress superposition and friction reduction. In this work, the stress reduction $\Delta\sigma$ caused by the large-amplitude vibration field is described as a combined effect of vibration softening and stress superposition as shown in Fig. 6(a), which can be expressed as

$$\Delta\sigma = \Delta\sigma_{su} + \Delta\sigma_{so} \tag{2}$$

where $\Delta\sigma_{so}$ and $\Delta\sigma_{su}$ is the stress reduction due to vibration softening and stress superposition respectively. As an important factor related to the stress reduction in the MV-ISF,

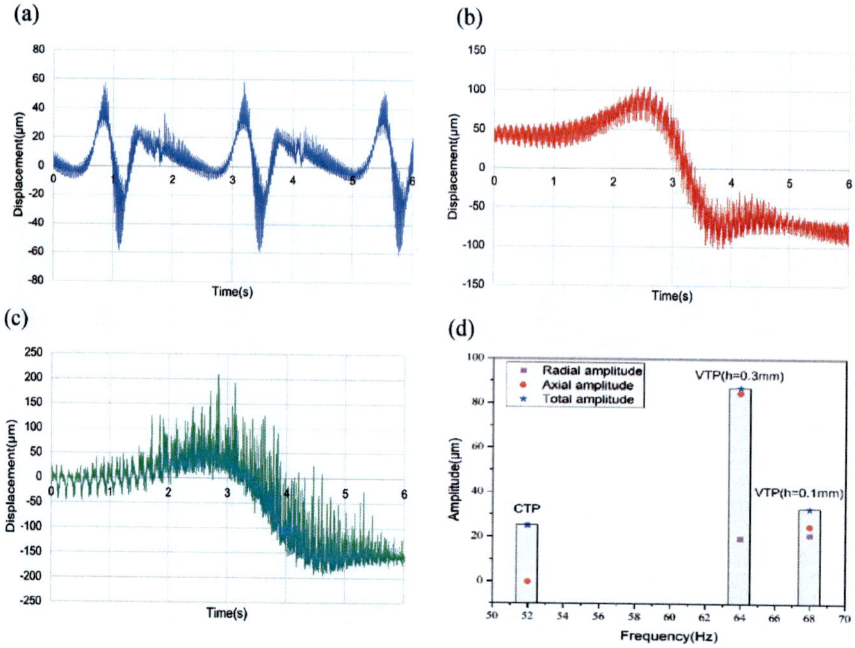

Fig. 5. Vibration measurements by using different tool paths: (a) T2 & CTP; (b) T2 & VTP $h = 0.1$ mm; (c) T2 & VTP $h = 0.3$ mm; (d) Vibration amplitude and frequency.

the stress superposition maintains an exponential relationship with vibration amplitude as investigated by Wang et al. [17] as

$$\Delta\sigma_{su} = -K\xi^n \tag{3}$$

where K and n are the hardening exponent and material strength coefficient of the material. Based on thermal activation theory, the stress reduction due to the stress softening is considered to be related to the vibration energy density [17]

$$\Delta\sigma_{so} = -\beta M\hat{\tau}(E/\hat{\tau})^m \tag{4}$$

where $\hat{\tau}$ is the mechanical threshold of the sheet material and E is the vibration energy density; β and m are the material parameters obtained from experiments, and M is the Taylor factor, which is taken as 3.06. It is considered that the vibration energy density E in the MV-ISF process maintains a positive relationship with the vibration amplitude and frequency. When the ellipsoidal tool rotates, its vibration energy density along radial direction can be calculated by

$$E_h = J\omega^2 = \frac{1}{5}\xi_h^2\omega^2\rho \tag{5}$$

The axial vibration energy exerted by the VTP can be obtained by

$$E_v = \frac{1}{2}\xi_v^2\omega^2\rho \tag{6}$$

where ξ_v and ξ_h are vibration amplitude along axial and radial direction, respectively; while ω and ρ are the vibration frequency and density of the material. The total energy density caused by vibration field in the MV-ISF process can be obtained by

$$E = E_v + E_h = (\frac{1}{2}\xi_v^2 + \frac{1}{5}\xi_h^2)\omega^2\rho \tag{7}$$

Figure 6 shows the calculated stress reduction using different tool paths by the developed analytical model in Eqs. (3) and (4). The values of vibration amplitude and frequency used in the analytical model are from the experimental results in Fig. 5(d). The material parameters of AA5251 are: $\rho = 2700$ kg/m^3, $\hat{\tau} = 138$ MPa, $\beta = 15$, m = 0.5, K = 389, n = 0.228. Higher stress reduction contributes to the improvement of formability. As shown in Fig. 6, the stress superposition effect shows no obvious difference between processes, but the calculated stress reduction due to vibration softening increases with the increase of the axial distance h and vibration energy in the VTP, which is consistent with the improvement of fracture strain shown in Fig. 4. Therefore, the analysis of stress reduction indicates that the improved formability of the MV-ISF is due to the effect of the axial vibration field and the increasing of vibration energy by the VTP.

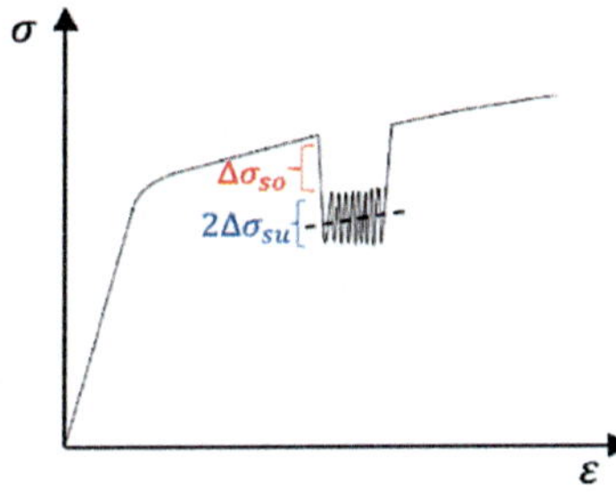

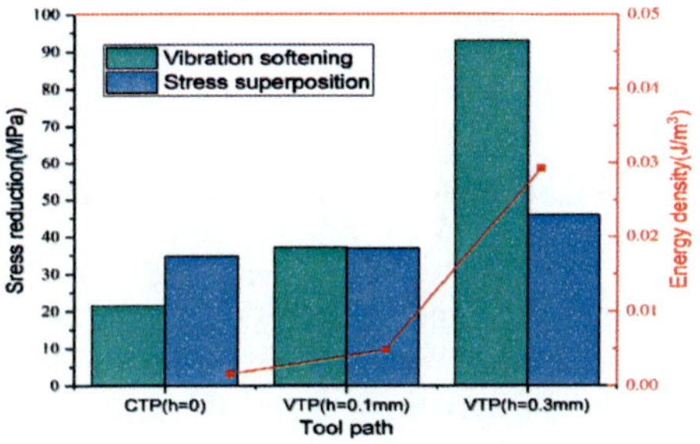

(a) Stress reduction under vibration field (b) Calculated stress reduction

Fig. 6. Comparison of stress reduction by using different ISF processes.

5 Finite Element Analysis

In order to understand the mechanism leading to the improved geometrical accuracy in the MV-ISF, finite element (FE) models of forming the conic part using different tool paths, shown in Fig. 1, are developed by using FE software ABAQUS to simulate the MV-ISF process. AA5251 material and shell elements using five integral points through thickness are selected in the simulation models using von Mises yield criterion. The coefficient of friction is assumed to be 0.05 in the simulation, taken from Eyckens et al. [18]. For a conventional ISF process, the sheet deformation is plane strain state [19], but it is essential to confirm whether the MV-ISF changes the deformation state. Figure 7(a) shows the strain paths obtained from both MV-ISF and CISF processes. It is clear that the minor strain is small when compared with the major strain in the MV-ISF. Therefore the deformation using the VTP and ellipsoidal headed tool in the MV-ISF is still in the plane strain state. The MV-ISF process can increase the fracture strain through introducing the vertical vibration field without changing the deformation state.

The simulated result in Fig. 7(b) shows that the stress triaxiality at contact area decreases cyclically under the effect of the vibration field, and the fluctuation of stress triaxiality is related to the vibration amplitude. The greater variation ranges of stress triaxiality in the MV-ISF provides a valid explanation for the improved formability of the material in the MV-ISF when compared with that in the CISF.

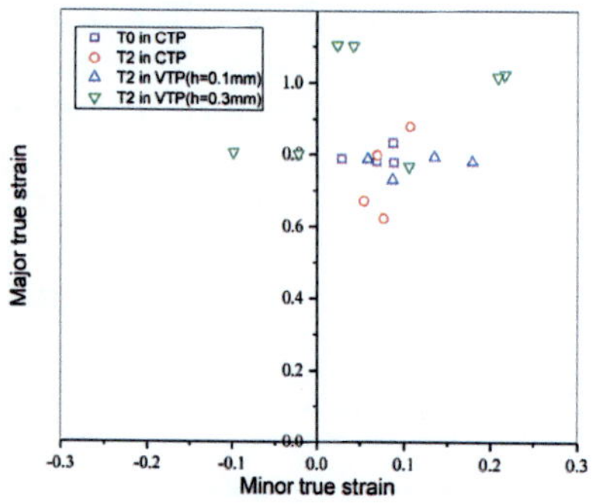

(a) strain path

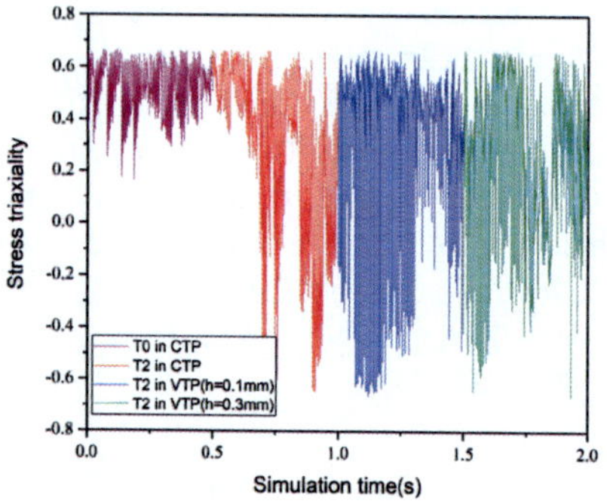

(b) stress triaxiality at the contact area

Fig. 7. Simulation results using different processing methods CISF and MV-ISF.

Figure 8 compares the residual von Mises stress distribution under CISF and MV-ISF. The average residual stress using the VTP is obviously smaller than that using the CTP, and the maximum residual stress by the VTP is about 580 MPa, which is smaller than that by the CTP of 650 MPa. It is shown that the decreased residual stress using the VTP is caused by the axial vibration field, and the internal residual stress is released during cyclic loading and unloading in the MV-ISF. Through releasing the residual stress, the springback of a formed part decreases which contributes to the improvement of geometrical accuracy. Based on the above simulated results of the residual stress, it can be concluded that the MV-ISF process can improve the geometrical accuracy of the formed part as shown in Fig. 2 as a result of releasing the residual stress due to vibration.

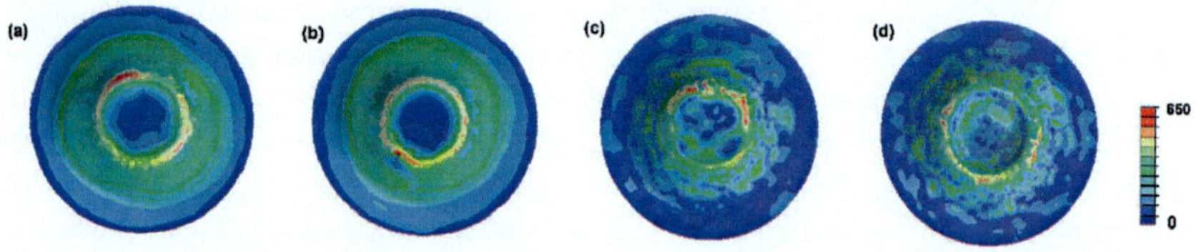

Fig. 8. Finite element simulation of residual stress distribution (MPa): (a) T0 & CTP; (b) T2 & CTP; (c) T2 & VTP $h = 0.1$ mm; (d) T2 & VTP $h = 0.3$ mm.

6 Conclusions

In this study, a new MV-ISF process is proposed by introducing axial and radial vibration fields in ISF through employing a vibrated tool path design combined with the use of the ellipsoidal headed tool. Experiments using AA5251 and AA6082, analytical modelling and finite element simulations have been conducted to investigate the improvement of the new process in geometrical accuracy, surface quality and formability. The main conclusions are:

- The MV-ISF process can significantly improve the fracture strain through increasing the vibration energy and decreasing the stress triaxiality under the effect of the exerted axial vibration field. This new vibrating tool path design provides a simple and feasible solution for improving the material formability.
- The new MV-ISF shows its advantage in decreasing the geometrical deviation of the formed part by releasing the residual stress due to the effect of the vibrated tool path.
- Increasing the axial distance in the vibrated tool path of the MV-ISF improves geometrical accuracy, surface quality and formability of formed parts.

Acknowledgments. The authors wish to acknowledge the funding support received from the UK EPSRC (Engineering and Physical Sciences Research Council) through project grants EP/W010089/1 and EP/T005254/1.

References

1. Vahdati, M., Mahdavinejad, R., Amini, S.: Investigation of the ultrasonic vibration effect in incremental sheet metal forming process. Proc. Inst. Mech. Eng. Pt. B J. Eng. Manuf. **231**, 971–982 (2015)
2. Mingshun, Y., Lang, B., Yunbo, L., et al.: Research on the radial accuracy of ultrasonic VA single point incremental forming parts. Int. J. Aerosp. Eng. **2019**, 1–9 (2019). https://doi.org/10.1155/2019/9809815
3. Cheng, Z., Li, Y., Li, J., et al.: Ultrasonic assisted incremental sheet forming: constitutive modeling and deformation analysis. J. Mater. Process. Tech. **299**, 117365 (2022)
4. Long, Y., Li, Y., Sun, J., Ille, I., Li, J., Twiefel, J.: Effects of process parameters on force reduction and temperature variation during ultrasonic assisted incremental sheet forming process. Int. J. Adv. Manuf. Technol. **97**(1–4), 13–24 (2018). https://doi.org/10.1007/s00170-018-1886-0
5. Bai, L., Li, Y., Yang, M., et al.: Modeling and analysis of single point incremental forming force with static pressure support and ultrasonic vibration. Materials **12**, 1899 (2019)
6. Amini, S., Hosseinpour Gollo, A., Paktinat, H.: An investigation of conventional and ultrasonic-assisted incremental forming of annealed AA1050 sheet. Int. J. Adv. Manuf. Technol. **90**, 1569–1578 (2016)
7. Yang, M., Bai, L., Li, Y., et al.: Influences of vibration parameters on formability of 1060 aluminum sheet processed by ultrasonic VA single point incremental forming. Adv. Mater. Sci. Eng. **2019**, 1–12 (2019). https://doi.org/10.1155/2019/8405438
8. Amini, S., Nazari, F., Baraheni, M., et al.: Investigating the effect of rotation speed and ultrasonic vibrations in the incremental forming process. Int. J. Adv. Design Manuf. Technol. **11**, 91–97 (2018)

9. Li, Y., Chen, X., Sun, J., et al.: Effects of ultrasonic vibration on deformation mechanism of incremental point-forming process. Procedia Eng. **207**, 777–782 (2017)
10. Lu, B., Li, Z., Long, H., et al.: Microstructure refinement by tool rotation-induced vibration in incremental sheet forming. Procedia Eng. **207**, 795–800 (2017). https://doi.org/10.1016/j.proeng.2017.10.831
11. Filice, L., Fratini, L., Micari, F.: Analysis of material formability in incremental forming. CIRP Ann. Manuf. Technol. **51**, 199–202 (2002)
12. Wang, J., Li, L., Zhou, P., Wang, X., Sun, S.: Improving formability of sheet metals in incremental forming by equal diameter spiral tool path. Int. J. Adv. Manuf. Technol. **101**(1–4), 225–234 (2018). https://doi.org/10.1007/s00170-018-2911-z
13. Wen, T., Chen, X., Zheng, J., et al.: Multi-directional incremental sheet forming-a novel methodology for flexibly producing thin-walled parts. Int. J. Adv. Manuf. Technol. **91**, 1909–1919 (2017)
14. Lu, B., Chen, J., Ou, H., et al.: Feature-based tool path generation approach for incremental sheet forming process. J. Mater. Process. Technol. **213**, 1221–1233 (2013)
15. Chang, Z., Huang, W., Chen, J.: A new tool path with point contact and its effect on incremental sheet forming process. Int. J. Adv. Manuf. Technol. **110**(5–6), 1515–1525 (2020). https://doi.org/10.1007/s00170-020-05960-0
16. Lu, B., Fang, Y., Xu, D.K., et al.: Mechanism investigation of friction-related effects in single point incremental forming using a developed oblique roller-ball tool. Int. J. Mach. Tools Manuf. **85**, 14–29 (2014)
17. Wang, C.J., Liu, Y., Guo, B., et al.: Acoustic softening and stress superposition in ultrasonic vibration assisted uniaxial tension of copper foil: experiments and modeling. Mater. Des. **112**, 246–253 (2016)
18. Eyckens, P., Belkassem, B., Henrard, C., et al.: Strain evolution in the single point incremental forming process: digital image correlation measurement and finite element prediction. Int J. Mater. Form. **4**, 55–71 (2011)
19. Silva, M., Skjoedt, M., Martins, P., et al.: Revisiting the fundamentals of single point incremental forming by means of membrane analysis. Int. J. Mach. Tools Manuf. **48**, 73–83 (2008)

Anisotropic-Asymmetric Hardening Characterization of BCC/FCC/HCP Metals: Experiments and Modeling

Yanshan Lou[1](✉), Chong Zhang[1], and Jeong Whan Yoon[2]

[1] Xi'an Jiaotong University, Xi'an, China
ys.lou@xjtu.edu.cn
[2] KAIST, Daejeon, South Korea

Abstract. This research investigates the anisotropic-asymmetric hardening behaviors of different types of metals by experiments and analytical modeling. Experiments are conducted at shear, uniaxial tension, plane strain tension and equibiaxial tension states to characterize the anisotropic hardening behaviors of aluminum alloy 7075-T6 and magnesium alloy AZ31. The strain hardening behaviors are compared between different loading directions and stress states. An obvious anisotropic-differential hardening evolution is observed for AZ31. The anisotropic-asymmetric hardening behaviors are characterized by a newly proposed stress invariants-based function with isotropic and anisotropic forms. A newly linear transformation tensor is introduced based on that of Barlat et al. (1991) to extend the isotropic stress invariants-based function to an anisotropic form. The comparison shows that the anisotropic-asymmetric hardening behaviors are precisely modeled by the proposed yield function under different stress states of shear, uniaxial tension, plane strain tension and equibiaxial tension, and different loading directions. The extended anisotropic form is recommended for strongly anisotropic materials such as AZ31 due to its less computationally cost while maintaining high accuracy.

Keywords: Anisotropic hardening · Strength differential effect · Yield surface evolution

1 Introduction

Sheet metal experiences various stress states during the stamping process includes uniaxial tension/compression, plane strain tension, equibiaxial tension and shear, etc. [1]. The strength modelling for different stress states is important especially for lightweight materials. Many lightweight materials such as advanced high-strength quenching & partitioning (Q&P) steel [2, 3], and magnesium alloy [4, 5] have complex strain hardening behavior between different stress states due to their complicated mechanical behavior.

Anisotropy is an inherent characteristic for rolled materials, which refers to the phenomenon where the material exhibits different properties or performance in different directions. It is influenced by various factors during the manufacturing process, such

K. Mocellin et al. (Eds.): ICTP 2023, LNME, pp. 613–621, 2024.
https://doi.org/10.1007/978-3-031-40920-2_63

as the crystal grain structure of the material, processing methods, stress state, etc. The strength difference between different stress states results in variations in the evolution of the hardening law, also known as differential hardening. Most traditional yield functions focus on the anisotropy instead of the strength difference between different stress states, including the classic Hill48 yield function [6], eight anisotropic parameters function Yld2000-2d [7], eighteen anisotropic parameters Yld2004-18p [8], anisotropic Drucker [9], reduced Yld2004 function [10] and so on. The above yield functions are based on the principal stress form, the stress-invariants based yield functions also have developed. The Drucker [11] yield function considers both the second and third invariants on yielding. Spitzig and Richmond [12] considered the pressure effect by adding the first stress invariant I1. Cazacu and Barlat [13] modified the Drucker yield function to an asymmetric form to account for strength differential (SD) effect. Lou et al. [14] coupled the pressure effect with the Yld2000-2d function to model both anisotropy and SD effect for sheet metals. Yoon et al. [15] proposed a general yield function that considers all three stress invariants for HCP model. Furthermore, Lou et al. [16] coupled the pressure effect in the Drucker function. Hu et al. [17] proposed a normalized stress invariant-based yield function in a polynomial form. The analytical approach has been developed for modelling of hardening evolution of different loading conditions, such as Stoughton-Yoon'2009 [18], CQN [19], Park et al. [20], Hou et al. [2], Hu et al. [21, 22], Chen et al. [23], Wu et al. [24], Lou and Yoon [25], Zhang and Lou [26] etc. Concexity is a very serious problem for these evolving yield function [27, 28]. All the yield functions above greatly improve the modelling accuracy of plastic deformation in metal forming processes. However, the stress states considered by the above functions are generally uniaxial tension, equibiaxial tension and uniaxial compression. Shear and plane strain tension are rarely considered together in analytical modelling.

In this study, a general yield function based on all three stress invariants is given to model the differential strain hardening behaviours under different stress states. It can reduce to other popular stress invariants-based functions such as Drucker-Prager, Cazacu-Barlat'2004, Yoon'2014 and pDrucker functions. Then the isotropic general yield function is extended to the analytical anisotropic form. The anisotropic parameters are analytically calculated so that the proposed yield function can model both differential hardening at various stress states and anisotropic hardening. The proposed yield function is applied to characterize the strain hardening behaviour under distinct stress states along different loading directions for materials AA7075 T6 and AZ31 in our previous study.

2 Experiments

Experiments are conducted in shear, uniaxial tension, plane strain tension and equibiaxial tension states to characterize the anisotropic hardening behaviors of AA7075-T6 and AZ31. Different specimens were tested include dogbone, notched, shear and bulge specimens under quasi-static loading conditions. The obtained equivalent Mises stress-equivalent plastic strain (EPS) curves are plotted in Fig. 1 (a) for AA7075 T6 and (b) for AZ31. The error bars displayed in Fig. 1 indicate a good repeatability of each test. For AA7075 T6, the strength difference under different stress states is about 8%, which is much larger than the 2% anisotropy difference as observed. The relationship between these four flow curves does not change much with respect to plastic strain, which indicates that the yield surface does not change severely for AA7075 T6 at different strains. The hardening evolution of AZ31 under different stress states is quite different from AA7075 T6. An obvious different hardening evolution between different stress states is observed. The shear flow curve is the lowest when the strain is less than 0.075 but rises to be higher than the uniaxial tensile flow curves when EPS > 0.09. In addition, the strength anisotropy of AZ31 is about 9%, which is much higher than AA7075 T6. However, the maximum difference of the flow curves under these four stress states is larger than 30%, which is much more apparent than the strength anisotropy of AZ31. The stress-strain curves are interpolated by the Hockett-Sherby law in Eq. (1) and hardening coefficients for the tested materials under different stress states are summarized in Table 1.

$$\sigma = A - (A - B)exp\left(-C\bar{\varepsilon}^{pn}\right) \tag{1}$$

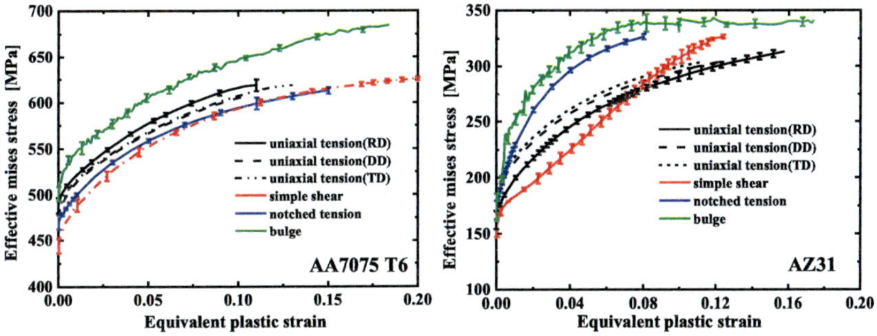

(a) (b)

Fig. 1. The Mises stress-EPS curves under the four different stress states: (a) AA7075 T6; (b) AZ31.

Table 1. The calibrated coefficients of the Hockett-Sherby hardening law as Eq. (1) for the tested materials under different stress states.

Materials	Stress states	coefficients			
		A [MPa]	*B* [MPa]	*C*	*n*
AA7075 T6	UT00	697	493	5.9	0.80
	UT45	960	473	1.23	0.57
	UT90	741	480	3.4	0.67
	SS45	640	451	11.8	0.88

(*continued*)

Table 1. (*continued*)

Materials	Stress states	coefficients			
		A [MPa]	B [MPa]	C	n
	PST90	640	470	8.8	0.83
	EBT	782	507	3.4	0.69
AZ31	UT00	349	165	6.6	0.76
	UT45	330	184	9.2	0.81
	UT90	327	193	12	0.88
	SS45	362	178	61.9	1.59
	PST45	343	171	18.8	0.83
	EBT	357.5	166.3	11.9	0.66

3 Constitutive Models

The universal yield function [29] has the isotropic form as Eq. (2) based on (I1, J2, J3) form and Eq. (3) based on $(\eta, \xi, \overline{\sigma}_{VM})$ form.

$$f(\sigma_{ij}) = a\left\{bI_1 + \left[\left(J_2^3 - cJ_3^2\right)^{1/2} - dJ_3\right]^{1/3}\right\} \tag{2}$$

$$f(\sigma_{ij}) = a\left\{3b\eta + \frac{1}{3}\left[\left(27 - 4c\xi^2\right)^{1/2} - 2d\xi\right]^{1/3}\right\}\overline{\sigma}_{VM} \tag{3}$$

where parameter b considers the pressure effect, parameter c adjusts the curvature of the yield surface as plotted in Fig. 1(a) for FCC and BCC metals, parameter d considers the SD effect for pressure-insensitive materials as plotted in Fig. 1(b) for HCP metals.

The isotropic parameters of the universal yield function are analytically solved to model differential hardening. The four isotropic parameters are calibrated by four critical stress states from uniaxial compression to equibiaxial tension including uniaxial compression, shear, uniaxial tension, plane strain and equibiaxial tension states. Their corresponding (ξ, η) are $(-1, -1/3)$, $(0, 0)$, $(1, 1/3)$, $(0, 1/\sqrt{3})$ and $(-1, 2/3)$, respectively. To model four of the five critical stress states, there are four different methods to calibrate the four parameters *a, b, c,* and *d* .

To model the anisotropic hardening, the $\overline{\sigma}_{VM}$ term in Eq. (3) is replaced by an introduced anisotropic $\overline{\sigma}'$. Accordingly, the analytical anisotropic form of the proposed is presented below

$$f(\sigma_{ij}) = a\left\{3b\eta + \frac{1}{3}\left[\left(27 - 4c\xi^2\right)^{1/2} - 2d\xi\right]^{1/3}\right\}\overline{\sigma}' \tag{4}$$

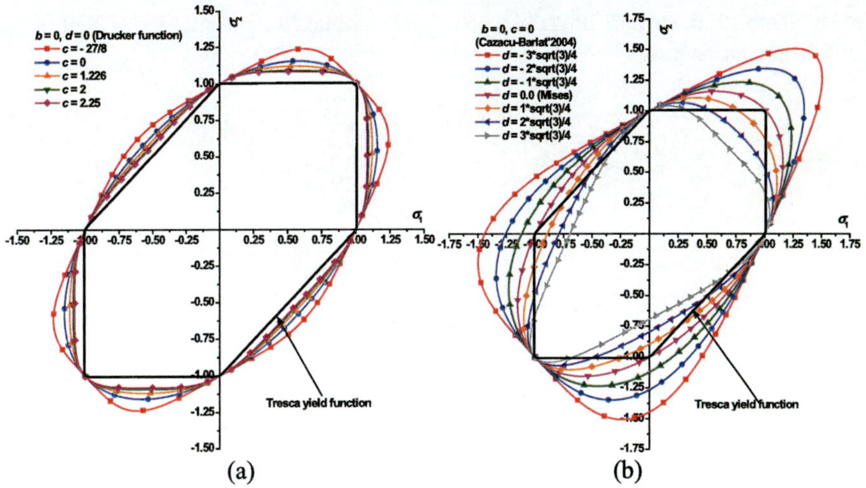

Fig. 2. Influence of the parameters c and d of the proposed yield function on yielding: (a) parameter c to modify the yield surface curvature (reduce to the Drucker yield criterion); (b) parameter d to consider the SD effect (reduce to the Cazacu-Barlat'2004 yield criterion).

Where $\overline{\sigma}'$ is the von Mises stress of the linear transformed stress tensor $\boldsymbol{s}'$ which incorporates anisotropic parameters as

$$\overline{\sigma}' = \sqrt{\frac{3}{2}\boldsymbol{s}'\boldsymbol{s}'} \ with\ \boldsymbol{s}' = \boldsymbol{L}'\boldsymbol{\sigma} \tag{5}$$

Where

$$\boldsymbol{L}' = \frac{1}{3}\begin{bmatrix} c_1' + c_2' & -c_1' & -c_2' & 0 & 0 & 0 \\ -c_1' & 2c_1' & -c_1' & 0 & 0 & 0 \\ -c_2' & -c_1' & c_1' + c_2' & 0 & 0 & 0 \\ 0 & 0 & 0 & 3c_3' & 0 & 0 \\ 0 & 0 & 0 & 0 & 3c_4' & 0 \\ 0 & 0 & 0 & 0 & 0 & 3c_5' \end{bmatrix}, \boldsymbol{\sigma} = \begin{bmatrix} \boldsymbol{\sigma}_{xx} \\ \boldsymbol{\sigma}_{yy} \\ \boldsymbol{\sigma}_{zz} \\ \boldsymbol{\sigma}_{yz} \\ \boldsymbol{\sigma}_{zx} \\ \boldsymbol{\sigma}_{xy} \end{bmatrix} \tag{6}$$

where c_i' are anisotropic parameters under spatial stress state. The linear transformation tensor $\boldsymbol{L}'$ is reduced from that of Barlat et al. [30]. When c_i' are all equal to 1.0, the isotropic form is recovered. Parameters c_1', c_2' and c_5' are calibrated to model the anisotropic behaviour of uniaxial tension along RD σ_0, DD σ_{45}, and TD σ_{90}. The rest two parameters c_3' and c_4' are related with the through-thickness shear properties.

To obtain the analytic anisotropic form of the proposed yield function, the anisotropic parameters of c_1', c_2' and c_3' are calculated from the $\sigma_{UT0}(\overline{\lambda})$, $\sigma_{UT90}(\overline{\lambda})$ and $\sigma_{UT45}(\overline{\lambda})$ flow curves, and differential parameters $(\overline{\lambda})$, $b(\overline{\lambda})$, $c(\overline{\lambda})$ and $d(\overline{\lambda})$ are calculated from hardening curves of different stress states. $\overline{\lambda}$ is the effective plastic strain. Here lists one of the calibration methods as in Eq. (7) for modelling differential hardening between shear, plane strain tension, uniaxial tension and equibiaxial tension states as well as the anisotropic hardening.

$$\begin{cases} c_1'(\bar{\lambda}) = \dfrac{\sigma_{UT_0}(\bar{\lambda})}{\sigma_{UT_{90}}(\bar{\lambda})} \\ c_2'(\bar{\lambda}) = \dfrac{\sqrt{12-3\left(c_1'(\bar{\lambda})\right)^2-\left(c_1'(\bar{\lambda})\right)}}{2} \\ c_5'(\bar{\lambda}) = \sqrt{\dfrac{4\left(\frac{\sigma_{UT_0}(\bar{\lambda})}{\sigma_{UT_{45}}(\bar{\lambda})}\right)^2-1}{3}} \end{cases} \quad \begin{cases} a(\bar{\lambda}) = \dfrac{\sqrt{3}\sigma_{UT_0}(\bar{\lambda})}{\tilde{\sigma}'_{SS_\theta}} \\ b(\bar{\lambda}) = \frac{1}{3}\left(\dfrac{\tilde{\sigma}'_{SS_\theta}(\bar{\lambda})}{\tilde{\sigma}'_{PST_\theta}(\bar{\lambda})}-1\right) \\ d(\bar{\lambda}) = \frac{27}{4}\left[\left(\dfrac{\sigma_{UT_0}(\bar{\lambda})}{a\tilde{\sigma}'_{EBT}(\bar{\lambda})}-2b\right)^3-\left(\frac{1}{a}-b\right)^3\right] \\ c(\bar{\lambda}) = \dfrac{27-\left[27\left(\frac{1}{a}-b\right)^3+2d\right]^2}{4} \end{cases} \tag{7}$$

where UT_θ, PST_θ, SS_θ and UC_θ indicate uniaxial tension, plane strain tension, simple shear and uniaxial compression states along loading direction θ with respect to RD. The calculation of $\overline{\sigma}'_{SS_\theta}$, $\overline{\sigma}'_{PST_\theta}$ and $\overline{\sigma}'_{UC_\theta}$ are depends on the loading angle θ.

4 Applications

The anisotropic and differential parameters evolution of the proposed anisotropic function is calculated by Eq. (7) as shown in Fig. 3 for AZ31. It indicates that the anisotropic and differential hardening is very strong for AZ31 since the parameters evolution is apparent for AZ31. The parameter evolution of AZ31 is apparent both for anisotropic and differential hardening parameters. The strongly anisotropic and differential hardening of AZ31 is expected to be modelled by the proposed anisotropic-asymmetric yield function in Eq. (4).

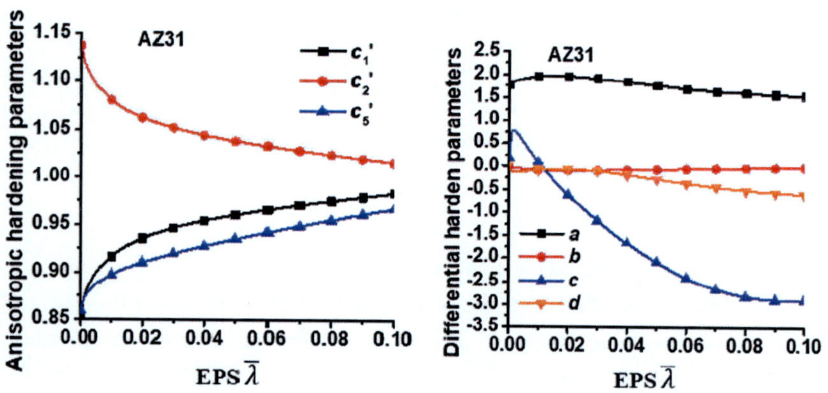

Fig. 3. The evolution of the four parameters a, b, c and d of the proposed yield function with respect to EPS for AZ31.

Then the anisotropic form of the proposed yield function is calibrated to model the anisotropy of test materials and is compared with the isotropic form as shown in Fig. 4

(a) for AA7075 T6 and (b) for AZ31. It shows the anisotropic form of the proposed function can model the strength under uniaxial tension along RD and TD precisely, but the isotropic form cannot. The yield surface differences of AA7075 T6 between the isotropic and anisotropic forms are quite small since AA7075 T6 is a weakly anisotropic material. The predicted uniaxial tension stress along different loading directions with respect to RD for AZ31 are shown in Fig. 5. There is a good fit between the prediction and experiments for uniaxial tensile strength along RD, DD and TD under different EPS values.

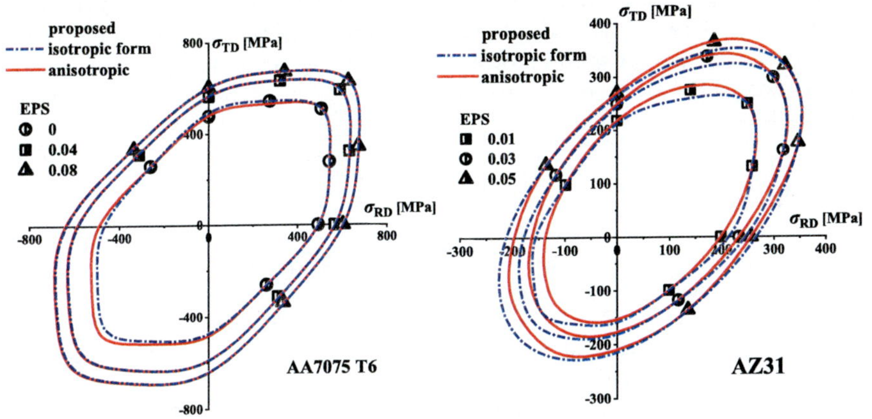

Fig. 4. Modelling of yield surface evolution by the proposed yield function with isotropic and anisotropic forms for (a) AA7075 T6 and (b) AZ31.

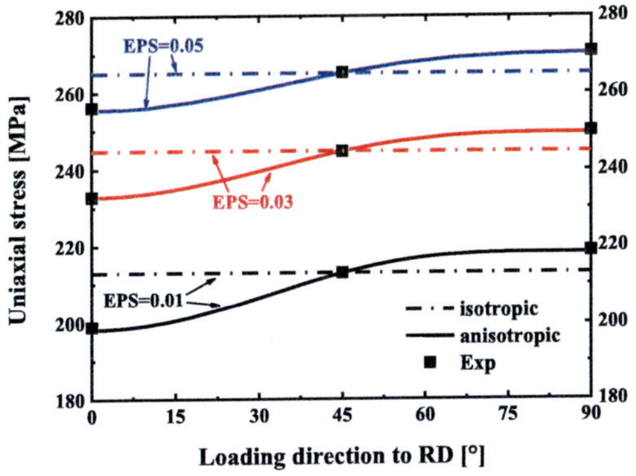

Fig. 5. Anisotropic strength evolution of AZ31 under uniaxial tension.

5 Conclusions

The isotropic form of the universal yield function is extended by linear transformation tensors with ease to take the anisotropy into account for enhanced numerical modelling of plastic deformation in metal forming. The anisotropic parameters are analytically calculated so that the proposed yield function can model both differential hardening at various stress states and anisotropic hardening along different loading directions. The extended anisotropic form is applied to characterize the strain hardening behaviour under distinct stress states for AA7075 T6 and AZ31. The results show that the proposed function precisely models the differential and anisotropic hardening for the tested metals under various stress states. It should be noted that the performance of the proposed yield function is not evaluated for other tests that are not used for calibration such as uniaxial compression test. For strongly anisotropic materials such as AZ31, the proposed analytical form is recommended due to its less computationally cost.

References

1. Brosius, A., Küsters, N., Lenzen, M.: New method for stress determination based on digital image correlation data. CIRP Ann. **67**, 269–272 (2018)
2. Hou, Y., Min, J., Stoughton, T.B., Lin, J., Carsley, J.E., Carlson, B.E.: A non-quadratic pressure-sensitive constitutive model under non-associated flow rule with anisotropic hardening: modeling and validation. Int. J. Plast. **135**, 102808 (2020)
3. Hou, Y., Min, J., Guo, N., Shen, Y., Lin, J.: Evolving asymmetric yield surfaces of quenching and partitioning steels: characterization and modeling. J. Mater. Process. Technol. **290**, 116979 (2021)
4. Lou, X.Y., Li, M., Boger, R.K., Agnew, S.R., Wagoner, R.H.: Hardening evolution of AZ31B Mg sheet. Int. J. Plast. **23**, 44–86 (2007)
5. Andar, M.O., Kuwabara, T., Steglich, D.: Material modeling of AZ31 Mg sheet considering variation of r-values and asymmetry of the yield locus. Mater. Sci. Eng. **549**, 82–92 (2012)
6. Hill, R.: A theory of the yielding and plastic flow of anisotropic metals. Proc. R Soc. London Ser. A Math. Phys. Sci. **193**, 281–297 (1948)
7. Barlat, F., Brem, J.C., Yoon, J.W., Chung, K., Dick, R.E., Lege, D.J.: Plane stress yield function for aluminum alloy sheets—part 1: theory. Int. J. Plast. **19**, 1297–1319 (2003)
8. Barlat, F., Aretz, H., Yoon, J.W., Karabin, M.E., Brem, J.C., Dick, R.E.: Linear transformation-based anisotropic yield functions. Int. J. Plast. **21**, 1009–1039 (2005)
9. Lou, Y., Yoon, J.W.: Anisotropic yield function based on stress invariants for BCC and FCC metals and its extension to ductile fracture criterion. Int. J. Plast. **101**, 125–155 (2018)
10. Lou, Y., Zhang, S., Yoon, J.W.: A reduced Yld 2004 function for modeling of anisotropic plastic deformation of metals under triaxial loading. Int. J. Mech. Sci. **161**, 105027 (2019)
11. Drucker, D.C.: Relation of experiments to mathematical theories of plasticity. J. Appl. Mech. **16**, 349–357 (1949)
12. Spitzig, W.A., Richmond, O.: The effect of pressure on the flow stress of metals. Acta Metall. **32**, 457–463 (1984)
13. Cazacu, O., Barlat, F.: A criterion for description of anisotropy and yield differential effects in pressure-insensitive metals. Int. J. Plast. **20**, 2027–2045 (2004)
14. Lou, Y., Huh, H., Yoon, J.W.: Consideration of strength differential effect in sheet metals with symmetric yield functions. Int. J. Mech. Sci. **66**, 214–223 (2013)

15. Yoon, J.W., Lou, Y., Yoon, J., Glazoff, M.V.: Asymmetric yield function based on the stress invariants for pressure sensitive metals. Int. J. Plast. **56**, 184–202 (2014)
16. Lou, Y., Zhang, S., Yoon, J.W.: Strength modeling of sheet metals from shear to plane strain tension. Int. J. Plast. **134**, 102813 (2020)
17. Hu, Q., Li, X., Han, X., Li, H., Chen, J.: A normalized stress invariant-based yield criterion: modeling and validation. Int. J. Plast. **99**, 248–273 (2017)
18. Stoughton, T.B., Yoon, J.W.: Anisotropic hardening and non-associated flow in proportional loading of sheet metals. Int. J. Plast. **25**, 1777–1817 (2009)
19. Lee, E.H., Stoughton, T.B., Yoon, J.W.: A yield criterion through coupling of quadratic and non-quadratic functions for anisotropic hardening with non-associated flow rule. Int. J. Plast. **99**, 120–143 (2017)
20. Park, N., Stoughton, T.B., Yoon, J.W.: A criterion for general description of anisotropic hardening considering strength differential effect with non-associated flow rule. Int. J. Plast. **121**, 76–100 (2019)
21. Hu, Q., Yoon, J.W.: Analytical description of an asymmetric yield function (Yoon2014) by considering anisotropic hardening under non-associated flow rule. Int. J. Plast. **140**, 102978 (2021)
22. Hu, Q., Yoon, J.W., Manopulo, N., Hora, P.: A coupled yield criterion for anisotropic hardening with analytical description under associated flow rule: modeling and validation. Int. J. Plast. **136**, 102882 (2021)
23. Chen, Z., Wang, Y., Lou, Y.: User-friendly anisotropic hardening function with non-associated flow rule under the proportional loadings for BCC and FCC metals. Mech. Mater. **165**, 104190 (2022)
24. Wu, P., Lou, Y., Chen, Q., Ning, H.: Modeling of temperature-and stress state-dependent yield and fracture behaviors for Mg-Gd-Y alloy. Int. J. Mech. Sci. **229**, 107506 (2022)
25. Lou, Y., Yoon, J.W.: Lode-dependent anisotropic-asymmetric yield function for isotropic and anisotropic hardening of pressure-insensitive materials. Part I: Quadratic function under non-associated flow rule. Int. J. Plast. (2023). https://doi.org/10.1016/j.ijplas.2023.103647
26. Zhang, C., Lou, Y.: Characterization and modelling of evolving plasticity behaviour up to fracture for FCC and BCC metals. J. Mater. Process. Technol. **317**, 117997 (2023)
27. Wang, J., Han, M., Zhang, C., Rayhan, H.M.A., Li, X., Lou, Y.: Anisotropic hardening of TRIP780 steel sheet: experiments and analytical modeling. Material **16**, 1414 (2023)
28. Lou, Y., Zhang, C., Wu, P., Yoon, J.W.: Geometry-inspired numerical convex analysis (GINCA) of yield functions under isotropic and anisotropic hardening. Submitted for review
29. Lou, Y., Zhang, C., Zhang, S., Yoon, J.W.: A general yield function with differential and anisotropic hardening for strength modelling under various stress states with non-associated flow rule. Int. J. Plast. **158**, 103414 (2022)
30. Barlat, F., Lege, D.J., Brem, J.C.: A six-component yield function for anisotropic materials. Int. J. Plast. **7**, 693–712 (1991)

Highly Efficient Characterization of the Dynamic Anisotropic Plasticity for Sheet Metals from the Heterogeneous High-Speed Impact Inertial Fields via the Virtual Fields Method

Jiawei Fu(✉), Zefei Yang, Yahui Cai, Jun Luo, and Lehua Qi

Northwestern Polytechnical University, Xi'an, China
jiawei.fu@nwpu.edu.cn

Abstract. In this paper, the virtual field method (VFM) is used to identify the dynamic anisotropic plasticity of the sheet metal subjected to highspeed impact. First, the selected dynamic anisotropic plasticity constitutive models and the principle of the VFM-based characterization method are introduced. Then, a heterogeneous highspeed impact testing prototype using the double-notched specimen configuration is proposed and the virtual inertial impact test carried out. The simulated strain and acceleration field data are extracted and analyzed, from which the target dynamic anisotropic plasticity constitutive parameters are retrieved. The results show that using the current identification scheme the anisotropic yielding and the rate-dependent hardening parameters can be simultaneously identified from the heterogeneous inertial kinetic fields with a single impact test. The proposed identification method is exempt from the one-dimensional stress wave propagation and homogeneous deformation state preconditions and can significantly simplify the identification process of the dynamic anisotropic plasticity of sheet metals.

Keywords: Sheet Metal · Dynamic Anisotropic Plasticity · High Strain Rate · Constitutive Parameter Identification · Virtual Field Method

1 Introduction

Sheet metals are essential structural materials in industry. Due to the evolution of crystallographic structure during forming process, sheet metals usually possess high anisotropic plasticity in yielding and hardening properties. For many sheet metals, the hardening behavior also depends on the rate of deformation. Metals usually tend to exhibit higher strengths and lower ductility when strain rate is increased. An accurate characterization of the dynamic anisotropic plasticity of sheet metals is a prerequisite for further purposes such as improving forming process of parts and predicting material behaviors under complex loading conditions. Classic dynamic testing methods such as the split Hopkinson pressure bar (SHPB) can provide reasonable stress-strain response of materials under high strain rates. However, they also suffer from limitations such as the

K. Mocellin et al. (Eds.): ICTP 2023, LNME, pp. 622–629, 2024.
https://doi.org/10.1007/978-3-031-40920-2_64

uniform deformation state and one-dimensional stress wave propagation preconditions, which increase the difficulty and complexity of the identification of dynamic anisotropic plasticity.

In recent years, the rapid development of ultra-highspeed imaging and full-field deformation measurement techniques have significantly promoted the characterization of dynamic mechanical properties. It allows the instantaneous non-uniform deformation field of sample surface to be recorded, and then material constitutive parameters to be identified from the heterogeneous kinetic fields using an inverse problem solution such as the finite element model updating (FEMU) [1], and the virtual field method (VFM) [2], etc. Based on the VFM, the so-called image-based inertial impact (IBII) tests have been developed and successfully applied to identify the linear elasticity [3, 4], isotropic elasto-plasticity [5], and visco-plasticity of materials under dynamic loads [6, 7].

In this paper, the VFM is further extended to the dynamic anisotropic plasticity of sheet metals undergoing high strain rates. First, the rate-dependent anisotropic plasticity constitutive models and principle of the VFM are introduced. Then, the virtual heterogeneous inertial impact test is carried out and the resulting dynamic field data exported. Finally, the input target constitutive parameters are retrieved from the full-field data and the identification results analyzed.

2 Methodology

2.1 The Virtual Fields Method

The basic idea of VFM is to establish the equilibrium equation of internal virtual work (IVW), external virtual work (EVW) and acceleration virtual work (AVW). For dynamic loadings, it is difficult to measure the dynamic load $\boldsymbol{T}$ on the boundary in practical experiments. In order to simplify the virtual work balance equation, one can select proper virtual displacement field $\boldsymbol{u}^*$ to offset the EVW caused by $\boldsymbol{T}$. Thus, the simplified virtual work balance equation can be expressed as:

$$-\int_V \boldsymbol{\sigma} : \boldsymbol{\varepsilon}^* dV = \int_V \rho \boldsymbol{a} \cdot \boldsymbol{u}^* dV \tag{1}$$

where, V is the object volume, $\boldsymbol{\sigma}$ is the Cauchy stress tensor, ρ is the density, $\boldsymbol{a}$ is the acceleration, $\boldsymbol{u}^*$ is the defined virtual displacement, and the virtual strain $\boldsymbol{\varepsilon}^*$ is obtained by the derivative of the virtual displacement $\boldsymbol{u}^*$.

For the nonlinear case of plasticity and plane stress state, a cost function between the IVW and AVW can be defined as:

$$\mathbb{C}(\boldsymbol{X}) = \sum_{i=1}^{N_i} \sum_{j=1}^{N_j} \left(\int_S \boldsymbol{\sigma}(\boldsymbol{X}, \boldsymbol{\varepsilon}) : \boldsymbol{\varepsilon}^* dS + \rho \int_S \boldsymbol{a} \cdot \boldsymbol{u}^* dS \right)^2 \tag{2}$$

where, S is the surface area, N_j is the total number of steps of the deformation process, j represents the j-th loading step. N_i is the total number of the defined independent virtual fields, and i the i-th virtual field. For anisotropic plasticity, multiple groups of virtual fields should be selected to improve the convergence and accuracy of the solution. The unknown parameters $\boldsymbol{X}$ can be obtained by minimizing $\mathbb{C}(\boldsymbol{X})$ using appropriate optimization algorithms.

2.2 Constitutive Models

Hill1948 yield criterion is chosen to describe the anisotropic yield behavior. In the plane stress state, the expression of Hill1948 anisotropic yield function is:

$$F(\boldsymbol{\sigma}, \varepsilon_p) = \sqrt{(H+G)\sigma_{xx}^2 + (H+F)\sigma_{yy}^2 - 2H\sigma_{xx}\sigma_{yy} + 2N\sigma_{xy}^2} - \sigma_s(\varepsilon_p) = 0 \quad (3)$$

where, $\boldsymbol{\sigma}$ is Cauchy stress tensor, ε_p is the equivalent plastic strain. σ_s is the current flow stress. H, G, F and N are the anisotropic yield parameters.

The current flow stress is a function of equivalent plastic strain where the Johnson-Cook type rate-dependent term is adopted, namely

$$\sigma_s(\varepsilon_p) = K(\varepsilon_0 + \varepsilon_p)^n\left(1 + C\ln\left(\dot{\varepsilon}_p / \dot{\varepsilon}_0\right)\right) \quad (4)$$

where, K, ε_0 and n are strain hardening parameters. C is the strengthening coefficient of strain rate. $\dot{\varepsilon}_p$ is the equivalent plastic strain rate. $\dot{\varepsilon}_0$ is the reference strain rate.

3 Numerical Validation of the Identification Method

3.1 Virtual Impact Test by Finite Element Simulation

In this study, a conventional desktop (Intel Core i5-6300HQ CPU, 8.00 GB RAM, 64-bit operating system) was used for calculation. The virtual inertial impact test prototype was established in Abaqus Explicit. As shown in Fig. 1, a double-notch specimen configuration was determined, which can provide heterogeneous stress states under unidirectional impact loading. The material parameters were somehow arbitrarily chosen according to the 2024 aluminum alloy. The test configurations are given in Table 1. During the test, the left boundary of the double-notched sample was fixed, and the right boundary subjected to uniformly varying pressure pulses. The evolution of the pulse over time is approximately a trapezoid with the temporal shape of the rising and falling edges specified using the "Smooth Step" option in Abaqus. The total duration of the pressure pulse is 30 μs.

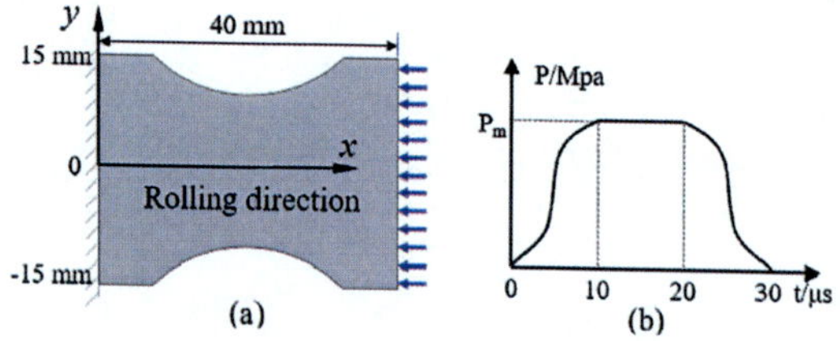

Fig. 1. (a) Schematic of the double-notched inertial impact loading, and (b) Temporal evolution of the applied pressure pulse.

Table 1. Test configurations of the virtual impact.

Specimen shape parameters	Length	40 mm	
	Width	30 mm	
	Thickness	4 mm	
Specimen material parameters	Density	2700 kg/m^3	
	Young's modulus	70 GPa	
	Poisson's ratio	0.3	
	Hill1948 yield criterion	H	0.4
		G	0.6
		F	0.45
		N	0.8
	Strain hardening	K	700 MPa
		ε_0	0.02
		n	0.22
	Strain rate hardening	C	0.008
		$\dot{\varepsilon}_0$	1
Testing parameters	Element type	CPS3	
	Mesh size	0.5 mm	
	Peak pressure	290 MPa	
	Damping coefficient	0.06	
	Interframe time	0.2 μs	
	Time steps	128	

3.2 Results of the Simulated Virtual Impact Test

The acceleration and strain field data were extracted at an interval of 0.2 μs from Abaqus. Figure 2 shows the acceleration distributions at different time steps. As can be seen, both acceleration components reach 10^6 m/s^2 under the current impact velocity. Figure 3 shows the normal strain and shear strain maps at the same time steps. It can be seen that the strain distributions are highly heterogeneous due to the double-notched configuration. Considerable shear strains are obtained in the notched regions. In order to illustrate the stress state, the elemental stress values are plotted in the two-dimensional normalized yield surface, as shown in Fig. 4. In this figure, the outer elliptic contour represents the yield surface where the shear stress is equal to zero, and the origin of the coordinate system represents the pure shear state. For the double-notched specimen, the stress points reasonably cover the internal contours of the ellipse, indicating a rich shear stress state.

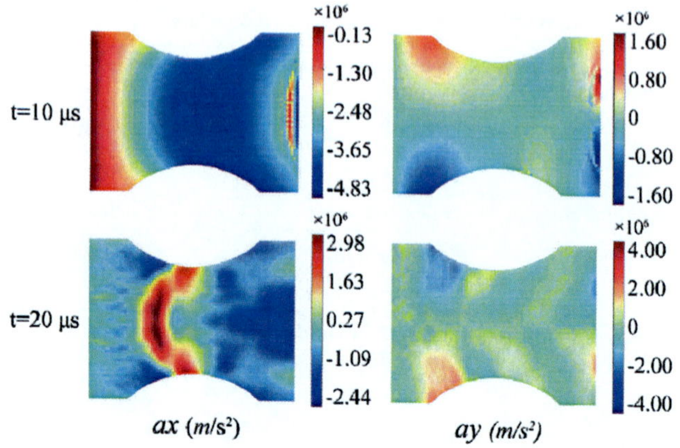

Fig. 2. Acceleration maps at the time steps of 10 μs and 20 μs.

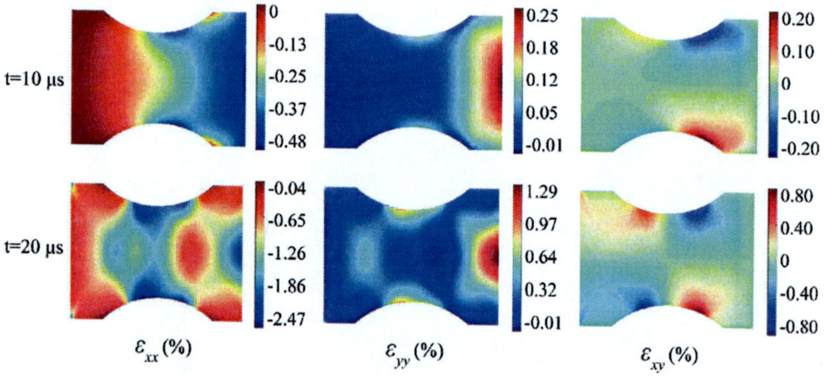

Fig. 3. Strain maps at the time steps of 10 μs and 20 μs.

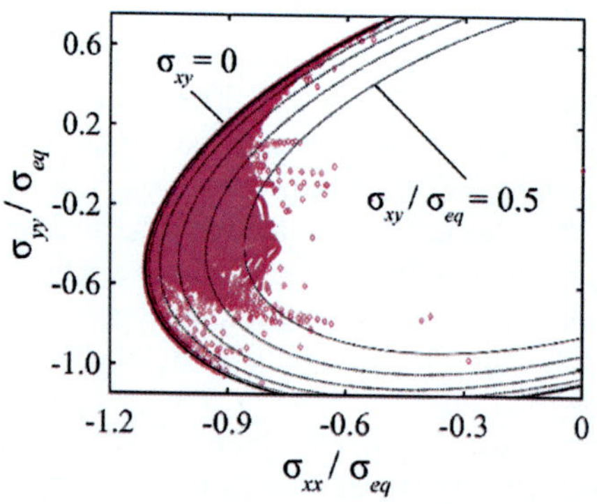

Fig. 4. The elemental stress distribution in the 2D view normalized yield surface

3.3 Set of Virtual Displacement Fields

Since the anisotropic and rate-dependent plasticity constitutive models are highly non-linear, multiple sets of virtual fields should be used to improve the convergence of the minimization process. Here, two sets of virtual fields are defined manually by a trial-and-error process.

$$
\begin{aligned}
VF1: &\quad u_x^* = x(x-l),\ u_y^* = 0 \\
&\quad \varepsilon_{xx}^* = 2x - l,\ \varepsilon_{yy}^* = 0,\ \gamma_{xy}^* = 0 \\
VF2: &\quad u_x^* = \frac{l}{2\pi}\sin\left(\frac{2\pi x}{l}\right),\ u_y^* = 0 \\
&\quad \varepsilon_{xx}^* = \cos\left(\frac{2\pi x}{l}\right),\ \varepsilon_{yy}^* = 0,\ \gamma_{xy}^* = 0
\end{aligned} \tag{5}
$$

3.4 Identification Results

The exported full-field kinetic data are input into the proposed identification program, and different initial values were selected to start the optimization. It was found that due to the large number of constitutive parameters the stability of the global identification results is poor, as shown in Table 2, two sets of initial estimates converge at different local minima. In this case, the strain hardening parameters are set as predetermined values to improve the stability. This is a reasonable means as in practice one can easily determine these values first from a quasi-static test. In Table 3, the results obtained for the anisotropic yield and rate-dependent hardening are shown. One can see that stable identification results now are obtained from the single heterogeneous inertial impact test. The maximum error is seen for the shear-related anisotropic yield parameter N with an error of ~6%, which is satisfactory at this stage. The calculation time for each initial estimate is no more than 10 min, verifying its high efficiency.

Table 2. Identification results of the global anisotropic plasticity constitutive parameters.

Direction	R.V	K	ε_0	n	C	H	F	N
		700	0.02	0.22	0.008	0.4	0.45	0.8
0°	I.E.1	800	0.01	0.1	0.005	0.5	0.5	0.5
	I.R.1	725	0.0183	0.219	0.00469	0.386	0.431	0.933
	R.E.1 (%)	−3.58%	8.36%	0.20%	41.3%	3.58%	4.26%	−16.7%

(continued)

Table 2. (*continued*)

Direction	R.V	K	ε_0	n	C	H	F	N
		700	0.02	0.22	0.008	0.4	0.45	0.8
	I.E.2	400	0.03	0.2	0.02	0.2	0.7	1.0
	I.R.2	644	0.0125	0.176	0.00571	0.385	0.436	0.794
	R.E.2 (%)	7.97%	37.3%	19.9%	28.6%	3.80%	3.09%	0.72%

(Note: R.V - Reference Values, I.E - Initial Estimates, I.R - Identified Results, R.E - Relative Errors to the References)

Table 3. Identification results of the anisotropic plasticity constitutive parameters when the Swift hardening parameters are pre-constrained.

Direction	R.V	K	ε_0	n	C	H	F	N
		700	0.02	0.22	0.008	0.4	0.45	0.8
0°	I.E.1	\	\	\	0.005	0.5	0.5	0.5
	I.R.1	\	\	\	0.00807	0.395	0.444	0.849
	R.E.1 (%)	\	\	\	−0.84%	1.22%	1.41%	−6.14%
	I.E.2	\	\	\	0.02	0.2	0.7	1.0
	I.R.2	\	\	\	0.00807	0.395	0.444	0.849
	R.E.2 (%)	\	\	\	−0.83%	1.17%	1.36%	−6.10%

4 Conclusions

In this paper, the dynamic anisotropic plasticity of the selected aluminum alloy sheet were characterized by the heterogeneous virtual impact test and the VFM. Using the proposed double-notched impact configuration, heterogeneous kinetic fields can be obtained, from which the anisotropic yield and strain rate strengthening parameters can be simultaneously identified from a single impact test. The current work has numerically verified the feasibility of applying VFM to dynamic anisotropic plasticity of sheet metals, providing sensitivity studies for future experimental applications.

References

1. Peirs, J., Verleysen, P., Van Paepegem, W., et al.: Determining the stress-strain behaviour at large strains from high strain rate tensile and shear experiments. Int. J. Impact Eng. **38**(5), 406–415 (2011)
2. Pierron, F., Grediac, M.: The Virtual Fields Method: Extracting Constitutive Mechanical Parameters from Full-Field Deformation Measurements. Springer, New York (2012)

3. Jiawei, F., Kaiyu, Z., Xiangfan, N., et al.: Inertia-based identification of elastic anisotropic properties for materials undergoing dynamic loadings using the virtual fields method and heterogeneous impact tests. Mater. Des. **203,** 109594 (2021)
4. Pierron, F., Zhu, H., Siviour, C.: Beyond Hopkinson's bar. Philos. Trans. **372**(2023), 20130195 (2014)
5. Fletcher, L., Davis, F., Dreuilhe, S., et al.: High strain rate elasto-plasticity identification using the image-based inertial impact (IBII) test part 1: error quantification. Strain **57**(2), e12375 (2021)
6. Bouda, P., Langrand, B., Notta-Cuvier, D., Markiewicz, E., Pierron, F.: A computational approach to design new tests for viscoplasticity characterization at high strain-rates. Comput. Mech. **64**(6), 1639–1654 (2019). https://doi.org/10.1007/s00466-019-01742-y
7. Fourest, T., Bouda, P., Fletcher, L., et al.: Image-Based inertial impact test for characterisation of strain rate dependency of Ti6Al4V titanium alloy. Exp. Mech. **60**(2), 235–248 (2020)

Experimental Investigation on the Anisotropy of Wrought Magnesium Alloy by Complete Evolution of Equivalent Plastic Work Contours in "σ_{xx}-σ_{yy}" Space

Baodong Shi(✉), Hui Zhao, and Jiaqing Jiang

National Engineering Research Center for Equipment and Technology of Cold Strip Rolling, School of Mechanical Engineering, Yanshan University, Qinhuangdao 066004, People's Republic of China

baodong.shi@ysu.edu.cn

Abstract. Focusing on non-proportional loading condition during forming process of HCP metals, the equivalent plastic work contours of AZ31 Mg alloy are investigated experimentally under combined biaxial test. Strong anisotropy including initial strength differential effect and subsequent distortional hardening are detected. The underlying deformation mechanisms are discussed based on microstructure observation. It is found that the specimen undergoes the main plastic deformation and high stress state along the direction with larger strain ratio. Under the combined load of tension and compression, the c-axis change of the grains in the sample is mainly the deflection along the normal direction, which results in the deformation dominated mainly by dislocation slip. The difference in deformation mechanism leads to different evolution of equivalent plastic work contour in four quadrants.

Keywords: AZ31 Mg alloy · Non-proportional loading · Equivalent plastic work contour · Anisotropy · Texture evolution

1 Introduction

Magnesium alloys are widely used in automotive, aerospace and 3C fields due to their low density and high specific strength, which can achieve lightweight, reduce CO_2 emissions and improve fuel economy [1–3]. However, due to the close-packed hexagonal (HCP) structure of magnesium alloys, the independent slip systems that can be activated at room temperature are very limited, resulting in poor ductility and formability. In addition, the polarity of twins and the strong {0001} initial texture leads to strong anisotropic mechanical properties and tension-compression asymmetry of wrought Mg alloys. Therefore, the plastic deformation characterization of magnesium alloys needs to be considered from different perspectives [4, 5]. Generally, the sheet is subjected to complex multi-axial loading during forming process (such as drawing and extrusion) [6]. Thus, it is of great practical significance to clarify the anisotropic mechanical behavior of wrought magnesium alloy sheets under combined stress state.

K. Mocellin et al. (Eds.): ICTP 2023, LNME, pp. 630–636, 2024.
https://doi.org/10.1007/978-3-031-40920-2_65

At present, more and more attention has been paid on the microstructural evolution and mechanical response of magnesium alloy under multiaxial stress state. Fu et al. studied the deformation mechanism and different work hardening behaviors of AZ31 magnesium alloy sheet under biaxial loading, and the results showed that compression twin and non-basal <a> slip dominate at high strain [7]. Yoon et al. clarified the effect of forming temperature on the formability of Mg-3Al-1Zn sheet under biaxial stress state and found that the temperature improves the ductility of the sheets, while the type of initial texture does not change under biaxial tension [8]. Xia et al. studied the microstructure evolution and deformation compatibility of AZ31 magnesium alloy under biaxial compression stress and indicated that the geometrically necessary dislocation density was lower due to higher activity of tensile twins in the biaxial compression state [9]. Li et al. carried out the non-proportional multi-axial cyclic loading of extruded AZ31 magnesium alloy by performing strain and stress-controlled axial torsion cycle tests under different loading paths, showing that the ratcheting strain is highly sensitive to the loading path [10]. However, the material is always in a complex stress state in practical applications (such as the coexistence of tensile stress and compressive stress). Therefore, it is necessary to study the deformation mechanism of magnesium alloy under non-proportional tension-compression combined loading. Based on the aforementioned research work, the mechanical response, microstructure evolution and texture evolution of AZ31 magnesium alloy sheet under different stress states were studied by setting proportional and non-proportional biaxial loading.

2 Experiment

2.1 Material and Settings

The material used in the experiment is 8mm thick AZ31 magnesium alloy hot rolled plate. The multi-axial loading experiment was carried out on MTS biaxial testing machine. The corresponding cruciform sample is designed according to the equipment, and the style and size are shown in Fig. 1a. The sample is processed by wire cutting, the central area is thinned to 2 mm by milling, and each arm of the sample is tightened by fixture, to increase the deformation of the central area. The biaxial loading experimental data are recorded by the force sensor on the MTS biaxial testing machine, as shown in Fig. 1b. The displacement and strain field obtained in the whole biaxial loading process are recorded by digital image correlation (DIC). The proportional loading tests with displacement control and different displacement ratio along the rolling direction (RD) and transverse direction (TD) are performed. The displacement ratios of RD and TD are 1:1, 1:−1, −1: −1, −1:1, −1:2, −1:−2, −2:1, −2:−1, respectively. Three samples were tested in the same loading condition for its repeatability.

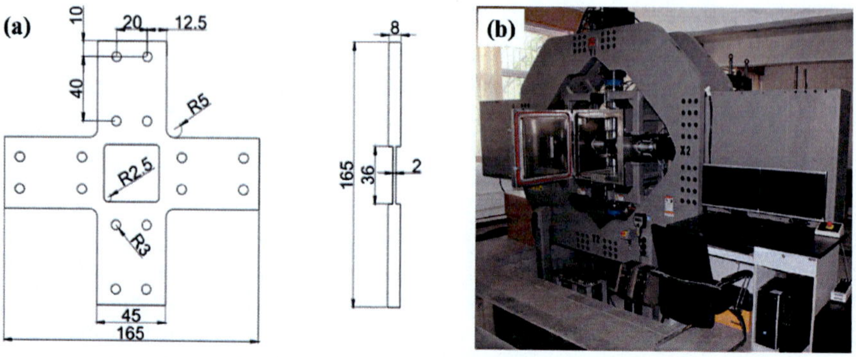

Fig. 1. (a) is the size of specimen, (b) is the MTS biaxial loading test machine.

2.2 Microstructure Characterization

The microstructure of the samples was observed by optical microscope (KEYENCE). Before observation, the samples were polished with a series of SiCs paper and etched with acetic acid solution. The texture of the sample in the RD-TD plane was detected by X-ray diffraction of Cu Kα radiation. The (0001) and (10−10) polar graphs are calculated using the MTEX algorithm developed by Hielscher and Schaeben [11]. The macroscopic textures of the samples before and after deformation were obtained by D-Max2500 XRD diffractometer.

3 Results and Discussion

3.1 Mechanical Responses Under Uniaxial and Biaxial Loadings

The true stress–strain curves for biaxial tensile tests with different displacement ratios are shown in Fig. 2. From Fig. 2(i), it is observed that the AZ31 Mg alloy exhibits strong anisotropy and SD effects in uniaxial loading along different directions. And under biaxial loading, the anisotropic behavior of AZ31 Mg alloy is also pronounced, as shown in Figs. 2(a) and 2(d). Compared with uniaxial loading, the maximum strain in biaxial loading test is relatively small. For non-proportional biaxial loading, the specimen undergoes mainly plastic deformation and high stress state in the direction of large displacement ratio, as shown in Fig. 2(e)–(h). It is found that in the third quadrant under biaxial loading, the strain of yielding in the RD direction gradually decreases in the displacement ratio of −1: −2, −1: −1, −2: −1. In the third quadrant of stress space, with the increase of the displacement ratio, the yield strength in the RD direction gradually increases ($\sigma_{-1:-2} < \sigma_{-1:-1} < \sigma_{-2:-1}$), while the yield strength in the TD direction changes is opposite.

During the biaxial loading process, some specimens exhibit a negative strain, as shown in Fig. 2(h). It is found that the negative strain is generated during biaxial compression. The displacement rate in the direction of negative strain is lower, which causes the strain in that direction influenced by the other direction and results in strain turbulence.

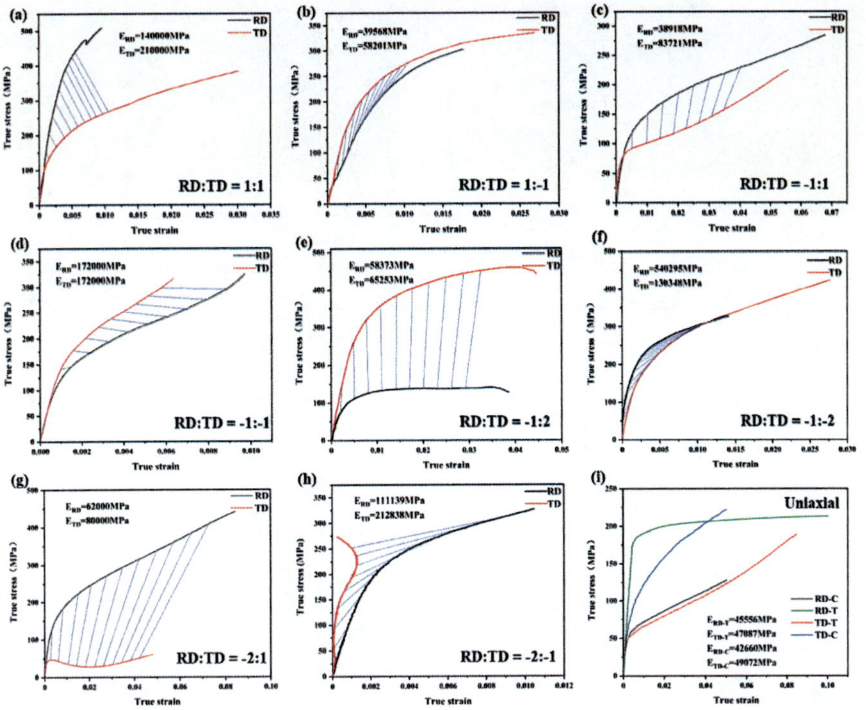

Fig. 2. (a)–(h) are the stress-strain responses for biaxial loading, and (i) are the results of uniaxial loading under the same experimental conditions.

3.2 Microstructure and Texture Analysis

The microstructure and texture evolutions of the samples before and after deformation are shown in Fig. 3. It is found that the strength of the {0001} texture is significantly enhanced. At the same time, no obvious twins are detected. Combined with the texture results, it is found that the deformation mechanism is mainly <a> basal slip and <a> prismatic slip under biaxial loading (including biaxial tension, compression and composite loading). Slip results in the c-axis orientation of the grains closer to the ND direction. Therefore, under combined tension-compression loading, the change of the c-axis of the grains is mainly reflected in the deflection along the ND direction, rather than the change of the c-axis length, which makes it difficult to generate twins.

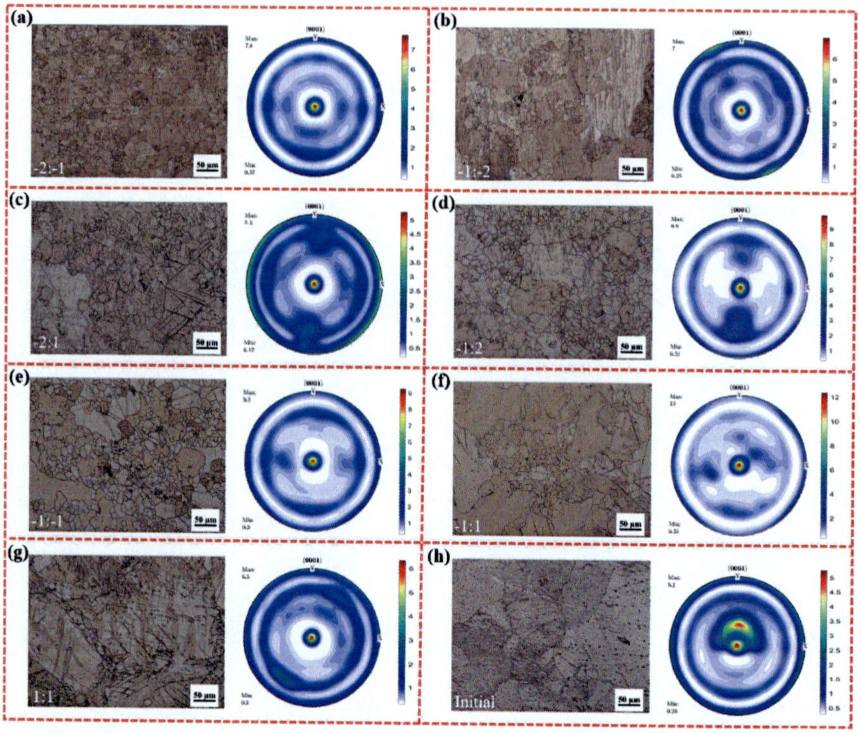

Fig. 3. The microstructure and texture of the specimens under different loading modes. Loading modes: (a) –2:–1; (b) –1:–2; (c) –2:1; (d) –1:2; (e) –1:–1; (f) –1:1; (g) 1:1; (h) initial microstructure

3.3 Complete Evolution of Equivalent Plastic Work Contours in "σ_{xx}-σ_{yy}" Space

The evolution of equivalent plastic work contours during uniaxial loading and biaxial loading with different paths are shown in Fig. 4. The von-Mises equivalent plastic strain is calculated from $\varepsilon_p = 0.053$ to $\varepsilon_p = 2.45$, as shown in Fig. 4(a). With the increase of the plastic strain, a concave evolution of the equivalent plastic work contour is detected, as shown in Fig. 4(b). Since the deformation mechanisms of the samples are similar in the second and fourth quadrants, which are dominated by dislocation slip, the concavity of the equivalent plastic work contour is not obvious. However, there is a big difference between the equivalent plastic work contour in the first and third quadrants. During the biaxial tensile loading process, the deformation mechanism is dominated by slip, and the tension twin is difficult to start. In contrast, under the biaxial compression loading, the RD-TD plane is compressed in both directions, thus the tensile twin is easily activated. Therefore, the difference of deformation mechanism leads to the strong anisotropic evolution of the equivalent plastic work contours.

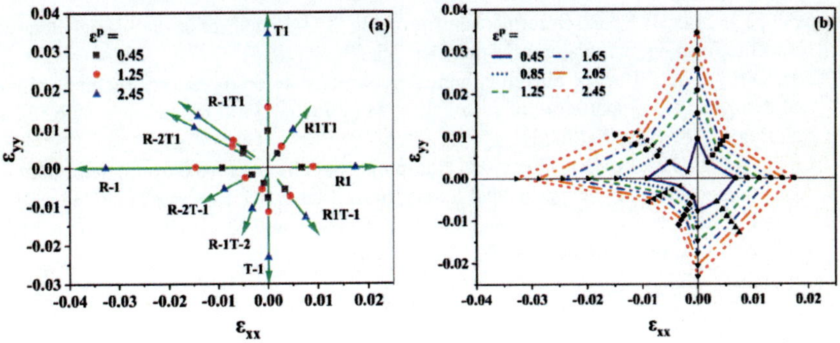

Fig. 4. (a) Loading paths in "σ_{xx}-σ_{yy}" space, (b) the complete evolution of the equivalent plastic work contours.

4 Conclusions

In this paper, the microstructure and macro texture of AZ31 magnesium alloy sheet are analyzed by multi-axis loading test, and the evolution process of the equivalent plastic work surface of "σ_{xx}-σ_{yy}" plane is further discussed. The conclusions are listed as follows:

1. The biaxial combined loading does not change the texture type of AZ31 magnesium alloy rolled sheet.
2. In biaxial loading, the direction with larger strain rate is the main direction of plastic deformation, and the other direction plays a role in coordinating deformation.
3. The difference in deformation mechanism leads to different trends of equivalent plastic work contour in four quadrants.

References

1. Chaudry, U.M., Tekumalla, S., Gupta, M., Jun, T.S., Hamad, K.: Designing highly ductile magnesium alloys: current status and future challenges. Crit. Rev. Solid State Mater. Sci. **47**, 194–281 (2022)
2. Wei, Q.M., et al.: Insights from the MEDE program: an overview of microstructure–property linkages in the dynamic behaviors of magnesium alloys. Mech. Mater. **163**, 104084 (2021)
3. Song, J.F., Chen, J., Xiong, X.M., Peng, X.D., Chen, D.L., Pan, F.S.: Research advances of magnesium and magnesium alloys worldwide in 2021. J. Magnes. Alloy. **10**, 863–898 (2022)
4. Trang, T.T.T., et al.: Designing a magnesium alloy with high strength and high formability. Nat. Commun. **9**, 2522 (2018)
5. Liu, B.Y., et al.: Large plasticity in magnesium mediated by pyramidal dislocations. Science **365**, 73–75 (2019)
6. Shi, B.D., Yang, C., Peng, Y., Zhang, F.H., Pan, F.S.: Anisotropy of wrought magnesium alloys: a focused overview. J. Magnes. Alloy. **10**(6), 1476–1510 (2022)
7. Fu, Y.J., Cheng, Y., Cui, Y., Xin, Y.C., Shi, S.W., Chen, G.: Deformation mechanisms and differential work hardening behavior of AZ31 magnesium alloy during biaxial deformation. J. Magnes. Alloy. **10**(2), 478–491 (2022)

8. Yoon, J.H., Lee, Y.S.: Fracture mechanism of Mg–3Al–1Zn sheet at the biaxial state with respect to forming temperatures. Mater. Des. **55**, 43–49 (2014)
9. Xia, D.B., et al.: Effect of biaxial compressive stress state on the microstructure evolution and deformation compatibility of rolled sheet Mg alloy AZ31 at room temperature. Mater. Sci. Eng. A **789**, 139599 (2020)
10. Li, H., Kang, G.Z., Liu, Y.J., Jiang, H.: Non-proportionally multiaxial cyclic deformation of AZ31 magnesium alloy: experimental observations. Mater. Sci. Eng. A **671**, 70–81 (2016)
11. Hielscher, R., Schaeben, H.: A novel pole figure inversion method: specification of the MTEX algorithm. J. Appl. Crystallogr. **41**(6), 1024 (2008)

On the Numerical Assessment of Failure in Stretch-Flanging by SPIF Using Equivalent Strain Versus Triaxiality Diagrams

José Andrés López-Fernández(✉), Gabriel Centeno, and Carpóforo Vallellano

Department of Mechanical and Manufacturing Engineering, University of Seville, Seville, Spain
{jlopez85,gaceba,carpofor}@us.es

Abstract. In the last few years, there has been a growing interest in the sheet metal forming community for analyzing the enhancement of formability attained in non-proportional forming processes such as incremental sheet forming (ISF), and especially in its dieless variant single point incremental forming (SPIF). To this regard, this analysis using classical forming limit diagrams (FLD) do not provide enough information for understanding the conditions upon which this postponed failure is attained. As an alternative, some recent studies made use of equivalent strain versus stress triaxiality diagrams, which proved to be more suitable for evaluating the non-proportional strain paths that lead to failure in SPIF.

On the other hand, flanged parts are commonly used in the aircraft industry to provide stiffness to the manufactured components as well as a target area for the assembly to other metallic parts. To this regard, the authors have recently carried out research works aiming investigating the formability and failure modes of AA2024-T3 sheet 1.2 mm thickness in stretch and shrink flanging by SPIF. These studies were carried out combining experimentation and numerical simulation using Finite Elements (FE), with the aim of evaluating and assessing the different modes of failure that occurred.

In this context, the present contribution consists on a numerical evaluation of failure by using a FE modeling of the stretch flanging process in LS-DYNA. To this purpose, Barlat's anisotropy is considered for the assessment of the forming limit at fracture within the material equivalent strain versus stress triaxiality diagram. The resulting numerical model, calibrated using principal strains experimental results, allows predicting the sheet material fracture and the mode of failure attained, which can be either failure by fracture at the flange edge or at the flange corner depending on the set of process parameters.

Keywords: Forming limit diagram (FLD) · Single point incremental forming (SPIF) · Stretch flanging

1 Introduction

Single point incremental forming (SPIF) is a promising manufacturing technique first introduced in the 1990s as a low-cost alternative for creating complex geometries and shapes in sheet metal components. This technology offers several advantages, including

K. Mocellin et al. (Eds.): ICTP 2023, LNME, pp. 637–645, 2024.
https://doi.org/10.1007/978-3-031-40920-2_66

adaptability, fast prototyping, and cost-efficient production by eliminating the need for costly specialized tools and equipment.

Despite its relatively recent development, SPIF has already found applications in a range of industries and technologies. It has shown promise in many areas, including the manufacturing of complex and high-precision components, particularly in aerospace and automotive industries. Additionally, the ability of SPIF to form materials without dedicated dies has led to its use in creating customized parts and prototypes quickly and inexpensively.

One of the main advantages of SPIF is its ability to increase formability, allowing the forming of materials above the conventional forming limit. While the traditional analysis of formability in flanging by SPIF has been done in terms of principal strains within the FLD [1], recent studies in both hole flanging [2, 3] and stretch flanging [4] by SPIF suggest that a stress-based approach provides a more comprehensive explanation of formability. These studies have emphasized the important role of stress triaxiality in voids growth and have shown that the lower stress triaxiality produced by SPIF, compared to press forming, helps to reduce voids growth [6] and avoid fracture.

The aim of this research paper is to study the different kinds of fracture present in stretch flanges produced through SPIF using a stress-based analyses. The study will use numerical models based on finite element methods to obtain stress data, which will be analyzed in terms of stress triaxiality and equivalent strain.

2 Experimental Procedures

2.1 Tensile Properties and Formability

The material used in this study is a 1,2 mm thick AA2024-T3 aluminum sheet. The material characterization was performed through tensile testing [7] and Lankford testing [8]. The results are summarized in Table 1, which includes the yield stress (σ_Y), ultimate tensile strength (σ_{UTS}), modulus of elasticity (E) and Lankford coefficients of anisotropy (r_θ) for the longitudinal, transverse, and diagonal directions. In addition, Fig. 1a provides the stress-strain curve for each of the three tested directions. The average tensile behavior along the rolling direction is represented by Eq. 1 corresponding to the Swift stress-strain expression. Further details regarding the calculation of these properties can be found in reference [6].

$$\overline{\sigma} = 742.36(0.025 + \varepsilon_p)^{0.235} \tag{1}$$

The material's formability was evaluated using the Forming Limit Diagram (FLD) shown in Fig. 1.b. This FLD was obtained through Nakazima tests [9] performed at constant speed of 1 mm/s on an Ericksen testing machine, for five specimen geometries. The software Aramis®, based on Digital Image Correlation (DIC), was used to obtain the strain evolution of the different specimens, which did not present signs of localized necking. Instead, the specimens showed uniform deformation of the material until ductile fracture occurred. As a result, the limiting strains in the FLD are represented using the Fracture Forming Limit (FFL) curve, as it is usual for a material without necking. To determine the fracture points, the minor and major strains from DIC were recorded until instants prior to fracture, obtaining proportional strain paths. Then the fracture points were determined measuring the sheet thickness at the crack to obtain the thickness strain, and applying volume constancy. A more exhaustive description of this process is detailed in reference [4].

Table 1. Tensile and anisotropy properties of AA2024-T3 sheet for longitudinal, transversal and rolling direction.

Direction	σ_Y (MPa)	σ_{UTS} (MPa)	E (GPa)	r_θ
0°	336	445	69.4	0.76
45°	306	427	67.1	0.95
90°	319	443	68.2	0.54

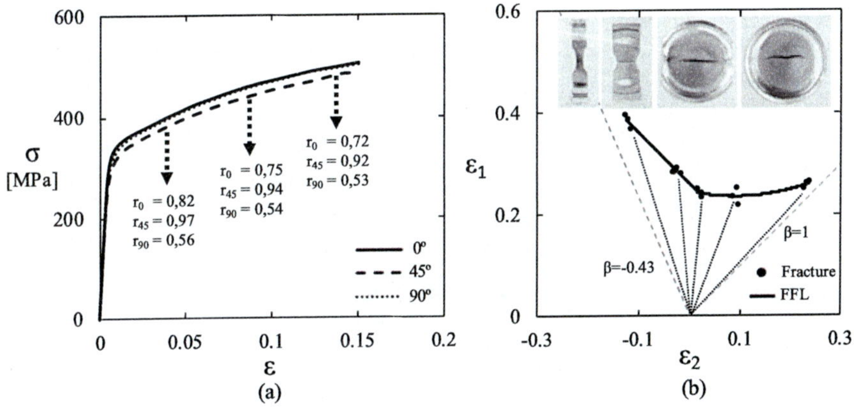

Fig. 1. (a) Stress-strain curve of AA2024-T3 sheet at different orientation with respect to the rolling direction. (b) FLD of the material represented within the principal strain space.

2.2 Stretch Flanging by SPIF

The flanging experiments were performed on an EMCO VMC200 three-axes milling machine using a flanging setup as shown in Fig. 2. The setup consisted of a lower die with 45 mm radius and a sheet holder to avoid bulges. To reduce any damage to the flange, a layer of PTFE was placed between the sheet and the dies.

The flanging process involved a hemispheric tool that was cnc-controlled and followed circular arcs in the horizontal plane and downward displacements in the z-axis. It was important to maintained a gap between the sheet and the die in order to prevent the ironing of the flange.

The experiments include the geometrical parameters of the flange initial width w_0 and initial length l_0, while the process parameters die radius, tool diameter, spindle speed, and step down were kept constant. The mentioned parameters are represented in Fig. 2a. To perform a principal strains measurement, a circle grid was previously electro-etched on the sheet surfaces as shown in Fig. 3. Once the flanging process was finished, the software ARGUS® based on Circle Grid Analysis (CGA) was utilized to determine the principal strains' distribution and validate the finite element model.

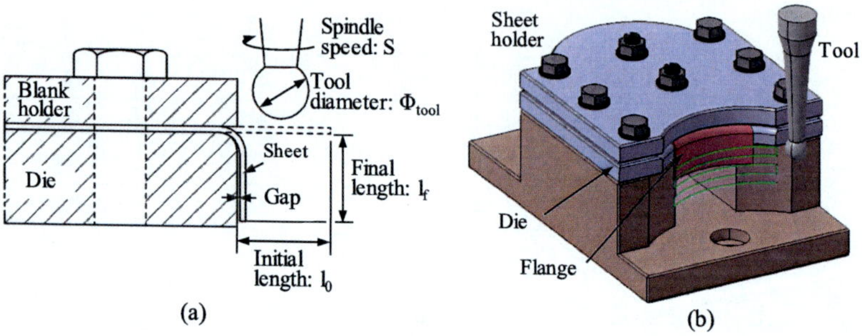

Fig. 2. (a) Elements and process parameters of the stretch flanging setup. (b) CAD model of the stretch flanging setup for SPIF.

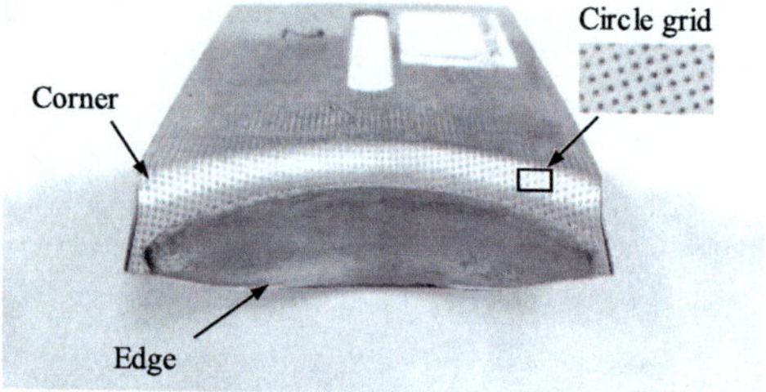

Fig. 3. Flange performed by SPIF.

3 Finite Element Modelling

This study involves a numerical model used to analyze the SPIF flanging process. The model was created using Ansys LS-DYNA®, a widely used finite element software for dynamic analysis. The mesh of the model consisted of shell elements with an element size of 1 mm (type SHELL163 in LS-DYNA®). The forming tool, die and sheet holder were represented by rigid bodies and Coulomb friction $\mu = 0.01$ was set for contact surfaces. The tool movement was identical to the actual SPIF process described in Sect. 2.2, i.e. arcs in the horizontal plane defined by 300 discrete points. The velocity of the tool was increased with respect to the real process by a factor of 1000 and mass scaling by a factor of 5 was used. The sheet was modeled using the Barlat-89 elastic plastic model for anisotropic behaviour, the stress-strain curve defined in Eq. 1 and the anisotropy coefficients presented in Table 1. The computer used had an 8 CPUs processor of 3 GHz and 32 Gb of RAM. Figure 4 shows the elements of the FE model and the major strain contour for a deformed flange. The details of the validation of the model was detailed in a previous research about stretch flanging [4].

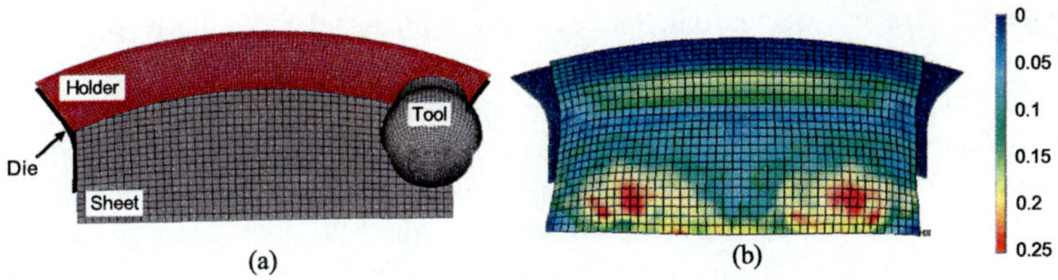

Fig. 4. Finite element model for a flange with $w_0 = 45$ mm and $l_0 = 20$ mm. (a) Undeformed mesh. (b) Major strain contour on the back surface of a deformed flange.

4 Results

4.1 Experimental Results

The experimental plan is presented in Table 2, which summarizes the results for flanges of different initial widths w_0 and lengths l_0 performed by SPIF. The process parameters were kept constant throughout the set of experiments and include the die radius R_{die} of 45 mm, spindle speed S of 1000 rpm, tool diameter ϕ_{tool} of 12 mm, and step down Δ_z of 0.4 mm.

Table 2. Set of stretch flanges performed by SPIF.

	w_0 [mm]				
l_o [mm]	*36*	*45*	*54*	*63*	*72*
20			S	S	**S**[3]
25	S	S	**F(C)**[1]	F(C)	**F(E)2**
30	S	S			

The process window shows two modes of failure in the flanges, fracture at the edge F(E) and fracture at the corner F(C), whereas some of the flanges were successful S and did not have fracture. On the one hand, the F(E) mode of failure was caused by stretching at the edge of the flange, which occurred on the back surface. On the other hand, F(C) was caused by stretching due to friction with the tool at the corner of the flange, which occurred on the front surface.

4.2 Critical Points Analysis

With the aim of studying the failure modes, the flanges are analyzed using FE. Figure 5 depicts the numerical major strain contours of a successful flange of $w_0 = 45$ mm and $l_0 = 25$ mm. Figure 5a shows the major strain concentrated near the edge of the flange on the back surface, namely critical points. Due to the high levels of major strain, this zone is where fracture occurred in flanges with fracture at edge. Similarly, Fig. 5b shows the major strain on the front surface concentrated in the corner of the flange. This is the critical point for flanges with fracture a corner. The experimental and numerical results show that the fracture occurred either at the edge or at the corner depending on the levels of major strain. Furthermore, experimental strain results from a recent study [4] show that SPIF allows to obtain principal strains above the FFL, and points out that this is not a good indicator of fracture in SPIF. Instead, this study performs a stress analysis to analyze both modes of failure. This will provide a more comprehensive understanding of fracture in flanging by SPIF.

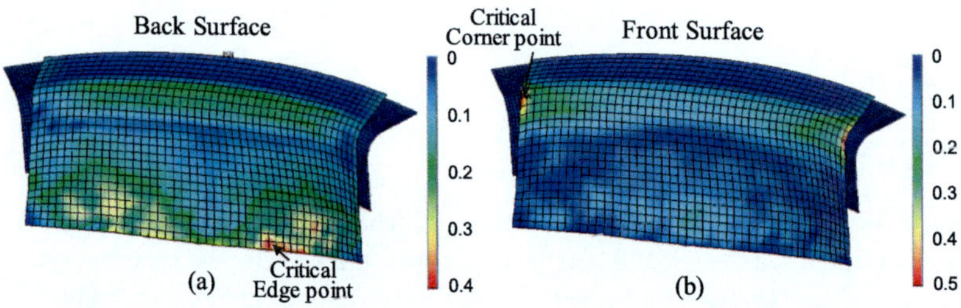

Fig. 5. Major strain contours for a flange of $w_0 = 45$ mm and $l_0 = 20$ mm. (a) Back surface. (b) Front surface.

4.3 Analysis Within the $\bar{\eta}$-ε_{eqv} Space

A stress approach is adopted to study the formability of flanges. The elements of the analysis are shown in Fig. 6a and 6b, which include the numerical evolution of some critical points corresponding to flanges marked in bold in Table 1.

The FFL was transformed from the FLD to the field of stress triaxiality. The evolution of the fracture points was obtained by numerical simulation of the Nakazima tests described in Sect. 2, modeled using Barlat's 89 plasticity criterion. The strain paths of the Nakazima simulations were found to be proportional, i.e. constant stress triaxiality ratio, as it is usual for this standard test. The end of each numerical test was set when the major strain reaches the fracture major strain of its corresponding experiment.

Figure 6a and 6b also presents the analysis of three flanges represented in Table 2: a flange with fracture at corner (Case 1), a flange with fracture at edge (Case 2) and a flange without fracture (Case 3). In Fig. 6a, the evolutions of the failed flanges at the corner and at the edge, respectively, is shown, using the label 'x' to mark the fracture onset. Additionally, Fig. 6b presents the evolution of points without fracture: the corner and the edge point of the successful flange (Case 3) respectively, and the corner point of Case

2. It can be seen that failed and successful flanges have lower average stress triaxiality and higher equivalent strain levels than the fracture points obtained in Nakazima tests. For the strain paths in SPIF are not proportional, the fracture locus within the $\overline{\eta}$-ε_{eqv} space is different to that obtained by Nakazima tests. In the literature, this phenomenon is attributed to the local and incremental character of SPIF.

Some classical ductile fracture criteria (DFC) are based either on the effect of maximum stress [10] or stress triaxiality [6, 11, 12] on voids evolution, but neglect void nucleation and coalescence. These criteria predict failure exclusively in mode I, and have problems to predict fracture near equibiaxial strain ($\eta = 0.66$). More recent DFCs such as Mohr-Coulomb [13] and Lou [14] are used to predict fracture in mode II, which is highly influenced by voids coalescence (related to shear stresses) and is usually observed within the range $-1/3 < \eta < 1/3$. However, these two DFC do not consider the existence of fracture in mode I for values of stress triaxiality $\eta < 1/3$.

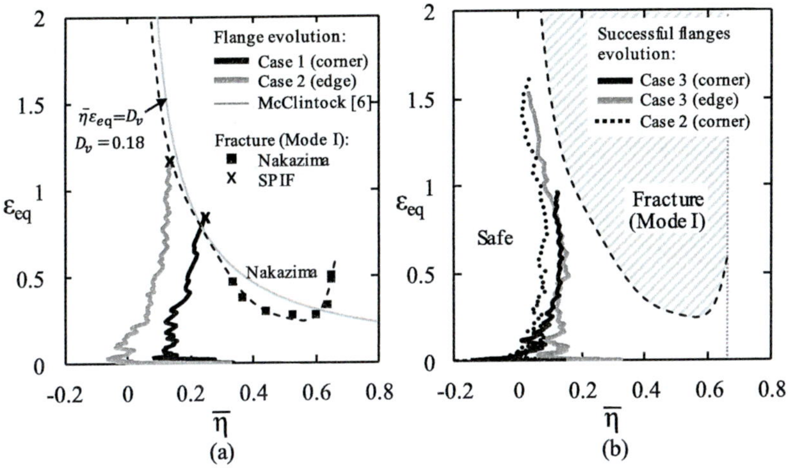

Fig. 6. Analysis of stretch flanges within the average stress triaxiality vs. equivalent strain diagram. Evolution of (a) fracture and (b) successful (non-fracture) points.

According to the results presented in Fig. 6 and previously studied in reference [5], fracture in flanging by SPIF occurs in mode I at levels of average stress triaxiality $\overline{\eta} < 0.33$. Note that in the case of SPIF, the averaged variable is used, as the instantaneous stress triaxiality oscillates due to the non-proportional loading. According to the experimental-numerical study carried out, a general DFC should predict fracture mode I for both proportional loading (Nakazima tests) and SPIF. As shown in Fig. 6a, the DFC of McClintock [6], which is based only in voids growth, and adjusted using SPIF and Nakazima fracture points, does not successfully predict the fracture points for Nakazima tests. A suitable fracture curve has to be continuous, smooth, and must be represented in the $\overline{\eta}$-ε_{eqv} space. In this study, and in absence of an analytic criterion which is still under refinement, an experimental-numerical curve is proposed.

On that basis, Fig. 6a and 6b show the fracture limit resulting from the interpolation of the results obtained in Nakazima tests and in SPIF respectively. In this regards, Fig. 6a shows the fracture line obtained for SPIF analyzed for two different modes of failure, F(E) and F(C), and considering the fracture points of Nakazima tests. As it is shown in Fig. 6b, the evolution of the edge and the corner of the successful flange (Case 3) and the corner of the flange failed by F(E) are below the fracture line proposed. It must be pointed out that this is a preliminary study and more incremental tests need to be carried out to propose a more accurate fracture experimental curve.

5 Conclusions

The aim of this research is to analyze fracture in stretch flanging by SPIF of AA2024-T3. To this purpose, the process is studied numerically by finite elements and analyzed in terms of stresses. From this study, the following conclusions are drawn:

- The average stress triaxiality levels in SPIF are notably lower than the FFL, which is likely related to lower levels voids growth.
- Fracture in SPIF is not well predicted by classical DFCs. Some of them predict fracture mode II for stress triaxiality levels $\overline{\eta} < 0.33$ whereas others do not adjust to the fracture locus obtained for SPIF and Nakazima experimental tests simultaneously.
- A mode I fracture limit curve for stretch flanging by SPIF in the $\overline{\eta} - \varepsilon_{eqv}$ space has been obtained using experimental and numerical data from Nakazima and SPIF tests.

Acknowledgements. The authors acknowledge the Spanish Government for funding this research line through the Major Grants PID2021-125934OB-I00 and TED2021-131153B-C22.

References

1. McAnulty, T., Jeswiet, J., Doolan, M.: Formability in SPIF: a comparative analysis of the state of the art. CIRP J. Man Sci. Tech. **16**, 43–54 (2017)
2. Mirnia, M.J., Shamsari, M.: Numerical prediction of failure in SPIF using a phenomenological ductile fracture criterion. J. Mat. Proc. Tech. **244**(8), 17–43 (2017)
3. Martínez-Donaire, A.J., Borrego, M., Morales-Palma, D., Centeno, G., Vallellano, C.: Analysis of the influence of stress triaxiality on formability of hole-flanging by single-stage SPIF. Int. J. Mech. Sci. **151**, 76–84 (2019)
4. Lopez-Fernandez, J.A., Centeno, G., Martinez-Donaire, A.J., Morales-Palma, D., Vallellano, C.: Stretch-flanging of AA2024-T3 sheet by single-stage SPIF. Thin-Walled Struct. **160**, 107338 (2021)
5. Lopez-Fernandez, J.A., Centeno, G., Vallellano, C.: Wrinkling in shrink flanging by single point incremental forming. Int. J. Mech. Sci. **240**, 107930 (2023)
6. McClintock, F.A.: A criterion for ductile fracture by the growth of holes. J. Appl. Mech. Trans. ASME **35**, 363–371 (1968)
7. ASTM E8/E8M–09: S. T. Methods for Tension Testing of Metallic Materials (2009)
8. ASTM E 517–00. Standard Method for Plastic Strain Ratio of Sheet Metal (2000)
9. ISO 12004-2:2008. Metallic materials-sheet and strip determination of forming limit curves, part 2: determination of forming limit curves in the laboratory (2008)
10. Cockcroft, M.G., Latham, D.J.: Ductility and the workability of metals. J. Inst. Met. **96**, 33–39 (1968)

11. Brozzo, P., DeLuca, B., Rendina, R.: A new method for the prediction of formability in metal sheets. In: Proceedings of the 7th Conference of IDDRG on Sheet Metal Forming and Formability (1972)
12. Oyane, M., Sato, T., Okimoto, K., Shima, S.: Criteria for ductile fracture and their applications. J. Mech. Work. Technol. **4**, 65–81 (1980)
13. Bao, Y.B., Wierzbicki, T.: On fracture locus in the equivalent strain and stress triaxiality space. Int. J. Mech. Sci. **46**, 81–98 (2004)
14. Lou, Y., Huh, H., Lim, S., Pack, K.: New ductile fracture criterion for prediction of fracture forming limit diagrams of sheet metals. Int. J. Sol. Struct. **49**, 3605–3615 (2012)

Texture and Microstructure After Roll-Bonding of an Fe-Al Multilaminate

Guillaume Hanon[1(✉)], Loïc Malet[2], and Laurent Delannay[1]

[1] iMMC, UCLouvain, 1348 Louvain-la-Neuve, Belgium
{guillaume.hanon,laurent.delannay}@uclouvain.be
[2] 4MAT, Université Libre de Bruxelles, 1050 Bruxelles, Belgium
loic.malet@ulb.be

Abstract. A 5-layer laminate composite was produced by roll-bonding at room temperature of a stacking of ARMCO steel and commercial purity aluminum alloy (AA1050). The large rolling draught led to heterogeneous deformation across the thickness of the layered aggregate, which was investigated using finite element modeling. The evolution of microstructure and crystallographic texture was probed using EBSD. Intense localized shear inside the AA1050 layers gave rise to continuous recrystallization, whereas a rolling texture developed in the adjacent iron layer. Both were reproduced using crystal plasticity modeling.

Keywords: Crystal plasticity · Laminate composite · Finite elements

1 Introduction

Roll bonding of metal sheets with contrasted mechanical properties may result in high-performance layered composites e.g. [1,2]. The plastic anisotropy of such composite metal sheet depends on the crystallographic texture development of its constituents, which in turn is influenced by internal stresses building up due to the prescribed co-deformation of adjacent layers [1]. The roll-bonding process typically requires a large reduction (∼50%) of the sheet thicknesses in a single rolling pass, thus involving significant microstructural changes.

If the diameter of the rolls is either much smaller or much larger than the sheet thickness, the rolled sheet undergoes heterogeneous deformation even if it is made of a single material. Both the roll-gap geometry and the friction of the sheet with the rolls may give rise to significant shear parallel to the rolling plane, resulting in a gradient of texture across the sheet thickness [3–8]. The deformation process fulfils orthotropic symmetry only along the sheet mid-thickness

The present work investigates across-the-thickness strain heterogeneity within roll-bonded multi-laminates made of commercially pure iron and aluminium. The preparation of samples and their characterisation via EBSD are presented in Sect. 2. Experimental observations are then compared to numerical simulations. This involves both the finite element modelling of a rolling pass and

K. Mocellin et al. (Eds.): ICTP 2023, LNME, pp. 646–653, 2024.
https://doi.org/10.1007/978-3-031-40920-2_67

crystal plasticity predictions of the texture development (Sect. 3). Results are discussed before reaching the conclusion.

2 Preparation and Characterisation of Roll-Bonded Samples

Two common metal grades were selected for the present work. With a yield stress of 75 MPa and a UTS of 120 MPa the 1050 aluminium alloy sheets were about three times softer than the sheets made of ARMCO steel ($\sigma_{yield} = 250\,\text{MPa}$ and $\sigma_{UTS} = 300\,\text{MPa}$). Before being stacked and roll-bonded, both metals had equiaxed grains and a very weak crystallographic textures.

The roll-bonding proceeded in two steps, using a laboratory cold-rolling mill. The diameter of the rolls was 112 mm. The first rolling pass was applied to a stacking of two 0.4 mm-thick AA1050 sheets separated by a single 0.8 mm-thick iron sheet. The total thickness was reduced to 0.62 mm, which corresponds to 61% thickness reduction. Then, two 0.2 mm-thick iron layers were added on either side of the 3-layer composite sheet. The second rolling pass reduced the total thickness to 0.62 mm, which corresponds to an extra 39% thickness reduction.

All to-be-bonded surfaces were preliminary degreased in Acetone and brushed with a 0.3 mm diameter wire-brush in order to remove a the oxide layer. Between the two rolling passes, the 3-layer composite was held at 500 °C during 30 min, which allowed annealing the AA1050 sheet.

Due to the strength contrast, it was a real challenge to prevent relative motion of the sheets at the entrance of the roll gap. Successful bonding was achieved by preliminary folding of the AA1050 sheet around the iron (pass 1) or the iron sheet around the three-layer stacking (pass 2). Figure 1 illustrates the successive folding steps before pass 1.

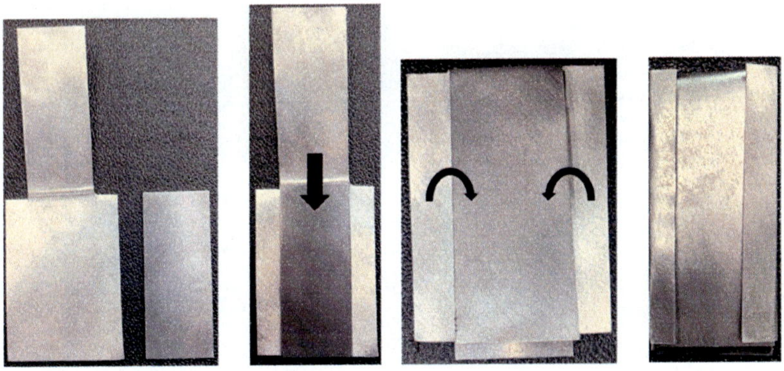

Fig. 1. Before the first rolling pass, the aluminium sheet is folded around the iron sheet to prevent relative motion, which would impede bonding.

Optical microscopy observation of the sheet longitudinal cross-section (Fig. 2) confirmed balanced co-deformation of the two metals and uniform thickness reduction in all layers. A thin oxide film at the interface (not fully eliminated after the inter-pass heating) did not impede successful bonding.

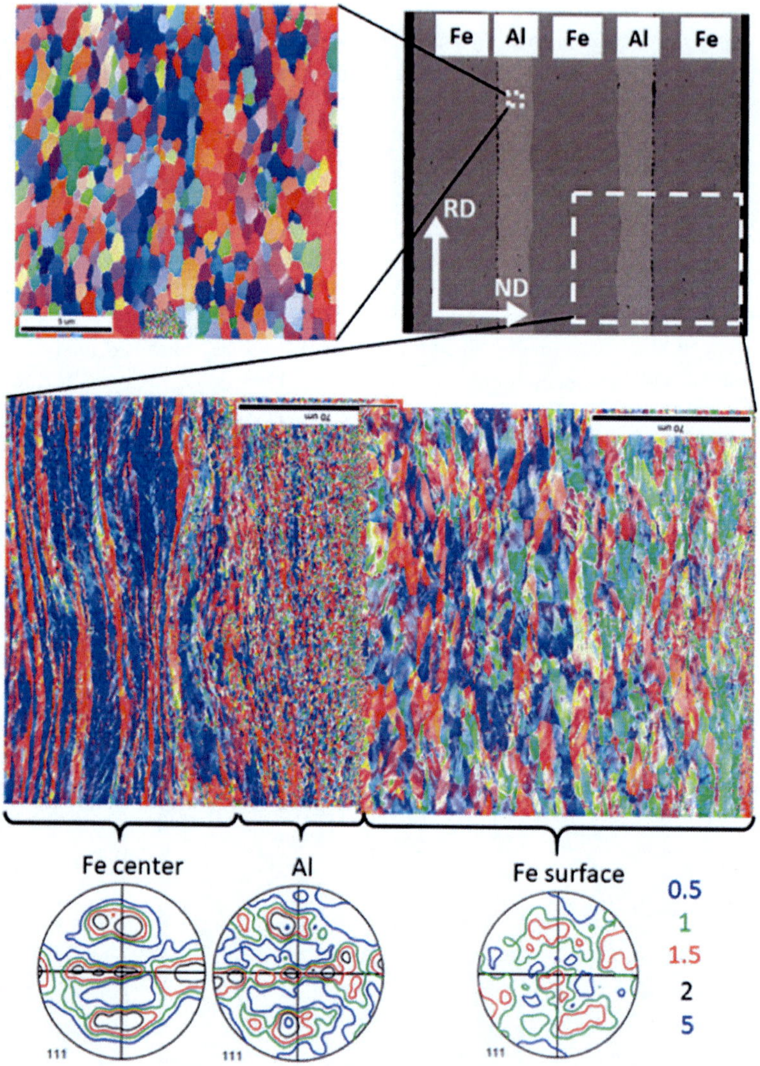

Fig. 2. Microstructural characterization based on optical microscopy and EBSD. The coloring of orientation maps is according to the ND-IPF. The ⟨111⟩ pole figures show the textures developed in the Fe and AA1050 layers.

The crystallographic texture was probed within each layer using electron back scatter diffraction (EBSD). This required prior polishing using SiC up to grid 1200 followed by diamond suspension up to 1 μ and OPS. The orientation maps are colored according to the inverse pole figure (IPF) of the normal to the rolling plane (ND). The iron layer located at the sheet mid-thickness (left on the figure) shows elongated grains and a texture which is typical of cold rolling. The pole figure of the external layer made of iron does not show a clear texture. This layer was added after the first rolling pass and it was hence less deformed than the central layers. The aluminum layer shows a refined equiaxed microstructure. The high indexation rate in the close-up EBSD scan is indicative of continuous recrystallization [9].

3 Numerical Modeling

Finite element modeling allowed simulating the heterogeneous deformation of the sheets inside the roll gap. The Abaqus/Standard (*d'Assault Systèmes*, France) commercial code was used. As illustrated in Fig. 3, the simulations were performed in 2D, assuming plane strain, and the rolling mill was considered as infinitely stiff. Based on [10], the friction coefficient was set equal to 0.3.

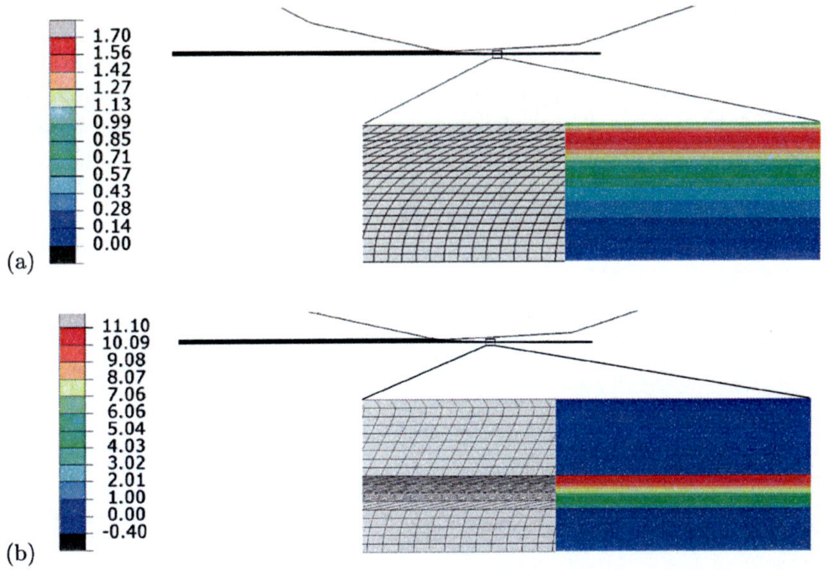

Fig. 3. Heterogeneity of shear (RD-ND component of the deformation gradient tensor) across the half-thickness of the rolled shear. The broken line on the top is the (cylindrical!) rolling mill having a diameter ~100 times larger than the sheet thickness. The half-thickness shows (a) only Fe (b) one AL1050 layer and two Fe layers.

The sheet thickness was reduced from 1.02 mm to 0.62 mm, corresponding to the second rolling pass in the experiment. Conforming to the laboratory experiment, the rolling draught was large, i.e. the contact length (between roll and sheet) was more than ten times larger than the (aggregate) sheet thickness. The half sheet-thickness was represented by 31896 quadrilateral first order elements.

Two variants of the simulation were carried out. In the first case, we assigned to all finite elements the Armco Iron properties (thus corresponding to traditional rolling). In the second case, a layer of finite elements were assigned properties corresponding to AA1050. The laminated structure then corresponded to the 5-layer samples which had been roll-bonded in the laboratory. The bonding itself was not simulated: the interface between the two materials was tied from the onset of the simulation. In the FE simulation, the material response was considered isotropic (J_2 elasto-plasticity) with a power-law hardening law. As shown in Fig. 3, when the aluminium layer is present, the shear deformation due to the contact with the rolls tends to be concentrated within the aluminium layer instead of spreading across the whole sheet thickness.

Crystal plasticity (CP) theory then allowed predicting the development of texture corresponding to the strain heterogeneity predicted by FE across the sheet thickness. Dislocation slip occurred along $\{111\}\langle 110\rangle$ slip systems in the AA1050 layer (fcc), whereas the $\{110\}\langle 111\rangle$ and $\{112\}\langle 111\rangle$ systems were active in Fe (bcc). The power-law strain hardening maintained an identical critical resolved shear stress (CRSS) on all slip systems. The initial random texture was represented by 1000 grains which were considered to interact in pairs according to the ALAMEL model [11–13]. In Fig. 4, the simulated textures are presented in the form of $\langle 111\rangle$ poles figures similar to those presented in Fig. 2. The surface layer made of iron was disregarded as the experimental texture was very weak. In order to check the effect of intense localized shear in AA1050, the simulation of the AA1050 texture was performed twice: either with or without RD-ND shear superimposed to the 75% thickness reduction under plane strain.

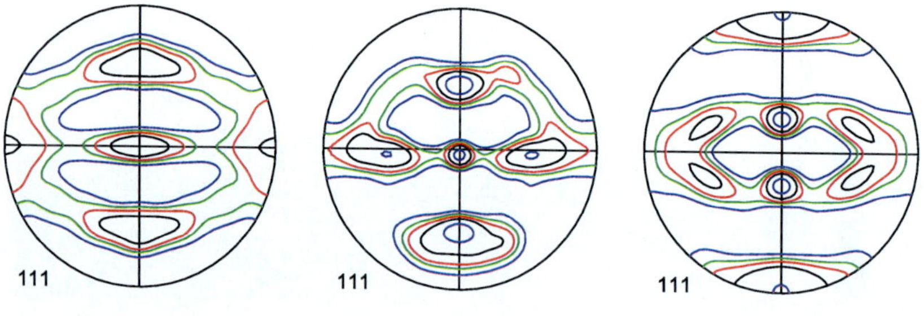

(a) Fe (centre) (b) AA1050 (c) AA1050 (no shear)

Fig. 4. $\langle 111\rangle$ poles figures predicted in the central Fe layer (a) and in the AL1050 layer when either using the strain predicted by FE (b) or suppressing the RD-ND shear (c). The contour levels are 0.5/1/1.5/2/5.

The orientation distribution function (ODF) allows a more accurate comparison of the predicted textures to the experimental ones. After rolling of cubic metals, the most relevant information lies in the $\varphi_2 = 45°$ section of the ODF. Figure 5 confirms that crystal plasticity theory provides valid predictions of the texture development in the central Fe layer and in the AA1050 layer on the condition that localized shear within the latter layer is accounted for.

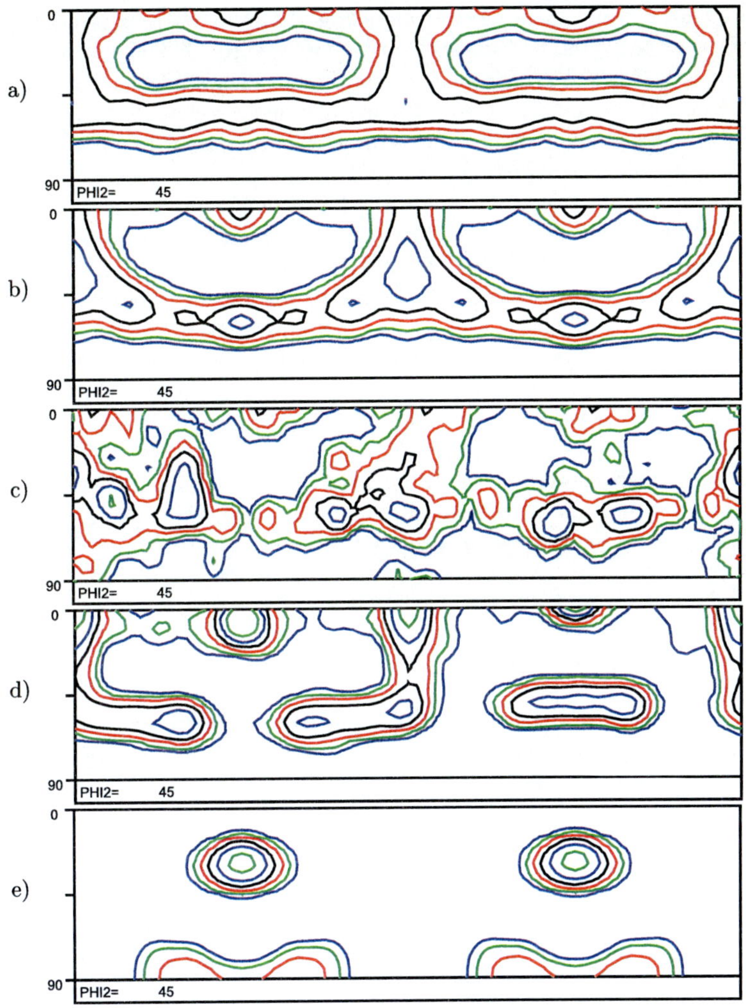

Fig. 5. $\varphi_2 = 45°$ section of the ODF allowing a quantitative assessment of the experimental (a) and predicted (b) textures in the central Fe layer. The experimental (c) texure in AA1050 is compared to predictions which either use the strain predicted by FE (d) or suppress the RD-ND shear (e). The contour levels are 0.5/1/2/4/8.

4 Discussion

Orientation mapping (based on EBSD measurements) is a powerful experimental technique which allowed us to probe both the texture and the microstructure in the different layers of the multilaminate (Fig. 2). However, the technique does not provide direct measurements of the microscopic strain field (unless it is combined with DIC e.g. [14]).

The present work demonstrates the possibility to assist the interpretation of EBSD by relying on numerical modeling (FE and CP). It is only through the combined use of all techniques that we are able to confirm that the AA1050 layer underwent intense shear, causing continuous recrystallization as well as the development of an unconventional texture after cold rolling.

Numerical models also provide estimates of the level of internal stresses, which are at the origin of damage, as well as the plastic anisotropy at the polycrystalline scale. Investigating this in the Fe-Al multilaminate is still ongoing.

It is noteworthy that CP modeling was carried out here in post-processing of the FE modeling which had assumed isotropic mechanical response. This sounds valid because the strain path inside the roll gap is not likely to depend much on plastic anisotropy. When modeling other metal forming processes, such as press forming of metal sheets, a direct micro-to-macro modeling is advisable e.g. [15].

5 Conclusions

- Folding one sheet around the other prior to roll-bonding allowed achieving the production of a 5-layer multilaminate made of ARMCO iron and AA1050.
- The central Fe layer developed a conventional cold rolling texture whereas texture development was slower in the surface Fe layer.
- EBSD measurements in the AA10150 layer revealed a highly refined and equiaxed microstructure, which is indicative of continuous recrystallization.
- According to FE simulations, the soft AA1050 layer hosted intense shear parallel to the rolling plane.
- The comparison of experimental textures to the predictions of CP modeling allowed us to convincingly link the EBSD observation of the different layers to the through-thickness strain heterogeneity predicted by FE modeling.

Acknowledgements. The authors thank Dr. Fengxiang Lin for her advice in the interpretation of EBSD maps. Financial support for this work was provided by the Fonds de la Recherche Scientifique de Belgique (F.R.S.-FNRS). Laurent Delannay is mandated by FSR-FNRS. Computational resources were provided by the Consortium des Équipements de Calcul Intensif (CÉCI), funded by FRS-FNRS under Grant No. 2.5020.11 and by the Walloon Region.

References

1. Nizolek, T., Mara, N.A., Beyerlein, I.J., Avallone, J.T., Scott, J.E., Pollock, T.M.: Processing and deformation behavior of bulk Cu-Nb nanolaminates. Metall. Microstruct. Anal. **3**, 470–476 (2014)

2. Kim, K.J., et al.: Formability of AA5182/polypropylene/AA5182 sandwich sheets. J. Mater. Process. Technol. **139**(1–3), 1–7 (2003)
3. Schoenfeld, S.E., Asaro, R.J.: Through thickness texture gradients in rolled polycrystalline alloys. Int. J. Mech. Sci. **38**(6), 661–683 (1996)
4. Engler, O., Tomé, C.N., Huh, M.Y.: A study of through-thickness texture gradients in rolled sheets. Metall. Mater. Trans. A. **31**(9), 2299–2315 (2000)
5. Tamimi, S., Gracio, J.J., Lopes, A.B., Ahzi, S., Barlat, F.: Asymmetric rolling of interstitial free steel sheets: microstructural evolution and mechanical properties. J. Manuf. Process. **31**, 583–592 (2018)
6. Sidor, J., Petrov, R.H., Kestens, L.: Texture control in aluminum sheets by conventional and asymmetric rolling. Compr. Mater. Process. **3**, 447–498 (2014)
7. Mishin, O.V., Bay, B., Jensen, D.J.: Through-thickness texture gradients in cold-rolled aluminum. Metall. Mater. Trans. A **31**, 1653–1662 (2000)
8. Zhang, Y.B., Mishin, O.V., Godfrey, A.: Analysis of through-thickness heterogeneities of microstructure and texture in nickel after accumulative roll bonding. J. Mater. Sci. **49**, 287–293 (2014)
9. Humphreys, F.J., Hatherly, M.: Recrystallization and Related Annealing Phenomena. Elsevier, Amsterdam (2012)
10. Tieu, A.K., Liu, Y.J.: Friction variation in the cold-rolling process. Tribol. Int. **37**(2), 177–183 (2004)
11. Delannay, L., Mishin, O.V.: Crystal plasticity modeling of the through-thickness texture heterogeneity in heavily rolled aluminum. In: Key Engineering Materials, vol. 554, pp. 1189–1194. Trans Tech Publications Ltd. (2013)
12. Dancette, S., Delannay, L., Renard, K., Melchior, M.A., Jacques, P.J.: Crystal plasticity modeling of texture development and hardening in TWIP steels. Acta Mater. **60**, 2135–2145 (2012)
13. Delannay, L., Tacq, J., Bardel, D., Seefeldt, M.: Direct micro-to-macro modelling of the cold rolling of pearlitic steel. In: MATEC Web of Conferences, vol. 80, p. 02008 (2016)
14. Lin, F.X., Shi, Q., Delannay, L.: Microscopic heterogeneity of plastic strain and lattice rotation in partially recrystallized copper polycrystals. Int. J. Solids Struct. **184**, 167–177 (2020)
15. Delannay, L., Beringhier, M., Chastel, Y., Loge, R.E.: Simulation of cup-drawing based on crystal plasticity applied to reduced grain samplings. Mater. Sci. Forum **495–497**, 1639–1644 (2005)

Thermal Ratcheting of Uranium Simulated with a Thermo-Elasto-Visco-Plastic Polycrystal Model

Carlos N. Tomé[1](✉) and Youngung Jeong[2]

[1] MST Division – Los Alamos National Laboratory, Los Alamos, NM 87545, USA
tome@lanl.gov

[2] School of Materials Science and Engineering, Changwon National University, Changwon, Republic of Korea

Abstract. Uranium aggregates accumulate large plastic deformation during thermal cycling (ratcheting). This process is the result of highly anisotropic single crystal properties, texture, and a complex coupling between thermal, elastic and plastic mechanisms of deformation. Here we present a thermo-elastic-viscous crystal plasticity model and apply it to the prediction of thermal ratcheting in clock-rolled uranium. The model captures the plastic relaxation of intergranular stress induced by thermal cycling and leads to an understanding of the coupling of the deformation mechanisms in uranium. Results are compared with ratcheting measurements.

Keywords: uranium · thermal ratcheting · crystal plasticity · thermo-elastic-viscous modeling

1 Introduction

Uranium is a metal with orthorhombic crystal symmetry. It has been known since the 1940s that, when subjected to thermal cycling in the absence of applied stress, textured uranium polycrystals accumulate large plastic deformation, in a process known as thermal ratcheting [1–3]. The phenomenon has been correctly attributed to the high degree of anisotropy in the crystal coefficients of thermal expansion (CTE), and consequent reaction stresses induced in the grains by the surrounding medium when temperature changes take place. Those stresses are sufficiently large to trigger non-recoverable plastic deformation in the grains.

Quantitative modeling of this phenomenon requires to use a polycrystal model that can account for texture and can couple the effects of plastic, elastic, and thermal strains. Except for a simplified model proposed recently by Calhoun et al. [4] no rigorous simulation of this process has been reported. In this paper, we add thermal effects to a recently developed incremental elasto-visco-plastic model (ΔEVPSC) [5, 6] and apply it to simulate the ratcheting of uranium. In what follows we address the most relevant features of the models used and refer the reader to the appropriate papers for detailed developments.

K. Mocellin et al. (Eds.): ICTP 2023, LNME, pp. 654–661, 2024.
https://doi.org/10.1007/978-3-031-40920-2_68

2 Thermo-Elastic-Visco-Plastic Self-Consistent Polycrystal Model

In this work, the strain rate of the grain includes contributions from elasticity, visco-plasticity, and thermal expansion (or contraction):

$$\dot{\boldsymbol{\varepsilon}} = \dot{\boldsymbol{\varepsilon}}^{el} + \dot{\boldsymbol{\varepsilon}}^{vp} + \dot{\boldsymbol{\varepsilon}}^{th} \tag{1}$$

The elastic strain rate follows from the linear Hooke's law using the single crystal compliance $\mathbb{M}^{el}$ and stress rate $\dot{\boldsymbol{\sigma}}$ such that $\dot{\boldsymbol{\varepsilon}}^{el} = \mathbb{M}^{el} : \dot{\boldsymbol{\sigma}}$. The thermal expansion (or contraction) rate is given by the product of the temperature rate of change times the coefficients of thermal expansion $\boldsymbol{\alpha}$ such that $\dot{\boldsymbol{\varepsilon}}^{th} = \boldsymbol{\alpha}\dot{T}$. The visco-plastic strain rate $\dot{\boldsymbol{\varepsilon}}^{vp}$ is a result of shear rate in slip and twinning systems $\dot{\gamma}^s$ such that $\dot{\boldsymbol{\varepsilon}}^{vp} = \sum_s \mathbb{m}^s \dot{\gamma}^s$ where $\mathbb{m}^s$ denotes the Schmid tensor. Applying a forward Euler integration to the stress and temperature rates gives an incremental form of Eq. (1):

$$\dot{\boldsymbol{\varepsilon}}_{ij} = \dot{\gamma}_0 \sum_s m^s_{ij} \left(\frac{\mathbf{m}^s : (\boldsymbol{\sigma}_{(k)} + \Delta\boldsymbol{\sigma})}{\boldsymbol{\tau}^s} \right)^n + \frac{(M^{el} : \Delta\boldsymbol{\sigma})_{ij}}{\Delta t} + \boldsymbol{\alpha}_{ij} \frac{\Delta T}{\Delta t} \tag{2}$$

where n, τ^s, and $\dot{\gamma}_0$ denote the exponent of the visco-plastic power law, the critical resolved shear stress (CRSS), and the reference shear rate, respectively. $\Delta\boldsymbol{\sigma}$ and ΔT denote the increment of stress tensor and temperature, respectively, that take place during the time increment Δt. Equation (2) is solved for the (k + 1) time increment that begins at time t and ends at time $t+\Delta t$, during which the stress is integrated as $\boldsymbol{\sigma}_{(k+1)} = \boldsymbol{\sigma}_{(k)} + \Delta\boldsymbol{\sigma}$. To utilize the Eshelby's solution Eq. (2) is written in a pseudo-linear form:

$$\dot{\boldsymbol{\varepsilon}} = \mathbb{M}^{evp} : \Delta\boldsymbol{\sigma} + \dot{\boldsymbol{\varepsilon}}^{evp,0} + \left(\frac{\alpha \Delta T}{\Delta t} \right) \tag{3}$$

where $\mathbb{M}^{evp} = \left[\mathbb{M}^{vp,tg}_{(\sigma_{(n)}+\Delta\sigma)} + \frac{\mathbb{M}^{el}}{\Delta t}\right]$ is the elasto-visco-plastic compliance and $\dot{\boldsymbol{\varepsilon}}^{evp,0} = \left[\mathbb{M}^{vp,sec}_{(\sigma_{(n)}+\Delta\sigma)} - \mathbb{M}^{vp,tg}_{(\sigma_{(n)}+\Delta\sigma)}\right] : \Delta\boldsymbol{\sigma} + \mathbb{M}^{vp,sec}_{(\sigma_{(n)}+\Delta\sigma)} : \boldsymbol{\sigma}_{(n)}$ is the back-extrapolated strain rate. The last term in Eq. (3) is the incremental form of the thermal strain rate. A law similar to Eq. (3) is proposed for the response of the aggregate:

$$\dot{\overline{\boldsymbol{\varepsilon}}} = \overline{\mathbb{M}} : \Delta\overline{\boldsymbol{\sigma}} + \dot{\overline{\boldsymbol{\varepsilon}}}^0 \tag{4}$$

The algebraic form of these equations is the same as for the Eshelby problem and so the Eshelby solution applies, with elastic strain and stress replaced by total strain rate and stress increment. By assuming that each grain is an elasto-visco-plastic (EVP) inclusion interacting with the effective EVP medium that represents the polycrystal, a set of equations is obtained that have to be solved self-consistently to obtain the EVP and the back extrapolated tensors of the polycrystal [5]. The advantage of this procedure compared with previous EVP models [7] is that the inconsistency of solving the elastic and viscous inclusion-medium interaction separately is avoided.

In the current study, the CRSS is described as a function of the evolving densities of dislocations in each slip system. The hardening model is based on the Kocks-Mecking rule [10]. The derivation of the associated equations are described in detail in [8, 9,

11], but space limitations prevent us from reproducing it here. The rate of change in the density of forest dislocation population $\dot{\rho}^s_{for}$ is the result of generation $\dot{\rho}^s_{gen}$ and recombination $\dot{\rho}^s_{rec}$ mechanisms such that:

$$\dot{\rho}^s_{for} = \dot{\rho}^s_{gen} - \dot{\rho}^s_{rec} = k_1^\alpha \sqrt{\rho^s_{for}} \dot{\gamma}^s - k_2^\alpha \rho^s_{for} \dot{\gamma}^s \tag{5}$$

The factor k_1 controls dislocation generation and the factor k_2, function of temperature and rate, controls the recombination rate [11]. A fraction $q^s \dot{\rho}^s_{rec}$ of the recombined dislocations contributes to the formation of substructure (i.e., dislocation cell walls), while the remaining fraction $(1 - q^s)\dot{\rho}^s_{rec}$ annihilates. The forest and substructure dislocation densities contribute to the CRSS of each slip system s:

$$\tau^s = \tau_0^\alpha + \tau^s_{for} + \tau^s_{sub} \tag{6}$$

where τ_0^α is the Peierls friction stress and its dependence with temperature is described by:

$$\tau_0^\alpha = A_0^\alpha + A_1^\alpha \exp\left(-\frac{T}{A_2^\alpha}\right) \tag{7}$$

A_0^α, A_1^α and A_2^α are empirical parameters specific for each slip mode. In this study the hardening parameters of the Δ EVPSC model are calibrated using mechanical tests done at different temperatures reported in [4, 8, 12, 13]. Of the three terms in Eq. (6) the Peierls stress τ_0^α is the most relevant one for the thermal cycling simulation because the dislocations generated during the lower temperature portion of the cycle tend to annihilate during the higher temperature portion of the cycle. Consequently, τ^s_{for} and τ^s_{sub} remain low, and the CRSS in the slip systems does not deviate much from τ_0^α during thermal cycling. In addition, twinning activation is minor during the thermal cycling simulation. For the sake of brevity, the fitting of the hardening parameters for slip and twin systems is not going to be described here and will be reported in [14].

3 Material Characterization

The main objective of the current study is to understand the experimentally observed thermal ratcheting observed in Uranium by utilizing the elasto-visco-plastic self-consistent polycrystal model. Experimental observations of thermal ratcheting reported by Kelman [15] and by Chiswik [16] are used here for comparison when assessing the predicted growth rate of uranium under temperature cycles. Unfortunately, reference [15] reports only a loosely defined extrusion texture of the uranium rod with a (001) component along the radial direction of the rod and a (010) component along the axial direction. For fitting the plastic parameters of the different slip and twinning modes we utilize recent texture and flow curves for α-Uranium measured for various loading conditions [8, 9, 12]. The experimental data reported in these studies originated from the same depleted uranium sample. Because texture, pedigree and thermal treatment of our Uranium differ from Chiswik's, only a semi-quantitative comparison to experiments is possible.

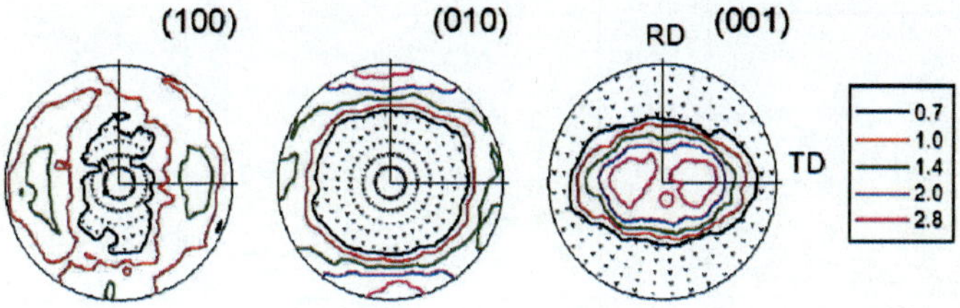

Fig. 1. (100), (010) and (001) pole figures of clock-rolled α-uranium [8]. The sample axes 1, 2 and 3 (out-of-plane) correspond to rolling, transverse, and normal directions. For comparison purposes they can be loosely correlated with the axial, radial and tangent directions of the extruded uranium rods reported by Chiswik [16].

The rolling and annealing processing route used is described in [12]. It results on aggregates with an average grain size of 25 microns and a rolling texture represented here by means of 8000 discrete orientations (Fig. 1).

The orthorhombic uranium crystal possesses 9 independent elastic constants C_{ijkl} and 3 independent thermal expansion moduli $\alpha_{ij}^{th} = \alpha_i \delta_{ij}$ In this study their temperature dependence was described using a 2nd degree polynomial fit to the elastic moduli of single crystal uranium reported by Fisher [17] (Fig. 2a) and to the coefficients of thermal expansion (CTE) reported by Korn [18] (Fig. 2b).

The twin and slip systems identified in previous work were made available in the simulation. They are (010)[100] wall slip (mult 1), (001)[100] floor slip (mult 1), {110}<1–10> chimney slip (mult 2), {021} <1–12> (mult 4) roof slip, {130} <3–10> twin (mult 2), {172} <3–12> twin (mult 4). The evolution of their associated Peierls stress τ_0^{α} with temperature is depicted in Fig. 2c. Of those, only floor, roof and chimney showed meaningful activity during thermal cycling.

The temperature range reported by Kelman is between 300 K and 800 K, similar to the one used by Chiswik of 20 °C and 550 °C. According to Fig. 2b, the cooling from 800 K to 300 K of single crystal uranium will induce about 1.5% negative strain along axes 1 and 3 of the crystal, and much smaller but positive strain along axis 2. While in the single crystal this strain is reversible, in the aggregate it represents a significant misfit by comparison to the average aggregate strains, which induces significant stress and causes irreversible plastic relaxation.

4 Thermal Contraction During Cooling

The anisotropic thermal contraction of polycrystalline uranium induced by cooling from 800 K to 300 K is shown in Fig. 3. To demonstrate the effect of visco-plasticity, additional simulations were performed setting all the critical resolved shear stresses to a high value of 10^3 GPa. Doing so suppresses the contribution from visco-plasticity and emulates the predictions of the thermo-elastic self-consistent (TESC) model reported in [19]. While both ΔEVPSC and TESC models lead to similar anisotropic contraction induced by cooling, the results shown in Fig. 3 imply that the macroscopic CTEs $(\partial \bar{\varepsilon}_{ij}/\partial T)$ are significantly affected by intergranular plasticity. Plasticity enhances the anisotropy in the

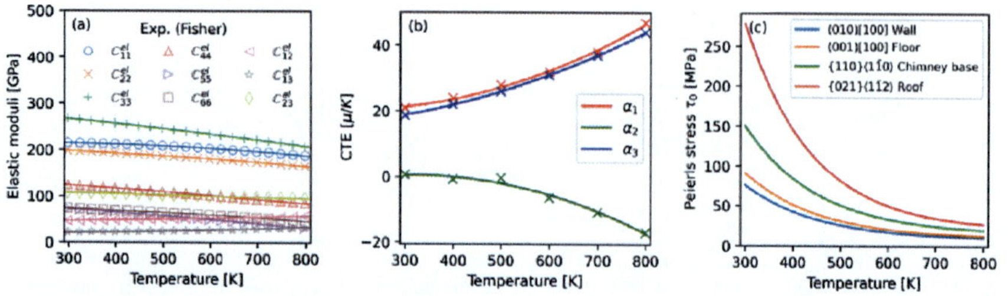

Fig. 2. Temperature dependence of elastic, thermal and plastic properties of a-uranium single crystal. (a) elastic constants; (b) coefficients of thermal expansion; (c) Peierls stress τ_0.

dimensional change of Uranium in that the contraction along RD (i.e., $\overline{\varepsilon}_{11}$) decreases while the contraction along TD and ND ($\overline{\varepsilon}_{22}$ and $\overline{\varepsilon}_{33}$) is enhanced by comparison with the TESC results.

It is also worth comparing this polycrystal contraction to that of the single crystal According to the CTEs depicted in Fig. 2b a negative and large contraction will take place along [100] and [001], while along [010] an extension, albeit small, is expected. On the other hand, according to the results in Fig. 3 the interaction between grains leads to a contraction of the textured aggregate in all three directions, albeit reduced by comparison with the single crystal.

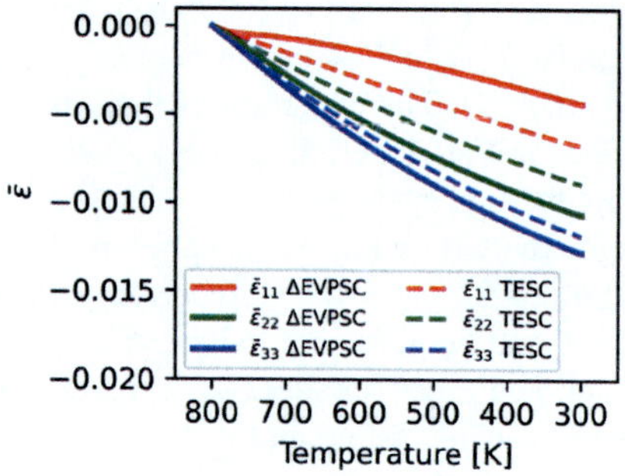

Fig. 3. Strain evolution during cooling from 800 K to 300 K of a -uranium polycrystal predicted by the current ΔEVPSC model and by a surrogate thermo-elastic model (TESC).

5 Thermal Ratcheting Simulations

The thermal ratchetting effects were experimentally reported in [15, 16]. In the Kelman experiments [15] the uranium rods were subjected to thermal cycles between 100 °C and 550 °C, which corresponds to 373 K and 823 K, respectively. In the ΔEVPSC simulation results, we use the reported heating and cooling rates of Kelman, which were (450 K)/(5 min) and (−450 K)/(22.5 min), respectively, for a full cycle duration of 27.5 min. Due to duration of the cycle, visco-plastic relaxation (and likely static dislocation annealing not included in the model) takes place during the high temperature portion. Figure 4 shows the strain evolution resulting from the thermal cycle simulations together with some of the experimental data reported by Chiswik [16]. The latter author tested a number of uranium rods that were prepared under different thermo-mechanical processing routes which resulted in a wide dispersion of texture and grain sizes. As mentioned previously, differences in texture and microstructure between their and our Uranium samples preclude making a rigorous quantitative comparison. However, the magnitude of the ratcheting strains predicted after 100 cycles is about 7% and comparable to the experimental. The coupling of the elastic, thermal and plastic strains via the thermally induced stress is captured and clearly elucidated in this work.

The relative slip and elastic contributions during each cycle is depicted in Fig. 5. It indicates that chimney slip (orange) and to a lesser extent roof slip (green) are the main sources of plastic relaxation, while floor slip (blue) contributes at most 5%. The contribution of twinning is negligible, which is consistent with EPSC simulation results of Calhoun et al. [20]. At the beginning of the heating ramp the relative activity of plastic systems is negligible until about 600 K and becomes relevant between 600 K and 800 K as a consequence of increasing intergranular stresses and decreasing CRSS with temperature. The peaking followed by a dropping is due to nearly instantaneous relaxation of the accumulated stress before reaching 800 K. During the cooling process there is a build-up of intergranular stress which get relaxed by slip (viscous creep) on the plastic systems. As temperatures approaches 400 K floor and roof slips become inactive while chimney slip is still active, albeit it contributes less than 10% deformation. Elasticity is the main contributor to the total strain increment below 400 K.

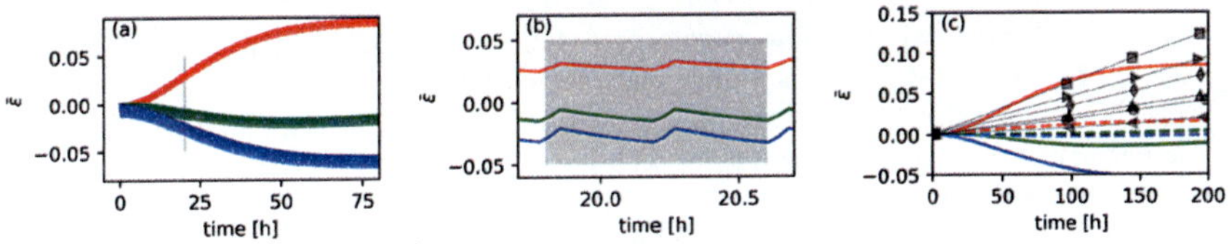

Fig. 4. Evolution of strain components $\bar{\varepsilon}_{11}$ (red), $\bar{\varepsilon}_{22}$ (green) and $\bar{\varepsilon}_{33}$ (blue) along RD, TD and ND during thermal cycling between 300 K and 800 K following the experimental condition reported in Kelman [15]. (a) overall evolution for 100 thermal cycles; (b) enlarged view of the shaded region in (a); (c) comparison of ΔEVPSC predictions using both, textured and un-textured uranium polycrystal. The symbols correspond to experimental axial strains (akin to our RD strain) reported by Chiswik [16] for bars with different thermomechanical treatments.

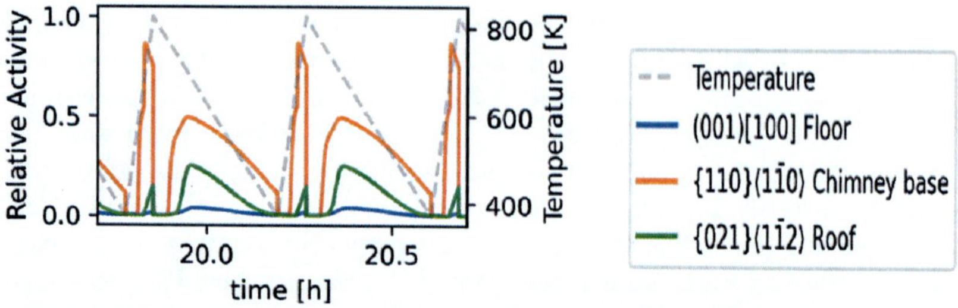

Fig. 5. Relative activity of deformation modes predicted by ΔEVPSC corresponding to two cycles in the time interval 19.5 to 21 h. The activity is defined as the percentage of shear rate on the slip planes contributed by each plastic mode. The elastic component (not plotted) contributes the balance activity.

References

1. Boas, W., Honeycombe, R.W.K.: Thermal fatigue of metals. Nature **153**, 494–495 (1944)
2. Boas, W., Honeycombe, R.W.K.: The plastic deformation of non-cubic metals by heating and cooling. Proc. R Soc. Lond. A Math. Phys. Sci. **186**, 57–71 (1946)
3. Boas, W., Honeycombe, R.W.K.: The anisotropy of thermal expansion as a cause of deformation in metals and alloys. Proc. R Soc. Lond. A Math. Phys. Sci. **188**, 427–439 (1947)
4. Calhoun, C.A., Garlea, E., Mulay, R.P., Sisneros, T.A., Agnew, S.R.: Investigation of the effect of thermal residual stresses on deformation of α-uranium through neutron diffraction measurements and crystal plasticity modeling. Acta Mater. **85**, 168–179 (2015)
5. Jeong, Y., Tomé, C.N.: An efficient elasto-visco-plastic self-consistent formulation: application to steel subjected to loading path changes. Int. J. Plast. **135**, 102812 (2020)
6. Jeong, Y., Jeon, B., Tomé, C.N.: Finite element analysis using an incremental elasto-visco-plastic self-consistent polycrystal model: FE simulations on Zr and low-carbon steel subjected to bending, stress-relaxation, and unloading. Int. J. Plast. **147**, 103110 (2021)
7. Wang, H., Wu, P.D., Tomé, C.N., Huang, Y.: A finite strain elastic-viscoplastic self-consistent model for polycrystalline materials. J. Mech. Phys. Solids **58**, 594–612 (2010)
8. McCabe, R.J., Capolungo, L., Marshall, P.E., Cady, C.M., Tomé, C.N.: Deformation of wrought uranium: experiments and modeling. Acta Mater. **58**, 5447–5459 (2010)
9. Knezevic, M., et al.: Modeling mechanical response and texture evolution of α-uranium as a function of strain rate and temperature using polycrystal plasticity. Int. J. Plast. **43**, 70–84 (2013)
10. Kocks, U.F., Mecking, H.: Physics and phenomenology of strain hardening: the FCC case. Prog. Mater. Sci. **48**, 171–273 (2003)
11. Beyerlein, I.J., Tomé, C.N.: A dislocation-based constitutive law for pure Zr including temperature effects. Int. J. Plast. **24**, 867–895 (2008)
12. Brown, D.W., et al.: Temperature and direction dependence of internal strain and texture evolution during deformation of uranium. Mater. Sci. Eng. A **512**, 67–75 (2009)
13. Calhoun, C.A., Garlea, E., Sisneros, T.A., Agnew, S.R.: In-situ neutron diffraction characterization of temperature dependence deformation in α-uranium. J. Nucl. Mater. **502**, 60–67 (2018)
14. Jeong, Y., Tomé, C.N.: under preparation
15. Kelman, L.R.: Dimensional Changes in Uranium Under Thermal Cycling, Argonne, IL (United States) (1949)

16. Chiswik, H.H.: The plastic deformation of uranium on thermal cycling. Trans. Am. Soc. Metals **49**, 623–654 (1957)
17. Fisher, E.S.: Temperature dependence of the elastic moduli in alpha uranium single crystals, part iv (298° to 923° K). J. Nucl. Mater. **18**, 39–54 (1966)
18. Korn, J.: An investigation of the cones of zero expansion and thermal coefficients for single crystal of α-uranium. J. Nucl. Mater. **71**, 320–326 (1978)
19. Tomé, C.N., Lebensohn, R.A.: Polycrystal thermo-elasticity revisited: theory and applications. Comptes Rendus Mécanique **348**, 877–891 (2021)
20. Calhoun, C.A., Wollmershauser, J.A., Brown, D.W., Mulay, R.P., Garlea, E., Agnew, S.R.: Thermal residual strains in depleted α-U. Scripta Mater. **69**, 566–569 (2013)

Interpretation of the Unloading Non-linearity in Dual-Phase 980 Steel Using an Elasto-Visco-Plastic Self-consistent Polycrystal Model

Bohye Jeon[1] and Youngung Jeong[2](✉)

[1] Department of Materials Convergence and System Engineering, Changwon National University, Changwon, Republic of Korea
[2] School of Materials Science and Engineering, Changwon National University, Changwon, Republic of Korea
yjeong@changwon.ac.kr

Abstract. A dual-phase 980 steel sample consisting of hard martensitic and ductile ferritic phases was investigated in terms of its nonlinear unloading behavior and the Bauschinger effect. The microstructural characters obtained from a set of electron back-scattered diffraction scans were used to analyze the crystallographic texture for each phase. The incremental elasto-visco-plastic self-consistent model was applied with an empirical Voce hardening law. Neither backstress nor the effect of geometrically necessary dislocation was accounted for. Despite the simplicity of this baseline Δ EVPSC model, the nonlinear unloading and the Bauschinger effect were captured. It seems that the contrast in the critical resolved shear stress of the two phases suffices to induce earlier and selective re-yielding of ductile ferrite.

Keywords: crystal plasticity · nonlinearity · unloading · bauschinger effect

1 Introduction

Dual-phase (DP) steel has been widely utilized in the manufacturing of various automotive components owing to its well-balanced combination of elongation and strength. Ever growing demand for lightweight vehicles encourages steelmakers to increase the volume fraction of the martensitic phase in DP steels. However, the increase in strength inevitably leads to an increase in springback. It has been reported in the literature that there exist many material characteristics that may affect the springback [1]. Among various factors, the non-linear behavior during elastic unloading has recently attracted attention in the literature, which needs to be accurately described to improve the predictive accuracy of springback [2, 3]. Recently, Li and Wagoner [4] demonstrated that their dissipative bow-out model can capture most of the non-linear characteristics observed during loading-unloading-loading (LUL) experiments of various steel and magnesium alloys. Their model can potentially quantify the contribution of dislocation bow-out mechanism and use the quantifiable contribution to the overall strain, which can be also experimentally observed as shown in [5].

K. Mocellin et al. (Eds.): ICTP 2023, LNME, pp. 662–670, 2024.
https://doi.org/10.1007/978-3-031-40920-2_69

While the dissipative bow-out modeling framework reported therein is promising, the model did not explicitly consider the heterogeneity in stress (and strain) responses of grains. The extension to account for heterogeneity can be realized by incorporating grain-scale physics into the polycrystal plasticity framework [6, 7].

The main objective of the current work is to investigate if the recently developed elasto-visco-plastic self-consistent model (denoted hereinafter as Δ EVPSC) [8, 9], in which unified elasto-visco-plastic medium represents the polycrystalline aggregate, is worth for such extension. To that end, the baseline behavior of Δ EVPSC using an empirical Voce hardening model without further accounting for, say, the dislocation bow-out mechanism, should be examined to demonstrate the limits in the baseline model.

There has been increasing use of similar self-consistent polycrystal formulations to link the micro-mechanisms to macroscopic behaviors by accounting for crystallographic backstress [10–12] or the geometrically necessary dislocation (GND) associated inter- and intra-granular stresses [6, 13, 14], and the dissipative characteristics in anelastic behavior [4]. The apparent advantage of the Δ EVPSC model is that it is already interfaced with a commercial finite element (FE) software [9] and can simulate, for example, bending and springback FE simulations [15]. Also, the visco-plastic formulation has an apparent advantage over an elasto-plastic formulation since it can more adequately describe stress relaxation and unloading behavior [15, 16].

In this study, we attempt to address the below questions:

- Can this baseline Δ EVPSC model capture some of the important characteristics regarding unloading non-linearity without advanced treatment?
- If not, what should be further improved?

2 Model Calculation

2.1 Elasto-Visco-Plastic Self-consistent Model

In the Δ EVPSC model, derived from the visco-plastic self-consistent (VPSC) formulation [17], the constitutive behavior of a single crystal is described through the equation below power-law:

$$\dot{\boldsymbol{\varepsilon}} = \dot{\gamma}_0 \sum_s \boldsymbol{m}^s \left\{ \frac{\left| \boldsymbol{m}^s : \left(\boldsymbol{\sigma}_{(k)} + \Delta\boldsymbol{\sigma} \right) \right|}{\tau^s} \right\}^n sgn\left(\boldsymbol{m}^s : \left(\boldsymbol{\sigma}_{(k)} + \Delta\boldsymbol{\sigma} \right) \right) + \left(\frac{\mathbb{M}^{el} : \Delta\boldsymbol{\sigma}}{\Delta t} \right) \tag{1}$$

The symbols $\dot{\boldsymbol{\varepsilon}}$, $\Delta\boldsymbol{\sigma}$, Δt, $\mathbb{M}^{el}$, $\boldsymbol{m}^s$, and τ^s denote the grain strain rate tensor, the incremental stress tensor, the time increment, the elastic modulus tensor, the Schmid tensor, and the critical resolved shear stress (CRSS), respectively. The superscript s denotes that the associated quantities are of each slip system, and the exponent n is the inverse of strain-rate sensitivity. The subscript (k) is appended to the stress tensor $\boldsymbol{\sigma}$ to denote that the stress state of grain converged in the previous step is used. The self-consistent homogenization is applied to obtain the averaged response of the polycrystal [8, 9, 17].

The evolution of dislocation density as well as the effect of the crystallographic backstress is already implemented and can be chosen as the hardening law [10–12] in the numerical code of Δ EVPSC. However, as mentioned in Introduction, only the baseline behavior of the Δ EVPSC model is investigated in this work. To that end, a simple empirical Voce hardening model was chosen, which requires only four different parameters in each slip mode. In it, CRSS τ is described as a function of total accumulative slip shear Γ via:

$$\tau_{(\Gamma)} = \tau_0 + (\tau_1 + \theta_1\Gamma)\left\{1 - \exp\left(-\frac{\Gamma\theta_0}{\tau_1}\right)\right\}. \tag{2}$$

For the current DP980 steel sample, two slip modes, i.e., $\langle 111\rangle\{1\bar{1}0\}$ and $\langle 111\rangle\{11\bar{2}\}$ consisting of 24 slip systems, were considered for each phase. For the sake of simplicity, the same set of parameters was used regardless of the slip mode during parameter calibration. The same simplification was often used for body-centered steels before, which led to reasonable predictive accuracy of a similar model [18, 19]. Likewise, the same elastic moduli were used for both phases to keep the model simple. The four parameters were calibrated by fitting the uniaxial tension flow stress-strain curves. Also, the relative ratio of ferrite and martensite phases was manually determined such that the lattice strain data reported in [20]. The resulting hardening parameters of the two phases differ significantly in that the initial CRSS (represented by τ_0) of martensite (800 MPa) is much higher than that of ferrite (270 MPa). The validity of the chosen parameters shown in Table 1 was checked in terms of the additional diffraction strains reported in the same reference [20], and the comparison with the experimental flow stress-strain curves and R-value. Detailed analysis of the accuracy of the characterized model is beyond the scope of the current study and is not the main purpose of this study. Therefore, neither the experimental stress-strain curves nor the comparison with the in-situ diffraction strain is reported in this study. Nevertheless, the important results discussed in later sections would not change even if the parameters were more finely adjusted provided that the two phases have significantly different values of CRSS.

Table 1. Voce hardening parameters [MPa] and elastic moduli [GPa]

Slip systems	Phase	τ_0	τ_1	θ_0	θ_1
$\langle 111\rangle\{1\bar{1}0\}$, $\langle 111\rangle\{11\bar{2}\}$	Ferrite	270	50	10	0
	Martensite	800	200	2000	0
Elastic moduli for both phases					
$\mathbb{C}_{11}$	$\mathbb{C}_{12}$		$\mathbb{C}_{44}$		
220	132		99		

3 Experiments and Results

3.1 DP980 Steel Microstructure

The DP980 steel sheet produced by POSCO was investigated in this study, and its initial thickness was 1.4 mm. The EBSD (electron back-scattered diffraction) data were used to obtain the initial crystallographic distribution for each constituent phase as well as the phase volume fraction. Two characteristics were utilized to identify the phases:

1. The EBSD result shows a bimodular distribution in the grain size: small martensitic islands are surrounded by large ferritic grains.
2. The martensitic grains show a distinctively high value of kernel angular misorientation (KAM) in comparison with that of ferrite – see e.g. [12].

A certain grain is identified as martensite if the below criteria are met simultaneously:

a. Grain size greater than 10 $\mu m.$
b. Grain-averaged value of KAM (i.e., GAM) smaller than 0.5 °.

Otherwise, the grain is identified as ferrite (Fig. 1).

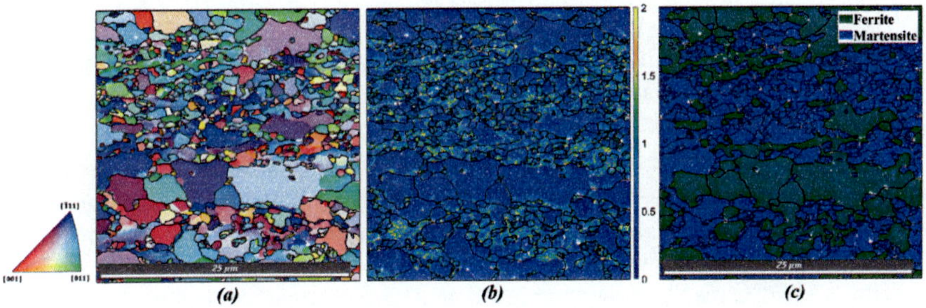

Fig. 1. (a) Inverse pole figure maps (RD), (b) kernel angular misorientation maps, and (c) phase maps of the as-received DP980 sample [degree]. Three EBSD maps obtained at two different depths in the thickness direction are stacked together.

The resulting volume fraction of ferrite to martensite (i.e., 6:4) is in good agreement with other similar DP steel reported in the literature [21]. Admittedly, the criteria applied to phase identification were rather arbitrary and were not further validated. Nonetheless, the resulting EBSD data suffice to provide the initial crystallographic texture required for the subsequent crystal plasticity simulations. Throughout the current study, each phase of the DP980 polycrystal was represented by 1000 discrete orientations, which were analyzed using MTEX [22].

3.2 Loading-Unloading-Reloading Simulation

The loading-unloading-loading (LUL) test along the rolling direction was simulated for the DP steel and is shown in Fig. 2a. In it, three different LUL curves are depicted, which represent the stress-strain responses of the overall polycrystal, ferritic, and martensitic phases, respectively. In addition, the stress-strain curve resulting from a tension-compression simulation is shown in Fig. 2b. Note that in Fig. 2b the stress of each phase shown in Fig. 2a is plotted against the strain of each phase. In the literature, a different plotting convention is often applied, in which the phase stress is plotted against the polycrystal-average strain. Following such convention, the same data in Fig. 2 are reused in Fig. 3b by plotting the phase stresses against the polycrystal-averaged strain not phase-average strain. It should be noted that Fig. 3b alone cannot clearly visualize phase-specific responses as the polycrystal-average strain responses therein contain the contributions from both phases.

The phase-specific stress and strain responses were used in Fig. 2. For example, the average strain of the ferritic phase was calculated as the weighted average of only the constituent grains such that

$$\langle \boldsymbol{\varepsilon} \rangle^{\text{ferrite}} = \frac{\sum_{g \in f} w^g \boldsymbol{\varepsilon}^g}{\sum_{g \in f} w^g} \tag{3}$$

where the symbol $\Sigma_{g \in f}$ indicates that the summation was performed only for ferritic grains, and w^g denotes the weight of grain g. The same method was applied to obtain the phase-specific stresses.

In Fig. 3a, the simulation result with only one LUL event (red) is compared with the tension-compression stress-strain curve (gray dashed line). The phase-specific stress vs. polycrystal strain curves, denoted by blue and green lines respectively for martensite and ferrite, are shown in Fig. 3b.

The Δ EVPSC results shown in Figs. 2 and 3 are summarized below:

- At the full unloads (i.e., at $\langle \boldsymbol{\sigma} \rangle^{polycrystal} = 0$), the martensitic phase remains tensile in average, while the ferritic phase is compressed in average (see Figs. 2a, b, and 3b).
- The LUL loops of martensite show the linear relationship between the stress and strain, which implies that the martensite remains elastic during the unloading and reloading event. In contrast, the LUL loop of the ferrite deviates from the linearity towards the end of macroscopic unloading, which suggests that the ferrite re-yields at the full unloading state of the polycrystal (see Figs. 2a and b).
- The closed loop (of the overall polycrystal) formed by the unloading and the loading legs was captured by the model, and its width increased with plastic deformation (see Fig. 2a).
- The predicted shape of the closed loop was *not* symmetric (see Figs. 2b and 3b) as opposed to the experimental and modeling results demonstrated in [4, 16] and deviated from the typical lens shape.
- The baseline Δ EVPSC model captured the Bauschinger effect – see the gray dashed line (denoted as TC polycrystal) in Figs. 2b, 3a, and b.

4 Discussion

Despite that the simple Voce hardening model with only four parameters was used for each phase, the baseline Δ EVPSC model could capture a few important behaviors: a) the formation of closed loops; b) the increasing width of the loop; and c) the Bauschinger effect.

The early re-yielding of ferrite seems responsible for the non-linearity near the fully unloaded state of the polycrystal. As evident in Fig. 2 the ferrite tends to re-yield even at unloaded state due. Also, the phase-specific stress, at which ferrite begins to re-yield decreased with plastic deformation thus leading to the increasing loop width. This behavior captured by the model is explained below:

i. Since the martensite has much higher CRSS (higher τ_0) and the hardening capability (higher τ_1), the stiffness of HEM (Homogeneous Equivalent Medium) is more strongly affected by the slip resistance of martensitic grains than ferrite.
ii. As plastic deformation increases, the flow stress of ferrite quickly saturates to its maximum circa 700 MPa (Fig. 2a) while that of martensite keeps on increasing due to the higher hardening capability.
iii. The HEM becomes increasingly harder with plastic deformation, while the stiffness of soft ferrite tends to saturate at the strain of ~0.02.
iv. As plastic deformation increases, the HEM hardens and more strongly compresses the ferrite, which yields even at the fully unloaded state.
v. The yielding of ferrite led to the nonlinearity contribution to the LUL loop and the Bauschinger effect.

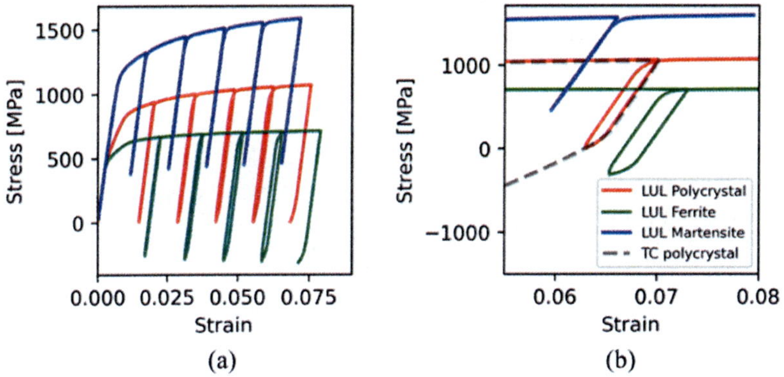

Fig. 2. Stress-strain curves resulting from the loading-unloading-loading and the monotonic uniaxial tension simulation are shown for each phase of the DP980 polycrystal and the overall polycrystal.

In the previous study using the Δ EVPSC on a single-phase mild steel [15], the unloading non-linearity was not captured even if the crystallographic backstress was accounted for together with advanced hardening modeling to describe the path-dependent evolution in the dislocation density. That the same Δ EVPSC modeling framework with a simple Voce hardening law successfully captured the mentioned characteristics implies that the effect of stress and strain heterogeneity (induced mainly by the drastic contrast in CRSS of soft ferrite and hard martensite) may play a significant role as analyzed above. Furthermore, the latter suggests that the backstress is not, at least for the current DP980 steel, a sole contributing factor to the unloading non-linearity and for the earlier yielding when reversely loaded.

Nevertheless, the LUL curves captured by the current baseline Δ EVPSC do not completely reflect the experimentally observed behavior. For example, while the shape of Δ EVPSC-predicted LUL loops is asymmetric, the experimental results suggest that they should be symmetric [4, 16]. It is expected that incorporating the advanced micro-mechanical models as in [4, 6, 13, 14] may improve the predictive accuracy.

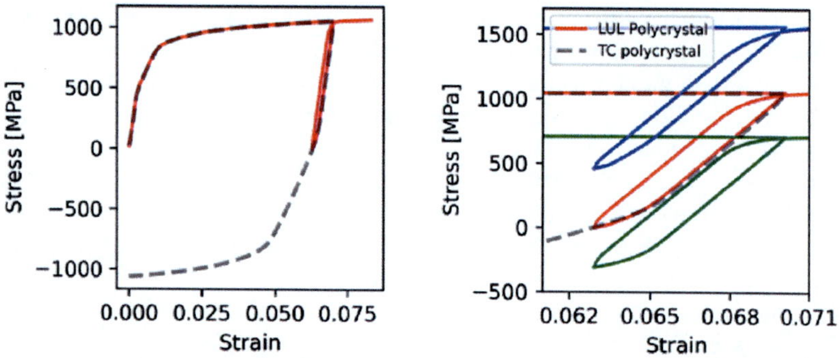

Fig. 3. Stress-strain curves resulting from the loading-unloading-loading and the tension-compression simulation of the DP980 polycrystal.

5 Conclusions

In this study, the baseline Δ EVPSC model with a simple Voce hardening law was applied to study the non-linear unloading of a DP980 steel sample. Uniaxial tension flow stress-strain curve, the diffraction lattice strain available in the literature, and the initial EBSD map were used to calibrate the empirical Voce hardening model employed in the baseline Δ EVPSC model without explicitly accounting for the back-stress. Despite of the simplicity, a few important characteristics were successfully captured in that the non-linearity between the flow stress and the strain during loading-unloading-loading sequence was well described by the model and is attributed to a large difference between the ferrite and the martensite phases in terms of the critical resolved shear stress. Moreover, the Bauschinger effect was reasonably captured despite that no explicit back-stress was accounted for. This result suggests that the HEM-based self-consistent polycrystal formulation would provide a promising modeling framework to describe mechanical behaviors under complex loading scenarios. As a future work, the model incorporating advanced micro-mechanical considerations such as the dissipative dislocation bow-out mechanism and the crystallographic backstress is required. Also, the model should be verified in comparison with experimental behavior.

Acknowledgment. This research was supported by Changwon National University in 2023–2024.

References

1. Wagoner, R.H., Lim, H., Lee, M.-G.: Advanced Issues in springback. Int. J. Plast. **45**, 3–20 (2013). https://doi.org/10.1016/j.ijplas.2012.08.006
2. Morestin, F., Boivin, M.: On the necessity of taking into account the variation in the young modulus with plastic strain in elastic-plastic software. Nucl. Eng. Des. (1996). https://doi.org/10.1016/0029-5493(95)01123-4
3. Sun, L., Wagoner, R.H.: Complex unloading behavior: Nature of the deformation and its consistent constitutive representation. Int. J. Plast. **27**, 1126–1144 (2011). https://doi.org/10.1016/j.ijplas.2010.12.003
4. Li, D., Wagoner, R.H.: The nature of yielding and anelasticity in metals. Acta Mater. **206**, 116625 (2021). https://doi.org/10.1016/j.actamat.2021.116625
5. Kupke, L., Madej, P.D., Hodgson, M., Weiss, experimental in-situ verification of the unloading mechanics of dual phase steels. Mater. Sci. Eng. A **760**, 134–140 (2019). https://doi.org/10.1016/j.msea.2019.05.091
6. Zhou, G., et al.: A predictive strain-gradient model with no undetermined constants or length scales. J. Mech. Phys. Solids. **145**, 104178 (2020). https://doi.org/10.1016/j.jmps.2020.104178
7. Bong, H.J., Lim, H., Lee, M.-G., Fullwood, D.T., Homer, E.R., Wagoner, R.H.: An RVE procedure for micromechanical prediction of mechanical behavior of dual-phase steel. Mater. Sci. Eng., A **695**, 101–111 (2017). https://doi.org/10.1016/j.msea.2017.04.032
8. Jeong, Y., Tomé, C.N.: An efficient elasto-visco-plastic self-consistent formulation: application to steel subjected to loading path changes. Int. J. Plast. **135**, 102812 (2020). https://doi.org/10.1016/j.ijplas.2020.102812
9. Jeong, Y., Jeon, B., Tomé, C.N.: Finite element analysis using an incremental elasto-visco-plastic self-consistent polycrystal model: FE simulations on Zr and low-carbon steel subjected to bending, stress-relaxation, and unloading. Int. J. Plast. **147**, 103110 (2021). https://doi.org/10.1016/j.ijplas.2021.103110
10. Wen, W., et al.: Mechanical behavior of low carbon steel subjected to strain path changes: experiments and modeling. Acta Mater. **111**, 305–314 (2016). https://doi.org/10.1016/j.actamat.2016.03.075
11. Jeong, Y., Barlat, F., Tomé, C.N., Wen, W.: A comparative study between micro- and macro-mechanical constitutive models developed for complex loading scenarios. Int. J. Plast. **93**, 212–228 (2017). https://doi.org/10.1016/j.ijplas.2016.07.015
12. Kim, H., Barlat, F., Lee, Y., Bin Zaman, S., Lee, C.S., Jeong, Y.: A crystal plasticity model for describing the anisotropic hardening behavior of steel sheets during strain-path changes. Int. J. Plast. **111**, 85–106 (2018). https://doi.org/10.1016/j.ijplas.2018.07.010
13. Daroju, S., Kuwabara, T., Knezevic, M.: Experimental characterization and crystal plasticity modeling of dual-phase steels subjected to strain path reversals. Mech. Mater. **168**, 104293 (2022). https://doi.org/10.1016/j.mechmat.2022.104293
14. Zecevic, M., Korkolis, Y.P., Kuwabara, T., Knezevic, M.: Dual-phase steel sheets under cyclic tension–compression to large strains: experiments and crystal plasticity modeling. J. Mech. Phys. Solids. (2016). https://doi.org/10.1016/j.jmps.2016.07.003
15. Joo, M., et al.: A crystal plasticity finite element analysis on the effect of prestrain on springback. Int. J. Mech. Sci. **237**, 107796 (2023). https://doi.org/10.1016/j.ijmecsci.2022.107796

16. Deng, N., Kuwabara, T., Korkolis, Y.P.: On the non-linear unloading behavior of a biaxially loaded dual-phase steel sheet. Int. J. Mech. Sci. **138–139**, 383–397 (2018). https://doi.org/10.1016/j.ijmecsci.2018.02.015
17. Lebensohn, R.A., Tomé, C.N.: A self-consistent anisotropic approach for the simulation of plastic deformation and texture development of polycrystals: application to zirconium alloys. Acta Metall. Mater. **41**, 2611–2624 (1993). https://doi.org/10.1016/0956-7151(93)90130-K
18. Jeong, Y., Iadicola, M.A., Gnäupel-Herold, T., Creuziger, A.: Multiaxial constitutive behavior of an interstitial-free steel: measurements through X-ray and digital image correlation. Acta Mater. **112**, 84–93 (2016). https://doi.org/10.1016/j.actamat.2016.04.013
19. Jeong, Y., Gnäupel-Herold, T., Barlat, F., Iadicola, M., Creuziger, A., Lee, M.-G.: Evaluation of biaxial flow stress based on elasto-viscoplastic self-consistent analysis of X-ray diffraction measurements. Int. J. Plast. **66**, 103–118 (2015). https://doi.org/10.1016/j.ijplas.2014.06.009
20. Vitzthum, S., et al.: In-situ analysis of the elastic-plastic characteristics of high strength dual-phase steel. Mater. Sci. Eng. A **857**, 144097 (2022). https://doi.org/10.1016/j.msea.2022.144097
21. Abedini, O., Behroozi, M., Marashi, P., Ranjbarnodeh, E., Pouranvari, M.: Intercritical heat treatment temperature dependence of mechanical properties and corrosion resistance of dual phase steel. Mater. Res. **22** (2019). https://doi.org/10.1590/1980-5373-mr-2017-0969
22. Bachmann, F., Hielscher, R., Schaeben, H.: Texture analysis with MTEX- free and open source software toolbox. In: Solid State Phenomena, pp. 63–68 (2010). https://doi.org/10.4028/www.scientific.net/SSP.160.63

Modeling the Effect of Backstress on Springback Predictions in AA 6016-T4 as a Function of Pre-strain

Dane Sargeant[1], Md Zahidul Sarkar[2], Rishabh Sharma[1], Marko Knezevic[1], David Fullwood[2], and Michael Miles[1](✉)

[1] Brigham Young University, Provo, UT 84663, USA
mmiles@byu.edu
[2] University of New Hampshire, Durham 03824, NH, UK

Abstract. This paper explores springback levels in AA6016-T4 sheets that were first pre-strained to various levels in uniaxial, plane-strain, and biaxial tension, then subjected to a pure bending operation and released. Finite element modeling of the pre-strain and bending steps was performed using both isotropic and elasto-plastic self-consistent (EPSC) crystal plasticity approaches. Because the EPSC model incorporates backstresses informed by GND content, the springback predictions were far more accurate than those of the isotropic model. In the unstrained base material, the EPSC predictions had a maximum error of 7% versus experiment, compared to 49% for the isotropic approach. The EPSC model accurately predicted the transition from springforward to springback as a function of applied pre-strain, but only when the backtress term was included in the hardening law. The backstress contribution was also found to be increasing influential in predicting springback as pre-strain levels increased, especially in the case of a uniaxial tension pre-strain.

Keywords: Aluminum Alloys · Sheet Metal · Springback · Pre-Stain · Strain Path Change

1 Introduction

Stamped parts for automotive body assembly often require multi-step forming operations to create complicated geometries. During the stamping process, interactions arise between hardening processes and the evolution of internal stresses that vary with strain path, and strongly influence backstresses that cause springback during unloading [1]. As the strain path can often be very different as a function of location within a complex part, it is critical that combined hardening and internal stress behavior be characterized using both macro and micro approaches to fully understand its impact on springback.

The springback behavior in high strength 6xxx or 7xxx series aluminum can be understood by considering the precipitate-based microstructures in these alloys, and associated dislocation substructures [2–5]. Long-range internal stress development is associated with the presence of geometrically necessary dislocations (GNDs) that form to maintain strain compatibility during deformation [6], that accumulate in pileups at

K. Mocellin et al. (Eds.): ICTP 2023, LNME, pp. 671–681, 2024.
https://doi.org/10.1007/978-3-031-40920-2_70

obstacles such as grain boundaries (GBs) or precipitates [7], or that are associated with substructures within grains [8]. Backstresses build up in the strain gradient regions, due to the interaction between dislocations of the same polarity and contribute to the cumulative hardening behavior of the alloy by opposing the applied resolved shear stress necessary for dislocation slip. However, they also assist dislocation slip upon reversal, or change in strain path, causing nonlinear unloading and softening and/or lower subsequent yield stresses.

In a recent study in AA6016-T4 sheet, GND content was measured by high resolution electron backscatter diffraction (HREBSD), after pre-straining the material in biaxial, plane-strain and uniaxial tension. Backstress levels estimated using these experimentally measured GND were shown to have a significant effect on the elasto-plastic transition and plastic response under subsequent uniaxial tension deformation [9]. Additional studies that included cyclic deformation cycles and various pre-strain conditions in advanced high strength steel (AHSS) alloys and a 7xxx series aluminum showed a decrement in elastic modulus values upon unloading-reloading, and an increase in springback with greater pre-strain levels [10–12]. The importance of including the inelastic recovery, i.e., the drop in elastic modulus after the plastic deformation of the metal alloy, as an important parameter in the prediction of springback has been highlighted previously [11, 12]. Several anisotropic models account for a dynamic modulus in their kinematic hardening law to improve springback simulation predictions [9].

Though these anisotropic models can capture deformation response and springback that occur during and after forming, when correctly calibrated to reflect a particular material and deformation path, they do not accurately capture the effects of strain path changes on nonlinear unloading response if backstress is not incorporated into the hardening law. In a prior study, Yue et al. [12] reported a springback prediction error of approximately 17.5% for an 11% uniaxial pre-strain. Liao et al. [14] performed twist springback simulations after applying uniaxial pre-strain and found that their simulations had an error of 15% and 12% in the rolling (RD) and transverse (TD) directions, respectively. In a similar study performed by Chen et al. [15], errors of 28.6% and 14.6% along the RD and the TD directions were observed due to an over-estimation of elastic modulus values in the model.

In this paper, an EPSC model is modified to incorporate a phenomenological backstress approach, along with a dislocation density-based hardening law, to predict springback in AA6016-T4 sheet material that is pre-strained along different strain paths and then subjected to a pure bending operation. It builds on EPSC models that have been shown to accurately predict the microstructural and mechanical response to strain path changes, including non-linear unloading, residual stress fields, the Bauschinger effect, hardening rates, and texture of the material [10–14].

2 Experimental Methods

2.1 Bending/Springback Experiments

Springback experiments were carried out by bending specimens on a 100-ton Accurpress press-brake, using standard tooling with an 88° bend angle (Fig. 1). Initial trials were performed on unstrained AA6016-T4 sheets, with the bend axis perpendicular to the rolling direction (RD), the transverse direction (TD) and 45° to the RD using the bottom bending tooling. All experiments were performed on specimens with the same geometry, having dimensions of 63.5 mm x 63.5 mm. Subsequent experiments were then performed on specimens that were pre-strained to varying amounts of effective strain (6%, 15%, 20%) relative to the RD direction. Prior to performing a bend experiment, mineral oil was applied to the specimen, as well as to the punch and die set, to minimize friction. To validate bottom bending results, each specimen was tracked via high-contrast HDR video. The punch displacement in which the material and v-die maintain full contact were captured. Thickness measurements were taken after bottom bending experiments over five points along the specimen and compared to the unstrained material to ensure no thinning (coining) occurred at the bottom of the punch stroke. The included angle of the specimen, before and after unloading, was used to evaluate springback. So any deviation with respect to the 88° die angle represents either springforward (included angle of specimen less than 88°) or springback (included angle of specimen greater than 88°).

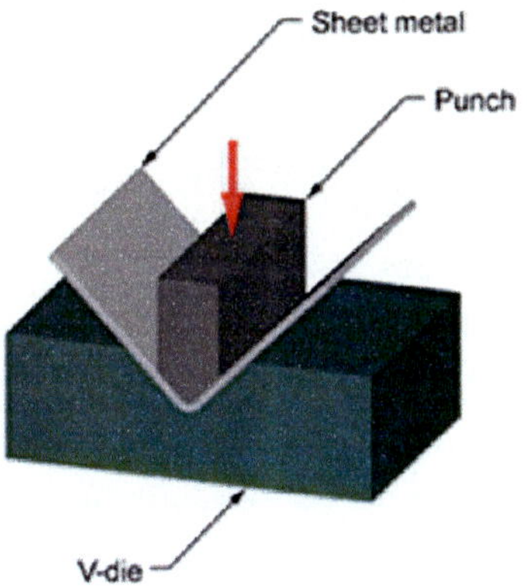

Fig. 1. Punch and die configuration for bending/springback experiments. The punch/die bend angle is 88° and the punch tip radius is 0.8 mm.

2.2 Pre-strain Experiments

Specimens were pre-strained in uniaxial, plane-strain, and biaxial tension prior to bendang/springback experiments. Specimen dimensions are shown in Fig. 2 for each strain path. Standard Marcinak tooling, with carrier blank of the same alloy and thickness between the sheet and the tooling, was used to impart plane-strain and biaxial tension pre-strains, while a large tensile specimen was used for uniaxial pre-strains.

Figure 3a shows the Marcinak tooling used to apply the plane-strain and biaxial tension pre-strains, using the specimens from Fig. 2a and 2b. An example of pre-strain level measured by digital image correlation (DIC) for a biaxial pre-strain of 15% effective strain, is shown in Fig. 3b. The pre-strains for uniaxial tension, performed with the specimen shown in Fig. 2c, were done using an Instron test frame, where specimen dimensions were measured to confirm each pre-strain level.

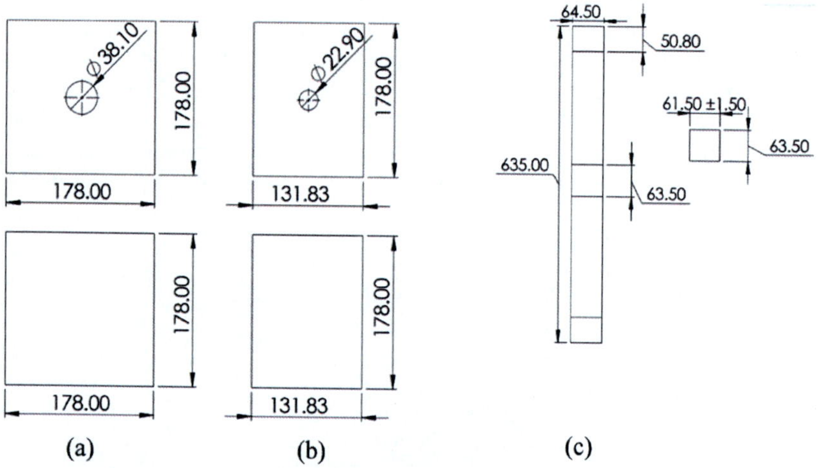

Fig. 2. Specimens used for pre-strain experiments for (a) biaxial tension, (b) plane-strain tension, and (c) uniaxial tension.

After pre-straining, square specimens with dimension 63.5 mm x 63.5 mm were cut from Marcinak specimens and used for bending/springback experiments (as depicted in Fig. 1). In the case of the uniaxial pre-strain, the specimen width started at about 64.5 mm and narrowed slightly to between 60 mm and 63 mm after testing; this small difference in width compared to the plane-strain and biaxial tension specimens was not considered significant in terms of the bending results for these thin sheets.

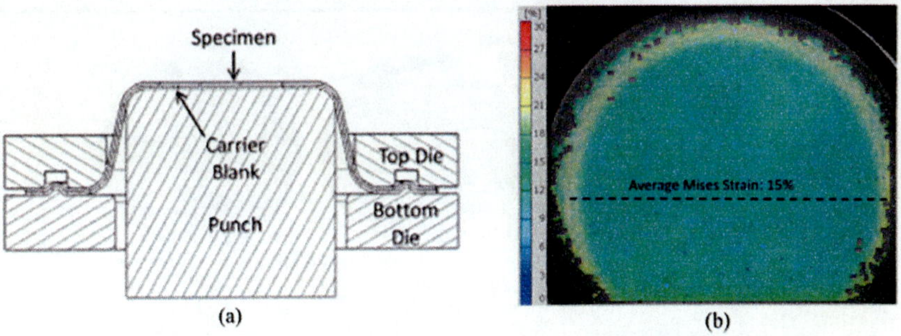

Fig. 3. (a) Marciniak tooling for biaxial and plane-strain tension testing, and (b) DIC strain map for a 15% pre-strained biaxial-tension specimen. Bend specimens were cut from the center of the flat portion of the specimens, after applying different levels of plane-strain and biaxial pre-strain.

2.3 Material Characterization

The AA6016-T4 alloy used for the experimental work had a 1 mm thickness and was provided by Commonwealth Rolled Products. Stress-strain curves produced by uniaxial tension tests provided material properties necessary for initial characterization. Specimens were cut from the sheet along the rolling direction (RD), transverse direction (TD), and 45° to the RD of the sheet according to the ASTM E8 specification. The specimens were pulled at a 1.5 mm/min crosshead displacement rate and an extensometer was used for strain measurement (Table 1).

Table 1. Base material tensile properties for AA 6016-T4 material.

Orientation	Young's modulus (GPa)	True failure strain	Yield strength (MPa)	Tensile strength (GPa)
RD	70.5	0.270	142	344
45° to RD	70.8	0.245	140	321
TD	71.7	0.249	140	326

Additional characterization was carried out to generate the texture data necessary for the EPSC simulations. Base material specimens were prepared for electron backscatter diffraction (EBSD) and then scanned to obtain data for approximately 1000 grains, which were subsequently compacted to 100 grains for each orientation [15]. The compaction is a necessary step to decrease the overall time of each simulation; as can be seen in Fig. 4, there is no visible difference between the original and the compacted pole figures, for the unstrained base material (100% accurate to L = 16 GSH expansion).

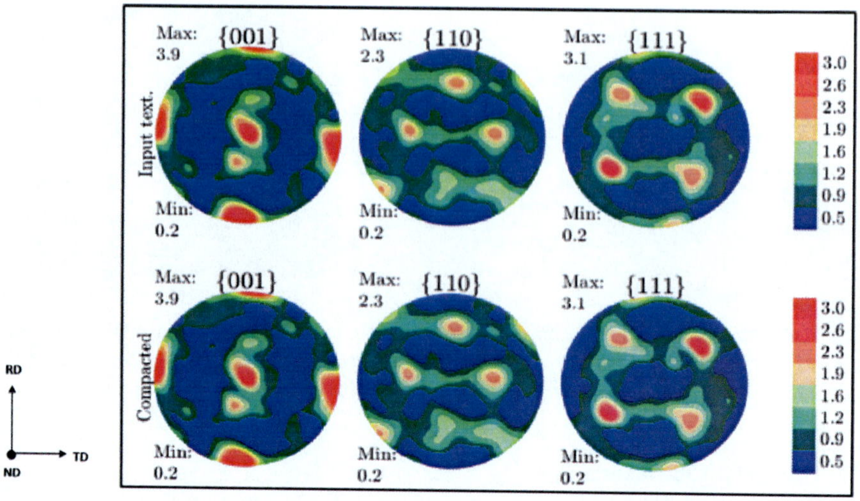

Fig. 4. Original texture for AA6016-T4 comprising ~ 1000 grains (top), and texture data compacted to 100 grains (bottom) for EPSC simulations.

3 Modeling Approach

The simulations were performed using an EPSC polycrystal plasticity model that was developed by Zecevic and Knezevic [14] and has been applied to the modeling of plastic deformation under different strain paths and for cyclic loading of 6016 Al alloy [16, 17]. Key details of the model, including the strain path sensitive hardening law with backstress evolution, and associated latent hardening parameters shown in Table 2, can be found in these prior papers. It was found that calibration of the backstress for AA 6016-T4 corresponded to a dislocation density that was reasonable, when considering both GND measurements done by HR-EBSD for this alloy, as well as XRD measurements that were performed in other studies for similar alloys [17]. As such, the parameters shown in Tables 3 and 4 were retained for the current work.

Table 2. Latent hardening parameters.

a_0	a_1	a_2	a_3	a_4	a_5
0.068	0.068	0.0454	0.625	0.137	0.122

In this paper, most EPSC simulations were performed with the slip system backstress component included in the hardening law, but modeling without backstress was also performed for comparison. In addition, some standard J2 plasticity simulations were run with Abaqus standard (the latter was only used for unstrained base material bending and springback simulations). For cases where a pre-strain was desired, the appropriate boundary conditions were applied to the sheet model in order to reach various plastic effective strain levels, prior to simulating the bending/springback step.

Table 3. Fitting parameters used for the evolution of slip resistance.

τ_0 [MPa]	$k_1[m^{-1}]$	g	D [MPa]	$\rho^s_{for}\left[m^{-2}\right]$
25	1.38 x10^8	0.09	400	4.1e12

Table 4. Fitting parameters used for the evolution of slip system backstress.

τ^{sat}_{bs} [MPa]	ν	γ_b	A
5	560	0.001	1

4 Results and Discussion

A comparison of the J2 plasticity simulation results with the EPSC predictions is shown in Fig. 5, with respect to experimental measurements. This comparison is for springback angle predictions after the bending operation with the base material, ie. no prestrain applied prior to bending/springback. For the base material, all specimens exhibited springforward, where the included angle after unloading is smaller than that of the die (88°).

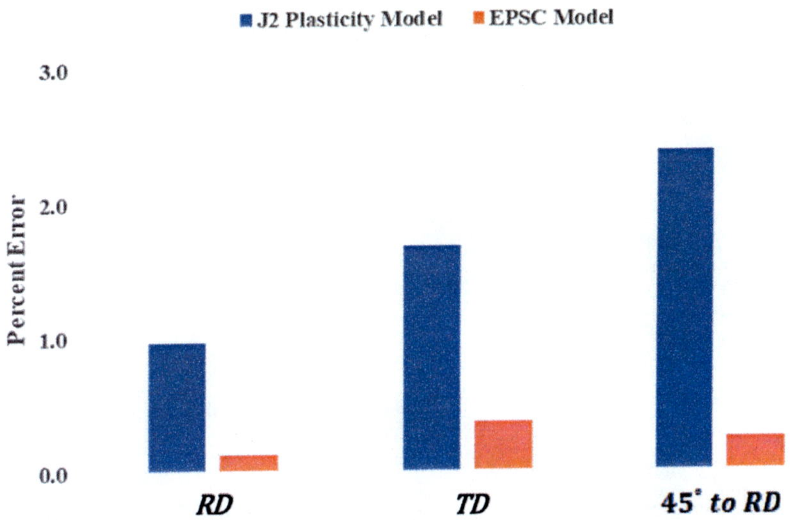

Fig. 5. Percent error of J2 and EPSC model predictions of springback angle, with respect to experiment.

While the J2 isotropic modeling approach captured the general spring-forward behavior after bending, it provided overestimates of spring-forward values. Overestimation of springback/spring-forward in aluminum alloys has been documented in prior work, when isotropic modeling approaches were employed [18]. The predicted values from the J2 isotropic model differed with experiment by as much as 2.4% (2.0°), for the 45° to RD specimen. By contrast, the EPSC predictions that included a backtress component within the hardening law, representing kinematic hardening behavior, were close to the experimental values; at worst, within 0.4% (0.3°) for the TD specimen. Percent error terms were calculated with respect to the included angle of the bend specimen, after unloading.

Given the superior accuracy of the EPSC model for the unstrained base material predictions, further work was done comparing EPSC predictions with experiments where pre-strain was first applied to the sheets, along different strain paths, before performing bending experiments and springback measurements. Correlation between EPSC springback predictions and experiment, for all pre-strained specimens, is shown in Fig. 6. Pre-strains along uniaxial, plane-strain, and biaxial tension strain paths were applied to the sheet for three levels of effective strain: 6%, 15%, and 20%. The strength of the correlation is very good, at 91%, even if the simulated angles produced a trendline with a slope a bit greater than 45°, indicating a mild overprediction of springback by the model.

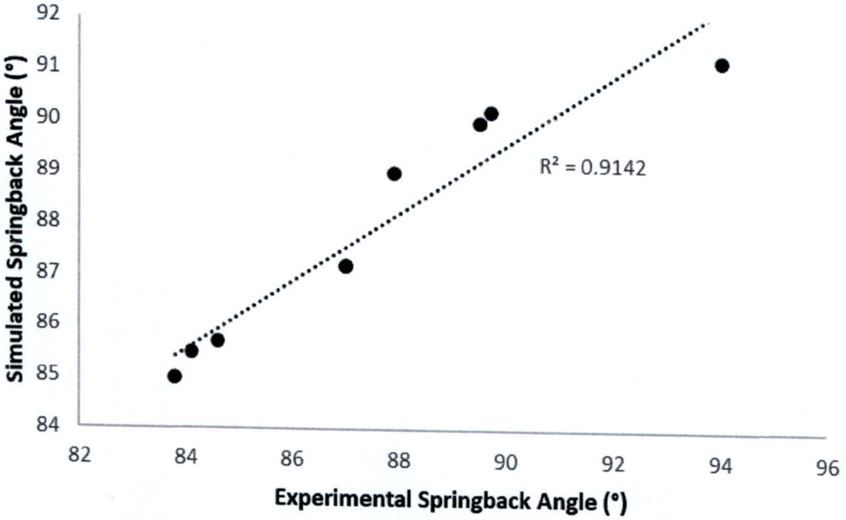

Fig. 6. Correlation between model predictions and experiments, for all pre-strain paths (uniaxial, plane-strain, biaxial) and levels (6%, 15%, and 20%).

Given the relatively small absolute values of the experimental springback angles, the differences between simulation and experiment appear to be reasonable and consistent, and most likely within tolerance when considering manufacturing assembly requirements.

The benefit of modeling backstress in the hardening law was also assessed by performing EPSC simulations both with, and without, this term. A comparison of both cases, against experimental springback angles, is shown in Fig. 7. The predictions are

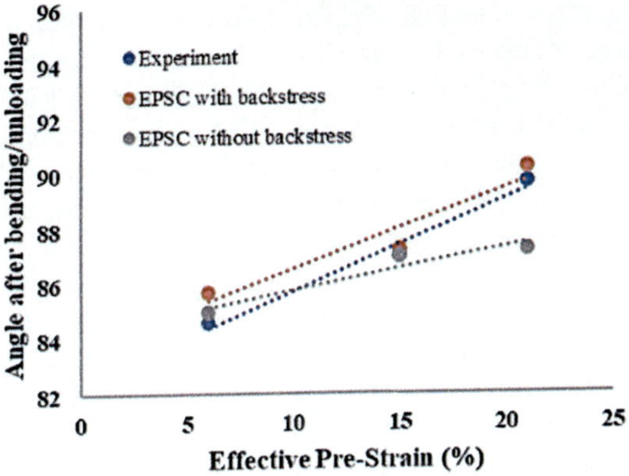

(a)

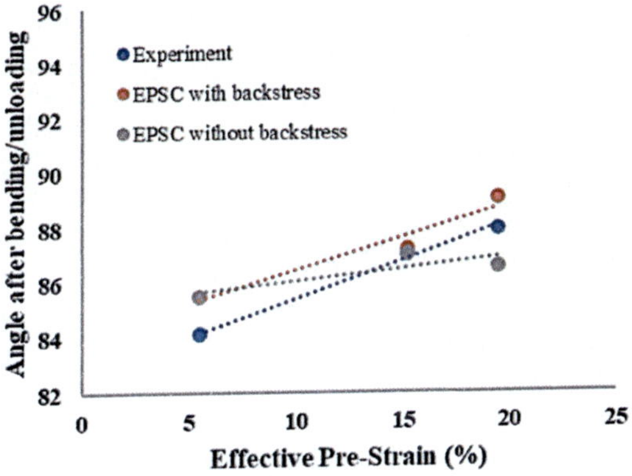

(b)

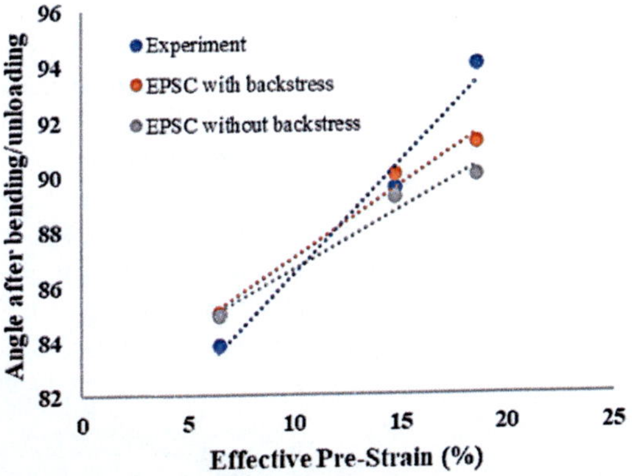

(c)

Fig. 7. Spring-forward/springback measurement comparison for sheets pre-strained in (a) uniaxial tension, (b) plane-strain tension, and (c) biaxial tension. EPSC simulations with, and without, backstress are compared to experiment.

clearly more accurate when including the backstress contribution in the EPSC simulation, and the influence of the backtress term on computation accuracy grows as effective strains increase. This is especially true for the uniaxial and plane-strain tension pre-strains, where the EPSC predictions with backstress provide trendlines parallel to the experimental data, while the EPSC predictions without backstress have trendlines that diverge from the experimental ones. While the springforward (negative angle) to springback (positive angle) transition was predicted by both models for the biaxial pre-strain experiments, the model without backstress did not accurately predict this transition for uniaxial and plane-strain tension pre-strains, as can be seen in Fig. 6, where best fit lines diverge at high levels of pre-strain. EPSC predictions without backstress do not accurately capture the relationship between pre-strain type/level and springback angle after bending, especially in the case of uniaxial tension and plane strain tension. In these last cases, the transition from springforward to springback is not predicted when backstress is not modeled, i.e. predicted specimen angles are less than 88° (springforward) and do exceed 88° (springback) at higher pre-strain levels. This is especially true for uniaxial and plane-strain tension pre-strains. When backstress is included in the model hardening law, the springforward to springback transition is predicted reasonably well.

5 Summary and Conclusions

Bending and springback experiments were performed on 1 mm thick AA 6016-T4 sheets, where a punch formed the sheets into an 88° v-block die, then released the material, allowing it to springback. Careful measurements documented springback angles for the base material specimens oriented along the RD, TD, and 45 deg directions (bend axis perpendicular to each of these directions). Additional experiments were conducted where pre-strains were first imparted to the sheet specimens, using Marcinak tooling to apply plane-strain and biaxial tension pre-strains and a large tensile specimen to impart uniaxial tension pre-strains. Specimens cut from these pre-strained sheets were then used for bending/springback experiments, performed as described earlier. A crystal plasticity EPSC code with the ability to model the directional evolution of dislocation density and backstress during plastic deformation was employed to simulate the bending/springback experiments, for both the base material and for pre-strained specimens.

It was found that the EPSC model with backstress provided accurate predictions of experimental spring-forward angles in the base material, with an error of 0.4% (0.3°) in the worst case, while the isotropic model overpredicted springforward angles by as much as 2.4% (2°). The EPSC model was also accurate in predicting the move from spring-forward (negative angle) to springback (positive angle), as a function of applied pre-strain; but this was not the case when the backtress term was excluded from the model. The effect of backstress was also seen to be increasingly influential to prediction accuracy for greater levels of pre-strain, especially in the case of a uniaxial tension pre-strain.

References

1. Boers, S.H.A., et al.: Experimental characterization and model identification of directional hardening effects in metals for complex strain path changes. Int. J. Solids Struct. **47**(10), 1361–1374 (2010)
2. El-Madhoun, Y., Mohamed, A., Bassim, M.: Cyclic stress–strain response and dislocation structures in polycrystalline aluminum. Mater. Sci. Eng. A **359**(1–2), 220–227 (2003)
3. Moan, G., Embury, J.: A study of the bauschinger effect in Al-Cu alloys. Acta Metall. **27**(5), 903–914 (1979)
4. Stoltz, R., Pelloux, R.: Cyclic deformation and bauschinger effect in Al-Cu-Mg alloys. Scr. Metall. **8**(3), 269–275 (1974)
5. Stoltz, R.E., Pelloux, R.M.: The Bauschinger effect in precipitation strengthened aluminum alloys. Metall. Trans. A **7**(8), 1295–1306 (1976)
6. Ashby, M.F.: The deformation of plastically non-homogeneous materials. Phil. Mag. **21**, 399–424 (1970)
7. Shen, Z., Wagoner, R.H., Clark, W.A.T.: Dislocation and grain boundary interactions in metals. Acta Metall. **36**(12), 3231–3242 (1988)
8. Field, D.P., et al.: Analysis of local orientation gradients in deformed single crystals. Ultramicroscopy **103**, 33–39 (2005)
9. Wagoner, R.H., Lim, H., Lee, M.-G.: Advanced Issues in springback. Int. J. Plast **45**, 3–20 (2013)
10. Neil, C.J., et al.: Modeling lattice strain evolution at finite strains and experimental verification for copper and stainless steel using in situ neutron diffraction. Int. J. Plast **26**(12), 1772–1791 (2010)
11. Wollmershauser, J.A., Clausen, B., Agnew, S.R.: A slip system-based kinematic hardening model application to in situ neutron diffraction of cyclic deformation of austenitic stainless steel. Int. J. Fatigue **36**(1), 181–193 (2012)
12. Zecevic, M., Knezevic, M.: A dislocation density based elasto-plastic self-consistent model for the prediction of cyclic deformation: application to AA6022-T4. Int. J. Plast **72**, 200–217 (2015)
13. Zecevic, M., et al.: Dual-phase steel sheets under cyclic tension–compression to large strains: experiments and crystal plasticity modeling. J. Mech. Phys. Solids **96**, 65–87 (2016)
14. Zecevic, M., Knezevic, M.: An implicit formulation of the elasto-plastic self-consistent polycrystal plasticity model and its implementation in implicit finite elements. Mech. Mater. **136**, 103065 (2019)
15. Eghtesad, A., Barrett, T.J., Knezevic, M.: Compact reconstruction of orientation distributions using generalized spherical harmonics to advance large-scale crystal plasticity modeling: verification using cubic, hexagonal, and orthorhombic polycrystals. Acta Mater. **155**, 418–432 (2018)
16. Daroju, S., et al.: Experimental characterization and crystal plasticity modeling for predicting load reversals in AA6016-T4 and AA7021-T79. Int. J. Plast **153**, 103292 (2022)
17. Sharma, R., et al.: Multi-strain path deformation behavior of AA6016-T4: experiments and crystal plasticity modeling. Int. J. Solids Struct. **244**, 111536 (2022)
18. Li, K.P., Carden, W.P., Wagoner, R.H.: Simulation of springback. Int. J. Mech. Sci. **44**(1), 103–122 (2002)

Honorary Symposium for Prof. Lang Li Hui

Prof. Lihui Lang and His Contribution to Metal Forming Technology

Yong Li, Meng Zhang, Yi Xiao, Dongdong Yan, and Xiaoqiang Li(✉)

School of Mechanical Engineering and Automation, Beihang University, Beijing 100191, China
lixiaoqiang@buaa.edu.cn

Abstract. Professor Lihui Lang is an internationally renowned scholar in the field of metal forming technology. He was deputy director of Youth Work Committee of China Society for Technology of Plasticity, editorial board members of many scientific journals and chair of many international conferences. During his more than 20 years' research in theories, technologies and equipment in hydroforming and powder hot isostatic pressing field, he has published about 10 textbooks and monographs, more than 300 academic papers, obtained more than 40 authorized invention patents, and cultivated nearly 100 doctoral and master students. Based on these contributions, he won international and national awards 3 times, including the Thomas Stephen Group Prize and second prize of Chinese National Award for Technological Inventions.

Keywords: Metal forming · Hydroforming · Powder hot isostatic pressing

1 Introduction

Lihui Lang is a professor in the School of Mechanical Engineering and Automation of Beihang University, who died in Beijing on June 28, 2022 with only 52 years old. He is an internationally renowned scholar in the field of metal forming technology. After obtaining a doctorate degree from Harbin Institute of Technology in 1998, he served as an associate professor at Aalborg University in Denmark from 2001 to 2004 and a professor at Beihang University in China from 2004. He was deputy director of Youth Work Committee of China Society for Technology of Plasticity, editorial board members of many international journals (e.g., International journal of Machine Tools and Manufacture) and Chinese journals (e.g. Forging Technology and Journal of Plastic Engineering), and chair of many international conferences (e.g., The 3rd International Conference on Advanced Composite Materials, ICACM 2020).

During his more than 20 years academic and education life in metal forming field, he has published many textbooks and monographs, such as "Forging Handbook (Stamping Volume)", "Advanced Hydroforming Technology for Sheet Metal" and "CAE Design and Application of Sheet Metal Forming - Based on DYNAFORM", more than 300 academic papers, obtained more than 40 authorized invention patents, cultivated nearly 100 doctoral and master students and won international and national awards 3 times,

K. Mocellin et al. (Eds.): ICTP 2023, LNME, pp. 685–690, 2024.
https://doi.org/10.1007/978-3-031-40920-2_71

including the Thomas Stephen Group Prize and second prize of Chinese National Award for Technological Inventions.

Prof. Lang's research focused on investigation and development of theories, technologies and equipment for hydroforming process of sheet metals, and he is also one of the pioneers in the world to utilize hydroforming process to manufacture fiber metal laminates. His research outcome directly supported the equipment development of hydroforming technology and their applications in aerospace and car manufacturing field. In the last 10 years, he started a new research direction of powder hot isostatic pressing (HIP) of titanium alloy parts, provided new thoughts in the process optimization and equipment development for potential aerospace products.

2 Research on Metal Forming Technology and Applications

2.1 Hydroforming of Lightweight Metals

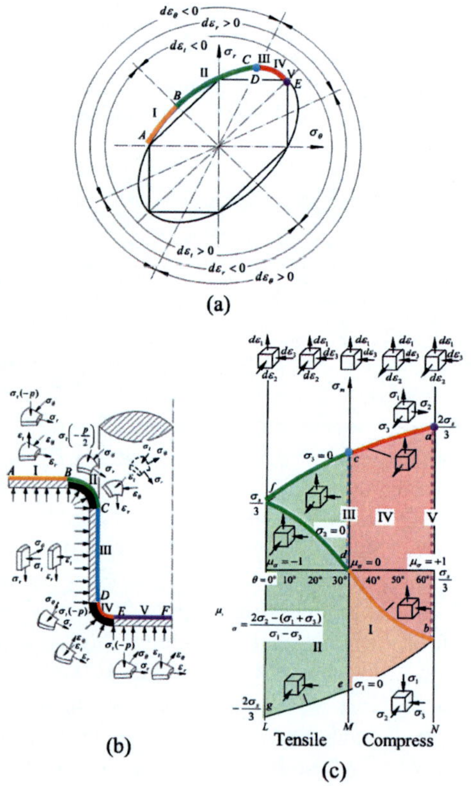

Fig. 1. Stress distribution of typical deformation zone of sheet hydroforming. (a) Distribution of cup part along the yield surface of Tresca and Mises plane stress, (b) typical stress-strain state of cup part in hydroforming considering normal stress and (c) distribution of stress in plane formed by Mises yield surface

Some research has been documented the accuracy and useful of plan-stress in sheet forming, while it may not suitable for hydroforming, especial in high pressure forming cases, due to the existence of normal stress cannot be ignored. Prof Lang presented the theoretical formula of springback in hydroforming with normal stress [1]. Compared with plane stress, as shown in Fig. 1 (a), it is added that the influence of the internal stress on the opposite side of the third direction stress leads to the change of stress distribution interval, which improves the calculation accuracy [2]. Moreover, the introduction of normal stress also helped to explain the higher forming limited curve benefits from hydraulic liquid [3]. In order to verify the above theory and obtained the exact material property in hydroforming, tube bulging and sheet bulging test platform were developed by Prof. Lang respectively [4]. In addition, based on these theoretical and technical outcome, Prof. Lang led his team to develop a various set of hydroforming equipment and solved many process problems in hydroforming of complex thin-walled structures. His research has significantly promoted the widely applications of hydroforming on complex thin-walled parts in aerospace industry, as shown in Fig. 2.

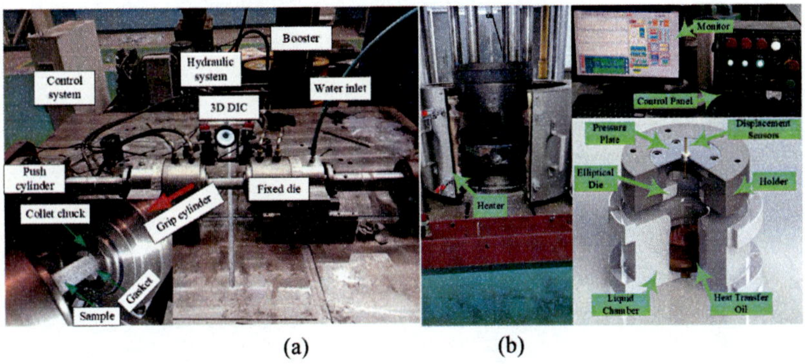

Fig. 2. Hydro-bulging test platform. (a) Tube hydro-bulging test platform and (b) sheet hydro-bulging test platform

2.2 Hydroforming of Fiber Metal Laminates (FMLs)

The FMLs traditional forming technology mostly are roll bending, laser forming and shot peening technology. However, the disadvantage of traditional forming technology is that they can only form simple parts with large size and small curvature. It is difficult to form complex structural parts. Hydroforming technology was initiated by professor Lang to manufacture FMLs parts [5], and successfully realized the forming and curing of complex structural parts.

Typical FMLs hydroforming technology uses liquid as a flexible punch or die to realize the formation of FMLs. First, the metal materials in the FMLs are subject to surface phosphoric acid anodizing treatment. Then the treated metal is paved with glass fiber prepreg to prepare semi-cured FMLs. By controlling the cavity pressure, blank holding force and pre-bulging pressure [6] to manufacture the FMLs parts. The schematic diagram of hydroforming of FMLs is shown in Fig. 3 (a) [7]. FMLs hydroforming technology can reduce the die cost and improves the application range of FMLs.

Prof Lang presented the theoretical formula of the relationship between zonal stress $\sigma^i{}_z(\theta)$, length of the three axis of ellipsoid a, b and h ($a > b > h$), thickness of layer t^i and friction coefficient of Γ^i as shown in Fig. 3 (b). The mechanism helps us understand why fracture always occurs at the polar point, and can reduce the stress at the polar point, prevent fracture and improve the forming limit by adjusting the tangential cohesion. Since 2015, Prof. Lang led the group to carry out a large number of numerical simulation and forming tests, as shown in Fig. 3 (c), which has been successfully applied to a certain type of aircraft.

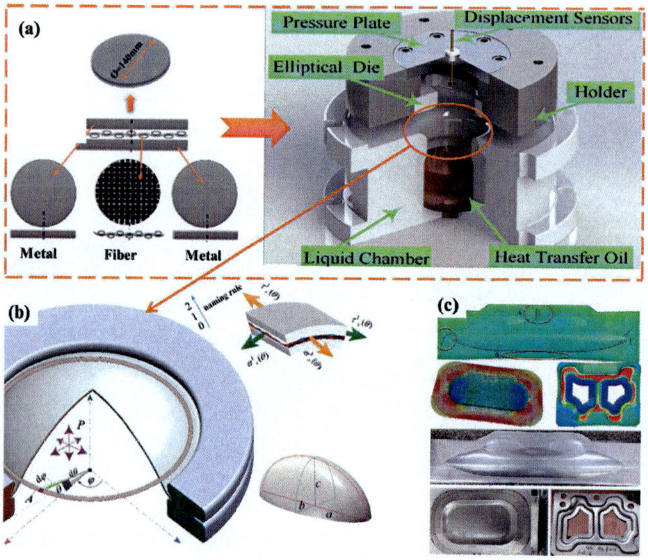

Fig. 3. (a) Schematic diagram of FMLs hydroforming technology, (b) uniform pressure loading test (UPLT) FMLs stress analysis, (b) FMLs assembly, (c) typical complex structural parts

2.3 Powder Hot Isostatic Pressing of Titanium Alloys

In the past ten years, Prof. Lang began to pay attention to hot isostatic pressing (HIP) technology to provide new thoughts for potential aerospace products. As shown in Fig. 4, he has carried out systematic research (including theory, numerical simulation, experiment, equipment, and application, etc.) to solve the scientific problems of the powder hot isostatic pressing technology, in which typical aerospace materials are involved, such as titanium alloys, aluminum alloys, superalloys, and tungsten alloys, etc. He participated in the design of large-scale HIP equipment, in which he has done a lot of work on scheme formulation and structural design [8]. He studied the densification behavior of powder during hot isostatic pressing, and the plastic flow rule and constitutive model were established, which was of valuable reference to the development of the powder densification theory [9]. He proposed an inverse modeling for the numerical simulation of powder forming based on the genetic algorithm, by which the constitutive parameters of powder materials could be conveniently and effectively calculated [10]. Based on a mass of experiments, he put forward appropriate powder hot isostatic pressing process systems and specifications for various materials [11, 12]. As a result, he led his team to form a variety of high-performance complex parts, which has effectively promoted the application of hot isostatic pressing in aerospace products.

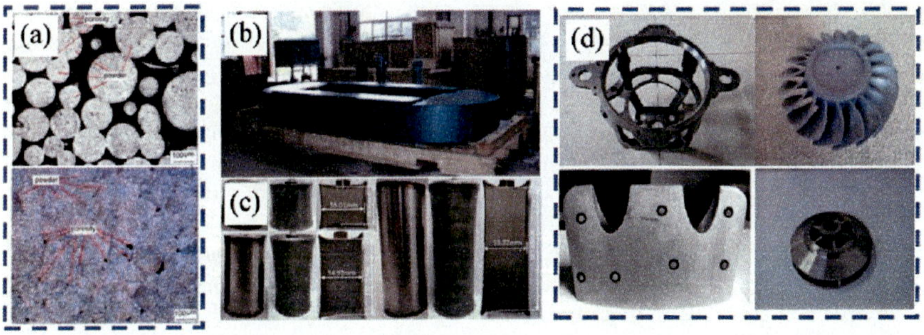

Fig. 4. Prof. Lang's contributions to powder hot isostatic pressing: (a) densification behavior; (b) equipment; (c) deformation behavior; (d) applications.

3 Summary

This paper is written in the memory of Professor Lihui Lang, who is an internationally renowned scholar in the field of metal forming technology. During his more than 20 years' research and teaching life, Prof. Lang's focused on investigation and development of theories, technologies and equipment for hydroforming process of sheet metals, as well as fiber metal laminates. The research outcome directly supported the equipment development of hydroforming technology and their applications in aerospace and car manufacturing field in China. Based on these outcomes, he has published about 10 textbooks and monographs, more than 300 academic papers, obtained more than 40 authorized invention patents, and cultivated nearly 100 doctoral and master students. His spirit in research set a good example for us and would stimulate our further research and exploring in the metal forming field.

References

1. Zhang, Q., Lang, L., Wang, Y., Sun, Z.: Theoretical investigation on the springback behavior of AA7B04 sheet in hydraulic bulge process. Int. J. Adv. Manuf. Technol. **87**(9–12), 2861–2871 (2016). https://doi.org/10.1007/s00170-016-8689-y
2. Sun, Z., Lang, L.: Effect of stress distribution on springback in hydroforming process. Int. J. Adv. Manuf. Technol. **93**(5–8), 2773–2782 (2017). https://doi.org/10.1007/s00170-017-0670-x

3. Liu, B., et al.: Forming characteristic of sheet hydroforming under the influence of through-thickness normal stress. J. Mater. Process. Technol. **212**(9), 1875–1884 (2012)
4. Wang, Y., et al.: Flow property of AA2B06 sheet using overlapping elliptical bulge test. Trans. Nonferrous Met. Soc. China **26**(8), 2179–2187 (2016)
5. Liu, S., Lang, L., Guan, S., et al.: Investigation into composites property effect on the forming limits of multi-layer hybrid sheets using hydroforming technology. Appl. Compos. Mater. **26**(1), 205–217 (2019)
6. Li, L., Lang, L., Hamza, B., et al.: Effect of hydroforming process on the formability of fiber metal laminates using semi-cured preparation. Int. J. Adv. Manuf. Technol. **107**(9), 3909–3920 (2020)
7. Meng, Z., Mirza, H., Zhang, Y., et al.: Investigation into a comprehensive forming limit curve considering multi-parameters effects for semi-cured FMLs through uniform pressure loading test. Int. J. Damage Mech. **32**, 1281893270 (2022)
8. Li, H.L., Lang, L.H., Zhang, J.Y., Yang, H.: Cost optimization method of large-scale pre-stressed wire winded framework on multiple-island genetic algorithm. Chin. J. Aeronaut. **24**(5), 673–680 (2011)
9. Xue, Y., Lang, L.H., Bu, G.L., Li, L.: Densification modeling of titanium alloy powder during hot isostatic pressing. Sci. Sinter. **43**(3), 247–260 (2011)
10. Duan, W., Lang, L.H., Wang, G., Huang, X.N., Liu, J.J.: Inverse method of powder material parameters based on numerical simulation and genetic algorithm. Forging Stamping Technol. **41**(12), 131–137 (2016). (in Chinese)
11. Huang, X.N., Lang, L.H., Wang, G., Alexandrov, S.: Effect of powder size on microstructure and mechanical properties of 2A12Al compacts fabricated by hot isostatic pressing. Adv. Mater. Sci. Eng. **2018**, 1989754 (2018)
12. Huang, X.N., Lang, L.H., Wang, G.: Effect of HIP post-treatment on the HIPed Ti6Al4V powder compacts. Powder Metall. **62**(1), 8–14 (2019)

A Study of Internal Defects in Flexible Medium Hydroforming of Complex Structures of Fiber Metal Laminates

Dongdong Yan[1](✉), Yong Li[1], Chiye Zhang[2], Sanmin Zhang[1], and Lihui Lang[1]

[1] School of Mechanical Engineering and Automation, Beihang University, Beijing, China
ydd2007@buaa.edu.cn

[2] Academy of Opto-Electronic, China Electronic Technology Group Corporation, Beijing, China

Abstract. A flexible medium hydroforming and curing (FMHC) process for fiber metal laminates (FMLs) was proposed and utilized to manufacture FMLs parts with complex structures. The formation mechanisms of wrinkling defects in the FMHC process of complex FMLs structure were studied through both numerical simulation and experimental methods. The results show that the compressive stress will occur at the T-shaped cavity when forming complex FMLs parts by FMHC process, which will cause wrinkling defects. To further eliminate the wrinkling, the steel-plate assisted flexible medium hydroforming and curing (SPAFMHC) process was proposed and validated. Industrial CT scanning technology was used to further analyze the distribution of internal defects in the FMLs parts formed by SPAFMHC. The results show that the defects are mainly concentrated in the corner of the ∞-shaped cavity, and the internal defects of the FMLs parts account for 4.7% of the total area of the part; The defects position was cut and observed. It is found that internal porosity and delamination are the main forms of internal defects.

Keywords: Flexible Medium Hydroforming · Fiber Metal Laminates · Complex Structure · Defects

1 Introduction

Fiber metal laminates (FMLs), as a lightweight hybrid composite material, have the advantages of high specific strength, high specific stiffness, high damage tolerance, and fatigue resistance compared with metal materials [1, 2]. FMLs has been widely concerned by researchers. However, due to the limitations of forming technology, at present, only simple structural parts with large size and small curvature characteristics can be manufactured, which limits the use and promotion of FMLs. It is necessary to research on the forming technology of FMLs parts with complex structures [3].

The traditional FMLs forming technology mainly include roll bending process [4, 5], laser forming process [6] and shot peening process [7, 8]. In recent years, more and more researchers have begun to explore the possibility of manufacturing FMLs parts using hydroforming process. Flexible medium hydroforming process is an advanced forming

K. Mocellin et al. (Eds.): ICTP 2023, LNME, pp. 691–698, 2024.
https://doi.org/10.1007/978-3-031-40920-2_72

technology, which can be used to form complex structural parts of sheet metal. It has the characteristics of good forming surface quality, high forming accuracy and excellent economy [9, 10]. Liu et al. [11] studied the influence rule of forming limit and fracture form of FMLs hemispherical parts formed by hydroforming process. Li et al. [12] studied the influence factors of wall thickness reduction rate, stress and strain distribution and fracture mode of semi-cured FMLs cup shaped parts formed by hydroforming process. Blala et al. [13] studied the influence of parameters such as blank holder force, cavity pressure, die radius, blank holder gap and stamping speed on the wall thickness reduction rate and wrinkling when forming semi-cured FMLs cup parts with hydroforming process. Although researchers have carried out a lot of research on the manufacturing of FMLs parts by hydroforming process, but the main focus is on the basic tests such as cup parts and free bulging [14], and the research that aims to manufacture complex structure parts is relatively rare, so it is necessary to further study the possibility of manufacturing FMLs parts with complex structures by hydroforming process.

In this paper, a new hydroforming process was proposed for manufacturing FMLs parts with complex structures. The wrinkling defect and formation mechanism of FMLs complex parts manufactured by conventional and new hydroforming processes were compared by numerical simulation and forming test. The CT scanning of FMLs complex parts was carried out, and the distribution and formation mechanism of internal porosity and delamination defects of FMLs complex structure after forming and curing were analyzed.

2 Material and Method

2.1 Material

The FMLs used in the test are sandwich structures, and the laminates are composed of two layers of aluminum plates and one layer of glass fiber prepreg, as shown in Fig. 1a. The aluminum is 2A12-T4 plate with thickness of 0.5 mm, and the material properties are shown in Table 1; The fiber layer material is the plain woven glass fiber prepreg WP-9011 purchased from Weihai Guangwei Composite Material Co., Ltd., Weihai, China. The glass fiber prepreg thickness is 0.2 mm. Table 1 provides the prepreg properties. The structure of aluminum alloy plate and glass fiber prepreg blank are shown in Fig. 1b, with a dimension of 380 mm in length and 300 mm in width.

Table 1. Mechanical property of 2A12-T4, WP-9011 and steel.

Material	Yield strength	Tensile strength	Elastic modulus	Fracture strain
2A12-T4	306 ~ 315 MPa	467 ~ 472 MPa	68 GPa	19%
WP-9011		450 MPa	18 GPa	
steel	151 MPa	302 MPa	20.4 GPa	43%

2.2 Forming Method and Die Design

The complex structure of the part is shown in Fig. 2a, with a length of 340 mm and a width of 254 mm. The deep drawing ratio of stamping forming can be calculated by Eq. (1). D_R is the deep drawing ratio, A_{blank} is the area of the blank, A_{punch} is the forming area. The number of A_{blank} is 92123.71 mm^2 exported from CATIA, the number of A_{punch} is 15571.15 mm^2 exported from CATIA, the number of D_R is 2.4. Using stamping form process it is hard to form a structure with a deep draw ratio of 2.4.

$$D_R = \sqrt{\frac{A_{blank}}{A_{punch}}} \tag{1}$$

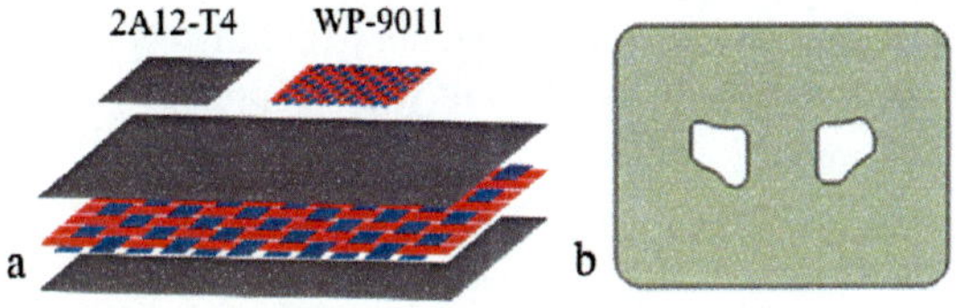

Fig. 1. (a) Schematic diagram of FMLs structure, (b) the blank structure.

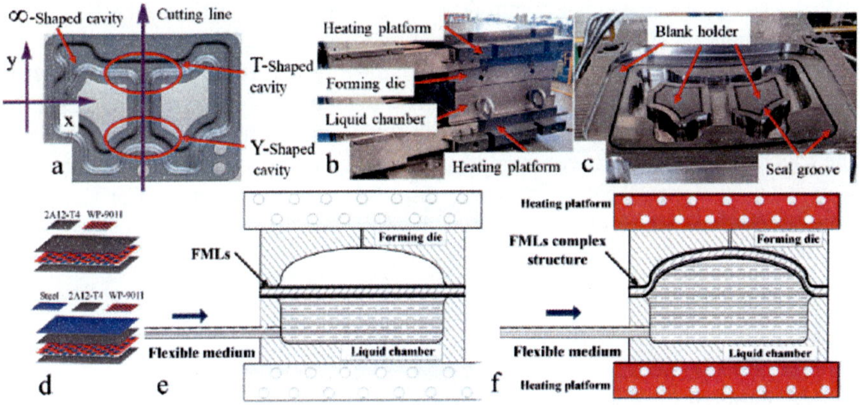

Fig. 2. (a) Complex structure part, (b) forming and curing integrated die, (c) liquid chamber seal groove, (d) schematic diagram of FMLs laminates and auxiliary FMLs laminates with steel plates, (e) hydroforming process, (f) curing process of FMLs laminates after forming.

The flexible medium hydroforming process is used to form this part. Active flexible medium hydroforming process use single action hydraulic press to manufacturing parts. The requirements for forming equipment are low. This paper adopts the active flexible medium hydroforming process to form this complex structural part. The forming schematic diagram is shown in Fig. 2d, e, f. In this paper, two kinds of hydroforming process are performed. One is flexible medium hydroforming and curing (FMHC) process and the other one is a steel-plate assisted flexible medium hydroforming and curing

(SPAFMHC) processes. The flexible medium is Changcheng L-QB300 heat transfer fluid. The forming temperature is at room temperature. As shown in Fig. 2c, sealing rubber rings are installed on the surface of the liquid chamber to ensure that the resin does not overflow during the forming process.

In order to meet the process requirement of forming and curing integration, the forming device is divided into two parts: heating platform and forming die, as shown in Fig. 2b. Compared with FMHC process, SPAFMHC process adds a layer of steel plate between FMLs and forming die. Due to the good plasticity of the steel plate and the large friction coefficient between the steel plate and the FMLs, the wrinkling defect of the FMLs can be effectively suppressed during the forming process. The thickness of the steel plate in the SPAFMHC process is 0.8 mm. As the cavity pressure and blank holder force are the key factor affecting the quality of hydroforming process. Hence in this paper, the progressive loading method is used to load the blank holder force and cavity pressure. As shown in Fig. 3a, the loading path of blank holder force and cavity pressure is used in the FMHC process and SPAFMHC process.

2.3 Numerical Simulation Model

ABAQUS explicit solver is used to perform the numerical simulation of FMHC process and SPAFMHC process respectively. The blank holder and die are rigid bodies, and the aluminum plate and glass fiber prepreg are set as deformable bodies. The mesh type is S4R, and the mesh size is 2.5 mm. The friction coefficient between the FMLs and the forming die is set to be 0.15[15], and the interlaminar tangential action coefficient between the glass fiber prepreg and the aluminum alloy sheet is 0.1. In the SPAFMHC process the steel plate material is set as a deformable body. The steel plate material properties are shown in Table 1. The mesh type is S4R, the mesh size is 2.5 mm. The friction coefficient between steel plate and aluminum plate is 0.2, and the friction coefficient between steel plate and forming die is 0.15. The Hill yield criterion is selected as the yield criterion of 2A12-T4 and steel. The failure criterion for the fiber layer is the maximum stress criterion.

3 Results and Discussion

3.1 Influence of Forming Process on Wrinkling Defect of FMLs Part

The comparison between the numerical simulation and the forming test results of the FMHC process are shown in Fig. 3b, c. It shows that the wrinkling defect appears at the T-shaped cavity. As is shown in Fig. 3b, the stress in the Y-axis direction of the part is mainly tensile stress, but there is a large compressive stress in the T-shaped cavity.

In order to further study the causes of wrinkling defects, the stress nephogram along the Y-axis direction of the part under the pressure of 3 MPa, 9 MPa and 18 MPa liquid chamber are analyzed, as shown in Fig. 4. The cutting line (shown in Fig. 2a) deformation displacement are extracted in the numerical simulation results, and the deformation displacement curve is obtained as shown in Fig. 4d. With the increase of the cavity pressure, the deformation displacement of the FMLs becomes larger, and Y direction cavity gradually sticks to the die. The wrinkling defect at the T-shaped cavity becomes larger and larger with the increase of the pressure in the liquid chamber. The closer to the T-shaped cavity, the more difficult it is for the FMLs to stick to the die.

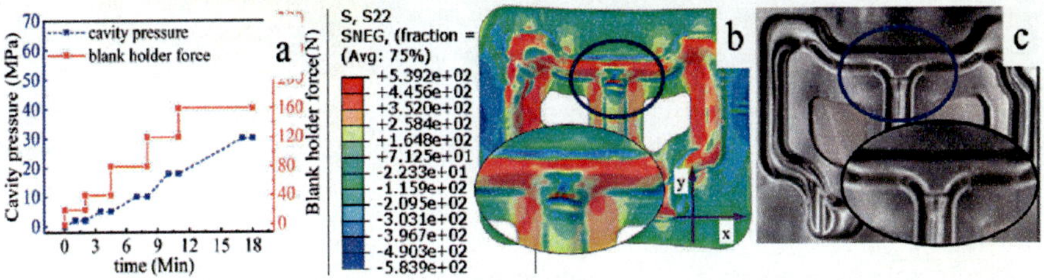

Fig. 3. (a) Blank holder force and cavity pressure loading path of FMHC process and SPAFMHC process, (b) simulation results of FMLs complex structures by FMHC process, (c) FMLs complex part formed by FMHC process.

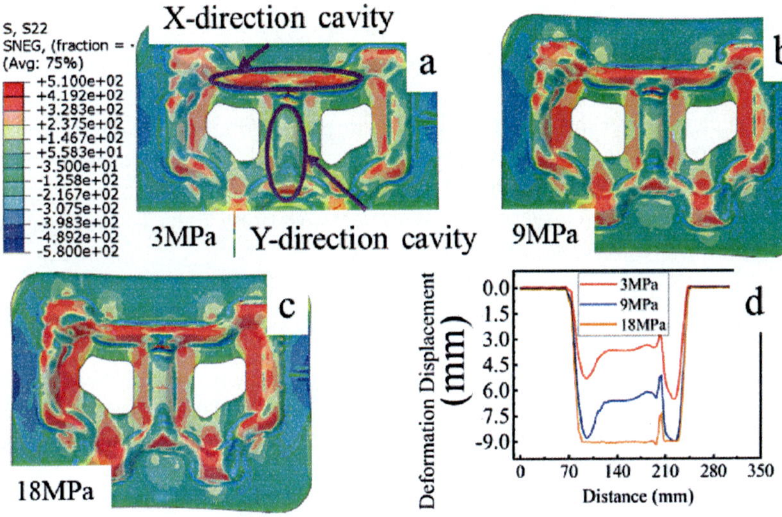

Fig. 4. (a) Stress nephogram of 3 MPa cavity pressure, (b) stress nephogram of 9 MPa cavity pressure, (c) stress nephogram of 18 MPa cavity pressure, (d) the deformation displacement of cutting line with the change of cavity pressure.

As shown in Fig. 4, with the increase of cavity pressure, the X-direction cavity is mainly subject to tensile stress, and the Y-direction cavity subject to compressive stress, and there is wrinkling defect in the Y-direction cavity. With the increase of the cavity pressure, the tensile stress in the X direction cavity gradually increases, and the compressive stress in the Y-direction cavity gradually increases. It can be seen that the wrinkling defect at the position of the T-shaped cavity is caused by the gradual increase of the compressive stress on the Y direction cavity.

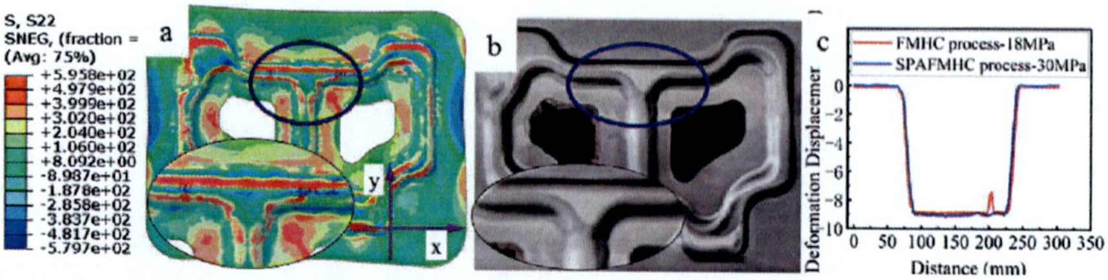

Fig. 5. (a) Simulation results of FMLs complex structures by SPAFMHC process, (b) FMLs complex part formed by SPAFMHC process; (c) comparison of deformation displacement of cutting line under FMHC process and SPAFMHC process.

The comparison between the numerical simulation and the forming test results of the SPAFMHC process is shown in Fig. 5. The comparison between Fig. 3b and Fig. 5a shows that the compressive stress at the T-shaped cavity of the FMLs complex part formed by the SPAFMHC process is significantly reduced. The SPAFMHC process has realized the good forming of the FMLs complex part as is shown Fig. 5b. As shown in Fig. 5c, the numerical simulation result of deformation displacement at cutting line (shown in Fig. 2a) of FMHC process and SPAFMHC process is compared. It can be seen that the wrinkling defect at the T-shaped cavity of the FMLs formed by the SPAFMHC process is obviously eliminated, and the FMLs at this position are well attached to the die, which improves the forming quality of the part. During the SPAFMHC forming process, the steel plate and FMLs is always contact with each other. The friction between the steel plate and FMLs can inhibiting the formation of wrinkling defects. Comparing Fig. 4c and Fig. 5a, it can be seen that the compressive stress at the T-shaped cavity is significantly reduced by using SPAFMHC process.

3.2 Internal Defect Formation Mechanism of FMLs Complex Structure

The FMHC process cannot form qualified components, so analyzing the internal defects of part formed by the FMHC process is meaningless. Internal defect analysis is conducted on SPAFMHC process formed part. Industrial CT scanning technology is used to scan the part manufactured by SPAFMHC process, and the results is shown in Fig. 6a. The result shows that the defect coverage area is about 3788.69 mm^2, the surface area of FMLs is 79049.64 mm^2, and the defect area accounts for 4.7% of the total part area. The internal defects are mainly concentrated in the bottom fillet area of the ∞ cavity. The defect position of the part is cutted for further observation. It is found that the main types of internal defects of the part are delamination and porosity defects. The delamination defect between the metal layer and the fiber layer are shown in Fig. 6b and Fig. 6e. The incompatible deformation between the fiber layer and the metal layer result in delamination defects between the fiber layer and the metal layer, during the forming process of T-shaped cavity and fillets. As shown in Fig. 6c and Fig. 6d, there are porosity defects in the part. Because when forming complex structures, with the plasticity deformation of metal and fiber sliding the resin is hard to fill the surrounding space, resulting in the internal porosity.

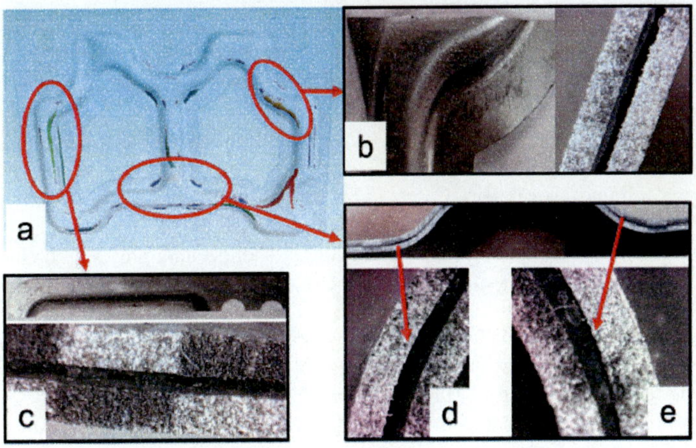

Fig. 6. (a) Internal defect distribution scaned by CT, (b) delamination defect, (c) porosity defect, (d) porosity defect, (e) delamination defect.

4 Conclusion

This paper proposes a FMLs forming and curing technology, which realizes the integrated manufacturing of forming and curing of FMLs parts with complex structure, and analyzes the distribution law and formation mechanism of internal defects of FMLs part after forming and curing. The main conclusions are as follows:

(1) The wrinkling defect of the T-shaped cavity is mainly caused by the compressive stress in the Y direction. The SPAFMHC process can effectively reduce the compressive stress in the Y direction of the T-shaped cavity. An FMLs part has been successfully manufacture through SPAFMHC process;
(2) The main types of defects are delamination and porosity defects. Delamination defects are mainly caused by the incompatible deformation of fibers and metals; With the plasticity deformation of metal and fiber sliding, the resin is hard to supplement to plastic deformation zone, resulting in the internal porosity at surrounding corners.

The complicate part of FMLs can be manufactured by SPAFMHC process with no wrinkling defects. But there have some porosity and delamination defects in the part. The future research can be focus on how to reduce the internal defects.

References

1. Asundi, A., Choi, A.Y.N.: Fiber metal laminates: an advanced material for future aircraft. J. Mater. Process. Technol. **63**(1), 384–394 (1997)
2. Sinmazcelik, T., Avcu, E., Bora, M., et al.: A review: Fibre metal laminates, background, bonding types and applied test methods. Mater. Des. Mater Des. **32**, 3671–3685 (2011)
3. Blala, H., Lang, L.H., Li, L., et al.: Investigation on the effect of blank holder gap in the hydroforming of cylindrical cups, made of fiber metal laminate. Int. J. Adv. Manuf. Technol. **108**(9–10), 2727–2740 (2020)

4. Palkowski, H., Stanic, V., Carradò, A.: Multilayer roll-bonded sandwich: processing, mechanical performance, and bioactive behavior. JOM **64**(4), 514–519 (2012)
5. Sinke, J.: Manufacturing of GLARE Parts and Structures. Appl. Compos. Mater. **10**(4), 293–305 (2003)
6. Edwardson, S., Dearden, G., Watkins, K, et al.: Laser forming of fibre metal laminates. Lasers in Eng. **15**(3), 233–255 (2005)
7. Russig, C., Bambach, M., Hirt, G., et al.: Shot peen forming of fiber metal laminates on the example of GLARE®. Int. J. Mater. Form. **7**(4), 425–438 (2014)
8. Mylonas, G., Labeas, G.: Numerical modelling of shot peening process and corresponding products: residual stress, surface roughness and cold work prediction. Surf. Coat. Technol. **205**, 4480–4494 (2011)
9. Sadegh-yazdi, M., Bakhshi-Jooybari, M., Shakeri, M., et al.: Optimization of pressure paths in hydrodynamic deep drawing assisted by radial pressure with inward flowing liquid using a hybrid method. Int. J. Adv. Manuf. Technol. **97**(5), 2587–2601 (2018)
10. Ghaforian Nosrati, H., Gerdooei, M., Falahati, N.M.: Experimental and numerical study on formability in tube bulging: a comparison between hydroforming and rubber pad forming. Mater. Manuf. Processes **32**(12), 1353–1359 (2017)
11. Liu, S.C., Lang, L.H., Guan, S.W., et al.: Investigation into composites property effect on the forming limits of multi-layer hybrid sheets using hydroforming technology. Appl. Compos. Mater. **26**(1), 205–217 (2019)
12. Li, L., Lang, L.H., Hamza, B., et al.: Effect of hydroforming process on the formability of fiber metal laminates using semi-cured preparation. Int. J. Adv. Manuf. Technol. **107**(9–10), 3909–3920 (2020)
13. Blala, H., Lang, L.H., Khan, S., et al.: An analysis of process parameters in the hydroforming of a hemispherical dome made of fiber metal laminate. Appl. Compos. Mater. **28**(3), 685–704 (2021)
14. Tabasum, M.N., Lang, L.H., Mirza, H.A., et al.: Numerical and experimental investigations on the effects of variable cavity pressure on the formability of GLARE using hydromechanical deep drawing. Int. J. Adv. Manuf. Technol. **119**(9–10), 6091–6101 (2022)
15. Blala, H., Lang, L., Khan, S., Alexandrov, S.: Experimental and numerical investigation of fiber metal laminate forming behavior using a variable blank holder force. Prod. Eng. Res. Devel. **14**(4), 509–522 (2020). https://doi.org/10.1007/s11740-020-00974-9

Diffusion Bonding Process and Interface Performance of 2198 Al-Li Alloy

Senbao Jiang[1], Zhiting Yang[2], Tianle Li[2], Zengyu Wang[2], Yue Yan[1], Yao Chen[1], and Xifeng Li[2(✉)]

[1] Chengdu Aircraft Industrial (Group) Co., Ltd., Chengdu 610092, China
[2] Department of Plasticity Technology, School of Materials Science and Engineering, Shanghai Jiao Tong University, Shanghai 200030, China
lixifeng@sjtu.edu.cn

Abstract. The diffusion bonding process can take full advantage of 2198 Al-Li alloy to produce high-efficiency integral structures. However, the dense and stable oxide layer on the surface to be bonded seriously hinders the diffusion bonding process. The suitable surface treatment method was selected to remove the oxide layer. Furthermore, the effects of bonding temperature, time and pressure on interface performance were studied. Optical microscopy (OM) and scanning electron microscope (SEM) are used to observe the bonded interface and fracture surface after the shear test. The maximum shear strength and strength ratio reach up to 131 MPa and 96%, respectively. In general, shear fracture surfaces account for the majority of bonded interfaces, which show ductile areas with dimples and brittle regions with quasi-cleavage behavior. This study indicates 2198 Al-Li alloy can be well bonded by suitable surface treatment and process parameters.

Keywords: 2198 Al-Li alloy · Diffusion bonding · Interface performance

1 Introduction

In recent years, Al-Li (Aluminum-Lithium) alloys have been given increasing attention for their use in weight and stiffness-critical structures serviced in aircraft, aerospace, and military industries due to their better properties, such as high specific strength and excellent strength-ductility than those of common Al alloys [1]. For the 3rd generation Al-Li alloys including 2195, 2050, 2297, 2198, etc., some micro-alloy elements like Mg, Mn, Zn, and Ti are attempted to add into Al-Li alloys in order to improve the final mechanical properties [2–4]. 2198 Al-Li alloy with Cu element of 2.9–3.4wt.% and Li element 0.9%–1.1wt.% is developed to replace 2024 and 2524 in aircraft structures in which damage tolerance is a critical design factor to meet the demand of high cycle fatigue and fatigue endurance limit regimes [5]. Fracture behavior and anisotropic deformation of 2198 Al-Li alloy subjected to different heat treatments were studied by experimental observation and numerical simulation [6, 7].

K. Mocellin et al. (Eds.): ICTP 2023, LNME, pp. 699–706, 2024.
https://doi.org/10.1007/978-3-031-40920-2_73

Apart from alloy composition design, the welding process is also a key method to favor the manufacturing of components with superior mechanical properties by replacing riveting and fastening procedures. Diffusion bonding, an advanced joining process, is often applied to manufacture components of ceramics and metals with small low-stress creep deformation under the action of a time of elevated temperature and pressure [8, 9]. Diffusion bonding of the Al alloys is always a tough task due to the oxide film characterized by chemically stubborn at the substrate surface which acts as the barrier of atoms interdiffusion [10]. To date, several methods have been attempted to improve the bonded quality of Al alloys, such as increased surface roughness [11], transient liquid phase with Cu as interlayer [12], and vacuum evaporation coating [13]. Jiang et al. [14] fabricated the hollow double-layer structure of Al-Li 5A90 alloy by superplastic forming/diffusion bonding process, and investigated the characteristics and mechanism with respect to forming and bonding.

The 3rd generation Al-Li alloys are becoming increasingly competitive in the aircraft industry, while the study on diffusion bonding of this type of Al-Li alloy has been rarely reported up to now. In this study, we report the role of bonded parameters including temperature, time, and pressure on the quality of joints and shear performance of 2198 Al-Li alloy.

2 Experimental Procedure

The 2198 Al-Li alloy plate with a thickness of 3.1 mm was thick and has a chemical composition of 3.46 Cu, 0.93 Li, 0.29 Mg, 0.03 Mn, 0.23 Zn, and the balance Ti in weight percent (wt.%). The plate surfaces were mechanically ground using P600 sandpaper followed by ultrasonic cleaning in an acetone bath. Diffusion bonding experiments were carried out in a hot-pressing furnace with a vacuum of 3×10^{-3}Pa at a heating rate of 10 °C/min and furnace cooling to room temperature. In order to study the effect of temperature on bonding, a group of samples were bonded at 475 °C, 500 °C, 525 °C, and 550 °C for 45 min and under a pressure of 15 MPa. A group of samples was heated to 500 °C for 45 min under various pressure including 5 MPa, 10 MPa, 15 MPa, and 20 MPa to observe the pressure action. In addition, a group of samples was bonded at 500 °C for different times of 30 min, 45 min, 60 min, and 90 min under 15 MPa.

Interface bonding observation for the bonded samples was performed with optical microscopy (OM) and scanning electron microscope (SEM). The fracture surfaces of the bonded samples were analyzed by SEM. Bonded interfaces for OM and SEM examinations were etched using a 1.5 ml hydrochloric acid (HCl), 1ml hydrofluoric acid (HF), 2.5 ml nitric acid (HNO_3), and 95 ml pure water (H_2O) solution for 10 s. Compression shear tests along the bonded interface were conducted on a universal test machine at a speed of 1 mm/min at room temperature. Shear stress can be calculated as F/A, where F refers to the measured force during the test, and A refers to the area that affords this force. Shear strength indicates the maximum shear stress during compression.

3 Results and Discussion

3.1 Bonded Pressure Action

Figure 1a–d shows the morphology of bonded interface at 500 °C for 45 min under different pressures. The bonded interface becomes increasingly indistinguishable due to enhanced joining with increasing pressure. There is no void in Fig. 1d, which indicates that 2198 Al-Li alloy can be completely bonded at 500 °C for 45 min under 20 MPa. It is found that several precipitate particles are formed at the bonded interface and substrate in Fig. 1e and f. Nayan et al. [15] pointed out that these precipitates are likely to be incoherent θ′, β′ (Al_3Zr), and T_B (Al_7Cu_4Li) precipitates in the study of microstructural evolution of Al-Li alloy 2198-T8 processed by friction stir weld.

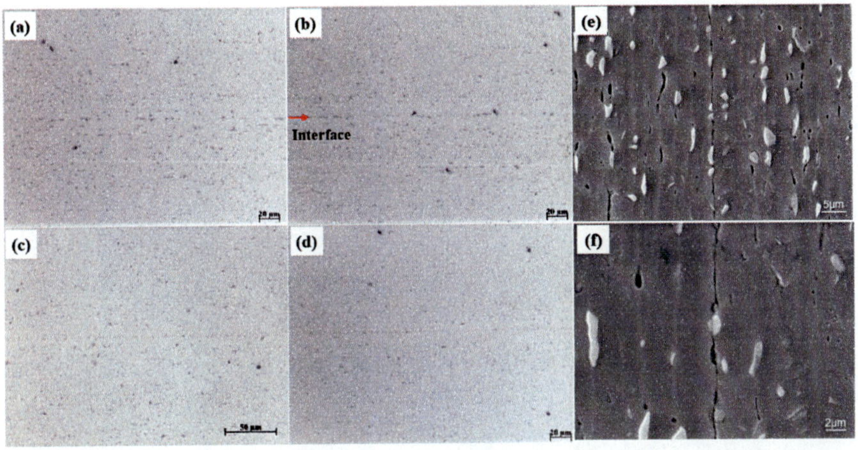

Fig. 1. OM (a–d) and SEM (e, f) observations on interface bonding and microstructure of samples bonded at 500 °C for 45 min under different pressures: (a) 5 MPa, (b) 10 MPa, (c) 15 MPa, (d) 20 MPa, (e, f) precipitate particles.

Figure 2 shows the shear strength and the corresponding strength ratio of bonded interfaces with different pressures. Strength ratio refers to a ratio of shear strength of bonded samples to shear strength of the monolithic Al plate subjected to the consistent heat treatment. Shear strength of the bonded interfaces increases from 113 MPa to 131 MPa with the increasing pressure, and strength ratio also increases from 85% to 96%.

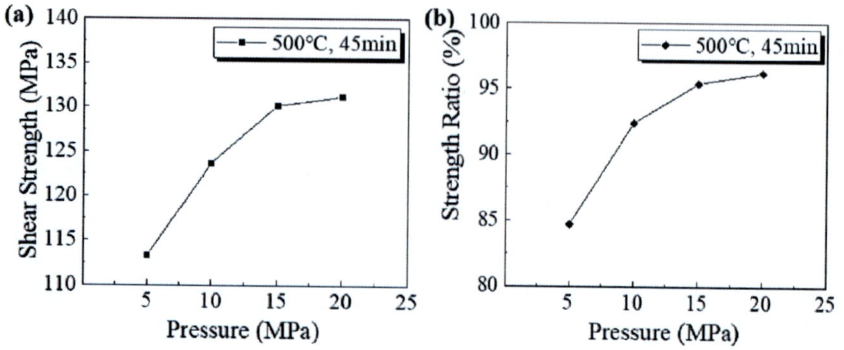

Fig. 2. Shear performance of bonded interfaces with various pressures: (a) shear strength, (b) strength ratio.

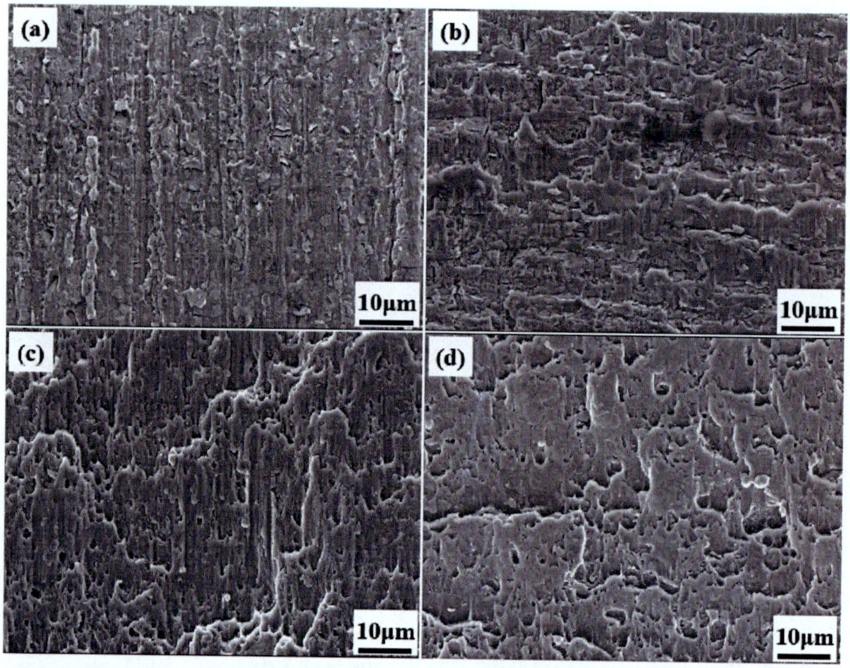

Fig. 3. SEM observation on fracture surfaces of samples bonded under different pressures after shear: (a) 5 MPa, (b) 10 MPa, (c) 15 MPa, (d) 20 MPa.

From Fig. 3a, shear fracture surface under 5 MPa is uneven compared with other samples. A fact that dimples are small and shallow can be attributed to the several retained unbonded areas. There are some ductile areas with dimples and brittle regions with quasi-cleavage behavior in Fig. 3b–d, which indicates the bonded interfaces of the samples under the pressure ranging from 10 MPa to 20 MPa present a mixed fracture pattern of ductile and brittle types.

3.2 Bonded Temperature Action

The bonding interface microstructure of samples bonded at different temperatures for 45 min under 15 MPa is observed by OM in Fig. 4. For samples bonded at 475 °C, long and narrow voids are distributed along the bonding interface, and only a few zones are completely bonded. There are no retained voids at the bonding interface of other samples due to the more active atom interdiffusion with the increasing temperature.

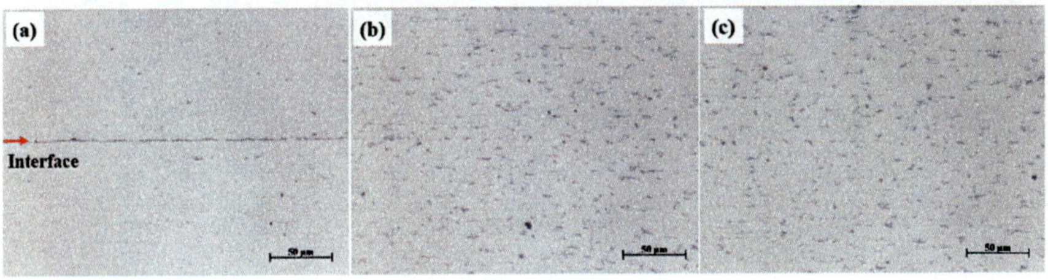

Fig. 4. OM observations on interface bonding of samples bonded at different temperatures for 45 min under 15 MPa: (a) 475 °C, (b) 525 °C, (c) 550 °C.

In Fig. 5, shear strength and the corresponding strength ratio of bonded interface present a consistent change trend with various temperatures. The maximum strength of bonded interface up to 127 MPa is obtained at 500 °C, and strength becomes decreased with further increasing temperature. There is no positive correlation between bonding ratio and shear strength. Another important factor affecting the shear strength is the microstructure of the bond interface, the increased temperature also causes grain growth and coarsening, which directly results in a gradual decrease in strength.

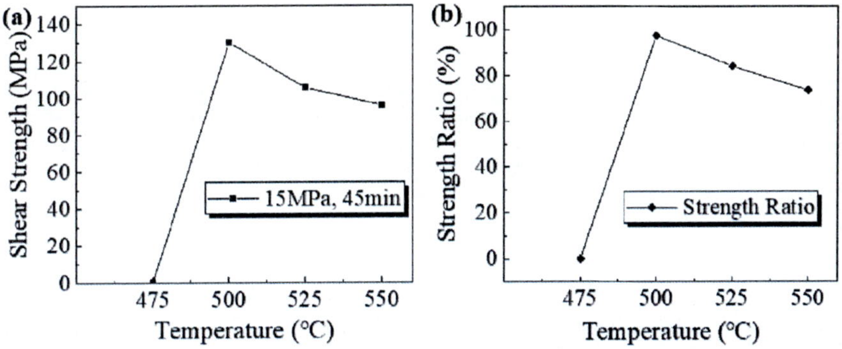

Fig. 5. Shear performance of bonded interfaces with various temperatures: (a) shear strength, (b) shear ratio.

The shear fracture surface of the interface obtained at 475 °C is very smooth in Fig. 6a, which can be attributed to the massive area without effective metallurgical bonding. The appearance of shear dimples with typical characteristics of a sharp shape at the end has obvious directivity consistent with the action direction of the shear loading. Fracture surface of samples bonded at 525 °C and 550 °C also present a mixed fracture pattern of ductile and brittle types.

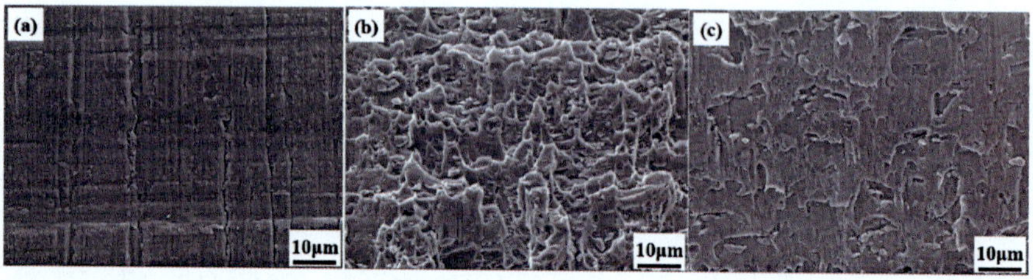

Fig. 6. SEM observation on fracture surfaces of samples bonded at different temperatures after shear: (a) 475 °C, (b) 525 °C, (c) 550 °C.

3.3 Bonded Time Action

In Fig. 7, It can be seen that the width of the bonded interface becomes narrower with increasing time, and the extension of bonded time also promotes the formation and growth of precipitate particles. Low stress can drive this alloy to creep activation at high temperature. The longer the creep time is, the more volume of alloy migrates to the voids. In addition, the movement of atoms towards the voids under the function of the stress concentration at the edge of the voids, which promotes the shrinking and even disappearing of retained voids. Figure 8a shows the relationship between shear strength and bonded time, and the relationship between strength ratio and bonded time is shown in Fig. 8b. The shear strength of the bonded interface increases at first and then decreases with increasing time.

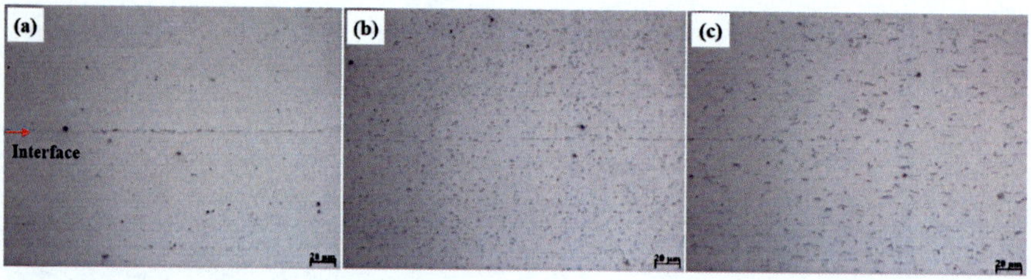

Fig. 7. OM observations on interface bonding of samples bonded at 500 °C under 15 MPa for different times: (a) 30 min, (b) 60 min, (c) 90 min.

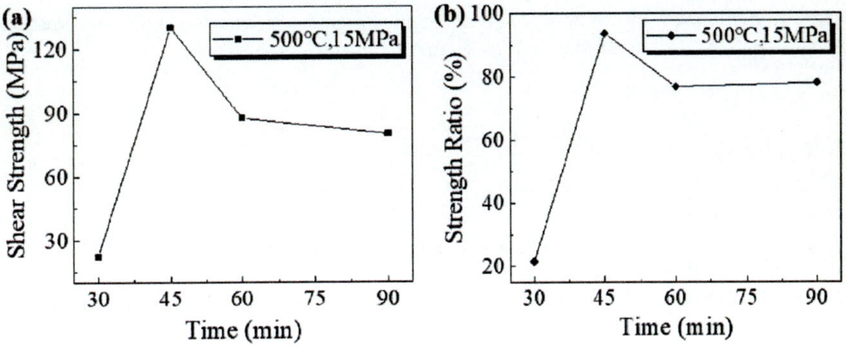

Fig. 8. Shear performance of bonded interfaces with various times: (a) shear strength, (b) shear ratio.

Figure 9 shows the fracture characteristics of samples bonded for various times ranging from 30 min to 90 min. For 30 min sample, the fracture surface consists of some parallel scratches along the shear direction which are traces left by mechanical friction. The shear fracture surface is flat showing obvious brittle fracture characteristics. For 60 min and 90 min samples, the fracture surfaces of the bonded interfaces still show the characteristics of ductile areas with dimples and brittle regions with quasi-cleavage behavior. In short, the crucial function of process parameters to interface bonding in sequence is temperature, time, and pressure.

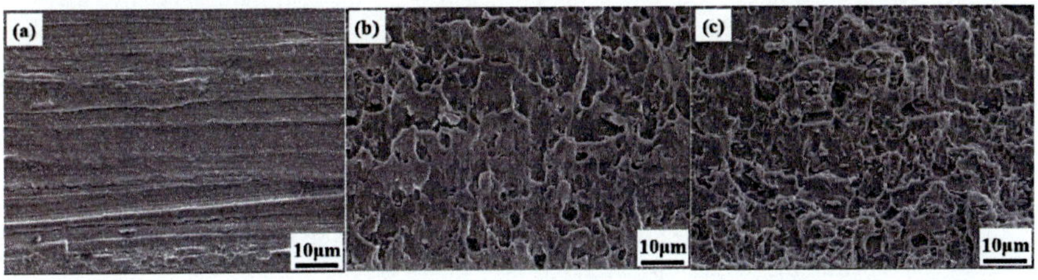

Fig. 9. SEM observation on fracture surfaces of samples bonded for different times after shear: (a) 30 min, (b) 60 min, (c) 90 min.

4 Conclusion

Interface bonding and shear performance of 2198 Al-Li alloy processed by diffusion bonding with changed temperature, time, and pressure are investigated. This Al-Li alloy can be completely metallurgical bonded at 500 °C for 45 min under 20 MPa. In this process parameters, shear strength and strength ratio reach up to 131 MPa and 96%, respectively. In general, shear fracture surfaces the majority of bonded interfaces show ductile areas with dimples and brittle regions with quasi-cleavage behavior.

Acknowledgments. This work was sponsored by Program of Shanghai Academic Research Leader (No. 22XD1421600). We are grateful for the financial support.

References

1. Abd El-Aty, A., Xu, Y., Guo, X.Z., Zhang, S.H., Ma, Y., Chen, D.Y.: Strengthening mechanisms, deformation behavior, and anisotropic mechanical properties of Al-Li alloys: a review. J. Adv. Res. **10**, 49–67 (2018)
2. Li, J.F., Liu, P.L., Chen, Y.L., Zhang, X.H., Zheng, Z.Q.: Microstructure and mechanical properties of Mg, Ag and Zn multi-microalloyed Al-(3.2-3.8)Cu-(1.0-1.4)Li alloys. Trans. Nonferrous Met. Soc. China **25**, 2103–2112 (2015)
3. Dursun, T., Soutis, C.: Recent developments in advanced aircraft aluminium alloys. Mater. Des. **56**, 862–871 (2014)
4. Li, J.Y., Pan, Y., Zhao, D.J., Lü, S.L., Wu, S.S., Guo, W.: Development of a novel high strength and toughness Al-Cu-Li alloy casting billet with a new process. Mater. Sci. Eng. A **854**, 143827 (2022)
5. Alexopoulos, N.D., Migklis, E., Stylianos, A., Myriounis, D.P.: Fatigue behavior of the aeronautical Al-Li (2198) aluminum alloy under constant amplitude loading. Int. J. Fatigue **56**, 95–105 (2013)
6. Chen, J.Q., Madi, Y., Morgeneyer, T.F., Besson, J.: Plastic flow and ductile rupture of a 2198 Al-Cu-Li aluminum alloy. Comput. Mater. Sci. **50**, 1365–1371 (2011)
7. Steglich, D., Wafai, H., Besson, J.: Interaction between anisotropic plastic deformation and damage evolution in Al 2198 sheet metal. Eng. Fract. Mech. **77**, 3501–3518 (2010)
8. Kovacevic, S., Pan, R., Sekulic, D.P., Mesarovic, S.D.: Interfacial energy as the driving force for diffusion bonding of ceramics. Acta Mater. **186**, 405–414 (2020)
9. Li, T.L., Zhong, L.X., Wu, H.P., An, D.Y., Li, X.F., Chen, J.: Microstructure evolution and fatigue crack growth of diffusion bonded Ti-6Al-4V titanium alloy. J. Alloy. Compd. **918**, 165816 (2022)
10. Lathashankar, B., Tejaswini, G.C., Suresh, R., Swamy, N.H.S.: Advancements in diffusion bonding of aluminium and its alloys: a comprehensive review of similar and dissimilar joints. Adv. Mater. Process. Technol. **8**, 1–19 (2022)
11. Zuruzi, A.S., Li, H., Dong, G.: Effects of surface roughness on the diffusion bonding of Al alloy 6061 in air. Mater. Sci. Eng. A **270**, 244–248 (1999)
12. Nami, H., Halvaee, A., Adgi, H.: Transient liquid phase diffusion bonding of Al/Mg2Si metal matrix composite. Mater. Des. **32**, 3957–3965 (2011)
13. Ureña, A., Gómez De Salazar, J.M., Quiñones, J., Martín, J.J.: TEM characterization of diffusion bonding of superplastic 8090 Al-Li alloy. Scripta Mater. **34**, 617–623 (1996)
14. Jiang, S.S., Jia, Y., Lu, Z., Shi, C.C., Zhang, K.F.: Superplastic Forming/diffusion bonding without interlayer of 5A90 Al-Li alloy hollow double-layer structure. J. Mater. Eng. Perform. **26**, 4265–4273 (2017)
15. Nayan, N., et al.: Microstructure and tensile response of friction stir welded Al-Cu-Li (AA2198-T8) alloy. Mater. Charact. **159**, 110002 (2020)

Enhanced Mechanical Properties of High Temperature Titanium Alloy Components by Fast Gas Forming with In-Die Quenching

Wentao Chen[1,3], Kexin Dang[3], Kehuan Wang[1,2,3](✉), and Gang Liu[1,2,3]

[1] National Key Laboratory for Precision Hot Processing of Metals, Harbin Institute of Technology, Harbin 150001, China
wangkehuan@hit.edu.cn

[2] Institute of High Pressure Fluid Forming, Harbin Institute of Technology, Harbin 150001, China

[3] School of Materials Science and Engineering, Harbin Institute of Technology, Harbin 150001, China

Abstract. In this paper, Fast Gas Forming with in-die Quenching (FGFQ) of TC31 thin-walled components was developed to integrate gas forming and heat treatment together. This forming process consists of rapid resistance heating, fast gas forming, and in-die quenching. The heating rate and pressurization rate can reach up to 30 °C/s and 20 MPa/s respectively. The whole forming cycle takes less than one minute, which can significantly improve the forming efficiency. The component possesses the best mechanical properties at the forming temperature of 1040 °C and the pressure of 15 MPa. The ultimate tensile strength of the component could reach up to 1292 and 837 MPa respectively at room temperature and 650 °C, which are improved by 18.2% and 29.8% compared with the initial state. Moreover, the elongation of 11.2% and 16.5% could also remain. The rapid resistance heating and fast gas forming can effectively prevent prior β grains from coarsening during the forming process, which results in the formation of high-density fine martensite and therefore enhance the post-form strength. Also, the refinement of the martensite contributes to the great ductility.

Keywords: High temperature titanium alloy · Fast gas forming with in-die quenching · Resistance heating · Strengthening mechanism

1 Introduction

Titanium alloys have been widely used in aerospace industries due to their high specific strength, outstanding corrosion resistance, and great creep resistance. With the development of high-speed aircraft, the requirement for complex-shaped components with closed cross-sections made of high temperature titanium alloys is gradually increasing. Hot metal gas forming (HMGF) process was developed to form such components [1]. To further promote the forming efficiency, non-isothermal methods were conducted on HMGF, where the temperature of the die was lower than the tube blank. Cold dies were

K. Mocellin et al. (Eds.): ICTP 2023, LNME, pp. 707–713, 2024.
https://doi.org/10.1007/978-3-031-40920-2_74

used in HMGF to form aluminum alloy and high-strength steel tubes in these years [2, 3]. Meanwhile, hot stamping with cold dies was also conducted to demonstrate the feasibility of titanium alloy components [4]. With the continuous improvement of forming efficiency, how to improve the mechanical properties of the formed components is equally important.

Strengthening and toughening have become crucial orientations to improve the mechanical properties of conventional metallic materials. However, they are always contradicting each other [5]. So, exploring a novel technique to achieve a great combination of strength and ductility is of great significance. Grain refinement is a typical method to improve the strength of titanium alloys, due to the significant grain boundary or α/β interface strengthening [6]. Compared with conventional equiaxed phase, lamellar martensitic microstructure has great strength with acceptable high temperature properties. Martensite of titanium alloys is usually produced by solution treatment in $\alpha + \beta$ or single β phase region followed by water quenching. A great combination of strength and ductility can be obtained by solution treatments below the β transus temperature. While solution treatments above the β transus temperature lead to rapid grain growth of prior β grains, which causes severe loss in strength and ductility [7]. Therefore, it is necessary to develop a new method to solve the above problems. Rapid resistance heating can effectively avoid prior β grains from coarsening and result in great mechanical properties. However, the combination of HMGF and rapid resistance heating has not yet been reported to form and strengthen high temperature titanium alloys.

In present work, Fast Gas Forming with in-die Quenching (FGFQ) for high temperature titanium alloys was studied, which could effectively promote forming efficiency of closed cross-sections titanium alloy components. The tube blank was heated by current in tens of seconds, and the pressurization was up to 20 MPa/s. The influence of forming parameters on forming accuracy, microstructure evolution, post-form mechanical properties, and the strengthening mechanism of the components were also illustrated.

2 Materials and Experimental Method

2.1 Material

A TC31 high temperature titanium alloy with a chemical composition of Ti-6Al-3Sn-3Zr-3Nb-3Mo-1W-0.2Si (in wt.%) was utilized in the present study. The as-received alloy was a hot rolled sheet with a thickness of 1 mm. The initial microstructure of the sheet is shown in Fig. 1, which consisted of equiaxed α phase, β phase, and secondary α phase. The fraction of β phase was less than 12.6%. The β transus temperature was 1025 °C according to optical microscopy. The TC31 titanium alloy tubes used in this paper were fabricated by a hot U-O forming process combined with laser beam welding [8]. The microstructure changed little during the U-O forming process because of the relatively low forming temperature of 800 °C and short forming time of 15 min. The tubes were 260 mm in length and 40 mm in diameter with a radius-thickness ratio of 40.

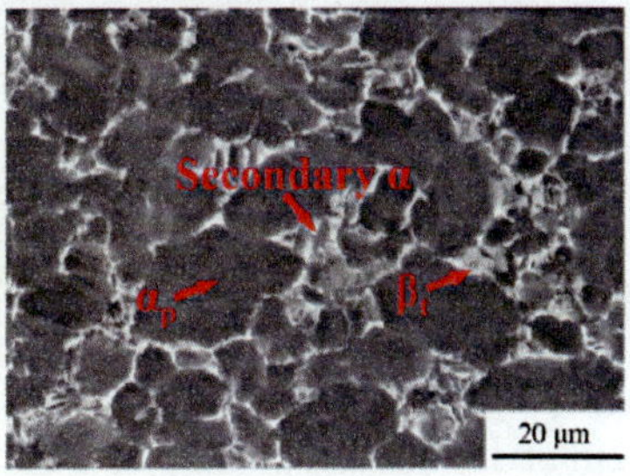

Fig. 1. Microstructure of the as-received TC31 titanium alloy sheet.

2.2 Principle of FGFQ

The schematic diagram of the forming device used in the study is shown in Fig. 2, which has an ultimate pressure of 70 MPa and a maximum pressurization rate of 40 MPa/s. The diameter at the bulging zone is 56 mm with an expansion ratio of 40%. The device is mainly composed of a gas-filling system, a current heating system, a die-closing and punch-sealing system, and an integrated control system. More information could be found in our previous publication [9]. Before the experiments, the surface of the tubes was polished to remove the oxide layer. BN high-temperature lubricant was sprayed on the bulging area of the tubes to reduce the friction and slow down the oxidation. At the early stage of the experiment, three tubes were heated with a current density of 10 A/mm^2, and the die was kept at room temperature. The heating device would automatically shut down when the temperature of the tubes reached 995, 1040, and 1080 °C respectively. Then the gas with a pressure of 15 MPa was filled into the tubes immediately under a fast pressurization rate of 20 MPa/s. The tubes would deform and contact the die within 1 s. Meanwhile, the temperature of the formed components would decrease sharply with a cooling rate of 200 °C/s. The components would be taken out after a 10 s pressure holding.

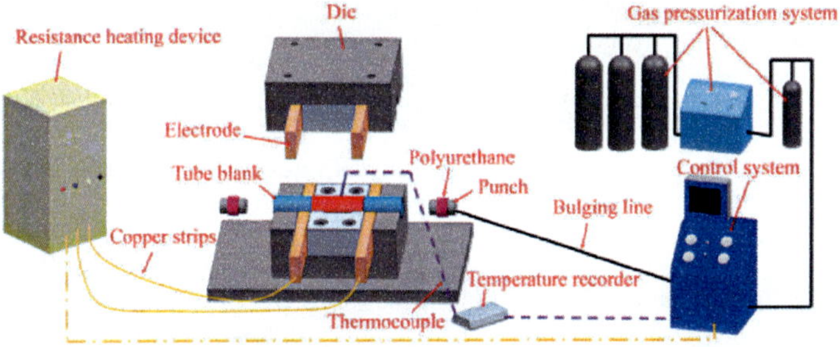

Fig. 2. Schematic diagram of the experimental device [9].

2.3 Microstructure and Property Characterization

To study the microstructure evolution of the TC31 titanium alloy during FGFQ, the microstructure was characterized by SEM. The samples were first grinding by silicate grit paper with 400–1500 grit numbers and then electropolished for 60 s with a mixed solution of 60 vol% CH_3OH, 34 vol% $CH_3(CH_2)_3OH$, and 6 vol% $HClO_4$ at −25 °C with a potential of 30 V and current of 0.9 A. The samples were chemically etched for 15 s with a solution of 80 vol% H_2O, 13 vol% HNO_3, and 7 vol% HF. Dog-bone-shaped samples with a gauge length of 15 mm and a width of 2 mm were machined by wire cutting from the bulging area of the tubes. Uniaxial tensile tests at room temperature and 650 °C were conducted on a Shimadzu AG-X plus electronic universal material testing machine with a strain rate of 10^{-3}/s. At least 3 samples were prepared for each condition to ensure repeatability.

3 Results and Discussion

The formed components at different temperatures are shown in Fig. 3. It can be seen that all the components fitted with the die perfectly due to the high forming temperature and forming pressure. Therefore, the influence of the forming temperature on the forming accuracy can be ignored at the temperature range of 995–1080 °C.

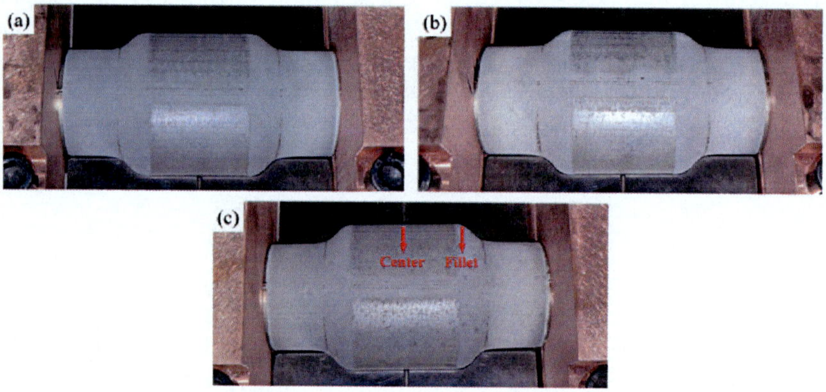

Fig. 3. Effect of forming temperature on formed parts of variable-diameter tubes: (a) 995 °C, (b) 1040 °C, (c) 1080 °C.

Samples cut from the center and fillet of the tubes were used for the microstructure observation as shown in Fig. 3c. The microstructure of components at different forming temperatures and positions is shown in Fig. 4. It is obvious that the microstructure of the center was almost the same as that of the fillet. Because the forming time was too short to cause an obvious temperature gradient, which was beneficial to obtain uniform microstructure in the bulging area of the variable-diameter tubes.

As shown in Fig. 4a, when the forming temperature was 995 °C, there was still much remained α_p phase. Because the rapid resistance heating can effectively improve the β transus temperature of TC31 titanium alloy. The microstructure consisted of dominating fine martensite, equiaxed α phase, and remained β phase. The fraction of the α phase was 41.7% and 46.5% respectively at the center and fillet of the variable-diameter tube in Fig. 4a and d. Meanwhile, the average grain size of the α phase was quite large.

When the forming temperature reached 1040 °C, The fraction of the α phase was 4.3% and 7.6% respectively at the center and fillet of the variable-diameter tube in Fig. 4b and e. The temperature was close to the β transus temperature, so a large amount of phase transformation occurred during the resistance heating. The grain size of the α phase was lower than that of 995 °C. At the forming temperature of 1080 °C, a fully martensitic microstructure was formed during FGFQ as shown in Fig. 4c and f.

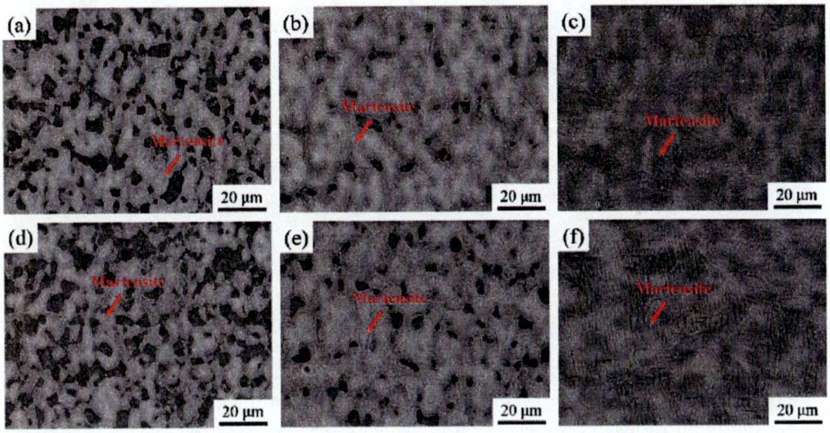

Fig. 4. Microstructure of variable-diameter tubes at different forming temperatures: (a) 995 °C at the center, (b) 1040 °C at the center, (c) 1080 °C at the center, (d) 995 °C at fillet, (e) 1040 °C at fillet, (f) 1080 °C at fillet.

Room temperature mechanical properties of the formed components of different forming conditions are presented in Fig. 5a. It is obvious that the ultimate tensile strength (UTS) of the formed components increased but the elongation decreased gradually with the forming temperature increasing from 995 to 1080 °C. After the tubes were formed at the temperature of 995, 1040, and 1080 °C, the UTS was 1226, 1292, and 1341 MPa. Compared with the initial state (1093 MPa), the UTS increased by 12.2%, 18.2%, and 22.7%. While the elongation decreased to 13.1%, 11.2%, and 6.0%, respectively from 14.1% of the initial state. When the forming temperature was 1040 °C, the tube had great strength and acceptable elongation. Therefore, the component had the best comprehensive performance at the forming temperature of 1040 °C.

TC31 titanium alloy is a newly developed high temperature titanium alloy utilized at 650 °C, so mechanical properties under the service temperature have a decisive influence on the application prospect of this alloy. The tensile properties of TC31 alloy at 650 °C are shown in Fig. 5b. It can be seen that the strength of the formed components was promoted sharply after FGFQ, but the elongation decreased with the increasing forming temperature. The UTS and elongation of the initial alloy were 645 MPa and 41.8% respectively. When the forming temperature was 995, 1040, and 1080 °C, the UTS was 735, 837, and 864 MPa, and the elongation was 22.0%, 16.5%, and 7.9% respectively. The component also had the best comprehensive high temperature properties at the forming temperature of 1040 °C, whose UTS increased by 29.8% compared with the initial state.

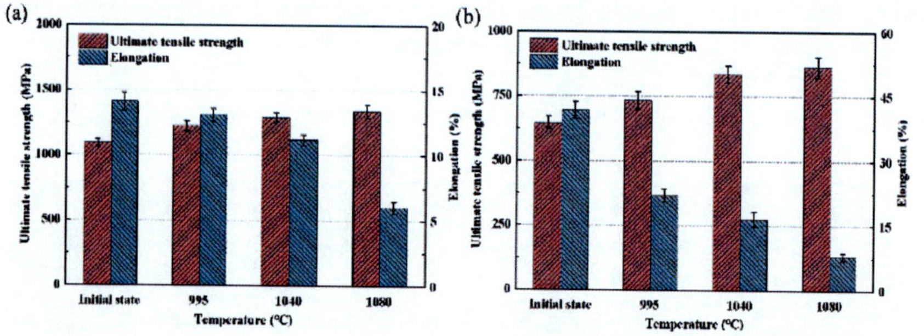

Fig. 5. Effect of the forming temperature on tensile properties: (a) room temperature, (b) 650 °C.

4 Conclusion

In this paper, the appropriate parameters for FGFQ of TC31 titanium alloy were obtained, which realized the efficient gas forming of TC31 thin-wall tubes. Meanwhile, the formed variable-diameter tubes expressed excellent mechanical properties at room temperature and service temperature (650 °C). The main conclusions are as follows:

1) The microstructure of the formed TC31 variable-diameter tubes by FGFQ consisted of dominating fine martensite, equiaxed α phase, and remained β phase with forming temperature increasing from 995 to 1040 °C. Meanwhile, the fraction of the α_p phase decreased from 41.7% to 4.3% at the center, and from 46.5% to 7.6% at the fillet. At the forming temperature of 1080 °C, a fully martensitic microstructure was formed during FGFQ.
2) When the forming temperature was 1040 °C, the variable-diameter tube had the best comprehensive properties. Compared with the initial state, the UTS at room temperature increased from 1093 to 1292 MPa by 18.2%, and the elongation remained 11.2%. The variable-diameter tube also had excellent properties at 650 °C. The UTS increased from 645 to 837 MPa by 29.8%, and the elongation was 16.5%.
3) The rapid resistance heating and fast gas forming can effectively prevent prior β grains from coarsening during the forming process, which results in the formation of high-density fine martensite and therefore enhances the post-form strength. Also, the refinement of the martensite contributes to the great ductility.

Acknowledgments. This research was funded by the Program of National Natural Science Foundation of China (No. U1937204 and No. 51905124), Heilongjiang Provincial Natural Science Foundation of China (No. LH2021E058), and China Postdoctoral Science Foundation (2021T140153).

References

1. Wang, K.H., et al.: High-efficiency forming processes for complex thin-walled titanium alloys components: state-of-the-art and perspectives. Int. J. Extreme Manuf. **2**, 032001 (2020)
2. Maeno, T., Mori, K., Adachi, K.: Gas forming of ultra-high strength steel hollow part using air filled into sealed tube and resistance heating. J. Mater. Process. Technol. **214**, 97–105 (2014)
3. Maeno, T., Mori, K., Fujimoto, K.: Hot gas bulging of sealed aluminium alloy tube using resistance heating. Manuf. Rev. **1**, 5 (2014)
4. Wang, K.H., et al.: Enhanced formability and forming efficiency for two-phase titanium alloys by Fast light Alloys Stamping Technology (FAST). Mater. Des. **194**, 108948 (2020)
5. Xu, X.F., et al.: Ti-6Al-4V alloy strengthening via instantaneous phase transformation induced by electropulsing. J. Alloy. Compd. **899**, 163303 (2022)
6. Zhang, Y.J., Fang, S.M., Wang, Y.M., Zhang, D.L.: Enhancement of tensile strength and ductility of titanium with a bimodal structure. Mater. Sci. Eng. A **803**, 140701 (2021)
7. Zou, Z.Y., Simonelli, M., Katrib, J., Dimitrakis, G., Hague, R.: Refinement of the grain structure of additive manufactured titanium alloys via epitaxial recrystallization enabled by rapid heat treatment. Scripta Mater. **180**, 66–70 (2020)
8. Wang, K.H., Liu, G., Zhao, J., Wang, J.L., Yuan, S.J.: Formability and microstructure evolution for hot gas forming of laser-welded TA15 titanium alloy tubes. Mater. Des. **91**, 269–277 (2016)
9. Dang, K.X., Wang, K.H., Chen, W.T., Liu, G.: Study on fast gas forming with in-die quenching for titanium alloys and the strengthening mechanisms of the components. J. Market. Res. **18**, 3916–3932 (2022)

Author Index

K. Mocellin et al. (Eds.): ICTP 2023, LNME, pp. 715–718, 2024.
https://doi.org/10.1007/978-3-031-40920-2